Graduate Texts in Contemporary Physics

Series Editors:

R. Stephen Berry
Joseph L. Birman
Mark P. Silverman
H. Eugene Stanley
Mikhail Voloshin

T0140511

Graduate Texts in Contemporary Physics

(continued after index)

V. Parameswaran Nair

Quantum Field Theory
A Modern Perspective

With 100 Illustrations

 Springer

V. Parameswaran Nair
Physics Department
City College
Convent Avenue & 138th Street
New York, NY 10031
USA

Series Editors

R. Stephen Berry
Department of Chemistry
University of Chicago
Chicago, IL 60637
USA

Joseph L. Birman
Department of Physics
City College of CUNY
New York, NY 10031
USA

Mark P. Silverman
Department of Physics
Trinity College
Hartford, CT 06106
USA

H. Eugene Stanley
Center for Polymer Studies
Physics Department
Boston University
Boston, MA 02215
USA

Mikhail Voloshin
Theoretical Physics Institute
Tate Laboratory of Physics
The University of Minnesota
Minneapolis, MN 55455
USA

On the cover: The pinching contribution to the interaction between fermions. See page 213 for discussion.

ISBN 978-1-4419-1946-5   e-ISBN 978-0-387-25098-4

Library of Congress Cataloging-in-Publication Data
Nair, V. P.
    Topics in quantum field theory / V.P. Nair.
      p. cm.
    Includes bibliographical references and index.

    1. Quantum field theory.   I. Title.
QC174.45.N32   2004
530.14′3—dc22   2004049910

Printed on acid-free paper

9 8 7 6 5 4 3 2 1

springeronline.com

To the memory of my parents
Velayudhan and Gowrikutty Nair

Preface

Quantum field theory, which started with Dirac's work shortly after the discovery of quantum mechanics, has produced an impressive and important array of results. Quantum electrodynamics, with its extremely accurate and well-tested predictions, and the standard model of electroweak and chromodynamic (nuclear) forces are examples of successful theories. Field theory has also been applied to a variety of phenomena in condensed matter physics, including superconductivity, superfluidity and the quantum Hall effect. The concept of the renormalization group has given us a new perspective on field theory in general and on critical phenomena in particular. At this stage, a strong case can be made that quantum field theory is the mathematical and intellectual framework for describing and understanding all physical phenomena, except possibly for quantum gravity.

This also means that quantum field theory has by now evolved into such a vast subject, with many subtopics and many ramifications, that it is impossible for any book to capture much of it within a reasonable length. While there is a common core set of topics, every book on field theory is ultimately illustrating facets of the subject which the author finds interesting and fascinating. This book is no exception; it presents my view of certain topics in field theory loosely knit together and it grew out of courses on field theory and particle physics which I have taught at Columbia University and the City College of the CUNY.

The first few chapters, up to Chapter 12, contain material which generally goes into any course on quantum field theory although there are a few nuances of presentation which the reader may find to be different from other books. This first part of the book can be used for a general course on field theory, omitting, perhaps, the last three sections in Chapter 3, the last two in Chapter 8 and sections 6 and 7 in Chapter 10. The remaining chapters cover some of the more modern developments over the last three decades, involving topological and geometrical features. The introduction given to the mathematical basis of this part of the discussion is necessarily brief, and these chapters should be accompanied by books on the relevant mathematical topics as indicated in the bibliography. I have also concentrated on developments pertinent to a better understanding of the standard model. There is no discussion of supersymmetry, supergravity, developments in field theory inspired by string theory, etc.. There is also no detailed discussion of the renormalization group either. Each of these topics would require a book in its own right to do justice to the topic. This book has generally followed the tenor of my courses, referring the students to more detailed treatments for many specific topics. Hence this is only a portal to so many more topics of detailed and ongoing research. I have also mainly cited the references pertinent to the discussion in the text, referring the reader to the many books which have been cited to get a more comprehensive perspective on the literature and the historical development of the subject.

I have had a number of helpers in preparing this book. I express my appreciation to the many collaborators I have had in my research over the years; they have all contributed, to varying extents, to my understanding of field theory. First of all, I thank a number of students who have made suggestions, particularly Yasuhiro Abe and Hailong Li, who read through certain chapters. Among friends and collaborators, Rashmi Ray and George Thompson read through many chapters and made suggestions and corrections, my special thanks to them. Finally and most of all, I thank my wife and long term collaborator in research, Dimitra Karabali, for help in preparing many of these chapters.

New York   *V. Parameswaran Nair*
May 2004   City College of the CUNY

Contents

1 Results in Relativistic Quantum Mechanics

1.1 Conventions

Summation over repeated tensor indices is assumed. Greek letters μ, ν, etc., are used for spacetime indices taking values $0, 1, 2, 3$, while lowercase Roman letters are used for spatial indices and take values $1, 2, 3$.

The Minkowski metric is denoted by $\eta_{\mu\nu}$. It has components $\eta_{00} = 1$, $\eta_{ij} = -\delta_{ij}$, $\eta_{0i} = 0$. We also use the abbreviation $\partial_\mu = \frac{\partial}{\partial x^\mu}$. The scalar product of four-vectors A_μ and B_ν is $A \cdot B = A_0 B_0 - A_i B_i$. Such products between momenta and positions appear often in exponentials; we then write it simply as px. It is understood that this is $p_0 x_0 - \boldsymbol{p} \cdot \boldsymbol{x}$, where the boldface indicates three-dimensional vectors.

The Levi-Civita symbol ϵ^{ijk} is antisymmetric under exchange of any two indices, and $\epsilon^{123} = 1$. $\epsilon^{\mu\nu\alpha\beta}$ is similarly defined with $\epsilon^{0123} = 1$.

Two spacetime points x, y are spacelike separated if $(x - y)^2 < 0$. This means that the spatial separation is more than the distance which can be traversed by light for the time-separation $|x^0 - y^0|$.

∂ is also used to denote the boundary of a spatial or spacetime region; i.e., ∂V and $\partial \Sigma$ are the boundaries of V and Σ, respectively.

We will now give a resumé of results from relativistic quantum mechanics. They are merely stated here, a proper derivation of these results can be obtained from most books on relativistic quantum mechanics.

1.2 Spin-zero particle

We consider particles to be in a cubical box of volume $V = L^3$, with the limit $V \to \infty$ taken at the end of the calculation. The single particle wave functions for a particle of momentum $\boldsymbol{k}$ can be taken as

$$u_k(x) = \frac{e^{-ikx}}{\sqrt{2\omega_k V}} \tag{1.1}$$

where $\omega_k = \sqrt{\boldsymbol{k} \cdot \boldsymbol{k} + m^2}$. We choose periodic boundary conditions for the spatial coordinates, i.e., $u_k(x + L) = u_k(x)$ for translation by L along any spatial direction; therefore the values of $\boldsymbol{k}$ are given by

$$k_i = \frac{2\pi n_i}{L} \qquad (1.2)$$

(n_1, n_2, n_3) are integers. The wave functions $u_k(x)$ obey the orthonormality relation

$$\int_V d^3x \, [u_k^*(i\partial_0 u_{k'}) - (i\partial_0 u_k^*)u_{k'}] = \delta_{k,k'} \qquad (1.3)$$

where $\delta_{k,k'}$ denotes the Kronecker δ's of the corresponding values of n_i's, i.e.,

$$\delta_{k,k'} = \delta_{n_1,n_1'} \delta_{n_2,n_2'} \delta_{n_3,n_3'} \qquad (1.4)$$

In the limit of $V \to \infty$, we have

$$\delta_{k,k'} \to \frac{(2\pi)^3}{V} \, \delta^{(3)}(k - k') \qquad (1.5)$$

$$\sum_k \to \int V \frac{d^3k}{(2\pi)^3} \qquad (1.6)$$

The completeness condition for the momentum eigenstates $|k\rangle$ can be written as

$$\int |k\rangle \frac{d^3k}{(2\pi)^3} \frac{1}{2k_0} \langle k| = 1 \qquad (1.7)$$

where $k_0 = \omega_k$.

The wave functions u_k are obviously solutions of the equation

$$i\frac{\partial u_k}{\partial t} = \sqrt{-\nabla^2 + m^2} \, u_k \qquad (1.8)$$

The differential operator on the right-hand side is not a local operator; it has to be understood in the sense of

$$\sqrt{-\nabla^2 + m^2} f(x) \equiv \int \frac{d^3k}{(2\pi)^3} \, e^{ik\cdot x} \sqrt{k^2 + m^2} f(k) \qquad (1.9)$$

where

$$f(x) = \int \frac{d^3k}{(2\pi)^3} e^{ik\cdot x} f(k) \qquad (1.10)$$

One can define a local differential equation for the u_k's; it is the Klein-Gordon equation

$$(\Box + m^2)u(x) = 0 \qquad (1.11)$$

where $\Box$ is the d'Alembertian operator, $\Box = \partial_\mu \partial^\mu = (\partial_0)^2 - \nabla^2$.

One can take the Klein-Gordon equation as the basic defining equation for the spinless particle and construct $u_k(x)$ as solutions to it. The inner product is then determined by the requirement that it be preserved under time-evolution according to the Klein-Gordon equation. The inner product for functions u, v obeying the Klein-Gordon equation is thus given by

$$(u|v) = \int d^3x \ [u^*i\partial_0 v - i\partial_0 u^* \ v] \tag{1.12}$$

The time-derivative of this gives

$$\begin{aligned}
\partial_0(u|v) &= \int d^3x \ [\partial_0 u^* i\partial_0 v - i\partial_0 u^* \ \partial_0 v + iu^* \partial_0^2 v - i\partial_0^2 u^* v] \\
&= \int d^3x \ [iu^*(\nabla^2 - m^2)v - i(\nabla^2 - m^2)u^* v] \\
&= \int d^3x \ \nabla \cdot [u^* i\nabla v - i\nabla u^* \ v] \\
&= \oint_{\partial V} dS \cdot [u^* i\nabla v - i\nabla u^* \ v] \\
&= 0 \tag{1.13}
\end{aligned}$$

The last equality follows from the periodic boundary conditions. We see that this inner product is preserved by time-evolution according to the Klein-Gordon equation; this is the reason that (1.12) is the correct choice and (1.3) is the correct form of the orthonormality condition to be used for this case.

1.3 Dirac equation

The basic variables are $\Psi_r(x)$, $r = 1, 2, 3, 4$, which can be thought of as a column vector. Each $\Psi_r(x)$ is a complex function of space and time. The Dirac equation is given by

$$(-i\gamma^\mu_{rs}\partial_\mu + m\delta_{rs})\Psi_s(x) = 0 \tag{1.14}$$

This can be written in a matrix notation as

$$(-i\gamma^\mu\partial_\mu + m\mathbf{1})\Psi(x) = 0 \tag{1.15}$$

Here $\mathbf{1}$ denotes the identity matrix , $\mathbf{1} = \delta_{rs}$. γ^μ are four matrices obeying the anticommutation rules, or the Clifford algebra relations,

$$\gamma^\mu\gamma^\nu + \gamma^\nu\gamma^\mu = 2\eta^{\mu\nu}\mathbf{1} \tag{1.16}$$

One set of matrices satisfying these relations is given by

$$\gamma^0 = \begin{pmatrix} 1 & 0 \\ 0 & -1 \end{pmatrix}, \qquad \gamma^i = \begin{pmatrix} 0 & \sigma^i \\ -\sigma^i & 0 \end{pmatrix} \tag{1.17}$$

The identity in the above expression for γ^0 is the 2×2-identity matrix. The gamma matrices are 4×4-matrices. σ^i are the Pauli matrices.

$$\sigma^1 = \begin{pmatrix} 0 & 1 \\ 1 & 0 \end{pmatrix}, \qquad \sigma^2 = \begin{pmatrix} 0 & -i \\ i & 0 \end{pmatrix}, \qquad \sigma^3 = \begin{pmatrix} 1 & 0 \\ 0 & -1 \end{pmatrix} \tag{1.18}$$

Clearly, a similarity transform of the above set of γ's will also obey the Clifford algebra. The fundamental theorem on Clifford algebras states that the only irreducible representation of the γ-matrices is given by the above set, up to a similarity transformation.

The Lagrangian for the Dirac equation is

$$\mathcal{L} = \bar{\Psi}(i\gamma \cdot \partial - m)\Psi \tag{1.19}$$

$\bar{\Psi}$ is related to the conjugate of Ψ as

$$\bar{\Psi} = \Psi^{\dagger}\gamma^0 \tag{1.20}$$

The Lorentz transformation of the Dirac spinor is given by

$$\Psi'(x) = S\,\Psi(L^{-1}x) \tag{1.21}$$

where $x'^{\mu} = (L)^{\mu}_{\nu}x^{\nu}$ is the Lorentz transformation of the coordinates. Infinitesimally, $x'^{\mu} \approx x^{\mu} + \omega^{\mu}_{\nu}x^{\nu}$, where $\omega^{\mu\nu} = -\omega^{\nu\mu}$ are the parameters of the Lorentz transformation. The transformation of the spinors is then given by

$$\Psi'(x) \approx \left(1 - \frac{i}{2}\omega^{\mu\nu}M_{\mu\nu}\right)\Psi(x) \tag{1.22}$$

$$M_{\mu\nu} = i(x_{\mu}\partial_{\nu} - x_{\nu}\partial_{\mu}) + S_{\mu\nu} \tag{1.23}$$

$S_{\mu\nu}$ is the spin term in $M_{\mu\nu}$,

$$S_{\mu\nu} = -\frac{1}{4i}[\gamma_{\mu}, \gamma_{\nu}] \tag{1.24}$$

By evaluating $S_{12} = S_3$, one can check that Ψ corresponds to spin $\frac{1}{2}$. Some further details on relativistic transformations are given in the appendix.

There are two types of plane wave solutions, those with $p_0 = \sqrt{\mathbf{p}^2 + m^2} \equiv E_p$ and those with $p_0 = -E_p = -\sqrt{\mathbf{p}^2 + m^2}$. They can be written as

$$\Psi(x) = u_r(p)\,e^{-ipx} = u_r(p)\,e^{-iEx^0 + i\mathbf{p}\cdot\mathbf{x}} \tag{1.25}$$

for the positive-energy solutions and

$$\Psi(x) = v_r(p)\,e^{ipx} = v_r(p)\,e^{iEx^0 - i\mathbf{p}\cdot\mathbf{x}} \tag{1.26}$$

for the negative-energy solutions. In these equations we have written the signs explicitly in the exponentials, so that p_0 in px is E for both cases.

The spinors $u_r(p)$, $v_r(p)$, $r = 1, 2$, are given by

$$u_r(p) = B(p)w_r, \qquad v_r(p) = B(p)\tilde{w}_r \tag{1.27}$$

where

$$w_1 = \begin{pmatrix} 1 \\ 0 \\ 0 \\ 0 \end{pmatrix}, \quad w_2 = \begin{pmatrix} 0 \\ 1 \\ 0 \\ 0 \end{pmatrix}, \quad \tilde{w}_1 = \begin{pmatrix} 0 \\ 0 \\ 1 \\ 0 \end{pmatrix}, \quad \tilde{w}_2 = \begin{pmatrix} 0 \\ 0 \\ 0 \\ 1 \end{pmatrix} \quad (1.28)$$

and

$$B(p) = \begin{pmatrix} \sqrt{\dfrac{E+m}{2m}} & \dfrac{\sigma \cdot p}{\sqrt{2m(E+m)}} \\ \dfrac{\sigma \cdot p}{\sqrt{2m(E+m)}} & \sqrt{\dfrac{E+m}{2m}} \end{pmatrix} \quad (1.29)$$

Here $E = \sqrt{p^2 + m^2}$ and we have used the representation for the gamma matrices given earlier.

It is easily seen that $B(p)$ is the boost transformation which takes us from the rest frame of the particle to the frame in which it has velocity $v^i = p^i/E$. From the Lorentz transformation properties, it is clear that $\Psi^\dagger \Psi$ is not Lorentz invariant. So we have chosen a Lorentz invariant normalization for the wave functions

$$\bar{u}_r(p)u_s(p) = \delta_{rs}, \qquad \bar{v}_r(p)v_s(p) = -\delta_{rs} \quad (1.30)$$

Using the definition of $B(p)$, we can establish the properties

$$\sum_r u_r(p)\bar{u}_r(p) = \frac{(\gamma \cdot p + m)}{2m}, \qquad \sum_r v_r(p)\bar{v}_r(p) = \frac{(\gamma \cdot p - m)}{2m} \quad (1.31)$$

The completeness relation for the solutions is expressed by

$$\sum_r u_r(p)\bar{u}_r(p) - v_r(p)\bar{v}_r(p) = 1 \quad (1.32)$$

Further

$$\bar{u}_r(p)\gamma^\mu u_s(p) = \frac{p^\mu}{m}\delta_{rs} = \bar{v}_r(p)\gamma^\mu v_s(p) \quad (1.33)$$

$$u_r^\dagger(p)v_s(p) = v_r^\dagger(p)u_s(-p) = 0 \quad (1.34)$$

The chirality matrix γ_5 is defined by

$$\begin{aligned} \gamma_5 &= i\gamma^0\gamma^1\gamma^2\gamma^3 \\ &= \frac{i}{4!}\epsilon_{\mu\nu\alpha\beta}\gamma^\mu\gamma^\nu\gamma^\alpha\gamma^\beta \end{aligned} \quad (1.35)$$

In the explicit representation of γ-matrices given above

$$\gamma_5 = \begin{pmatrix} 0 & 1 \\ 1 & 0 \end{pmatrix} \quad (1.36)$$

Another useful representation is

$$\gamma^0 = \begin{pmatrix} 0 & 1 \\ 1 & 0 \end{pmatrix}, \quad \gamma^i = \begin{pmatrix} 0 & \sigma^i \\ -\sigma^i & 0 \end{pmatrix}$$

$$\gamma^5 = \begin{pmatrix} -1 & 0 \\ 0 & 1 \end{pmatrix} \tag{1.37}$$

The left and right chirality projections are defined by

$$\Psi_L = \frac{1}{2}(1 + \gamma_5)\Psi, \qquad \Psi_R = \frac{1}{2}(1 - \gamma_5)\Psi \tag{1.38}$$

They correspond to eigenstates of γ_5 with eigenvalues ± 1, respectively.

References

1. Most of the results in this chapter are standard and can be found in almost any book on advanced quantum mechanics. A detailed book is W. Greiner, *Relativistic Quantum Mechanics: Wave Equations*, Springer-Verlag, 3rd edition (2000).
2. The basic theorem on representation of Clifford algebras is given in many of the general references, specifically, S. S. Schweber, *An Introduction to Relativistic Quantum Field Theory*, Harper and Row, New York (1961) and J. M. Jauch and F. Rohrlich, *The Theory of Photons and Electrons*, Springer-Verlag (1955 & 1976), to name just two. For an interesting discussion of spinors, see Appendix D of Michael Stone, *The Physics of Quantum Fields*, Springer-Verlag (2000).

2 The Construction of Fields

2.1 The correspondence of particles and fields

Ordinary point-particle quantum mechanics can deal with the quantum description of a many-body system in terms of a many-body wave function. However, there are many situations where the number of particles is not conserved, e.g., the β-decay of the neutron, $n \rightarrow p + e + \bar{\nu}_e$. There are also situations like $e^+e^- \rightarrow 2\gamma$ where the number of particles of a given species is not conserved, even though the number of particles of all types taken together is conserved. In order to discuss such processes, the usual formalism of many-body quantum mechanics, with wave functions for fixed numbers of particles, has to be augmented by including the possibility of creation and annihilation of particles via interactions. The resulting formalism is quantum field theory.

In many situations such as atomic and condensed matter physics, a nonrelativistic description will suffice. But for most applications in particle physics relativistic effects are important. Relativity necessarily brings in the possibility of conversion of mass into energy and vice versa, i.e., the creation and annihilation of particles. Relativistic many-body quantum mechanics necessarily becomes quantum field theory. Our goal is to develop the essentials of quantum field theory.

Quite apart from the question of creation and annihilation of particles, there is another reason to discuss quantized fields. We know of a classical field which is fundamental in physics, viz., the electromagnetic field. Analyses by Bohr and Rosenfeld show that there are difficulties in having a quantum description of various charged particle phenomena such as those that occur in atomic physics while retaining a classical description of the electromagnetic field. One has to quantize the electromagnetic field; this is independent of any many-particle interpretation that might emerge from quantization. Similar arguments can be made for quantizing the dynamics of other fields also.

There are two complementary approaches to field theory. One can postulate fields as the basic dynamical variables, discuss their quantum mechanics by diagonalization of the Hamiltonian operator, etc., and show that the result can be interpreted in many-particle terms. Alternatively, one can start with point-particles as the basic objects of interest and derive or construct the field operator as an efficient way of organizing the many-particle states.

We shall begin with the latter approach. We shall end up constructing a field operator for each type or species of particles. Properties of the particle will be captured in the transformation laws of the field operator under rotations, Lorentz transformations, etc. The one-to-one correspondence of species of particles and fields is exemplified by the following table.

Particle	Field
Spin-zero bosons	$\phi(\boldsymbol{x}, t)$, ϕ is a real scalar field
Charged spin-zero bosons	$\phi(\boldsymbol{x}, t)$, ϕ is a complex scalar field
Photons (spin-1, massless bosons)	$A_\mu(\boldsymbol{x}, t)$, real vector field (Electromagnetic vector potential)
Spin-$\frac{1}{2}$ fermions ($e^{\pm}$, quarks, etc.)	$\psi_r(\boldsymbol{x}, t)$, a spinor field

The simplest case to describe is the theory of neutral spin-zero bosons, so we shall begin with this.

2.2 Spin-zero bosons: construction of the field operator

We consider noninteracting spin-zero uncharged bosons of mass m. The wave function $u_k(x)$ for a single particle of four-momentum k_μ was given in Chapter 1. With the box normalization,

$$u_k(x) = \frac{e^{-ikx}}{\sqrt{2k_0 V}} \qquad (2.1)$$

The states of the system can evidently be represented as follows.

$|0\rangle$ = vacuum state, state with no particles.

$|1_k\rangle = |k\rangle$ = one-particle state of momentum $\boldsymbol{k}$, energy $k_0 = \sqrt{\boldsymbol{k}^2 + m^2} = \omega_k$.

$|1_{k_1}, 1_{k_2}\rangle = |k_1, k_2\rangle$ = two-particle state, with one particle of momentum $\boldsymbol{k}_1$ and one particle of momentum $\boldsymbol{k}_2$, with corresponding energies.

$|n_{k_1}, n_{k_2}, \ldots\rangle$ = many-particle state, with n_{k_1} particles of momentum $\boldsymbol{k}_1$, n_{k_2} particles of momentum $\boldsymbol{k}_2$, etc.

We now introduce operators which connect states with different numbers of particles. It is sufficient to concentrate on states $|0\rangle, |1_k\rangle, |2_k\rangle, \ldots |n_k\rangle$ with a fixed value of $\boldsymbol{k}$, introduce the connecting operators and then generalize to all $\boldsymbol{k}$. We thus define a particle annihilation operator a_k by

$$a_k|n_k\rangle = \alpha_n|n_k - 1\rangle \tag{2.2}$$

Since the vacuum has no particles, we require

$$a_k|0\rangle = 0 \tag{2.3}$$

The many-particle states are orthonormal, i.e.,

$$\langle 0|0\rangle = \langle 1_k|1_k\rangle = \langle 2_k|2_k\rangle = \ldots = 1 \tag{2.4}$$

$$\langle n_k|n_k'\rangle = 0, \qquad n_k \neq n_k' \tag{2.5}$$

From (2.2), we can then write, omitting the subscripts k for a while,

$$\langle n - 1|a|n\rangle = \alpha_n \tag{2.6}$$

Since $\langle \psi|A\phi\rangle = \langle A^\dagger\psi|\phi\rangle$ for an operator A, (2.6) gives

$$\langle a^\dagger(n - 1)|n\rangle = \alpha_n \tag{2.7}$$

This shows, with the orthogonality (2.5), that $a^\dagger|n-1\rangle$ must be proportional to $|n\rangle$. Thus $a^\dagger$ is a particle creation operator and we may write, from (2.7),

$$a^\dagger|n\rangle = \alpha_{n+1}^*|n + 1\rangle \tag{2.8}$$

The operators $aa^\dagger$ and $a^\dagger a$ are diagonal on the states. We have

$$a^\dagger a|n\rangle = |\alpha_n|^2 \, |n\rangle \tag{2.9}$$

Further, $a^\dagger a|0\rangle = 0$ using (2.3); thus $\alpha_0 = 0$.

The only quantum number characterizing the state $|n\rangle$, since we are look-ing at a fixed value of k, is the number of particles n. We shall thus identify $a^\dagger a$ as the number operator, i.e., the operator which counts the number of particles; this is the simplest choice and gives $\alpha_n = \sqrt{n}$. (An irrelevant phase is set to one.) Notice that $aa^\dagger$, the other diagonal operator, is not a suitable definiton of the number operator, since $\langle 0|aa^\dagger|0\rangle = 1$. With the identification of $a^\dagger a$ as the number operator, we have

$$a|n\rangle = \sqrt{n} \, |n - 1\rangle, \qquad a^\dagger|n\rangle = \sqrt{n + 1} \, |n + 1\rangle \tag{2.10}$$

These properties of a, $a^\dagger$ may be summarized by the commutation rules

$$[a, a] = 0, \qquad [a^\dagger, a^\dagger] = 0, \qquad [a, a^\dagger] = 1 \tag{2.11}$$

In fact, these commutation rules serve as the definitions of the operators a, $a^\dagger$. With the definiton of the vacuum by $a|0\rangle = 0$, $\langle 0|0\rangle = 1$, we can recursively build up all the states.

So far we have discussed one value of k. We can generalize the above discussion to all values of k by introducing a sequence of creation and anni-hilation operators with each pair being labeled by k. Thus we write

$$a_k | n_{k_1}, n_{k_2}, \ldots, n_k, \ldots \rangle = \sqrt{n_k} \; | n_{k_1}, n_{k_2}, \ldots, (n-1)_k, \ldots \rangle$$
$$a_k^\dagger | n_{k_1}, n_{k_2}, \ldots, n_k, \ldots \rangle = \sqrt{n_k + 1} \; | n_{k_1}, n_{k_2}, \ldots, (n+1)_k, \ldots \rangle$$

$$(2.12)$$

with the commutation rules

$$[a_k, a_l] = 0, \qquad [a_k^\dagger, a_l^\dagger] = 0, \qquad [a_k, a_l^\dagger] = \delta_{kl} \qquad (2.13)$$

Our discussion has so far concentrated on the abstract states, labeled by the momenta. It is possible to represent the above results in terms of the wave functions (2.1). We can actually combine the operators a_k, $a_k^\dagger$ with the one-particle wave functions $u_k(x)$ and *define* a field operator $\phi(x)$ by

$$\phi(x) = \sum_k \left[a_k \, u_k(x) \; + \; a_k^\dagger \, u_k^*(x) \right] \qquad (2.14)$$

Since u_k and u_k^* obey the Klein-Gordon equation, we see that $\phi(x)$ obeys the Klein-Gordon equation, viz.,

$$(\Box + m^2)\phi(x) = 0 \qquad (2.15)$$

As we noticed in Chapter 1, the wave functions actually obey the equation

$$i\frac{\partial}{\partial t} u_k = \sqrt{-\nabla^2 + m^2} \; u_k \qquad (2.16)$$

The operator $\sqrt{-\nabla^2 + m^2}$ is not a local operator. Since we would like to keep the theory as local as possible, we choose the second-order form of the equation. One may also wonder why we could not define a field operator just by the combination $\sum_k a_k u_k$ or its hermitian conjugate. The reason is that, once we decide on the Klein-Gordon equation rather than its first order version (2.16), the complete set of solutions include both the positive and negative frequency functions, i.e., both $u_k(x)$ and $u_k^*(x)$. Combining these together as in (2.14), we can reverse the roles of (2.14) and (2.15). We can postulate (2.15) as the fundamental equation for $\phi(x)$, and then the expansion of $\phi(x)$ in a complete set of solutions will give us (2.14). The coefficients of the mode expansion, viz., a_k, $a_k^\dagger$ are then taken as operators satisfying (2.13). This leads to a reconstruction of the many-particle description, but with the field $\phi(x)$ as the fundamental dynamical object. Notice that the negative frequency solutions, which are difficult to be interpreted as wave functions in one-particle quantum mechanics, now naturally emerge as being associated with the creation operators.

In terms of the field operator $\phi(x)$, the many-particle wave function for a state $| n_{k_1}, n_{k_2} \ldots \rangle$ may be written, up to a normalization factor, as

$$\Psi(x_1, x_2 \ldots x_N) = \langle 0 | \phi(x_1)\phi(x_2) \ldots | n_{k_1}, n_{k_2}, \ldots \rangle \qquad (2.17)$$

where $N = n_{k_1} + n_{k_2} + \ldots$. From the fact that the a_k's commute among themselves, we see that the wave function $\Psi(x_1, x_2 \ldots x_N)$ is symmetric under exchange of the positions of particles. The particles characterized by the commutation rules (2.13) are thus bosons.

To recapitulate, we have seen that we can introduce creation and annihilation operators on the Hilbert space of many-particle states. They obey the commutation rules (2.13); the field operator $\phi(x)$ is constructed out of these and obeys the Klein-Gordon equation. Conversely, one can postulate the field $\phi(x)$ as obeying the Klein-Gordon equation; expansion of $\phi(x)$ in a complete set of solutions gives (2.14). The amplitudes or coefficients of this expansion can then be taken as operators obeying (2.13). One can then recover the many-particle interpretation.

The field operator $\phi(x)$ is a scalar; it is hermitian and so, corresponds, classically to a scalar field which is real. The particles described by this field are bosons.

2.3 Lagrangian and Hamiltonian

The field operator $\phi(x)$ obeys the equation of motion

$$(\Box + m^2)\phi = 0 \tag{2.18}$$

If $\phi(x)$ were not an operator but an ordinary c-number field $\varphi(x)$, we could write down a Lagrangian and an action such that the corresponding variational equation (or extremization condition) is the Klein-Gordon equation (2.18). Such a Lagrangian is given by

$$\mathcal{L} = \tfrac{1}{2}\left[(\partial_\mu \varphi \partial^\mu \varphi) - m^2 \varphi^2\right] \tag{2.19}$$

with the action, for a spacetime volume Σ,

$$S = \int_\Sigma d^4x \, \mathcal{L} \tag{2.20}$$

The equation of motion can be derived as the condition satisfied by the fields which extremize the action S with fixed boundary values for the fields; i.e., as the condition $\delta S = 0$. We find

$$\delta S = \int_\Sigma d^4x \, \left[-(\Box + m^2)\varphi\right]\delta\varphi + \oint_{\partial\Sigma} d\sigma^\mu \, (\partial_\mu \varphi)\delta\varphi \tag{2.21}$$

We consider variations with the value of φ fixed on the boundary $\partial\Sigma$ of Σ. i.e., $\delta\varphi = 0$ on $\partial\Sigma$ and the extremization of the action gives the equations of motion

$$(\Box + m^2)\varphi = 0 \tag{2.22}$$

since $\delta\varphi$ is arbitrary in the interior of Σ.

Notice that the Lagrangian $\mathcal{L}$ is a Lorentz scalar. If we write the action as

$$S = \int dt\, d^3x\, \left[\tfrac{1}{2}(\partial_0\varphi)^2 - \tfrac{1}{2}\{(\nabla\varphi)^2 + m^2\varphi^2\}\right] \tag{2.23}$$

we see that it has the standard form $\int dt\,(T - U)$, with the kinetic energy $T = \int d^3x\, \tfrac{1}{2}(\partial_0\varphi)^2$ and potential energy $U = \int d^3x\, \tfrac{1}{2}[(\nabla\varphi)^2 + m^2\varphi^2]$. The Hamiltonian is given by

$$H = T + U = \int d^3x\, \tfrac{1}{2}\left[(\partial_0\varphi)^2 + (\nabla\varphi)^2 + m^2\varphi^2\right] \tag{2.24}$$

If we now replace the c-number field φ by the field operator $\phi(x)$, we get a Hamiltonian operator

$$H = \int d^3x\, \tfrac{1}{2}\left[(\partial_0\phi)^2 + (\nabla\phi)^2 + m^2\phi^2\right] \tag{2.25}$$

Use of the mode expansion (2.14) for $\phi(x)$ gives

$$H = \sum_k \omega_k\, a_k^\dagger a_k + \sum_k \tfrac{1}{2}\omega_k \tag{2.26}$$

where $\omega_k = k_0 = \sqrt{k^2 + m^2}$. Acting on the many-particle states, $a_k^\dagger a_k$ is the number of particles of momentum k, and thus H in (2.26) gives the energy of the state, except for the additional term $\sum_k \tfrac{1}{2}\omega_k$. This term is the energy of the vacuum state and is referred to as the zero-point energy. It arises because of the ambiguity of ordering of operators. The c-number expression (2.24) does not specify the ordering of a_k's and $a_k^\dagger$'s when we replace φ by the operator ϕ. We have to drop the zero-point term in (2.26) and define the Hamiltonian operator as

$$H = \sum_k \omega_k\, a_k^\dagger a_k \tag{2.27}$$

to obtain agreement with the many-particle description. Actually there are more fundamental reasons to subtract out the zero-point term as we have done. This has to do with the Lorentz invariance of the vacuum, as will be explained later. For the moment, we may take it as part of the rule of quantization, i.e., in replacing φ by the operator ϕ, we must choose the ordering of operators such that the vacuum energy is zero.

Analogous to the definition of the Hamiltonian, we can define a momentum operator

$$P_i = \sum_k k_i\, a_k^\dagger a_k \tag{2.28}$$

which can be checked to give the total momentum of a many-particle state.

The Lagrangian has essentially all the information about the theory; it gives the equations of motion, operators such as the Hamiltonian and momentum, the commutation rules, as we shall see later, and is a succinct way of specifying interactions, incorporating symmetries, etc. It will play a major role in all of what follows.

2.4 Functional derivatives

A mathematical notion which is very useful to all of our discussion is that of the functional derivative. The action S is a functional of the field $\varphi(x)$, i.e., its value depends on the specific function $\varphi(x)$ we use to evaluate it. More concretely, we may specify $\varphi(x)$ by an expansion in terms of a complete set of functions $f_n(x)$ as

$$\varphi(x) = \sum_n c_n \, f_n(x) \tag{2.29}$$

We can specify the function $\varphi(x)$ by giving the set of values $\{c_n\}$. One set of values $\{c_n\}$ gives one function, a different set $\{c_n'\}$ will give a different function and so on. Thus variation of the functional form of $\varphi(x)$ is achieved by variation of the c_n's; i.e.,

$$\varphi(x) \, + \, \delta\varphi(x) \, = \sum_n \, (c_n + \delta c_n) f_n(x) \tag{2.30}$$

$$\delta\varphi(x) = \sum_n \, \delta c_n \, f_n(x) \tag{2.31}$$

A functional, i.e., a quantity that depends on the functional form of another quantity $\varphi(x)$, can be written generically as

$$I[\varphi] = \int_\Sigma d^4x \, \rho(\varphi, \partial\varphi, \ldots) \tag{2.32}$$

For most of the applications in our discussions, we shall only need the variations of functionals like $I[\varphi]$ when we change φ in the interior of Σ, keeping the values of φ on the boundary fixed. This means that we can evaluate the variation of $I[\varphi]$ by carrying out partial integrations if necessary, using $\delta\varphi = 0$ on $\partial\Sigma$. The variation can then be brought to the form

$$\delta I[\varphi] = \int_\Sigma d^4x \, \sigma(x)\delta\varphi(x) \tag{2.33}$$

The functional derivative $\frac{\delta I}{\delta\varphi(x)}$ is then defined as $\sigma(x)$, the coefficient of $\delta\varphi(x)$. For example,

$$\frac{\delta \varphi(x)}{\delta \varphi(y)} = \delta^{(4)}(x - y)$$

$$\frac{\delta}{\delta \varphi(x)} \int_{\Sigma} d^4 y \, \varphi^2(y) = 2\varphi(x)$$

$$\frac{\delta}{\delta \varphi(x)} \int_{\Sigma} d^4 x \, (\partial \varphi)^2 = 2(-\Box \varphi(x)) \tag{2.34}$$

From (2.21,2.22), we see that

$$\frac{\delta S}{\delta \varphi(x)} = -(\Box + m^2)\varphi(x) \tag{2.35}$$

and the equation of motion is just $\frac{\delta S}{\delta \varphi} = 0$.

We shall now express a little more precisely the ideas of functional variations and derivatives. $\varphi(x)$ is real-valued, so let us define a space which is the set of all real-valued functions from the spacetime region Σ to $\mathbf{R}$, the real numbers. Since we shall be considering functionals like the action, which involve integrals of φ^2 and $(\partial \varphi)^2$, we require further that the functions we consider satisfy

$$\int_{\Sigma} d^4 x \, \varphi^2 < \infty, \qquad \int_{\Sigma} d^4 x \, (\partial \varphi)^2 < \infty \tag{2.36}$$

We may thus specify the function space $\mathcal{F}$ as

$$\mathcal{F} = \{\text{set of all } \varphi \text{'s such that } \varphi : \Sigma \to \mathbf{R},$$
$$\text{with the finiteness conditions (2.36)}\} \tag{2.37}$$

Elements of $\mathcal{F}$ are functions; if desired, one can also define a mode expansion which furnishes a basis for $\mathcal{F}$. A functional like the action is simply a map from $\mathcal{F}$ into the real numbers; i.e., it is a real-valued function on $\mathcal{F}$. The functional derivative is thus the usual notion of derivative applied to this function. Of course, the function space $\mathcal{F}$ is infinite-dimensional, since in general we need an infinite number of functions $f_n(x)$ to obtain a basis; as a result, one has to be careful about the convergence of sums and integrals.

The conditions (2.36) are relevant for the problem of the scalar field. In different physical situations, the conditions defining a suitable function space may be different. Likewise, the functions may not always be real-valued. In any case, it is clear that one can, in a way analogous to what we have done, define a suitable function space and functional derivatives.

2.5 The field operator for fermions

The wave functions for free spin-$\frac{1}{2}$ particles have been given in Chapter 1 as the solutions of the Dirac equation. We shall now introduce the creation and annihilation operators. Annihilation and creation operators for the particle are denoted by $a_{p,r}$ and $a^\dagger_{p,r}$, and those for the antiparticle are denoted by $b_{p,r}$ and $b^\dagger_{p,r}$. (r labels the spin states.) The important difference with the spin-zero case is that spin-$\frac{1}{2}$ particles are fermions. (This is part of a general result, which tells us that integral values of spin correspond to bosons and half-odd-integral values of spin to fermions. This "spin-statistics theorem" will be discussed later.) *For fermions, we have the exclusion principle; there cannot be double occupancy of any state.* Consider a fixed value of momentum and fixed spin state. Dropping indices for the moment, the states are $|0\rangle$, $c\,|1\rangle = a^\dagger|0\rangle$, where c is a normalization factor and $|2\rangle = (a^\dagger)^2|0\rangle \equiv 0$. Since there cannot be a two-particle occupancy of the state, we need $(a^\dagger)^2 = 0$, $(b^\dagger)^2 = 0$, which also gives

$$a^2 = 0, \qquad b^2 = 0 \tag{2.38}$$

The vacuum state or the state of no-particles $|0\rangle$ obeys

$$a|0\rangle = b|0\rangle = 0 \tag{2.39}$$

We can define $a^\dagger a$ as the particle number operator, as before. This leads to

$$a^\dagger a|0\rangle = 0, \qquad a^\dagger a|1\rangle = |1\rangle \tag{2.40}$$

This shows that $a|1\rangle = (1/c)|0\rangle$ and the above equation, along with this, gives $|c|^2 = 1$ from the orthonormality of states. We also have the results $\langle 0|aa^\dagger|0\rangle = |c|^2$ and $\langle 1|aa^\dagger|1\rangle = 0$. The combination $aa^\dagger + a^\dagger a$ is thus equal to one, on both the states $|0\rangle$ and $|1\rangle$. We shall thus use the anti-commutation rules

$$a^2 = 0, \qquad (a^\dagger)^2 = 0, \qquad aa^\dagger + a^\dagger a = 1 \tag{2.41}$$

for the operators a, $a^\dagger$, and similarly for the antiparticle operators. Notice that it is inconsistent to impose a rule like $aa^\dagger - a^\dagger a = constant$. The generalization of the rules (2.41) with momentum and spin labels is

$$a_{p,r}a^\dagger_{k,s} + a^\dagger_{k,s}a_{p,r} = \delta_{rs}\delta_{p,k}$$
$$b_{p,r}b^\dagger_{k,s} + b^\dagger_{k,s}b_{p,r} = \delta_{rs}\delta_{p,k}$$
$$a_{p,r}a_{k,s} + a_{k,s}a_{p,r} = 0, \qquad a^\dagger_{p,r}a^\dagger_{k,s} + a^\dagger_{k,s}a^\dagger_{p,r} = 0 \tag{2.42}$$
$$b_{p,r}b_{k,s} + b_{k,s}b_{p,r} = 0, \qquad b^\dagger_{p,r}b^\dagger_{k,s} + b^\dagger_{k,s}b^\dagger_{p,r} = 0$$
$$a_{p,r}b_{k,s} + b_{k,s}a_{p,r} = 0, \qquad a_{p,r}b^\dagger_{k,s} + b^\dagger_{k,s}a_{p,r} = 0$$
$$a^\dagger_{p,r}b_{k,s} + b_{k,s}a^\dagger_{p,r} = 0, \qquad a^\dagger_{p,r}b^\dagger_{k,s} + b^\dagger_{k,s}a^\dagger_{p,r} = 0$$

It can also be checked that, starting from these rules and defining the vacuum state by $a_{p,r}|0\rangle = b_{p,r}|0\rangle = 0$, we can recursively obtain all the multiparticle states of the fermions.

We now combine these operators with the one-particle wave functions to construct the fermion field operator. We can combine $u_r(p)\, e^{-ipx}$ with $a_{p,r}$. The solution $v_r(p)\, e^{ipx}$ has an exponential e^{iEt}, indicating that it must be interpreted as the conjugate wave function, corresponding to creation of particles. It must be combined with a creation operator. However, we cannot use $a^\dagger_{p,r}$; if we do, the combination $a_{p,r} u_r(p)\, e^{-ipx} + a^\dagger_{p,r} v_r(p)\, e^{ipx}$ does not have definite fermion number or charge, since one term annihilates particles (a process with a change of -1 for fermion number) and the other term creates them (a process with a change of $+1$ for fermion number). We must thus use $b^\dagger_{p,r}$; this is consistent since annihilating particles and creating antiparticles change charge or fermion number by the same amount. The field operator is thus given by

$$\psi(x) = \sum_{p,r} \sqrt{\frac{m}{E_p V}} \left[a_{p,r}\, u_r(p)e^{-ipx} + b^\dagger_{p,r} v_r(p)e^{ipx}\right]$$

$$\bar\psi(x) = \sum_{p,r} \sqrt{\frac{m}{E_p V}} \left[a^\dagger_{p,r}\, \bar u_r(p)e^{ipx} + b_{p,r} \bar v_r(p)e^{-ipx}\right] \qquad (2.43)$$

These obey the equations

$$(i\gamma \cdot \partial - m)\psi = 0, \qquad -i\partial_\mu \bar\psi\gamma^\mu - m\bar\psi = 0 \qquad (2.44)$$

We have used the complete set of solutions to the equations (2.44); one may therefore think of (2.44) as the starting point. Writing the general mode expansion for the fields ψ and $\bar\psi$, one can interpret the coefficients as operators obeying the anti-commutation rules (2.42) and thus recover the many-particle picture.

References

1. The formalism of creation and annihilation operators for particles goes back to Dirac's 1927 paper on the absorption and emission of radiation. Anticommutation rules were introduced by Jordan and Wigner in 1929. These have become such staple fare of physics, and even chemistry where they have been used for reaction kinetics, that citing original articles is somewhat irrelevant in a book which does not claim to trace the historical development of the subject. For the historical development of the subject, see S. S. Schweber, *QED and the Men Who Made It*, Princeton University Press (1994). Many of the original papers are easily accessible in the reprint collection, J. Schwinger, *Selected Papers in Quantum Electrodynamics*, Dover Publications, Inc. (1958).
2. The Bohr-Rosenfeld analyses are in N. Bohr and L. Rosenfeld, Kgl. Danske. Vidensk. Selsk. Mat-Fys. Medd, **12**, No. 8, (1933); Phys. Rev. **78**, 794 (1950).

3 Canonical Quantization

3.1 Lagrangian, phase space, and Poisson brackets

In this chapter we develop the essentials of canonical quantization. Instead of constructing fields in terms of particle wave functions, we consider fields as the fundamental dynamical variables and discuss how to obtain a quantum theory of fields.

We shall first consider bosonic fields. The fields will be denoted by $\varphi_r(x)$. The index r or part of it may be a spacetime index for vector and tensor fields; it can also be an internal index labeling the number of independent fields. The Lagrangian $\mathcal{L}$ is a scalar function of $\varphi_r(x)$ and its spacetime derivatives. We shall assume that the equations of motion are at most second order in the time-derivatives. Correspondingly, $\mathcal{L}$ involves at most $(\partial_0\varphi)^2$. This is the most relevant case. If the equations of motion involve higher-order time-derivatives of the fields, there are usually unphysical ghost modes (modes which have negative norm in the quantum theory). (There is a generalization of the canonical formalism for theories with higher than first-order derivatives in time; this is due to Ostrogradskii.) Higher powers of $(\partial_0\varphi)$ also generally lead to difficulties in quantization and do not seem to be relevant for any realistic situation. We shall not discuss these situations further.

Since the Lagrangian has at most the square of $(\partial_0\varphi)$, we expect, based on Lorentz invariance, that $\mathcal{L}$ is at most quadratic in space-derivatives as well. (There are some topological Lagrangians with one time-derivative and several different space-derivatives of fields. We will not consider them here; some examples are briefly discussed in Chapter 20 which describes geometric quantization.) The action in a spacetime volume Σ can be written as

$$\mathcal{S} = \int_\Sigma d^4x \, \mathcal{L}(\varphi_r, \partial_\mu\varphi_r) \tag{3.1}$$

The spacetime region will be taken to be of the form $V \times [t_f, t_i]$, where V is a spatial region. The equations of motion are given by the variational principle, viz., the classical trajectory $\varphi_r(\boldsymbol{x}, t)$, which connects specified initial and final field configurations $\varphi_r(\boldsymbol{x}, t_i)$ and $\varphi_r(\boldsymbol{x}, t_f)$ at times t_i and t_f, extremizes the action. In other words, we can vary the action with respect to $\varphi(\boldsymbol{x}, t)$ for $t_i < t < t_f$ and set $\delta\mathcal{S}$ to zero to obtain the equations of motion. Explicitly

$$\delta \mathcal{L} = \frac{\partial \mathcal{L}}{\partial \varphi_r} \delta \varphi_r + \frac{\partial \mathcal{L}}{\partial (\partial_\mu \varphi_r)} \partial_\mu \delta \varphi_r$$

$$= \left[\frac{\partial \mathcal{L}}{\partial \varphi_r} - \frac{\partial}{\partial x^\mu} \frac{\partial \mathcal{L}}{\partial (\partial_\mu \varphi_r)} \right] \delta \varphi_r + \frac{\partial}{\partial x^\mu} \left(\frac{\partial \mathcal{L}}{\partial (\partial_\mu \varphi_r)} \delta \varphi_r \right) \quad (3.2)$$

(Summation over the repeated index, in this case r, is assumed as usual.) When we integrate the variation of $\mathcal{L}$ over the spacetime region Σ to obtain δS, the second term in (3.2), being a total divergence, becomes a surface integral over $\partial \Sigma$. Since we fix the initial and final field configurations $\varphi_r(\boldsymbol{x}, t_i)$ and $\varphi_r(\boldsymbol{x}, t_f)$, $\delta \varphi_r = 0$ at t_i, t_f. Further, we assume that either $\delta \varphi_r$ or $\frac{\partial \mathcal{L}}{\partial (\partial_i \varphi_r)}$ vanishes at the spatial boundary ∂V. Eventually, we are interested in the limit of large spatial volumes; this condition is physically quite reasonable in this case; alternatively, we could require periodic boundary conditions for the spatial directions. Either way the surface integral is zero and

$$\delta S = \int_\Sigma d^4 x \left[\frac{\partial \mathcal{L}}{\partial \varphi_r} - \frac{\partial}{\partial x^\mu} \frac{\partial \mathcal{L}}{\partial (\partial_\mu \varphi_r)} \right] \delta \varphi_r \quad (3.3)$$

The extremization condition $\delta S = 0$ now yields the equations of motion, since $\delta \varphi_r$ is arbitrary, as

$$\frac{\partial \mathcal{L}}{\partial \varphi_r} - \frac{\partial}{\partial x^\mu} \frac{\partial \mathcal{L}}{\partial (\partial_\mu \varphi_r)} = 0 \quad (3.4)$$

We now consider more general variations of fields, with $\delta \varphi_r$ not zero at t_i or t_f. The total divergence term in (3.2) integrates out to $\Theta(t_f) - \Theta(t_i)$, where

$$\Theta(t) = \int_V d^3 x \frac{\partial \mathcal{L}}{\partial (\partial_0 \varphi_r)} \delta \varphi_r \quad (3.5)$$

This quantity Θ is called the canonical one-form.

In the variation of the action when using the variational principle, we specify the initial and final values of the field configurations. Since there is then a unique classical trajectory, we may say that the initial and final values label the classical trajectories. The set of all classical trajectories is defined to be the phase space of the theory. Alternatively, we can specify the classical trajectories by the initial data for the equations of motion rather than initial and final values for the field. Since our equations are second order in time-derivatives, the initial data are clearly $\varphi_r(\boldsymbol{x}, t)$ and $\partial_0 \varphi_r(\boldsymbol{x}, t)$, at some starting time t. It will be more convenient for the formalism to use

$$\pi_r(\boldsymbol{x}, t) = \frac{\partial \mathcal{L}}{\partial (\partial_0 \varphi_r)} \quad (3.6)$$

rather than $\partial_0 \varphi_r$. The phase space for a set of scalar fields is thus equivalent to the set $\{\pi_r(\boldsymbol{x}), \varphi_r(\boldsymbol{x})\}$ (for all $\boldsymbol{x}$) which is used to label the classical trajectories. The phase space for a field theory is obviously infinite-dimensional. π_r is called the canonical momentum conjugate to φ_r.

The canonical one-form Θ can be written as

$$\Theta = \int_V d^3x \; \pi_r \delta\varphi_r \tag{3.7}$$

(The name is due to the fact that this is a differential one-form on the phase space, as will be explained in Chapter 20.) We will denote the phase space variables (coordinates on the phase space) by $\xi^i(\boldsymbol{x})$ for a general dynamical system, which could be more general than a scalar field theory. The canonical one-form Θ is identified from the surface term in the variation of the action and has the general form

$$\Theta = \int d^3x \; A_i(\xi, \boldsymbol{x}) \; \delta\xi^i(\boldsymbol{x}) \tag{3.8}$$

where A_i could depend on ξ. (For the scalar field $\xi^i = (\pi_r, \varphi_r)$ and $A_i = (\pi_r, 0)$.) Given Θ, we define

$$\Omega_{ij}(\boldsymbol{x}, \boldsymbol{x}') = \frac{\delta}{\delta\xi^i(\boldsymbol{x})} A_j(\boldsymbol{x}') - \frac{\delta}{\delta\xi^j(\boldsymbol{x}')} A_i(\boldsymbol{x})$$
$$= \partial_I A_J - \partial_J A_I = -\Omega_{ji}(\boldsymbol{x}', \boldsymbol{x}) \tag{3.9}$$

where in the last line, we have introduced the composite indices $I = (i, \boldsymbol{x})$ and $J = (j, \boldsymbol{x}')$ and $\partial_I = \delta/\delta\xi^i(\boldsymbol{x})$ to avoid clutter in the notation. Ω is called the *symplectic structure* or the canonical two-form. (It can be considered as a differential form on the space of fields and their time-derivatives.) Just as the metric tensor defines the basic geometric structure for any spacetime, Ω defines the basic geometric structure of the phase space. Notice that from the definition of Ω, we have the Bianchi identity

$$\partial_I \Omega_{JK} + \partial_J \Omega_{KI} + \partial_K \Omega_{IJ} = 0 \tag{3.10}$$

A concept of central importance in canonical quantization is that of a canonical transformation and the generator associated with it. Let $\xi^i \to \xi^i + a^i(\xi)$ be an infinitesimal transformation of the canonical variables. This transformation is called canonical if it preserves the canonical structure Ω. The change in Ω arises from two sources, firstly due to the ξ-dependence of the components Ω_{IJ} and secondly due to the fact that Ω_{IJ} transforms under change of phase space coordinate frames. (Ω_{IJ} transforms as a covariant rank-two tensor under change of coordinates.) The total change is

$$\delta\Omega_{IJ} = \partial_I \alpha_J - \partial_J \alpha_I$$
$$\delta\Omega_{ij}(\boldsymbol{x}, \boldsymbol{x}') = \left[\frac{\delta}{\delta\xi^i(\boldsymbol{x})} \alpha_j(\boldsymbol{x}') - \frac{\delta}{\delta\xi^j(\boldsymbol{x}')} \alpha_i(\boldsymbol{x}) \right] \tag{3.11}$$

where

$$\alpha_I = a^K \Omega_{KI}$$

$$\alpha_i(\boldsymbol{x}) = \int_V d^3x' \, a^k(\boldsymbol{x}') \Omega_{ki}(\boldsymbol{x}', \boldsymbol{x}) \tag{3.12}$$

(In these equations, we have expanded out the composite notation to show how it works out.) From (3.11, 3.12), we see that the transformation $\xi^i \rightarrow \xi^i + a^i(\xi)$ will preserve Ω and hence be a canonical transformation, if

$$\alpha_I \equiv a^K \Omega_{KI} = -\partial_I \, G \tag{3.13}$$

for some function G of the phase space variables. G so defined is called the generator of the canonical transformation. (Equation (3.13) is a necessary and sufficient condition locally on the phase space. If the phase space has nontrivial topology, the vanishing of $\delta\Omega$ may have more general solutions. Even though locally all solutions look like (3.13), G may not exist globally on the phase space. We shall return to the case of nontrivial topology in later chapters.)

If we add a total divergence $\partial_\mu F^\mu$ to the Lagrangian, the equations of motion do not change, but Θ changes as $\Theta \rightarrow \Theta + \delta \int d^3x \, F^0$. This is of the form (3.13) with $A_I \rightarrow A_I + \partial_I \int F^0$ and hence Ω is unchanged. Thus the addition of total derivatives to a Lagrangian is an example of a canonical transformation.

The inverse of Ω is defined by $(\Omega^{-1})^{IJ} \Omega_{JK} = \delta^I_K$ which expands out as

$$\int_V d^3x' (\Omega^{-1})^{ij}(\boldsymbol{x}, \boldsymbol{x}') \Omega_{jk}(\boldsymbol{x}', \boldsymbol{x}'') = \delta^i_k \, \delta^{(3)}(x - x'') \tag{3.14}$$

As will be clear from the following discussion, it is important to have an invertible Ω_{IJ}. If Ω is not invertible, the Lagrangian is said to be singular. There are many interesting cases, e.g., theories with gauge symmetries, where it is not possible to define an invertible Ω in terms of the obvious field variables. One has to define a nonsingular Ω in such cases, by suitable elimination of redundant degrees of freedom. (A gauge theory is an example of this; the redundant variables are eliminated by the procedure of gauge-fixing.)

Using the inverse of Ω, we can rewrite (3.13) with an Ω^{-1} on the right-hand side as

$$a^I = (\Omega^{-1})^{IJ} \partial_J G \tag{3.15}$$

The discussion from equation (3.11) to (3.15) shows that to every infinitesimal canonical transformation, modulo the topological issues mentioned above, we can associate a function G on the phase space, and conversely, given any function G we can associate to it an infinitesimal canonical transformation.

The change of any function under the transformation $\xi^I \rightarrow \xi^I + a^I$ is given by the action of the functional differential operator $V_a = a^I \partial_I$. The commutator of two such transformations is given by

$$[V_a, V_b] = \left(a^J \partial_J b^I - b^J \partial_J a^I \right) \partial_I \tag{3.16}$$

Let F, G be the functions associated, via (3.15), with a^I and b^I, respectively. We then find

$$a^J \partial_J b^I - b^J \partial_J a^I = \left[(\Omega^{-1})^{JL} \partial_J (\Omega^{-1})^{IK} - (\Omega^{-1})^{JK} \partial_J (\Omega^{-1})^{IL} \right] \partial_L F\, \partial_K G$$
$$+ (\Omega^{-1})^{JL} (\Omega^{-1})^{IK} \left[\partial_L F \partial_J \partial_K G - \partial_J \partial_K F \partial_L G \right] \tag{3.17}$$

$$(\Omega^{-1})^{JL} \partial_J (\Omega^{-1})^{IK} - (K \leftrightarrow L) = (\Omega^{-1})^{IM} \, (\Omega^{-1})^{JL} (\Omega^{-1})^{NK}$$
$$\times \left[\partial_N \Omega_{MJ} + \partial_J \Omega_{NM} \right]$$
$$= -(\Omega^{-1})^{IM} \, (\Omega^{-1})^{JL} (\Omega^{-1})^{NK} \partial_M \Omega_{JN}$$
$$= (\Omega^{-1})^{IM} \partial_M (\Omega^{-1})^{KL} \tag{3.18}$$

where we have used the identity (3.10). This can now be used to simplify (3.17) as

$$a^J \partial_J b^I - b^J \partial_J a^I = (\Omega^{-1})^{IM} \partial_M \left[(\Omega^{-1})^{KL} \partial_K G \partial_L F \right] \tag{3.19}$$

This shows that if $a \leftrightarrow F$, $b \leftrightarrow G$, the commutator of the corresponding infinitesimal transformations corresponds to a function $-\{F, G\}$, where

$$\{F, G\} = (\Omega^{-1})^{IJ} \partial_I F \partial_J G$$
$$= \int d^3x\, d^3x' \, (\Omega^{-1})^{ij} (\boldsymbol{x}, \boldsymbol{x}') \frac{\delta F}{\delta \xi^i(\boldsymbol{x})} \frac{\delta G}{\delta \xi^j(\boldsymbol{x}')} \tag{3.20}$$

The function $\{F, G\}$ is called the Poisson bracket of the functions F and G. It arises naturally in the composition of canonical transformations. For the ξ^I's themselves, we find $\{\xi^I, \xi^J\} = (\Omega^{-1})^{IJ}$ or

$$\{\xi^i(\boldsymbol{x}), \xi^j(\boldsymbol{x}')\} = (\Omega^{-1})^{ij}(\boldsymbol{x}, \boldsymbol{x}') \tag{3.21}$$

Notice also that if Θ has the simple form (3.7), the Poisson bracket of two functions $F(\pi, \varphi)$, $G(\pi, \varphi)$ of the phase space variables becomes

$$\{F, G\} = \int_V d^3x \left[\frac{\delta F}{\delta \varphi_r} \frac{\delta G}{\delta \pi_r} - \frac{\delta F}{\delta \pi_r} \frac{\delta G}{\delta \varphi_r} \right] \tag{3.22}$$

Comparing equation (3.13) with the definition of Poisson brackets, we see that it is equivalent to

$$a^I \equiv \delta \xi^I = \{\xi^I, G\} \tag{3.23}$$

In fact, this equation may be taken as the definition of the generator. Conversely, for any function G on the phase space, the transformations on ξ^I defined by (3.23), i.e., Poisson brackets with G, are canonical. Notice that for the simple case of $\xi^i = (\pi_r, \varphi_r)$, (3.23) is equivalent to

$$\delta \varphi_r(\boldsymbol{x}) = \frac{\delta G}{\delta \pi_r(\boldsymbol{x})}, \qquad \delta \pi_r(\boldsymbol{x}) = -\frac{\delta G}{\delta \varphi_r(\boldsymbol{x})} \tag{3.24}$$

More generally, the change of any function F under a canonical transformation generated by G is given by

$$\delta F = a^I \partial_I F = (\Omega^{-1})^{IJ} \partial_J G \partial_I F$$
$$= \{F, G\} \tag{3.25}$$

These equations show why Poisson brackets are important. The change of any variable, so long as it is canonical, is given by the Poisson bracket of the variable with the generating function for the transformation.

We now find the generators of some important canonical transformations.

1. *Change of $\varphi_r(\boldsymbol{x})$.*
   For $\varphi_r \to \varphi_r + a_r(\boldsymbol{x})$, $\pi_r \to \pi_r$,

$$G = \int_V d^3x \, a_r(\boldsymbol{x}) \pi_r(\boldsymbol{x}) \tag{3.26}$$

2. *Change of $\pi_r(\boldsymbol{x})$.*
   For $\varphi_r \to \varphi_r$, $\pi_r \to \pi_r + a_r(\boldsymbol{x})$,

$$G = - \int_V d^3x \, a_r(\boldsymbol{x}) \varphi_r(\boldsymbol{x}) \tag{3.27}$$

3. *Space translations.*
   For $x^i \to x^i + a^i$, a^i being constants, $\delta\varphi_r = a^i \partial_i \varphi_r$, $\delta\pi_r = a^i \partial_i \pi_r$ and

$$G = \int_V d^3x \, a^i \partial_i \varphi_r \pi_r = a^i P_i \tag{3.28}$$

$$P_i = \int_V d^3x \, \partial_i \varphi_r \pi_r \tag{3.29}$$

The generator of space translations, P_i, is the momentum of the system.

4. *Time translations.*
   The generator of time translations is the Hamiltonian $H(\pi, \varphi)$; this is the definition of the Hamiltonian. From (3.24), this means that the equations of motion should be of the form

$$\partial_0 \varphi_r = \frac{\delta H}{\delta \pi_r}, \qquad \partial_0 \pi_r = -\frac{\delta H}{\delta \varphi_r} \tag{3.30}$$

One can easily see that

$$H = \int_V d^3x \, (\pi_r \partial_0 \varphi_r - \mathcal{L}) \tag{3.31}$$

The easiest way to check this is to use (3.31) to write the action as

$$S = \int d^4x \ \pi_r \partial_0 \varphi_r - \int dt \ H \tag{3.32}$$

and then use the variational principle to write the equations of motion. The equations of motion so obtained are seen to be (3.30), showing the consistency of (3.31) as the generator of time translations.

The Hamiltonian and momentum components can be expressed in terms of an energy-momentum tensor $T_{\mu\nu}$ defined by

$$T_{\mu\nu} = \partial_\mu \varphi_r \frac{\partial \mathcal{L}}{\partial(\partial^\nu \varphi_r)} - \eta_{\mu\nu} \mathcal{L} + \partial^\alpha B_{\alpha\mu\nu} \tag{3.33}$$

where $B_{\alpha\mu\nu}$ is related to spin contributions. (We discuss this a little later in this chapter.) In terms of $T_{\mu\nu}$,

$$P_\mu = (H, P_i) = \int_V d^3x \ T_{\mu 0} \tag{3.34}$$

Notice that the tensor $B_{\alpha\mu\nu}$ does not contribute to the expressions for P_μ.

5. *Lorentz transformations.*

For Lorentz transformations, $\delta x^\mu = \omega^{\mu\nu} x_\nu$. The generator of Lorentz transformations can be checked to be

$$M_{\mu\nu} = \int_V d^3x \ (x_\mu T_{\nu 0} - x_\nu T_{\mu 0}) \tag{3.35}$$

3.2 Rules of quantization

As with any quantum mechanical system, the states are represented by vectors (actually rays) in a Hilbert space $\mathcal{H}$. The scalar product $\langle \varphi | \alpha \rangle = \Psi_\alpha[\varphi]$ is the wave function of the state $|\alpha\rangle$ in a φ-diagonal representation; it is the probability amplitude for finding the field configuration $\varphi(x)$ in the state $|\alpha\rangle$.

Observables are represented by linear hermitian operators on $\mathcal{H}$. Fields are in general linear operators on $\mathcal{H}$, not necessarily always hermitian or observable. We have the operator $\phi_r(x, t)$ corresponding to $\varphi_r(x, t)$ and the operator $\pi_r(x, t)$ corresponding to the canonical momentum.

The change of any operator F under any infinitesimal unitary transformation of the Hilbert space is given by

$$i \ \delta F = FG - GF = [F, G] \tag{3.36}$$

where G is the generator of the transformation; it is a hermitian operator. If we were to start directly with the quantum theory, we can regard this as the basic postulate. The fact that observables are linear hermitian operators follow from this because observations or measurements correspond to infinitesimal unitary transformations of the Hilbert space.

However, in starting from a classical theory and quantizing it, we need a rule relating the operator structure to the classical phase space structure. The basic rule is that, in passing to the quantum theory, *canonical transformations should be represented as unitary transformations on the Hilbert space*. The generator of the unitary transformation is obtained by replacing the fields in the classical canonical generator by the corresponding operators. (This replacement rule has ambiguities of ordering of operators; e.g., classically, $\pi_r \varphi_r$ and $\varphi_r \pi_r$ are the same, but the corresponding quantum versions $\pi_r \phi_r$ and $\phi_r \pi_r$ are not the same, since ϕ_r and π_r do not necessarily commute. The correct ordering for the quantum theory can sometimes be understood on grounds of desirable symmetries. There is no general rule.)

Comparing the rule (3.25) for the change of a function under a canonical transformation with the rule (3.36) for the change of an operator under a unitary transformation, we see that $-i[F, G]$ should behave as the Poisson bracket $\{F, G\}$ in going to the classical limit. Therefore the commutator algebra of the operators, apart from ordering problems mentioned above, will be isomorphic to the Poisson bracket algebra of the corresponding classical functions.

The finite version of (3.36) is

$$F' = e^{iG} \, F \, e^{-iG} \tag{3.37}$$

The transformation law for states is given by

$$|\alpha'\rangle = e^{iG}|\alpha\rangle \tag{3.38}$$

Equations (3.37) and (3.38) say that classical canonical transformations are realized as unitary transformations in the quantum theory.

Many useful results follow from (3.36) to (3.38). From the generators (3.26) and (3.27) of changes in φ_r and π_r, we find, using (3.36),

$$[\phi_r(\boldsymbol{x}, t), \phi_s(\boldsymbol{x}', t)] = 0$$
$$[\pi_r(\boldsymbol{x}, t), \pi_s(\boldsymbol{x}', t)] = 0$$
$$[\phi_r(\boldsymbol{x}, t), \pi_s(\boldsymbol{x}', t)] = i \, \delta_{rs} \, \delta^{(3)}(x - x') \tag{3.39}$$

These give us the basic commutation rules, sometimes called the canonical commutation rules, to be imposed on the operators of the theory. (More generally, we would have $[\xi^i(\boldsymbol{x}, t), \xi^j(\boldsymbol{x}', t)] = i(\Omega^{-1})^{ij}(\boldsymbol{x}, \boldsymbol{x}')$.)

The generator of time-translations is the Hamiltonian and we get from (3.36)

$$i\frac{\partial F}{\partial t} = [F, H] \tag{3.40}$$

This is the quantum equation of motion, called the Heisenberg equation of motion.

Using the canonical commutation rules, one can also work out the commutator algebra of various operators of interest. For example, using expressions

(3.34,3.35) and replacing the fields and their canonical momenta by operators, we get the operators P_μ, $M_{\mu\nu}$, which give the action of the Poincaré transformations on any quantity in the quantum theory as in (3.36). In particular, using the canonical commutation rules, one can check that these operators obey the Poincaré algebra commutation relations given in the appendix.

3.3 Quantization of a free scalar field

We now apply the rules of quantization to obtain the theory of a free scalar field φ. The Lagrangian is

$$\mathcal{L} = \tfrac{1}{2}\left[(\partial\varphi)^2 - m^2\varphi^2\right] \tag{3.41}$$

In the quantum theory, the field becomes an operator $\phi(\boldsymbol{x}, t)$. The canonical momentum is $\pi(\boldsymbol{x}, t) = \partial_0\phi(\boldsymbol{x}, t)$. The Hamiltonian is

$$H = \int d^3x \, \tfrac{1}{2}\left[\pi^2 + (\nabla\phi)^2 + m^2\phi^2\right] \tag{3.42}$$

The basic commutation rules are

$$[\phi(\boldsymbol{x}, t), \phi(\boldsymbol{x}', t)] = 0$$
$$[\pi(\boldsymbol{x}, t), \pi(\boldsymbol{x}', t)] = 0$$
$$[\phi(\boldsymbol{x}, t), \pi(\boldsymbol{x}', t)] = i\delta^{(3)}(x - x') \tag{3.43}$$

The Heisenberg equation of motion becomes, using (3.42,3.43),

$$(\Box + m^2)\phi = 0 \tag{3.44}$$

The field operator obeys the Klein-Gordon equation.

Since ϕ commutes with itself, it is possible to choose a ϕ-diagonal representation where

$$\phi|\varphi\rangle = \varphi(\boldsymbol{x})|\varphi\rangle \tag{3.45}$$

Here $\varphi(\boldsymbol{x})$ is some c-number field configuration which is the eigenvalue for $\phi(\boldsymbol{x}, t)$. In this case, we can write $\pi(\boldsymbol{x}) = -i\delta/\delta\varphi(\boldsymbol{x})$. This is the analog of the Schrödinger representation. We can in fact understand the theory by writing the Schrödinger equation, which would be a functional differential equation in this case, and solving it for the eigenstates of the Hamiltonian. However, the diagonalization of the Hamiltonian is most easily done in another representation where we solve the equation of motion (3.44). (Evidently, we are also using the Heisenberg picture where operators evolve with time.) The solutions are obviously plane waves. Choosing a normalization as we have done in Chapter 1, we can thus write the general solution to (3.44) as

$$\phi(x) = \sum_k [a_k u_k(x) + a_k^\dagger u_k^*(x)] \tag{3.46}$$

where

$$u_k(x) = \frac{e^{-ikx}}{\sqrt{2\omega_k \, V}} \tag{3.47}$$

($\omega_k = \sqrt{k^2 + m^2}$.) (Notice that the u_k, u_k^* appear here merely as mode functions for the expansion of a general solution of the equation of motion.) The fact that we have an operator is accounted for by considering the coefficients of the expansion a_k, $a_k^\dagger$ to be operators. Notice that since we have a real field classically, we need a hermitian field operator and so the coefficient of $u_k^*(x)$ in (3.46) must be the hermitian conjugate of a_k. By using the orthogonality property of the $u_k(x)$, $u_k^*(x)$ we have

$$a_k = \int d^3x \, u_k^*(x)(\omega_k \, \phi + i\pi), \qquad a_k^\dagger = \int d^3x \, u_k(x)(\omega_k \, \phi - i\pi) \tag{3.48}$$

With these expressions, we can obtain the commutation rules for a_k, $a_k^\dagger$ using the fundamental commutation rules (3.43). We find

$$\begin{aligned} \left[a_k, a_l\right] &= 0 \\ \left[a_k^\dagger, a_l^\dagger\right] &= 0 \\ \left[a_k, a_l^\dagger\right] &= \delta_{kl} \end{aligned} \tag{3.49}$$

The commutation rules for a_k, $a_k^\dagger$ are the same as for the creation and annihilation operators. These rules were obtained in Chapter 2 by considerations of the many-particle states. Here they emerge as the fundamental rules of quantization for the field $\phi(\boldsymbol{x}, t)$, which is the dynamical degree of freedom.

The mode expansion for the canonical momentum π is obtained from the mode expansion (3.46) for ϕ as $\partial_0 \phi$. We can then evaluate the Hamiltonian as

$$H = \sum_k \tfrac{1}{2}\omega_k(a_k a_k^\dagger + a_k^\dagger a_k) \; = \; \textstyle\sum_k [\omega_k a_k^\dagger a_k + \tfrac{1}{2}\omega_k] \tag{3.50}$$

Similarly, the momentum operator P_i is

$$P_i \equiv \int \partial_i \phi \pi \; = \; \sum_k \tfrac{1}{2}k_i(a_k a_k^\dagger + a_k^\dagger a_k) \; = \; \textstyle\sum_k k_i a_k^\dagger a_k \tag{3.51}$$

(We have used the commutation rules and $\sum_k k_i = 0$ to simplify the expressions. Strictly speaking, such expressions have to be defined by regulating the sum, which can be done by defining partial sums over N modes and then taking the limit $N \to \infty$ eventually. For the momentum operator, we are using a reflection symmetric way of doing this, so that the contribution due to $\boldsymbol{k}$ is cancelled by the contribution due to $-\boldsymbol{k}$.)

We are now in a position to interpret these results. Apart from the constant $\frac{1}{2}\omega_k$-term, the Hamiltonian involves the positive operator $a^\dagger a$. This is positive since $\langle\alpha|a^\dagger a|\alpha\rangle = \sum_\beta \langle\alpha|a^\dagger|\beta\rangle\langle\beta|a|\alpha\rangle = \sum_\beta |\langle\beta|a|\alpha\rangle|^2 \geq 0$. This can vanish only for a state obeying $a|\alpha\rangle = 0$. The lowest energy state, identified as the vacuum state and denoted $|0\rangle$, can thus be defined by

$$a_k|0\rangle = 0 \tag{3.52}$$

We see that the vacuum state has energy equal to $\sum_k \frac{1}{2}\omega_k$. This is an (infinite) constant contribution to the energy and is a result of the ordering ambiguity mentioned earlier. The classical expression does not tell us whether we must use $a_k^\dagger a_k$ or $\frac{1}{2}(a_k^\dagger a_k + a_k a_k^\dagger)$. Actually the correct quantum operator should be $\omega_k a_k^\dagger a_k$ so that the vacuum has zero energy. This can be seen as follows. The operators P_μ, $M_{\mu\nu}$ obey the Poincaré algebra. In particular we have the relation

$$[K_i, P_j] = i\, \delta_{ij} H \tag{3.53}$$

If we have a unitary realization of the Lorentz transformations and if the vacuum state is invariant under Lorentz transformations, so that different observers see the vacuum in exactly the same way, we have $K_i|0\rangle = 0$, $\langle 0|K_i = 0$; the vacuum expectation value of (3.53) then shows that we must have $\langle 0|H|0\rangle = 0$. This implies that $H = \sum_k \omega_k a_k^\dagger a_k$ is the correct expression. Thus the requirement of Lorentz invariance of the vacuum can be used to choose the correct ordering of operators in this case. Similar arguments can be made for the momentum; the correct expression is $P_i = \sum_k k_i a_k^\dagger a_k$. (For relativistic field theory, the requirement of invariance of the vacuum is physically reasonable. In situations where we do not have Lorentz invariance, e.g., in special laboratory settings with conducting surfaces or when we do not have flat Minkowski space as in the neighborhood of a gravitating body, the vacuum energy, or more precisely, the ground state energy, is important and can lead to physical effects such as the Casimir effect or Hawking radiation.) From now on we will consider the correctly ordered expressions $H = \sum_k \omega_k a_k^\dagger a_k$ and $P_i = \sum_k k_i a_k^\dagger a_k$.

The vacuum state has $H|0\rangle = 0$, $P_i|0\rangle = 0$. Consider now $a_k^\dagger|0\rangle$. We have

$$H\, a_k^\dagger|0\rangle = \omega_k\, a_k^\dagger|0\rangle, \qquad P_i\, a_k^\dagger|0\rangle = k_i\, a_k^\dagger|0\rangle \tag{3.54}$$

This state has momentum k_i and energy $\omega_k = \sqrt{k^2 + m^2}$. The relationship between energy and momentum is what we expect for a relativistic point-particle of mass m, and so we can identify $a_k^\dagger|0\rangle$ as a one-particle state of momentum k_i. Higher states can be obtained by the application of a string of $a^\dagger$'s to the vacuum state. An arbitrary state

$$|n_{k_1}, n_{k_2}...\rangle = \frac{(a_{k_1}^\dagger)^{n_{k_1}}}{\sqrt{n_{k_1}!}} \frac{(a_{k_2}^\dagger)^{n_{k_2}}}{\sqrt{n_{k_2}!}}...|0\rangle \tag{3.55}$$

can be seen, by evaluation of H and P_i to be a multiparticle state with n_{k_1} particles of momentum k_1 (and corresponding energies), n_{k_2} particles of momentum k_2, etc. The $\sqrt{n_k!}$ factors are needed for normalization. One can also compute the angular momentum of these states and show that they are spin-zero particles. The states (3.55) give the full Hilbert space. In this version, when the states are constructed from the vacuum by the application of creation operators, the full Hilbert space also called a Fock space.

The N-particle wave function for an N-body state can be defined, up to a normalization factor, as

$$\Psi(x_1, x_2, ...x_n) = \langle 0|\phi(x_1)\phi(x_2)...\phi(x_N)|N\rangle \tag{3.56}$$

where $|N\rangle$ is the N-particle state as in (3.55). For one- and two-particle states,

$$\Psi(x) = u_k(x), \qquad \Psi(x_1, x_2) = u_{k_1}(x_1)u_{k_2}(x_2) + u_{k_2}(x_1)u_{k_1}(x_2) \tag{3.57}$$

The two-particle wave function is symmetric under exchange of particles, due to the fact that a_k's commute. This shows that the particles described by the scalar field are bosons.

In conclusion, through quantization of the scalar field, we have obtained a description of spin-zero bosons. We have recovered the many-particle theory starting from fields as the basic dynamical variables, complementing our construction of the field operator from the many-particle approach.

3.4 Quantization of the Dirac field

The Lagrangian for the Dirac field is

$$\mathcal{L} = \bar{\psi}(i\gamma \cdot \partial - m)\psi \tag{3.58}$$

The momentum canonically conjugate to ψ is given by

$$\pi = i\psi^\dagger \tag{3.59}$$

One may expect that the commutation rule is of the form $[\psi(x), \psi^\dagger(x')] = \delta^{(3)}(x - x')$, but we shall see shortly that one has to use anticommutators for the Dirac theory.

The Hamiltonian operator is given by

$$H = \int d^3x \; \psi^\dagger(i\gamma^0\gamma^i\partial_i + m\gamma^0)\psi \tag{3.60}$$

From our discussion of the plane wave solutions of the Dirac equation, we can write the general solution as

$$\psi(x) = \sum_{p,r} \sqrt{\frac{m}{E_pV}} \left[a_{p,r}u_r(p)e^{-ipx} + c_{p,r}v_r(p)e^{ipx}\right]$$

$$\bar{\psi}(x) = \sum_{p,r} \sqrt{\frac{m}{E_pV}} \left[a^\dagger_{p,r}\bar{u}_r(p)e^{ipx} + c^\dagger_{p,r}\bar{v}_r(p)e^{-ipx}\right] \tag{3.61}$$

where $E_p = \sqrt{p^2 + m^2}$. (We follow the convention of using E_p for fermions, rather than ω_p.) The normalization factors in (3.61) are chosen for later simplifications. The coefficients of the plane wave expansion, viz., $a_{p,r}$, $a^\dagger_{p,r}$, $c_{p,r}$, $c^\dagger_{p,r}$ are operators in the quantum theory. Using this expansion and the orthonormality properties of the u and v-spinors given in Chapter 1,

$$H = \sum_{p,r} E_p(a^\dagger_{p,r} a_{p,r} - c^\dagger_{p,r} c_{p,r}) \qquad (3.62)$$

If we use the canonical commutation rules for ψ and $\psi^\dagger$, we find that a, $a^\dagger$ and c, $c^\dagger$ obey the commutation rules for the creation and annihilation operators. Equation (3.62) then shows the difficulty of using commutation rules. The Hamiltonian is not positive; there are states of negative energy. The way to avoid this is to use anticommutation rules. First we redefine

$$c_{p,r} = b^\dagger_{p,r}, \qquad c^\dagger_{p,r} = b_{p,r} \qquad (3.63)$$

If we further assume the anticommutation rules $b^\dagger_{p,r} b_{k,s} + b_{k,s} b^\dagger_{p,r} = \delta_{rs}\delta_{p,k}$, the Hamiltonian can be written as

$$H = \sum_{p,r} E_p(a^\dagger_{p,r} a_{p,r} + b^\dagger_{p,r} b_{p,r}) - \sum_p 2E_p \qquad (3.64)$$

The change of sign for the second term is due to the anticommutation property. (Since the fields ψ and $\bar\psi$ involve sums over $a_{p,r}$, $c_{p,r}$ and $a^\dagger_{p,r}$, $c^\dagger_{p,r}$, we must take $a_{p,r}$, $a^\dagger_{p,r}$ to have anticommutation rules as well, to have commutation rules for the fields consistent with various physical requirements.) We can now define the vacuum state by $a_{p,r}|0\rangle = b_{p,r}|0\rangle = 0$. The vacuum energy (or the zero-point energy) $-\sum_p 2E_p$ has the opposite sign to what we found for the scalar field. The magnitude per mode is actually the same, $\frac{1}{2}E_p$ for each of the two spin states of the positive energy solutions and for the two spin states of the negative energy solutions. We shall redefine the Hamiltonian by subtracting out the vacuum energy, for the same reasons as before, viz., Lorentz invariance of the vacuum. The corrected Hamiltonian then reads

$$H = \sum_{p,r} E_p(a^\dagger_{p,r} a_{p,r} + b^\dagger_{p,r} b_{p,r}) \qquad (3.65)$$

With the interpretation of $a_{p,r}, b_{p,r}$ as annihilation operators and the vacuum defined by $a_{p,r}|0\rangle = b_{p,r}|0\rangle = 0$, we see that H is always positive.

The anticommutation rules can be formulated as follows.

$$\begin{aligned}
\{\psi(\mathbf{x},t), \psi(\mathbf{x}',t\} &= 0, \\
\{\pi(\mathbf{x},t), \pi(\mathbf{x}',t)\} &= 0 \\
\{\psi(\mathbf{x},t), \pi(\mathbf{x}',t)\} &= i\,\delta^{(3)}(x - x')\,\mathbf{1}
\end{aligned} \qquad (3.66)$$

(The spinor labels are not explicitly shown; the term $\mathbf{1}$ on the right-hand side refers to the identity for spinor labels. Also recall that π is $i\psi^\dagger$, so that these anticommutation rules can be rewritten in terms of ψ, $\psi^\dagger$.) We use the standard abbreviation $AB + BA = \{A, B\}$. Using the mode expansion for the fields, the commutation rules can be obtained in terms of the operators a, $a^\dagger$, b, $b^\dagger$ as

$$\{a_{p,r}, a_{k,s}\} = \{a_{p,r}, b_{k,s}\} = 0$$
$$\{a_{p,r}, b^\dagger_{k,s}\} = \{a^\dagger_{p,r}, b_{k,s}\} = 0 \tag{3.67}$$
$$\{a_{p,r}, a^\dagger_{k,s}\} = \delta_{rs}\delta_{p,k}$$
$$\{b_{p,r}, b^\dagger_{k,s}\} = \delta_{rs}\delta_{p,k} \tag{3.68}$$

The hermitian conjugates of the relations (3.67) hold as well, although we do not display them here. These rules are the same as what we obtained in the discussion of fermions and subsequent construction of the field operator in Chapter 2.

The momentum operator can be evaluated as

$$P_i = \int d^3x \; \psi^\dagger(-i\,\partial_i)\psi = \sum_{p,r} p_i \left[a^\dagger_{p,r}a_{p,r} + b^\dagger_{p,r}b_{p,r}\right] \tag{3.69}$$

We also define a charge or fermion number operator by

$$Q = \int d^3x \; \psi^\dagger\psi = \sum_{p,r} \left[a^\dagger_{p,r}a_{p,r} - b^\dagger_{p,r}b_{p,r}\right] \tag{3.70}$$

(We have chosen an ordering of operators in Q which makes it zero on the vacuum.)

We can now study the states. The vacuum $|0\rangle$ has zero energy, momentum, and charge. The next set of states are $a^\dagger_{p,r}|0\rangle$ and $b^\dagger_{p,r}|0\rangle$. These have energy E_p and momentum p_i. Since $E_p = \sqrt{p^2 + m^2}$, we see that these can be interpreted as one-particle states of momentum p_i and mass m. The label r gives the spin states; these are spin-$\frac{1}{2}$ particles. We have seen this in terms of the one-particle wave functions in Chapter 1. It follows in our present discussion by noting that $\langle 0|\psi(x)a^\dagger_{p,r}|0\rangle$ and $\langle 0|\bar\psi(x)b^\dagger_{p,r}|0\rangle$ are the one-particle wave functions discussed in that chapter. This result can also be checked by direct calculation of the angular momentum. The states $a^\dagger_{p,r}|0\rangle$, $b^\dagger_{p,r}|0\rangle$ have charges $+1$ and -1, respectively. We can thus interpret these as the states of a single particle and its antiparticle. Evidently, $a^\dagger_{p,r}$ is a particle creation operator, $b^\dagger_{p,r}$ is an antiparticle creation operator; their hermitian conjugates are the corresponding annihilation operators.

Multiparticle states can be obtained by applying a string of creation operators to the vacuum state. Because of the anticommutation rules, we have $a^\dagger_{p,r}a^\dagger_{p,r} = 0$. Thus we cannot have more than one particle for every value of p, r. This is the exclusion principle. Also we see that the two-particle wave function given by

$$\Psi_{p,r;k,s}(x_1,x_2) = \langle 0|\psi(x_1)\psi(x_2)|p,r;k,s\rangle$$
$$= -\sqrt{\frac{m}{E_pV}\frac{m}{E_kV}}\left(e^{-ipx_1}e^{-ikx_2} - e^{-ipx_2}e^{-ikx_1}\right)u_r(p)u_s(k)$$

$$(3.71)$$

is antisymmetric under exchange of particles. From these two results we can see that the Dirac field describes fermions.

The use of the anticommutation rule was necessary to avoid states of negative energy. One may wonder how this ties in with the general idea of observables being generators of unitary transformations in the quantum theory since unitary transformations, infinitesimally, lead to commutation operations.

Consider, for example, the Heisenberg equation of motion,

$$i\frac{\partial\psi}{\partial t} = [\psi, H] \tag{3.72}$$

The Hamiltonian is of the form $\psi^\dagger h\psi$, $h = i\gamma^0\gamma^i\partial_i + m\gamma^0$. Commutators of the form $[A, BC]$ can be written out in two ways, either as $[A,B]C + B[A,C]$, which is useful in evaluating commutators for a theory of bosons or as $\{A,B\}C - B\{A,C\}$, which can be used for fermionic theories where the basic rules are anticommutation rules. Thus we can have commutation operations at the level of operators which are quadratic in (or generally even powers of) the basic field variables. Hermitian operators involving even powers of the fermionic field operators generate unitary transformations in the quantum theory. We have consistency if we require that all observables involve even powers of the fermionic field operators. With this condition, our quantization for fermions is consistent with the general rule of quantization (3.36).

The spinor field must be quantized by anticommutation rules, that is, as fermions obeying the exclusion principle in order to have a positive Hamiltonian. This is a special case of the more general *spin-statistics theorem*, which states that

1. Quantization of half-odd integer spin fields or spinors using commutation rules will lead to states of negative energy.
2. Quantization of integer spin fields using anticommutation rules will lead to states of negative norm, which do not, therefore, admit a probabilistic interpretation, or, in a different version, to lack of Lorentz covariance.

The consequence is that spin-$\frac{1}{2}$, -$\frac{3}{2}$, -$\frac{5}{2}$,... particles must be fermions while spin-zero, -1, -2, ... particles must be bosons. Originally proved for relativistic theories, this result has been improved over the years. Recently, there have been attempts to prove such a spin-statistics theorem based only on certain general topological arguments and the existence of antiparticles.

3.5 Symmetries and conservation laws

A symmetry of a classical field theory is a transformation $\varphi \to \varphi'$ under which the Lagrangian changes at most by a total divergence. A total divergence integrates to a surface term in the action and so does not change the equations of motion. Preserving the Lagrangian up to a total divergence is a sufficient but not necessary condition for a symmetry of the equations of motion. It is possible to have a symmetry of the equations of motion which is not a symmetry of the Lagrangian (even up to total divergence). The simplest example is nonrelativistic free particle motion described by

$$\mathcal{L} = \frac{1}{2}m\frac{dx^i}{dt}\frac{dx^i}{dt}$$

$$\frac{d^2 x^i}{dt^2} = 0 \tag{3.73}$$

We see that any general linear transformation $x^i \to M^i_j x^j$, where M is an invertible constant matrix is a symmetry of the equations of motion, but only those M's which are orthogonal, viz., $M^T M = 1$ preserve the Lagrangian. However, in anticipation of the quantum theory, we shall be interested only in symmetries which are canonical transformations; such symmetries preserve the Lagrangian up to a total divergence.

There are many discrete symmetries of interest in physics such as parity and time-reversal. We shall postpone their discussion for now and consider continuous symmetries. For continuous symmetries, the changes in the fields are specified by a continuous set of parameters. (We consider global symmetries for which the parameters are constants, i.e., independent of spacetime. Local symmetries for which the parameters can be spacetime dependent will be discussed in Chapter 10 where we introduce gauge theories.) For a continuous symmetry it is possible to consider infinitesimal transformations which are very close to the identity transformation. i.e., we can write $\varphi_r \to \varphi'_r = \varphi_r + \epsilon^A \tilde{\varphi}_{Ar}$. ϵ^A are the infinitesimal parameters of the transformation; $\tilde{\varphi}_{Ar}$ is defined by the change in φ_r. Being a symmetry, we must have

$$\delta\mathcal{L} = \partial_\mu(\epsilon^A K^\mu_A) \tag{3.74}$$

for some K^μ_A.

Consider a general change $\varphi_r \to \varphi_r + \delta\varphi_r$. We have

$$\delta\mathcal{L} = \left[\frac{\partial\mathcal{L}}{\partial\varphi_r} - \frac{\partial}{\partial x^\mu}\frac{\partial\mathcal{L}}{\partial(\partial_\mu\varphi_r)}\right]\delta\varphi_r + \frac{\partial}{\partial x^\mu}\left(\frac{\partial\mathcal{L}}{\partial(\partial_\mu\varphi_r)}\delta\varphi_r\right) \tag{3.75}$$

We define a current associated to a symmetry transformation by

$$J^\mu_A = \frac{\partial\mathcal{L}}{\partial(\partial_\mu\varphi_r)}\tilde{\varphi}_{Ar} - K^\mu_A \tag{3.76}$$

For a symmetry, using (3.74, 3.76), we can write (3.75) as

$$\left[\frac{\partial \mathcal{L}}{\partial \varphi_r} - \frac{\partial}{\partial x^\mu} \frac{\partial \mathcal{L}}{\partial(\partial_\mu \varphi_r)}\right] \delta\varphi_r + \epsilon^A \partial_\mu J_A^\mu = 0 \qquad (3.77)$$

We see that J_A^μ evaluated along the classical trajectories is conserved; i.e., $\partial_\mu J_A^\mu = 0$, if the fields obey the equations of motion. Notice that (3.76) defines J_A^μ even for field configurations which do not obey the equations of motion.

We have obtained the result that for every continuous symmetry of the theory, there is a current which is conserved by the time evolution of the fields given by the equations of motion. This result is known as Noether's theorem.

Integrating $\partial_\mu J_A^\mu = 0$ over all space, we see that the charge

$$Q_A = \int d^3x \, J_A^0 \qquad (3.78)$$

is preserved in time, i.e., $\frac{dQ}{dt} = 0$. (This is true if the surface term $\oint \boldsymbol{J}_A \cdot d\boldsymbol{S}$ at the spatial boundary is zero; otherwise, this surface integral tells us the rate at which charge is flowing out of the volume under consideration.)

We now consider some examples illustrating this result.

1. Consider a complex scalar field φ with the Lagrangian

$$\mathcal{L} = \partial_\mu \varphi^* \partial^\mu \varphi - m^2 \varphi^* \varphi \qquad (3.79)$$

The transformation $\varphi \to \varphi' = e^{i\theta}\varphi$ is evidently a symmetry, i.e., $\mathcal{L}(\varphi') = \mathcal{L}(\varphi)$, for constant θ. Thus

$$\delta\varphi = i\theta\varphi, \qquad \delta\varphi^* = -i\theta\varphi^*, \qquad K^\mu = 0 \qquad (3.80)$$

The current is given by

$$J^\mu = -i\left[\varphi^*(\partial^\mu \varphi) - (\partial^\mu \varphi^*)\varphi\right] \qquad (3.81)$$

One can easily check that this current is conserved, using the equations of motion. The charge is given by

$$Q = -i \int d^3x \, (\varphi^* \pi^* - \pi\varphi) \qquad (3.82)$$

π and π^* are the canonical momenta for φ and φ^* respectively. θQ is the generator of the symmetry transformation; i.e., for any operator F built up of ϕ, ϕ^* and the corresponding canonical momenta,

$$i\delta F = [F, \theta Q] \qquad (3.83)$$

2. Consider the Dirac theory with the Lagrangian

$$\mathcal{L} = \bar{\psi}(i\gamma \cdot \partial - m)\psi \tag{3.84}$$

The transformation $\psi \to \psi' = e^{-i\theta}\psi$, $\bar{\psi} \to \bar{\psi}' = e^{i\theta}\bar{\psi}$ is a symmetry with $\mathcal{L}(\psi', \bar{\psi}') = \mathcal{L}(\psi, \bar{\psi})$. The current is given by

$$J^\mu = \bar{\psi}\gamma^\mu\psi \tag{3.85}$$

The corresponding charge is

$$Q = \int d^3x \; \psi^\dagger\psi \tag{3.86}$$

This charge is the fermion number charge we introduced in the last section for Dirac particles. From the anticommutation rules for the fields

$$[\psi(\boldsymbol{x}, t), \theta Q] = \theta\psi(\boldsymbol{x}, t), \qquad [\bar{\psi}(\boldsymbol{x}, t), \theta Q] = -\bar{\psi}\theta \tag{3.87}$$

showing that θQ is the generator of the transformations $\psi \to e^{-i\theta}\psi$, $\bar{\psi} \to e^{i\theta}\bar{\psi}$.

3.6 The energy-momentum tensor

Symmetries we have considered in the examples given above leave the Lagrangian unaltered. Generally, in the case of spacetime symmetries, the Lagrangian changes by a total divergence; there is a nonzero K_A^μ. The spacetime symmetries of interest to us are translations and Lorentz transformations which lead to the conservation of energy, momentum, and angular momentum. Rather than use the formula for the current, we work out once again the derivation of the conservation laws.

We consider the transformation $x^\mu \to x^\mu + \xi^\mu$, with $\xi^\mu = a^\mu + \omega^{\mu\nu}x_\nu$, which corresponds to a constant translation by a^μ and an infinitesimal Lorentz transformation with parameters $\omega^{\mu\nu}$; the vector ξ^μ obeys $\partial_\mu\xi^\mu = 0$. The change in the fields is given by

$$\delta\varphi_r = \xi^\mu\partial_\mu\varphi_r - \frac{i}{2}\omega^{\mu\nu}(S_{\mu\nu}\varphi)_r$$

$$\delta\partial_\mu\varphi_r = \xi^\mu\partial_\mu\varphi_r + \partial_\mu\xi^\nu\partial_\nu\varphi_r - \frac{i}{2}\omega^{\alpha\beta}(S_{\alpha\beta}\varphi)_r \tag{3.88}$$

Here $S_{\mu\nu}$ are the spin matrices, whose explicit form is given in Chapter 1 and the appendix for spin-$\frac{1}{2}$ and spin-1 cases. In (3.88), they are understood to be in the representation to which the fields φ_r belong.

The Lagrangian is a scalar function of x^μ; the change in $\mathcal{L}$ is thus given by

$$\delta\mathcal{L} = \xi^\mu\frac{\partial\mathcal{L}}{\partial x^\mu} = \partial_\mu(\xi^\mu\mathcal{L}) \tag{3.89}$$

The Lagrangian does not have explicit dependence on the coordinates; the dependence on x^μ is through the fields and their derivatives. So we can also write, as in equation (3.75),

$$\delta \mathcal{L} = \mathcal{E}_r \delta \varphi_r + \partial_\nu \left(\delta \varphi_r \frac{\partial \mathcal{L}}{\partial(\partial_\nu \varphi_r)} \right) \tag{3.90}$$

$$\mathcal{E}_r = \frac{\partial \mathcal{L}}{\partial \varphi_r} - \frac{\partial}{\partial x^\nu} \left(\frac{\partial \mathcal{L}}{\partial(\partial_\nu \varphi_r)} \right) \tag{3.91}$$

Combining equations (3.88 - 3.91), we have

$$\frac{\partial}{\partial x^\nu} \left[\left(\xi^\mu \partial_\mu \varphi_r \frac{\partial \mathcal{L}}{\partial(\partial_\nu \varphi_r)} \right) - \xi^\nu \mathcal{L} - \frac{i}{2} (\omega^{\alpha\beta} S_{\alpha\beta} \varphi)_r \frac{\partial \mathcal{L}}{\partial(\partial_\nu \varphi_r)} \right] = -\mathcal{E}_r \delta \varphi_r \tag{3.92}$$

We see that the quantity in the square brackets on the left-hand side is conserved for fields which obey the equations of motion, viz., when $\mathcal{E}_r = 0$. The current for translations is given by

$$t_{\mu\nu} = \partial_\mu \varphi_r \frac{\partial \mathcal{L}}{\partial(\partial^\nu \varphi_r)} - \eta_{\mu\nu} \mathcal{L} \tag{3.93}$$

(This is not quite the energy-momentum tensor, which is why we use the lowercase letter.) There is some arbitrariness in defining $t_{\mu\nu}$ from the conservation condition; one can add a term like $\partial^\alpha B_{\mu\alpha\nu}$, where $B_{\mu\alpha\nu}$ is antisymmetric in α and ν, to $t_{\mu\nu}$. This does not affect the conservation condition. A specific choice for $B_{\mu\alpha\nu}$ as a function of the fields and their derivatives will be made below, motivated by symmetry properties.

The four-momentum of the system is given by

$$P_\mu = \int d^3x \; t_{\mu 0} \tag{3.94}$$

From equation (3.92), the current for Lorentz transformations, viz., the density for angular momentum, is defined by

$$\tfrac{1}{2} \omega^{\mu\alpha} \mathcal{M}_{\mu\alpha\nu} = \omega^{\mu\alpha} \left[x_\alpha t_{\mu\nu} - \tfrac{i}{2} (S_{\mu\alpha} \varphi)_r \frac{\partial \mathcal{L}}{\partial(\partial_\nu \varphi_r)} \right] \tag{3.95}$$

Since only the product of $\omega^{\mu\alpha}$ and $\mathcal{M}$ is defined, there is some freedom in the identification of the density $\mathcal{M}_{\mu\alpha\nu}$. We now define

$$B_{\mu\alpha\nu} = \frac{i}{2} \left[(S_{\mu\alpha} \varphi)_r \frac{\partial \mathcal{L}}{\partial(\partial_\nu \varphi_r)} - (S_{\mu\nu} \varphi)_r \frac{\partial \mathcal{L}}{\partial(\partial_\alpha \varphi_r)} - (S_{\alpha\nu} \varphi)_r \frac{\partial \mathcal{L}}{\partial(\partial_\mu \varphi_r)} \right] \tag{3.96}$$

Notice that $B_{\mu\alpha\nu}$ is antisymmetric in α and ν. The angular momentum density, consistent with (3.95), is defined as

$$\mathcal{M}_{\mu\alpha\nu} = x_\alpha t_{\mu\nu} - x_\mu t_{\alpha\nu} - (B_{\mu\alpha\nu} - B_{\alpha\mu\nu}) \tag{3.97}$$

The energy-momentum tensor is now defined as

$$T_{\mu\nu} = t_{\mu\nu} + \partial^\alpha B_{\mu\alpha\nu} \tag{3.98}$$

From the antisymmetry of $B_{\mu\alpha\nu}$ and the conservation of $t_{\mu\nu}$, it follows that $T_{\mu\nu}$ is conserved. Further, the four-momentum can be written as

$$P_\mu = \int d^3x \; T_{\mu 0} \tag{3.99}$$

since the term involving B gives a surface integral at spatial infinity, which is zero for fields which vanish there or obey appropriate periodic boundary conditions. We can also write the angular momentum density as

$$\mathcal{M}_{\mu\alpha\nu} = x_\alpha T_{\mu\nu} - x_\mu T_{\alpha\nu} - \partial^\beta (x_\alpha B_{\mu\beta\nu} - x_\mu B_{\alpha\beta\nu}) \tag{3.100}$$

The angular momenta and boost generators are given by

$$M_{\mu\alpha} = \int d^3x \; \mathcal{M}_{\mu\alpha 0} = \int d^3x \; (x_\alpha T_{\mu 0} - x_\mu T_{\alpha 0}) \tag{3.101}$$

The divergence term in (3.100) leads to a surface integral at spatial infinity and does not contribute in $M_{\mu\alpha}$ for B's which vanish sufficiently fast. Therefore we may also define the angular momentum density as

$$\mathcal{M}_{\mu\alpha\nu} = x_\alpha T_{\mu\nu} - x_\mu T_{\alpha\nu} \tag{3.102}$$

From the conservation of the angular momentum density, it follows that $T_{\mu\nu}$ is a symmetric tensor, when the fields obey the equations of motion. Our choice of $B_{\mu\alpha\nu}$, which is the ambiguity in defining the currents from the conservation equation (3.92), is motivated by this symmetry property. The tensor $B_{\mu\alpha\nu}$ is sometimes referred to as the Belinfante tensor. In working out the expression for the tensor $T_{\mu\nu}$ from (3.98), one may encounter, depending on the specific theory, terms which are proportional to $\mathcal{E}_r$. Classically such terms are clearly irrelevant; quantum theoretically, one needs further physical requirements to define such terms; these physical requirements are related to the renormalization of the energy-momentum tensor.

3.7 The electromagnetic field

An example of a field with spin for which the above construction of the symmetric energy-momentum tensor can be applied is the electromagnetic field. The basic field variable is the electromagnetic vector potential $A_\mu(x)$. The field strengths are given by the tensor $F_{\mu\nu} = \partial_\mu A_\nu - \partial_\nu A_\mu$. The electric field E_i and the magnetic field B_i are given by $E_i = F_{0i}$ and $B_i = \frac{1}{2}\epsilon_{ijk}F^{jk}$. The Lagrangian is given by

$$\mathcal{L} = -\tfrac{1}{4} F^{\mu\nu} F_{\mu\nu} \tag{3.103}$$

This gives

$$\frac{\partial \mathcal{L}}{\partial(\partial_\nu A_\alpha)} = F^{\alpha\nu}$$

$$t_{\mu\nu} = -\partial_\mu A_\alpha F_\nu{}^\alpha + \tfrac{1}{4} \eta_{\mu\nu} F^2 \tag{3.104}$$

The Belinfante tensor is given by

$$B_{\mu\alpha\nu} = -A_\mu F_{\alpha\nu} \tag{3.105}$$

and the symmetric energy-momentum tensor is easily seen to be

$$T_{\mu\nu} = -F_{\mu\alpha} F_\nu{}^\alpha + \tfrac{1}{4} \eta_{\mu\nu} F^2 \tag{3.106}$$

3.8 The energy-momentum tensor and general relativity

We have seen that the energy-momentum tensor $T_{\mu\nu}$ is symmetric. For fields with spin, this is achieved after the addition of the Belinfante term to $t_{\mu\nu}$. This term arises from the rotation of coordinate frames which are needed for defining the components of a vector or tensor, or generally for the components of any field with spin. General coordinate transformations, as encountered in the general theory of relativity, include these frame rotations, and therefore we have another way to derive the energy-momentum tensor. We first make the action invariant under general coordinate transformations by including the metric tensor to carry out contractions of spacetime indices and to make the integration measure invariant. In this expression, the frame rotations are compensated by the change of the metric components. The energy-momentum tensor can thus be obtained by varying the action with respect to the metric components. Specifically, the formula is easily seen to be

$$\delta S = \frac{1}{2} \int \sqrt{-g}\, d^4 x\, T_{\mu\nu}\, \delta g^{\mu\nu} \tag{3.107}$$

In this expression, $g = \det(g_{\mu\nu})$ and $g^{\mu\nu}$ is the inverse to $g_{\mu\nu}$, i.e., $g^{\mu\nu} g_{\nu\alpha} = \delta^\mu_\alpha$. Since the metric tensor $g_{\mu\nu}$ is symmetric, the energy-momentum tensor so derived is automatically symmetric and coincides with the symmetric energy-momentum tensor obtained via the Belinfante addition. After identifying $T_{\mu\nu}$, we may set $g_{\mu\nu} = \eta_{\mu\nu}$ to obtain the result for flat spacetime. The derivation of the equations of motion for gravity from an action principle involves the variation of the action with respect to the metric. Thus the above formula is also just what we expect from the fact that the energy-momentum tensor for matter fields acts as the source for the gravitational field.

As an example, the action for the electromagnetic field is easily covariantized as

$$S = -\frac{1}{4} \int \sqrt{-g} \, d^4x \, g^{\mu\alpha} g^{\nu\beta} F_{\mu\nu} F_{\alpha\beta} \tag{3.108}$$

Using $\delta\sqrt{-g} = -\frac{1}{2}\sqrt{-g}\, g_{\mu\nu}\delta g^{\mu\nu}$, we find

$$
\begin{aligned}
T_{\mu\nu} &= -F_{\mu\alpha} F_\nu^\alpha + \frac{1}{4} g_{\mu\nu} F^2 \\
&\to -F_{\mu\alpha} F_\nu^\alpha + \frac{1}{4} \eta_{\mu\nu} F^2
\end{aligned}
\tag{3.109}
$$

Similarly, for the scalar field we find

$$S = \int \sqrt{-g} \, d^4x \left[\frac{1}{2} g^{\mu\nu} \partial_\mu\varphi \partial_\nu\varphi - \frac{1}{2} m^2\varphi^2 \right] \tag{3.110}$$

which leads to

$$T_{\mu\nu} = \partial_\mu\varphi\partial_\nu\varphi - \frac{1}{2}\eta_{\mu\nu}\left[(\partial\varphi)^2 - m^2\varphi^2 \right] \tag{3.111}$$

These expressions coincide with those which were derived previously in a more tedious fashion.

3.9 Light-cone quantization of a scalar field

A simple example which illustrates the use of the symplectic structure $\Omega_{ij}(\boldsymbol{x}, \boldsymbol{x}')$ is the light-cone quantization of a scalar field. We consider a real scalar field φ with a Lagrangian

$$\mathcal{L} = \tfrac{1}{2}(\partial\varphi)^2 - U(\varphi) \tag{3.112}$$

where $U(\varphi) = \tfrac{1}{2}m^2\varphi^2 + V(\varphi)$.

We now introduce light-cone coordinates, corresponding to a light-cone in the (z, t)-direction as

$$
\begin{aligned}
u &= \frac{1}{\sqrt{2}}(z + t) \\
v &= \frac{1}{\sqrt{2}}(z - t)
\end{aligned}
\tag{3.113}
$$

Instead of considering evolution of the fields in time t, we can consider evolution in one of the the light-cone coordinates, say, u. The other light-cone coordinate v and the two coordinates $x^T = x, y$ transverse to the light-cone parametrize the equal-u hypersurfaces. Field configurations $\varphi(u, v, x, y)$ at fixed values of u, i.e., real-valued functions of v, x, y, characterize the trajectories. They form the phase space of the theory. The action can be written as

$$\mathcal{S} = -\int du\, dv\, d^2x^T \left[\partial_u\varphi\partial_v\varphi + \tfrac{1}{2}(\partial_T\varphi)^2 + U(\varphi)\right] \tag{3.114}$$

A naive definition of the canonical momentum π as $\frac{\partial \mathcal{L}}{\partial(\partial_u\varphi)}$ gives $\pi = -\partial_v\varphi$, which is not independent of $\varphi(v, x^T)$. Such a definition is therefore not very useful. However, from the variation of the action $\mathcal{S}$, we can identify the canonical one-form Θ as

$$\Theta = \int dv\, d^2x^T \left(-\partial_v\varphi\, \delta\varphi\right) \tag{3.115}$$

The symplectic two-form is thus given by

$$\Omega(v, x^T, v', x'^T) = -2\, \partial_v\delta(v - v')\delta^{(2)}(x^T - x'^T) \tag{3.116}$$

We need the inverse of Ω. Writing

$$\delta(v - v')\, \delta^{(2)}(x^T - x'^T) = \int \frac{d^3p}{(2\pi)^3} \exp\left(-ip_u(v - v') - ip^T \cdot (x^T - x'^T)\right) \tag{3.117}$$

we see that

$$\begin{aligned}
\Omega^{-1}(v, x^T, v', x'^T) &= \frac{1}{2} \int \frac{d^3p}{(2\pi)^3} \frac{1}{ip_u} \exp(-ip_u(v - v') - ip^T \cdot (x^T - x'^T)) \\
&= -\frac{1}{4}\epsilon(v - v')\, \delta^{(2)}(x^T - x'^T)
\end{aligned} \tag{3.118}$$

Here $\epsilon(v - v')$ is the signature function, equal to 1 for $v > 0$ and equal to -1 for $v < 0$.

The phase space is thus given by field configurations $\varphi(v, x^T)$ with the Poisson brackets

$$\{\varphi(u, v, x^T), \varphi(u, v', x'^T)\} = -\frac{1}{4}\epsilon(v - v')\, \delta^{(2)}(x^T - x'^T) \tag{3.119}$$

The Hamiltonian for u-evolution is given by

$$H = \int dv\, d^2x^T \left[\tfrac{1}{2}(\partial_T\varphi)^2 + U(\varphi)\right] \tag{3.120}$$

The Hamiltonian equations of motion are easily checked using the Poisson brackets (3.119).

Quantization is achieved by replacing φ by an operator ϕ with commutation rules given by i-times the Poisson bracket.

3.10 Conformal invariance of Maxwell equations

Translations and Lorentz transformations (or Poincaré transformations) are not the only kind of spacetime symmetries possible. An example of another spacetime symmetry of interest is conformal symmetry. The free or sourceless Maxwell equations are invariant under conformal transformations.

Lorentz transformations are of interest because they arise as isometries of the Minkowski space. (We give a brief discussion of isometries and the Killing equation in the appendix.) The change of the metric or distance function

$$ds^2 = g_{\mu\nu}dx^\mu dx^\nu \tag{3.121}$$

under an infinitesimal transformation $x^\mu \to x^\mu + \xi^\mu$ is given by

$$\delta g_{\mu\nu} = g_{\alpha\nu}\frac{\partial\xi^\alpha}{\partial x^\mu} + g_{\mu\alpha}\frac{\partial\xi^\alpha}{\partial x^\nu} + \xi^\alpha\frac{\partial g_{\mu\nu}}{\partial x^\alpha} \tag{3.122}$$

Setting this to zero, we get the Killing equation. For any given metric, the solutions of the Killing equation give the isometries. For the Minkowski metric which is constant, the Killing equation becomes

$$\partial_\mu\xi_\nu + \partial_\nu\xi_\mu = 0 \tag{3.123}$$

The most general solution, as noted in the appendix, is given by $\xi_\mu = a_\mu + \omega_{\mu\nu}x^\nu$; i.e., the general isometries for the Minkowski space are translations and Lorentz transformations (including rotations). This is why these transformations are important for field theories on Minkowski spacetime.

Conformal transformations preserve the metric up to a scale factor; i.e., the change in the metric tensor is given by $\lambda g_{\mu\nu}$ where λ is a scalar. In this case, $ds^2 \to (1+\lambda)ds^2$. The propagation of light rays is given by $ds^2 = 0$ and this condition is preserved by conformal transformations. For this and many other reasons, conformal transformations are also important in physics. From (3.122), we see that conformal transformations are given by

$$\xi^\alpha\partial_\alpha g_{\mu\nu} + g_{\alpha\nu}\partial_\mu\xi^\alpha + g_{\mu\alpha}\partial_\nu\xi^\alpha = \lambda g_{\mu\nu} \tag{3.124}$$

This is the conformal Killing equation. For the Minkowski metric this simplifies to

$$\partial_\mu\xi_\nu + \partial_\nu\xi_\mu = \lambda\eta_{\mu\nu} \tag{3.125}$$

The contraction of this with $\eta^{\mu\nu}$ gives $\lambda = \frac{1}{2}\partial_\mu\xi^\mu$. The conformal Killing equation becomes

$$\partial_\mu\xi_\nu + \partial_\nu\xi_\mu - \frac{1}{2}(\partial_\alpha\xi^\alpha)\,\eta_{\mu\nu} = 0 \tag{3.126}$$

The most general solution is given by

$$\xi_\mu = a_\mu + \omega_{\mu\nu}x^\nu + b^\nu(x^2\eta_{\mu\nu} - 2x_\mu x_\nu) + \epsilon\,x_\mu \tag{3.127}$$

for constant a_μ, $\omega_{\mu\nu}$, b^ν, ϵ. The first two sets of parameters correspond to translations and Lorentz transformations as before. The transformations corresponding to the parameters b^ν are called special conformal transformations and the transformation corresponding to ϵ is called a dilatation. The special conformal transformations may be understood as follows. Define an inverted coordinate $y^\mu = (x^\mu/x^2)$, $x \neq 0$. An infinitesimal translation of y^μ as given by $y^\mu \to y^\mu + b^\mu$ can be easily checked to be the special conformal transformation when it is expressed in terms of x^μ.

Now consider the change of A_μ and $F_{\mu\nu}$ under a coordinate transformation $x^\mu \to x^\mu + \xi^\mu$, viz.,

$$\delta A_\mu = \xi^\alpha F_{\alpha\mu} + \partial_\mu(A_\alpha \xi^\alpha)$$
$$\delta F_{\mu\nu} = \xi^\alpha \partial_\alpha F_{\mu\nu} + F_{\alpha\nu}\partial_\mu \xi^\alpha + F_{\mu\alpha}\partial_\nu \xi^\alpha \tag{3.128}$$

The change in the Lagrangian (3.103) is given by

$$\delta \mathcal{L} = \partial_\mu(\mathcal{L}\xi^\mu) - \frac{1}{2}F^{\mu\alpha}F^\nu_\alpha \left[\partial_\mu \xi_\nu + \partial_\nu \xi_\mu - \frac{1}{2}(\partial \cdot \xi)\eta_{\mu\nu}\right] \tag{3.129}$$

For a conformal transformation which obeys (3.126), we see that the change in the Lagrangian is a total divergence. The equations of motion, which are the sourceless Maxwell equations, therefore are invariant. In other words, the sourceless Maxwell theory has conformal symmetry at the classical level.

The conserved current for these symmetries may be obtained using (3.75). Since $K^\mu = \mathcal{L}\xi^\mu$, we find

$$\left(\frac{\partial \mathcal{L}}{\partial(\partial_\mu \varphi_r)}\delta\varphi_r\right) - K^\mu = \xi^\alpha \left[-F^{\mu\nu}F_{\alpha\nu} - \mathcal{L}\delta^\mu_\alpha\right] - F^{\mu\nu}\partial_\nu(A \cdot \xi)$$
$$= \xi^\alpha T^\mu_\alpha - F^{\mu\nu}\partial_\nu(A \cdot \xi) \tag{3.130}$$

Equation (3.75) becomes

$$\partial_\mu \left[\xi^\alpha T^\mu_\alpha - F^{\mu\nu}\partial_\nu(A \cdot \xi)\right] \approx 0 \tag{3.131}$$

where the sign $\approx$ indicates that equality holds when the equations of motion are used. Notice that $\partial_\mu[F^{\mu\nu}\partial_\nu(A \cdot \xi)] = 0$ by itself upon using the equations of motion for $F^{\mu\nu}$, so that we have the conservation law

$$\partial_\mu[\xi^\alpha T^\mu_\alpha] \approx 0 \tag{3.132}$$

The conserved currents for various transformations may be obtained by substituting (3.127) in $\xi^\alpha T^\mu_\alpha$ and taking the coefficients of the parameters of the transformations. For example, the dilatation current is given by

$$J^\mu = x^\alpha T^\mu_\alpha \tag{3.133}$$

The change $\partial_\mu(A \cdot \xi)$ is in the form of a gauge transformation; it is not a change in the physical field configurations and can be dropped from these considerations. This is why the term $F^{\mu\nu}\partial_\nu(A \cdot \xi)$ may be removed from the expression for the current. It also follows from (3.107) and (3.108) that $T^\mu_\mu = 0$, since $\delta S = 0$ for $g_{\mu\nu} \to \lambda g_{\mu\nu}$.

References

1. The canonical formalism when the Lagrangian has time-derivatives which are higher than second order, alluded to in text, is due to M. Ostrogradskii, Mem. Act. St.Petersburg, **VI 4**, 385 (1850).

2. The result that the surface term at the final time-slice in the variation of the action gives the canonical one-form is, in essence, an old result going back to nineteenth century work on analytical mechanics. In the context of quantum field theory, it is also the basis of Schwinger's quantum action principle. For some modern references, see V. Guillemin and S. Sternberg, *Symplectic Techniques in Physics*, Cambridge University Press (1990); J. Schwinger, *Phys.Rev.* 82, 914 (1951); C. Črnkovic and E. Witten, in *Three Hundred Years of Gravitation*, S.W. Hawking and W. Israel (eds.), Cambridge University Press (1987); G.J. Zuckerman, in *Mathematical Aspects of String Theory*, S.T. Yau (ed.), World Scientific (1987).

3. Casimir effect and Hawking radiation were briefly mentioned in text. Casimir effect is covered in detail in K.A. Milton, *The Casimir Effect*, World Scientific Pub. Co. (2001).

4. Hawking radiation and many related effects are discussed in V.P. Frolov and I.D. Novikov, *Black Hole Physics: Basic Concepts and New Developments*, Kluwer Academic Publishers (1998).

5. The original spin-statistics theorem is due to Pauli, with later more general approaches due to many others. In the context of relativistic quantum field theory, a good general reference is R.F. Streater and A.S. Wightman, *PCT, Spin and Statistics and All That*, W.A. Benjamin, Inc. (1964).

6. For the topological approach to spin-statistics for point particles, see R.P. Feynman, "The Reason for Antiparticles" in *Elementary Particles and the Laws of Physics*, Oxford University Press (1987); R.D. Tscheuschner, Int. J. Theor. Phys. **28**, 1269 (1989); A.P. Balachandran *et al*, Mod. Phys. Lett. **A5**, 1574 (1990). This is related to earlier work on spin-statistics for solitons by a number of people, some of which can be traced from the last reference quoted.

7. A nice discussion of the spin-statistics theorem in the framework of path integrals is K. Fujikawa, Int. J. Mod. Phys. **A16**, 4025 (2001).

8. For the canonical procedure on symmetrization of energy-momentum tensor, see F. Belinfante, Physica **6**, 887 (1939); *ibid.* **7**, 305 (1940). The symmetric energy-momentum tensor via variation of the metric is discussed in many books on relativity, for example, S. Weinberg, *Gravitation and Cosmology*, Wiley Text Books (1972); L. Landau and E.M. Lifshitz, *Classical Theory of Fields*, Butterworth-Heinemann, 4th edition (1980).

9. The classical conformal invariance of Maxwell equations goes back to E. Cunningham, Proc. Lond. Math. Soc. **8**, 77 (1910); H. Bateman, Proc. Lond. Math. Soc. **8**, 223 (1910).

4 Commutators and Propagators

4.1 Scalar field propagators

As a prelude to the discussion of interactions we calculate commutators and propagators.

Consider the theory of the scalar field first. The field $\phi(x)$ has the mode expansion

$$\phi(x) = \sum_k a_k u_k(x) + a_k^\dagger u_k^*(x) \tag{4.1}$$

where

$$u_k = \frac{e^{-ikx}}{\sqrt{2\omega_k V}} \tag{4.2}$$

($\omega_k = \sqrt{k^2 + m^2}$). The commutator $[\phi(x), \phi(y)]$ can be directly calculated as

$$
\begin{aligned}
[\phi(x), \phi(y)] &= \sum_k u_k(x) u_k^*(y) - u_k^*(x) u_k(y) \\
&= \int \frac{d^3k}{(2\pi)^3} \frac{1}{2\omega_k} \left(e^{-ik(x-y)} - e^{ik(x-y)} \right)
\end{aligned} \tag{4.3}
$$

in the limit $V \to \infty$. We can rewrite this as

$$[\phi(x), \phi(y)] = \int_C \frac{d^4k}{(2\pi)^4} \left(\frac{-i}{k^2 - m^2} \right) e^{-ik(x-y)} \equiv \Delta(x, y) \tag{4.4}$$

where the contour is shown below.

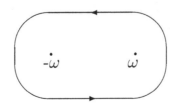

Fig 4.1. Contour for the commutator

Equation (4.4) is to be interpreted as follows. We carry out the k_0-integration first, as a contour integral in the complex k_0-plane, along the contour shown. The contour encloses both the poles of function $(k^2 - m^2)^{-1}$, which are at $k_0 = \pm \omega_k$. Equation (4.4) then reproduces the expression in (4.3).

The commutator $\Delta(x, y)$ is a Lorentz-invariant function of the proper distance $(x - y)^2$. If $x^0 = y^0$, the commutator vanishes since the fields commute at equal times. Since it is Lorentz invariant, it will thus vanish for all points which are Lorentz transforms of $(x^0, \boldsymbol{x})$, $(x^0, \boldsymbol{y})$, i.e., for all spacelike separations of the points x, y or equivalently for $(x - y)^2 < 0$.

$$[\phi(x), \phi(y)] = 0, \qquad\qquad (x - y)^2 < 0 \qquad (4.5)$$

ϕ is a hermitian operator and qualifies as an observable. The fact that the fields commute at spacelike separations tells us that it is possible to measure ϕ at two points with no uncertainties if the two points are spacelike separated. This is a reflection of the fact that, as with any signals, the disturbances due to the measurement process cannot travel faster than light.

Notice that if we use only the u_k's in defining a field operator, say, $\chi(x) = \sum_k a_k u_k(x)$, then we do not have $[\chi(x) + \chi^\dagger(x), i(\chi(y) - \chi^\dagger(y))] = 0$ for spacelike separation of x, y. Thus we cannot interpret arbitrary hermitian combinations of χ and $\chi^\dagger$ as measurable quantities and be consistent with relativity. This is another reason that $\phi(x) = \sum_k a_k u_k(x) + a_k^\dagger u_k^*(x)$ is the appropriate field operator in the relativistic theory.

Green's functions for the Klein-Gordon operator will be important for the discussion of interactions. We define them generically by

$$(\Box + m^2)\, \mathcal{G}(x, y) = -i\delta^{(4)}(x - y) \qquad (4.6)$$

The solution can be written as

$$\mathcal{G}(x, y) = \int \frac{d^4 k}{(2\pi)^4} \left(\frac{i}{k^2 - m^2} \right) e^{-ik(x-y)} \qquad (4.7)$$

The choice of contour in dealing with the singularities at $k_0 = \pm \omega_k$ will determine the type of Green's function we have. The contours for the advanced and retarded functions are as shown below.

Consider the retarded contour C_R. For $x^0 > y^0$, we must complete the contour in the lower half-plane so that the exponential $e^{-|Im k_0|(x^0 - y^0)}$ will guarantee that the large semicircle will not contribute. We thus get $\mathcal{G}(x, y) = G_R(x, y)$,

$$G_R(x, y) = \int \frac{d^3 k}{(2\pi)^3} \frac{1}{2\omega_k} (e^{-ik(x-y)} - e^{ik(x-y)}), \qquad\qquad x^0 > y^0$$

$$= 0, \qquad\qquad\qquad\qquad x^0 < y^0 \qquad (4.8)$$

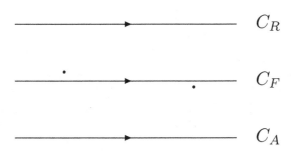

Fig 4.2. Contours for retarded, advanced and Feynman propagators

In other words,
$$G_R(x,y) = \theta(x^0 - y^0)[\phi(x), \phi(y)] \tag{4.9}$$
where $\theta(x^0 - y^0)$ is the step function. A similar result, $G_A(x,y) = \theta(y^0 - x^0)\Delta(y,x)$, can be obtained for the advanced Green's function.

Another quantity of interest is the Feynman propagator. It is defined by
$$G(x,y) = \langle 0|T\ \phi(x)\phi(y)|0\rangle \tag{4.10}$$
where time-ordering, denoted by T, is defined by
$$T\phi(x)\phi(y) = \theta(x^0 - y^0)\phi(x)\phi(y) + \theta(y^0 - x^0)\phi(y)\phi(x) \tag{4.11}$$

The T-symbol orders the operators on which it acts and rearranges them such that the operator with the latest time-argument is at the left, the one with the next latest time-argument comes next, and so on , with the operator with the earliest time-argument at the right. Using the expansion for $\phi(x)$, we find

$$
\begin{aligned}
G(x,y) &= \sum_k \left[\theta(x^0 - y^0)u_k(x)u_k^*(y) + \theta(y^0 - x^0)u_k(y)u_k^*(x)\right] \\
&= \int \frac{d^3k}{(2\pi)^3}\frac{1}{2\omega_k}\left[\theta(x^0 - y^0)e^{-ik(x-y)} + \theta(y^0 - x^0)e^{ik(x-y)}\right]
\end{aligned}
\tag{4.12}
$$

In terms of contour integrals, we can write
$$G(x,y) = \int \frac{d^4k}{(2\pi)^4}\left(\frac{i}{k^2 - m^2 + i\epsilon}\right)e^{-ik(x-y)} \tag{4.13}$$

The contour is the real axis (and a suitable completion in the upper or lower half-plane). ϵ is a small positive number, with $\epsilon \to 0$ eventually. It shifts the poles as shown. The pole at $k_0 = \omega_k$ contributes for $x^0 > y^0$; the pole at $k_0 = -\omega_k$ contributes for $y^0 > x^0$.

The propagator obeys the equation

$$(\Box + m^2 - i\epsilon)G(x,y) = -i\delta^{(4)}(x - y) \tag{4.14}$$

so that it is a Green's function in the mathematical sense. $G(x,y)$ is the inverse of $i(\Box + m^2 - i\epsilon)$.

The propagator can be interpreted as the amplitude for particle propagation. To see this, let us define a function $K(x,y)$ as follows. For $x^0 > y^0$, $K(x,y)$ is the amplitude for the particle to propagate from $\boldsymbol{y}$ to $\boldsymbol{x}$ in time $(x^0 - y^0)$. For $y^0 > x^0$, since x^0 is the earlier time, the particle must propagate from $\boldsymbol{x}$ to $\boldsymbol{y}$; i.e., $K(x,y)$ is the amplitude for this propagation. In terms of x-diagonal states in quantum mechanics, we can write

$$\begin{aligned} K(x,y) &= \langle \boldsymbol{x}, x^0 | \boldsymbol{y}, y^0 \rangle, & x^0 > y^0 \\ &= \langle \boldsymbol{y}, y^0 | \boldsymbol{x}, x^0 \rangle, & y^0 > x^0 \end{aligned} \tag{4.15}$$

Thus

$$K(x,y) = \theta(x^0 - y^0)\langle \boldsymbol{x}|e^{-iH(x^0-y^0)}|\boldsymbol{y}\rangle + \theta(y^0 - x^0)\langle \boldsymbol{y}|e^{-iH(y^0-x^0)}|\boldsymbol{x}\rangle \tag{4.16}$$

For the relativistic particle, $H = \sqrt{\boldsymbol{p}^2 + m^2}$. We can insert momentum eigenstates $|k\rangle$ using the completeness relation (1.7) from Chapter 1, where $p|k\rangle = k|k\rangle$ and $\langle \boldsymbol{x}|k\rangle = \exp(i\boldsymbol{k}\cdot\boldsymbol{x})$. The eigenvalues of H are ω_k and we can evaluate the above expression to find $K(x,y) = G(x,y)$.

The propagator can thus be interpreted as the function which, for $x^0 > y^0$ gives the amplitude for particle-propagation from y to x, and for $y^0 > x^0$ gives the amplitude for particle-propagation from x to y. (As mentioned before, we should use both $u_k(x)$ and $u_k^*(x)$ to define the field $\phi(x)$; this is the only combination which is physical in the sense of having $[\phi(x),\phi(y)] = 0$ for spacelike separated x, y. Thus the definition of the propagator in terms of the field $\phi(x)$ will include both cases, $x^0 > y^0$ and $x^0 < y^0$.) The interpretation of the propagator given here generalizes to N-point Green's functions, as we shall see later.

The analyticity of the integrand in expression (4.13) tells us that it is possible to deform the contour of integration for the propagator to lie along the imaginary k_0-axis. There is no crossing of poles of the integrand in this deformation. We may thus write

$$G(x,y) = \int_{-i\infty}^{i\infty} dk_0 \int \frac{d^3k}{(2\pi)^4} \left(\frac{i}{k^2 - m^2} \right) e^{-ik(x-y)} \tag{4.17}$$

Introducing $k_0 = ik_4$ and $x^0 = ix^4$, this equation can be written as

$$G(x,y) = G_E(x,y)\Big]_{x^4=-ix^0, y^4=-iy^0} \tag{4.18}$$

$$G_E(x,y) = \int \frac{d^4k}{(2\pi)^4} \left(\frac{1}{k^2 + m^2} \right) e^{ik(x-y)} \tag{4.19}$$

where we integrate over the real line for all k's including k_4 and $k^2 = \boldsymbol{k} \cdot \boldsymbol{k} + k_4^2$, $k(x-y) = \boldsymbol{k} \cdot (\boldsymbol{x} - \boldsymbol{y}) + k_4(x^4 - y^4)$. The metric used in (4.19) is thus the standard Euclidean one. The propagator can be considered as the analytic continuation of the Euclidean Green's function $G_E(x,y)$ to imaginary values of x^4, y^4. $G_E(x,y)$ obeys the equation

$$(-\Box_E + m^2)G_E(x,y) = \delta^{(4)}(x-y) \tag{4.20}$$

Thus $G_E(x,y)$ is the inverse to the operator $(-\Box_E + m^2)$.

The propagator we have defined describes the probability amplitude for the propagation of a single particle. This is clearly a quantity of physical interest, for if the particle undergoes interactions either with an external field or with other particles during the course of its propagation from $(y^0, \boldsymbol{y})$ to $(x^0, \boldsymbol{x})$, this will affect the probability amplitude for the propagation. The calculation of the propagator $\langle 0|T(\phi(x)\phi(y))|0\rangle$, suitably generalized to include interactions will then capture the effect of the interactions. Scattering amplitudes, for example, will be directly given by the propagator.

A generalization of the propagator to the many-particle case can be easily made. The quantities of interest are the N-point functions defined by

$$G(x_1, x_2, \cdots, x_N) = \langle 0|T(\phi(x_1)\phi(x_2)\cdots\phi(x_N))|0\rangle \tag{4.21}$$

As an example, consider the 4-point function $G(x_1, x_2, x_3, x_4)$. In the limit of $x_1^0, x_2^0, x_3^0 \to \infty$ and $x_4^0 \to -\infty$, we have one-particle in the far past and three particles in the far future, corresponding to the process of a particle decaying into three others. $G(x_1, x_2, x_3, x_4)$, with these assignments of time-labels, gives the probability amplitude for such a process. Likewise, in the limit of $x_1^0, x_2^0 \to \infty$ and $x_3^0, x_4^0 \to -\infty$, we have two particles in the far past and two particles in the far future, and the corresponding $G(x_1, x_2, x_3, x_4)$ gives the amplitude for two-particle scattering. Similarly, the N-point functions give amplitudes for a variety of physical processes. It is clear that the N-point functions are the quantities of interest.

A succinct way to describe all the N-point functions is to collect them together into a generating functional $Z[J]$ defined by

$$Z[J] = \sum_N \frac{1}{N!} \int d^4x_1 d^4x_2 \cdots d^4x_N \, G(x_1, x_2, \cdots, x_N)J(x_1)J(x_2)\cdots J(x_N)$$

$$= \langle 0|T\left[\exp\left(\int J\phi\right)\right]|0\rangle \tag{4.22}$$

where $J(x)$ is an arbitrary function of the spacetime coordinates x. J is not an operator. It is often referred to as a source function. By expanding $Z[J]$ in powers of J we can easily recover all the N-point functions as the coefficients of the expansion. Alternatively we may write

$$G(x_1, x_2, \cdots, x_N) = \frac{\delta}{\delta J(x_1)} \frac{\delta}{\delta J(x_2)} \cdots \frac{\delta}{\delta J(x_N)} \, Z[J]\Bigg|_{J=0} \tag{4.23}$$

We now derive an equation obeyed by $Z[J]$. The basic ingredients for this calculation will be the operator equations of motion and the canonical equal time commutation rules. We start with the quantity

$$\frac{\delta Z[J]}{\delta J(x)} = \sum_N \frac{1}{N!} \int J(y_1)J(y_2)\cdots J(y_N)\langle 0|T\phi(x)\phi(y_1)\phi(y_2)\cdots\phi(y_N)|0\rangle$$

$$= \langle 0|T\phi(x)e^{\int J\phi}|0\rangle \tag{4.24}$$

Consider applying $\Box_x$ on this quantity. The space-derivatives go through the time-ordering symbol and act on $\phi(x)$. The time-derivatives can produce extra terms. This can be seen as follows. From the definition of time-ordering, we can write

$$\langle 0|T\phi(x)e^{\int J\phi}|0\rangle = \langle 0|e^{\int_{t_{M-1}}^{t_M} dy^0 d^3y J\phi}e^{\int_{t_{M-2}}^{t_{M-1}} dy^0 d^3y J\phi}\cdots e^{\int_{x^0}^{t_1} dy^0 d^3y J\phi}\phi(x)$$

$$\times e^{\int_{t_1'}^{x^0} dy^0 d^3y J\phi}e^{\int_{t_2'}^{t_1'} dy^0 d^3y J\phi}\cdots e^{\int_{t_M'}^{t_{M-1}'} dy^0 d^3y J\phi}|0\rangle$$

$$= \langle 0|Te^{\int_{x^0}^{\infty} dy^0 d^3y J\phi}\phi(x)Te^{\int_{-\infty}^{x^0} dy^0 d^3y J\phi}|0\rangle$$

$$= \langle 0|P(\infty,x^0)\phi(x)P(x^0,-\infty)|0\rangle \tag{4.25}$$

where in the first equation we have divided the time interval into $2M$ intervals, labeled by t_i, t_i'. Eventually, $M\to\infty$, with the intervals shrinking to zero, as in the usual definition of the integral. The second step isolates the x^0-dependence in the appropriate order. We have also made the notation compact by defining

$$P(z^0,y^0) = T\exp\left(\int_{y^0}^{z^0} d^4x\, J(x)\phi(x)\right) \tag{4.26}$$

This has the property

$$\frac{\partial}{\partial z^0}P(z^0,y^0) = \int d^3x\, J(z^0,\boldsymbol{x})\phi(z^0,\boldsymbol{x})\, P(z^0,y^0)$$

$$\frac{\partial}{\partial y^0}P(z^0,y^0) = -P(z^0,y^0)\int d^3x\, J(y^0,\boldsymbol{x})\phi(y^0,\boldsymbol{x}) \tag{4.27}$$

Taking the time-derivative of (4.25) and using (4.27), we get

$$\frac{\partial}{\partial x^0}\langle 0|T\phi(x)e^{\int J\phi}|0\rangle = \langle 0|P(\infty,x^0)\frac{\partial\phi(x)}{\partial x^0}P(x^0,-\infty)|0\rangle$$

$$+ \int d^3y J(x^0,\boldsymbol{y})\langle 0|P(\infty,x^0)\Big[\phi(x)\phi(x^0,\boldsymbol{y})$$

$$- \phi(x^0,\boldsymbol{y})\phi(x)\Big]P(x^0,-\infty)|0\rangle \tag{4.28}$$

The equal time commutator $[\phi(x^0, \boldsymbol{x}), \phi(x^0, \boldsymbol{y})]$ is the extra term we have; this is zero by the canonical commutation rules and so we may write

$$\frac{\partial}{\partial x^0}\langle 0|T\phi(x)e^{\int J\phi}|0\rangle = \langle 0|P(\infty, x^0)\frac{\partial\phi(x)}{\partial x^0}P(x^0, -\infty)|0\rangle \tag{4.29}$$

In a similar fashion, the second time-derivative becomes

$$\frac{\partial^2}{(\partial x^0)^2}\langle 0|T\phi(x)e^{\int J\phi}|0\rangle = \langle 0|P(\infty, x^0)\frac{\partial^2\phi(x)}{(\partial x^0)^2}P(x^0, -\infty)|0\rangle$$
$$+ \int d^3y J(x^0, \boldsymbol{y})\langle 0|P(\infty, x^0)[\dot\phi(x^0, \boldsymbol{x}), \phi(x^0, \boldsymbol{y})]$$
$$P(x^0, -\infty)|0\rangle \tag{4.30}$$

The extra term $[\dot\phi(x^0, \boldsymbol{x}), \phi(x^0, \boldsymbol{y})]$ is $-i\delta^{(3)}(x - y)$. Using this, we find

$$\frac{\partial^2}{(\partial x^0)^2}\langle 0|T\phi(x)e^{\int J\phi}|0\rangle = \langle 0|P(\infty, x^0)\frac{\partial^2\phi(x)}{(\partial x^0)^2}P(x^0, -\infty)|0\rangle$$
$$- iJ(x)\langle 0|Te^{\int J\phi}|0\rangle$$
$$= \langle 0|T\left[\frac{\partial^2\phi(x)}{(\partial x^0)^2}e^{\int J\phi}\right]|0\rangle - iJ(x)Z[J] \tag{4.31}$$

Using this equation we find

$$(\Box_x + m^2)\frac{\delta Z[J]}{\delta J(x)} = \langle 0|T\left[(\Box_x + m^2)\phi(x)e^{\int J\phi}\right] - iJ(x)Z[J] \tag{4.32}$$

So far we have used the canonical commutation rules to simplify the above expression. The operator equation of motion $(\Box + m^2)\phi = 0$ for the free scalar field can now be used to obtain the equation

$$(\Box_x + m^2)\frac{\delta Z[J]}{\delta J(x)} = -iJ(x)Z[J] \tag{4.33}$$

This is the equation of motion for $Z[J]$. All the N-point functions may be obtained by solving this functional equation. The solution is actually quite easy to write down; it is given by

$$Z[J] = \mathcal{N}\exp\left[\frac{1}{2}\int d^4x d^4y\ J(x)G(x, y)J(y)\right] \tag{4.34}$$

where $\mathcal{N}$ is some quantity independent of J. The definition of $Z[J] = \langle 0|T\exp(\int J\phi)|0\rangle$ shows that we must have $Z[0] = 1$. This fixes the normalization factor $\mathcal{N}$ to be 1. Expression (4.34) is easily verified to be a solution to (4.33) provided $(\Box_x + m^2)G(x, y) = -i\delta^{(4)}(x - y)$. A priori, there are many

Green's functions $G(x, y)$ which could be used here. However, from the solution (4.34), and using the definition of $Z[J]$ expanded to quadratic order in the J's, we find that $\langle 0|T(\phi(x)\phi(y)|0\rangle = G(x, y)$. This identifies $G(x, y)$ in (4.34) as the Feynman propagator, or equivalently the boundary conditions for inverting the operator $(\square + m^2)$ have been specified.

The N-point functions $G(x_1, \cdots, x_N)$ can be written down explicitly by using (4.23) and (4.34) as

$$G(x_1, \cdots, x_{2n}) = \sum_P G(x_{i_1}, x_{i_2})G(x_{i_3}, x_{i_4}) \cdots G(x_{i_{2n-1}}, x_{i_{2n}}) \qquad (4.35)$$

where the sum is over all pairings of the coordinate labels $x_1, x_2, \cdots, x_{2n}$.

4.2 Propagator for fermions

The field operator for spin-$\frac{1}{2}$ particles (fermions) is given by

$$\psi(x) = \sum_{p,r} \sqrt{\frac{m}{E_p V}} \left[a_{p,r}\, u_r(p)e^{-ipx} + b^\dagger_{p,r} v_r(p)e^{ipx} \right] \qquad (4.36)$$

$$\bar{\psi}(x) = \sum_{p,r} \sqrt{\frac{m}{E_p V}} \left[a^\dagger_{p,r}\bar{u}_r(p)e^{ipx} + b_{p,r}\bar{v}_r(p)e^{-ipx} \right] \qquad (4.37)$$

where the creation and annihilation operators obey anticommutation rules. In the case of the scalar field, the time-ordered product $T[\phi(x)\phi(y)]$ was defined as $\phi(x)\phi(y)$ for $x^0 > y^0$ and $\phi(y)\phi(x)$ for $y^0 > x^0$. We get the same expression as $x^0 \to y^0$ from above or below, because the ϕ's commute at equal time. In the case of fermions, we have anticommutation rules at equal times, and, in order to get the same expression as $x^0 \to y^0$ from either side, we must define

$$T\,\psi(x)\psi(y) = \begin{cases} \psi(x)\psi(y), & x^0 > y^0 \\ -\psi(y)\psi(x) & y^0 > x^0 \end{cases}$$

For ψ and $\bar{\psi}$, we have similarly

$$T\,\psi(x)\bar{\psi}(y) = \begin{cases} \psi(x)\bar{\psi}(y), & x^0 > y^0 \\ -\bar{\psi}(y)\psi(x), & y^0 > x^0 \end{cases} \qquad (4.38)$$

The propagator is defined by

$$\begin{aligned} S(x, y) &= \langle 0|T\psi(x)\bar{\psi}(y)|0\rangle \\ &= \sum_{p,s} \left(\frac{m}{E_p V} \right) \left[\theta(x^0 - y^0)u_{sp}\bar{u}_{sp}e^{-ip(x-y)} \right. \\ &\qquad\qquad \left. - \theta(y^0 - x^0)v_{sp}\bar{v}_{sp}e^{ip(x-y)} \right] \end{aligned}$$

$$= \sum_p \left(\frac{1}{2E_p V} \right) \Big[\theta(x^0 - y^0)(\gamma \cdot p + m)e^{-ip(x-y)}$$

$$- \theta(y^0 - x^0)(\gamma \cdot p - m)e^{ip(x-y)} \Big]$$

$$= \int \frac{d^3p}{(2\pi)^3} \frac{1}{2E_p} \Big[\theta(x^0 - y^0)(\gamma \cdot p + m)e^{-ip(x-y)}$$

$$- \theta(y^0 - x^0)(\gamma \cdot p - m)e^{ip(x-y)} \Big]$$

$$(4.39)$$

(as $V \to \infty$). We have used the properties

$$\sum_{p,s} u_{sp} \bar{u}_{sp} = \frac{(\gamma \cdot p + m)}{2m}$$

$$\sum_{p,s} v_{sp} \bar{v}_{sp} = \frac{(\gamma \cdot p - m)}{2m} \qquad (4.40)$$

The propagator (4.39) can be written as

$$S(x,y) = \langle 0|T\psi(x)\bar{\psi}(y)|0\rangle$$

$$= i \int \frac{d^4p}{(2\pi)^4} \frac{\gamma \cdot p + m}{p^2 - m^2 + i\epsilon} e^{-ip(x-y)}$$

$$= (i\gamma \cdot \partial + m)G(x,y) \qquad (4.41)$$

We also have $\langle 0|T\psi(x)\psi(y)|0\rangle = \langle 0|T\bar{\psi}(x)\bar{\psi}(y)|0\rangle = 0$.

The fermion propagator (4.41) is easily seen to obey the equation

$$(i\gamma \cdot \partial - m)S(x,y) = i\delta^{(4)}(x - y) \qquad (4.42)$$

The propagator is thus the inverse of the operator $(i\gamma \cdot \partial - m)$.

4.3 Grassman variables and generating functional for fermions

In defining multiparticle propagators for fermions, it is again useful to collect them together into a generating functional. Recall that the important characteristic of fermions is that they obey the exclusion principle; there cannot be double occupancy of states. This is encoded in the anticommutation rules obeyed by fermionic fields. The definition of time-ordering for fermionic operators, viz., (4.38) also reflects this. Appropriate source functions (the analogs of J) for collecting together the multiparticle propagators and the functional differentiation with respect to them is then provided by anticommuting c-number functions or Grassman variables. A Grassman number η has the property

$$\eta^2 = 0, \qquad\qquad \eta \neq 0 \qquad\qquad (4.43)$$

In other words, a Grassman number is nilpotent. One could, if desired, give an explicit realization of η as

$$\eta = p \begin{pmatrix} 0 & 1 \\ 0 & 0 \end{pmatrix} \qquad\qquad (4.44)$$

where p is a real number. When there are many Grassman variables, this can lead to considerable notational complexity. Fortunately, for our purpose, such explicit realizations will not be necessary; the algebraic structure is all we need. We can describe a number of Grassman variables η_i, $i = 1, 2, ..., N$ by

$$\eta_i \eta_j + \eta_j \eta_i = 0 \qquad\qquad (4.45)$$

More generally, one can also consider Grassman-valued functions $\eta(x)$. (x may be thought of as a continuous version of the index i.) For Grassman-valued functions, we have

$$\eta(x)\eta(y) + \eta(y)\eta(x) = 0 \qquad\qquad (4.46)$$

The product of two Grassman numbers (or generally an even number of such numbers) behaves like an ordinary c-number function or bosonic variable in its commutation properties.

One can also define functions of a Grassman variable. These are defined by a Taylor series expansion in the Grassman variable around zero. The series is always finite for a finite number of Grassman variables since the square of a Grassman variable is zero. For example, for functions of a single Grassman variable we may write

$$f(x, \eta) = f_0(x) + \eta f_1(x) \qquad\qquad (4.47)$$

or if we have N variables η_i, $i = 1, 2, ..., N$,

$$\begin{aligned}
f(x, \eta) &= f_0(x) + \sum_i f_i(x)\eta_i + \frac{1}{2!}\sum_{ij} f_{ij}(x)\eta_i\eta_j + \cdots \\
&\quad + \frac{1}{N!}\sum_{i_1 i_2 ... i_N} f_{i_1 i_2 ... i_N}(x)\eta_{i_1}\eta_{i_2}...\eta_{i_N}
\end{aligned} \qquad (4.48)$$

Notice that $f_{i_1 i_2 ... i_k}$ must be antisymmetric under exchange of any two indices since the product of the η's which multiplies it has this property.

One can define differentiation of such functions by making a variation of η_i, bringing the resulting $\delta\eta_i$ to, say, the left end by making use of the antisymmetry property on the indices and then defining the coefficient of $\delta\eta_i$ as the derivative. We find

$$\frac{\partial f}{\partial \eta_i} = f(x)_i + \sum_{i_2} f_{i i_2}\eta_{i_2} + ... + \frac{1}{(N-1)!}\sum_{i_2 ... i_N} f(x)_{i i_2 ... i_N}\eta_{i_2}...\eta_{i_N} \qquad (4.49)$$

Differentiation has the property

$$\frac{\partial}{\partial \eta_i} \frac{\partial}{\partial \eta_j} = -\frac{\partial}{\partial \eta_j} \frac{\partial}{\partial \eta_i} \tag{4.50}$$

Functional differentiation can likewise be defined by a power series expansion in $\eta(x)$ and obeys the rule

$$\frac{\partial}{\partial \eta(x)} \frac{\partial}{\partial \eta(y)} = -\frac{\partial}{\partial \eta(y)} \frac{\partial}{\partial \eta(x)} \tag{4.51}$$

The notion of Grassman numbers and variables may not be as intuitive or as natural as our notion of real and complex numbers, owing to their nilpotent nature. Nevertheless, the above rules for Grassman variables form a consistent set of algebraic rules for doing calculations and this suffices for the purpose we have in mind.

We now return to fermionic fields. The propagator for fermions is given by $S(x, y) = \langle 0|T\psi(x)\bar{\psi}(y)|0\rangle$, which obeys (4.42). The many-body fermion propagators can be discussed by introducing a generating functional

$$Z[\eta, \bar{\eta}] = \langle 0|T \exp\left(\int \bar{\eta}\psi + \bar{\psi}\eta\right)|0\rangle \tag{4.52}$$

where we have introduced Grassman-valued functions $\eta, \bar{\eta}$ in a way analogous to J for bosonic fields; namely, one can calculate the N-point functions for fermions by differentiating $Z[\eta, \bar{\eta}]$ an appropriate number of times and then setting $\eta, \bar{\eta}$ to zero. The sources $\eta, \bar{\eta}$ are spinors and have to be Grassman-valued. We take them to anticommute with $\psi, \bar{\psi}$ as well. The advantage of introducing such sources is that we can write

$$\begin{aligned} T\bar{\eta}(x)\psi(x)\bar{\psi}(y)\eta(y) &= \bar{\eta}(x)\psi(x)\bar{\psi}(y)\eta(y), & x^0 > y^0 \\ &= \bar{\psi}(y)\eta(y)\bar{\eta}(x)\psi(x), & y^0 > x^0 \end{aligned} \tag{4.53}$$

Thus the combinations $\bar{\eta}(x)\psi(x)$ and $\bar{\psi}(y)\eta(y)$ behave like bosonic operators for time-ordering. The additional minus sign which was introduced for the time-ordering of fermionic operators is taken account of by the Grassman-valued sources. We can now write

$$\frac{\delta Z}{\delta \bar{\eta}(x)} = \langle 0|T\psi(x)e^{\int \bar{\eta}\psi + \bar{\psi}\eta}|0\rangle \tag{4.54}$$

Acting on this with the operator $(i\gamma \cdot \partial - m)$, we get

$$(i\gamma \cdot \partial - m)\frac{\delta Z}{\delta \bar{\eta}(x)} = \langle 0|T(i\gamma \cdot \partial - m)\psi(x)e^{\int \bar{\eta}\psi + \bar{\psi}\eta}|0\rangle +$$

$$i\gamma^0 \int d^3y \langle 0|T[\psi(x), (\bar{\eta}\psi(y) + \bar{\psi}\eta(y))] e^{\int \bar{\eta}\psi + \bar{\psi}\eta}|0\rangle \tag{4.55}$$

The first term on the right-hand side vanishes by the Dirac equation for ψ. The commutator in the second term, which is an equal time commutator, can be evaluated by the canonical anticommutation rules and gives $\gamma^0 \eta \delta^{(3)}(x-y)$. The above equation then simplifies to

$$(i\gamma \cdot \partial - m)\frac{\delta Z}{\delta \bar{\eta}(x)} = i\eta(x)Z \tag{4.56}$$

The solution to this equation is given by

$$Z[\eta, \bar{\eta}] = \exp\left[\int d^4x d^4y \; \bar{\eta}(x)S(x,y)\eta(y)\right] \tag{4.57}$$

where we have chosen the normalization to agree with $Z[0,0] = 1$.

References

1. Commutativity of fields at spacelike separations is an important physical requirement. It can be considered an axiom of field theory in the Wightman formulation, see R.F. Streater and A.S. Wightman, *PCT, Spin and Statistics and All That*, W.A. Benjamin, Inc. (1964). The relation between causal propagation of fields and antiparticles is also discussed in R.P. Feynman, "The Reason for Antiparticles" in *Elementary Particles and the Laws of Physics*, Oxford University Press (1987).
2. The importance of the Feynman propagator was realized by Stückelberg, Feynman and Dyson, see J. Schwinger, *Selected Papers in Quantum Electrodynamics*, Dover Publications, Inc. (1958); S. S. Schweber, *QED and the Men Who Made It*, Princeton University Press (1994).
3. For Grassmann variables applied to field theory, the original reference is F.A. Berezin, *The Method of Second Quantization*, Academic Press (1966). Grassmann variables have been used in Statistical Mechanics for evaluation of partition functions, some scattering problems, etc. For an interesting solution of the Ising model using Grassmann variables, see S. Samuel, J. Math. Phys. **19**, 1438 (1978).

5 Interactions and the S-matrix

5.1 A general formula for the S-matrix

In this chapter we begin the description of interacting field theories. The case of the scalar field theory with an interaction of the form ϕ^4 will be treated as an example. The formalism is, of course, more general and is easily extended to any polynomial interaction.

The Lagrangian for the theory we are considering is thus

$$\mathcal{L} = \tfrac{1}{2}\left[(\partial\phi)^2 - m^2\phi^2\right] - \lambda\phi^4 \tag{5.1}$$

Here λ is a constant; it is a measure of the strength of the interaction and is referred to as the coupling constant. The equation of motion is given by

$$(\Box_x + m^2)\phi(x) + 4\lambda\phi^3(x) = 0 \tag{5.2}$$

In the quantum theory, $\phi(x)$ is an operator on a Hilbert space and the above equation is an operator equation of motion. In addition to the equation of motion, we also have the canonical commutation rules

$$\begin{aligned}
\left[\phi(x^0, \boldsymbol{x}), \phi(x^0, \boldsymbol{y})\right] &= 0 \\
\left[\pi(x^0, \boldsymbol{x}), \phi(x^0, \boldsymbol{y})\right] &= -i\,\delta^{(3)}(x - y) \\
\left[\pi(x^0, \boldsymbol{x}), \pi(x^0, \boldsymbol{y})\right] &= 0
\end{aligned} \tag{5.3}$$

where $\pi(x^0, \boldsymbol{x}) = \partial_0\phi(x^0, \boldsymbol{x})$. If the interaction term $4\lambda\phi^3$ is set to zero, then $\phi(x)$ is a free scalar field and one gets the standard many-particle description. In this case, the commutation rules and the equation of motion show that $\phi(x)$ can be written as $\phi(x) = \sum a_k u_k(x) + a_k^\dagger u_k^*(x)$, where $a_k, a_k^\dagger$ represent annihilation and creation operators for the particles. Since the equation for ϕ is linear, the notion of what a single particle state is does not change with time. In the case with an interaction term, the equation of motion is nonlinear and we see that if we start with $a_k^\dagger$, then because of the nonlinear term, we can get $a^{\dagger 3}, a^\dagger a^3, ..$ terms. Thus a state $a_k^\dagger|0\rangle$ can evolve into $a_k^{\dagger 3}|0\rangle$ for example. This would describe the decay of a one-particle state into a three-particle state. The evolution of $a_k^\dagger a_l^\dagger|0\rangle$ into $a_p^\dagger a_q^\dagger|0\rangle$ would describe the two-particle scattering with the momenta as shown. All these processes will be generically referred to as scattering. The basic quantity of interest to us is the scattering amplitude or the transition amplitude for such a process.

As we have discussed before, the amplitude for such processes can be obtained from the N-point functions

$$G(x_1, x_2, \cdots, x_N) = \langle 0|T(\phi(x_1)\phi(x_2) \cdots \phi(x_N))|0\rangle \qquad (5.4)$$

by taking the time labels to $\pm\infty$ in a way appropriate to the process of interest. Therefore as a first step in calculating the scattering amplitudes, we shall derive an equation for such functions in the interacting case. The generating functional for the N-point functions was defined as

$$Z[J] = \langle 0|T \exp\left[\int d^4x \ J(x)\phi(x)\right]|0\rangle \qquad (5.5)$$

In the last chapter, we also obtained the equation (4.32),

$$(\Box_x + m^2 - i\epsilon)\frac{\delta Z[J]}{\delta J(x)} = \langle 0|T\left[(\Box_x + m^2)\phi(x)e^{\int J\phi}\right]|0\rangle - iJ(x)Z[J] \quad (5.6)$$

We have put in the $i\epsilon$ explicitly to specify that the Green's function to be used is the Feynman propagator $G(x,y)$. In the discussion for the free case, the equation of motion was $(\Box + m^2)\phi(x) = 0$ and this was used to simplify the above equation. The only difference in the interacting case is that the equation of motion is different. In fact, using (5.2), we get the equation for the generating functional in the interacting case as

$$(\Box_x + m^2 - i\epsilon)\frac{\delta Z[J]}{\delta J(x)} + \langle 0|T\left[4\lambda\phi^3(x)e^{\int J\phi}\right]|0\rangle = -iJ(x)Z[J] \qquad (5.7)$$

Since

$$\frac{\delta}{\delta J(x_1)}\frac{\delta}{\delta J(x_2)}\frac{\delta}{\delta J(x_3)}Z[J] = \langle 0|T\left[\phi(x_1)\phi(x_2)\phi(x_3)e^{\int J\phi}\right]|0\rangle \qquad (5.8)$$

we may write (5.7) as

$$(\Box_x + m^2 - i\epsilon)\frac{\delta Z[J]}{\delta J(x)} + 4\lambda\left(\frac{\delta}{\delta J(x)}\right)^3 Z[J] = -iJ(x)Z[J] \qquad (5.9)$$

In the free case the solution to this equation was given as

$$Z_0[J] = \exp\left[\frac{1}{2}\int d^4x d^4y \ J(x)G(x,y)J(y)\right] \qquad (5.10)$$

The solution to (5.9) is then given by

$$Z[J] = \mathcal{N}\exp\left[-i\lambda\int d^4x \ \left(\frac{\delta}{\delta J(x)}\right)^4\right] Z_0[J] \qquad (5.11)$$

$\mathcal{N}$ is a normalization factor. The above solution is easily checked as follows.

$$
\begin{aligned}
(\Box_x + m^2 - i\epsilon)\frac{\delta Z[J]}{\delta J(x)} &= \mathcal{N}e^{-i\lambda\int(\delta/\delta J)^4}(\Box_x + m^2 - i\epsilon)\frac{\delta Z_0[J]}{\delta J(x)} \\
&= \mathcal{N}e^{-i\lambda\int(\delta/\delta J)^4}(-iJ(x)Z_0[J]) \\
&= e^{-i\lambda\int(\delta/\delta J)^4}(-iJ(x))e^{i\lambda\int(\delta/\delta J)^4}\ Z[J] \\
&= -4\lambda\left(\frac{\delta}{\delta J(x)}\right)^3 Z[J] - iJ(x)\ Z[J]
\end{aligned}
$$

$$\tag{5.12}$$

where we have used the identity

$$
\begin{aligned}
e^{-i\lambda\int(\delta/\delta J)^4}(-iJ(x))e^{i\lambda\int(\delta/\delta J)^4} &= -iJ(x) - \lambda[(\int \delta/\delta J)^4, J(x)] \\
&= -iJ(x) - 4\lambda\left(\frac{\delta}{\delta J(x)}\right)^3 \tag{5.13}
\end{aligned}
$$

The solution (5.11) can be brought to a more useful form by using the following identity.

$$
F\left[\frac{\delta}{\delta\varphi}\right]G[\varphi]e^{\int J\varphi}\bigg|_{\varphi=0} = G\left[\frac{\delta}{\delta J}\right]F[J] \tag{5.14}
$$

for functionals F, G which can be expanded in a power series. Here φ is also an arbitrary function of the spacetime coordinates; it is not the field operator. This identity can be checked as follows. We may write the left-hand side of (5.14) as

$$
\begin{aligned}
F\left[\frac{\delta}{\delta\varphi}\right]G[\varphi]\ e^{\int J\varphi}\bigg|_{\varphi=0} &= F\left[\frac{\delta}{\delta\varphi}\right]G\left[\frac{\delta}{\delta J}\right]e^{\int J\varphi}\bigg|_{\varphi=0} \\
&= G\left[\frac{\delta}{\delta J}\right]F\left[\frac{\delta}{\delta\varphi}\right]e^{\int J\varphi}\bigg|_{\varphi=0} \\
&= G\left[\frac{\delta}{\delta J}\right]F[J]\,e^{\int J\varphi}\bigg|_{\varphi=0} \\
&= G\left[\frac{\delta}{\delta J}\right]F[J] \tag{5.15}
\end{aligned}
$$

Using (5.14), we can write (5.11) as

$$
Z[J] = \mathcal{N}e^{\frac{1}{2}\int G\delta\delta}\ e^{-i\int\lambda\varphi^4+\int J\varphi}\bigg|_{\varphi=0} \tag{5.16}
$$

where we have used obvious abbreviations like

$$\int G\delta\delta = \int d^4x d^4y \, G(x,y) \, \frac{\delta}{\delta\varphi(x)} \, \frac{\delta}{\delta\varphi(y)}$$

This expression can be further simplified by bringing the $J\varphi$-term to the left end. We can do this by

$$e^{\frac{1}{2}\int G\delta\delta} e^{\int J\varphi} = e^{\int J\varphi} \left[e^{-\int J\varphi} e^{\frac{1}{2}\int G\delta\delta} e^{\int J\varphi} \right]$$

$$= e^{\int J\varphi} \exp\left[e^{-\int J\varphi} (\frac{1}{2} \int G\delta\delta) e^{\int J\varphi} \right]$$

$$= e^{\int J\varphi} \exp\left[\frac{1}{2} \int G\delta\delta + [\frac{1}{2} \int G\delta\delta, \int J\varphi] \right.$$

$$\left. + \frac{1}{2!} [[\frac{1}{2} \int G\delta\delta, \int J\varphi], \int J\varphi] \right]$$

$$= e^{\int J\varphi} \exp\left[\frac{1}{2} \int G\delta\delta + \int JG\delta + \frac{1}{2} \int JGJ \right] \quad (5.17)$$

The solution (5.16) can now be written as

$$Z[J] = e^{\frac{1}{2}\int JGJ} \left[e^{\int JG\delta} \mathcal{F}[\varphi] \right]_{\varphi=0}$$

$$\mathcal{F}[\varphi] = \mathcal{N} e^{\frac{1}{2}\int G\delta\delta} e^{-i\lambda \int \varphi^4} \quad (5.18)$$

In arriving at (5.18) we have not used any perturbative approximation. The form (5.18) is, however, well suited to a perturbative expansion, which can be obtained by expanding the exponential $\exp(-i\lambda \int \varphi^4)$ in powers of λ.

The quantity $\exp(-i\lambda \int \varphi^4)$ is, of course, $\exp(iS_{int})$, where $S_{int} = -\lambda \int \varphi^4$ is the interaction part of the action. We can thus generalize the above formulae to any polynomial type of interaction as follows.

$$Z[J] = e^{\frac{1}{2}\int JGJ} \left[e^{\int JG\delta} \mathcal{F}[\varphi] \right]_{\varphi=0}$$

$$\mathcal{F}[\varphi] = \mathcal{N} \exp\left(\frac{1}{2} \int G\delta\delta \right) \exp\left(iS_{int}[\varphi] \right) \quad (5.19)$$

The normalization factor $\mathcal{N}$ is fixed by requiring $Z[0] = 1$, which is equivalent to

$$\mathcal{N} \left[e^{\frac{1}{2}\int G\delta\delta} e^{iS_{int}[\varphi]} \right]_{\varphi=0} = 1 \quad (5.20)$$

We now turn to the scattering amplitude. Consider first the part of $Z[J]$ given by $e^{\int JG\delta} \mathcal{F}[\varphi]$. The contribution of this term to the N-point function is obtained as

$$G(x_1, x_2, \cdots, x_N) = \int d^4 z_1 d^4 z_2 \cdots d^4 z_N \; G(x_1, z_1) G(x_2, z_2) \cdots$$

$$G(x_N, z_N) \; V(z_1, z_2, ..., z_N) \qquad (5.21)$$

$$V(z_1, z_2, ..., z_N) = \frac{\delta}{\delta\varphi(z_1)} \frac{\delta}{\delta\varphi(z_2)} \cdots \frac{\delta}{\delta\varphi(z_N)} \mathcal{F}[\varphi] \bigg]_{\varphi=0} \qquad (5.22)$$

$V(z_1, z_2, ..., z_N)$ is often referred to as a vertex function.

Consider $x_1^0, x_2^0, ..., x_n^0 \to -\infty$ and $x_{n+1}^0, ..., x_N^0 \to \infty$. If the vertex function does not extend to infinity, i.e., if it has compact support, we may then take $x_1^0, x_2^0, ..., x_n^0 < z_1^0, z_2^0, ..., z_n^0$ and $x_{n+1}^0, ..., x_N^0 > z_{n+1}^0, ..., z_N^0$. The propagators $G(x, z)$ can then be replaced by their expressions for the appropriate time-ordering. We then find

$$G(x_1, x_2, \cdots, x_N) = \int d^4 z_1 d^4 z_2 \cdots d^4 z_N \sum_{k_i} u_{k_1}(z_1) u_{k_1}^*(x_1) u_{k_2}(z_2) u_{k_2}^*(x_2)...$$

$$u_{k_{n+1}}(x_{n+1}) u_{k_{n+1}}^*(z_{n+1})... \; V(z_1, z_2, ..., z_N) \qquad (5.23)$$

Along the lines of our general interpretation of the N-point functions as probability amplitudes, this has the interpretation as the probability amplitude for the propagation of particles introduced at $x_1, x_2, ..., x_n$ to be observed as particles at $x_{n+1}, ..., x_N$. This is indeed the quantity of physical interest, but it is expressed in a basis where the particle positions are specified. For most scattering situations, we specify the momenta. The corresponding amplitude can be obtained by projecting out the momentum eigenstates by appropriate Fourier transformation using the orthonormality relation (1.3) from Chapter 1. The result is

$$S(k_1, k_2, ..., k_n \to k_{n+1}, ..., k_N) = \int d^4 z_1 d^4 z_2 \cdots d^4 z_N \; u_{k_1}(z_1) u_{k_2}(z_2)...$$

$$u_{k_{n+1}}^*(z_{n+1})...u_{k_N}^*(z_N) \; V(z_1, z_2, ..., z_N)$$

$$= \int d^4 z_1 d^4 z_2 \cdots d^4 z_N \; u_{k_1}(z_1) u_{k_2}(z_2)...$$

$$u_{k_{n+1}}^*(z_{n+1})...u_{k_N}^*(z_N)$$

$$\times \frac{\delta}{\delta\varphi(z_1)} \frac{\delta}{\delta\varphi(z_2)} \cdots \frac{\delta}{\delta\varphi(z_N)} \mathcal{F}[\varphi] \bigg]_{\varphi=0}$$

$$(5.24)$$

$S(k_1, k_2, ..., k_n \to k_{n+1}, ..., k_N)$ is the amplitude for the scattering process with the momenta indicated. It can be thought of as the matrix element

$$S(k_1, k_2, ..., k_n \to k_{n+1}, ..., k_N) = \langle k_{n+1}, ..., k_N | \hat{S} | k_1, k_2, ..., k_n \rangle$$

of an operator $\hat{S}$ which is appropriately called the scattering operator or the S-operator. Equations (5.22,5.24) show that this matrix element is determined by the functional $\mathcal{F}[\varphi]$; it is thus appropriate to refer to $\mathcal{F}[\varphi]$ as the S-matrix functional. The functional differentiations in (5.24) show that the S-matrix element is obtained by replacing the φ's in $\mathcal{F}[\varphi]$ by the one-particle wave functions, u_k for the incoming particles and u_k^* for the outgoing particles.

So far we have not included the effect of the terms which arise from $\exp(\frac{1}{2}\int JGJ)$ in $Z[J]$. This can lead to terms in $G(x_1, x_2, \cdots, x_N)$ of the form $G(x_1, x_2)G(x_3, ..., x_N)$ and similar product forms where a number of free propagators are multiplied by a many-particle propagator with less than N particles. From the factorization of the propagators in terms of u_k, u_k^*, it is clear that these correspond to processes where some of the particles do not participate in the interaction but just fly by; they are propagators disconnected from the main part of the scattering process. Thus contributions from $\exp(\frac{1}{2}\int JGJ)$ describe subscattering processes and are not of great interest. (The nontrivial scattering contribution in such terms is taken account of at a lower order in J.)

Notice that from (5.21) we may write

$$V(x_1, x_2, ..., x_N) = \prod_j i(\Box_{x_j} + m^2)G(x_1, x_2, \cdots, x_N) \qquad (5.25)$$

where we have used the relation $i(\Box_x + m^2)G(x, y) = \delta^{(4)}(x - y)$. We may thus write the scattering amplitude as

$$S(k_1, k_2, ..., k_n \to k_{n+1}, ..., k_N) = \prod_{j=1}^{n} \int_x u_{k_j}(x_j)i(\Box_{x_j} + m^2) \times$$

$$\prod_{r=n+1}^{N} \int_y u_{k_r}^*(x_r)i(\Box_{x_r} + m^2)\ G(x_1, ..., x_N)$$

$$(5.26)$$

This result is often known as the reduction formula. (The operators $i(\Box_{x_j} + m^2)$ cancel poles in the propagators and hence, in this formula, partial integration of the derivatives such that they act on the one-particle wave functions is not justified. Alternatively, the wave functions may be taken to be arbitrary, to be set to being solutions of the free wave equation only at the end.)

By substituting $J = i(\Box + m^2 - i\epsilon)\varphi$ in equation (5.19) and doing some partial integrations, we see that the S-matrix functional can be written as

$$\mathcal{F}[\varphi] = e^{-\frac{i}{2}\int \varphi(\Box + m^2)\varphi}\ Z[i(\Box + m^2)\varphi] \qquad (5.27)$$

i.e., it is just $Z[J]$ evaluated for the choice $J = i(\Box + m^2 - i\epsilon)\varphi$, apart from a trivial factor of $\exp(-\frac{i}{2}\int \varphi(\Box + m^2)\varphi)$.

5.2 Wick's theorem

We now want to do the perturbative expansion of the S-matrix functional. In doing this, we encounter terms of the form $\exp(\int \frac{1}{2} G \delta \delta)\, \varphi(1)\varphi(2) \cdots \varphi(N)$ where $\varphi(i)$ stands for $\varphi(x_i)$.

Consider the action of one power of $\int \frac{1}{2} G \delta \delta$ on the fields φ.

$$
\frac{1}{2} \int_{x,y} G(x,y) \frac{\delta}{\delta\varphi(x)} \frac{\delta}{\delta\varphi(y)}\, \varphi(1)\varphi(2) = \frac{1}{2} \left[\int_x G(x,1) \frac{\delta}{\delta\varphi(x)} \varphi(2) \right.
$$
$$
\left. + \int_x G(x,2) \frac{\delta}{\delta\varphi(x)} \varphi(1) \right]
$$
$$
= G(1,2) \tag{5.28}
$$

The operator $\int \frac{1}{2} G \delta \delta$ replaces the pair of fields by its propagator. This is known as Wick contraction. When applied on $\varphi(1)\varphi(2) \cdots \varphi(N)$ we get

$$
\int \tfrac{1}{2} G \delta \delta\, \varphi(1)\varphi(2) \cdots \varphi(N) = \sum_{i<j} G(i,j)\varphi(1) \cdots \varphi(i-1)\varphi(i+1) \cdots
$$
$$
\varphi(j-1)\varphi(j+1) \cdots \varphi(N) \tag{5.29}
$$

Notice that we again get the sum of all pairings or Wick contractions on the right-hand side, with one propagator only. There is no other numerical factor. The $G(i,j)$ term arises in two ways: when the first functional derivative acts on $\varphi(i)$, the second on $\varphi(j)$, and when the first functional derivative acts on $\varphi(j)$ and the second on $\varphi(i)$. This removes the factor of $\frac{1}{2}$.

Now consider the term quadratic in the G's in the expansion of $\exp(\int \frac{1}{2} G \delta \delta)$. We find

$$
\frac{1}{2!} \int \tfrac{1}{2} G \delta \delta \int \tfrac{1}{2} G \delta \delta\, \varphi(1)\varphi(2)\varphi(3)\varphi(4) = G(1,2)G(3,4) + G(1,3)G(2,4)
$$
$$
+ G(1,4)G(2,3) \tag{5.30}
$$

We again get the sum of all Wick contractions. The pair $(1,2)$ can occur from the action of the first $\int \frac{1}{2} G \delta \delta$ or the second; this removes the $\frac{1}{2!}$. Applied on $\varphi(1)\varphi(2) \cdots \varphi(N)$, we get the sum of all terms with two Wick contractions with no other numerical factors. Continuing in this way, we get the formula

$$
e^{\int \frac{1}{2} G \delta \delta}\, \varphi(1)\varphi(2) \cdots \varphi(N) = \varphi(1)\varphi(2) \cdots \varphi(N)
$$
$$
+ \sum \text{terms with 1 Wick contraction each}
$$
$$
+ \sum \text{terms with 2 Wick contractions each}
$$
$$
+ \sum \text{terms with 3 Wick contractions each}
$$
$$
+ \cdots \tag{5.31}
$$

This result is known as Wick's theorem. It gives a simple rule to carry through the differentiations that we encounter in using the formula for $\mathcal{F}[\varphi]$. We shall refer to $\exp(\int \frac{1}{2} G\delta\delta)$ as the Wick contraction operator.

5.3 Perturbative expansion of the S-matrix

We now consider the perturbative expansion of the S-matrix functional as given by (5.18). This is easily done by expansion of the exponential in powers of λ.

1. *Zeroth order in λ*

   We get $Z[J] = Z_0[J]$ and there is no nontrivial scattering.

2. *First order in λ*

   For the first order term we get

$$\mathcal{N}^{-1}\mathcal{F}^{(1)} = e^{\frac{1}{2}\int G\delta\delta}(-i\lambda \int \varphi^4)$$

$$= -i\lambda \int \varphi^4 - i6\lambda \int \varphi(x)^2 G(x,x)$$

$$- i3\lambda \int G(x,x)G(x,x) \tag{5.32}$$

We shall analyze each of these terms separately. The first term leads to a vertex function with four points and hence to processes with four external particles. This can describe a decay process $1 \to 3$, a $2 \to 2$-scattering, or a $3 \to 1$ process. The amplitudes are easily written down by replacing the φ's by the wave functions as in (5.24). For the $1 \to 3$ decay process, we have

$$S(k \to p_1, p_2, p_3) = (-i\lambda)4! \int d^4x \, u_k(x)u_{p_1}^*(x)u_{p_2}^*(x)u_{p_3}^*(x)$$

$$= (-i\lambda)4! \frac{(2\pi)^4\delta^{(4)}(k - p_1 - p_2 - p_3)}{\sqrt{(2\omega_k V)(2\omega_{p_1} V)(2\omega_{p_2} V)(2\omega_{p_3} V)}} \tag{5.33}$$

We can represent this diagrammatically as shown below. This diagram, known as a Feynman diagram, not only gives an intuitive picture of the process involved, but it is also a mnemonic for the mathematical expression for the scattering amplitude. We associate a factor of $(1/\sqrt{2\omega V})$ with each external line; the vertex carries a factor of $(-i\lambda)4!$ and an energy-momentum conservation δ-function $(2\pi)^4 \, \delta^{(4)}(k - p_1 - p_2 - p_3)$. Formula (5.33) can then be written down by taking the product of such factors. (Since all the particles involved have the same mass, this particular process is forbidden by energy conservation. The amplitude can be written down as in (5.33), but the δ-function has no support and vanishes. The same rules for associating mathematical expressions with a diagram are general and apply to situations for which the amplitude does not vanish.)

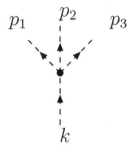

Fig 5.1. $1 \to 3$ decay process

For the $2 \to 2$ scattering process, we have

$$S(k_1, k_2 \to p_1, p_2) = (-i\lambda)4!\frac{(2\pi)^4\delta^{(4)}(k_1 + k_2 - p_1 - p_2)}{\sqrt{(2\omega_{k_1}V)(2\omega_{k_2}V)(2\omega_{p_1}V)(2\omega_{p_2}V)}} \qquad (5.34)$$

We can represent this diagrammatically as follows. This describes two-particle scattering to the lowest order in the coupling constant λ.

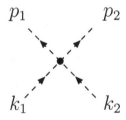

Fig 5.2. $2 \to 2$ scattering process

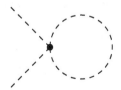

Fig 5.3. Scalar self-energy or mass correction

The term $-i6\lambda G(x,x)\varphi(x)^2$ can only lead to processes with two external particles. It gives a correction to the propagation of a single particle and contains information about corrections to the mass of the particle due to its self-interactions. Representing the propagator $G(x,y)$ as a line from x to y, we can diagrammatically represent this term as shown in figure 5.3. The proper treatment of this term requires ideas about renormalization and will be postponed for now.

Fig 5.4. The first order vacuum process

The term $-i3\lambda G(x,x)G(x,x)$ can be diagrammatically represented as in figure 5.4. It is a pure vacuum correction, i.e., changes $\langle 0|0\rangle$; but, of course, it is canceled by our choice of normalization $\mathcal{N}$. In fact

$$\mathcal{N} = 1 + i3\lambda \int d^4x \; G(x,x)G(x,x) + \mathcal{O}(\lambda^2) \tag{5.35}$$

(There can be situations where the vacuum diagrams can be important physically. If we consider field theory in the presence of external fields, vacuum or ground state (which now includes the external field) can decay or undergo other interesting changes. These can be described by the vacuum diagrams. Likewise, for fields at finite temperature, the vacuum diagrams are important temperature-dependent corrections to the partition function.)

2. *Second order in* λ

The various terms of order λ^2 are given by

$$\mathcal{N}^{-1}\mathcal{F}^{(2)} = e^{\frac{1}{2}\int G\delta\delta} \; \frac{(-i\lambda)^2}{2!} \int d^4x d^4y \; \varphi^4(x)\varphi^4(y) \tag{5.36}$$

This leads to many terms which are repetitions of first-order processes. Thus we get terms which can be diagrammatically represented as

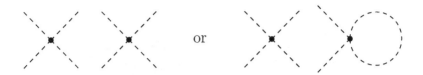

Fig 5.5. Disconnected diagrams, repetitions of first order processes

or

Fig 5.6. Another disconnected diagram, repetition of first order processes

Genuinely new terms arise when there is a propagator $G(x, y)$ connecting points x and y. The simplest such term is

$$\frac{(-i\lambda)^2}{2!} 4^2 \int \varphi^3(x) G(x, y) \varphi^3(y) \tag{5.37}$$

Fig 5.7. A contribution to $3 \to 3$ scattering process

This term includes many processes, such as $3 \to 3$ scattering, $2 \to 4$ scattering, $1 \to 5$ decay, etc. As an example consider $3 \to 3$ scattering $k_1, k_2, k_3 \to p_1, p_2, p_3$. The amplitude is given by

$$S(k_i, p_i) = (-i\lambda 4!)^2 \int u_{k_1}(x)u_{k_2}(x)u_{k_2}(x)G(x,y)u_{p_1}^*(y)u_{p_2}^*(y)u_{p_3}^*(y)$$

$$+ \text{permutations}$$

$$= (-i\lambda 4!)^2 \frac{i}{(k_1 + k_2 + k_3)^2 - m^2 + i\epsilon} \frac{(2\pi)^2 \delta^{(4)}(\sum_i k_i - \sum_i p_i)}{\prod_i \sqrt{2\omega_{k_i} V}\sqrt{2\omega_{p_i} V}}$$

$$+ \text{permutations}$$

$$(5.38)$$

Diagrammatically this can be repesented as shown in figure 5.7. If we make a rule of representing each internal line or propagator in momentum space by $(i/(k^2 - m^2 + i\epsilon))$, then the mathematical expression can be written down at once from the diagram. The conservation of four-momentum at each vertex gives the correct momentum (in this case $k_1 + k_2 + k_3$) to the propagator. Various other processes contained in (5.37) can be treated similarly.

The terms with two propagators connecting x, y are contained in

$$\frac{(-i\lambda)^2}{2} \frac{(4 \times 3)^2}{2} \int d^4x d^4y\; \varphi^2(x)\varphi^2(y)G(x,y)G(x,y) \qquad (5.39)$$

This can give the next order ($\mathcal{O}(\lambda^2)$ corrections) to $2 \to 2$ scattering, $1 \to 3$ decay, etc. For example, for the scattering $k_1, k_2 \to p_1, p_2$, we find

$$\frac{(-i\lambda 4!)^2}{2} \int \Big[\; u_{k_1}(x)u_{k_2}(x)G(x,y)G(x,y)u_{p_1}^*(y)u_{p_2}^*(y)$$

$$+ u_{k_1}(x)u_{p_1}^*(x)G(x,y)G(x,y)u_{k_2}(y)u_{p_2}^*(y)$$

$$+ u_{k_1}(x)u_{p_2}^*(x)G(x,y)G(x,y)u_{k_2}(y)u_{p_1}^*(y)\Big] \quad (5.40)$$

These terms are obtained by different assignment of momenta to the basic process represented diagrammatically by figure 5.8.

Fig 5.8. $\mathcal{O}(\lambda^2)$ correction to $2 \to 2$ scattering

The factor of $\frac{1}{2}$ in (5.40) follows from straightforward functional differentiation, but it can also be understood as arising from the symmetry of the diagram under the exchange of the two propagators. Putting in $G(x, y)$ and integrating, the first of the terms in (5.40) becomes

$$\frac{(-i\lambda 4!)^2}{2} \frac{(2\pi)^4 \delta^{(4)}(k_1 + k_2 - p_1 - p_2)}{\sqrt{(2\omega_{k_1} V)(2\omega_{k_2} V)(2\omega_{p_1} V)(2\omega_{p_2} V)}}$$

$$\times \int \frac{d^4 q}{(2\pi)^4} \frac{i}{(k_1 + k_2 + q)^2 - m^2 + i\epsilon} \frac{i}{q^2 - m^2 + i\epsilon} \quad (5.41)$$

We find a new rule that we must integrate over the loop momentum q with the measure $(d^4 q / (2\pi)^4)$. This concludes our discussion of the basic diagrammatic rules.

It is clear that the expansion of $\mathcal{F}[\varphi]$ can be carried out to any order and that at each order, the terms in $\mathcal{F}[\varphi]$ describe a variety of processes.

5.4 Decay rates and cross sections

The S-matrix functional gives the amplitude for a process via (5.24). The transition amplitude, we have seen, has the form

$$S = \left(\prod_i \frac{1}{\sqrt{2\omega_i V}} \right) (2\pi)^4 \delta^{(4)}\left(\sum k - \sum p \right) \mathcal{M} \quad (5.42)$$

where the product is over all external particles and $\mathcal{M}$ is an invariant matrix element.

Decay Rate

Consider the decay of a particle of momentum k into n particles of momenta p_i. The S-matrix element has the form

$$S = \frac{1}{\sqrt{2\omega_k V}} \prod_i^n \frac{1}{\sqrt{2\omega_i V}} (2\pi)^4 \delta^{(4)}\left(k - \sum p \right) \mathcal{M} \quad (5.43)$$

The square of this amplitude gives the decay probability. The square of the δ-function is ambiguous. This problem arises because we have considered infinite spacetime volume. We must interpret it as

$$\int d^4 x e^{i(\sum k - \sum p)x} \int d^4 y e^{i(\sum k - \sum p)y} = (2\pi)^4 \delta^{(4)}\left(\sum k - \sum p \right) \int d^4 y$$

$$= (2\pi)^4 \delta^{(4)}\left(\sum k - \sum p \right) V\tau \quad (5.44)$$

where τ is the range of time-integration, taken finite for now; we shall take $\tau \to \infty$ eventually. The absolute square of (5.43), keeping in mind (5.44), gives

$$|S|^2 = \frac{1}{2\omega_k V} \prod_i^n \frac{1}{2\omega_i V} (2\pi)^4 \delta^{(4)}(\sum k - \sum p) V\tau \, |\mathcal{M}|^2 \qquad (5.45)$$

The decay rate is thus given by

$$\frac{|S|^2}{\tau} = \frac{1}{2\omega_k} \prod_i^n \frac{1}{2\omega_i V} (2\pi)^4 \delta^{(4)}(k - \sum p) \, |\mathcal{M}|^2 \qquad (5.46)$$

This is the decay rate into specified sharp values of the final momenta (of the outgoing particles). In practice, we are interested in scattering into a small range $d^3 p_1, d^3 p_2, \ldots$ of each final momentum around mean values $p_1, p_2, \ldots$. This can be obtained by summing over all states with this range of momenta, viz., $(V d^3 p_i/(2\pi)^3$ for each p_i. Thus the decay rate into the specified ranges of final mometa is given by

$$d\Gamma = \frac{1}{2\omega_k} |\mathcal{M}|^2 (2\pi)^4 \delta^{(4)}(k - \sum p) \prod_i \frac{d^3 p_i}{2\omega_{p_i} (2\pi)^3} \qquad (5.47)$$

The total decay rate is given by integration over all final momenta as

$$\Gamma = \int d\Gamma \qquad (5.48)$$

The lifetime of the particle is given by $1/\Gamma$. Notice that $d\Gamma$ and Γ are invariant except for the factor $(1/2\omega_k)$. This factor gives them the correct Lorentz-transformation property and gives the time-dilation effect for lifetimes of fast-moving particles.

Cross sections

We now consider a $2 \to n$ scattering process with momenta $k_1, k_2 \to p_1, p_2, \ldots, p_n$. The amplitude has the form

$$S = \frac{1}{\sqrt{2\omega_{k_1} V}} \frac{1}{\sqrt{2\omega_{k_2} V}} \prod_i^n \frac{1}{\sqrt{2\omega_i V}} (2\pi)^4 \delta^{(4)}(k_1 + k_2 - \sum p) \, \mathcal{M} \qquad (5.49)$$

Taking the absolute square using (5.44) as before, we get the rate for the process as

$$\frac{|S|^2}{\tau} = \frac{1}{2\omega_{k_1}} \frac{1}{2\omega_{k_2} V} \prod_i^n \frac{1}{2\omega_i V} (2\pi)^4 \delta^{(4)}(k_1 + k_2 - \sum p) \, |\mathcal{M}|^2 \qquad (5.50)$$

We must divide this by the flux to obtain the cross section. The flux can be computed as follows. Consider the collision of two particles of masses m_1, m_2, and momenta k_1, k_2. If one of the particles is at rest, say, $k_1 = (m_1, 0, 0, 0)$, the flux due to the other particle is given by

$$F = \frac{1}{V} \frac{|k_2|}{\omega_{k_2}} \tag{5.51}$$

since $(|k_2|/\omega_{k_2})$ is the speed of the second particle and $(1/V)$ is the density of the particle for the plane waves we have. We can generalize this to any frame as follows. Since $k_1^2 = \omega_{k_1}^2 - \boldsymbol{k_1}^2 = m_1^2, k_2^2 = \omega_{k_2}^2 - \boldsymbol{k_2}^2 = m_1^2$, the only kinematic invariant is $k_1 \cdot k_2$. We have $(k_1 \cdot k_2)^2 - m_1^2 m_2^2 = m_1^2(\omega_{k_2}^2 - m_2^2) = m_1^2 \boldsymbol{k_2}^2$ if $k_1 = (m_1, 0, 0, 0)$. Thus the generalization of (5.51), preserving the necessary $1 \leftrightarrow 2$ symmetry, is

$$F = \frac{1}{V} \frac{\sqrt{(k_1 \cdot k_2)^2 - m_1^2 m_2^2}}{\omega_{k_1} \omega_{k_2}} \tag{5.52}$$

Dividing the rate (5.50) by this flux gives the cross section for scattering into sharp values of the final momenta. For a range of momenta of dispersion $d^3 p_i$ around the p_i, we get the differential cross section

$$d\sigma = |\mathcal{M}|^2 \frac{(2\pi)^4 \delta^{(4)}(k_1 + k_2 - \sum p)}{4\sqrt{(k_1 \cdot k_2)^2 - m^4}} \prod_i \frac{d^3 p_i}{2\omega_{p_i}(2\pi)^3} \tag{5.53}$$

where we have set $m_1 = m_2 = m$, which is the case for us. Notice that $d\sigma$ is completely Lorentz invariant. Also factors of V anf τ have canceled out, so we can take the limits $V, \tau \to \infty$.

5.5 Generalization to other fields

The generalization of the S-matrix functional to the case of many types of fields in interaction is quite straightforward. Notice that the formula (5.19) involves the propagator of free particles and the interaction part of the Lagrangian. Thus, once we know the propagators and the interaction Lagrangian, we can immediately write down the expression for the S-matrix functional.

As an example, consider a theory with two types of bosonic fields, denoted by φ and ξ, of masses μ_1 and μ_2, respectively, and two types of fermionic fields denoted by P and N with masses m_1 and m_2, respectively. The various propagators are then given by

$$G_1(x, y) = G(x, y, \mu_1) = \int \frac{d^4 k}{(2\pi)^4} \left(\frac{i}{k^2 - \mu_1^2 + i\epsilon} \right) e^{-ik(x-y)}$$

$$G_2(x, y) = G(x, y, \mu_2) = \int \frac{d^4 k}{(2\pi)^4} \left(\frac{i}{k^2 - \mu_2^2 + i\epsilon} \right) e^{-ik(x-y)}$$

$$S_1(x, y) = S(x, y, m_1) = i \int \frac{d^4 p}{(2\pi)^4} \frac{\gamma \cdot p + m_1}{p^2 - m_1^2 + i\epsilon} e^{-ip(x-y)}$$

$$S_2(x, y) = S(x, y, m_2) = i \int \frac{d^4 p}{(2\pi)^4} \frac{\gamma \cdot p + m_2}{p^2 - m_2^2 + i\epsilon} e^{-ip(x-y)}$$

$$\tag{5.54}$$

The S-matrix functional can then be written as

$$
\mathcal{F}[\varphi, \xi, P, N] = \mathcal{N} \exp\left[\frac{1}{2}\int \frac{\delta}{\delta\varphi(x)} G(x, y, \mu_1) \frac{\delta}{\delta\varphi(y)}\right]
$$

$$
\times \exp\left[\frac{1}{2}\int \frac{\delta}{\delta\xi(x)} G(x, y, \mu_2) \frac{\delta}{\delta\xi(y)}\right]
$$

$$
\times \exp\left[-\int \frac{\delta}{\delta P_r(x)} S_{rs}(x, y, m_1) \frac{\delta}{\delta\bar{P}_s(y)}\right]
$$

$$
\times \exp\left[-\int \frac{\delta}{\delta N_r(x)} S_{rs}(x, y, m_2) \frac{\delta}{\delta\bar{N}_s(y)}\right] \; e^{iS_{int}[\varphi, \xi, P, N]}
$$

$$(5.55)$$

where we have used the same symbols φ, ξ, P, N to denote the (operator-valued) fields and the corresponding c-number functions which appear in the expression for the S-matrix functional.

The one-particle wave functions to be used in the calculation of the scattering amplitudes are the scalar and spin-$\frac{1}{2}$ wave functions given before, with the appropriate masses.

Another interesting set of examples is given by the theory of nonrelativistic particles which can be considered as the theory of the Schrödinger field. The Lagrangian for the free theory is

$$
\mathcal{L} = i\psi^* \frac{\partial\psi}{\partial t} + \psi^* \frac{\nabla^2}{2m} \psi
$$

$$(5.56)$$

It is possible to quantize this field either using commutation rules giving the theory of nonrelativistic bosons or anticommutation rules giving the theory of nonrelativistic fermions. Consider the fermionic case as an example. The anticommutation rules, obtained via the canonical one-form, are

$$
\{ \psi(x^0, \boldsymbol{x}), \ \psi(x^0, \boldsymbol{y}) \} = 0
$$
$$
\{\psi^\dagger(x^0, \boldsymbol{x}), \psi^\dagger(x^0, \boldsymbol{y}) \} = 0
$$
$$
\{ \psi(x^0, \boldsymbol{x}), \psi^\dagger(x^0, \boldsymbol{y}) \} = \delta^{(3)}(x - y)
$$

$$(5.57)$$

The one-particle wave functions are given by

$$
u_k(x) = \frac{1}{\sqrt{V}} \; e^{-iE_k x^0 + i\boldsymbol{k}\cdot\boldsymbol{x}}
$$

$$(5.58)$$

where $E_k = k^2/2m$. The mode expansions for the fields are given by

$$
\psi(x) = \sum_k a_k u_k(x)
$$
$$
\psi^\dagger(x) = \sum_k a_k^\dagger u_k^*(x)
$$

$$(5.59)$$

where

$$\{a_k, a_l\} = \{a_k^\dagger, a_l^\dagger\} = 0$$
$$\{a_k, a_l^\dagger\} = \delta_{kl} \tag{5.60}$$

The propagator is given by

$$S(x, y) = \int \frac{d^4k}{(2\pi)^4} \frac{i}{k^0 - E_k + i\epsilon} e^{-ik^0(x^0 - y^0) + i\mathbf{k} \cdot \mathbf{x}} \tag{5.61}$$

$S(x, y)$ is the inverse to $\partial_0 - i\nabla^2/2m$.

The S-matrix functional is given by

$$\mathcal{F}[\varphi] = \mathcal{N} \exp\left[-\int \frac{\delta}{\delta\psi(x)} S(x, y) \frac{\delta}{\delta\psi^*(y)} \right] e^{iS_{int}[\psi, \psi^*]} \tag{5.62}$$

where ψ, ψ^* are Grassman-valued functions now.

The functional differentiations involved in evaluating the S-matrix functional can be carried out by using the rule of Wick's theorem given earlier. Wick's theorem for fermions works just as in the case of bosons with possible extra signs from moving the Grassman variables around to bring them to the correct order to be identified with the propagator. For example, for fermion fields $\psi, \bar{\psi}$,

$$\mathcal{W} \, \psi(1)\bar{\psi}(2) = \psi(1)\bar{\psi}(2) + S(1, 2) \tag{5.63}$$

where we have denoted by $\mathcal{W}$ the Wick contraction operator

$$\exp\left[-\int \frac{\delta}{\delta\psi(x)} S(x, y) \frac{\delta}{\delta\bar{\psi}(y)} \right] \equiv \mathcal{W} \tag{5.64}$$

For the product of fields in the order $\bar{\psi}(2)\psi(1)$, we will get the propagator $-S(1, 2)$. We may think of this as writing $\bar{\psi}(2)\psi(1) = -\psi(1)\bar{\psi}(2)$ first, so that the fields are in the order in which they appear in the propagator and then identifying the pair with the propagator. By extension we then find

$$\begin{aligned}
\mathcal{W}\bar{\psi}(1)\psi(2)\bar{\psi}(3)\psi(4) = {} &\bar{\psi}(1)\psi(2)\bar{\psi}(3)\psi(4) - S(2, 1)\bar{\psi}(3)\psi(4) \\
&-\psi(2)\bar{\psi}(3)S(4, 1) + \bar{\psi}(1)\psi(4)S(2, 3) \\
&-\bar{\psi}(1)\psi(2)S(4, 3) \\
&+S(2, 1)S(4, 3) - S(2, 3)S(4, 1) \tag{5.65}
\end{aligned}$$

We see that we get all possible pairings with coefficients equal to ± 1 depending on the number of transpositions of the Grassman-valued fields. This kind of rule extends easily to higher numbers of fields.

5.6 Operator formula for the N-point functions

We now obtain a formula for the N-point functions expressed in terms of an operator expectation value. We start with the formulae (5.10, 5.11), which give

$$Z_0[J] = \exp\left[\tfrac{1}{2}\int d^4x d^4y\ J(x)G(x,y)J(y)\right]$$

$$Z[J] = \mathcal{N}\exp\left[-i\lambda\int d^4x\ \left(\frac{\delta}{\delta J(x)}\right)^4\right] Z_0[J] \qquad (5.66)$$

Since $Z_0[J]$ is the generating functional for the N-point functions of a free scalar field, we can write, using the definition (5.5),

$$Z_0[J] = \langle 0|T \exp\left[\int d^4x\ J(x)\phi_{in}(x)\right]|0\rangle \qquad (5.67)$$

where $\phi_{in}(x)$ is a *free field*. In other words, it has the expansion

$$\phi_{in}(x) = \sum_k a_k u_k(x) + a_k^\dagger u_k^*(x) \qquad (5.68)$$

(The operator $\phi_{in}(x)$ is called the "in-field" because the incoming states can be contructed in terms of its action on the vacuum state.) Using the expression (5.67) for $Z_0[J]$ in (5.66), and doing the functional derivatives inside the matrix element, we can write $Z[J]$ as

$$\begin{aligned} Z[J] &= \mathcal{N}\langle 0|T e^{\int J\phi_{in}} e^{-i\lambda\int \phi_{in}^4}|0\rangle \\ &= \mathcal{N}\langle 0|T e^{\int J\phi_{in}} e^{iS_{int}(\phi_{in})}|0\rangle \end{aligned} \qquad (5.69)$$

By functionally differentiating N times with respect to J and setting J to zero we obtain

$$G(x_1, x_2, ..., x_N) = \langle 0|T\ \phi_{in}(x_1)\phi_{in}(x_2)...\phi_{in}(x_N)e^{iS_{int}(\phi_{in})}|0\rangle \qquad (5.70)$$

This gives the Green's functions of the interacting theory in terms of free field expectation values. Evidently a similar formula is obtained for more general situations and more general interaction terms in the Lagrangian.

We can also write the scattering operator in terms of the in-fields ϕ_{in}. In the above formula, we have arbitrary Green's functions expressed in terms of ϕ_{in}. We now take a set of time-labels $x_1^0, x_2^0, ..., x_n^0$ to $-\infty$ and $x_{n+1}^0, ..., x_N^0$ to $+\infty$. In this limit, by virtue of the time-ordering, we can write (5.70) as

$$G(x_1, x_2, ..., x_N) = \langle 0|T\ \phi_{in}(x_{n+1})...\phi_{in}(x_N)T\ e^{iS_{int}(\phi_{in})}\phi_{in}(x_1)...\phi_{in}(x_n)|0\rangle \qquad (5.71)$$

where we assume that the fields in the interaction term do not extend to $\pm\infty$ in time. Since $G(x_1, x_2, ..., x_N)$ represents the scattering amplitude (albeit in terms of positions rather than momenta), we see that the scattering operator is given by

$$\widehat{S} = T \; e^{iS_{int}(\phi_{in})} = U(\infty, -\infty)$$

$$U(x^0, y^0) = T \; \exp \left[i \int_{y^0}^{x^0} dx^0 \int d^3x \; \mathcal{L}_{int}(\phi_{in}(x)) \right] \qquad (5.72)$$

This is a very simple formula: the S-operator is given by the exponential of the interaction part of the action with the *free field operator* ϕ_{in} substituted for the field ϕ. The S-operator as defined by the above formula is unitary. The operator U defined above obeys the properties

$$U(x^0, y^0)U(y^0, z^0) = U(x^0, z^0)$$
$$U^\dagger(x^0, y^0) = U(y^0, x^0) = U^{-1}(x^0, y^0) \qquad (5.73)$$

for $x^0 \geq y^0 \geq z^0$.

One can also define, at this point, the *interacting field operator* ϕ by

$$\phi(x) = \phi_{in}(x) + i \int_y G_R(x, y) \; \rho(y) \qquad (5.74)$$

where $G_R(x, y)$ is the retarded Green's function of (4.9) and ρ is defined to be

$$\rho(x) = -i\widehat{S}^{-1} \frac{\delta \widehat{S}}{\delta \phi_{in}(x)} \qquad (5.75)$$

The functional derivative with respect to ϕ_{in} is understood as follows. We shift ϕ_{in} in $\widehat{S}$ by a c-number (non-operator) function $\delta f(x)$, i.e., $\phi_{in} \to \phi_{in} + \delta f$. The derivative is then the coefficient of δf in the variation of the operator. In other words, the derivative is defined by

$$\delta \widehat{S} \equiv \int_x \frac{\delta \widehat{S}}{\delta \phi_{in}(x)} \delta f(x) \qquad (5.76)$$

Since the retarded Green's function obeys the equation $(\Box + m^2)G_R(x, y) = -i\delta^{(4)}(x - y)$, we find that

$$(\Box + m^2)\phi(x) = \rho(x) \qquad (5.77)$$

Conversely, (5.74) may be thought of as the solution of the above equation with the retarded boundary conditions, which is appropriate if $\phi \to \phi_{in}$ as $x^0 \to -\infty$. In principle, we can express ρ in terms of ϕ rather than ϕ_{in} by using (5.74). This will give an equation of motion expressed entirely in terms of ϕ. We shall now see that this agrees with the Heisenberg equation of motion, so that ϕ as defined by (5.74) can be taken as the Heisenberg field operator itself. First of all, a solution to ϕ as defined by (5.74) is given by

$$\phi(x) = U(-\infty, x^0)\phi_{in}(x)U(x^0, -\infty) \tag{5.78}$$

We can check this by rewriting it as

$$\phi(x) = \phi_{in}(x) + U(-\infty, x^0)\left[\phi_{in}(x), U(x^0, -\infty)\right] \tag{5.79}$$

Since $U(x^0, -\infty)$ is defined in terms of the free field ϕ_{in}, the commutator can be evaluated as

$$\left[\phi_{in}(x), U(x^0, -\infty)\right] = i\int_{-\infty}^{x^0} d^4y\, U(x^0, y^0)\Delta(x, y)\left(\frac{\delta S_{int}(\phi_{in})}{\delta\phi_{in}}\right)_y U(y^0, -\infty)$$

$$= i\int d^4y\, G_R(x, y)U(x^0, y^0)\frac{\delta S_{int}(\phi_{in})}{\delta\phi_{in}(y)}U(y^0, -\infty) \tag{5.80}$$

using $G_R(x, y) = \theta(x^0 - y^0)\Delta(x, y)$. Equation (5.79) can thus be written, by combining U's using (5.73), as

$$\phi(x) = \phi_{in}(x) + i\int_y G_R(x, y)U(-\infty, y^0)\frac{\delta S_{int}(\phi_{in})}{\delta\phi_{in}(y)}U(y^0, -\infty)$$

$$= \phi_{in}(x) + i\int_y G_R(x, y)\frac{\delta S_{int}(\phi)}{\delta\phi(y)} \tag{5.81}$$

where we have used (5.78) again in the last step. Now, from the definition of $U(x^0, y^0)$ in (5.72), we see that

$$\rho(x) = -i\widehat{S}^{-1}\frac{\delta\widehat{S}}{\delta\phi_{in}(x)} = U(-\infty, x^0)\frac{\delta S_{int}(\phi_{in})}{\delta\phi_{in}}U(x^0, -\infty) \tag{5.82}$$

Comparison of this equation with (5.81) shows that (5.78) is indeed a solution to (5.74). Further, from (5.74) and (5.78) we see that ρ can be expressed in terms of the interacting field ϕ as

$$\rho(x) \equiv -i\widehat{S}^{-1}\frac{\delta\widehat{S}}{\delta\phi_{in}(x)}$$

$$= \frac{\delta S_{int}(\phi)}{\delta\phi(x)} \tag{5.83}$$

Consequently, (5.77) is the Heisenberg equation of motion for a field with the action $S = \int\frac{1}{2}[(\partial\phi)^2 - m^2\phi^2] + S_{int}(\phi)$. In other words, ϕ as defined by (5.74) can be identified with the operator ϕ we started with in this chapter.

Finally, notice that, since ϕ_{in} is a free field, we can write the N-point functions of this field as

$$\langle 0| T \ \phi_{in}(x_1)\phi_{in}(x_2)...\phi_{in}(x_N)|0\rangle = \frac{\delta}{\delta J(x_1)} \frac{\delta}{\delta J(x_2)} \cdots \frac{\delta}{\delta J(x_N)} e^{\frac{1}{2}\int JGJ} \Bigg]_{J=0}$$

$$= e^{\frac{1}{2}\int G\delta\delta} \ \varphi(x_1)\varphi(x_2)...\varphi(x_N) \Bigg]_{\varphi=0}$$

$$(5.84)$$

where we have used (5.67) and (5.16), with the interaction term set to zero. Using this formula for the N-point functions of ϕ_{in} and equation (5.19), we find that we can write the S-matrix functional as

$$\mathcal{F}[\varphi] = \mathcal{N}\langle 0|T \ \exp\left[i\mathcal{S}_{int}(\phi_{in} + \varphi)\right]|0\rangle \qquad (5.85)$$

References

1. The idea of the S-matrix is due to W. Heisenberg, *Zeits. für Phys.* **120**, 513, 673 (1943). Covariant techniques for the perturbative calculation of the S-matrix were developed by J. Schwinger, S. Tomonaga, R.P. Feynman, F. Dyson and others. For original papers, see J. Schwinger, *Selected Papers in Quantum Electrodynamics*, Dover Publications, Inc. (1958).
2. The use of the Heisenberg picture for the S-matrix, which is how we start in this chapter, is due to C.N. Yang and D. Feldman, Phys. Rev. **79**, 972 (1950). It led to the more complete LSZ formulation of H. Lehmann, K. Symanzik and W. Zimmerman, Nuovo Cimento, **1**, 1425 (1955); *ibid.* **2**, 425 (1955); *ibid.* **6**, 319 (1957). The reduction formula and the in-fields were introduced in this work.
3. The operator formula for the N-point functions is due to M. Gell-Mann and F. Low, Phys. Rev. **84**, 350 (1951).
4. An S-matrix functional has been defined in I. Ya Aref'eva, A.A. Slavnov and L.D. Faddeev, Theor. Math. Phys. **21**, 1165 (1974). In this context, see also L.D. Faddeev and A.A. Slavnov, *Gauge Fields, Introduction to the Quantum Theory*, Benjamin-Cummings, MA (1980).

6 The Electromagnetic Field

6.1 Quantization and photons

The action for the electromagnetic field is given by

$$S = \int -\frac{1}{4}F_{\mu\nu}F^{\mu\nu} - A_\mu J^\mu = \int \tfrac{1}{2}(E^2 - B^2) - A_0 J_0 + A_i J_i \qquad (6.1)$$

where

$$F_{\mu\nu} = \partial_\mu A_\nu - \partial_\nu A_\mu$$
$$F_{0i} = E_i, \qquad F_{ij} = \epsilon_{ijk}B_k \qquad (6.2)$$

A_μ is the four-vector potential. The equations of motion for this action, obtained by varying the A_μ, are the Maxwell equations

$$\partial_\mu F^{\mu\nu} = J^\nu \qquad (6.3)$$

which can be written out in terms of the components as

$$\partial_i E_i = J_0 \qquad (6.4)$$
$$\partial_0 E_i + \epsilon_{ijk}\partial_j B_k = J_i \qquad (6.5)$$

Notice that, since $\partial_\mu\partial_\nu F^{\mu\nu} = 0$, for consistency, the current to which A_μ couples, viz., J^μ, must be conserved. In other words

$$\partial_\mu J^\mu = 0 \qquad (6.6)$$

In considering the quantization of this field, we must take account of two related facts. Notice that if we have two vector potentials A_μ and $A'_\mu = A_\mu + \partial_\mu\theta$ for some scalar function θ, the corresponding electric and magnetic fields are unchanged, i.e., $F_{\mu\nu}(A) = F_{\mu\nu}(A')$. Thus we have a redundancy of variables in using A_μ to describe the electromagnetic field. We will need to eliminate the redundant degrees of freedom before we can apply the rules of quantization. We also notice that equation (6.4) cannot be realized as a Heisenberg equation of motion. Heisenberg equations of motion are of the form $\partial_0 C =$ something, for an operator C. Equation (6.5) shows that the equation of motion for A_i is second order in time-derivatives. Thus, $E_i = \partial_0 A_i - \partial_i A_0$ and A_i form the set of phase-space variables at a fixed time. Their initial values must be specified whereupon their time-evolution is given by (6.5). In terms of such phase-space variables, (6.4) has no time-derivatives and hence it cannot be obtained as a Heisenberg equation of motion.

The elimination of the redundant variables can be done as follows. First of all, we can choose $A_0 = 0$. For, if it is not zero, we use the physically equivalent set $A'_0 = A_0 + \partial_0\theta$, $A'_i = A_i + \partial_i\theta$ and choose θ such that $\partial_0\theta = -A_0$ to get $A'_0 = 0$ and some nonzero A'_i, which we can rename as A_i. We can then split A_i as

$$A_i = A_i^T + \partial_i f \tag{6.7}$$

where A_i^T is 'transverse'; i.e., it obeys the condition $\partial_i A_i^T = 0$. This leads to the relation $E_i = \partial_0 A_i^T + \partial_i(\partial_0 f)$. Equation (6.4) now tells us that $\partial_i\partial_i(\partial_0 f) = J_0$, or

$$\partial_0 f = \int d^3 y \; G_C(\boldsymbol{x} - \boldsymbol{y}) J_0(x^0, \boldsymbol{y})$$

$$\partial_i\partial_i G_C(\boldsymbol{x} - \boldsymbol{y}) = \delta^{(3)}(x - y) \tag{6.8}$$

$$G_C(\boldsymbol{x} - \boldsymbol{y}) = -\frac{1}{4\pi}\frac{1}{|\boldsymbol{x} - \boldsymbol{y}|}$$

$G_C(\boldsymbol{x} - \boldsymbol{y})$ is the Coulomb Green's function. $B_i(A) = B_i(A^T)$ since $\partial_i f$ has vanishing curl. The $A_i J_i$-term can be simplified as

$$\int d^4x \; A_i J_i = \int d^4x (A_i^T J_i + \partial_i f J_i) = \int d^4x (A_i^T J_i - f\partial_i J_i)$$

$$= \int d^4x (A_i^T J_i - f\partial_0 J_0) = \int d^4x (A_i^T J_i + \partial_0 f J_0)$$

$$= \int d^4x \; A_i^T J_i$$

$$+ \int dx^0 d^3x d^3y \; J_0(x^0, \boldsymbol{x}) G_C(\boldsymbol{x} - \boldsymbol{y}) J_0(x^0, \boldsymbol{y}) \tag{6.9}$$

Using equations (6.7) to (6.9) in the action (6.1), we find, upon partial integrations,

$$\mathcal{S} = \int d^4x \; \frac{1}{2} \left[\partial_0 A_j^T \partial_0 A_j^T - \partial_i A_j^T \partial_i A_j^T \right]$$

$$+ \int d^4x d^4y \; \frac{1}{2} J_0(x) G_C(\boldsymbol{x} - \boldsymbol{y}) \delta(x^0 - y^0) J_0(y) + \int d^4x \; A_i^T J_i$$

$$\tag{6.10}$$

We have included a δ-function in the time-variables for the term involving J_0's to write the integration measure in a covariant form. We see that (6.10) is equivalent to the action for two massless fields corresponding to the two transverse directions or polarizations of A_i^T. (The combination of conditions, $A_0 = 0$ and $\nabla \cdot A = 0$, which our physical fields A_i^T obey, is called the radiation gauge.) In the absence of the current J^μ, the equation of motion is

$$\Box\, A_i^T = 0 \tag{6.11}$$

which is the same as (6.5). The general solution to this equation can be written as

$$A_i^T(x) = \sum_{k\lambda} a_{k\lambda} e_i^{(\lambda)} u_k(x) + a_{k\lambda}^\dagger e_i^{(\lambda)} u_k^*(x)$$

$$u_k(x) = \frac{1}{\sqrt{2\omega_k V}} e^{-ikx} \tag{6.12}$$

where $\omega_k = \sqrt{k^2}$ and kx in the exponent is, as usual, $k_0 x_0 - k_i x_i$. $e_i^{(\lambda)}$, $\lambda = 1,2$ are unit vectors in the two independent directions transverse to k_i (or ∂_i), so as to be consistent with the condition $\partial_i A_i^T = 0$. For most purposes, we do not need an explicit form for these; one choice, if an explicit form is needed, is

$$e^{(1)} = \frac{1}{\sqrt{k_1^2 + k_2^2}} (k_2, -k_1, 0)$$

$$e^{(2)} = \frac{1}{\sqrt{k_1^2 + k_2^2}\sqrt{k \cdot k}} (k_1 k_3, k_2 k_3, -(k_1^2 + k_2^2)) \tag{6.13}$$

Along with $e_i^{(3)} = k_i/\sqrt{k \cdot k}$, these form an orthonormal triad of unit vectors. We also have

$$\sum_{\lambda=1,2} e_i^{(\lambda)} e_j^{(\lambda)} = \left(\delta_{ij} - \frac{k_i k_j}{k \cdot k} \right) \tag{6.14}$$

This can be seen as follows. The right-hand side must be a symmetric tensor P_{ij} with $k_i P_{ij} = 0$, $k_j P_{ij} = 0$ since $k_i e_i = 0$. Thus P_{ij} must be proportional to $(\delta_{ij} - k_i k_j/k \cdot k)$. The proportionality constant is seen to be 1 by evaluating the trace.

In the quantum theory, the coefficients $a_{k\lambda}, a_{k\lambda}^\dagger$ in (6.12) are operators. The canonical commutation rules are obtained from (6.10) via the canonical one-form. Equivalently, notice that the action is essentially that of two copies of a scalar field (even though the transformation properties of the fields are different) and hence the commutation rules can be easily seen to be

$$\left[a_{k\lambda}, a_{k'\lambda'} \right] = 0$$

$$\left[a_{k\lambda}^\dagger, a_{k'\lambda'}^\dagger \right] = 0 \tag{6.15}$$

$$\left[a_{k\lambda}, a_{k'\lambda'}^\dagger \right] = \delta_{kk'} \delta_{\lambda\lambda'}$$

Since the free theory without currents mimics two scalar fields, the Hamiltonian for the free case is

$$H = \sum_{k\lambda} \omega_k a_{k\lambda}^\dagger a_{k\lambda} \tag{6.16}$$

As in the case of the scalar field, Lorentz invariance requires that we choose an ordering of operators such that there is no zero point energy or vacuum energy.

The vacuum state obeys $a_{k\lambda}|0\rangle = 0$. One can build up many-particle states as we have done in detail for the scalar field. The particles in this case are massless (since $\omega_k = \sqrt{\boldsymbol{k} \cdot \boldsymbol{k}}$) and come in two polarizations orthogonal to $\boldsymbol{k}$. They are photons. $a_{k\lambda}^\dagger$ is the creation operator for a photon of wavevector or momentum $\boldsymbol{k}$ and polarization λ.

The propagator can be obtained as

$$\begin{aligned} D_{ij}(x,y) &= \langle 0|T\, A_i^T(x)A_j^T(y)|0\rangle \\ &= \int \frac{d^4k}{(2\pi)^4} \left(\delta_{ij} - \frac{k_i k_j}{\boldsymbol{k}\cdot\boldsymbol{k}}\right) \frac{i}{k^2 + i\epsilon} e^{-ik(x-y)} \end{aligned} \tag{6.17}$$

This is not manifestly covariant as it stands. If we start from an interaction of the form $A_\mu J^\mu$, then the interaction part of the action, after elimination of redundant variables, is

$$S_{int} = \int d^4x d^4y\, \frac{1}{2} J_0(x) G_C(\boldsymbol{x}-\boldsymbol{y})\delta(x^0 - y^0) J_0(y) + \int d^4x\, A_i^T J_i \tag{6.18}$$

The S-matrix functional can be written down as

$$\mathcal{F}(A) = \exp\left[\frac{1}{2}\int D_{ij}(x,y)\frac{\delta}{\delta A_i^T(x)}\frac{\delta}{\delta A_j^T(y)}\right] e^{iS_{int}} \tag{6.19}$$

The first term which involves a photon propagator is the term quadratic in the currents. The quadratic terms are given by

$$\begin{aligned} \mathcal{F}^{(2)} &= i\int d^4x d^4y\, \frac{1}{2} J_0(x) G_C(\boldsymbol{x}-\boldsymbol{y})\delta(x^0 - y^0) J_0(y) \\ &\quad - \frac{1}{2}\int d^4x d^4y\, J_i(x) D_{ij}(x,y) J_j(y) \end{aligned} \tag{6.20}$$

The term involving the $k_i k_j / \boldsymbol{k}\cdot\boldsymbol{k}$ part of the propagator may be written as

$$\begin{aligned} \int_{x,y} J_i(x) J_j(y) \int_k \frac{k_i k_j}{\boldsymbol{k}\cdot\boldsymbol{k}} \frac{e^{-ik(x-y)}}{k^2+i\epsilon} &= \int_{x,y} \partial_i J_i(x)\partial_j J_j(y) \int_k \frac{1}{\boldsymbol{k}\cdot\boldsymbol{k}} \frac{e^{-ik(x-y)}}{k^2+i\epsilon} \\ &= \int_{x,y} \partial_0 J_0(x)\partial_0 J_0(y) \int_k \frac{1}{\boldsymbol{k}\cdot\boldsymbol{k}} \frac{e^{-ik(x-y)}}{k^2+i\epsilon} \\ &= \int_{x,y} J_0(x) J_0(y) \int_k \frac{k_0^2}{\boldsymbol{k}\cdot\boldsymbol{k}} \frac{e^{-ik(x-y)}}{k^2+i\epsilon} \end{aligned}$$

$$\tag{6.21}$$

where we have used the conservation of the current as in (6.6). Noting that $G_C(\boldsymbol{x} - \boldsymbol{y})$ is the Fourier transform of $-(1/\boldsymbol{k} \cdot \boldsymbol{k})$, we can combine (6.20,6.21) to get

$$\mathcal{F}^{(2)} = \frac{1}{2} \int d^4x d^4y \left[J_0(x)J_0(y) \int_k \frac{i}{k^2 + i\epsilon} e^{-ik(x-y)} \right.$$
$$\left. - J_i(x)J_j(y) \int_k \frac{i}{k^2 + i\epsilon} e^{-ik(x-y)} \right]$$
$$= \frac{1}{2} \int d^4x d^4y \; J^\mu(x) D_{\mu\nu}(x,y) J^\nu(y) \tag{6.22}$$

$$D_{\mu\nu}(x,y) = \eta_{\mu\nu} \int \frac{d^4k}{(2\pi)^4} \frac{i}{k^2 + i\epsilon} e^{-ik(x-y)} \tag{6.23}$$

Thus the propagator, when applied to conserved currents, can be taken as the covariant propagator given by $D_{\mu\nu}(x, y)$. (Even in this form, one has the freedom of adding a term proportional to $\partial_\mu \partial_\nu$ to $\eta_{\mu\nu}$, since J^μ is conserved. This will also correspond to the freedom of gauge choice; it is discussed in more detail in Chapter 10.) The above result is exactly what we get if we apply the covariant Wick contraction on a covariant interaction term, viz.,

$$\mathcal{F}^{(2)} = -\frac{1}{2} \int_{x,y} D_{\mu\nu}(x,y) \frac{\delta}{\delta A_\mu(x)} \frac{\delta}{\delta A_\nu(y)} \frac{i^2}{2!} \int A_\mu J^\mu \int A_\nu J^\nu \tag{6.24}$$

For all terms in the S-matrix functional involving photon propagators, a similar simplification can be done. (We do not discuss this in detail here because these issues become much simpler and clearer once the functional integral is introduced.) For incoming and outgoing photons, we may still write the covariant form of the interaction with the understanding that the polarization vectors vanish for the time-components and are transverse to $\boldsymbol{k}$ for the space components. The S-matrix functional may therefore be taken as

$$\mathcal{F} = \exp\left(-\frac{1}{2} \int_{x,y} D_{\mu\nu}(x,y) \frac{\delta}{\delta A_\mu(x)} \frac{\delta}{\delta A_\nu(y)} \right) e^{i \int A_\mu J^\mu} \tag{6.25}$$

6.2 Interaction with charged particles

The coupling of the electromagnetic field to charged particles follows the "minimal coupling" principle. This may be formulated as follows. We have seen that in describing the electromagnetic field using the gauge potential A_μ, there is redundancy of variables. Both A_μ and $A'_\mu = A_\mu + \partial_\mu \theta$ describe the same physical situation since the electric and magnetic fields are the same. The change $A_\mu \to A_\mu + \partial_\mu \theta$ is a gauge transformation and one can say that the Maxwell equations are invariant with respect to gauge transformations. This invariance principle is the key to electromagnetic interactions. Coupling to charged fields must respect this invariance.

Consider a complex scalar field; the free Lagrangian is

$$\mathcal{L} = \partial_\mu \phi \partial^\mu \phi^* - m^2 \phi \phi^* \tag{6.26}$$

This theory has invariance under the transformation $\phi \to e^{-iQ\theta}\phi$, where θ is independent of x^μ. In other words, $\mathcal{L}(e^{-iQ\theta}\phi) = \mathcal{L}(\phi)$; we have already seen that this leads to a conserved current; Q is the charge carried by a single particle as given by this current. This symmetry is now made local; i.e., it is extended to the case where θ is considered to be a function of x^μ. Since $\partial_\mu(e^{-iQ\theta}\phi) = e^{-iQ\theta}(\partial_\mu - iQ\partial_\mu\theta)\phi$, clearly the terms in Lagrangian (6.26) with derivatives of the fields are not invariant. However, using A_μ, we can form the combination

$$D_\mu \phi = \partial_\mu \phi + iQ A_\mu \phi \tag{6.27}$$

We then have

$$D_\mu(A')\phi' = e^{-iQ\theta} \, D_\mu(A)\phi \tag{6.28}$$
$$A'_\mu = A_\mu + \partial_\mu\theta$$
$$\phi' = e^{-iQ\theta}\phi \tag{6.29}$$

In other words, $D_\mu\phi$ transforms homogeneously (or covariantly) even for local x-dependent θ if we combine the phase transformation of ϕ and the gauge transformation of A_μ. In fact, the phase transformation of ϕ can be identified as the gauge transformation for the matter field ϕ. $D_\mu\phi$ is called the covariant derivative. Since it transforms covariantly, we see that if we replace the derivative $\partial_\mu\phi$ in the Lagrangian (6.26) by $D_\mu\phi$, we get a Lagrangian which has invariance under the gauge transformations (6.29) and has interactions coupling ϕ, ϕ^* to the electromagnetic field. This is the *gauge principle*, which can be summarized as follows.

> Charged fields have Lagrangians which have invariance under constant phase transformations of the fields. This symmetry is made local by replacing all derivatives in the Lagrangian by the covariant derivatives. The resulting Lagrangian is gauge invariant and incorporates coupling to the electromagnetic field.

The application of this principle gives us, for the charged scalar field,

$$\mathcal{L} = D_\mu\phi D^\mu\phi^* - m^2\phi\phi^*$$
$$= \partial_\mu\phi\partial^\mu\phi^* - m^2\phi\phi^* + \mathcal{L}_{int}$$
$$\mathcal{L}_{int} = -iQA^\mu\left(\phi^*\partial_\mu\phi - \partial_\mu\phi^* \, \phi\right) + Q^2 A^\mu A_\mu \phi\phi^* \tag{6.30}$$

The S-matrix functional is then given by

$$\mathcal{F} = \exp\left[\int_{x_1,x_2} G(x_1,x_2)\frac{\delta}{\delta\varphi(x_1)}\frac{\delta}{\delta\varphi^*(x_2)}\right] \times$$
$$\exp\left(-\frac{1}{2}\int_{x,y} D_{\mu\nu}(x,y)\frac{\delta}{\delta A_\mu(x)}\frac{\delta}{\delta A_\nu(y)}\right) e^{iS_{int}} \tag{6.31}$$

We have used the covariant propagator for the photon, even though the use of the covariant propagator was justified only for the coupling $A_\mu J^\mu$ where J^μ is a conserved current. The use of the covariant propagator is actually correct. Rather than justify it at this stage, we shall simply assume it. The correctness of (6.31) will emerge naturally once we have defined functional integrals.

Some comments on the validity of the gauge principle are appropriate here. It is valid for all fundamental charged particles. Indeed, it is the defining principle for electromagnetic interactions. (And a suitable generalization gives other interactions as well.) It also applies to composite charged particles for the coupling of long-wavelength modes of A_μ, viz., those modes whose wavelengths are large compared to the size of the composite particle. For shorter-wavelength A_μ-modes coupling to composite particles, the effective interaction has to be derived from the details of the composite nature. An example is the neutron, which has overall charge neutrality but does have a magnetic moment. It has a coupling to the electromagnetic field given by $c\bar{N}\gamma_\mu\gamma_\nu N F^{\mu\nu}$, where N is the neutron field and the constant c is to be determined from the fact that the neutron is a bound state of three fundamental quarks. Clearly this is not of the form of a minimal coupling. Another example is the neutral π-meson which decays electromagnetically to two photons. The interaction Lagrangian is effectively

$$\mathcal{L}_{int} = \frac{e^2}{4\pi^2 f_\pi} \boldsymbol{E} \cdot \boldsymbol{B} \, \phi \tag{6.32}$$

where f_π is the so-called pion decay constant and ϕ represents the pion field. This is again not of the minimal type. It can be derived from the fact that the pion is made up of a quark and an antiquark which couple minimally to the electromagnetic field. This will be derived later using anomaly considerations.

6.3 Quantum electrodynamics (QED)

Quantum electrodynamics usually refers to the theory of electrons, positrons, and the electromagnetic field in interaction. Electrons and positrons can be described by a Dirac spinor field with the Lagrangian

$$\mathcal{L} = \bar{\psi}(i\gamma \cdot \partial - m)\psi \tag{6.33}$$

The Lagrangian obviously has invariance under the constant phase transformation $\psi \to e^{ie\theta}\psi$. The covariant derivative can be written as

$$D_\mu\psi = \partial_\mu\psi - ieA_\mu\psi \tag{6.34}$$

This corresponds to $Q = -e$, which is appropriate for interpreting ψ as corresponding to the annihilation of electrons of charge $-e$ and creation of positrons of charge e. The gauge principle now gives for the interacting theory

$$\mathcal{L} = \bar{\psi}(i\gamma \cdot D - m)\psi$$
$$= \bar{\psi}(i\gamma \cdot \partial - m)\psi + e\bar{\psi}\gamma^\mu \psi A_\mu$$
$$\mathcal{L}_{int} = e\bar{\psi}\gamma^\mu \psi A_\mu \tag{6.35}$$

Notice that the interaction part of the Lagrangian has the form of $A^\mu J_\mu$. The S-matrix functional for quantum electrodynamics (QED) may thus be written as

$$\mathcal{F} = \exp\left(-\frac{1}{2}\int_{x,y} D_{\mu\nu}(x,y)\frac{\delta}{\delta A_\mu(x)}\frac{\delta}{\delta A_\nu(y)}\right) \times$$
$$\exp\left[-\int \frac{\delta}{\delta\psi_r(x)}S_{rs}(x,y)\frac{\delta}{\delta\bar{\psi}_s(y)}\right]e^{ie\int A_\mu\bar{\psi}\gamma^\mu\psi} \tag{6.36}$$

One can also write down the operator formula for the S-matrix for QED as

$$\hat{S} = T\,\exp\left(ie\int A_{\mu\ in}\bar{\psi}_{in}\gamma^\mu\psi_{in}\right) \tag{6.37}$$

where the in-field $A_{\mu\ in}$ has the mode expansion given by (6.12) and propagator given by (6.23). The fermion fields have the expansions

$$\psi_{in}(x) = \sum_{p,r}\sqrt{\frac{m}{E_p V}}\left[a_{p,r}\,u_r(p)e^{-ipx} + b^\dagger_{p,r}v_r(p)e^{ipx}\right] \tag{6.38}$$

$$\bar{\psi}_{in}(x) = \sum_{p,r}\sqrt{\frac{m}{E_p V}}\left[a^\dagger_{p,r}\bar{u}_r(p)e^{ipx} + b_{p,r}\bar{v}_r(p)e^{-ipx}\right] \tag{6.39}$$

and the propagator given by (4.41) as

$$S(x,y) = i\int \frac{d^4p}{(2\pi)^4}\frac{\gamma \cdot p + m}{p^2 - m^2 + i\epsilon}e^{-ip(x-y)} \tag{6.40}$$

References

1. The quantization of the electromagnetic field in the noncovariant way given in text goes back to Dirac's original work; see J. Schwinger, *Selected Papers in Quantum Electrodynamics*, Dover Publications, Inc. (1958). Covariant operator quantization methods were developed by E. Fermi, Lincei Ren. **9**, 881 (1929); Rev. Mod. Phys. **4**, 87 (1932); S.N. Gupta, Proc. Roy. Soc. **63** 681 (1950); *ibid.* **64**, 850 (1951); K. Bleuler, Helv. Phys. Acta, **23**, 567 (1950); K.Bleuler and W. Heitler, Prog. Theor. Phys. **5**, 600 (1950). We do not discuss these because the Faddeev-Popov and BRST formalisms introduced in Chapter 10 give a more efficient approach.

7 Examples of Scattering Processes

7.1 Photon-scalar charged particle scattering

Compton scattering, historically, refers to photon-electron scattering. We consider here the same type of process, except that we have scalar charged particles instead of electrons. This may be referred to as scalar Compton scattering. The S-matrix functional for charged particles coupled to photons is given by

$$\mathcal{F} = \exp\left[\int_{x_1,x_2} G(x_1,x_2)\frac{\delta}{\delta\varphi(x_1)}\frac{\delta}{\delta\varphi^*(x_2)}\right] \times$$
$$\exp\left(-\frac{1}{2}\int_{x,y} D_{\mu\nu}(x,y)\frac{\delta}{\delta A_\mu(x)}\frac{\delta}{\delta A_\nu(y)}\right)e^{iS_{int}}$$
$$\mathcal{L}_{int} = -ieA^\mu\left(\varphi^*\partial_\mu\varphi - \partial_\mu\varphi^* \varphi\right) + e^2 A^\mu A_\mu \varphi\varphi^* \qquad (7.1)$$

To the lowest nontrivial order, viz., to second order in e, the relevant term in $\mathcal{F}$ is given by

$$\mathcal{F} = e^2 \int A^\mu(x)A^\nu(y)\left(\varphi^*(x)\overleftrightarrow{\partial}_\mu G(x,y)\overleftrightarrow{\partial}_\nu \varphi(y)\right) + ie^2 \int A^2\varphi^*\varphi \qquad (7.2)$$

where $f\overleftrightarrow{\partial}_\mu g = f\partial_\mu g - \partial_\mu f\ g$. Denoting the incoming and outgoing photon momenta by k and k' respectively, and by p and p' for the charged particle, the amplitude, as given by the general formula (5.24) is seen to be

$$\mathcal{A} = \frac{(2\pi)^4\delta^{(4)}(p+k-p'-k')}{V^2\sqrt{2\omega 2\omega' 2E_p 2E_{p'}}}\mathcal{M}$$
$$\mathcal{M} = ie^2\left[2e^{(\lambda)}(k)\cdot e^{(\lambda')}(k') - \frac{(2p'-k)\cdot e^{(\lambda)}(k)(2p-k')\cdot e^{(\lambda')}(k')}{(p-k')^2 - m^2 + i\epsilon}\right.$$
$$\left. - \frac{(2p'+k')\cdot e^{(\lambda')}(k')(2p+k)\cdot e^{(\lambda)}(k)}{(p+k)^2 - m^2 + i\epsilon}\right] \qquad (7.3)$$

Since $p^2 - m^2 = 0$, $k^2 = 0$, etc., we have $(p-k')^2 - m^2 = p^2 - m^2 + k'^2 - 2p\cdot k' = -2p\cdot k'$, etc. Further since $k\cdot e^{(\lambda)}(k) = 0$, $k'\cdot e^{(\lambda')}(k') = 0$, we can simplify the above expression to

$$\mathcal{M} = 2ie^2 \left[e^{(\lambda)} \cdot e^{(\lambda')} + \frac{p \cdot e^{(\lambda')} \, p' \cdot e^{(\lambda)}}{p \cdot k'} - \frac{p \cdot e^{(\lambda)} \, p' \cdot e^{(\lambda')}}{p \cdot k} \right] \tag{7.4}$$

where we use $e^{(\lambda)}, e^{(\lambda')}$ for $e^\lambda(k), e^{\lambda'}(k')$. The cross section is especially simple to evaluate in the rest frame of the incoming charged particle. In this case, $p = (m, 0, 0, 0)$ and hence $p \cdot e^{(\lambda)} = p \cdot e^{(\lambda')} = 0$ since the polarizations do not have time-components. We then find for the differential rate for scattering into a range of final momenta

$$\frac{|\mathcal{A}|^2}{\tau} = 4e^4 \frac{(2\pi)^4 \delta^{(4)}(p + k - p' - k')}{16\omega\omega' E_p E_{p'} V} (e^{(\lambda)} \cdot e^{(\lambda')})^2 \frac{d^3 p'}{(2\pi)^3} \frac{d^3 k'}{(2\pi)^3} \tag{7.5}$$

The flux is given by $F = \sqrt{(p \cdot k)^2}/V E_p \omega = 1/V$. Thus the differential cross section is

$$d\sigma = \frac{e^4}{4} \frac{(2\pi)^4 \delta^{(4)}(p + k - p' - k')}{\omega\omega' m E'} (e^{(\lambda)} \cdot e^{(\lambda')})^2 \frac{d^3 p'}{(2\pi)^3} \frac{d^3 k'}{(2\pi)^3} \tag{7.6}$$

Here $E' = E_{p'}$. The p'-integration is trivially done, identifying p' as $p' = p + k - k' = k - k'$. Also define the scattering angle θ by $k \cdot k' = \omega\omega' \cos\theta$. Then

$$d\sigma = \left(\frac{e^2}{4\pi} \right)^2 \frac{\omega'(e^{(\lambda)} \cdot e^{(\lambda')})^2}{\omega m E'} \delta(m + \omega - \omega' - E') d\omega' d\Omega \tag{7.7}$$

The remaining integral can also be done easily. The argument of the δ-function depends on ω' directly and through $E' = \sqrt{p'^2 + m^2} = \sqrt{(k - k')^2 + m^2} = \sqrt{\omega^2 + \omega'^2 - 2\omega\omega' \cos\theta + m^2}$. Thus,

$$\frac{d(\omega' + E')}{d\omega'} = \frac{E' + \omega' - \omega\cos\theta}{E'} \tag{7.8}$$

Carrying out the ω'-integration, we then get

$$d\sigma = \frac{\alpha^2}{m} \frac{\omega'}{\omega} \frac{(e^{(\lambda)} \cdot e^{(\lambda')})^2}{(E' + \omega' - \omega\cos\theta)} d\Omega \tag{7.9}$$

where $\alpha = e^2/4\pi$ is the fine structure constant and ω' is given by $m + \omega = E' + \omega'$. We may write this as $(m + \omega - \omega')^2 = E'^2 = \omega^2 + \omega'^2 - 2\omega\omega' \cos\theta + m^2$ or

$$\frac{\omega'}{\omega} = \frac{m}{m + \omega(1 - \cos\theta)} \tag{7.10}$$

From this result, we also find that $E' + \omega' - \omega\cos\theta = m(\omega/\omega')$. Thus

$$d\sigma = \frac{\alpha^2}{m^2} \left(\frac{\omega'}{\omega} \right)^2 (e^{(\lambda)} \cdot e^{(\lambda')})^2 d\Omega \tag{7.11}$$

This is the cross section for specific polarizations of the incoming and outgoing photons. If we are considering an initially unpolarized beam of photons, we should average over the initial polarizations. Likewise, if the final polarizations are not measured separately, but only the cross section for all final polarizations is considered, we must sum over all final polarizations. In this case

$$\left(\frac{d\sigma}{d\Omega}\right)_{unpol} = \frac{1}{2} \sum_{\lambda} \sum_{\lambda'} \left(\frac{d\sigma}{d\Omega}\right)$$

$$= \frac{\alpha^2}{2m^2} \left(\frac{\omega'}{\omega}\right)^2 (1 + \cos^2 \theta) \qquad (7.12)$$

where we have used the result

$$\sum_{\lambda} e_i^{\lambda}(k) e_j^{\lambda}(k) = \left(\delta_{ij} - \frac{k_i k_j}{k^2}\right) \qquad (7.13)$$

and a similar result for the final polarizations.

For the total cross section, we integrate over all angles to get

$$\sigma = \frac{\pi\alpha^2}{m^2} \left[\frac{4}{1+2x} + \frac{4}{x^2} - \frac{2(1+x)}{x^3} \log(1+2x)\right] \qquad (7.14)$$

where $x = \omega/m$. For small x, we can expand the logarithm to obtain

$$\sigma \approx \frac{8\pi\alpha^2}{3m^2} \qquad (7.15)$$

which agrees with the classical Thomson scattering cross section. For $\omega \gg m$ or large x,

$$\sigma \approx \frac{2\pi\alpha^2}{xm^2} \qquad (7.16)$$

7.2 Electron scattering in an external Coulomb field

This is also known as Mott scattering. The interaction Lagrangian for electrons in an external electromagnetic field is given by

$$\mathcal{L}_{int} = e\bar{\psi}\gamma^{\mu}\psi A_{\mu}^{ext} \qquad (7.17)$$

The S-matrix functional is given by

$$\mathcal{F} = \exp\left[-\int \frac{\delta}{\delta\psi_r(x)} S_{rs}(x,y) \frac{\delta}{\delta\bar{\psi}_s(y)}\right] e^{ie\int A_{\mu}^{ext}\bar{\psi}\gamma^{\mu}\psi} \qquad (7.18)$$

For one incoming electron of momentum p and spin state α and one outgoing electron of momentum p' and spin state β, the amplitude is given, to lowest order in the coupling e, as

$$\mathcal{A} = ie\frac{m}{V\sqrt{EE'}}\bar{u}_{\beta p'}\gamma^{\mu}u_{\alpha p}\int d^4x A_{\mu}^{ext}(x)e^{-i(p-p')x} \tag{7.19}$$

Consider an external screened Coulomb potential

$$A_0 = \frac{Ze}{r}e^{-ar}, \qquad\qquad A_i = 0 \tag{7.20}$$

We consider the screened potential to avoid difficulties of integration in (7.19). At the end, we can take the limit $a \to 0$. Using

$$\int d^3x e^{-i\boldsymbol{q}\cdot\boldsymbol{x}}\frac{Ze}{r}e^{-ar} = \frac{4\pi Ze}{q^2 + a^2} \tag{7.21}$$

we find, in the limit $a \to 0$

$$\mathcal{A} = ie\frac{m}{EV}\frac{4\pi Ze}{q^2}\bar{u}_{\beta p'}\gamma^0 u_{\alpha p}2\pi\delta(E - E') \tag{7.22}$$

where $\boldsymbol{q} = \boldsymbol{p} - \boldsymbol{p}'$. Squaring and using the trick of replacing one factor of the δ-function by the total time τ, we get

$$\frac{|\mathcal{A}|^2}{\tau} = \frac{(4\pi Ze^2)^2}{q^4}\frac{m^2}{E^2V^2}|\bar{u}_{\beta p'}\gamma^0 u_{\alpha p}|^2 2\pi\delta(E - E') \tag{7.23}$$

Since the source of the potential is at rest, the flux is given by $F = |\boldsymbol{p}|/EV$. We then get for the cross section

$$\begin{aligned}
d\sigma &= \frac{|\mathcal{A}|^2}{\tau}\frac{EV}{|\boldsymbol{p}|}V\frac{d^3p'}{(2\pi)^3}\\
&= \frac{4(Ze^2)^2m^2}{q^4E|\boldsymbol{p}|}|\bar{u}_{\beta p'}\gamma^0 u_{\alpha p}|^2\delta(E - E')p'^2dp'd\Omega
\end{aligned} \tag{7.24}$$

We must have $|\boldsymbol{p}| = |\boldsymbol{p}'|$ because of the $\delta(E-E')$. Further $E'dE' = p'dp'$ from $E' = \sqrt{p'^2 + m^2}$. Keeping these in mind, the integration over the magnitude of p' can be done trivially because of the δ-function and gives

$$\begin{aligned}
\frac{d\sigma}{d\Omega} &= \frac{4(Ze^2)^2m^2}{q^4}|\bar{u}_{p'}^{\beta}\gamma^0 u_p^{\alpha}|^2\\
&= \frac{(Ze^2)^2m^2}{4|\boldsymbol{p}|^4\sin^4(\theta/2)}|\bar{u}_{p'}^{\beta}\gamma^0 u_p^{\alpha}|^2
\end{aligned} \tag{7.25}$$

where we have used $q^2 = \boldsymbol{p}\cdot\boldsymbol{p}+\boldsymbol{p}'\cdot\boldsymbol{p}'-2\boldsymbol{p}\cdot\boldsymbol{p}' = 2\boldsymbol{p}\cdot\boldsymbol{p}(1-\cos\theta) = 4|\boldsymbol{p}|^2\sin^2(\theta/2)$. The angle θ between the incoming direction and the outgoing direction for the electron is called the scattering angle.

The above result is for specific choices of initial spin (polarized incoming beam) and specific spin values for the scattered electrons. If we have an unpolarized beam and if there is no experimental discrimination of final spin values, we must average over initial spins and sum over all final spins. This involves the quantity

$$
\frac{1}{2} \sum_{\alpha=1,2} \sum_{\beta=1,2} |\bar{u}_{\beta p'} \gamma^0 u_{\alpha p}|^2 = \frac{1}{2} \sum_{\alpha=1,2} \sum_{\beta=1,2} \bar{u}_{\beta p'} \gamma^0 u_{\alpha p} \bar{u}_{\alpha p} \gamma^0 u_{\beta p'}
$$

$$
= \frac{1}{2} \sum_{\alpha,\beta} \mathrm{Tr} \left[u_{\alpha p} \bar{u}_{\alpha p} \gamma^0 u_{\beta p'} \bar{u}_{\beta p'} \gamma^0 \right]
$$

$$
= \frac{1}{8m^2} \mathrm{Tr} \left[(m + \gamma \cdot p) \gamma^0 (m + \gamma \cdot p') \gamma^0 \right]
$$

$$
= \frac{1}{8m^2} \left[4m^2 + 4 p_\mu p'_\nu (\eta_{\mu 0} \eta_{\nu 0} - \eta_{\mu\nu} \eta_{00} + \eta_{\nu 0} \eta_{\mu 0}) \right]
$$

$$
= \frac{1}{2m^2} \left[E^2 + m^2 + \boldsymbol{p} \cdot \boldsymbol{p}' \right]
$$

$$
= \frac{E^2}{m^2} \left[1 - v^2 \sin^2(\theta/2) \right] \tag{7.26}
$$

where $v = |\boldsymbol{p}|/E$ is the velocity of the incoming electron. Using this calculation, we find for the unpolarized cross section

$$
\frac{d\sigma}{d\Omega} = \frac{(Ze^2)^2}{4E^2 v^4 \sin^4(\theta/2)} \left(1 - v^2 \sin^2(\theta/2) \right) \tag{7.27}
$$

For $v \ll 1$ this gives the classical Rutherford scattering cross section. The term $[1 - v^2 \sin^2(\theta/2)]$ arises because of the spin of the electron. (In the rest frame of the electron, the source is moving. The electron thus feels a magnetic field with which the spin can interact.)

The cross section diverges at $\theta = 0$. This is related to the infinite range of the Coulomb potential as seen from (7.21). This divergence can be consistently eliminated by including the finite angular resolution of the detector.

7.3 Slow neutron scattering from a medium

Scattering of neutrons from crystals and liquids is a useful technique for studying the properties of such materials. This is a problem which can be discussed using standard many-particle quantum mechanics, but it is interesting to apply field theory to it. The interaction between the neutron and the nuclei can be represented by

$$
\mathcal{L}_{int} = \lambda \bar{N} N \bar{P} P \tag{7.28}
$$

where P represents the nucleus which may be bound in a crystal, for example. The S-matrix element is obtained by replacing the c-number fields in the S-matrix functional by the wave functions of the particles. So far we have used free wave functions for the incoming and outgoing particles. But in the present case, the wave function for the nucleus is not free; in the case of scattering from a crystal, the nucleus or the positive ion is bound to the crystal and the wave function must be the corresponding one. Denoting this by $\Psi(x)$, we find for the amplitude, to lowest order in λ,

$$\mathcal{A} = \frac{i\lambda}{V} \int e^{-iqx} \Psi_f^*(x) \Psi_i(x) \tag{7.29}$$

where $q = k - k'$, k, k' are the incoming and outgoing momenta of the neutron and the subscripts i, f on the Ψ's refer to the initial and final states for the nucleus. We have used nonrelativistic wave functions for the neutron.

Consider first the case of a crystal lattice. In this case, the wave function of the nucleus corresponds to an oscillating ion at a lattice site. Denoting the position of the lattice point by $\boldsymbol{a}_n$ we have

$$\Psi(x) = e^{-iEx^0} \psi(\boldsymbol{x} - \boldsymbol{a}_n) \tag{7.30}$$

E is the energy of the state of the crystal. Using this, we can write

$$\int d^3x \, e^{i\boldsymbol{q}\cdot\boldsymbol{x}} \, \Psi_f^*(x) \Psi_i(x)$$

$$= e^{i(E_f - E_i)x^0} \int d^3x \, \psi_f^*(\boldsymbol{x} - \boldsymbol{a}_n)\psi_i(\boldsymbol{x} - \boldsymbol{a}_n)e^{i\boldsymbol{q}\cdot(\boldsymbol{x}-\boldsymbol{a}_n)}e^{i\boldsymbol{q}\cdot\boldsymbol{a}_n}$$

$$= e^{i(E_f - E_i)x^0} \langle f|e^{i\boldsymbol{q}\cdot\boldsymbol{\xi}_n}|i\rangle e^{i\boldsymbol{q}\cdot\boldsymbol{a}_n} \tag{7.31}$$

$\boldsymbol{\xi}_n$ is the position operator for the particle (ion) at the lattice site $\boldsymbol{a}_n$. Since the neutron can scatter from any one of the ions at the various lattice points, we have, for the total amplitude

$$\mathcal{A} = \frac{i\lambda}{V} 2\pi\delta(q_0 + E_i - E_f) \sum_n \langle f|e^{i\boldsymbol{q}\cdot\boldsymbol{\xi}_n}|i\rangle e^{i\boldsymbol{q}\cdot\boldsymbol{a}_n} \tag{7.32}$$

The incoming flux is k/mV, and so the cross section becomes

$$d\sigma = \frac{\lambda^2 m}{k} \sum_{nm} 2\pi\delta(q_0 + E_i - E_f)\langle i|e^{-i\boldsymbol{q}\cdot\boldsymbol{\xi}_m}|f\rangle\langle f|e^{i\boldsymbol{q}\cdot\boldsymbol{\xi}_n}|i\rangle e^{i\boldsymbol{q}\cdot(\boldsymbol{a}_n-\boldsymbol{a}_m)} \frac{d^3k'}{(2\pi)^3} \tag{7.33}$$

The final state of the crystal is seldom discriminated in any scattering, so the quantity of interest is the cross section for scattering into all final states of the crystal; we should sum over all final states. We have

$$\sum_f 2\pi\delta(q_0 + E_i - E_f)\langle i|e^{-i\boldsymbol{q}\cdot\boldsymbol{\xi}_m}|f\rangle\langle f|e^{i\boldsymbol{q}\cdot\boldsymbol{\xi}_n}|i\rangle$$

$$= \sum_f \int dz^0 e^{-i(q_0+E_i-E_f)z^0} \langle i|e^{-i\boldsymbol{q}\cdot\boldsymbol{\xi}_m}|f\rangle\langle f|e^{i\boldsymbol{q}\cdot\boldsymbol{\xi}_n}|i\rangle$$

$$= \int dz^0 e^{-iq_0 z^0} \sum_f \langle i|e^{-i\boldsymbol{q}\cdot\boldsymbol{\xi}_m}|f\rangle\langle f|e^{iHz^0}e^{i\boldsymbol{q}\cdot\boldsymbol{\xi}_n}e^{-iHz^0}|i\rangle$$

$$= \int dz^0 e^{-iq_0 z^0} \langle i|e^{-i\boldsymbol{q}\cdot\boldsymbol{\xi}_m(0)}e^{i\boldsymbol{q}\cdot\boldsymbol{\xi}_n(z^0)}|i\rangle \qquad (7.34)$$

where we have used the completeness relation $\sum_f |f\rangle\langle f| = 1$ and $\boldsymbol{\xi}_n(z^0) = e^{iHz^0}\boldsymbol{\xi}_n e^{-iHz^0}$ denotes the Heisenberg time-dependent version of the position operator at the indicated site. Finally, the initial state is seldom prepared to be in any fixed pure state. Rather, one has thermal excitations of the lattice vibrations. This can be taken care of by summing over all initial states with a probability c_i for state $|i\rangle$ or equivalently by taking the average with a density matrix $\rho = \sum_i c_i |i\rangle\langle i|$. Using (7.34) and averaging, we get

$$d\sigma = \frac{\lambda^2 m}{k} \int dz^0 e^{-iq_0 z^0} \sum_{nm} \mathrm{Tr}\left[\rho e^{-i\boldsymbol{q}\cdot\boldsymbol{\xi}_m(0)}e^{i\boldsymbol{q}\cdot\boldsymbol{\xi}_n(z^0)}\right] e^{i\boldsymbol{q}\cdot(\boldsymbol{a}_n-\boldsymbol{a}_m)}\frac{d^3k'}{(2\pi)^3}$$

$$(7.35)$$

The summation over the lattice sites with the factor $e^{i\boldsymbol{q}\cdot(\boldsymbol{a}_n-\boldsymbol{a}_m)}$ leads to sharp peaks at values of the momentum transfer corresponding to the reciprocal lattice vectors. At these values of momentum transfer, the factors $e^{i\boldsymbol{q}\cdot(\boldsymbol{a}_n-\boldsymbol{a}_m)}$ are all equal to 1 and we have complete constructive interference. The factor $\mathrm{Tr}\left[\rho e^{-i\boldsymbol{q}\cdot\boldsymbol{\xi}_m(0)}e^{i\boldsymbol{q}\cdot\boldsymbol{\xi}_n(z^0)}\right]$ suppresses the peaks somewhat and gives the so-called Debye-Waller factor. It also, due to the time-dependence, contains information about processes where phonons are absorbed and emitted during the scattering. Further simplification of (7.35) will depend on the specific context.

Consider now another medium, say, a liquid or gas. Notice that we can write

$$\Psi_f^*(x)\Psi_i(x) = \langle f|\bar{P}P(x)|i\rangle \qquad (7.36)$$

which follows from writing out the operator P in terms of a mode expansion with Ψ's as the one-particle wave functions. $\bar{P}P$ is the density operator for the nuclei which we will denote by $J_0(x)$. (It is after all the time-component of a number density current.) We then find

$$|\mathcal{A}|^2 = \frac{\lambda^2}{V^2} \int d^4x d^4y\; e^{-iq(x-y)}\langle i|J_0(y)|f\rangle\langle f|J_0(x)|i\rangle \qquad (7.37)$$

Summation over final states gives the quantity $\langle i|J_0(y)J_0(x)|i\rangle$. We can simplify this as follows.

$$\begin{aligned}
\langle i|J_0(y)J_0(x)|i\rangle &= \langle i|e^{iPy}J_0(0)e^{-iP(y-x)}J_0(0)e^{-iPx}|i\rangle \\
&= \langle i|J_0(0)e^{iP(x-y)}J_0(0)e^{-ip_i(x-y)}|i\rangle \\
&= \langle i|J_0(0)e^{iP(x-y)}J_0(0)e^{-iP(x-y)}|i\rangle \\
&= \langle i|J_0(0)J_0(x-y)|i\rangle \qquad (7.38)
\end{aligned}$$

where we have used the energy-momentum operator $P = (H, \boldsymbol{P})$ to translate the arguments of the J_0's to the indicated points. (This P is not to be confused with the field operator for the nucleus.) p_i is the eigenvalue of energy-momentum for the initial state $|i\rangle$. Using this result and changing variables from x, y to $x - y, y$ we can do the integration over y in (7.37). The cross section can thus be written as

$$d\sigma = \frac{\lambda^2 m}{k} S(q) \frac{d^3 k'}{(2\pi)^3} \tag{7.39}$$

$$S(q) = \int d^4x \; e^{-iqx} \langle i|J_0(0)J_0(x)|i\rangle \tag{7.40}$$

$S(q)$ is the two-point function for the density operator. We see that it can be measured by neutron scattering.

In some situations, such as slow neutron scattering which is essentially elastic, it is a good approximation to take J_0 to be independent of time. In this case

$$S(q) = 2\pi\delta(k_0 - k_0')S(\boldsymbol{q})$$
$$S(\boldsymbol{q}) = \int d^3x \; e^{i\boldsymbol{q}\cdot\boldsymbol{x}} \langle i|J_0(0)J_0(\boldsymbol{x})|i\rangle \tag{7.41}$$

$S(\boldsymbol{q})$ is called the structure factor.

7.4 Compton scattering

We now consider Compton scattering with spin-$\frac{1}{2}$ charged particles, for example, the scattering of photons by electrons or positrons. It is described by the QED S-matrix as given in (6.36). The relevant term of the S-matrix functional is

$$\mathcal{F} = \exp\left[-\int \frac{\delta}{\delta\psi_r(x)} S_{rs}(x, y) \frac{\delta}{\delta\bar{\psi}_s(y)}\right] e^{ie \int A_\mu \bar{\psi}\gamma^\mu \psi}$$
$$= (ie)^2 \int \bar{\psi}(x)\gamma^\mu S(x, y)\gamma^\nu \psi(y) A_\mu(x) A_\nu(y) \tag{7.42}$$

Upon replacing the $\psi, \bar{\psi}$ and the A's by the one-particle wave functions according to (a generalization of the) formula (5.24) we get the amplitude for the process as

$$A = \sqrt{\frac{m^2}{EVE'V2\omega V2\omega'V}} \; (2\pi)^4 \delta^{(4)}(p + k - p' - k') \; \mathcal{M}$$

$$\mathcal{M} = -ie^2\left[\bar{u}_{\beta p'}\gamma \cdot e^{(\lambda')} \frac{1}{\gamma \cdot (p + k) - m}\gamma \cdot e^{(\lambda)} u_{\alpha p}\right.$$

$$\left. + \bar{u}_{\beta p'}\gamma \cdot e^{(\lambda)} \frac{1}{\gamma \cdot (p - k') - m}\gamma \cdot e^{(\lambda')} u_{\alpha p}\right] \tag{7.43}$$

where we have denoted the incoming and outgoing electron momenta by p, p', respectively, and the corresponding photon momenta by k, k'. Subscripts α, β on the spinors refer to the spins of the incoming and outgoing electrons and $e^{(\lambda)}, e^{(\lambda')}$ refer to the corresponding photon polarizations. The kinematics of the process remain as in the discussion of photon scattering with a scalar charged particle. If we consider the rest frame of the initial electron, then the flux is $1/V$ and we get

$$d\sigma = \frac{1}{16\pi^2} \left(\frac{\omega'}{\omega} \right)^2 |\mathcal{M}|^2 d\Omega \tag{7.44}$$

where, as before,

$$\omega' = \frac{m\omega}{m + \omega(1 - \cos\theta)} \tag{7.45}$$

There is no significant difference with the scalar case until this point, except of course that $\mathcal{M}$ is different.

The cross section (7.44) applies to fixed incoming polarizations for the photon and electron and for scattering into specified final polarizations. For unpolarized incoming electrons and where final polarizations are not measured, the quantity of interest is the above expression averaged over initial spins of the electron and summed over final spins. The result can be written as

$$d\sigma = \alpha^2 \left(\frac{\omega'}{\omega} \right)^2 \frac{1}{2} \sum_{\alpha,\beta} |\mathcal{N}|^2 d\Omega \tag{7.46}$$

$$\mathcal{N} = \bar{u}_{\beta p'} \mathcal{O}_1^{\mu\nu} u_{\alpha p} e_\mu^{(\lambda')} e_\nu^{(\lambda)} + \bar{u}_{\beta p'} \mathcal{O}_2^{\mu\nu} u_{\alpha p} e_\mu^{(\lambda)} e_\nu^{(\lambda')}$$

$$\mathcal{O}_1^{\mu\nu} = \gamma^\mu \frac{1}{\gamma \cdot (p+k) - m} \gamma^\nu$$

$$\mathcal{O}_2^{\mu\nu} = \gamma^\mu \frac{1}{\gamma \cdot (p-k') - m} \gamma^\nu$$

The square of $\mathcal{N}$ will involve four types of terms. In general, when we square a term like $\bar{u}_\beta \mathcal{O} u_\alpha$ we get

$$\sum_{\alpha,\beta} \bar{u}_{\beta p'} \mathcal{O} u_{\alpha p} u_{\alpha p}^\dagger \mathcal{O}^\dagger \gamma^0 u_{\beta p'} = \sum_{\alpha\beta} \bar{u}_{\beta p'} \mathcal{O} u_{\alpha p} \bar{u}_{\alpha p} \mathcal{O} u_{\beta p'}$$

$$= \text{Tr} \left[\mathcal{O} \frac{(m + \gamma \cdot p)}{2m} \mathcal{O} \frac{(m + \gamma \cdot p')}{2m} \right] \tag{7.47}$$

where we have used the fact that $\gamma^0 \mathcal{O}^\dagger \gamma^0 = \mathcal{O}$ for the combinations of γ-matrices we have, which follows from the fact that γ^i are antihermitian and anticommute with γ^0 which is also hermitian. We thus get

$$4m^2 \sum |\mathcal{N}|^2 = \text{Tr} \left[e^{(\lambda')} \cdot \mathcal{O}_1 \cdot e^{(\lambda)} (m + \gamma \cdot p) e^{(\lambda)} \cdot \mathcal{O}_1 \cdot e^{(\lambda')} (m + \gamma \cdot p') \right]$$

$$+ \mathrm{Tr}\left[e^{(\lambda)} \cdot \mathcal{O}_2 \cdot e^{(\lambda')}(m + \gamma \cdot p)e^{(\lambda')} \cdot \mathcal{O}_2 \cdot e^{(\lambda)}(m + \gamma \cdot p')\right]$$

$$+ \mathrm{Tr}\left[e^{(\lambda')} \cdot \mathcal{O}_1 \cdot e^{(\lambda)}(m + \gamma \cdot p)e^{(\lambda')} \cdot \mathcal{O}_2 \cdot e^{(\lambda)}(m + \gamma \cdot p')\right]$$

$$+ \mathrm{Tr}\left[e^{(\lambda)} \cdot \mathcal{O}_2 \cdot e^{(\lambda')}(m + \gamma \cdot p)e^{(\lambda)} \cdot \mathcal{O}_1 \cdot e^{(\lambda')}(m + \gamma \cdot p')\right]$$

$$(7.48)$$

For the traces of the γ-matrices, we have the formulae

$$\mathrm{Tr}[\gamma^\mu \gamma^\nu] = 4\eta^{\mu\nu}$$
$$\mathrm{Tr}[\gamma^\mu \gamma^\nu \gamma^\alpha \gamma^\beta] = 4\left(\eta^{\mu\nu}\eta^{\alpha\beta} - \eta^{\mu\alpha}\eta^{\nu\beta} + \eta^{\mu\beta}\eta^{\nu\alpha}\right)$$
$$\mathrm{Tr}[\gamma^{\mu_1}\gamma^{\mu_2}\cdots\gamma^{\mu_{2n}}] = \eta^{\mu_1\mu_2}\mathrm{Tr}[\gamma^{\mu_3}\gamma^{\mu_4}\cdots\gamma^{\mu_{2n}}]$$
$$- \eta^{\mu_1\mu_3}\mathrm{Tr}[\gamma^{\mu_2}\gamma^{\mu_4}\cdots\gamma^{\mu_{2n}}]$$
$$+ \eta^{\mu_1\mu_4}\mathrm{Tr}[\gamma^{\mu_2}\gamma^{\mu_4}\gamma^{\mu_5}\cdots\gamma^{\mu_{2n}}] - \cdots \quad (7.49)$$

These rules follow from the basic Clifford algebra relations $\gamma^\mu\gamma^\nu + \gamma^\nu\gamma^\mu = 2\eta^{\mu\nu}$. The last result is a recursion rule which can be used for higher numbers of γ-matrices to reduce such traces to lower ones. Using these rules, the relevant traces in (7.48) can be evaluated, although this is rather tedious. The result, in the rest frame of the incoming electron, with $p = (m, 0, 0, 0)$, becomes

$$\sum |\mathcal{N}|^2 = \frac{1}{2m^2}\left[\frac{\omega'}{\omega} + \frac{\omega}{\omega'} + 4(e^{(\lambda)} \cdot e^{(\lambda')})^2 - 2\right] \quad (7.50)$$

The differential scattering cross section (7.46) can now be written as

$$\frac{d\sigma}{d\Omega} = \frac{\alpha^2}{4m^2}\left(\frac{\omega'}{\omega}\right)^2\left[\frac{\omega'}{\omega} + \frac{\omega}{\omega'} + 4(e^{(\lambda)} \cdot e^{(\lambda')})^2 - 2\right] \quad (7.51)$$

This is the Klein-Nishina formula. In the nonrelativistic limit or when the photon energy is small, we have $\omega' \approx \omega$, and this reduces to the Thomson cross section

$$\frac{d\sigma}{d\Omega} \approx \frac{\alpha^2}{m^2}(e^{(\lambda)} \cdot e^{(\lambda')})^2 \quad (7.52)$$

The expression (7.51) applies to the case where the electrons are unpolarized, but the photons are polarized. We can get the cross section for all final polarizations of the photon and for unpolarized incoming photons by summing over final polarizations and averaging over the initial polarizations. The use of (7.13) then gives

$$\frac{1}{2}\sum(e^{(\lambda)} \cdot e^{(\lambda')})^2 = \frac{1}{2}(1 + \cos^2\theta)$$
$$\frac{1}{2}\sum\left[\frac{\omega'}{\omega} + \frac{\omega}{\omega'} - 2\right] = 2\left[\frac{\omega'}{\omega} + \frac{\omega}{\omega'} - 2\right] \quad (7.53)$$

Using this in (7.51), we find for the unpolarized cross section

$$\frac{d\sigma}{d\Omega} = \frac{\alpha^2}{2m^2} \left(\frac{\omega'}{\omega}\right)^2 \left[\frac{\omega'}{\omega} + \frac{\omega}{\omega'} - \sin^2\theta\right] \tag{7.54}$$

Integration over all angles gives the total cross section. In the nonrelativistic limit, it is the Thomson total cross section.

$$\sigma(\omega) \approx \sigma_{Th} = \frac{8\pi}{3}\frac{\alpha^2}{m^2} \tag{7.55}$$

More generally one has

$$\sigma(\omega) = \sigma_{Th} \, F(\omega/m) \tag{7.56}$$

where

$$
\begin{aligned}
F(\omega/m) &\to 1 & &\text{as} & \frac{\omega}{m} &\to 0 \\
&\to \frac{3m}{8\omega}\left[\log(2\omega/m) + \frac{1}{2} + \mathcal{O}\left(\frac{\log(\omega/m)}{(\omega/m)}\right)\right] & &\text{as} & \frac{\omega}{m} &\to \infty
\end{aligned}
\tag{7.57}
$$

The result that the cross section agrees with the Thomson cross section when the photon energy is low is the same as what we found for the scalar particle. Our calculation is only at the lowest nontrivial order in the coupling constant e and only for spins-0 and $-\frac{1}{2}$. This reduction to Thomson cross section is, however, universal. Based on the gauge invariance of the electromagnetic interactions, one can prove the following low-energy theorem. The exact cross section for Compton scattering reduces to the Thomson cross section as the photon energy ω goes to zero. Of course, we have $(Q^2/4\pi)$, instead of α, for a particle of charge Q in the formula for the cross section. This theorem is very general and holds even after including renormalization effects and contributions from other hadronic intermediate states and so on. It can be used to define what is meant by the charge of the particle, viz., we can say that the charge Q of a particle is defined by the formula

$$\left(\frac{Q^2}{4\pi}\right)^2 = \frac{3}{8\pi}m^2\sigma(\omega)\bigg]_{\omega\to 0} \tag{7.58}$$

where $\sigma(\omega)$ is the Compton scattering cross section. (For a related low-energy theorem, see Chapter 11.)

7.5 Decay of the π^0 meson

The π^0-meson (or the neutral pion) is a neutral particle made of a quark and an antiquark. It decays into two photons. The interaction Lagrangian responsible for this decay can be shown to be

$$\mathcal{L}_{int} = \frac{\alpha}{\pi f} \boldsymbol{E} \cdot \boldsymbol{B} \phi \tag{7.59}$$

where ϕ is the pion field. α is the fine structure constant, and $f = f_\pi \approx 93 MeV$ is the pion decay constant, which can be measured in the leptonic decays of the charged pions. The interaction (7.59) will be derived in the discussion of anomalies later.

We consider incoming pion momentum p and outgoing photons of momenta k_1, k_2 and polarizations $e^{(\lambda)}, e^{(\lambda')}$. The amplitude is obtained by replacing the fields by the one-particle wave functions in $e^{iS_{int}}$. We find

$$\mathcal{A} = \frac{(2\pi)^4 \delta^{(4)}(p - k_1 - k_2)}{\sqrt{2p_0 V 2\omega_1 V 2\omega_2 V}} \mathcal{M}$$

$$\mathcal{M} = \frac{i\alpha}{\pi f} \epsilon_{ijr}(k_{1i}\omega_2 - k_{2i}\omega_1)e_j^{(\lambda)} e_r^{(\lambda')} \tag{7.60}$$

The decay rate is given by $|\mathcal{A}|^2/\tau$, where we use the usual trick of replacing the square of the delta function by one delta function and a factor of $V\tau$. As regards final states, there is an additional factor of 2. This can be seen as follows. The final state has two photons, which are identical bosonic particles. The state $((k_1, e), (k_2, e'))$ is not distinguishable from $((k_2, e'), (k_1, e))$. In summing over final states, if we do an unrestricted sum over momenta and polarizations, there would be double counting. We can remove this by dividing by 2. Thus the total decay rate is given by

$$\Gamma = \int \frac{1}{2} \sum_{final} \frac{|\mathcal{A}|^2}{\tau}$$

$$= \int \frac{d^3 k_1}{(2\pi)^3} \frac{d^3 k_2}{(2\pi)^3} \frac{(2\pi)^4 \delta^{(4)}(p - k_1 - k_2)}{16 p_0 \omega_1 \omega_2} \sum_{\lambda, \lambda'} |\mathcal{M}|^2 \tag{7.61}$$

The k_2-integration is trivial because of the delta function, giving $k_2 = p - k_1$. In the following, we shall choose the rest frame of the pion, so that $p = 0, p_0 = m_\pi$ and hence $k \equiv k_1 = -k_2$ and $\omega \equiv \omega_1 = \omega_2$. We also have

$$\sum_{\lambda, \lambda'} |\mathcal{M}|^2 = \frac{4\alpha^2}{\pi^2 f^2} \sum \left[\omega \epsilon_{ijr} k_i e'_j e_r \right]^2$$

$$= \frac{8\alpha^2 \omega^4}{\pi^2 f^2} \tag{7.62}$$

The decay rate is thus given by

$$\Gamma = \int \frac{8\alpha^2 \omega^4}{\pi^2 f^2} \frac{2\pi \delta(m_\pi - 2\omega)}{16 m_\pi \omega^2} \frac{d^3 k}{(2\pi)^3}$$

$$= \frac{\alpha^2 m_\pi^3}{64 \pi^3 f^2} \tag{7.63}$$

The numerical value given by (7.63) is approximately 7.63 eV to be compared to the experimental value of (7.37 ± 1.5) eV, which is not bad as we have made some assumptions such as ignoring the composite nature of the pion.

7.6 Čerenkov radiation

The basic process involved here is single photon emission by a charged particle. Such a process is kinematically forbidden in vacuum, but in a medium it can happen if the velocity of the particle exceeds the velocity of light in the medium. While the dielectric constant itself arises from interactions of the photon with the charged particles in the medium, for the purpose of discussing Čerenkov radiation, we can have an effective description where the medium is assigned a dielectric constant $\epsilon(\omega)$. In this case, we have for free photons, $\boldsymbol{k}^2 = \epsilon(\omega)\omega^2$. Let p, p' denote the initial and final momenta of the particle and let k denote the momentum of the emitted photon. Conservation of energy-momentum gives

$$\boldsymbol{p} = \boldsymbol{p}' + \boldsymbol{k}$$
$$E = E' + \omega \tag{7.64}$$

Squaring the second of these and using the first, we find

$$\cos\theta = \frac{E\omega}{pk} + \frac{(\epsilon - 1)\omega^2}{2pk}$$
$$= \frac{1}{v\sqrt{\epsilon}} + \frac{(\epsilon - 1)\omega^2}{2pk} \tag{7.65}$$

where θ is defined by $\boldsymbol{p} \cdot \boldsymbol{k} = pk \cos\theta$ and $v = p/E$ is the velocity of the particle. As $\epsilon \to 1$, the right-hand side becomes larger than 1 with no solution for θ and the photon emission is kinematically forbidden. In the medium, light is emitted on a cone whose angle is given by (7.65). For $\epsilon \approx 1$, we can approximate the above relation by

$$\cos\theta = \frac{1}{v\sqrt{\epsilon}} \tag{7.66}$$

Of course, we need $v\sqrt{\epsilon} > 1$ for this to make sense. (Actually, the second term in (7.65) is a quantum effect and hence (7.66) is all that appears classically.)

We now turn to the dynamics. The photon Lagrangian, with $A_0 = 0$, has the form

$$\mathcal{L} = \frac{1}{2}\left[\epsilon\left(\frac{\partial \boldsymbol{A}}{\partial t}\right)^2 - \boldsymbol{B}^2\right] \tag{7.67}$$

The canonical momentum is $\Pi_i = \epsilon \dot{A}_i$. The expansion for A_i is now

$$A_i^T(x) = \sum_{k\lambda} a_{k\lambda} e_i^{(\lambda)} u_k(x) + a_{k\lambda}^\dagger e_i^{(\lambda)} u_k^*(x)$$

$$u_k(x) = \frac{1}{\sqrt{2\,\epsilon\,\omega_k V}} e^{-ikx} \tag{7.68}$$

where $a, a^\dagger$ have the usual commutation rules. The factor of $\sqrt{\epsilon}$ is required in the denominator of $u_k(x)$ so that Π, A have the standard commutation rules. The Hamiltonian can be written as

$$H = \int d^3x\, \frac{1}{2}\left[\frac{\Pi^2}{\epsilon} + B^2\right]$$

$$= \sum_k \omega a_{k\lambda}^\dagger a_{k\lambda} \tag{7.69}$$

We take the charged particle to be a scalar, so that the interaction term for one photon emission is

$$\mathcal{L}_{int} = -ie\phi^* \overset{\leftrightarrow}{\partial}_\mu \phi A^\mu + \cdots \tag{7.70}$$

The amplitude for single photon emission is thus given by

$$A = ie \int d^4x\, (p+p')^\mu e_\mu^{(\lambda)} \frac{e^{-ipx+ip'x+ikx}}{\sqrt{2\epsilon\omega V 2EV 2E'V}}$$

$$= -ie\frac{2p \cdot e^{(\lambda)}}{\sqrt{2\epsilon\omega V 2EV 2E'V}} (2\pi)^4 \delta^{(4)}(p - p' - k) \tag{7.71}$$

where we have used the fact that $p' \cdot e = (p+k) \cdot e = p \cdot e$ since $k \cdot e = 0$. The photon emission rate is given by

$$\frac{|A|^2}{\tau} = \frac{e^2}{2V^2}\frac{(p \cdot e)^2}{\epsilon\omega EE'} (2\pi)^4 \delta^{(4)}(p - p' - k) \tag{7.72}$$

The rate of energy loss is obtained by multiplying the above expression by ω since the energy loss per photon emission is ω. Putting in the final state summations, we find

$$\frac{dE}{dt} = \sum_\lambda \omega \frac{e^2}{2}\frac{(p \cdot e)^2}{\epsilon\omega EE'} (2\pi)^4 \delta^{(4)}(p - p' - k)\frac{d^3k}{(2\pi)^3}\frac{d^3p'}{(2\pi)^3}$$

$$= \alpha\frac{p^2}{EE'\epsilon}(1 - \cos^2\theta)\delta(E - \sqrt{(\boldsymbol{p}-\boldsymbol{k})^2 + m^2} - \omega)k^2 dk d(\cos\theta) \tag{7.73}$$

where we have carried out the integration over $\boldsymbol{p'}$ and also used the fact that

$$\sum_\lambda p \cdot e\, p \cdot e = \boldsymbol{p}^2 - \frac{(\boldsymbol{p} \cdot \boldsymbol{k})^2}{k^2}$$

$$= p^2(1 - \cos^2\theta) \tag{7.74}$$

We carry out the angular integration; this sets the value of θ to that given by (7.65). For the remaining integral over k, we then find

$$\frac{dE}{dt} = \alpha \int \frac{p}{E} \frac{kdk}{\epsilon} (1 - \cos^2 \theta) \tag{7.75}$$

Now, approximately,

$$1 - \cos^2 \theta \approx \left(1 - \frac{1}{v^2 \epsilon}\right)$$

$$kdk \approx \epsilon \omega d\omega \tag{7.76}$$

where we neglect $\partial \epsilon / \partial \omega$. Also, the final result is usually expressed as energy loss per unit length of the path; this can be done using $dx = vdt$. Combining these results

$$\frac{dE}{dx} = \alpha \int d\omega \; \omega \left(1 - \frac{1}{v^2 \epsilon}\right) \tag{7.77}$$

This is the Frank-Tamm result for energy loss by Čerenkov radiation.

7.7 Decay of the ρ-meson

The decay of the neutral ρ-meson into charged leptons can be modeled by a Lagrangian which incorporates mixing between the ρ and the photon. The Lagrangian for the ρ_μ, e^+, e^- and the photon is given by

$$\begin{aligned}
\mathcal{L} &= \mathcal{L}_0 + \mathcal{L}_{int} \\
\mathcal{L}_0 &= -\frac{1}{4} F_{\mu\nu} F^{\mu\nu} - \frac{1}{2} (\partial \cdot A)^2 - \frac{1}{2} \partial_\mu \rho_\nu \partial^\mu \rho^\nu + \frac{1}{2} M^2 \rho^2 \\
&\quad + \bar{\psi} (i\gamma \cdot \partial - m) \psi \\
\mathcal{L}_{int} &= \frac{e}{g} M^2 \rho \cdot A + e \bar{\psi} \gamma \cdot A \psi
\end{aligned} \tag{7.78}$$

with $\partial \cdot \rho = 0$. The $\rho - A$ term is consistent with the mixing of the ρ and the photon in the so-called vector dominance approach to ρ-meson interactions. The constant g has the numerical value, $g \approx 5$. The ρ-meson has many channels for its decay. The Lagrangian given above can describe the decay of a ρ-meson into an e^+e^--pair. The partial decay rate for this mode can thus be calculated using the Lagrangian (7.78).

The amplitude for the decay of the ρ of momentum p and polarization $\epsilon_\mu^{(\lambda)}$ into an e^+e^--pair of momenta k, k', is given, to the lowest order in the coupling e by

$$\mathcal{A} = \frac{(2\pi)^4 \delta^{(4)}(k + k' - p)}{\sqrt{2\omega_p V E_k V E_{k'} V}} \; i \frac{e^2}{g} M^2 m \; \epsilon_\mu^{(\lambda)} \frac{(\bar{u}_{\alpha k} \gamma^\mu u_{\beta k'})}{p^2} \tag{7.79}$$

The factor of $1/p^2$ arises from the photon propagator due to the Wick contraction of the A's in the expansion of $e^{iS_{int}}$. For this calculation, p^2 is equal to M^2. We will calculate the decay rate for an unpolarized ρ-meson; there are three polarization states and the average over the polarizations can be done using

$$\sum_\lambda \epsilon_\mu^{(\lambda)} \epsilon_\nu^{*(\lambda)} = -\eta_{\mu\nu} + \frac{p_\mu p_\nu}{M^2} \tag{7.80}$$

The decay rate for an unpolarized ρ-meson with a small range of final momenta and all possible final spins is given by

$$\Gamma = \frac{e^4}{g^2} m^2 \int \frac{1}{3} \sum_\lambda \sum_{\alpha,\beta} |\bar{u}_k \gamma \cdot \epsilon^{(\lambda)} u'_{k'}|^2 \frac{(2\pi)^4 \delta^{(4)}(k + k' - p)}{2\omega_p E_k E_{k'}} \frac{d^3k d^3k'}{(2\pi)^6} \tag{7.81}$$

Using the polarization summation formula (7.80), we find

$$\sum_\lambda \sum_{\alpha,\beta} |\bar{u}_k \gamma \cdot \epsilon^{(\lambda)} u'_{k'}|^2 = 6 + \frac{4\boldsymbol{k}^2}{m^2} \tag{7.82}$$

For a ρ-meson at rest, $p = (M, 0, 0, 0)$ and $\boldsymbol{k}' = -\boldsymbol{k}$, $E_{k'} = E_k$. In this case the decay rate is given by

$$\begin{aligned}
\Gamma &= \frac{e^4 m^2}{8\pi^2 g^2} \frac{1}{3} \int d^3k \left(6 + \frac{4\boldsymbol{k}^2}{m^2}\right) \frac{\delta(2E_k - M)}{M E_k^2} \\
&= \frac{4\pi\alpha^2}{3g^2} M \left(1 + \frac{2m^2}{M^2}\right) \sqrt{1 - \frac{4m^2}{M^2}}
\end{aligned} \tag{7.83}$$

where $\alpha = e^2/4\pi$. Since the electron mass is small compared to the mass of the ρ, we can simplify this to

$$\Gamma \approx \frac{4\pi\alpha^2}{3} \frac{M}{g^2} \tag{7.84}$$

References

1. The calculations of Compton scattering for scalar particles and for electrons are standard examples found in many books; so is Mott scattering. A book with the calculation of a number of QED processes is J. M. Jauch and F. Rohrlich, *The Theory of Photons and Electrons*, Springer-Verlag (1955 & 1976).
2. Neutron scattering and Čerenkov radiation are included to show how the field theory can offer a vantage point of view for some processes which are usually calculated by other techniques. For scattering from a crystal, see J. Callaway, *Quantum Theory of the Solid State*, Academic Press (1974). The structure factor, which can be measured by neutron scattering, is useful in some condensed matter contexts, for example, Feynman's theory of liquid Helium; see R.P. Feynman, *Statistical Mechanics*, Addison-Wesley Publishing Co. (1972).

3. The Frank-Tamm result on Čerenkov radiation is worked out classically in J.D. Jackson, *Classical Electrodynamics*, 3rd edition, Wiley Text Books (1998).

4. The decay of the neutral pion is included to illustrate how one may calculate with phenomenological Lagrangians. It was first calculated by J. Steinberger, Phys. Rev. **82**, 664 (1949). The understanding of pion decay, especially appreciating its role as a Goldstone boson, is in terms of anomalies, calculated by S. Adler, Phys. Rev. **177**, 2426 (1969); J. Bell and R. Jackiw, Nuov. Cim. **60A**, 47 (1969). The effective Lagrangian is derived from this symmetry point of view.

5. The dominant decay mode of the ρ-meson is into two pions, not the mode we discuss. This calculation is given as another example of the use of a phenomenological Lagrangian. For ρ-γ mixing and vector dominance, see J.J. Sakurai, *Currents and Mesons*, University of Chicago Press (1969).

8 Functional Integral Representations

8.1 Functional integration for bosonic fields

We have discussed the notion of functional differentiation earlier. Here we shall discuss the notion of functional integration. A certain function space $\mathcal{F}$ was defined in Chapter 2. For integration, we need a volume measure on the function space $\mathcal{F}$. We can do this by specifying a distance function or metric on $\mathcal{F}$. For a real scalar field $\varphi(x)$, the distance between the configurations $\varphi(x)$ and $\varphi(x) + \delta\varphi(x)$ may be taken as

$$ds^2 = \|\delta\varphi\|^2 = \int_\Sigma d^4x \; (\delta\varphi)^2 \tag{8.1}$$

If we expand $\varphi(x)$ in terms of a real orthonormal basis, i.e., $\varphi(x) = \sum_n c_n u_n(x)$, the coefficients c_n are real and we have

$$ds^2 = \|\delta\varphi\|^2 = \sum_n (\delta c_n)^2 \tag{8.2}$$

We see that the distance defined by (8.1) is Euclidean. This would be appropriate for real-valued scalar functions. The metric will in general depend on the nature of φ. For example, consider maps from the spacetime region Σ to the two-sphere S^2. Using (θ, α) as the coordinates of S^2 (where α is the azimuthal angle), the maps we are considering are $\theta(x), \alpha(x)$. On the corresponding function space, the metric should be of the form

$$ds^2 = \int_\Sigma d^4x \; [(\delta\theta)^2 + \sin^2\theta \; (\delta\alpha)^2] \tag{8.3}$$

Since there is a choice of c_n for each function, and each function corresponds to a point in $\mathcal{F}$, c_n can be taken as local coordinates on $\mathcal{F}$. Once we have a set of local coordinates c_n and a metric tensor g_{nm}, the volume element is given by

$$[d\varphi] = dV = \sqrt{\det g} \; dc_1 dc_2 \cdots \tag{8.4}$$

In general, since we have an infinite number of c_n's, we must specify the volume measure by a limiting procedure, i.e., define

$$dV^{(N)} = \sqrt{\det g^{(N)}}\, dc_1 dc_2 \cdots dc_N \tag{8.5}$$

where we consider N modes with a corresponding $(N \times N)$-matrix g_{nm}. Then

$$dV = \lim_{N \to \infty} dV^{(N)} \tag{8.6}$$

The finite mode version of the metric is referred to as a regularized metric. Generally, all metrics on function spaces have to be defined with proper regularization.

The integral of interest to us is a Gaussian integral, which we now evaluate. Consider

$$I = \int [d\varphi] \exp\left[-\frac{1}{2} \int_\Sigma d^4x d^4y\, \varphi(x) M(x,y)\varphi(y)\right] \tag{8.7}$$

Introducing a mode expansion in an orthonormal basis given by $\varphi(x) = \sum_n c_n u_n(x)$ as before, the exponent in the above equation can be written as

$$\int_\Sigma d^4x d^4y\, \varphi(x) M(x,y)\varphi(y) = \sum_{n,m} c_n M_{nm} c_m \tag{8.8}$$

where

$$M_{nm} = \int_\Sigma d^4x d^4y\, u_n(x) M(x,y) u_m(y) \tag{8.9}$$

We are interested in situations where M_{nm} is diagonalizable with eigenvalues which have a positive real part. We can then choose an orthonormal basis which diagonalizes M; i.e., we consider eigenfunctions $f_n(x)$ of $M(x,y)$ defined by

$$\int_\Sigma d^4y\, M(x,y) f_n(y) = \lambda_n f_n(x) \tag{8.10}$$

Expanding $\varphi(x)$ as $\varphi(x) = \sum_n a_n f_n(x)$, we have $\|\delta\varphi\|^2 = \sum_n (\delta a_n)^2$, $dV = \prod da_n$, so that

$$\begin{aligned}
I &= \lim_{N \to \infty} \int \prod_1^N [da_n] \exp\left[-\frac{1}{2}\sum_1^N a_n^2 \lambda_n\right] \\
&= \lim_{N \to \infty} \prod_1^N \sqrt{\frac{2\pi}{\lambda_n}} \\
&= \left[\det\left(\frac{M}{2\pi}\right)\right]^{-\frac{1}{2}}
\end{aligned} \tag{8.11}$$

The determinant should be defined by the limiting procedure of first evaluating $\det M^{(N)}$ and then taking $N \to \infty$. This method of truncating to a finite number of modes N and then taking the limit of $N \to \infty$ to define functional determinants, and functional integrals in general, is referred to as a regularization procedure. Since the result involves the determinant which is independent of the basis, it is clear that our choice of the diagonalizing basis $\{f_n(x)\}$ is not a restriction.

We now consider a related integral

$$I[J] = \int [d\varphi] \exp\left[-\frac{1}{2}\int_\Sigma d^4x d^4y\, \varphi(x)M(x,y)\varphi(y) + \int_\Sigma d^4x\, J(x)\varphi(x)\right] \tag{8.12}$$

This can be evaluated by completing the square in the exponent. We find

$$I[J] = \int [d\varphi] e^{-\frac{1}{2}\int \varphi M\varphi + \int J\varphi}$$

$$= \int [d\varphi] e^{-\frac{1}{2}\int (\varphi - JM^{-1})M(\varphi - M^{-1}J)} e^{\frac{1}{2}\int JM^{-1}J}$$

$$= \int [d\varphi] e^{-\frac{1}{2}\int \varphi M\varphi} e^{\frac{1}{2}\int JM^{-1}J}$$

$$= \left[\det\left(\frac{M}{2\pi}\right)\right]^{-\frac{1}{2}} \exp\left[\frac{1}{2}\int_\Sigma d^4x d^4y\, J(x)M^{-1}(x,y)J(y)\right] \tag{8.13}$$

where we have used the translational invariance of the measure to shift the variable from φ to $(\varphi - M^{-1}J)$.

We have considered real functions so far. For a complex scalar function $\varphi(x)$ we can write $\varphi(x) = (\varphi_1(x) + i\varphi_2(x))/\sqrt{2}$, where φ_1, φ_2 are real. Then the following result emerges from what we have done so far.

$$I[J, \bar{J}] = \int [d\varphi d\bar\varphi] e^{-\int \bar\varphi M\varphi + \int \bar{J}\varphi + \bar\varphi J}$$

$$= \left[\det\left(\frac{M}{2\pi}\right)\right]^{-1} e^{\int \bar{J}M^{-1}J} \tag{8.14}$$

Since there are two scalar fields, we get two copies of $[\det(M/2\pi)]^{-\frac{1}{2}}$.

8.2 Green's functions as functional integrals

The integral in (8.13) can be used to obtain a functional integral representation for the generating functional $Z[J]$ of the N-point functions for a scalar field. The free part of the action for a scalar field φ can be written as

$$\mathcal{S}_0 = \int d^4x\, \tfrac{1}{2}\left[(\partial\varphi\partial\varphi) - m^2\varphi^2\right]$$

$$= -\int d^4x\, \tfrac{1}{2}\varphi(x)(\Box + m^2)\varphi(x) \tag{8.15}$$

We need to consider the integration over all fields of $\exp(i\mathcal{S}_0)$. $\mathcal{S}_0$ is real and hence such an integrand is oscillatory. We shall therefore consider the integral of $\exp(iS_0 - \frac{1}{2}\epsilon \int \varphi^2)$; ϵ, which is a small positive real number, is introduced for convergence at large values of the field. In the end, it can be set to zero. We have

$$iS_0 - \frac{1}{2}\epsilon \int \varphi^2 = -\frac{1}{2}\int \varphi(x)M(x,y)\varphi(y)$$

$$M(x,y) = i(\Box_x + m^2 - i\epsilon)\delta^{(4)}(x-y) \tag{8.16}$$

If $G(x,y)$ is the Feynman propagator, we have

$$\int d^4z\ M(x,z)G(z,y) = \delta^{(4)}(x-y) \tag{8.17}$$

which shows that $G(x,y)$ is the inverse to $M(x,y)$. Direct application of (8.13) then gives

$$\int [d\varphi] \exp\left[iS_0 - \tfrac{1}{2}\epsilon \int \varphi^2 + \int J\varphi\right] = C\ \exp\left[\frac{1}{2}\int d^4x d^4y\ J(x)G(x,y)J(y)\right] \tag{8.18}$$

where the C stands for the determinant in (8.13). It is independent of J. The right-hand side, apart from this constant, is just the generating functional $Z_0[J]$ for a free scalar field theory. In other words, we have shown that

$$Z_0[J] = C^{-1} \int [d\varphi] \exp\left[iS_0 - \tfrac{1}{2}\epsilon \int \varphi^2 + \int J\varphi\right] \tag{8.19}$$

It is interesting in this context to note that the ϵ-term was introduced in the functional integral for convergence at large values of the field and at the same time it helps to pick out the Feynman contour for the Green's function, leading to the propagator in the formula (8.18). The choice of the Feynman contour is crucial for the integrand of the propagator to be continued to Euclidean space. Likewise, the functional integral may be defined in terms of the Euclidean action and then continued to Minkowski space and the convergence factor $i\epsilon$ then appears naturally. The reasons for $i\epsilon$ in the two contexts are clearly related.

Consider now the interacting field theory. For a φ^4-interaction, we have shown in Chapter 5 that the generating functional $Z[J]$ is given by

$$Z[J] = \mathcal{N} \exp\left[-i\lambda \int d^4x \left(\frac{\delta}{\delta J(x)}\right)^4\right] Z_0[J] \tag{8.20}$$

Using (8.19) for $Z_0[J]$ we find

$$Z[J] = \mathcal{N}' \exp\left[-i\lambda \int d^4x \left(\frac{\delta}{\delta J(x)}\right)^4\right] \int [d\varphi] \exp\left[iS_0 - \tfrac{1}{2}\epsilon \int \varphi^2 + \int J\varphi\right]$$

$$= \mathcal{N}' \int [d\varphi] \exp\left[iS_0 - i\lambda \int \varphi^4 - \tfrac{1}{2}\epsilon \int \varphi^2 + \int J\varphi\right]$$

$$= \mathcal{N}' \int [d\varphi] \exp\left[iS - \tfrac{1}{2}\epsilon \int \varphi^2 + \int J\varphi\right] \tag{8.21}$$

where

$$S = \int d^4x \ \left[\tfrac{1}{2}(\partial\varphi\partial\varphi) - \tfrac{1}{2}m^2\varphi^2 - \lambda\varphi^4\right] \tag{8.22}$$

is the full classical action for the interacting theory. The constant $\mathcal{N}'$ is determined by the requirement of $Z[0] = 1$.

We can generalize (8.21) to any polynomial type of interaction as follows.

$$Z[J] = \mathcal{N} \int [d\varphi] \exp\left[iS - \tfrac{1}{2}\epsilon \int \varphi^2 + \int J\varphi\right] \tag{8.23}$$

where S is the classical action for the interacting theory and the normalization factor is determined as

$$\mathcal{N}^{-1} = \int [d\varphi] \exp\left[iS - \tfrac{1}{2}\epsilon \int \varphi^2\right] \tag{8.24}$$

This result can also be directly derived as follows. We have seen in Chapter 6 that the operator equations of motion, for the φ^4-theory, are

$$(\Box_x + m^2)\phi(x) + 4\lambda\phi^3(x) = 0 \tag{8.25}$$

The canonical equal-time commutation rules are

$$\begin{aligned}
\left[\phi(x^0, \boldsymbol{x}), \phi(x^0, \boldsymbol{y})\right] &= 0 \\
\left[\pi(x^0, \boldsymbol{x}), \phi(x^0, \boldsymbol{y})\right] &= -i\, \delta^{(3)}(x - y) \\
\left[\pi(x^0, \boldsymbol{x}), \pi(x^0, \boldsymbol{y})\right] &= 0
\end{aligned} \tag{8.26}$$

where $\pi(x^0, \boldsymbol{x}) = \partial_0\phi(x^0, \boldsymbol{x})$. Based on these, we derived, in Chapter 5, the equation obeyed by $Z[J]$ as

$$(\Box_x + m^2 - i\epsilon)\frac{\delta Z[J]}{\delta J(x)} + 4\lambda \left(\frac{\delta}{\delta J(x)}\right)^3 Z[J] = -iJ(x)Z[J] \tag{8.27}$$

Instead of solving this, as we have done, in terms of the free theory after separating off the interaction term and obtaining (8.20), we will show how it arises in the context of functional integrals. Notice that we have the following identity:

$$\int [d\varphi] \frac{\delta}{\delta\varphi}\left[\exp\left[iS - \tfrac{1}{2}\epsilon \int \varphi^2 + \int J\varphi\right]\right] = 0 \tag{8.28}$$

The integrand in (8.28) is a total derivative and so we can do the integral and express it in terms of the values of the integrand at the boundary of φ-space, i.e., at large values of $|\varphi|$. The integrand vanishes for large $|\varphi|$, because of the $\exp(-\tfrac{1}{2}\epsilon \int \varphi^2)$-term and this leads to (8.28). Writing out the derivative in this equation and noticing that

$$F\left(\frac{\delta}{\delta J}\right)\int[d\varphi]\exp\left[i\mathcal{S} - \tfrac{1}{2}\epsilon\int\varphi^2 + \int J\varphi\right] =$$

$$\int[d\varphi]F(\varphi)\exp\left[i\mathcal{S} - \tfrac{1}{2}\epsilon\int\varphi^2 + \int J\varphi\right] \qquad (8.29)$$

we get

$$\left[(\Box_x + m^2 - i\epsilon)\frac{\delta}{\delta J(x)} + 4\lambda\left(\frac{\delta}{\delta J(x)}\right)^3 + iJ(x)\right]\times$$

$$\int[d\varphi]\ \exp\left[i\mathcal{S} - \tfrac{1}{2}\epsilon\int\varphi^2 + \int J\varphi\right] = 0$$

$$(8.30)$$

Comparing this with (8.27) we see that the functional integral of (8.23) gives the solution for $Z[J]$.

Equations (8.23, 8.24) are rather remarkable. They allow us to bypass a lot of the operator formalism of the quantum theory and go directly from the classical action for the theory to the generating functional for the N-point functions, and hence to the S-matrix. And all manipulations are "classical", i.e., just integrals. As we shall see in the next chapter, things are not so simple; the need for renormalization of various coupling parameters will require the use of a slightly different version (or rewriting) of the classical action in practical computations. But this modification is still rather minor and (8.23, 8.24) give a way of proceeding directly from the classical action to the quantum theoretic results for the N-point functions and the S-matrix.

8.3 Functional integration for fermionic variables

We must now discuss a functional integral representation which is suitable for fermions. The functional integral representation, not surprisingly, is provided by integrations over anticommuting c-number functions or Grassmann variables.

We start with a definition of integration over Grassmann variables. Consider first the case of one variable η. Since any function of η is of the form $f(x)_0 + f(x)_1\eta$, we have to define only $\int d\eta$ and $\int d\eta\,\eta$. The definition of integration will be formal, but will be consistent with the expected behaviour of definite integrals over the entire range of η. Consider $\int d\eta\,\eta$. Since our formal definition of integration should mimic integration over all η, this should be invariant under translations of η. Thus we require that

$$\int d\eta\ (\eta + \alpha) = \int d\eta\ \eta \qquad (8.31)$$

for any Grassmann number α. This tells us that $\int d\eta = 0$. $\int d\eta\,\eta$ need not be zero. We define it to be 1. Thus the rules of Grassmann integration are

$$\int d\eta = 0, \qquad\qquad \int d\eta\,\eta = 1 \qquad\qquad (8.32)$$

More generally, if we have N Grassmann variables η_i, we find, by repeated application of the above result that all integrals are zero except the one whose integrand involves the product of all η's. The only nonzero integral is

$$\int d\eta_N d\eta_{N-1}...d\eta_1 \ \ \eta_1\eta_2...\eta_N = 1 \qquad\qquad (8.33)$$

For a more general ordering of the η's, from the antisymmetry property of the η's under commutation, we get

$$\int [d\eta]\ \eta_{i_1}\eta_{i_2}...\eta_{i_N} = \epsilon_{i_1 i_2...i_N} \qquad\qquad (8.34)$$

where $[d\eta] = d\eta_N d\eta_{N-1}...d\eta_1$.

The integral of interest to us for field theory calculations will be a Gaussian integral. Consider N variables η_i and N variables $\bar{\eta}_i$, all mutually anti-commuting. ($\bar{\eta}_i$ are independent Grassmann variables, they are not the "conjugate" of η_i despite the notation.) We can now evaluate the integral

$$I = \int [d\eta d\bar{\eta}]\ \exp\left(-\sum_i \bar{\eta}_i M_{ij}\eta_j\right) \qquad\qquad (8.35)$$

The only term that contributes, by (8.33), should have N η's and N $\bar{\eta}$'s. Expanding the exponential and using (8.34)

$$I = (-1)^{\frac{1}{2}N(N+1)}\frac{1}{N!}\epsilon_{i_1 i_2...i_N}\epsilon_{j_1 j_2...j_N} M_{i_1 j_1} M_{i_2 j_2}...M_{i_N j_N}$$

$$= (-1)^{\frac{1}{2}N(N+1)}(\det M) \qquad\qquad (8.36)$$

The generalization to functions is given by

$$I = \int [d\eta d\bar{\eta}]\ \exp\left(-\int_{x,y} \bar{\eta}(x)M(x,y)\eta(y)\right)$$

$$= (\det M) \qquad\qquad (8.37)$$

We have ignored the overall sign since it will not be relevant for us. As in the bosonic case, the determinant has to be evaluated in a regularized way, i.e., it should be evaluated for a finite number of modes and then the limit of an infinite number of modes should be taken in an appropriate way.

Finally, consider the Grassmann-valued functions $\psi(x)$ and $\bar{\psi}(x)$ and $\eta(x)$, $\bar{\eta}(x)$ and the integral

$$I[\eta,\bar{\eta}] = \int [d\psi d\bar{\psi}]\exp\left(-\int_{x,y} \bar{\psi}(x)M(x,y)\psi(y) + \int_x (\bar{\eta}(x)\psi(x) + \bar{\psi}(x)\eta(x))\right)$$

$$= \int [d\psi d\bar{\psi}] \, \exp\left(-\int (\bar{\psi} - \bar{\eta}M^{-1})M(\psi - M^{-1}\eta)\right) \exp\left(\int \bar{\eta}M^{-1}\eta\right)$$

$$= (\det M) \exp\left(\int \bar{\eta}M^{-1}\eta\right) \tag{8.38}$$

These results can now be applied to fermionic fields. Many-fermion propagators have been discussed by introducing a generating functional

$$Z[\eta, \bar{\eta}] = \langle 0 | T \exp\left(\int \bar{\eta}\psi + \bar{\psi}\eta\right) | 0 \rangle \tag{8.39}$$

The sources $\eta, \bar{\eta}$ are spinors and are Grassmann-valued. They anticommute with $\psi, \bar{\psi}$ as well. The expression for $Z[\eta, \bar{\eta}]$ was given in Chapter 4 as

$$Z[\eta, \bar{\eta}] = \exp\left[\int d^4x d^4y \, \bar{\eta}(x)S(x,y)\eta(y)\right] \tag{8.40}$$

where we have chosen the normalization $Z[0,0] = 1$.

The integral (8.38) can now be applied to obtain a functional integral representation for $Z[\eta, \bar{\eta}]$. Taking $M(x,y) = (-i)(i\gamma \cdot \partial_x - m)\delta^{(4)}(x-y)$ the integral in (8.38) gives

$$I[\eta, \bar{\eta}] = \int [d\psi d\bar{\psi}] \, \exp\left(-\int_{x,y} \bar{\psi}(x)M(x,y)\psi(y) + \int_x (\bar{\eta}(x)\psi(x) + \bar{\psi}(x)\eta(x))\right)$$

$$= (\det M)Z[\eta, \bar{\eta}] \tag{8.41}$$

We can write

$$-\int_{x,y} \bar{\psi}(x)M(x,y)\psi(y) = i \int d^4x \, \bar{\psi}(i\gamma \cdot \partial - m)\psi$$

$$= iS[\psi, \bar{\psi}] \tag{8.42}$$

where $S[\psi, \bar{\psi}]$ is the action for the Dirac theory. The integral formula for $Z[\eta, \bar{\eta}]$ becomes

$$Z[\eta, \bar{\eta}] = \mathcal{N} \int [d\psi d\bar{\psi}] \, \exp\left[iS[\psi, \bar{\psi}] + \int d^4x \, (\bar{\eta}(x)\psi(x) + \bar{\psi}(x)\eta(x))\right] \tag{8.43}$$

The generalization of this result to interacting fermion theories is entirely straightforward. We find in all cases, involving both fermionic and bosonic fields, that the generating functional for the N-point functions may be written as

$$Z[J, \eta, \bar{\eta}] = \mathcal{N} \int [d\varphi d\psi d\bar{\psi}] \, \exp\Big[iS[\varphi, \psi, \bar{\psi}]$$

$$+ \int d^4x \, (J(x)\varphi(x) + \bar{\eta}(x)\psi(x) + \bar{\psi}(x)\eta(x))\Big] \tag{8.44}$$

where, in the integral, the fermionic variables are Grassmann valued and the bosonic fields are ordinary real- or complex-valued functions. The normalization $\mathcal{N}$ is to be fixed by the requirement of $Z[0,0,0] = 1$. Thus given the classical action for the fields, we can bypass most of the operator formalism and proceed to the generating functional for the N-point functions by making use of the above formula. In some sense, it may be taken as the definition of the quantum theory of the fields.

8.4 The S-matrix functional

Since we have a functional integral representation for $Z[J]$, it is clear that we have a similar representation for the S-matrix as well. In Chapter 5, for a scalar field theory, we obtained the formula

$$\mathcal{F}[\varphi] = \exp\left[-\frac{i}{2}\int \varphi(\Box + m^2)\varphi\right]\ Z[i(\Box + m^2)\varphi] \qquad (8.45)$$

We now use the functional integral (8.23) for $Z[J]$. The action can be split as

$$\mathcal{S}(\chi) = \int d^4x\ \left[\tfrac{1}{2}(\partial\chi)^2 - \tfrac{1}{2}m^2\chi^2\right] + \mathcal{S}_{int}(\chi)$$
$$= \mathcal{S}_0 + \mathcal{S}_{int} \qquad (8.46)$$

We complete the squares in $\mathcal{S}(\chi) + \int i\chi(\Box + m^2)\varphi$ to obtain

$$Z[i(\Box + m^2)\varphi] = \mathcal{N}\int [d\chi]\exp\Big[i\mathcal{S}_0(\chi - \varphi) - \tfrac{1}{2}\epsilon\int \chi^2$$
$$+ i\mathcal{S}_{int}(\chi) + \frac{i}{2}\int \varphi(\Box + m^2)\varphi\Big]$$
$$= \mathcal{N}\int [d\chi]\exp\Big[i\mathcal{S}_0(\chi) - \tfrac{1}{2}\epsilon\int \chi^2 + i\mathcal{S}_{int}(\chi + \varphi)$$
$$+ \frac{i}{2}\int \varphi(\Box + m^2)\varphi\Big]$$
$$= \exp\left[\frac{i}{2}\int \varphi(\Box + m^2)\varphi\right]\ \langle\ \exp(i\mathcal{S}_{int}(\chi + \varphi))\rangle_0$$

$$(8.47)$$

We made a shift of the variable of integration χ as $\chi \to \chi + \varphi$ in the second step. The angular brackets denote functional average over χ's with just the free part of the action; i.e.,

$$\langle\mathcal{O}\rangle_0 = \mathcal{N}\int [d\chi]\exp\left[i\mathcal{S}_0(\chi) - \tfrac{1}{2}\epsilon\int \chi^2\right]\ \mathcal{O} \qquad (8.48)$$

We have also omitted some terms of order ϵ which can be set to zero without affecting convergence of the integral. Using equation (8.47) in (8.45), we get

$$\mathcal{F}[\varphi] = \langle\, \exp\left[i\mathcal{S}_{int}(\chi + \varphi)\right] \rangle_0 \tag{8.49}$$

This equation relates $\mathcal{F}[\varphi]$ to the free average of the interaction term in the action $\mathcal{S}$. We can directly expand the exponential to generate the perturbation series for the S-matrix. Being a Gaussian average, the average of a product of χ's will factorize into the products of two-point functions $\langle \chi(x)\chi(y)\rangle_0$ with suitable symmetrizations. In the equation for the averages and the S-matrix, we have included the full normalization factor $\mathcal{N}$ for the interacting theory. One can use the normalization factor for the free theory to define the functional averages provided the S-matrix is properly normalized by the condition $\mathcal{F}[0] = 1$ at the end. Notice also that equation (8.49) is a way of rewriting the operator formula (5.85).

The above derivation, although given for the scalar field theory, can be easily extended to more general theories and one can write the S-matrix functional as

$$\mathcal{F}[\varphi, a_\mu, \psi, \bar{\psi}] = \langle \exp\left[i\mathcal{S}_{int}(\chi + \varphi, A_\mu + a_\mu, \Psi + \psi, \bar{\Psi} + \bar{\psi})\right] \rangle_0 \tag{8.50}$$

χ, A_μ, Ψ and $\bar{\Psi}$ are the fields which are integrated over with the free action in the measure to define the averages in this equation.

8.5 Euclidean integral, quantum electrodynamics, etc.

The propagators and the N-point functions can be obtained, as we have discussed in Chapter 4, as the Minkowski space continuation via $x^4 \to ix^0$ of the corresponding results in Euclidean space. A Euclidean space version of the functional integrals is very useful, especially for higher-order calculations. We define the Euclidean free field actions for a scalar field and a fermion field as

$$\mathcal{S}_{E0} = \int d^4x \left[\frac{1}{2}(\partial\varphi)^2 + \frac{1}{2}m^2\varphi^2 + \bar{\psi}(\gamma \cdot \partial + m)\psi\right] \tag{8.51}$$

where all scalar products are taken with the Euclidean metric $\delta_{\mu\nu}$ and the γ-matrices obey $\gamma_\mu\gamma_\nu + \gamma_\nu\gamma_\mu = 2\delta_{\mu\nu}$. The Euclidean version of (8.18, 8.41) is

$$\int [d\varphi d\psi d\bar{\psi}] \, \exp\left[-\mathcal{S}_{E0}(\varphi, \bar{\psi}, \psi) + \int d^4x \, (J\varphi + \bar{\eta}\psi + \bar{\psi}\eta)\right]$$
$$= C\exp\left[\int d^4x d^4y \, \left(\frac{1}{2}J(x)G_E(x, y)J(y) + \bar{\eta}(x)S_E(x, y)\eta(y)\right)\right] \tag{8.52}$$

where

$$G_E(x,y) = \int \frac{d^4p}{(2\pi)^4} \frac{1}{p^2 + m^2} e^{ip(x-y)}$$

$$S_E(x,y) = \int \frac{d^4p}{(2\pi)^4} \frac{1}{i\gamma \cdot p + m} e^{ip(x-y)} \tag{8.53}$$

We may write, more generally with interactions, for the generating functional Z in Minkowski space

$$Z[J,\bar{\eta},\eta] = Z_E[J,\bar{\eta},\eta]\Big|_{x^4 \to ix^0}$$

$$Z_E[J,\bar{\eta},\eta] = \mathcal{N} \int [d\varphi d\psi d\bar{\psi}] \, \exp\left[-\mathcal{S}_E(\varphi,\bar{\psi},\psi) + \int d^4x \, (J\varphi + \bar{\eta}\psi + \bar{\psi}\eta)\right] \tag{8.54}$$

Now we turn to quantum electrodynamics (QED). As we have shown before, the photon propagator, for calculations in QED, may be taken to have the covariant form given in equation (6.23), namely,

$$D_{\mu\nu}(x,y) = \eta_{\mu\nu} \int \frac{d^4k}{(2\pi)^4} \frac{i}{k^2 + i\epsilon} e^{-ik(x-y)} \tag{8.55}$$

The corresponding Euclidean propagator is given by

$$D_{E\mu\nu} = \delta_{\mu\nu} \int \frac{d^4k}{(2\pi)^4} \frac{1}{k^2} e^{ik(x-y)} \tag{8.56}$$

The Euclidean functional integral for QED is thus given by

$$Z_E[J,\bar{\eta},\eta] = \mathcal{N} \int [dA d\psi d\bar{\psi}] \, \exp\left[-\mathcal{S}_E + \int d^4x \, (A_\mu J^\mu + \bar{\eta}\psi + \bar{\psi}\eta)\right] \tag{8.57}$$

where

$$\mathcal{S}_E(A,\bar{\psi},\psi) = \int d^4x \left[\frac{1}{2}(\partial_\mu A_\nu \partial_\mu A_\nu) + \bar{\psi}(\gamma \cdot (\partial - ieA) + m)\psi\right]$$

$$= \int d^4x \left[\frac{1}{4}F_{\mu\nu}F^{\mu\nu} + \frac{1}{2}(\partial \cdot A)^2 + \bar{\psi}(\gamma \cdot (\partial - ieA) + m)\psi\right] \tag{8.58}$$

$\mathcal{N}$ is, as usual, fixed by the requirement $Z[0,0,0,] = 1$. The integration over A's in (8.57) is done with the standard Euclidean measure on the space of A's.

The general idea of the functional integral being given by the integration over the classical action, as discussed after equation (8.44), would suggest the use of the Maxwell action $\int \frac{1}{4}F^2$, or its Euclidean version, in the formula (8.57). The fact that one has to use the action as given in (8.58) has to do with gauge transformations. Since A_μ and its gauge transform $A_\mu + \partial_\mu\theta$

are physically equivalent, there is redundancy in the choice of field variables. For integration over the Maxwell action, one must then use a measure where this redundancy has been removed. The use of such a measure, which we shall discuss in more detail in the chapter on gauge theories, will lead to the above formulae for QED. In summation then, for all theories, the Euclidean functional integral for the generating functional can be written as the integral over $\exp[-S_E]$ with integration over the space of all physical, i.e., non-redundant, field configurations. (For QED there is a further averaging over gauges needed to get to the formula (8.57), see Chapter 10.)

8.6 Nonlinear sigma models

Nonlinear sigma models are an important class of field theory models. The functional integral for these theories has a nontrivial measure for integration over the fields. It is interesting to derive this following our general approach of using the operator equations of motion and the equal-time commutation rules.

Consider a general Riemannian space $\mathcal{M}$, which may have nonzero curvature, with coordinates φ^A and metric

$$ds^2 = G_{AB}d\varphi^A d\varphi^B \tag{8.59}$$

The metric tensor G_{AB} is in general a function of the coordinates φ^A. The action for a point particle moving on such a space can be taken as

$$S = \int d\tau \, \frac{1}{2} G_{AB} \frac{d\varphi^A}{d\tau} \frac{d\varphi^B}{d\tau} \tag{8.60}$$

where τ parametrizes the trajectory. The classical equations of motion are the geodesic equations, the classical trajectories are geodesics. As a generalization of this notion one can consider fields $\varphi^A(x)$ which give a map from the spacetime to the Riemannian space $\mathcal{M}$. This space $\mathcal{M}$ into which we are mapping from spacetime is often called the target space. The action is then taken as

$$S = \int \sqrt{-g}d^4x \, g^{\mu\nu} \frac{1}{2} G_{AB} \frac{\partial\varphi^A}{\partial x^\mu} \frac{\partial\varphi^B}{\partial x^\nu} \tag{8.61}$$

The action is determined by the metric of the target space. For generality, we have written $g^{\mu\nu}$ for the spacetime (inverse) metric. The formulation of the theory along the lines given here is applicable for the case of a general spacetime which may have nonzero curvature as well. One can also consider Euclidean signature for the spacetime. For flat Minkowski spacetime, which we consider for the rest of this section, $g^{\mu\nu} = \eta^{\mu\nu}$.

Field theories defined by the action (8.61) are called nonlinear sigma models or sometimes, they are called chiral models. (The reason for the name "sigma models" is historical; it arose in particle physics literature, in the context of chiral symmetry breaking, from a linear field theory model where there was a field which was designated by σ; the nonlinear model resulted from taking the mass of σ to be large compared to the momenta of interest.)

The classical equations of motion for this action are

$$- \frac{\partial}{\partial x^\mu} \left[G_{AB} \frac{\partial \varphi^B}{\partial x_\mu} \right] + \frac{1}{2} \frac{\partial G_{BC}}{\partial \varphi^A} \frac{\partial \varphi^B}{\partial x^\mu} \frac{\partial \varphi^C}{\partial x_\mu} = 0 \tag{8.62}$$

The solutions to this equation form a special class of maps from the spacetime to the target space $\mathcal{M}$ which are a generalization of the notion of the geodesic. Such maps are called harmonic maps in the mathematical literature.

The importance of this theory in physics has to do with Goldstone's theorem. Consider a field theory which has a continuous global symmetry corresponding to a Lie group G. (A symmetry is global if the parameters of the symmetry transformation are constant in spacetime.) If the vacuum state does not have this symmetry G, but is symmetric under a subgroup H, we say that the symmetry G is spontaneously broken down to H. A typical example is the Heisenberg ferromagnet for which the action has three-dimensional rotational symmetry, but the ground state, which has spontaneous magnetization along some direction, breaks this rotational symmetry. Goldstone's theorem tells us that there will be massless particles corresponding to each broken symmetry. There are $(dimG - dimH)$ such massless particles called Goldstone particles. Further, since the particles are massless, at low energies, where one does not have enough energy to excite massive particles, we essentially get the dynamics of the Goldstone particles interacting among themselves. The action for the Goldstone bosons is given by a nonlinear sigma model of the form (8.61) where the target space is the group coset space G/H. Thus nonlinear sigma models are important in all physical contexts where spontaneous breaking of continuous symmetries occurs. (The phenomenon of spontaneous symmetry breaking is discussed in detail in Chapter 12.)

In this section we will just consider the functional integral for the sigma model following our general derivation of the functional equation for $Z[J]$ which is defined in the usual manner as

$$Z[J] = \langle 0 | \, T \, \exp \left(\int J_A \varphi^A \right) | 0 \rangle$$

$$\equiv \langle \, T \, \exp \left(\int J_A \varphi^A \right) \, \rangle \tag{8.63}$$

The canonical commutation rules for the fields are

$$\begin{aligned}
&[\, \varphi^A(x^0, \boldsymbol{x}), \, \varphi^B(x^0, \boldsymbol{y}) \,] = 0 \\
&[\, \varphi^A(x^0, \boldsymbol{x}), \, \Pi^0_B(x^0, \boldsymbol{y}) \,] = i \, \delta^A_B \, \delta^{(3)}(x - y) \\
&[\, \Pi^0_A(x^0, \boldsymbol{x}), \, \Pi^0_B(x^0, \boldsymbol{y}) \,] = 0
\end{aligned} \tag{8.64}$$

where $\Pi_A^\mu = G_{AB}\partial^\mu\varphi^B$. We also define

$$F(x^0, y^0) = T \exp\left(\int_{y^0}^{x^0} dz^0 \int d^3z J_A(z)\varphi^A(z)\right) \qquad (8.65)$$

By direct differentiation we check that this obeys

$$\frac{\partial}{\partial x^0} F(x^0, y^0) = \left[\int d^3z J_A(x^0, \mathbf{z})\varphi^A(x^0, \mathbf{z})\right] F(x^0, y^0)$$

$$\frac{\partial}{\partial y^0} F(x^0, y^0) = -F(x^0, y^0)\left[\int d^3z J_A(y^0, \mathbf{z})\varphi^A(y^0, \mathbf{z})\right] \qquad (8.66)$$

Using this quantity we can write

$$\partial_\mu\langle T\Pi_A^\mu(x)e^{\int J\varphi}\rangle = \partial_\mu\langle F(\infty, x^0)\Pi_A^\mu(x)F(x^0, -\infty)\rangle$$

$$= \langle F(\infty, x^0)[\Pi_A^0(x), \int d^3z J_A(x^0, \mathbf{z})\varphi^A(x^0, \mathbf{z})] F(x^0, -\infty)\rangle$$

$$+ \langle T\partial_\mu\Pi_A^\mu e^{\int J\varphi}\rangle$$

$$= -iJ_A(x)Z[J] + \frac{1}{2}\langle T\frac{\partial G_{BC}}{\partial\varphi^A}\frac{\partial\varphi^B}{\partial x^\mu}\frac{\partial\varphi^C}{\partial x_\mu}e^{\int J\varphi}\rangle \qquad (8.67)$$

where in the first step we have used equations (8.66) and in the second step we have used the commutation rules (8.64) and the equation of motion (8.62) interpreted as an operator Heisenberg equation of motion. We now write

$$\langle TG_{AB}\partial^\mu\varphi^B e^{\int J\varphi}\rangle = \left[G_{AB}(\hat\varphi)\frac{\partial\hat\varphi^B}{\partial x_\mu}\right] Z[J] \qquad (8.68)$$

where $\hat\varphi^A = \delta/\delta J_A$. In taking the factor $\partial^\mu\varphi^B$ outside the vacuum expectation value and replacing φ^B by the functional derivative with respect to J_B, we may worry that there is an equal time commutator term due to the time-ordering. There is indeed such a term, but it involves the commutator of φ with φ and vanishes by (8.64). Equation (8.68) can be used to simplify the left-hand side of (8.67). We want to make a similar simplification of the right-hand side. We start with the expression

$$\frac{\partial}{\partial z^\mu}\langle T\frac{\partial G_{BC}}{\partial\varphi^A(x)}\partial^\mu\varphi^B(x)\varphi^C(z) e^{\int J\varphi}\rangle \equiv \frac{\partial}{\partial z^\mu}\langle T\,\mathcal{O}_{CA}^\mu(x)\,\varphi^C(z)\,e^{\int J\varphi}\rangle \qquad (8.69)$$

In simplifying this, we encounter commutators when we bring the z^0-derivative inside the time-ordering symbol. By writing out the time-ordered product as

$$T\,\mathcal{O}_{CA}^\mu(x)\varphi^C(z)\,e^{\int J\varphi}$$

$$= \left[F(\infty, x^0)\mathcal{O}_{CA}^\mu(x)F(x^0, z^0)\varphi^C(z)F(z^0, -\infty)\theta(x^0 - z^0)\right.$$

$$+F(\infty, z^0)\varphi^C(z)F(z^0, x^0)\mathcal{O}^\mu_{CA}(x)F(x^0, -\infty)\theta(z^0 - x^0)\Bigg]$$

$$(8.70)$$

and carrying out the differentiation using (8.66) we get

$$\frac{\partial}{\partial z^\mu}\langle T\,\mathcal{O}^\mu_{CA}(x)\varphi^C(z)\,e^{\int J\varphi}\rangle = \langle T\frac{\partial G_{BC}}{\partial\varphi^A}\frac{\partial\varphi^B}{\partial x_\mu}\frac{\partial\varphi^C}{\partial z^\mu}\,e^{\int J\varphi}\rangle$$

$$+i\langle\frac{\partial G_{BC}}{\partial\varphi^A}G^{BC}\delta^{(4)}(x-z)\,e^{\int J\varphi}\rangle$$

$$(8.71)$$

We have used the fact that

$$\frac{\partial G_{BC}}{\partial\varphi^A}[\partial^0\varphi^B(x^0,\boldsymbol{x}),\varphi^C(x^0,\boldsymbol{z})] = \frac{\partial G_{BC}}{\partial\varphi^A}G^{BD}[G_{DE}\partial^0\varphi^E(x^0,\boldsymbol{x}),\varphi^C(x^0,\boldsymbol{z})]$$

$$= \frac{\partial G_{BC}}{\partial\varphi^A}G^{BD}[\Pi^0_D(x^0,\boldsymbol{x}),\varphi^C(x^0,\boldsymbol{z})]$$

$$= -i\frac{\partial G_{BC}}{\partial\varphi^A}G^{BC}\delta^{(3)}(x-z)\qquad(8.72)$$

The idea now is to take $z \to x$ so that the first term on the right-hand side of (8.71) is what we want to simplify in (8.67). But this leads to the expression $\delta^{(4)}(0)$ which is the related to the volume of the momentum space. Thus we have to have a truncation of modes to do this entirely correctly. But we can proceed by writing this expression formally as follows. Notice that pointwise

$$\frac{\partial G_{BC}}{\partial\varphi^A}G^{BC} = \frac{\partial}{\partial\varphi^A}\mathrm{tr}\,\log G \qquad(8.73)$$

where the trace is over the indices B, C. The quantity $\log G$ may be regarded as an integral kernel with

$$\langle x|\log G|z\rangle = \delta^{(4)}(x-z)\,\log G \qquad(8.74)$$

so that by taking a functional trace as well as the trace over the indices B, C, we get

$$\mathrm{Tr}\,\log G = \delta^{(4)}(0)\,\mathrm{Tr}\,\log G \qquad(8.75)$$

where Tr denotes the functional trace as well. Using equations (8.71) to (8.75), we can now write

$$\frac{1}{2}\langle T\frac{\partial G_{BC}}{\partial\varphi^A}\frac{\partial\varphi^B}{\partial x^\mu}\frac{\partial\varphi^C}{\partial x_\mu}e^{\int J\varphi}\rangle\Bigg]_{z\to x} = \frac{1}{2}\frac{\partial}{\partial z^\mu}\langle T\,\mathcal{O}^\mu_{CA}(x)\varphi^C(z)\,e^{\int J\varphi}\rangle\Bigg]_{z\to x}$$

$$-\frac{i}{2}\langle\frac{\delta}{\delta\varphi^A(x)}(\mathrm{Tr}\log G)\,e^{\int J\varphi}\rangle \qquad(8.76)$$

Taking the remaining terms out of the expectation value as functional derivatives with respect to J does not cause problems. The commutators encountered along the way involve only φ's and vanish. In particular, we bring out the $\mathcal{O}^\mu_{CA}(x)\varphi^C(z)$ in the first term on the right-hand side as functional differential operators on $Z[J]$ and then carry out the differentiation with respect to z^μ and then take the limit $z \to x$. Using (8.68) and (8.76) in (8.67), we get

$$
\left[\frac{\partial}{\partial x^\mu} \left(G_{AB}(\varphi) \frac{\partial \varphi^B}{\partial x_\mu} \right) - \frac{1}{2} \frac{\partial G_{BC}}{\partial \varphi^A} \frac{\partial \varphi^B}{\partial x^\mu} \frac{\partial \varphi^C}{\partial x_\mu} + \frac{i}{2} \frac{\delta}{\delta \varphi^A(x)} \left(\operatorname{Tr} \log G \right) \right]_{\varphi \to \hat\varphi} Z[J]
$$
$$
= -i J_A(x) Z[J]
$$
(8.77)

In terms of the action (8.61) we can express this equation as

$$
\left[-\frac{\delta S}{\delta \varphi^A(x)} + \frac{i}{2} \frac{\delta}{\delta \varphi^A(x)} \operatorname{Tr} \log G \right]_{\varphi \to \hat\varphi} Z[J] = -i J_A(x) \, Z[J] \qquad (8.78)
$$

The integral of a total functional derivative will vanish with appropriate fall-off behavior for large φ, so we can write

$$
\int [d\varphi] \frac{\delta}{\delta \varphi^A} \exp \left(iS + \tfrac{1}{2} \operatorname{Tr} \log G + \int J\varphi \right) = 0 \qquad (8.79)
$$

Carrying out the differentiation leads to

$$
\int [d\varphi] \left[i \frac{\delta S}{\delta \varphi^A(x)} + \frac{1}{2} \frac{\delta}{\delta \varphi^A(x)} \operatorname{Tr} \log G + J_A(x) \right] e^{iS + \tfrac{1}{2} \operatorname{Tr} \log G + \int J\varphi} = 0
$$
(8.80)

Bringing out the φ's as differentiation with respect to J's in this equation, we get

$$
\left[-\frac{\delta S}{\delta \varphi^A(x)} + \frac{i}{2} \frac{\delta}{\delta \varphi^A(x)} \operatorname{Tr} \log G \right]_{\varphi \to \hat\varphi} \int [d\varphi] e^{iS + \tfrac{1}{2} \operatorname{Tr} \log G + \int J\varphi}
$$
$$
= -i J_A(x) \int [d\varphi] e^{iS + \tfrac{1}{2} \operatorname{Tr} \log G + \int J\varphi} \qquad (8.81)
$$

Comparison with (8.78) shows that we may solve for $Z[J]$ as a functional integral

$$
Z[J] = \mathcal{N} \int [d\varphi] \, \exp \left(iS + \tfrac{1}{2} \operatorname{Tr} \log G + \int J\varphi \right)
$$
$$
= \mathcal{N} \int [d\varphi] \sqrt{\det G} \, \exp \left(iS + \int J\varphi \right) \qquad (8.82)
$$

As usual, this result is up to a normalization factor $\mathcal{N}$.

On a Riemannian space with metric $G_{AB}d\varphi^A d\varphi^B$, the volume element is given as $\sqrt{\det G} \prod d\varphi$. This is indeed the volume that has emerged naturally from the above analysis for the functional measure for the sigma model. What needs to be done regarding the $\delta^{(4)}(0)$ that we encountered is also clear by now. We need to define a suitable regularization procedure to evaluate the functional determinant $\det G$ and then use it to evaluate the functional integral as well in a regulated way. As in the derivation for the φ^4-theory, the $i\epsilon$-term which has to be added to ensure the convergence of the functional integral at large values of φ^A (and which leads to the choice of Feynman contour for propagators) will give the required fall-off behavior for the validity of the vanishing of the integral of the total derivative in (8.79).

8.7 The connected Green's functions

In Chapter 5, we saw that the perturbative expansion of the Green's functions and the S-matrix can lead to connected and disconnected Feynman diagrams. For example at the second order, the diagram in figure 8.1 is a disconnected diagram with two connected pieces while the diagram in figure 8.2 is an example of a connected diagram. At higher orders, obviously, we would get large numbers of disconnected diagrams. We now show that there is a general relationship between the generating functional for Green's functions or the S-matrix functional and the generating functional for connected diagrams.

Fig 8.1. An example of a disconnected diagram

The generating functional $Z[J]$ can be expanded as

$$Z[J] = \sum_N \frac{1}{N!} \int d^4x_1...d^4x_N J(x_1)...J(x_N)G(x_1,...x_N) \qquad (8.83)$$

where

$$G(x_1,...x_N) = \mathcal{N} \int [d\varphi]e^{-S_E} \varphi(x_1)...\varphi(x_N)$$
$$\equiv \langle \varphi(x_1)...\varphi(x_N) \rangle \qquad (8.84)$$
$$\mathcal{N}^{-1} = \int [d\varphi]e^{-S_E} \qquad (8.85)$$

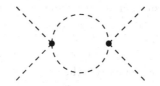

Fig 8.2. An example of a connected diagram

(We shall use Euclidean Green's functions for most of what follows.) The Green's functions are averages over the fields with a probability distribution given by $\mathcal{N}e^{-S_E}$. Disconnected Green's functions arise when an average like $\langle\varphi(x_1)...\varphi(x_N)\rangle$ factorizes as, for example, into $\langle\varphi(x_1)...\varphi(x_k)\rangle$ $\langle\varphi(x_{k+1})...\varphi(x_N)\rangle$. Let G^c denote connected Green's functions, viz., G_N^c is the fully connected Green' function with N fields. A general Green's function can be written as the sums of products of connected Green's functions. Let $G(x_1,...x_N)$ have n_1 factors of G_1^c, n_2 factors of G_2^c, ...,etc. The number of ways of factorizing $G(x_1,...x_N)$ in this fashion is the same as the number of ways of partitioning N particles with n_1 boxes with one particle each, n_2 boxes with two particles each and so on. For convenience of the argument, define $Q = \int d^4x J(x)\varphi(x)$. The expansion of $Z[J]$ involves the averages of products of Q's. Let G_N denote the average of N Q's. In the partition of G_N, it is clear that the exchange of the n_k boxes does not give a new partition; also exchange of the k particles in a box does not give a new partition. Thus the partition of G_N looks like

$$G_N = \sum_{\{n_k\}} N! \left[\frac{(G_1^c)^{n_1}}{n_1!} \frac{(G_2^c/2!)^{n_2}}{n_2!} ... \right] \tag{8.86}$$

subject to the condition $\sum_k n_k k = N$. Thus we may write

$$G_N = \sum_{\{n_k\}} N!\delta(N - \sum n_k k) \prod_s \frac{(G_s^c/s!)^{n_s}}{n_s!} \tag{8.87}$$

We now use this factored form in (8.83); the summation over N in (8.83) then effectively removes the constraint imposed by $\delta\left(N - \sum n_k k\right)$ and we find

$$Z[J] = \sum_N \delta\left(N - \sum n_k k\right) \sum_{\{n_k\}} \frac{(G_1^c)^{n_1}}{n_1!} \frac{(G_2^c/2!)^{n_2}}{n_2!} \frac{(G_3^c/3!)^{n_3}}{n_3!} ...$$

$$= \sum_{\{n_k\}} \frac{(G_1^c)^{n_1}}{n_1!} \frac{(G_2^c/2!)^{n_2}}{n_2!} \frac{(G_3^c/3!)^{n_3}}{n_3!} ...$$

$$= \exp\left(G_1^c\right) \exp\left(\frac{G_2^c}{2!}\right) ...$$

$$= \exp\left(W[J]\right) \tag{8.88}$$

where

$$W[J] = G_1^c + \frac{1}{2!}G_2^c + \dots$$

$$= \int d^4x J(x)G^c(x) + \frac{1}{2!}\int d^4x_1 d^4x_2 J(x_1)J(x_2)G^c(x_1, x_2) + \dots$$

$$= \sum_N \frac{1}{N!}\int\int d^4x_1 \dots d^4x_N J(x_1)\dots J(x_N)G^c(x_1, \dots x_N) \tag{8.89}$$

$W[J]$ is the generating functional for the connected Green's functions.

In the above result, we have used the normalized probability distribution $\mathcal{N}e^{-\mathcal{S}_E}$. Therefore there are no purely vacuum diagrams here; they are given by $\int[d\varphi]e^{-\mathcal{S}_E}$ and are removed by the normalization factor. A result similar to (8.89) holds for the vacuum diagrams as well. If calculations are done at a finite temperature and density, the average $\int[d\varphi]e^{-\mathcal{S}_E}$ represents the statistical partition function. Its dependence on control parameters such as temperature and chemical potential are of interest. (The background heat bath now plays the role of the vacuum state.) In this case also, there is a version of the result (8.89). This is best seen by writing

$$\mathcal{N}^{-1} \equiv \int[d\varphi]e^{-\mathcal{S}_E} = \langle e^{-\mathcal{S}_{int}}\rangle_0$$

$$= \sum_N \frac{1}{N!}\langle(-\mathcal{S}_{int})^N\rangle_0 \tag{8.90}$$

where the average is taken with the free action. Just as we considered the number of different ways of partitioning the product of Q's to get (8.87), we can now consider the number of different ways of distributing the $\mathcal{S}_{int}$'s to write, in terms of connected functions,

$$\mathcal{N}^{-1} = \sum_N \delta\left(N - \sum n_k k\right)\sum_{\{n_k\}} \frac{(\langle\mathcal{S}_{int}\rangle^c)^{n_1}}{n_1!}\frac{(\langle\mathcal{S}_{int}^2\rangle^c/2!)^{n_2}}{n_2!}\frac{(\langle\mathcal{S}_{int}^3\rangle^c/3!)^{n_3}}{n_3!}\dots$$

$$= \sum_{\{n_k\}} \frac{(\langle\mathcal{S}_{int}\rangle^c)^{n_1}}{n_1!}\frac{(\langle\mathcal{S}_{int}^2\rangle^c/2!)^{n_2}}{n_2!}\frac{(\langle\mathcal{S}_{int}^3\rangle^c/3!)^{n_3}}{n_3!}\dots$$

$$= \exp\left(\langle\mathcal{S}_{int}\rangle^c\right)\exp\left(\frac{\langle\mathcal{S}_{int}^2\rangle^c}{2!}\right)\dots$$

$$= \exp\left(W\right) \tag{8.91}$$

where the superscripts on the angular brackets again denote the connected vacuum diagrams with the indicated number of $\mathcal{S}_{int}$'s. W is the sum of all the connected vacuum diagrams. (Again, in the statistical context, by vacuum diagrams we mean processes involving scattering between particles in the heat bath or other similar thermal fluctuation effects, with no incoming or outgoing particles except those in the heat bath.)

8.8 The quantum effective action

Closely related to the generating functional for the connected Green's functions is the quantum effective action $\Gamma[\Phi]$. It is the Legendre transform of $W[J]$ defined by

$$\Gamma[\Phi] = \int J\Phi - W[J] \qquad (8.92)$$

where the relation between Φ and J is defined by the connecting relations

$$\frac{\delta\Gamma}{\delta\Phi(x)} = J(x)$$

$$\frac{\delta W}{\delta J(x)} = \Phi(x) \qquad (8.93)$$

These equations may be regarded as defining $\Gamma[\Phi]$ if we are given $W[J]$, or as defining $W[J]$ if we are given $\Gamma[\Phi]$. If $W[J]$ is given, we obtain J as a function of the free variable Φ by the second equation in (8.93); substituting this into (8.92), we obtain $\Gamma[\Phi]$. Conversely, given $\Gamma[\Phi]$, we can obtain Φ as a function of a variable J using the first equation in (8.93), and then use (8.92) to define $W[J]$. One can expand $\Gamma[\Phi]$ in powers of Φ as

$$\Gamma[\Phi] = \sum_N \frac{1}{N!} \int d^4x_1 d^4x_2 \cdots d^4x_N \ \Phi(x_1)\Phi(x_2)\cdots\Phi(x_N) \ \tilde{V}(x_1, x_2, ..., x_N)$$

$$(8.94)$$

The coefficients $\tilde{V}(x_1, x_2, ..., x_N)$ are easily checked to be vertex functions, namely, Green's functions with the external lines removed or amputated, as in (5.25). However, these are actually one-particle irreducible. In the graphical representation of vertices and Green's functions, a diagram is said to be one-particle irreducible (1PI) if it does not become disconnected upon cutting any single one-particle propagator; diagrams which become disconnected are reducible. $\Gamma[\Phi]$ generates all the 1PI-vertices.

From the connecting relations (8.93), we can write

$$\int d^4z \left[\frac{\delta^2\Gamma}{\delta\Phi(x)\delta\Phi(z)} \right] \tilde{G}(z, y) = \delta^{(4)}(x - y) \qquad (8.95)$$

where

$$\tilde{G}(x, y) = \frac{\delta^2 W}{\delta J(x)\delta J(y)} \qquad (8.96)$$

The result (8.95) follows since $\delta\Phi(x)/\delta J(y)$ and $\delta J(x)/\delta\Phi(y)$ are inverses of each other. Notice that we have not set $J = 0$ in (8.95) or (8.96). By differentiating the relation (8.95) many times with respect to $\Phi(x)$ and using the connecting relations (8.93), one can check that the coefficients $\tilde{V}(x_1, x_2, ..., x_N)$ in the expansion of Γ are indeed the 1PI-vertex functions.

We can obtain an equation of motion for Γ directly from the equation of motion for $Z[J]$. Using the functional integral representation

$$Z[J] = \mathcal{N} \int [d\varphi] \, \exp\left[-S_E(\varphi) + \int J\varphi\right] \tag{8.97}$$

we have the equation of motion

$$\left[\frac{\delta S_E}{\delta\varphi}\right]_{\varphi=\frac{\delta}{\delta J}} Z[J] = J \, Z[J] \tag{8.98}$$

In fact, it was from the Minkowski space version of this equation, viz., (5.9) or (8.30), that we obtained the functional representation for $Z[J]$. In the above equation, the left-hand side involves various powers of derivatives with respect to J acting on $Z[J]$. This can be simplified as follows. Using $W = \log Z$ in (8.93) and differentiating, we get

$$\frac{1}{Z}\frac{\delta^2 Z}{\delta J(x_1)\delta J(x_2)} = \frac{\delta\Phi(x_1)}{\delta J(x_2)} + \Phi(x_1)\Phi(x_2)$$
$$= \tilde{G}(x_1, x_2) + \Phi(x_1)\Phi(x_2)$$
$$= \hat{\phi}(x_1)\hat{\phi}(x_2) \cdot 1 \tag{8.99}$$

where

$$\hat{\phi}(x_1) = \Phi(x_1) + \int d^4 x_2 \, \tilde{G}(x_1, x_2)\frac{\delta}{\delta\Phi(x_2)} \tag{8.100}$$

Differentiating once more, we find

$$\frac{1}{Z}\frac{\delta^3 Z}{\delta J(x_1)\delta J(x_2)\delta J(x_3)} = \Phi(x_1)\Phi(x_2)\Phi(x_3) + \Phi(x_1)\tilde{G}(x_2, x_3)$$
$$+ \Phi(x_2)\tilde{G}(x_3, x_1) + \Phi(x_3)\tilde{G}(x_1, x_2)$$
$$+ \int_{x_4} \tilde{G}(x_1, x_4)\frac{\delta\tilde{G}(x_2, x_3)}{\delta\Phi(x_4)}$$
$$= \hat{\phi}(x_1)\hat{\phi}(x_2)\hat{\phi}(x_3) \cdot 1 \tag{8.101}$$

where we have also used (8.96). We see that differentiation with respect to J may be replaced by the action of $\hat{\phi}$. We can therefore write (8.98) as

$$\frac{\delta\Gamma}{\delta\Phi(x)} = \left[\frac{\delta S_E}{\delta\varphi(x)}\right]_{\varphi=\hat{\phi}} \cdot 1$$

$$\int d^4 z \left[\frac{\delta^2\Gamma}{\delta\Phi(x)\delta\Phi(z)}\right] \tilde{G}(z, y) = \delta^{(4)}(x - y) \tag{8.102}$$

We have also repeated (8.95). It can be interpreted as defining $\tilde{G}(x, y)$ in terms of Γ. With $\tilde{G}(x, y)$ given by this, the first equation in (8.102) is a nonlinear equation for $\Gamma[\Phi]$. Given a classical action S_E for the field, we can directly set up the equations (8.102). These equations can be considered as the fundamental equations for defining the quantum theory of the field. In this approach, $W[J]$ is a derived quantity given by (8.92) as

$$W[J] = \int J\Phi - \Gamma[\Phi]$$

$$J(x) = \frac{\delta\Gamma}{\delta\Phi(x)} \qquad (8.103)$$

The second equation is to be solved for J as a function of Φ. Equations (8.102), which may be taken as another definition of the quantum theory of the field, are a functional version of what are often referred to as the Schwinger-Dyson equations. They can be constructed in an analogous manner for any field theory.

It is useful to write out the Schwinger-Dyson equations for the simple example of a φ^4-theory in some more detail. The Euclidean action is

$$S_E = \int d^4x \left[\frac{1}{2}(\partial\varphi)^2 + \frac{1}{2}m^2\varphi^2 + \lambda\varphi^4 \right] \qquad (8.104)$$

For the first of equations (8.102) we find

$$\frac{\delta\Gamma}{\delta\Phi(x)} = K\Phi(x) + 4\lambda\Phi^3(x) + 12\lambda\Phi(x)\tilde{G}(x,x)$$

$$-4\lambda \int_{z_1,z_2,z_3} \tilde{G}(x,z_1)\tilde{G}(x,z_2)\tilde{G}(x,z_3)\tilde{V}_\Phi(z_1,z_2,z_3) \qquad (8.105)$$

where $K = (-\Box + m^2)$ and we have used the fact that we can write

$$\frac{\delta\tilde{G}(x,y)}{\delta\Phi(z)} = -\int_{z_1,z_2} \tilde{G}(x,z_1)\frac{\delta^3\Gamma}{\delta\Phi(z_1)\delta\Phi(z_2)\delta\Phi(z)}\tilde{G}(z_2,y)$$

$$= -\int_{z_1,z_2} \tilde{G}(x,z_1)\tilde{V}_\Phi(z_1,z_2,z)\tilde{G}(z_2,y)$$

$$\tilde{V}_\Phi(z_1,z_2,z) \equiv \frac{\delta^3\Gamma}{\delta\Phi(z_1)\delta\Phi(z_2)\delta\Phi(z)} \qquad (8.106)$$

which follows from the fact that $\tilde{G}(x,y)$ is inverse to $\delta^2\Gamma/\delta\Phi(y)\delta\Phi(z)$, according to the second of equations (8.102). The vertex functions $\tilde{V}_\Phi$ depend on Φ; when Φ is set to zero, they become the usual vertex functions of (8.94). Likewise, $\tilde{G}(x,y)$ becomes the exact propagator when $\Phi = 0$.

The second derivative of Γ is given by

$$\tilde{V}_\Phi(1,2) = K(1,2) + 12\lambda \left[\Phi(1)^2 + \tilde{G}(1,1) \right] \delta(1,2)$$

$$-12\lambda\Phi(1)\int_{3,4}\tilde{G}(1,3)\tilde{G}(1,4)\tilde{V}_\Phi(3,4,2)$$

$$+12\lambda\int_{3,\dots,7}\tilde{G}(1,3)\tilde{V}_\Phi(3,4,2)\tilde{G}(4,5)\tilde{G}(1,6)\tilde{G}(1,7)\tilde{V}_\Phi(5,6,7)$$

$$-4\lambda\int_{3,4,5}\tilde{G}(1,3)\tilde{G}(1,4)\tilde{G}(1,5)\tilde{V}_\Phi(3,4,5,2)\tag{8.107}$$

We use the simplified notation $\tilde{G}(1,2) = \tilde{G}(x_1,x_2)$, etc. and $K(1,2) = (-\square_1 + m^2)\delta(x_1,x_2)$. When Φ is set to zero, the three-point vertices vanish since the theory has symmetry under $\Phi \to -\Phi$. The second equation in (8.102) can then be written as

$$\int_3 V(1,3)G(3,2) = \delta(1,2)$$

$$V(1,2) = K(1,2) + \Sigma(1,2)\tag{8.108}$$

$$\Sigma(1,2) = 12\lambda G(1,1)\delta(1,2)$$

$$-4\lambda\int_{3,4,5}G(1,3)G(1,4)G(1,5)V(3,4,5,2)$$

The equation for $V(1,2)$ involves higher V's, the four-point one in this case. One can derive equations for the higher vertices by further differentiations of equations (8.105) or (8.107). This will involve still higher vertices, leading to a whole infinite chain of equations. One needs to truncate them at some stage to do useful calculations. For example, for the four-point vertex we find

$$V(1,2,3,4) = 4!\lambda\,\delta(1,2)\delta(2,3)\delta(2,4) + \cdots\tag{8.109}$$

If we truncate this by keeping only the first term, the term which is explicitly shown in (8.109), we get

$$\Sigma(1,2) \approx 12\lambda G(1,1)\delta(1,2) - 4!\,4\lambda^2 G(1,2)^3\tag{8.110}$$

This can be used in (8.108) to get a closed set of equations for the propagator $G(1,2)$.

The Schwinger-Dyson equations are a set of equations for the exact propagator and exact vertices. One can generate the perturbation expansion from them by expanding around the free propagator. Let $G_0(x,y)$ denote the free propagator; the subscript is to emphasize that this is for the free theory. We can then convert the equation for G into an integral equation as

$$G(1,2) = G_0(1,2) - \int_{3,4}G_0(1,3)\Sigma(3,4)G(4,2)\tag{8.111}$$

This equation, together with the equation for $\Sigma(1,2)$, can be used to generate a series expansion for G. If we use the approximation (8.109) and compare the resulting series with the standard perturbative expansion for G, we can see that the equations (8.108) amount to resummation of an infinite set of Feynman diagrams.

8.9 The S-matrix in terms of Γ

One can extend this point of view, namely, regarding the definition of Γ directly in terms of the classical action as the way to quantize the theory, to the calculation of the scattering matrix as well. We have seen in (5.27) that the S-matrix functional may be written as

$$\mathcal{F}[\varphi] = Z[i(\Box + m^2)\varphi] = e^{W[J]}\Big|_{J=i(\Box+m^2)\varphi} \qquad (8.112)$$

(There is also a trivial factor $e^{-\frac{i}{2}\int \varphi(\Box+m^2)\varphi}$ which we have not displayed since it is not important for what follows.) The S-matrix elements are obtained by replacing φ by the free one-particle wave functions as in (5.24). The free one-particle wave functions obey the condition $(\Box + m^2)\varphi = 0$, or $J = 0$. (This is to be done after the required number of differentiations with respect to J or φ.) Using the Minkowski space version of (8.92), we may write

$$W[J] = \int d^4x\, J\Phi + i\Gamma[\Phi] \qquad (8.113)$$

When J is set to zero, we have

$$\frac{\delta\Gamma}{\delta\Phi} = 0 \qquad (8.114)$$

and W becomes $i\Gamma[\Phi]$ evaluated on solutions of the equation (8.114). In other words, from equations (8.112, 8.114), we can write

$$\mathcal{F} = \hat{S} = \exp\Big(i\,\Gamma[\Phi]\Big)\Big|_{\frac{\delta\Gamma}{\delta\Phi}=0} \qquad (8.115)$$

The S-matrix is given by the quantum effective action evaluated on solutions of the equation (8.114). This relation gives a nonperturbative definition of the S-matrix. The solutions to the equation $\frac{\delta\Gamma}{\delta\Phi} = 0$ will be parametrized by some set of variables; this free data in the solutions are the quantities on which $\hat{S}$ depends. Perturbatively, the free data are the amplitudes a_k, a_k^* in the solution for φ written as $\varphi = \sum_k a_k u_k(x) + a_k^* u_k^*(x)$. Here a_k, a_k^* are viewed as c-number quantities. The amplitudes for specific processes are then obtained by differentiating $\hat{S}$ appropriately with respect to a_k, a_k^* and then setting them to zero. There are many applications of this definition of the S-matrix; for example, one can use it to derive recursion rules for scattering amplitudes whereby amplitudes with a certain number of external lines are generated recursively from amplitudes with lower numbers of external lines.

Equation (8.114) can be considered as the effective quantum equation of motion, namely, as a c-number equation which nevertheless captures the full effect of the quantum dynamics. We see that Γ is not only the generating functional for 1PI diagrams, it defines the quantum theory of the field entirely. By solving the equations of motion (extremization condition for Γ), we can define the S-matrix. One can also use it for analysis of nonperturbative aspects of the theory, for example, for analyzing nontrivial ground state properties. In most of the situations we have considered so far, the ground state of the sytem was the ground state of the free-field theory. While this is adequate for perturbation theory, there are many situations where the ground state is modified by the interaction. We will need a nonperturbative analysis to see if a new ground state is dynamically chosen or preferable. The calculation of Γ (nonperturbatively) and its subsequent extremization can answer this question.

8.10 The loop expansion

The diagrammatic expansion of the Green's functions or $W[J]$ leads to a diagrammatic expansion of the vertex functions in Γ. A systematic expansion procedure, which is useful both conceptually and for practical calculations, is given by expanding Γ in powers of $\hbar$.

The functional representation for $Z[J]$ is given by

$$Z[J] = \mathcal{N} \int [d\varphi] \; \exp\left[-\frac{1}{\hbar}S_E(\varphi) + \int J\varphi\right] \tag{8.116}$$

This is the same as equation (8.97), but we have now explicitly indicated where $\hbar$ appears. Since the propagator G is the inverse to K where $\frac{1}{2}\int \varphi K\varphi$ is the free part of the action, we see that, with $\hbar$ included, $G = \hbar K^{-1} \sim \hbar$. From (8.116), we also see that the vertices must go like $1/\hbar$. For any term in Γ represented as a Feynman diagram with V_i vertices of type i, I internal lines or propagators and E external lines, we have

$$\sum_i v_i V_i - 2I = E \tag{8.117}$$

where v_i is the valence of the vertex of type i; for example, for an interaction term $\lambda \varphi^4$, the valence is 4. The E-lines carry E external momenta. The I-momenta for the internal lines are constrained by the momentum conservation δ-functions at each vertex. One δ-function simply expresses overall conservation of momentum. There are thus $\sum_i V_i - 1$ δ-functions constraining the internal momenta. The unconstrained internal momenta are the loop momenta which are to be integrated over. Thus, for the number of loop momenta or loops L, we have

$$L = I - \left(\sum_I V_i - 1\right) \tag{8.118}$$

The number of powers of $\hbar$ is given by $I - \sum_i V_i$, which is equal to $L-1$ from the above equation. A term with L loops in its diagrammatic representation will go like $\hbar^{L-1}$. In this way, we see that the quantum effective action has an expansion of the form

$$\frac{1}{\hbar}\Gamma[\Phi] = \sum_{L=0}^{\infty} \hbar^{L-1} \, \Gamma^{(L)}[\Phi] \qquad (8.119)$$

with the $L=0$ term corresponding to the classical theory; this has no loops. Diagrams with no loops are called tree diagrams. The $\hbar$-expansion is a systematic way to classify and analyze the quantum corrections.

In the expansion (8.119), the term $\Gamma^{(0)}[\Phi]$ is in fact the classical action $S_E(\Phi)$. We can see this as follows. Using the definition of Γ in (8.116), with the factor of $\hbar$ inserted, we can write

$$
\begin{aligned}
\exp(-\frac{1}{\hbar}\Gamma[\Phi]) &= \mathcal{N} \int [d\varphi] \, \exp\left(-\frac{1}{\hbar}S_E(\varphi) + \int J(\varphi - \Phi) \right) \\
&= \mathcal{N} \int [d\varphi] \, \exp\left(-\frac{1}{\hbar}S_E(\varphi + \Phi) + \int J\varphi \right) \\
&= \mathcal{N} \int [d\varphi] \, \exp\left(-\frac{1}{\hbar}S_E(\varphi + \Phi) + \frac{1}{\hbar}\int \varphi\frac{\delta\Gamma}{\delta\Phi} \right)
\end{aligned}
\qquad (8.120)
$$

Using the expansion (8.119) and Taylor-expanding in powers of φ, we find

$$
\exp(-\sum_{L=0}^{\infty} \hbar^{L-1} \, \Gamma^{(L)}[\Phi]) = \mathcal{N} \exp(-\frac{1}{\hbar}S_E(\Phi)) \times
$$
$$
\int [d\varphi] \, \exp\left[\frac{1}{\hbar}\int \varphi\left(\frac{\delta\Gamma^{(0)}}{\delta\Phi} - \frac{\delta S_E}{\delta\Phi} \right) \right.
$$
$$
\left. -\frac{1}{2\hbar}\int \varphi\left(\frac{\delta^2 S_E}{\delta\varphi\delta\varphi} \right)_{\Phi} \varphi + \cdots \right]
\qquad (8.121)
$$

The first two terms in the $\hbar$-expansion give

$$
\begin{aligned}
\Gamma^{(0)}[\Phi] &= S_E(\Phi) \\
\Gamma^{(1)}[\Phi] &= \frac{1}{2}\log\det\left(\frac{\delta^2 S_E}{\delta\varphi(x)\delta\varphi(y)} \right)_{\Phi}
\end{aligned}
\qquad (8.122)
$$

The determinant involved is a functional determinant. For example, for the $\lambda\varphi^4$-theory we have

$$
\Gamma = \int \left[\frac{1}{2}(\partial\Phi)^2 + \frac{1}{2}m^2\Phi^2 + \lambda\Phi^4 \right]
$$
$$
+ \frac{\hbar}{2}\log\det(-\Box + m^2 + 12\lambda\Phi^2) + \mathcal{O}(\hbar^2) \qquad (8.123)
$$

The identity $\text{Tr}\log A = \log \det A$, which is certainly valid for any diagonalizable matrix A, and more generally for all matrices since diagonalizable matrices are dense, can be used to define and evaluate functional determinants. This gives

$$\frac{\hbar}{2}\log\det(-\square + m^2 + 12\lambda\Phi^2) = \frac{\hbar}{2}\text{Tr}\log(-\square + m^2 + 12\lambda\Phi^2)$$

$$= \frac{\hbar}{2}\int d^4x\langle x|\log(-\square + m^2 + 12\lambda\Phi^2)|y\rangle\Big]_{y\to x}$$
(8.124)

Expansion of this in powers of Φ will lead to a series of terms which correspond to one-loop diagrams with increasing numbers of external lines or Φ. A general formula for the result is difficult because Φ depends on x and one can have many derivatives of Φ appearing. If one is interested only in very slowly varying fields Φ, one can evaluate the determinant explicitly, neglecting all derivatives of Φ and treating it as a constant. This would also be the lowest-order term in an expansion of the determinant in powers of derivatives acting on Φ. The term in Γ corresponding to constant fields is called the effective potential; it was defined and evaluated to one-loop order by Coleman and Weinberg. For the one-loop correction to the effective potential, namely, the lowest-order term with no derivatives of Φ, we have

$$\frac{\hbar}{2}\int d^4x\langle x|\log(-\square + m^2 + 12\lambda\Phi^2)|y\rangle\Big]_{y\to x} = \frac{\hbar}{2}\int d^4x\int\frac{d^4k}{(2\pi)^4}\log(k^2 + s)$$

$$\equiv F(s)$$
(8.125)

where $s = m^2 + 12\lambda\Phi^2$.

$$\frac{\partial F(s)}{\partial s} = \frac{\hbar}{2}\int d^4x\int\frac{d^4k}{(2\pi)^4}\frac{1}{k^2 + s}$$

$$= \frac{\hbar}{2}\int d^4x\int\frac{d^3k}{(2\pi)^3}\frac{1}{2\sqrt{\boldsymbol{k}\cdot\boldsymbol{k} + s}}$$
(8.126)

Integrating, we find

$$F(s) = \int d^4x\int\frac{d^3k}{(2\pi)^3}\left[\frac{1}{2}\hbar\sqrt{\boldsymbol{k}\cdot\boldsymbol{k} + s} - \frac{1}{2}\hbar\sqrt{\boldsymbol{k}\cdot\boldsymbol{k} + m^2}\right]$$
(8.127)

The expression for F, apart from the volume integration, is of the form of the sum over zero point energies with s in place of m^2. We have chosen the constant of integration for the s-integration to be the zero-point energy for the free theory, so that $F(s)$ is zero without the Φ^4-interaction. The k-integration in (8.127) is divergent and the proper way to handle this is to evaluate it with a cutoff on the momentum and absorb potentially divergent terms into various parameters of the theory. This is the procedure of renormalization discussed in the next chapter. For now, we evaluate it with a cutoff, or an upper limit for $|\boldsymbol{k}|$, denoted by Λ and obtain

$$
\begin{aligned}
F(s) = \frac{\hbar}{32\pi^2} \int d^4x \Bigg[&24\lambda\Phi^2 \left(\Lambda^2 + \frac{m^2}{4} - \frac{m^2}{2}\log(4\Lambda^2/m^2) \right) \\
&+ (12\lambda\Phi^2)^2 \left(\frac{1}{4} - \frac{1}{2}\log(4\Lambda^2/m^2) \right) \\
&+ \frac{1}{2}(m^2 + 12\lambda\Phi^2)^2 \log\left(1 + \frac{12\lambda\Phi^2}{m^2} \right) \Bigg]
\end{aligned}
$$

$$(8.128)$$

For the sake of completeness, we will give here the renormalized form of the effective action, although details are discussed only later. To one-loop order, Γ is given by

$$
\begin{aligned}
\Gamma = \int &\left[\frac{1}{2}(\partial\Phi)^2 + \frac{1}{2}m^2\Phi^2 + \lambda\Phi^4 \right] \\
&+ \frac{\hbar}{64\pi^2} \left[(m^2 + 12\lambda\Phi^2)^2 \log\left(1 + \frac{12\lambda\Phi^2}{m^2} \right) - 12\lambda m^2\varphi^2 - 216\lambda^2\varphi^4 \right]
\end{aligned}
$$

$$(8.129)$$

As mentioned at the end of the last section, extremization of Γ can define the quantum theory. In (8.129), we have done a one-loop evaluation of Γ for slowly varying fields. Notice that when we set the variation of Γ to zero, the source J is zero and hence Φ is identical to $\langle 0|\phi|0\rangle$; it is the vacuum (or ground state) expectation value. If the ground state is translationally invariant, then the expectation value is a constant. Thus, the effective potential approximation, where gradients of Φ are neglected, is adequate to analyze the ground state expectation value of ϕ in the full theory, if we assume that the vacuum is translationally invariant. Our evaluation of the effective potential to one-loop order can be used as a first approximation in this endeavor.

References

1. Functional integration applied to quantum field theory is discussed in many books by now. The early work which is a standard reference, particularly for fermions, is F.A. Berezin, *The Method of Second Quantization*, Academic Press (1966). A very good book, with many early references, is J. Glimm and A. Jaffe, *Quantum Physics: A Functional Integral Point of View*, Springer-Verlag (1981 & 1987).
2. Systematic study of Euclidean Green's functions started with J. Schwinger, Proc. Nat. Acad. Sci. U.S.A, **44**, 956 (1958); Phys. Rev. **115**, 721 (1959); K. Symanzik, J. Math. Phys. **7**, 510 (1966); in *Local Quantum Theory*, R.

Jost (ed.), Academic Press (1969). The correspondence of Euclidean and Minkowski formulations is given in general form by K. Osterwalder and R. Schrader, Commun. Math. Phys. **31**, 83 (1973); *ibid.* **42**, 281 (1975).

3. The emergence of the correct Riemannian measure for nonlinear sigma models was given by I.S. Gerstein, R. Jackiw, B.W. Lee and S. Weinberg, Phys. Rev. **D3** 2486 (1971). In this context, see also J. Honerkamp and K. Meetz, Phys. Rev. **D3**, 1996 (1971); A. Slavnov, Nucl. Phys. B31, 301 (1971).

4. The importance of one-particle irreducible graphs was recognized by F. Dyson in his classic work on renormalization of QED, see J. Schwinger, *Selected Papers in Quantum Electrodynamics*, Dover Publications, Inc. (1958). The effective action Γ was defined by J. Goldstone, A. Salam and S. Weinberg, Phys. Rev. **127**, 965 (1962). It was obtained as the Legendre transform of W by G. Jona-Lasinio, Nuovo. Cim. **34**, 1790 (1964).

5. Schwinger-Dyson equations are derived in F. Dyson, Phys. Rev. **75**, 1736 (1949); J. Schwinger, Proc. Nat. Acad. Sci. U.S.A. **37**, 452 and 455 (1951).

6. The S-matrix functional, in relation to Γ, has been analyzed in L.D. Faddeev and A.A. Slavnov, *Gauge Fields, Introduction to the Quantum Theory*, Benjamin-Cummings, MA (1980); A. Jevicki and C. Lee, Phys. Rev. **D37** (1988) 1485; C. Kim and V.P. Nair, Phys. Rev. **D55**, 3851 (1997).

7. The effective potential and its one-loop calculation were given by S. Coleman and E. Weinberg, Phys. Rev. **D7**, 1888 (1973).

9 Renormalization

9.1 The general procedure of renormalization

We have already seen that the effective action Γ can be expanded in powers of $\hbar$ as $\Gamma = \sum_L \hbar^L \Gamma^{(L)}$ where $\Gamma^{(L)}$ generates the one-particle irreducible (1PI) or proper vertices with L loops in the Feynman diagram. The loop integrations correspond to the fact that the interaction can induce virtual transitions to various intermediate states. Alternatively they may be thought of as the interactions of the incoming fields with the quantum fluctuations of the field in the vacuum (which must exist since the field and its conjugate momentum do not commute). If the loop-momenta are allowed to become arbitrarily large, which is to say that if transitions to virtual states of arbitrarily large momenta can occur, some of the integrals can and do diverge. In effect, this means that the field theories we are discussing, with point-like interactions at short distances, are inadequate as descriptions of the physical world at very high momenta or at very short distances. We must consider these theories as valid only for momenta less than some very large value Λ. All loop-integrations are to be cut off in some fashion at this value Λ. The specific procedure for introducing a high momentum (or ultraviolet) cut-off in the theory is called a regulator or a regularization procedure. The resulting theory, which has an ultraviolet cut-off, and hence no divergent integrals, is called a regularized theory. The aim of quantum field theory is to provide a description of physical phenomena in terms of such regularized theories.

In using quantum field theory for practical calculations, we must therefore take account of the following points.

1. First of all, one needs a regulator which makes all loop integrals mathematically welldefined and finite. The calculated results with a regulator will depend on the cut-off Λ and hence on the specific regulator used. Unless we have a good reason to choose a specific regulator, this would lead to some ambiguity in the predictions of the theory even after a Lagrangian is chosen.

2. The calculated results such as the S-matrix will also depend on the parameters such as the coupling constants (generically denoted λ_0) and masses (denoted m_0) which appear in the Lagrangian. (These parameters in the Lagrangian are often called the bare parameters.) The physically

measured couplings and masses are not the parameters in the Lagrangian, since there are, in general, corrections to them due to interactions. The measured couplings and masses are thus functions of the bare couplings and masses and the cut-off Λ. These are calculable functions once a regulator is chosen. In interpreting the calculated results such as the S-matrix or the effective action Γ and in comparing them with experiments, we have to rewrite them in terms of the actual measured parameters λ and m.

The two issues above are related. The idea of renormalization is that one can absorb all the Λ-dependence or regulator dependence of the calculated results into the relation between the measured parameters λ, m and the bare parameters λ_0, m_0. In other words, the calculated results, when expressed in terms of the physically measured parameters, do not depend on the choice of the regulator. Thus after computation of the loop corrections and rewriting everything in terms of the measured parameters, we end up with the unambiguous predictions of the theory. The measured parameters λ, m are often called the renormalized parameters.

The transformation of the parameters can be done at the level of the starting Lagrangian itself. For example, for the scalar field theory with quartic interaction, we can write

$$\mathcal{L} = \left[\frac{1}{2}(\partial\chi)^2 + \frac{1}{2}m_0^2\,\chi^2\right] + \lambda_0\,\chi^4 \tag{9.1}$$

$$= Z_3\left[\frac{1}{2}(\partial\varphi)^2 + \frac{1}{2}(m^2 - \delta m^2)\varphi^2\right] + Z_1\,\lambda\,\varphi^4 \tag{9.2}$$

The transformation between the bare and renormalized quantities is explicitly given by

$$\begin{aligned}
\lambda_0 &= Z_1\,Z_3^{-2}\,\lambda \\
m_0^2 &= m^2 - \delta m^2 \\
\chi &= \sqrt{Z_3}\,\varphi
\end{aligned} \tag{9.3}$$

Z_1, Z_3 and δm^2 are functions of Λ and the renormalized parameters λ, m. It will become clear that a transformation of the fields will also be necessary; this is the reason for the factor Z_3. (This may be thought of as arising from corrections to the canonical structure of the theory.) The quantities Z_1, Z_3 and δm^2 are called the renormalization constants.

Since corrections arise due to loop diagrams (which carry powers of $\hbar$), we can write

$$Z_1 = 1 + \sum_1^\infty \hbar^L Z_1^{(L)}$$

$$Z_3 = 1 + \sum_1^\infty \hbar^L Z_3^{(L)}$$

$$\delta m^2 = \sum_{1}^{\infty} \hbar^L (\delta m^2)^{(L)} \tag{9.4}$$

The renormalized parameters are, by definition, the measured values of the coupling constants and masses. Therefore, there are no further corrections to them. This means that the renormalization constants must be such that they cancel out any corrections which may arise from the loop calculations. The strategy for perturbative calculations is then the following. We start with the Lagrangian (9.2) and calculate $\Gamma[\Phi]$ with some regulator. δm^2, Z_1 and Z_3 are then chosen so that in $\Gamma[\Phi]$ the mass is m^2, the Φ^4-coupling is λ and the normalization of the kinetic energy term is 1. $\Gamma[\Phi]$ should then have no terms which diverge as Λ becomes very large. The specific value of Λ is then immaterial except that it should be large enough so that terms of order $1/\Lambda$ can be ignored. From $\Gamma[\Phi]$ one can obtain $W[J]$, the generating functional for the connected Green's functions by the Legendre transformation (8.103). The S-matrix can then be constructed from this. (What we have described is one 'scheme' of renormalization. There is some freedom of Λ-independent redefinitions in relating the renormalized parameters to experimental measurements, leading to other 'schemes'. This is briefly discussed in the next section. Such schemes can be useful in some contexts, for example, when we have massless particles.)

We shall now work through the implementation of these ideas to one-loop order in the scalar field theory with quartic interaction. There are still many more features of this renormalization procedure which require further discussion, but we shall do that at a later stage.

9.2 One-loop renormalization for scalar field theory

Since we are not interested in vacuum diagrams, at least not at this point, the simplest one-loop correction we can calculate corresponds to two external lines and is given by the following diagram.

Fig 9.1. Scalar self-energy

The 1PI-diagrams and the corresponding vertex functions can be obtained from the general expression for the S-matrix functional given in (5.18). The Euclidean version of this is

$$\mathcal{F}[\varphi] = \mathcal{N} \exp\left(\frac{1}{2}\int G \frac{\delta}{\delta\varphi}\frac{\delta}{\delta\varphi}\right) \exp\left[-Z_1\lambda\int\varphi^4\right] \tag{9.5}$$

The mathematical expression corresponding to the Feynman diagram given above will have one power of the coupling constant and one Wick contraction. It is thus given by the term

$$\mathcal{F}[\varphi] \approx \mathcal{N}\left(\frac{1}{2}\int G\frac{\delta}{\delta\varphi}\frac{\delta}{\delta\varphi}\right)\left[-Z_1\lambda\int\varphi^4\right]$$
$$= -6Z_1\lambda\mathcal{N}\int d^4x\, G(x,x)\varphi^2(x) \tag{9.6}$$

The propagator is given by

$$G(x,y) = \frac{1}{Z_3}\int \frac{d^4k}{(2\pi)^4}\frac{e^{ik\cdot(x-y)}}{k^2+m^2-\delta m^2} \tag{9.7}$$

For this calculation, we will get one power of $\hbar$ from the propagator. Thus, to get the $\mathcal{O}(\hbar)$-term in the effective action, we only need Z_1, Z_3 and δm^2 to $\mathcal{O}(\hbar^0)$ in the propagator and in (9.6); i.e., we can take $Z_1 \approx 1$, $Z_3 \approx 1$, $\delta m^2 \approx 0$ on the right hand side. The term corresponding to (9.6) in $\Gamma[\Phi]$ is then

$$\Gamma_2^{(1)} = \left[12\lambda\int\frac{d^4p}{(2\pi)^4}\frac{1}{p^2+m^2}\right]\int d^4x\,\frac{1}{2}\Phi^2(x) \tag{9.8}$$

If the integration over the loop momentum p in this expression is unrestricted in range, we see that this integral will be quadratically divergent. We must interpret the theory as having a cut-off for the momentum integration, $p^2 \leq \Lambda^2$. We can evaluate the integral easily with this cut-off. Notice that the integral is spherically symmetric in four-dimensional p-space and so can be evaluated using spherical coordinates in p-space. We then find

$$\Gamma_2^{(1)} = \frac{12\lambda}{16\pi^2}\int_0^{\Lambda^2} ds\,\frac{s}{s+m^2}\int d^4x\,\frac{1}{2}\Phi^2(x)$$
$$= \frac{3\lambda}{4\pi^2}\left[\Lambda^2 - m^2\log\left(1+\frac{\Lambda^2}{m^2}\right)\right]\int d^4x\,\frac{1}{2}\Phi^2(x) \tag{9.9}$$

where we have used the fact that

$$\int_{angles} d^4p = 2\pi^2 p^3 dp = \pi^2 s\, ds \tag{9.10}$$

with $s = p^2$.

We now have to include this term in Γ and identify δm^2. This will be done after evaluating the correction to the four-point vertex also.

The one-loop correction to the φ^4-interaction can be represented by the Feynman diagram shown in figure 9.2. The mathematical expression for the this diagram must have two powers of λ and two Wick contractions. The relevant term in $\mathcal{F}[\varphi]$ is then

$$\mathcal{F}[\varphi] = \mathcal{N} \frac{1}{2!} \int \frac{1}{2} G \frac{\delta}{\delta\varphi} \frac{\delta}{\delta\varphi} \int \frac{1}{2} G \frac{\delta}{\delta\varphi} \frac{\delta}{\delta\varphi} \frac{Z_1^2\lambda^2}{2!} \int \varphi^4(x) \int \varphi^4(y)$$

$$= \mathcal{N}(6\lambda)^2 \int_{x,y} \varphi^2(x) G(x,y)^2 \varphi^2(y) \tag{9.11}$$

Fig 9.2. One-loop correction to φ^4-interaction

Once again, we will get one power of $\hbar$ from each of the propagators and so, as before, we can set $Z_1 \approx 1$, $Z_3 \approx 1$, $\delta m^2 \approx 0$ in evaluating this.

Using (9.7) for the propagators with $Z_1 \approx Z_3 \approx 1$, $\delta m^2 \approx 0$, and with a change of variables, we get

$$\Gamma_4^{(1)} = \int d^4x d^4y \ \Phi^2(x) \ V(x,y) \ \Phi^2(y) \tag{9.12}$$

where

$$V(x,y) = \int \frac{d^4k}{(2\pi)^4} e^{ik\cdot(x-y)} \ V(k)$$

$$V(k) = -(6\lambda)^2 \int \frac{d^4p}{(2\pi)^4} \frac{1}{(p^2+m^2)[(k-p)^2+m^2]} \tag{9.13}$$

The behaviour of the integrand at large p shows that this integral is logarithmically divergent. We evaluate $V(k)$ with a cut-off Λ as before. By using the Feynman integral representation

$$\frac{1}{AB} = \int_0^1 dv \frac{1}{[A(1-v)+Bv]^2} \tag{9.14}$$

we can combine the denominators of the two propagators to get

$$V(k) = -(6\lambda)^2 \int \frac{d^4p}{(2\pi)^4} \int_0^1 dv \frac{1}{[(p^2 + m^2)(1 - v) + ((k - p)^2 + m^2)v]^2}$$

$$= -(6\lambda)^2 \int \frac{d^4p}{(2\pi)^4} \int_0^1 dv \frac{1}{[((p - kv)^2 + m^2) + k^2v(1 - v)]^2} \qquad (9.15)$$

By a shift of variables $p \to p + kv$, we see that the integral will be spherically symmetric in the four-dimensional p-space. The corrections due to the change of the limits of integration will be negligible if the momentum k is small compared to Λ. In this case we have

$$V(k) = -\frac{(6\lambda)^2}{16\pi^2} \int_0^1 dv \int_0^{\Lambda^2} ds \frac{s}{[s + k^2v(1 - v) + m^2]^2}$$

$$= -\frac{9\lambda^2}{4\pi^2} \int_0^1 dv \left[\log \left(\frac{\Lambda^2}{k^2v(1 - v) + m^2} \right) - 1 \right] \qquad (9.16)$$

We have used (9.10) again. Since we have a parameter with the dimensions of mass in the theory, it is convenient to split the logarithm and write this as

$$V(k) = V(0) + \frac{9\lambda^2}{4\pi^2} \int_0^1 dv \, \log \left(1 + \frac{k^2v(1 - v)}{m^2} \right)$$

$$V(0) = -\frac{9\lambda^2}{4\pi^2} \left[\log \left(\frac{\Lambda^2}{m^2} \right) - 1 \right] \qquad (9.17)$$

We now turn to the choice of the renormalization constants. The effective action, with the one-loop corrections to the Φ^2-vertex and the Φ^4-vertex included is

$$\Gamma = \int Z_3 \left[\frac{1}{2}(\partial\Phi)^2 + \frac{1}{2}(m^2 - \delta m^2)\Phi^2 \right] + Z_1\lambda \int \Phi^4 + \hbar\Gamma_2^{(1)} + \hbar\Gamma_4^{(1)} + \cdots \qquad (9.18)$$

The term which corresponds to the Φ^2-vertex is given, to first order in $\hbar$, by

$$\Gamma_2 = \int \left[\frac{1}{2}(\partial\Phi)^2 + \frac{1}{2}m^2\Phi^2 \right] + \hbar Z_3^{(1)} \int \left[\frac{1}{2}(\partial\Phi)^2 + \frac{1}{2}m^2\Phi^2 \right]$$

$$+ \hbar \left\{ -(\delta m^2)^{(1)} + \frac{3\lambda}{4\pi^2} \left[\Lambda^2 - m^2 \log \left(1 + \frac{\Lambda^2}{m^2} \right) \right] \right\} \int \frac{1}{2}\Phi^2 \qquad (9.19)$$

The requirement that the normalization of the kinetic energy term should be 1 gives $Z_3^{(1)} = 0$. The mass is given by m^2 if we choose

$$(\delta m^2)^{(1)} = \frac{3\lambda}{4\pi^2} \left[\Lambda^2 - m^2 \log \left(1 + \frac{\Lambda^2}{m^2} \right) \right] \qquad (9.20)$$

The Φ^2-term in the effective action is then

$$\Gamma_2 = \int \frac{1}{2} \left[(\partial \Phi)^2 + m^2 \Phi^2 \right] \tag{9.21}$$

The term which corresponds to the Φ^4-vertex reads, to first order in $\hbar$,

$$\Gamma_4 = \lambda \int \Phi^4 + \hbar \left(\lambda Z_1^{(1)} + V(0) \right) \int \Phi^4$$

$$+ \hbar \int_{x,y} \Phi^2(x) \left[\int \frac{d^4k}{(2\pi)^4} e^{ik\cdot(x-y)} \ V^*(k) \right] \Phi^2(y)$$

$$V^*(k) = \left[\frac{9\lambda^2}{4\pi^2} \int_0^1 dv \ \log \left(1 + \frac{k^2 v(1-v)}{m^2} \right) \right] \tag{9.22}$$

The basic strategy of renormalization is to choose $Z_1^{(1)}$ such that the Λ-dependence of the effective action is canceled out. In the present case, we can choose $\lambda Z_1^{(1)} + V(0) = 0$. The Φ^4-term of the effective action is thus

$$\Gamma_4 = \lambda \int \Phi^4 + \hbar \int_{x,y} \Phi^2(x) \left[\int \frac{d^4k}{(2\pi)^4} e^{ik\cdot(x-y)} \ V^*(k) \right] \Phi^2(y) \tag{9.23}$$

The Λ-dependence is eliminated by choosing the renormalization constant Z_1 to first order in $\hbar$ as

$$Z_1 = 1 + \hbar \frac{9\lambda}{4\pi^2} \left[\log \left(\frac{\Lambda^2}{m^2} \right) - 1 \right] \tag{9.24}$$

With this choice, the Φ^4-term of the effective action can be written as

$$\Gamma_4 = \int V(x_1, x_2, x_3, x_4) \Phi(x_1) \Phi(x_2) \Phi(x_3) \Phi(x_4)$$

$$V(x_1, x_2, x_3, x_4) = \int \prod_i \frac{d^4k_i}{(2\pi)^2} e^{ik_i \cdot x_i} \ (2\pi)^4 \delta^{(4)} \left(\sum_i k_i \right) \ V_4(k_i)$$

$$V_4(k_i) = \lambda + \left[\frac{9\lambda^2}{4\pi^2} \int_0^1 dv \ \log \left(1 + \frac{(k_1 + k_2)^2 \ v(1-v)}{m^2} \right) \right] \tag{9.25}$$

We kept factors of $\hbar$ up to this point to see how the renormalization constants cancel out the Λ-dependence in a systematic expansion. From now $\hbar$ will be set to 1 again.

There are certain ambiguities in the way we have separated out the Λ-dependent part and the "finite" part (the part which is finite as $\Lambda \to \infty$). For example, we could split the logarithm in (9.17) using an arbitrary mass scale μ to obtain

$$V(k) = V(0) + \frac{9\lambda^2}{4\pi^2} \int_0^1 dv \ \log \left(\frac{m^2}{\mu^2} + \frac{k^2 v(1-v)}{\mu^2} \right)$$

$$V(0) = -\frac{9\lambda^2}{4\pi^2} \left[\log \left(\frac{\Lambda^2}{\mu^2} \right) - 1 \right] + \cdots \tag{9.26}$$

We also have a similar ambiguity in choosing Z_1; for example, we could choose $\lambda Z_1^{(1)} + V(0) = c\lambda$. The resulting $V^*(k)$ and Z_1 would be

$$V^*(k) = \frac{9\lambda^2}{4\pi^2} \int_0^1 dv \, \log\left(\frac{m^2}{\mu^2} + \frac{k^2 v(1-v)}{\mu^2}\right)$$

$$Z_1 = 1 + c + \frac{9\lambda}{4\pi^2}\left[\log\left(\frac{\Lambda^2}{\mu^2}\right) - 1\right] + \cdots \tag{9.27}$$

Notice that c can actually be absorbed into the definition of μ, so that there is only a one-parameter ambiguity corresponding to μ. The μ-dependence of $V^*(k)$, after eliminating $Z_1^{(1)}$ and $V(0)$, has to do with the meaning of the renormalized coupling constant. The coupling constant is a way of parametrizing the strength of the interaction and has to be determined via some scattering experiment. What is measured is the scattering amplitude at some momenta for the incoming particles. We can pick one value of these momenta and define the corresponding amplitude as the coupling constant. The scattering amplitudes at other momenta are then parametrized by this constant. Because of this μ-dependence, strictly speaking, we must write $\lambda(\mu)$ for the renormalized coupling constant. In our first way of defining the renormalization constant via equations (9.22, 9.23, 9.24), which corresponds to the special choice $\mu = m$, we see that $V^*(k) \to 0$ as $k \to 0$. As a result, for the four-point vertex V_4, we find

$$V_4(k_1, k_2, k_3, k_4)\Big|_{k_i=0} = \lambda \tag{9.28}$$

We can identify λ as the value of the four-point vertex function (or the four-particle scattering amplitude, apart from trivial kinematical factors) when the momenta of the particles involved goes to zero. This is the λ we have used in the action; it can be further specified as $\lambda(m)$. In our second way of defining Z_1, we find

$$V_4(k_i)\Big|_{k_i=0} = \lambda + \frac{9\lambda^2}{4\pi^2} \log\left(\frac{m^2}{\mu^2}\right)$$

$$= \lambda(\mu) \tag{9.29}$$

We see that the second choice corresponds to a different definition of the coupling constant. Physical quantities will be independent of this ambiguity. This is easily seen in the present case by writing the various expressions in terms of the physical amplitudes at chosen momenta. First we write $V^*(k)$ as

$$V^*(k) = \frac{9\lambda^2}{4\pi^2} \int_0^1 dv \, \log\left(\frac{m^2}{\mu^2} + \frac{k^2 v(1-v)}{\mu^2}\right)$$

$$= \frac{9\lambda^2}{4\pi^2} \log\left(\frac{m^2}{\mu^2}\right) + \frac{9\lambda^2}{4\pi^2} \int_0^1 dv \, \log\left(1 + \frac{k^2 v(1-v)}{m^2}\right) \tag{9.30}$$

We can then write V_4 as

$$V_4(k_i) = V_4(0) + \frac{9V_4(0)^2}{4\pi^2} \int_0^1 dv \, \log\left(1 + \frac{k^2 v(1-v)}{m^2}\right) + \cdots \qquad (9.31)$$

where $V_4(0)$ denotes $V_4(k_i)$ at zero momentum for the particles involved and $k = k_1 + k_2$. We have eliminated λ in favor of $V_4(0)$, to the order we have calculated. Equation (9.31) expresses the true prediction of the theory, giving $V_4(k_i)$ in terms of its value at some fixed choice of momenta, $k_i = 0$ in this case. In a massive theory, $\mu = m$ is a natural and convenient choice and we shall use this for the φ^4-theory from now on.

It is interesting to examine the low- and high-energy behavior of the effective action. By momentum conservation at the vertex, $k = k_1 + k_2$, so that, for low-energy processes, we can consider the expansion of $V^*(k)$ in powers of k/m. We find

$$V^*(k) = \frac{3\lambda^2}{8\pi^2} \frac{k^2}{m^2} + \cdots \qquad (9.32)$$

This shows that the first correction to the low-energy result, expressed in coordinate space, is a term of the form $(\partial_\mu \varphi^2)(\partial^\mu \varphi^2)$.

In the high energy limit we find

$$V^*(k) \approx \frac{9\lambda^2}{4\pi^2} \log\left(\frac{k^2}{m^2}\right) \qquad (9.33)$$

The corresponding scattering amplitude increases logarithmically with k^2.

The calculations given above illustrate the renormalization procedure. We see that, at least as far as the Φ^2- and Φ^4-vertices are concerned, all Λ-dependence has completely disappeared and we have well-defined one-loop corrections in terms of the experimentally measured parameters. To complete the renormalization procedure to the one-loop order, we must consider higher vertices also, for example, the Φ^6-vertex given by the diagram

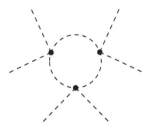

Fig 9.3. One-loop correction to the 6-point vertex function

In this case we have three propagators and one loop momentum p, so that the integral at large p has the form

$$I \sim \int d^4p \frac{1}{p^6} \tag{9.34}$$

There is no divergence at large Λ; the result is finite plus terms which are of order $\frac{1}{\Lambda^2}$. Vertices with higher number of Φ's are also finite as Λ becomes arbitrarily large. These vertices can all be calculated without additional restrictions or without introducing new renormalization constants. (Such vertices give further unambiguous and testable predictions about various types of processes.) We have thus carried out the renormalization of the theory to one-loop order.

The renormalization constant Z_3 is equal to 1 up to one-loop order. The correction to Z_3 starts at the two-loop order and arises from the diagram

Fig 9.4. Two-loop correction to two-point function

This can also give two-loop mass corrections. There are also two-loop corrections to the Φ^4-vertex from diagrams such as those in figure 9.5.

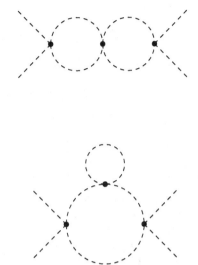

Fig 9.5. Two-loop corrections to the 4-point vertex function

We will not do a systematic calculation of the renormalization constants to two-loop order but will calculate $Z_3^{(2)}$ to illustrate how this can arise. The Φ^2-term corresponding to the two-loop diagram given is

$$\Gamma_2^{(2)} = \int_{x,y} \Phi(x)\, V(x,y)\, \Phi(y)$$

$$V(x,y) = -48\lambda^2 G^3(x,y)$$

$$= \int \frac{d^4k}{(2\pi)^4} e^{ik\cdot(x-y)}\, V(k)$$

$$V(k) = -48\lambda^2 \int \frac{d^4p}{(2\pi)^4}\frac{d^4q}{(2\pi)^4}\frac{1}{(p^2+m^2)(q^2+m^2)[(p+q-k)^2+m^2]}$$

$$(9.35)$$

The evaluation of the integral is rather involved, but the leading Λ-dependent term, which is the term relevant for Z_3, can be obtained without difficulty. The result is

$$\Gamma_2^{(2)} = \frac{1}{2}\int_{x,y}\partial\Phi(x)\left[\frac{3\lambda^2}{16\pi^4}\log\left(\frac{\Lambda^2}{m^2}\right)\delta^{(4)}(x-y) - F(x,y)\right]\partial\Phi(y)$$

$$F(x,y) = \int \frac{d^4k}{(2\pi)^4} e^{ik\cdot(x-y)}\left[\frac{3\lambda^2}{16\pi^4}\log\left(\frac{k^2}{m^2}\right)\right]$$

$$(9.36)$$

The terms in Γ which are of order $\hbar^2$ and involve two powers of Φ are

$$\Gamma_2 = Z_3^{(2)} \frac{1}{2} \left[(\partial \Phi)^2 + m^2 \Phi^2 \right]$$

$$+ \frac{1}{2} \int_{x,y} \partial \Phi(x) \left[\frac{3\lambda^2}{16\pi^4} \log \left(\frac{\Lambda^2}{m^2} \right) \delta^{(4)}(x-y) - F(x,y) \right] \partial \Phi(y) + \cdots$$

$$(9.37)$$

There are also two-loop corrections to the mass which we have not displayed above. From the coefficient of the term $\frac{1}{2}(\partial \Phi)^2$, we see that the choice of Z_3 which gives a Λ-independent Γ is

$$Z_3 = 1 - \frac{3\lambda^2}{16\pi^4} \log \left(\frac{\Lambda^2}{m^2} \right) + \cdots \tag{9.38}$$

9.3 The renormalized effective potential

In the last chapter, we calculated the one-loop correction to the effective potential. The effective action, with this term added, was

$$\Gamma = \int Z_3 \left[\frac{1}{2}(\partial \Phi)^2 + \frac{1}{2}(m^2 - \delta m^2)\Phi^2 \right] + Z_1 \lambda \Phi^4 + F(\Phi)$$

$$F(\Phi) = \frac{1}{32\pi^2} \int d^4x \left[24\lambda \Phi^2 \left(\Lambda^2 + \frac{m^2}{4} - \frac{m^2}{2} \log(4\Lambda^2/m^2) \right) \right.$$

$$+ (12\lambda \Phi^2)^2 \left(\frac{1}{4} - \frac{1}{2} \log(4\Lambda^2/m^2) \right)$$

$$\left. + \frac{1}{2}(m^2 + 12\lambda \Phi^2)^2 \log \left(1 + \frac{12\lambda \Phi^2}{m^2} \right) \right]$$

$$(9.39)$$

We have put in the Z-factors and δm^2. Renormalization can now be carried out by choosing these constants to cancel the potential divergences. When this is done, we get the Coleman-Weinberg potential as

$$\Gamma = \int \left[\frac{1}{2}(\partial \Phi)^2 + \frac{1}{2}m^2 \Phi^2 + \lambda \Phi^4 \right]$$

$$+ \frac{1}{64\pi^2} \left[(m^2 + 12\lambda \Phi^2)^2 \log \left(1 + \frac{12\lambda \Phi^2}{m^2} \right) - 12\lambda m^2 \Phi^2 - 216\lambda^2 \Phi^4 \right]$$

$$(9.40)$$

where

$$\delta m^2 = \frac{3\lambda}{2\pi^2} \left(\Lambda^2 + \frac{m^2}{2} - \frac{m^2}{2} \log(4\Lambda^2/m^2) \right) + \cdots$$

$$Z_1 = 1 - \frac{9\lambda}{2\pi^2} \left(1 - \frac{1}{2} \log(4\Lambda^2/m^2) \right) + \cdots \tag{9.41}$$

These renormalization constants differ from the set (9.20), (9.24) because of differences of regularization; notice, however, that the leading divergent terms have the same coefficients in (9.41), (9.20) and (9.24).

9.4 Power-counting rules

The renormalization procedure requires the introduction of the renormalization constants and then choosing them appropriately. Some of the questions which naturally arise at this stage are the following. Which terms or Feynman diagrams are potentially divergent and need regularization? How many renormalization constants do we need? How do we extend the one-loop procedure systematically to higher loops? The answer to these questions will require a more systematic analysis of the possible divergences of Feynman diagrams. Divergences arise from loop integrations. Propagators, since they behave like $1/k^2$ for bosons at high momenta and $1/k$ for fermions, can improve the convergence of the integral. We define the superficial degree of divergence δ of a 1PI-Feynman diagram with loop integrations as the number of positive powers of loop momenta minus the number of negative powers of loop momenta. In other words, each loop integration contributes $+4$ to δ; each propagator which carries a loop-momentum variable gives -2 if it is a bosonic propagator and -1 if it is a fermionic propagator. The superficial degree of divergence gives the highest powers of Λ which can arise from the evaluation of the integral with a cut-off Λ. It tells us which diagrams are potentially divergent. For a general Feynman diagram, for the φ^4-interaction, let E_B be the number of (bosonic) external lines and I_B be the number of (bosonic) internal lines. If V denotes the number of vertices, since each vertex has valency 4, we have $E_B = 4V - 2I_B$. With $\hbar$ for each propagator and $1/\hbar$ for each vertex, the number of powers of $\hbar$ which must be equal to the number of loops L is given by $L - 1 = I_B - V$. For a one-particle irreducible diagram, δ is given by $4L - 2I_B$ so that

$$\delta = 4(I_B - V + 1) - 2I_B = 4 - E_B \qquad (9.42)$$

δ is determined by the number of external lines only and so there are only a finite number of diagrams at a given loop order which can be potentially divergent, namely, diagrams with $E_B \leq 4$. Corrections to the φ^2-term have $\delta = 2$ since there are two external lines. The leading divergence is a quadratic divergence. There can be a subleading divergence which is logarithmic and which must carry two powers of external momenta for dimensional reasons. This would correspond to a contribution like $(\partial\varphi)^2$ appearing with a coefficient which behaves like $\log \Lambda$. We will therefore need a renormalization of the field as well as a mass renormalization. Corrections to φ^4 have $\delta = 0$ corresponding to a logarithmic divergence and will need an additional renormalization constant corresponding to coupling constant renormalization. In this theory we will not need any further renormalizations, since this exhausts all possible cases of $\delta \geq 0$. (Since the theory has symmetry under $\varphi \to -\varphi$, terms odd in φ do not appear in the perturbation theory.)

The superficial degree of divergence δ can be related to the dimension of the corresponding monomial of fields. We assign a dimension to the field φ. The action has to be dimensionless because, among other reasons, it is in the exponent of the functional integral. Since the kinetic term is $\int d^4x (\partial \varphi)^2$, assigning a mass dimension of -1 to the coordinate x and requiring that the action be dimensionless, we find that the dimension of φ must be 1. The dimensions of the coupling constants are determined from this so that all terms in the action remain dimensionless. (Equivalently, the Lagrangian must have dimension equal to 4.) The effective action is also dimensionless and we can use this to determine the dimensions of the coefficients of various possible terms in Γ. Thus, for example, the term $\int d^4x \; \varphi^2$ has dimension -2; the coefficient must have dimension 2. Since all other dimensionful parameters can be ignored compared to the cut-off Λ, we can conclude that the coefficient of this term must go like Λ^2. In other words, we expect a quadratic divergence for this term. Similarly, for the term $\int (\partial \varphi)^2$ the coefficient must be dimensionless; it has to be a logarithmic divergence. The coefficient of the φ^4-term can have logarithmic divergences by the same reasoning. Notice that in all these cases, δ is essentially 4 minus the dimension of the number of φ's involved.

While it can help to identify the potential divergences and the kind of renormalizations needed, the superficial degree of divergence is not the whole story. The actual degree of divergence may be different. A superficially convergent diagram may have subdivergences as in the following case, fig 9.6.

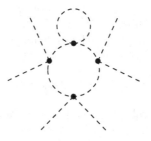

Fig 9.6. A two-loop correction to 6-point vertex function

One may also get entanglement or overlap of loop-integrations. It may happen that for one of the loop-integrations, the other loop momenta are to be treated as external momenta. Then, depending on how the loop-integration is done, the degree of divergence for the next set of loop-integrations may be different.

The way to deal with this in perturbation theory is to develop a systematic recursive procedure, based on the power-counting rules, with a subtraction procedure for separating out and eliminating the divergences. We shall take up this question later, after we consider the one-loop renormalization of QED.

9.5 One-loop renormalization of QED

The Euclidean action for QED was given in (8.58) as

$$
\mathcal{S}_E(A, \bar\psi, \psi) = \int d^4x \left[\frac{1}{2} (\partial_\mu A_\nu \partial_\mu A_\nu) + \bar\psi \left(\gamma \cdot (\partial - ieA) + m \right) \psi \right]
$$
$$
= \int d^4x \left[\frac{1}{4} F_{\mu\nu} F^{\mu\nu} + \frac{1}{2} (\partial \cdot A)^2 + \bar\psi \left(\gamma \cdot (\partial - ieA) + m \right) \psi \right]
$$
(9.43)

The free part of the action shows that ψ and $\bar\psi$ are fields of dimension $\frac{3}{2}$, while the photon field has dimension 1. The interaction Lagrangian is of dimension 4; e has no dimension. $\mathcal{L}_{int}$ shows that there are two fermions and one photon line at each vertex. For a Feynman diagram with E_F external fermion lines, E_B external photon lines, I_F fermion internal lines or propagators, I_B photon propagators and V vertices, we have

$$
E_F = 2(V - I_F)
$$
$$
E_B = V - 2I_B
$$
$$
L - 1 = I_F + I_B - V
$$
(9.44)

The superficial degree of divergence is given by

$$
\delta = 4L - I_F - 2I_B
$$
$$
= 4 - \frac{3}{2} E_F - E_B
$$
(9.45)

Notice that δ is again 4 minus the dimension of the fields involved.

The potentially divergent diagrams, which correspond to $\delta \geq 0$, are thus the following.

1) $\delta = 4, E_F = E_B = 0$.

These purely vacuum diagrams will be canceled by the normalization of the functional integral or equivalently by the requirement that $\langle 0|0 \rangle = 1$. (If we have a nontrivial background such as a gravitational field, diagrams with no external lines can be important. The relative vacuum contribution of different backgrounds is what is relevant and this is obtained by comparison of the vacuum diagrams with the backgrounds involved.)

2) $\delta = 3, E_F = 0,\ E_B = 1$

Diagrams with an odd number of external photon lines will vanish due to the charge conjugation invariance of QED; we shall discuss this symmetry later.

3) $\delta = 2, E_F = 0,\ E_B = 2$

These correspond to corrections to the photon propagator and are called vacuum polarization diagrams (or photon self-energy diagrams). Naively, the degree of divergence is 2, suggesting a quadratic divergence. A local photon mass term of the form $A_\mu(x)A^\mu(x)$ is not allowed by a combination of Lorentz invariance and gauge invariance. As a result, if we use a regulator which is gauge- and Lorentz-invariant, there will be no quadratic divergence. The lowest-dimension term which is gauge- and Lorentz-invariant is $F_{\mu\nu}F^{\mu\nu}$, which is of dimension 4. The coefficient being dimensionless, we expect only a logarithmic divergence for the vacuum polarization.

Although the $(\partial \cdot A)^2$-term breaks gauge invariance, these diagrams have fermion loops and, since the fermion part of the action has gauge invariance, a gauge-invariant regulator can be used.

4) $\delta = 1, E_F = 0,\ E_B = 3$

These diagrams vanish by charge conjugation invariance.

5) $\delta = 1, E_F = 2,\ E_B = 0$

These correspond to corrections to the fermion propagator and are called fermion self-energy diagrams. Naively, they are linearly divergent. But actually, because of the properties of Dirac γ-matrices, the divergence is only logarithmic. The theory with zero mass has a symmetry, the so-called chiral symmetry. The linear divergence, which would correspond to a mass term $\bar{\psi}(x)\psi(x)$, has to be zero to respect this symmetry. As a result, even in the massive theory, this term has to have a factor of m in its coefficient. The remainder is of zero dimension or at most a logarithmic divergence.

6) $\delta = 0, E_F = 0,\ E_B = 4$

This describes photon-photon scattering; the corresponding lowest-dimension monomial in the effective action is $A_\mu A^\mu A_\nu A^\nu$. Such a term is disallowed by gauge invariance, the lowest allowed term has four factors of $F_{\mu\nu}$, giving a dimension equal to 8. The coefficient has dimension -4, and hence this diagram is convergent if evaluated in a gauge-invariant manner.

7) $\delta = 0, E_F = 2,\ E_B = 1$

These correspond to corrections to the basic vertex $e\bar{\psi}\gamma^\mu\psi A_\mu$ and are called vertex corrections and ultimately are part of the charge renormalization. The divergence is logarithmic.

The discussion given above shows that there are three types of diagrams, namely, the photon and fermion self-energy diagrams and the vertex correction diagrams, which we must consider to work out the renormalization of QED. The photon self-energy correction can lead to a term like $F_{\mu\nu}F^{\mu\nu}$ in the effective action with a logarithmically divergent coefficient. The fermion self-energy can give terms like $m\bar{\psi}\psi$ and $\bar{\psi}\gamma \cdot \partial\psi$, both with logarithmically divergent coefficients. The vertex correction is again logarithmically divergent and gives a term of the form $\bar{\psi}\gamma^{\mu}\psi A_{\mu}$. We will need to introduce renormalization constants corresponding to these monomials. Since they are of the form of the terms in the action, this amounts to modifying the coefficients of the terms in the Lagrangian.

Interpreting the parameters in (9.43) as the bare parameters, we can rewrite it, in terms of the renormalized parameters, as

$$
S_E(A, \bar{\psi}, \psi) = \int d^4x \left[\frac{Z_3}{4}F_{\mu\nu}F^{\mu\nu} + \frac{\lambda}{2}(\partial \cdot A)^2 + Z_2\bar{\psi}\left(\gamma \cdot \partial + m - \delta m\right)\psi \right.
$$

$$
\left. -ieZ_1\bar{\psi}\gamma^{\mu}\psi A_{\mu} \right] \qquad (9.46)
$$

The renormalization constants $Z_1, Z_2, Z_3, \lambda, \delta m$ have the $\hbar$-expansion

$$
Z_i = 1 + \sum_1^{\infty} \hbar^L Z_i^{(L)}
$$

$$
\lambda = 1 + \sum_1^{\infty} \hbar^L \lambda^{(L)}
$$

$$
\delta m = \sum_1^{\infty} \hbar^L (\delta m)^{(L)} \qquad (9.47)
$$

for $i = 1, 2, 3$.

We now turn to the calculation of these constants to the first order in $\hbar$. This requires the evaluation of the one-loop contributions to the electron (fermion) self-energy, the vacuum polarization and the vertex correction. (Once again, $\hbar$ will be set to 1 in what follows, the terms of the same order in $\hbar$ will be collected together in the effective action.)

Electron self − energy

The electron self-energy diagram is given by

Fig 9.7. Electron self-energy

The corresponding contribution to Γ is given by

$$\Gamma^{(1)}_{\bar{\psi}\psi} = e^2 \int d^4x d^4y \frac{d^4k}{(2\pi)^4} \frac{d^4p}{(2\pi)^4} e^{i(k+p)\cdot(x-y)} \frac{1}{k^2} \bar{\psi}(x)\gamma^\mu \frac{1}{m+i\gamma\cdot p}\gamma_\mu \psi(y)$$

$$= \int_{x,y} \bar{\psi}(x)\Sigma(x,y)\psi(y) \tag{9.48}$$

where

$$\Sigma(x,y) = \int \frac{d^4p}{(2\pi)^4} e^{ip\cdot(x-y)} \Sigma(p)$$

$$\Sigma(p) = e^2 \int \frac{d^4k}{(2\pi)^4} \frac{1}{k^2} \left[\gamma^\mu \frac{1}{m+i\gamma\cdot(p-k)}\gamma_\mu\right] \tag{9.49}$$

We have set $Z_i \approx 1$, $\lambda \approx 1$, $\delta m \approx 0$ in the propagators and vertices in (9.49) since this is adequate to the order we are calculating. Combining denominators using the formula (9.14) and making a shift of the variable of integration, we find

$$\Sigma(p) = e^2 \int_0^1 dv \int \frac{d^4q}{(2\pi)^4} \left[\frac{2m(1+v)}{[q^2 + p^2v(1-v) + m^2v]^2} \right.$$

$$\left. +(m+i\gamma\cdot p) \frac{2(1-v)}{[q^2 + p^2v(1-v) + m^2v]^2} \right] \tag{9.50}$$

(As before, the change of the limits of integration gives a negligible correction if the cut-off Λ is very large compared to the momenta p.) The integral can be explicitly evaluated as

$$\int \frac{d^4q}{(2\pi)^4} \frac{1}{[q^2 + p^2v(1-v) + m^2v]^2} = \frac{1}{16\pi^2} \left[\log\left(\frac{\Lambda^2}{m^2}\right) - (1 + 2\log v) \right.$$

$$\left. + \log\left(\frac{m^2v^2}{m^2v^2 + K\,v(1-v)}\right) \right] \tag{9.51}$$

where $K = p^2 + m^2$. The self-energy contribution is now

$$\Sigma(p) = \Sigma_1 + (m + i\gamma \cdot p)\Sigma_2 + \Sigma^*$$

$$\Sigma_1 = \frac{3\alpha m}{4\pi} \log\left(\frac{\Lambda^2}{m^2}\right) + \frac{\alpha m}{2\pi}$$

$$\Sigma_2 = \left[\frac{\alpha}{4\pi} \log\left(\frac{\Lambda^2}{m^2}\right) + \frac{\alpha}{2\pi}\right]$$

$$\Sigma^* = -\frac{\alpha m}{2\pi} \int_0^1 dv(1+v) \log\left(\frac{m^2 v^2 + K\, v(1-v)}{m^2 v^2}\right)$$

$$- (m + i\gamma \cdot p) \frac{\alpha}{2\pi} \int_0^1 dv\ (1-v) \log\left(\frac{m^2 v^2 + K\, v(1-v)}{m^2 v^2}\right) \tag{9.52}$$

where $\alpha = e^2/4\pi$. (It is the fine-structure constant.) Notice that K, continued to Minkowski space, will vanish for free electrons of mass m; the last term Σ^* will thus vanish for free electrons. The contribution to Γ can be written as

$$\Gamma^{(1)}_{\bar\psi\psi} = \int_x \left[\Sigma_1\, \bar\psi\psi + \Sigma_2\, \bar\psi(\gamma \cdot \partial + m)\psi\right] + \int_{x,y} \bar\psi(x)\Sigma^*(x,y)\psi(y) \tag{9.53}$$

Vacuum polarization

The vacuum polarization or photon self-energy is given by the diagram

Fig 9.8. Vacuum polarization or photon self-energy

The corresponding mathematical expression is

$$\Gamma^{(1)}_{AA} = \frac{1}{2} \int_{x,y} A^\mu(x)\Pi_{\mu\nu}(x,y)A^\nu(y)$$

$$\Pi_{\mu\nu} = \int \frac{d^4 k}{(2\pi)^4} e^{ik\cdot(x-y)} \Pi_{\mu\nu}(k)$$

$$\Pi_{\mu\nu}(k) = -e^2 \int \frac{d^4 p}{(2\pi)^4} \frac{\mathrm{Tr}\left[\gamma_\mu(m - i\gamma \cdot (p+k))\gamma_\nu(m - i\gamma \cdot p)\right]}{(p^2 + m^2)[(p+k)^2 + m^2]} \tag{9.54}$$

The main difficulty in evaluating this expression is the question of the cut-off. The leading, Λ-dependent term in Γ_{AA} behaves like $\Lambda^2 A^\mu(x) A_\mu(x)$, a gauge-noninvariant mass like term for the photon. The reason for this is that a cut-off for the range of momentum integration is not gauge-invariant. One possibility is to introduce a gauge-invariant regulator which is a little more sophisticated than just a cut-off for the range of momentum. Dimensional regularization is one such regulator where the theory is first defined in spacetime dimension $4 - \epsilon$, $\epsilon \ll 1$, by analytic continuation. This isolates the potential divergences as poles in ϵ. The limit $\epsilon \to 0$ may be taken after the poles have been canceled by choice of the renormalization constants. This technique preserves gauge-invariance at all stages. Another, more simple-minded approach is to introduce one more renormalization constant, adding a term $\frac{1}{2}\delta\mu^2 A^2$ to the starting Lagrangian. A cut-off procedure can then be used, keeping the $\Lambda^2 A^2$-term generated by the vacuum polarization. At the end, we choose $\delta\mu^2$ so as to have gauge-invariance for Γ. In effect, this amounts to dropping the gauge-noninvariant part of $\Pi_{\mu\nu}$ in (9.54). We shall follow this approach here. (Since fermion action is gauge-invariant, it is completely consistent to require gauge invariance in the evaluation of this diagram.)

Gauge-invariance requires invariance under $A_\mu \to A_\mu + \partial_\mu\theta$, where θ is an arbitrary function of the spacetime variables. If $\Pi_{\mu\nu}$ obeys $k^\mu\Pi_{\mu\nu}(k) = 0$, in other words, if it is transverse to k^μ, this symmetry holds for the vacuum polarization contribution. We shall therefore separate out the transverse part of $\Pi_{\mu\nu}$ and set the nontransverse part to zero and then use a cut-off to evaluate the integral. Taking the trace of the γ-matrices in (9.54) and combining the denominators, we can write

$$\Pi_{\mu\nu}(k) = -4e^2 \int_0^1 dv \int \frac{d^4q}{(2\pi)^4} \frac{(m^2 + \frac{1}{2}q^2)\delta_{\mu\nu} + v(1-v)(2k_\mu k_\nu - k^2\delta_{\mu\nu})}{[q^2 + m^2 + k^2v(1-v)]^2}$$

(9.55)

where $q = p + kv$ and we have done a shift of the variable of integration. Terms odd in q have been dropped since they give zero upon angular integration. We have also used the fact that the integral of $q_\mu q_\nu$ is proportional to $\delta_{\mu\nu}$, i.e.,

$$\int \frac{d^4q}{(2\pi)^4} F(q^2)\, q_\mu q_\nu = \frac{\delta_{\mu\nu}}{4} \int \frac{d^4q}{(2\pi)^4} F(q^2)q^2$$

(9.56)

for a function of q^2. The transverse part of $\Pi_{\mu\nu}$ is given by

$$\Pi^T_{\mu\nu}(k) = \Pi_{\mu\nu} - \delta_{\mu\nu}\frac{k^\alpha k^\beta \Pi_{\alpha\beta}}{k^2}$$
$$= (k^2\delta_{\mu\nu} - k_\mu k_\nu)\, 8e^2 \int_0^1 dv \int \frac{d^4q}{(2\pi)^4} \frac{v(1-v)}{[q^2 + m^2 + k^2v(1-v)]^2}$$

(9.57)

The evaluation of the integral is now straightforward and gives

$$\Pi_{\mu\nu}^{T}(k) = (k^2 \delta_{\mu\nu} - k_\mu k_\nu) \left[\frac{\alpha}{3\pi} \left(\log\left(\frac{\Lambda^2}{m^2}\right) - 1 \right) \right.$$
$$\left. - \frac{2\alpha}{\pi} \int_0^1 dv \, v(1-v) \log\left(1 + \frac{k^2 v(1-v)}{m^2}\right) \right]$$

(9.58)

Using this expression, we find the contribution to Γ as

$$\Gamma_{AA}^{(1)} = \frac{1}{4} \int F_{\mu\nu} F^{\mu\nu} \left[\frac{\alpha}{3\pi} \left(\log\left(\frac{\Lambda^2}{m^2}\right) - 1 \right) \right]$$
$$+ \frac{1}{4} \int_{x,y} F_{\mu\nu}(x) \left[\int \frac{d^4 k}{(2\pi)^4} e^{ik\cdot(x-y)} \Pi^*(k) \right] F^{\mu\nu}(y)$$

(9.59)

where

$$\Pi^*(k) = -\frac{2\alpha}{\pi} \int_0^1 dv \, v(1-v) \, \log\left(1 + \frac{k^2 v(1-v)}{m^2}\right)$$

(9.60)

The vertex correction

The vertex correction is described by the diagram shown in figure 9.9. The contribution to Γ may be written as

$$\Gamma_{\bar\psi A\psi}^{(1)} = -ie \int \bar\psi(x) \Gamma_\mu(x,z,y) A^\mu(z) \psi(y)$$
$$\Gamma_\mu(x,z,y) = \int \frac{d^4 p}{(2\pi)^4} \frac{d^4 p'}{(2\pi)^4} e^{ip\cdot x} e^{-ip'\cdot y} e^{-i(p-p')\cdot z} \, \Gamma_\mu(p,p')$$
$$\Gamma_\mu(p,p') = -e^2 \int \frac{d^4 k}{(2\pi)^4} \frac{\gamma_\alpha[m - i\gamma\cdot(p-k)]\gamma_\mu[m - i\gamma\cdot(p'-k)]\gamma^\alpha}{k^2[(p-k)^2 + m^2][(p'-k)^2 + m^2]}$$

(9.61)

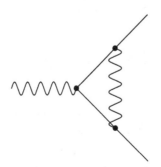

Fig 9.9. One-loop vertex correction

We will evaluate only the leading Λ-dependent term in (9.61). For this, since the high k-values are the important regime, the external momenta p, p' can be taken to be small compared to k. We then find

$$\Gamma_\mu \approx e^2 \int \frac{d^4k}{(2\pi)^4} \frac{\gamma_\alpha\gamma \cdot k\gamma_\mu\gamma \cdot k\gamma^\alpha}{k^2(k^2+m^2)^2} \tag{9.62}$$

The algebra of the Dirac γ-matrices gives $\gamma_\alpha\gamma \cdot k\gamma_\mu\gamma \cdot k\gamma^\alpha = 2\gamma_\mu k^2 - 4k_\mu k_\nu \gamma^\nu$. Further, since the integral involves only k^2, we can use (9.56). Equation (9.62) for Γ_μ now becomes

$$\Gamma_\mu = \gamma_\mu \, e^2 \int \frac{d^4k}{(2\pi)^4} \frac{1}{(k^2+m^2)^2} + \cdots$$

$$= \gamma_\mu \, \frac{\alpha}{4\pi} \left[\log\left(\frac{\Lambda^2}{m^2}\right) - 1 \right] + \cdots$$

$$\approx \gamma_\mu \, \frac{\alpha}{4\pi} \log\left(\frac{\Lambda^2}{m^2}\right) + \cdots \tag{9.63}$$

The Λ-dependent part of the $\bar\psi A\psi$-vertex in Γ is thus

$$\Gamma^{(1)}_{\bar\psi A\psi} = -ie \int \bar\psi\gamma \cdot A\psi \left[\frac{\alpha}{4\pi} \log\left(\frac{\Lambda^2}{m^2}\right) \right] \tag{9.64}$$

We are now in a position to collect the results together. The effective action, to first order in $\hbar$, can be written as

$$\Gamma = \int \frac{1}{4}F_{\mu\nu}F^{\mu\nu} + \frac{1}{2}(\partial \cdot A)^2 + \bar\psi(\gamma \cdot \partial + m)\psi - ie\bar\psi\gamma^\mu\psi A_\mu$$

$$\int + \frac{1}{4}F_{\mu\nu}F^{\mu\nu} \left[Z_3^{(1)} + \frac{\alpha}{3\pi}\left(\log\left(\frac{\Lambda^2}{m^2}\right) - 1\right) \right] + \lambda^{(1)}\frac{1}{2}(\partial \cdot A)^2$$

$$+ \left[Z_2^{(1)} + \frac{\alpha}{4\pi}\log\left(\frac{\Lambda^2}{m^2}\right) + \frac{\alpha}{2\pi} \right] \bar\psi(\gamma \cdot \partial + m)\psi$$

$$- \left[\delta m^{(1)} - \frac{3\alpha m}{4\pi}\log\left(\frac{\Lambda^2}{m^2}\right) - \frac{\alpha m}{2\pi} \right] \bar\psi\psi$$

$$- \left[Z_1^{(1)} + \frac{\alpha}{4\pi}\log\left(\frac{\Lambda^2}{m^2}\right) \right] ie\bar\psi\gamma^\mu\psi A_\mu$$

$$+ \Lambda - \text{independent terms} + \cdots \tag{9.65}$$

The Λ-dependent terms are absorbed into the definition of the physical parameters if we choose the renormalization constants as

$$Z_1 = 1 - \frac{\alpha}{4\pi}\log\left(\frac{\Lambda^2}{m^2}\right) + \cdots$$

$$Z_2 = 1 - \frac{\alpha}{4\pi}\log\left(\frac{\Lambda^2}{m^2}\right) - \frac{\alpha}{2\pi} + \cdots$$

$$Z_3 = 1 - \frac{\alpha}{3\pi} \left[\log \left(\frac{\Lambda^2}{m^2} \right) - 1 \right] + \cdots$$

$$\delta m = \frac{3\alpha m}{4\pi} \log \left(\frac{\Lambda^2}{m^2} \right) + \frac{\alpha m}{2\pi} + \cdots$$

$$\lambda = 1 + \cdots \tag{9.66}$$

where the ellipsis stand for terms of order $\hbar^2$. The resulting Γ, to first order in $\hbar$, is given by

$$\Gamma = \int \left(\frac{1}{4} F_{\mu\nu} F^{\mu\nu} + \frac{1}{2} (\partial \cdot A)^2 + \bar{\psi}(\gamma \cdot \partial + m)\psi - ie\bar{\psi}\gamma^\mu \psi A_\mu \right)$$

$$+ \int \bar{\psi} \Sigma^* \psi + \frac{1}{4} \int F_{\mu\nu} \Pi^* F^{\mu\nu} - ie \int \bar{\psi} \Gamma_\mu^* A^\mu \psi + \cdots \tag{9.67}$$

where Σ^* is given in (9.52), Π^* is given in (9.60), and Γ_μ^* denotes the finite, Λ-independent term of Γ_μ of (9.61). We have not explicitly calculated Γ_μ^* here, but it does appear in Γ. There are also all the one-particle irreducible vertices with more external particles which we have not calculated. They are convergent as $\Lambda \to \infty$ and do not affect the determination of the renormalization constants.

The finite corrections Σ^* and Π^* vanish when the fields obey the free field equations of motion with the correct masses; i.e., when $i\gamma \cdot p + m = 0$ for the electron and $k^2 = 0$ for the photon. These corrections do not affect the propagation of the free particles, in the perturbative expansion we are using. For a bound electron, the correction need not vanish and indeed gives a measurable effect, the Lamb shift, for example.

Notice also that, to the order that we have calculated,

$$Z_1 = Z_2 \tag{9.68}$$

We have only calculated the dominant Λ-dependent part for the vertex correction, so the comparison can only be made with the Λ-dependent part of Z_2. The relation (9.68) is a consequence of gauge invariance and is an example of a Ward-Takahashi (WT) identity. It gives gauge invariance for gauge transformations involving the physical (renormalized) charge e. Equation (9.68) and other related identities will be derived later.

More on vacuum polarization

We shall now consider the simplification of the finite correction (9.60) to photon propagation, the vacuum polarization effect. First consider the case of very large values of k^2, i.e., $k^2 \gg m^2$. In this case, we can approximate Π^* as

$$\Pi^*(k) \approx -\frac{\alpha}{3\pi} \log \left(\frac{k^2}{m^2} \right) \tag{9.69}$$

By solving the equations of motion for Γ with an external current J_μ, we see that the interaction is given by

$$W_2[J] = \frac{1}{2} \int J_\mu(x) \, D(x,y) \, J^\mu(y)$$

$$D^{-1}(k) = k^2 \left[1 - \frac{\alpha}{3\pi} \log\left(\frac{k^2}{m^2}\right) \right] \tag{9.70}$$

We see that the interaction is effectively stronger at high values of k^2, or at short distances. Vacuum polarization is thus a screening effect, the interaction at large separations is weaker. We also see that the true expansion parameter for QED is $\alpha \log(k^2/m^2)$ and not just α. The formula (9.70) actually shows a pole for $D(k)$ at $k^2 = m^2 \exp(3\pi/\alpha)$. This is known as a Landau pole. Clearly, perturbation theory will cease to be valid before one gets to these energies, so one cannot conclude anything definite about the pole from the calculation we have done. The pole is at best indicative of the fact that pertubative QED must be considered as an effective theory valid only for energy scales far below this value.

The result (9.70) can also be expressed as $\alpha D(k) = \alpha_{eff}(k)/k^2$ by defining an effective k-dependent charge

$$\alpha_{eff}(k) = \frac{\alpha}{1 - \frac{\alpha}{3\pi} \log\left(k^2/m^2\right)} \tag{9.71}$$

Consider now the low energy limit of vacuum polarization. This would be important for atomic systems, where the typical value of k is in the electron-volt range which is much smaller than the mass of the electron ($m_e \approx 0.51 MeV$). Expanding the logarithm in Π^*, we find

$$D(k) \approx \frac{1}{k^2 \left[1 - (\alpha k^2/15\pi m^2)\right]} \approx \frac{1}{k^2} + \frac{\alpha}{15\pi m^2} + \mathcal{O}(\alpha^2) \tag{9.72}$$

The interaction between two charges is modified by the second term. For example, the electrostatic interaction between an electron and an atomic nucleus of charge Ze is now given by

$$V(r) = -Ze^2 \left[\frac{1}{r} + \frac{\alpha}{15\pi m^2} \delta^{(3)}(r) \right] \tag{9.73}$$

The correction term to the standard Coulomb term is known as the Uehling potential. It gives a shift to the atomic energy levels; it contributes $-27MHz$ to the Lamb shift in Hydrogen and is thus a measurable effect.

Lamb shift

The $2S_{\frac{1}{2}} - 2P_{\frac{1}{2}}$ splitting of the energy levels of Hydrogen-like atoms is known as the Lamb shift. For the case of Hydrogen, two recent experimental values, due to Lundeen and Pipkin (LP) and Andrews and Newton (AN), are

$$E(2S_{\frac{1}{2}}) - E(2P_{\frac{1}{2}}) = 1057.893 \pm 0.020 MHz \quad (LP)$$

$$= 1057.862 \pm 0.020 MHz \quad (AN) \tag{9.74}$$

Theoretically, there are many contributions to this effect. The major contribution is from electron self-energy. While the finite correction Σ^* vanishes for a free particle, Σ^*(bound), calculated with bound-state propagators is not zero, and leads to shifts in the energy levels of a bound system. Calculating this self-energy correction to order α^2 and including other effects due to the anomalous magnetic moment of the electron, the finite size of the nucleus, nuclear recoil effects and the Uehling correction discussed above, the theoretical prediction becomes

$$E(2S_{\frac{1}{2}}) - E(2P_{\frac{1}{2}}) = 1057.864 \pm 0.014 MHz \tag{9.75}$$

This excellent agreement between theory and experiment may be taken as a confirmation of the self-energy and vacuum polarization effects.

Anomalous magnetic moment of the electron

The finite part of the vertex correction, which we have not calculated, predicts an anomalous magnetic moment for the electron. This can be obtained by isolating the term proportional to $[\gamma_\mu, \gamma_\nu](p-p')^\nu$ in (9.61) and evaluating it in the limit of low external momentum. The result can be expressed as a term $\bar{\psi}[\gamma_\mu, \gamma_\nu]\psi \, F^{\mu\nu}$ in the effective action. This is like a magnetic moment interaction and one can express this as a shift of the gyromagnetic ratio of the electron, $g \to 2(1 + \alpha/2\pi)$. In other words, the theory predicts that the gyromagnetic ratio of the electron should be different from the value 2 which is given by the one-particle Dirac equation; the electron has an anomalous magnetic moment. Including higher-order (upto $\mathcal{O}(\alpha^3)$) contributions, the result is

$$\frac{g-2}{2} = (1\ 159\ 652\ 140(5.3)(4.1)(27.1)) \times 10^{-12} \qquad \text{(theory)}$$

$$= (1\ 159\ 652\ 188.4(4.3) \pm 200) \times 10^{-12} \qquad \text{(experiment)}$$
$$\tag{9.76}$$

The numbers in brackets indicate the uncertainties, due to various sources, in the last decimal places.) This is one of the most accurately tested predictions of quantum electrodynamics.

9.6 Renormalization to higher orders

The one-loop calculations we have done so far show that if we start with the action of (9.46), namely,

$$\mathcal{S}_E(A, \bar{\psi}, \psi) = \int d^4x \left[\frac{Z_3}{4} F_{\mu\nu} F^{\mu\nu} + \frac{\lambda}{2}(\partial \cdot A)^2 + Z_2 \bar{\psi}(\gamma \cdot \partial + m - \delta m)\psi \right.$$

$$\left. -ieZ_1 \bar{\psi}\gamma^\mu \psi A_\mu \right] \tag{9.77}$$

and if we choose the renormalization constants as given in (9.66), then the effective action Γ is independent of the cut-off Λ when Λ becomes very large compared to the momenta involved. This is the renormalization procedure to one-loop order. We shall now discuss how this procedure can be extended to higher orders in $\hbar$.

It is convenient to rewrite the above action in the form

$$S_E(A, \bar{\psi}, \psi) = S_{cl}(A, \bar{\psi}, \psi) + S_C(A, \bar{\psi}, \psi) \tag{9.78}$$

where

$$S_{cl}(A, \bar{\psi}, \psi) = \int d^4x \left[\frac{1}{4} F_{\mu\nu} F^{\mu\nu} + \frac{1}{2} (\partial \cdot A)^2 + \bar{\psi} (\gamma \cdot \partial + m) \psi - ie\bar{\psi}\gamma^\mu \psi A_\mu \right]$$

$$S_C(A, \bar{\psi}, \psi) = \int d^4x \left[\frac{Z_3 - 1}{4} F_{\mu\nu} F^{\mu\nu} + \frac{\lambda - 1}{2} (\partial \cdot A)^2 \right.$$

$$+ (Z_2 - 1)\bar{\psi} (\gamma \cdot \partial + m) \psi - Z_2 \delta m \bar{\psi}\psi$$

$$\left. - ie(Z_1 - 1)\bar{\psi}\gamma^\mu \psi A_\mu \right] \tag{9.79}$$

S_C is the sum of the so-called counterterms. The counterterms have coefficients which depend on Λ and are canceled out in the effective action by the loop corrections.

As we have seen, for the calculation of the corrections at the L-th loop order, we need the renormalization constants (or counterterms) only to one order less in $\hbar$, namely, to order $\hbar^{L-1}$. The calculated loop corrections at the L-th order can then be used to fix the renormalization constants to the L-th order, and these in turn can be used for the next-order calculations. In other words, one can recursively carry through the renormalization procedure. The integrands are to be expanded in powers of the external momenta or combinations of these such as the mass-shell quantities like $p^2 + m^2$ and $i\gamma \cdot p + m$. The first few terms, up to and including the power corresponding to δ, the superficial degree of divergence, can have Λ-dependence. The remainder will be finite as Λ becomes very large.

There is another, related way of viewing this procedure. We can use just the action S_{cl} of (9.79) with no counterterms, and then, instead of potential divergences being canceled by counterterms, we can equivalently give a procedure to define a new integrand, the so-called renormalized integrand, which can be used for the loop calculations. The renormalized integrand is defined so that there will be no further divergences when the loop integrals are done. In other words, given the integrand $\mathcal{I}_G$ for a 1PI Feynman diagram G, which is constructed from S_{cl}, we replace $\mathcal{I}_G$ by a renormalized integrand R_G and then do the loop integrals. We then have to give a method of writing down R_G given $\mathcal{I}_G$. This will involve subtracting certain terms from $\mathcal{I}_G$; in comparison with the counterterm approach, the terms which are subtracted may be viewed as the conribution of the counterterms. This approach to renormalization is usually called the BPHZ method, after Bogoliubov, Parasiuk, Hepp and Zimmerman.

The Feynman diagram G may have subdiagrams which are divergent, even if G is superficially convergent. We have seen an example of this at the end of section 9.4 for the scalar field theory. (In the counterterm approach, some of the subdivergences are canceled by the lower-order counterterms.) Consider a particular term in Γ with a specific monomial of fields. Let $\bar{R}_G$ be the corresponding integrand at a given loop order obtained by including the subtractions due to all lower-order subdivergences. Since subdivergences are subtracted out, the only possibility of divergence in using $\bar{R}_G$ is from the whole diagram, namely, only if $\delta_G \geq 0$. Therefore, if $\delta_G < 0$, we define the renormalized intergrand R_G as

$$R_G = \bar{R}_G, \qquad\qquad \delta_G < 0 \qquad\qquad (9.80)$$

and if $\delta_G \geq 0$, we make a Taylor expansion in powers of the external momenta (or mass-shell quantities) and define

$$R_G = \bar{R}_G - T_G \bar{R}_G = (1 - T_G)\,\bar{R}_G, \qquad\qquad \delta_G \geq 0 \qquad\qquad (9.81)$$

T_G is the Taylor expansion of $\bar{R}_G$ in powers of external momenta (or mass-shell quantities) up to and including the order δ_G. (In comparison with the counterterm approach, $T_G\bar{R}_G$ gives the renormalization constants or the counterterms corresponding to the order of the calculation.)

We must now give the construction of $\bar{R}_G$ from $\mathcal{I}_G$. Suppose $\mathcal{I}_G$ has a divergent subdiagram γ. We can then write $\mathcal{I}_G = \mathcal{I}_{G/\gamma}\,\mathcal{I}_\gamma$. The contribution of the counterterms is $-T_\gamma \bar{R}_\gamma$, for γ, so for G we get $\mathcal{I}_{G/\gamma}(-T_\gamma\bar{R}_\gamma)$. Suppose γ_1, γ_2 are disjoint subdiagrams with renormalizations. (The diagrams are disjoint if they have no common line or vertex.) In this case, we can isolate the potential divergences as

$$\mathcal{I}_{G/\{\gamma_1,\gamma_2\}}(-T_{\gamma_1}\bar{R}_{\gamma_1})(-T_{\gamma_2}\bar{R}_{\gamma_2}) \qquad\qquad (9.82)$$

If the subdiagrams are not disjoint, we cannot write the counterterm contribution as above. For example, for the diagram in figure 9.10 we will have Z_1-counterterm contributions or lower-order subtractions of the type shown in figure 9.11. The **X**'s in the diagrams in figure 9.11 denote the one-loop vertex counterterms. The mathematical expression corresponding to this is of the form $T_{\gamma_1}\bar{R}_{\gamma_1} + T_{\gamma_2}\bar{R}_{\gamma_2}$, where γ_1, γ_2 are as shown in the diagram 9.12 below. This subtraction is clearly not of the form (9.82).

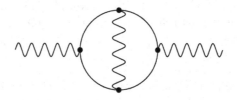

Fig 9.10. A two-loop correction to vacuum polarization

Fig 9.11. Z_1 counterterm contributions in calculating vacuum polarization
to $\mathcal{O}(\hbar^2)$

We can continue the series (9.82) to all disjoint subdiagrams. Once all lower-order subdivergences are subtracted out in this way, we get $\bar{R}_G$. Thus

$$\bar{R}_G = \mathcal{I}_G + \sum_{\gamma_1,\gamma_2,..} \mathcal{I}_{G/\{\gamma_1,\gamma_2,..\}} \prod_i (-T_{\gamma_i} \bar{R}_{\gamma_i}) \qquad (9.83)$$

This is Bogoliubov's recursion formula. The sum is over all disjoint families of subdiagrams, namely, those with $\gamma_i \cap \gamma_j = \emptyset$. From (9.83) and (9.81), we get R_G.

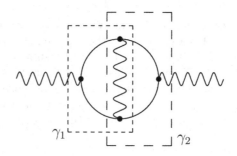

Fig 9.12. Subdiagrams for two-loop vacuum polarization

As an example of how this formula may be applied, consider the two-loop vertex correction in figure 9.13.

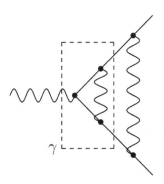

Fig 9.13. A contribution to two-loop vertex correction

The disjoint families are γ (and the null diagram $\emptyset$ and the full diagram G). The recursion formula gives

$$\bar{R}_G = \mathcal{I}_G + \mathcal{I}_{G/\gamma}(-T_\gamma \bar{R}_\gamma) = \mathcal{I}_G + \mathcal{I}_{G/\gamma}(-T_\gamma \mathcal{I}_\gamma)$$
$$= (1 - T_\gamma)\mathcal{I}_G$$
$$R_G = (1 - T_G)(1 - T_\gamma)\mathcal{I}_G \tag{9.84}$$

We have used the fact that $\bar{R}_\gamma = \mathcal{I}_\gamma$ since, at one-loop, there are no divergences from the previous order.

As another example, consider again the two-loop correction to vacuum polarization. In this case the disjoint families are γ_1 and γ_2 as shown, in addition to the null and full diagrams. We then have

$$\bar{R}_G = \mathcal{I}_G + \mathcal{I}_{G/\gamma_1}(-T_{\gamma_1}\bar{R}_{\gamma_1}) + \mathcal{I}_{G/\gamma_2}(-T_{\gamma_2}\bar{R}_{\gamma_2})$$
$$= (1 - T_{\gamma_1} - T_{\gamma_2})\,\mathcal{I}_G \tag{9.85}$$

where, once again, $\bar{R}_{\gamma_1} = \mathcal{I}_{\gamma_1}$ and $\bar{R}_{\gamma_2} = \mathcal{I}_{\gamma_2}$. The renormalized integrand is thus

$$R_G = (1 - T_G)\,(1 - T_{\gamma_1} - T_{\gamma_2})\,\mathcal{I}_G \tag{9.86}$$

Notice that $R_G \neq (1 - T_G)(1 - T_{\gamma_1})(1 - T_{\gamma_2})\mathcal{I}_G$. This is because of the overlapping nature of the two loop integrations.

In these formulae, we have a recursion rule; we must first determine $\bar{R}_{\gamma_i}$ and then use (9.83) to obtain $\bar{R}_G$. A solution to (9.83) and (9.81) directly in terms of $\mathcal{I}_G$ has been given by Zimmerman. This is the forest formula. A forest $\cup$ is defined as a family of subdiagrams with the following properties.

1. $\gamma_i \in \cup$ are proper superficially divergent diagrams.

2. γ_1, $\gamma_2 \in \cup \implies \gamma_1 \subset \gamma_2$ or $\gamma_2 \subset \gamma_1$ or $\gamma_1 \cap \gamma_2 = \emptyset$; i.e., they are either nonoverlapping or one is contained in the other.
3. $\cup$ may be empty.

The solution to (9.81) and (9.83) is then given by

$$R_G = \sum_{\cup} \prod_{\gamma \in \cup} (-T_\gamma) \mathcal{I}_G \tag{9.87}$$

This is Zimmerman's forest formula.

Consider the application of this formula to the two-loop vertex diagram. The forests are $\emptyset$, $\{\gamma\}$, $\{G\}$, $\{\gamma, G\}$. From the forest formula

$$\begin{aligned} R_G &= \mathcal{I}_G - T_\gamma \mathcal{I}_G - T_G \mathcal{I}_G + T_\gamma T_G \mathcal{I}_G \\ &= (1 - T_G)(1 - T_\gamma) \mathcal{I}_G \end{aligned} \tag{9.88}$$

which is in agreement with (9.84). Similarly, for the two-loop vacuum polarization diagram, the forests are $\emptyset$, $\{G\}$, $\{\gamma_1\}$, $\{\gamma_2\}$, $\{\gamma_1, G\}$, $\{\gamma_2, G\}$. The forest formula gives

$$\begin{aligned} R_G &= (1 - T_{\gamma_1} - T_{\gamma_2} - T_G + T_{\gamma_1} T_G + T_{\gamma_2} T_G) \, \mathcal{I}_G \\ &= (1 - T_G)(1 - T_{\gamma_1} - T_{\gamma_2}) \, \mathcal{I}_G \end{aligned} \tag{9.89}$$

This agrees with (9.86) as well.

9.7 Counterterms and renormalizability

We now turn to the renormalizability of QED. A field theory is said to be renormalizable if the renormalization procedure can be carried and a finite effective action Γ constructed such that it involves only a finite number of undetermined parameters (masses and coupling constants). These parameters are to be determined from experiments. If Γ involves an infinite number of undetermined parameters, we say the theory is nonrenormalizable. QED is a renormalizable theory.

For the purpose of doing loop calculations, the action for QED was written in the form (9.78), viz.,

$$S_E(A, \bar{\psi}, \psi) = S_{cl}(A, \bar{\psi}, \psi) + S_C(A, \bar{\psi}, \psi) \tag{9.90}$$

where

$$S_{cl}(A, \bar{\psi}, \psi) = \int d^4x \left[\frac{1}{4} F_{\mu\nu} F^{\mu\nu} + \frac{1}{2} (\partial \cdot A)^2 + \bar{\psi} (\gamma \cdot \partial + m) \psi - ie\bar{\psi}\gamma^\mu \psi A_\mu \right]$$

$$S_C(A, \bar{\psi}, \psi) = \int d^4x \left[\frac{Z_3 - 1}{4} F_{\mu\nu} F^{\mu\nu} + \frac{\lambda - 1}{2} (\partial \cdot A)^2 \right]$$

$$+(Z_2 - 1)\bar{\psi}\,(\gamma \cdot \partial + m)\,\psi - Z_2 \delta m \bar{\psi}\psi$$

$$\left. -ie(Z_1 - 1)\bar{\psi}\gamma^\mu\psi A_\mu \right] \tag{9.91}$$

$\mathcal{S}_C$ is the counterterm action. In renormalizing the theory, we choose the counterterms to cancel the divergences to get a finite effective action Γ. The counterterms are thus determined by the calculation of the regularized integrals which isolate potential divergences as $\Lambda \to \infty$. As a result, we have the following property.

If the classical theory has a symmetry which is preserved by the regulators, then the counterterm action (and hence the functional integral) will have the same symmetry.

An example of this is gauge-invariance in QED. The classical action has the gauge-invariance given by

$$\mathcal{S}_{cl}(A', \bar{\psi}', \psi') = \mathcal{S}_{cl}(A, \bar{\psi}, \psi)$$
$$A'_\mu = A_\mu + \partial_\mu\theta$$
$$\psi' = e^{ie\theta}\psi$$
$$\bar{\psi}' = e^{-ie\theta}\bar{\psi} \tag{9.92}$$

If we use a gauge-invariant regulator, we can expect $\mathcal{S}_C$ to have this symmetry. This leads immediately to $Z_1 = Z_2$. Thus, with a gauge-invariant regulator, one can restrict $\mathcal{S}_C$ to be as given in (9.91), but with Z_1 set equal to Z_2. (The functional integral involves a term $\frac{1}{2}(\partial \cdot A)^2$, which is clearly not gauge-invariant. However, this does not affect our argument because A_μ is coupled to a conserved current. This will become clearer when we discuss the Ward-Takahashi identities in more detail in Chapter 11.)

The counterterms one has to choose have the same structure in terms of fields and derivatives as the terms in the action. This can be understood as follows. Suppose there is a term of the form $\mathcal{O}$ which is a function of the fields and derivatives and which can be generated with a divergent coefficient $c(\Lambda)$ as a result of a loop claculation, up to some order. We must then have a term $Z\mathcal{O}$ in the action to cancel the divergent part of this. Putting the two together, we then have $(Z + c)\mathcal{O}$ in the effective action. The Λ-dependent terms in the combination $Z + c$ cancel out, but there is no a priori theoretical reason to say that $Z + c$ is zero; it could be any finite number λ_R, whose value must be taken from experiments. It should be viewed as another coupling constant in the theory. We can restate the above result as saying that there is a term $\lambda_0\mathcal{O}$ in the classical action, which we then write as $(\lambda_R + Z)\mathcal{O}$. This shows that the counterterms are of the same nature as the terms in the action. If a particular kind of term is forbidden by symmetry arguments and the regulator preserves that symmetry, then it will not be generated by loop calculations and so can be excluded consistently from the action and the counterterms.

In the case of QED, all terms in the action have dimensions less than or equal to zero. (We include the d^4x in the power counting.) From the superficial degree of divergence δ, we see that all the terms which can be generated with divergent coefficients are also of dimensions less than or equal to zero. This result is true for other field theories as well. In general, we have the result that an action of the form $\mathcal{S} = \sum_i \alpha_i \mathcal{O}_i(\varphi)$ where $dim\mathcal{O}_i \leq 0$ leads to counterterms with dimension ≤ 0. (At the level of the Lagrangian, local monomials of the field and its derivatives which have dimension ≤ 4 will lead to monomials of dimension ≤ 4.) Thus the most general combination of terms made up of the fields of interest which are of dimension ≤ 0 can be expected to be renormalizable. This means that after renormalization, the effective action will have a finite number of undetermined parameters or coupling constants α_i. Once the values of these parameters are taken from experiments, we can then make predictions using the functional integral. Thus, in the case of QED, we may expect a renormalizable action to be of the form

$$\mathcal{S} = \sum_i \alpha_i \mathcal{O}_i(A, \bar{\psi}, \psi) \tag{9.93}$$

where $\mathcal{O}_i$ are monomials of fields and derivatives which are of dimension ≤ 0. (Strictly speaking, the action must be somewhat more restricted, since a mass term for the photon can lead to bad ultraviolet behavior for the propagator. This is briefly discussed in the next chapter; for now, we ignore this complication, because we are imposing gauge-invariance anyway.) Without symmetries, this is the best we can say. For QED we have the symmetries

1. Lorentz-invariance
2. Gauge-invariance
3. Parity and charge conjugation

Since there exists a regulator respecting these symmetries, and if we use such a regulator, we can choose a more restricted action which has these symmetries. The most general action consistent with these symmetries and with terms of dimension ≤ 0 is (9.90) and so it can be a renormalizable theory.

So far we have used the superficial degree of divergence. The fact that our analysis holds in general is due to the following convergence theorem due to Weinberg.

If δ for a graph and all its subgraphs is < 0, the corresponding integral is absolutely convergent.

In building up the theory, one has to use this in conjunction with Bogoliubov's recursion formula. Consider, for example, the $\mathcal{O}(\hbar)$ terms. By considering graphs with $\delta \geq 0$, we see that all divergent terms are of dimension ≤ 0. At $\mathcal{O}(\hbar^2)$, consider $\bar{R}_G$. We have already subtracted out subdivergences, so the kind of divergences generated by $\bar{R}_G$, viz., $T_G\bar{R}_G$ are analyzed by considering just the superficial degree of divergence of G, since the subgraphs are convergent. Analyzing terms with $\delta(G) \geq 0$ shows that we again have only $dim \leq 0$ terms as potential divergences. Thus recursively, one is led to the result that

$$dim \leq 0 \text{ terms} \Longrightarrow \text{potential divergent terms of } dim \leq 0$$

This result combined with the symmetry argument can prove the perturbative renormalizability of QED.

If terms with $dim > 0$ are used, we lose renormalizability. For example, consider $\int d^4x(\bar{\psi}\psi)^2$ which is of dimension 2. This can lead to the graph shown in figure 9.14. The contribution of this diagram is divergent and requires counterterms of the type $\int d^4x(\bar{\psi}\psi)^3$. This means that we should also have a term of this type in the action with some arbitrary coupling constant. Such a term, in turn, leads to the graph shown in figure 9.15. The contribution of this diagram is also divergent and needs a counterterm of the type $\int d^4x(\bar{\psi}\psi)^6$. This leads to more divergent graphs in turn and we end up with an infinite set of counterterms and hence an infinite set of coupling constants in the theory. Thus, quite generally, the addition of a term of dimension > 0 in the action leads to a non-renormalizable theory.

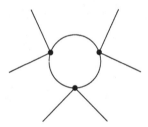

Fig 9.14. A one-loop 6-fermion vertex generated by $(\bar{\psi}\psi)^2$

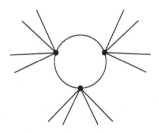

Fig 9.15. A graph generated by $(\bar{\psi}\psi)^3$

Some interesting examples of renormalizable theories are QED, nonabelian gauge theories, scalar field theory with a φ^4-interaction, and scalar QED with the action

$$\mathcal{S} = \int d^4x \left[|(\partial + ieA)\phi|^2 + m^2\phi^*\phi + \lambda(\phi^*\phi)^2 + \frac{1}{4}F^2 + \frac{1}{2}(\partial \cdot A)^2 \right] \quad (9.94)$$

Some prominent examples of non-renormalizable theories are the nonlinear sigma models, the $V - A$ current-current theory of weak interactions, and Einstein's theory of gravity.

The renormalization procedure can be and needs to be carried out in any theory. The difference between renormalizable and non-renormalizable theories is then a matter of how one interprets and uses the effective action. For a renormalizable theory, one has a finite number of coupling constants and other parameters and predictions are straightforward. In a non-renormalizable theory, Γ involves an infinite set of coupling constants. The action to be used in such a theory will have monomials $\mathcal{O}_i$ of the fields and derivatives of arbitrary dimension. The effect of a term of high dimension on low-energy processes is suppressed by powers of the energies involved. For example, consider a term like $\lambda \int d^4x(\bar{\psi}\psi)^2$. The coupling constant λ must have dimension -2. So let us write it as g/M^2 where g is dimensionless and M is some constant with the dimension of mass which is characteristic of the theory. This term can contribute to two-particle scattering. The total cross section for such scattering, in the lowest order, for example, will go like g^2p^2/M^4, just on dimensional grounds. Here p denotes some typical scale for the momenta of external particles. This becomes negligibly small when $p \ll M$. Thus, even though we need to know the value of g to make a prediction about the two-particle scattering in general, we can actually make a prediction for low energies (low momenta) without knowing g, if g is not abnormally large. The range of validity of this is determined by the value of M, which will be a characteristic mass scale of the theory. We can thus use a non-renormalizable theory sensibly, make predictions with it and so on, if we restrict the range of validity of our predictions to energy regimes which are low compared to the characteristic mass scale.

One can also understand and estimate this mass scale as follows. The cross section we have calculated grows with energy. But cross sections have to be bounded by unitarity of the S-matrix. The growing cross section will violate this bound at some energy. This gives a rough estimate of the energy beyond which we will need the value of g to make predictions. (An explicit calculation along these lines is given in Chapter 13.) Notice that if we introduce a new set of particles (and fields) at this stage, the Hilbert space is enlarged and the implementation of unitarity changes. Thus the breakdown of unitarity can also be interpreted, in some cases, as the signal of producing new particles as intermediate states in collisions. In turn, we may consider this as a signal of the need to augment the theory with new particles.

Ultimately, rather than having an infinite set of coupling constants, we might consider it desirable to introduce relations among them, based on some symmetry perhaps, and adding some new fields as well. This would leave us with a finite number of parameters again and would be equivalent to embedding the theory in a renormalizable theory which would then have a much larger range of validity.

As an example of how this might work out, consider the nonlinear sigma model which is used for describing low energy pion-nucleon dynamics.

$$S = -\frac{1}{2}f_\pi^2 \int d^4x \left(\text{Tr}\left[(U^{-1}\partial_\mu U)^2\right] + \bar{N}\gamma\cdot\partial N \right.$$

$$\left. + m\,\bar{N}\,U\,P_-N + \bar{N}\,U^\dagger\,P_+N \right)$$

$$P_\pm = \frac{1}{2}(1\pm\gamma_5) \tag{9.95}$$

where N is a two-component column vector, each component of it is a four-spinor; N_1 denotes the proton field, N_2 denotes the neutron field, and U is a (2×2)-matrix denoting the triplet of pions φ^a, $a = 1, 2, 3$, via $U = \exp(i\tau^a\varphi^a/\sqrt{2}\,f_\pi)$. τ^a are the Pauli matrices, connecting the proton and neutron, and thus generating the isospin transformations. f_π is a constant, called the pion decay constant. This theory is non-renormalizable, with interaction terms containing arbitrarily high powers of the pion field. One has to add an infinity of terms to (9.95) to make a renormalizable theory; the simplest such term would be

$$S_1 = c\int d^4x\,\text{Tr}\left[(U^{-1}\partial_\mu U)^4\right] \tag{9.96}$$

Pion-pion scattering shows that the theory (9.95) can be used up to approximately $4\pi f$ in energy. Around that value of energy, the contribution from the term (9.96) becomes important and one must add this as well. Terms of higher dimension than S_1 also become important at higher energies. Eventually, one needs so many new fields and so many terms that one has to seek a better theory to which (9.95) is a low-energy approximation. In the case of pions and nucleons, such a theory is quantum chromodynamics (QCD).

A similar situation holds for gravity as well. We can, and do, use Einstein's theory to make low-energy predictions. The energy scale involved is the Planck mass $M_P = (G_N)^{-\frac{1}{2}} \approx 10^{19} GeV$. ($G_N$ is Newton's constant.) All experiments to date are far below this energy and so we have not faced any serious discrepancy in the theory of gravity yet. Terms of dimension higher than the Einstein action, such as the square of the Riemann tensor, should be there in the theory of gravity. The hope is that there is some symmetry which will gather up all such higher-dimensional terms, an infinity of them, into an action with a finite number of parameters. One can then formulate gravity as a renormalizable theory, with possible infinities being restricted by the symmetry. We do not know what this is yet, but string theory may be such a theory.

A note on regulators

When we change regulators, the divergent terms change by finite amounts. Thus the counterterms change by finite amounts as well. The parameters, as they appear in the action, are also changed. But they are determined by experiments and so we can specify them in terms of scattering amplitudes at certain chosen momenta. If we express the cross sections in terms of the value of scattering amplitudes at the chosen momenta, then the parameters are eliminated (in favor of the amplitudes at the chosen momenta) and the results are independent of the regulator.

Given the relationship between the counterterms and the regulator, we can say that a choice of regulator is equivalent to a choice of (the coefficients of) the counterterms. This shows that there is nothing particularly important about symmetries in a regulator either. If a regulator exists which has the desired symmetry, we can still choose to use a regulator which does not respect this symmetry. In this case, we simply have to have counterterms which do not respect the symmetry. One can then impose desirable symmetries at the level of the renormalized Γ. For example, if we calculate the vacuum polarization graph in QED without a gauge-invariant regulator, we will find a term like $\Lambda^2 A^2$. This can be canceled by a counterterm $\mu^2 A^2$ by choosing μ^2 appropriately. This counterterm does not have gauge invariance; this, by itself, is not a problem. Gauge invariance is important for eliminating the unphysical polarizations of the photon and thereby getting a unitary theory. We must thus impose it on the renormalized theory, which will require the counterterm A^2 to cancel the A^2 term generated from vacuum polarization exactly. This is in fact the procedure we used in retaining only the transverse part of $\Pi^{\mu\nu}$ in the vacuum polarization. Thus, the use of a regulator which does not have gauge invariance is perfectly acceptable so long as we can impose gauge invariance on the renormalized action by suitably choosing the coefficients of the counterterms. Nevertheless, in practical calculations, especially with many external particles and loops, it is algebraically much simpler if we choose a gauge-invariant regulator. (We might also note that the experimental upper bound on the photon mass is very small, $< 2 \times 10^{-16} eV$.)

There are situations when there exists no regulator which has all the symmetries of the classical action. Typically, this happens when we have many symmetries and we cannot have a regulator which preserves all of them. In this case, it becomes impossible to impose all the symmetries at the level of the renormalized theory. We impose the most important ones, such as those required by unitarity, and use a Γ which breaks some of the other symmetries. The symmetries which are broken by the regularization process are said to be anomalous. This is discussed in more detail in Chapter 13.

9.8 RG equation for the scalar field

The fact that the Green's functions $G(x_1, x_2, ..., x_N)$ of the renormalized theory do not depend on the cut-off Λ leads to constraints on their asymptotic behavior. This can be expressed by an equation which tells us how the Green's functions behave under a scale change of the momenta or the coordinates. This is the renormalization group (RG) equation. The RG equation can be used to determine the asymptotic behavior of Green's functions. It has also been used to determine the critical exponents of various field-theory models in statistical mechanics. In this section, we will derive the RG equations for a scalar field theory.

The Green's functions for a massless scalar field theory are given by

$$G(x_1, x_2, ..., x_N) = \mathcal{N} \int [d\varphi] \ e^{-S} \ \varphi(x_1)\varphi(x_2) \cdots \varphi(x_N) \tag{9.97}$$

where

$$S = \int d^4x \ \left[\frac{1}{2}Z_3(\Lambda)(\partial\varphi)^2 + \lambda Z_1(\Lambda)\varphi^4\right] \tag{9.98}$$

The Green's functions so-defined are finite as $\Lambda \to \infty$. In terms of χ defined by $\varphi = Z_3^{-\frac{1}{2}}\chi$, we have

$$S = \int d^4x \ \left[\frac{1}{2}(\partial\chi)^2 + \lambda_0 \ \chi^4\right]$$
$$\lambda_0 = \lambda Z_1(\Lambda)Z_3^{-2}(\Lambda) \tag{9.99}$$

λ_0 depends on λ and Λ. We can invert this relation and write λ as a function of λ_0 and Λ; i.e., $\lambda = \lambda(\lambda_0, \Lambda)$. Since λ does not depend on the cut-off Λ, it must be that the Λ-dependence of λ_0 cancels out the explicit Λ-dependence of the function λ. In other words

$$\left(\frac{\partial\lambda}{\partial\lambda_0}\right)_\Lambda \frac{\partial\lambda_0}{\partial\log\Lambda} + \left(\frac{\partial\lambda}{\partial\log\Lambda}\right)_{\lambda_0} = 0 \tag{9.100}$$

In terms of the variables χ, the Green's functions can be expressed as

$$G(x_1, x_2, ..., x_N) = \tilde{\mathcal{N}} \ Z_3^{-N/2} \int [d\chi] \ e^{-S} \ \chi(x_1)\chi(x_2) \cdots \chi(x_N)$$
$$= Z_3^{-\frac{1}{2}N} \ \langle \ \chi(x_1)\chi(x_2) \cdots \chi(x_N) \ \rangle \tag{9.101}$$

where we have used the abbreviation

$$\langle \mathcal{O} \rangle = \tilde{\mathcal{N}} \int [d\chi] \, e^{-\mathcal{S}} \, \mathcal{O} \tag{9.102}$$

We now turn to the RG equations. The Λ-independence of the Green's functions gives

$$\frac{d}{d \log \Lambda} \, G(x_1, x_2, ..., x_N) = 0 \tag{9.103}$$

Using the expression for G as given in (9.101) and differentiating through, we find

$$-\frac{N}{2} \frac{d \log Z_3}{d \log \Lambda} \, G \, - Z_3^{-N/2} \langle \, \chi(x_1)\chi(x_2) \cdots \chi(x_N) \, \left(\frac{d\mathcal{S}}{d \log \Lambda} \right) \, \rangle = 0 \tag{9.104}$$

In order to simplify this further, we need the cut-off dependence of the action $\mathcal{S}$. This comes from two sources, one due to the cut-off dependence of λ_0 in the interaction term and the other in the integrals of the monomials of the fields. The latter point becomes clear if we write the expressions in momentum space. For example,

$$\mathcal{S}_0(\Lambda) \equiv \frac{1}{2} \int d^4x \, (\partial\chi)^2 = \frac{1}{2} \int_{p^2 \leq \Lambda^2} \frac{d^4p}{(2\pi)^4} \, \chi(-p) \, p^2 \, \chi(p) \tag{9.105}$$

The largest value of momentum is given by Λ, or equivalently, there is a short-distance cut-off for spatial separations; we need $|x - y| \geq 1/\Lambda$. If we replace Λ by $(1 + \sigma)\Lambda$, we must have $|x - y| \geq 1/(1 + \sigma)\Lambda$. The integral can then be related to the integral with cut-off Λ by the change of variables $x \rightarrow z = x \, (1 + \sigma), |z - z'| \geq 1/\Lambda$. Thus

$$\mathcal{S}_0((1 + \sigma)\Lambda) = \frac{1}{2} \int d^4x \, (\partial\chi)^2 \Big|_{(1+\sigma)\Lambda} = \frac{1}{2} \frac{1}{(1 + \sigma)^2} \int d^4z \, (\partial\chi)^2 \Big|_{z/1+\sigma}$$

$$= \frac{1}{2} \int d^4z \, (\partial\tilde{\chi})^2 \tag{9.106}$$

where

$$\tilde{\chi}(z) = \frac{1}{1 + \sigma} \, \chi(z/1 + \sigma)$$

$$\approx \chi(z) - \sigma \left(z \cdot \frac{\partial}{\partial z} + 1 \right) \chi(z)$$

$$\approx \chi - \sigma(D\chi) \tag{9.107}$$

The effect of the change of cut-off is thus to change the fields χ by $\sigma(D\chi)$ where

$$D\chi \equiv \left(z \cdot \frac{\partial}{\partial z} + 1 \right) \chi \tag{9.108}$$

gives the effect of a dilatation or scale change on the fields. The result (9.106) can thus be written as

$$-\frac{d\mathcal{S}_0}{d\log\Lambda} = \int d^4x\ (D\chi)\frac{\delta}{\delta\chi}\ \mathcal{S}_0 \tag{9.109}$$

For the interaction term, we have a similar expression and, in addition, we have the change due to the explicit Λ-dependence of λ_0.

$$-\frac{d\mathcal{S}_{int}}{d\log\Lambda} = \int d^4x\ (D\chi)\frac{\delta}{\delta\chi}\ \mathcal{S}_{int}\ -\ \frac{d\lambda_0}{d\log\Lambda}\int d^4x\ \chi^4 \tag{9.110}$$

Using these results in (9.104) we find

$$-\frac{N}{2}\frac{d\log Z_3}{d\log\Lambda}\ G + Z_3^{-N/2}\langle\ \chi(x_1)\chi(x_2)\cdots\chi(x_N)\ \left(\int(D\chi)\frac{\delta}{\delta\chi}\mathcal{S}\right)\ \rangle$$

$$-\frac{d\lambda_0}{d\log\Lambda}\ Z_3^{-N/2}\langle\ \chi(x_1)\chi(x_2)\cdots\chi(x_N)\int d^4x\ \chi^4\ \rangle =0 \tag{9.111}$$

A change of variables $\chi \to \chi' = \chi + \delta\chi$ in the functional integral leads to the identity

$$\int[d\chi]\ e^{-\mathcal{S}(\chi)}\ \mathcal{O}(\chi) = \int[d\chi']\ e^{-\mathcal{S}(\chi')}\ \mathcal{O}(\chi')$$

$$= \int[d\chi]\ e^{-\mathcal{S}(\chi)}\ \mathcal{O}(\chi) + \int[d\chi]\ e^{-\mathcal{S}(\chi)}\left[\delta\mathcal{O} - \mathcal{O}\ \delta\mathcal{S}\right] \tag{9.112}$$

In other words

$$\langle\delta\mathcal{O}\rangle\ -\ \langle\mathcal{O}\ \delta\mathcal{S}\rangle = 0 \tag{9.113}$$

Equation(9.111) can now be simplified as

$$\left[\sum x_i\cdot\frac{\partial}{\partial x_i}\ +N-\frac{N}{2}\frac{d\log Z_3}{d\log\Lambda}\right]\ G(x_1, x_2,\cdots, x_N)$$

$$-\frac{\partial\lambda_0}{\partial\log\Lambda}\ Z_3^{-N/2}\langle\ \chi(x_1)\chi(x_2)\cdots\chi(x_N)\int d^4x\ \chi^4\ \rangle = 0 \tag{9.114}$$

The term with the insertion of the interaction term $\int\chi^4$ can be obtained from the coupling constant dependence of the Green's function. Differentiating the expression (9.101) for the Green's function

$$\frac{\partial}{\partial\lambda_0}\ G = -\frac{N}{2}\left(\frac{\partial\log Z_3}{\partial\lambda_0}\right)_\Lambda\ G$$

$$-\ Z_3^{-N/2}\langle\chi(x_1)\chi(x_2)\cdots\chi(x_N)\int\chi^4\rangle \tag{9.115}$$

Here we consider Z_3 as a function of λ_0 and Λ. Further, writing

$$\frac{d \log Z_3}{d \log \Lambda} = \left(\frac{\partial \log Z_3}{\partial \lambda_0} \right)_\Lambda \frac{\partial \lambda_0}{\partial \log \Lambda} + \left(\frac{\partial \log Z_3}{\partial \log \Lambda} \right)_{\lambda_0}, \qquad (9.116)$$

equation (9.114) becomes

$$\left[\sum x_i \cdot \frac{\partial}{\partial x_i} + N(1 + \gamma) + \frac{\partial \lambda_0}{\partial \log \Lambda} \frac{\partial}{\partial \lambda_0} \right] G(x_1, x_2, \cdots, x_N) = 0 \quad (9.117)$$

where

$$\gamma = -\frac{1}{2} \left(\frac{\partial \log Z_3}{\partial \log \Lambda} \right)_{\lambda_0} \qquad (9.118)$$

Finally, we want to express this in terms of derivatives with respect to the renormalized coupling constant λ rather than λ_0. Since λ_0 is a function of λ as given by (9.99),

$$\begin{aligned}
\frac{\partial \lambda_0}{\partial \log \Lambda} \frac{\partial}{\partial \lambda_0} G &= \frac{\partial \lambda_0}{\partial \log \Lambda} \left(\frac{\partial \lambda}{\partial \lambda_0} \right)_\Lambda \frac{\partial G}{\partial \lambda} \\
&= - \left(\frac{\partial \lambda}{\partial \log \Lambda} \right)_{\lambda_0} \frac{\partial G}{\partial \lambda} \\
&= \beta(\lambda) \frac{\partial G}{\partial \lambda} \qquad (9.119)
\end{aligned}$$

where we have also used (9.100). Combining this with (9.117), we get the renormalization group (RG) equation

$$\left[\sum x_i \cdot \frac{\partial}{\partial x_i} + N(1 + \gamma) + \beta \frac{\partial}{\partial \lambda} \right] G(x_1, x_2, \cdots, x_N) = 0 \qquad (9.120)$$

The quantities β and γ in this equation are, once again,

$$\beta = - \left(\frac{\partial \lambda}{\partial \log \Lambda} \right)_{\lambda_0}$$

$$\gamma = -\frac{1}{2} \left(\frac{\partial \log Z_3}{\partial \log \Lambda} \right)_{\lambda_0} \qquad (9.121)$$

These are functions of the coupling constant λ.

For the φ^4 theory it is easy to calculate β and γ to the lowest order in perturbation theory using our explicit formulae for the renormalization constants. The results are

$$\beta(\lambda) = \frac{9\lambda^2}{2\pi^2} + \cdots$$

$$\gamma(\lambda) = \frac{3\lambda^2}{16\pi^4} + \cdots \qquad (9.122)$$

In the noninteracting theory, β and γ are zero and the RG-equation shows that the Green's functions behave as functions of dimension N, or equivalently, the fields φ have dimension 1. This is also clear from (9.108) and

(9.109), which show the effect of a dilatation on the fields. In the interacting theory, the Green's functions no longer have any scaling behavior in general. However, in the special case when $\beta = 0$, there is still scaling behavior; the dimension of N-point Green's function is then $N(1 + \gamma)$, not just N. For this reason, γ is referred to as the anomalous dimension of the field φ. β is a measure of the scaling violations due to the interactions; it is referred to as the β-function. Even though there is, in general, no scaling behavior for the Green's functions in the interacting theory, scaling behavior may occur at values of the coupling for which $\beta = 0$. In other words, the theory can be at a zero of the β-function. Such zeroes can occur at very high or at very low momenta, depending on the theory. In order to show how this can happen and how scaling behavior is obtained, we shall need to solve the RG-equation.

9.9 Solution to the RG equation and critical behavior

The solution is most easily obtained in terms of an effective coupling constant $\bar{\lambda}$, referred to as a running coupling constant. It is defined by

$$\frac{d\bar{\lambda}}{ds} = -\beta(\bar{\lambda}) \tag{9.123}$$

Here s is a scaling parameter, which scales the coordinates as $x \to e^s x$. We may write (9.123) as

$$s = -\int_\lambda^{\bar{\lambda}} \frac{du}{\beta(u)} \tag{9.124}$$

Notice that $\bar{\lambda}$ is obtained as a function of λ and s and we have chosen the initial condition $\bar{\lambda} = \lambda$ at $s = 0$. Differentiating both sides of (9.124) with respect to λ, we also find

$$\beta(\lambda) \frac{\partial \bar{\lambda}}{\partial \lambda} = \beta(\bar{\lambda}) \tag{9.125}$$

We may now write the solution of the RG equation as

$$G(e^s x_1, e^s x_2, ..., e^s x_N, \lambda) = G(x_1, x_2, ..., x_N, \bar{\lambda}) \, \exp \left(\int_\lambda^{\bar{\lambda}} \frac{\Delta(u)}{\beta(u)} du \right) \tag{9.126}$$

where $\Delta = N(1 + \gamma)$. One can check this solution by differentiating with respect to s. This will give the equation

$$\sum_i x_i \cdot \frac{\partial}{\partial x_i} G = -\beta(\bar{\lambda}) \frac{\partial G}{\partial \bar{\lambda}} - \Delta(\bar{\lambda}) G \tag{9.127}$$

By directly differentiating with respect to λ and using (9.125), we find

$$\beta(\lambda) \frac{\partial G}{\partial \lambda} = \beta(\bar{\lambda}) \frac{\partial G}{\partial \bar{\lambda}} + \Delta(\bar{\lambda}) G - \Delta(\lambda) G \tag{9.128}$$

The last two equations (9.127) and (9.128) verify that G given in (9.126) is indeed the solution of the RG equation.

The nature of the solution can be understood by first considering the zeroes of the β-function. Perturbatively, β is zero at $\lambda = 0$. In general, one may have other zeroes for the β-function. Let λ_* denote a zero of β, $\beta(\lambda_*) = 0$. The RG equation (9.120) can be trivially solved at this value of the coupling constant, by setting $\beta = 0$, as

$$G(e^s x_1, e^s x_2, ..., e^s x_N, \lambda_*) = \exp(-s\Delta(\lambda_*))\, G(x_1, x_2, ..., x_N, \lambda_*) \quad (9.129)$$

This shows that the Green's function has a simple scaling behavior with dimension $\Delta(\lambda_*) = N(1 + \gamma(\lambda_*))$. N is the expected canonical dimension of G since φ has dimension 1. γ evaluated at λ_* is the anomalous dimension at the chosen value of coupling. Notice that once we are at a zero of the β-function, the running coupling is a constant with respect to scaling and does not run anymore. Zeroes of the β-function are therefore called fixed points. In the neighborhood of a fixed point λ_*, β is given by $\beta \approx \beta'(\lambda_*)(\lambda - \lambda_*)$. From equation (9.123), $\bar{\lambda}$ will approach the fixed point λ_* as s increases if $\beta'(\lambda_*) > 0$. The Green's function is, in general, a function of differences of coordinates $x_i - x_j$, so that an increase in s corresponds to considering processes at larger separations of points, in other words, the infrared limit. For this reason, we say that such a fixed point is an infrared stable fixed point. The Green's functions or correlators of the theory show scaling behavior in the infrared limit; the theory is effectively at the value of the coupling constant λ_*, which is the infrared stable fixed point. Because of the scaling behavior, we say the theory is critical. On the other hand, if $\beta'(\lambda_*) < 0$, $\bar{\lambda}$ approaches the fixed point as s decreases or as we go to shorter and shorter separations. This gives an ultraviolet stable fixed point and the theory can be described by the critical theory at large momenta. The approach to such fixed points is also important; if the approach to scaling is slow, say, logarithmic, we never attain criticality; we have only asymptotic criticality.

The general solution (9.126) shows that we do not have scaling of Green's functions in general. Nevertheless, there is a simple rule to obtain the behavior of the Green's function under scaling. The effect of scaling the coordinates x is obtained by replacing the coupling constants in the Green's function by the running coupling constant $\bar{\lambda}$ and then there is an extra exponential factor due to the dimension of the function.

In perturbation theory, we always have a zero of the β-function when the coupling constants are zero, since β is obtained as a power series in the coupling constants. Near zero coupling, if β is positive, $\lambda = 0$ is an infrared stable fixed point; the theory tends to remain perturbative for processes at low momenta. This is indeed the case for the φ^4-theory we have considered. (The same property holds for QED as well.) In the ultraviolet region, we get a strongly coupled theory as the effective, or running, coupling grows. If β is negative, the free theory is obtained in the ultraviolet limit, for for processes at large momenta. This is in fact the case for nonabelian theories, if there are not too many matter fields. The high-energy limit of the theory is perturbative; the approach to the free theory is only logarithmic and so this case is referred to as asymptotic freedom. We will discuss asymptotic freedom in QCD in some more detail in the next chapter.

The renormalization group has been extensively applied to the study of critical behavior of field theories. At a second-order phase transition, there is scaling behavior for correlators and it is possible to calculate the critical exponents for the correlators using RG techniques. We will give only a very simple example of this here. Consider the φ^4-theory in three space dimensions. Such a theory can be argued to be a good description for the critical point of the three-dimensional Ising model. The simplest way to apply our analysis is to regard the theory as being defined in $4 - \epsilon$ dimensions. For the particular case of interest, $\epsilon = 1$, but the idea is to regard ϵ as an expansion parameter, setting it equal to 1 at the end. Then we can apply much of our analysis. The dimension of the field is now $1 - \frac{1}{2}\epsilon$. The interaction term $\int d^{4-\epsilon}x \, \varphi^4$ has a nonzero canonical dimension $-\epsilon$. By redoing the calculation in (9.110), we can see that the β function has an additional term $-\epsilon\lambda$, in addition to what is calculated in the theory at $4 - \epsilon$ dimensions.

$$\beta(\lambda) = \beta(\lambda)|_{4-\epsilon} - \epsilon\lambda \tag{9.130}$$

The perturbative calculation of β to the lowest order in λ and ϵ is

$$\beta(\lambda)|_{4-\epsilon} \approx \beta(\lambda)|_4 \approx \frac{9\lambda^2}{2\pi^2} \tag{9.131}$$

We now see that there is a fixed point or a zero of the β-function at

$$\lambda_* = \frac{2\pi^2}{9}\epsilon \tag{9.132}$$

For the anomalous dimension we find

$$\gamma(\lambda)|_{4-\epsilon} \approx \gamma(\lambda)|_4 \approx \frac{3\lambda^2}{16\pi^4}$$

$$\gamma(\lambda_*) \approx \frac{\epsilon^2}{108} \tag{9.133}$$

The solution of the RG equation then shows that there is scaling behavior for the correlators and φ has effective dimension $1 - \frac{1}{2}\epsilon + \gamma(\lambda_*) \approx \frac{1}{2} + \frac{1}{108}$. Since the Green's function depends on differences of coordinates, we obtain the behavior

$$G(x_1, x_2) \sim \frac{1}{|x_1 - x_2|^{1+\eta}} \tag{9.134}$$

where the critical exponent $\eta = \epsilon^2/54 = 1/54$ to the order we have calculated. Notice that the fixed-point value of the coupling and the exponent are close enough to zero that perturbation theory is adequate for the calculation presented here.

As a method of analyzing the three-dimensional theory, the ϵ-expansion can be carried to higher orders; it is an asymptotic expansion, and so, the actual values of critical exponents will not be improved by calculating to arbitrarily high orders, they are best approximated by a calculation to a certain optimal order which depends on the theory.

In principle, the correlators can be calculated by resummation of perturbation theory or by solving the Schwinger-Dyson equations of the theory. The RG method can thus be looked upon as an efficient way of solving the Schwinger-Dyson equations in the kinematic regime (either ultraviolet or infrared) of interest. There is another related and very useful point of view: If we are interested in the infrared regime, we may think of the RG technique as a way of incorporating the effect of modes of the field of high momenta to obtain an effective action for the low-energy theory.

References

1. The idea of renormalization is due to H.A. Kramers, Nuovo Cim. **15**, 108 (1938). Renormalization was extensively developed in the early days of quantum electrodynamics; for a historical account, see S. S. Schweber, *QED and the Men Who Made It*, Princeton University Press (1994). For many of the original papers, see J. Schwinger, *Selected Papers in Quantum Electrodynamics*, Dover Publications, Inc. (1958).
2. The one-loop effective potential is due to S. Coleman and E. Weinberg, Phys. Rev. **D7**, 1888 (1973).
3. Early proofs of the renormalizability of QED were carried out by F. Dyson, P.T. Matthews and A. Salam, J.C. Ward and others; for a good summary, see A.S. Wightman, in *Renormalization Theory*, G. Velo and A.S. Wightman (eds.), D. Reidel Publishing Co. (1976).
4. The Ward-Takahashi identity is due to J.C. Ward, Phys. Rev. **78**, 182 (1950); Y. Takahashi, Nuovo Cim. **6**, 371 (1957).
5. The Landau pole is discussed by L.D. Landau, in *Niels Bohr and the Development of Physics*, Pergamon Press (1955).
6. The full one-loop calculation of vacuum polarization is by J. Schwinger. For this and the Uehling limit, see Schwinger's reprint collection, in reference 1.
7. The first understanding of Lamb shift via the electron self-energy calculation is due to H.A. Bethe, Phys. Rev. **72**, 339 (1947). Calculations in the relativistic theory were done by N.M. Kroll and W. Lamb, Phys. Rev. **75**, 388 (1949); J.B. French and V. Weisskopf, Phys. Rev. **75**, 1240 (1949). This was elaborated upon, and redone, by many others. The general theoretical framework is given in G.W. Erickson and D.R. Yennie, Ann. Phys. **35**, 271 (1965). The theoretical value quoted is from P.J. Mohr, Phys. Rev. Lett. **34**, 1050 (1974).

8. The experimental values for Lamb shift come from S.R. Lundeen and F.M. Pipkin, Phys. Rev. Lett. **34**, 1368 (1975); D.A. Andrews and G. Newton, Phys. Rev. Lett. **37**, 1254 (1976).

9. The first correction to $g - 2$ was calculated by Schwinger. The latest theoretical situation is reviewed in the article by T. Kinoshita, in *Quantum Electrodynamics*, T. Kinoshita (ed.), World Scientific Pub. Co. (1990).

10. The experimental value quoted is from R.S. Van Dyck, Jr., P.B. Schwinberg and H.G. Dehmelt, Phys. Rev. Lett. **59**, 26 (1987).

11. For the BPHZ analysis, the original references are: N.N. Bogoliubov and O.S. Parasiuk, Doklady Acad. Nauk. SSR, **100**, 25, 429 (1955); Acta Math. **97**, 227 (1957); K. Hepp, Commun. Math. Phys. **2**, 301 (1966); W. Zimmerman, Commun. Math. Phys. **6**, 161 (1967); *ibid.* **11**, 1 (1969); *ibid.* **15**, 208 (1969). See also articles by J.H. Lowenstein and by W. Zimmerman, in the book by G. Velo and A.S. Wightman, reference 3. The article by Lowenstein also discusses the renormalization of composite operators.

12. The theorem on the asymptotic behaviour of Feynman integrals is due to S. Weinberg, Phys. Rev. **118**, 838 (1960).

13. The renormalization group has a long history, starting with M. Gell-Mann and F. Low, Phys. Rev. **95**, 1300 (1954); E.C.G. Stückelberg and A. Peterman, Helv. Phys. Acta **26**, 499 (1953). The general method of renormalization group is due to K.G. Wilson and F. Wegner. Some of the early papers are: K.G. Wilson, Phys. Rev. **B4** 3174, 3184 (1971); Phys. Rev. **D3**, 1818 (1971); *ibid.* **D6**. 419 (1972); F. Wegner, Phys. Rev. **B5**, 4529 (1972).

14. The RG equations we write are close to the Callan-Symanzik equations, C.G. Callan, Jr., Phys. Rev. **D2**, 1541 (1970); K. Symanzik, Commun. Math. Phys. **18**, 227 (1970).

15. The application of the RG method to the calculation of critical exponents is due to K.G. Wilson and M.E. Fisher, Phys. Rev. Lett. **28**, 240 (1972); K.G. Wilson, Phys. Rev. Lett. **28**, 548 (1972). This is a very mature subject by now; see D.J. Amit, *Field Theory, Renormalization Group and Critical Phenomena*, McGraw Hill (1978); J. Zinn-Justin, *Quantum Field Theory and Critical Phenomena*, Clarendon Press (1996).

16. There is a version of the RG equation, known as the exact RG equation, given in K.G. Wilson and J. Kogut, Phys. Rep. **12**, 75 (1974); K.G. Wilson, Rev. Mod. Phys. **47**, 773 (1975); F. Wegner and A. Houghton, Phys. Rev. **A8**, 401 (1972). We do not discuss this here, but it can be used to provide comparatively simple proofs of the renormalizability of certain theories, in addition to its many other uses; see J. Polchinski, Nucl. Phys. **B231**, 269 (1984).

10 Gauge Theories

10.1 The gauge principle

We have already discussed the principle of minimal coupling in electrodynamics and how it can be viewed as the requirement of gauge-invariance. The coupling of a fermion field ψ to the electromagnetic field A_μ is described by the Lagrangian

$$\mathcal{L} = \bar{\psi}\left[i\gamma^\mu(\partial_\mu - ieA_\mu) - m\right]\psi \tag{10.1}$$

It has the property of gauge-invariance, viz., it is invariant under the transformations

$$\psi \to \psi' = e^{ie\theta}\psi, \qquad \bar{\psi} \to \bar{\psi}' = \bar{\psi}e^{-ie\theta}$$
$$A_\mu \to A'_\mu = A_\mu + \partial_\mu\theta \tag{10.2}$$

where θ is an arbitrary function on spacetime $\mathcal{M}$. We can write the above set of transformations as

$$\psi' = g\psi, \qquad \bar{\psi}' = \bar{\psi}g^{-1}$$
$$eA'_\mu \equiv eA^g_\mu = g(eA_\mu)g^{-1} - i\partial_\mu gg^{-1} \tag{10.3}$$

where $g(x) = \exp(ie\theta(x))$ is a function on spacetime $\mathcal{M}$ taking values in the group $U(1)$. (We have absorbed e into the function θ.) In this chapter, we want to generalize the above principle of minimal coupling to groups other than $U(1)$.

Consider a set of fields ψ_i, $i = 1, 2, ..., N$, which can transform into each other under the action of the fundamental representation of $SU(N)$. (In other words, ψ_i are elements of an N-dimensional vector space, which can carry the fundamental representation of $SU(N)$.) Introduce a gauge transformation on ψ_i by

$$\psi'_i = g_{ij}(x)\,\psi_j \tag{10.4}$$

where g_{ij} is an element of $SU(N)$ in the fundamental representation; i.e., it is an $(N \times N)$ unitary matrix with determinant equal to one. g_{ij} depends on the spacetime points, and hence we have an $SU(N)$-valued function on spacetime. For any other representation R and a set of fields ϕ_α which carry this representation we can write

$$\phi'_\alpha = \mathcal{D}_{\alpha\beta}(g)\ \phi_\beta \tag{10.5}$$

where $\mathcal{D}_{\alpha\beta}(g)$ is the representative of the element g in the representation R. One can likewise write down transformation laws for any group G and any chosen representation.

Generalizing what we did earlier, we now define a "covariant" derivative on such ψ's. Consider $\partial_\mu\psi'$ given by

$$\partial_\mu\psi' = (\partial_\mu g)\psi + g(\partial_\mu\psi) \tag{10.6}$$

$\partial_\mu\psi$ does not transform covariantly as ψ does, but there is an extra term $\partial_\mu g$. We can define a derivative which transforms covariantly, denoted $D_\mu\psi$, given by

$$D_\mu\psi = (\partial_\mu + A_\mu)\ \psi \tag{10.7}$$

We introduce a matrix potential A_μ and choose its transformation law so as to cancel the inhomogeneous term in the transformation of $\partial_\mu\psi$. In other words, we require the covariance condition

$$D_\mu(A^g)(g\psi) = g\ (D_\mu(A)\psi) \tag{10.8}$$

From this it follows that the transformation law of A_μ is given by

$$A^g_\mu = gA_\mu g^{-1} - (\partial_\mu g)g^{-1} \tag{10.9}$$

The quantity $(\partial_\mu g)g^{-1}$ is Lie algebra valued; i.e., it has the form $-it^a\partial_\mu\theta^b e^a_b(\theta)$ for g of the form $g = \exp(-it^a\theta^a(x))$. t^a are matrix representatives of the generators of the Lie algebra of G and satisfy the commutation rules

$$[t^a, t^b] = if^{abc}t^c \tag{10.10}$$

t^a are hermitian matrices and f^{abc} are real constants; $a, b, c = 1, 2, ..., dimG$, where $dimG$ is the dimension of the group. Since the θ^a's are $dimG$ independent functions, we will need $dimG$ A_μ's in general; further, A_μ is Lie algebra-valued, since it has to be added to $\partial g g^{-1}$ as in (10.9). Thus we can write

$$A_\mu = -it^a A^a_\mu \tag{10.11}$$

where A^a_μ are real functions. We have absorbed the coupling constant e into A^a_μ; it can be restored at any stage by replacing A^a_μ by eA^a_μ. The gauge transformation law can be written, for infinitesimal transformation $g \approx 1 - it^a\theta^a$, $\theta \ll 1$, as

$$A^g_\mu \approx A_\mu + it^a \left(\partial_\mu\theta^a + f^{abc}A^b_\mu\theta^c\right) \tag{10.12}$$

The combination $\left(\partial_\mu\theta^a + f^{abc}A^b_\mu\theta^c\right)$ is the covariant derivative for the adjoint representation. The Lie algebra (10.10) is satisfied by the matrices $(T^a)_{bc} = -if^{abc}$ due to the Jacobi identity; this is the adjoint representation of the group. In this case, (10.7) can be simplified to

$$(D_\mu \theta)^a \equiv \partial_\mu \theta^a - i A_\mu^k (T^k)^{ab} \theta^b = \left(\partial_\mu \theta^a + f^{abc} A_\mu^b \theta^c \right) \qquad (10.13)$$

We may thus write the infinitesimal transformation law (10.12) as

$$A_\mu^g - A_\mu \approx i t^a \, (D_\mu \theta)^a \qquad (10.14)$$

The covariant derivative gives us a prescription for coupling A_μ's to any matter field. What we do is to replace all ordinary derivatives by the covariant derivatives. Of course, one still needs a Lagrangian for the A_μ's themselves; this is provided by the field strength. The field strength tensor $F_{\mu\nu}$ is defined by

$$\begin{aligned} F_{\mu\nu} &= D_\mu D_\nu - D_\nu D_\mu = [D_\mu, D_\nu] \\ &= -i t^a \left(\partial_\mu A_\nu^a - \partial_\nu A_\mu^a + f^{abc} A_\mu^b A_\nu^c \right) \\ &\equiv -i t^a F_{\mu\nu}^a \end{aligned} \qquad (10.15)$$

From the covariance of D_μ we find

$$F_{\mu\nu}(A^g)(g\psi) = [D_\mu(A^g), D_\nu(A^g)] \, (g\psi) = g \, [D_\mu, D_\nu] \, \psi = g F_{\mu\nu} \psi \qquad (10.16)$$

In other words,

$$F_{\mu\nu}(A^g) = g \, F_{\mu\nu} \, g^{-1} \qquad (10.17)$$

Thus $F_{\mu\nu}$ transforms covariantly. Using $F_{\mu\nu} = -i t^a F_{\mu\nu}^a$ and $g t^b g^{-1} = \mathcal{D}^{ab}(g) t^a$, we get

$$F_{\mu\nu}^a(A^g) = \mathcal{D}^{ab}(g) \, F_{\mu\nu}^b \qquad (10.18)$$

where $\mathcal{D}^{ab}(g)$ is the adjoint representation of g. We choose the normalization of the t^a-matrices to be given by $\mathrm{Tr}(t^a t^b) = \frac{1}{2} \delta^{ab}$. The adjoint representation of g can then be written as $\mathcal{D}^{ab}(g) = 2\mathrm{Tr}(t^a g t^b g^{-1})$.

One choice of a Lagrangian for A_μ is

$$\mathcal{L} = -\frac{1}{4e^2} F_{\mu\nu}^a F^{a\mu\nu} = \frac{1}{2e^2} \mathrm{Tr} \left(F_{\mu\nu} F^{\mu\nu} \right) \qquad (10.19)$$

From the covariant transformation law of $F_{\mu\nu}$, viz., $F_{\mu\nu}(A^g) = g F_{\mu\nu} g^{-1}$, it is clear that $\mathcal{L}$ is gauge-invariant. The Lagrangian (10.19) is called the Yang-Mills Lagrangian. It is the generalization of the Maxwell Lagrangian for the electromagnetic field. It is the simplest generalization for a general Lie group, but there are other Lagrangians possible, at least in special cases. For example, in three spacetime dimensions another possibility is the so-called Chern-Simons term

$$\mathcal{L} = -\frac{\kappa}{4\pi} \mathrm{Tr} \left(A_\mu \partial_\nu A_\alpha + \frac{2}{3} A_\mu A_\nu A_\alpha \right) \epsilon^{\mu\nu\alpha} \qquad (10.20)$$

where $\epsilon^{\mu\nu\alpha}$ is the Levi-Civita symbol in three dimensions. (This Lagrangian is actually not invariant under all gauge transformations. It changes by a total derivative under infinitesimal transformations; the action is invariant. There are also the so-called homotopically nontrivial transformations under which the action is not invariant, but e^{iS} (which is what is important in the quantum theory) is invariant for integral values of κ. Thus for these cases, one has a well-defined quantum theory.) At this stage we shall concentrate on the Yang-Mills type theories.

The equations of motion for a Yang-Mills field A_μ with coupling to matter fields is easily obtained from (10.19). $\delta F^a_{\mu\nu} = (D_\mu \delta A_\nu - D_\nu \delta A_\mu)^a$ which gives $\delta \mathcal{L} = (1/e^2) F^a_{\mu\nu}(D_\mu \delta A_\nu)$. The equations of motion are thus

$$(D_\mu F^\mu_\nu)^a + e^2 \frac{\delta \mathcal{S}_m}{\delta A^a_\nu} = 0 \tag{10.21}$$

where $\mathcal{S}_m$ is the matter part of the action, i.e., terms in the action other than the Yang-Mills term. Analogous to the identity $\partial_\alpha F_{\mu\nu} + cyclic = 0$ for electrodynamics, we have the Bianchi identity

$$D_\alpha F_{\mu\nu} + D_\nu F_{\alpha\mu} + D_\mu F_{\nu\alpha} = 0 \tag{10.22}$$

The Jacobi identity for commutators, viz., $[D_\mu, [D_\nu, D_\alpha]] + [D_\nu, [D_\alpha, D_\mu]] + [D_\alpha, [D_\mu, D_\nu]] = 0$, gives the above result directly.

The importance of the gauge principle is that it offers a uniform way to couple A_μ's to matter of different charges. We have already seen this in the case of electrodynamics. Thus in QED, for fields corresponding to particles of charge n, we have

$$D_\mu \psi = (\partial_\mu - in A_\mu)\psi$$
$$\psi' = e^{in\theta} \psi \tag{10.23}$$

For a nonabelian Lie group G, the analog of different charges would be different representations. Thus for a general representation R we have

$$D_\mu \psi = (\partial_\mu - it^a_R A^a_\mu)\psi$$
$$\psi' = \mathcal{D}_R(g) \psi \tag{10.24}$$

The A^a_μ are always the same, $dim G$ in number. t^a_R are the matrix representatives of the generators of the Lie algebra of G in the R-representation. There is only one coupling constant e for each simple group. For example, if we consider the group $SU(2)$ with matter fields in the representations of dimensions 2 and 3 (namely, for j-values $\frac{1}{2}$ and 1), we have

$$\psi_\alpha, \quad \alpha = 1, 2, \quad (t^a)_{\alpha\beta} = \left(\frac{\sigma^a}{2}\right)_{\alpha\beta}, \quad (D_\mu \psi)_\alpha = \partial_\mu \psi_\alpha - i\left(\frac{\sigma^a}{2} A^a_\mu \psi\right)_\alpha$$

$$\phi_b, \quad b = 1, 2, 3, \quad (t^a)_{bc} = -i\epsilon^a_{bc}, \quad (D_\mu \phi)^a = \partial_\mu \phi^a + \epsilon^{abc} A^b_\mu \phi^c$$

$$\tag{10.25}$$

10.2 Parallel transport

The matter fields ψ, ϕ, etc., can be considered as the components of a $dimR$-vector with respect to some chosen basis in a $dimR$-dimensional linear vector space V. In other words, we can write

$$\psi = \psi_i e_i \quad \in V_x \qquad (10.26)$$

where e_i are a set of basis vectors for V at the point x^μ. The gauge transformation $\psi'_i = g_{ij}\psi_j$ is equivalent to choosing a different basis for V. Since g depends on x^μ, the gauge invariance of the theory reflects the fact that physics is independent of our choice of basis (frame) at each point labeled by x^μ. We can choose frames independently at each spacetime point, which is, arguably, a sensible requirement even on a priori grounds. After all, it would be surprising if the *ad hoc* or conventional choice of a frame in this field space were to affect physical results. The fact that physics is independent of the local choice of frames is analogous to the case of the general theory of relativity, except that in the latter case V_x is not an arbitrary vector space, but the tangent space at x^μ to the spacetime manifold.

The field strength $F_{\mu\nu}$ is the gauge theory analog of the Riemann curvature $R^{ab}_{\mu\nu}$ and can also be understood as the angular deficit for parallel transport around small loops on the spacetime $\mathcal{M}$, with the qualification that the angular rotation is a frame rotation in the internal field space of the ψ's.

One can analyze parallel transport by considering covariantly constant ψ's. Consider

$$D_\mu\psi = (\partial_\mu + A_\mu)\psi = 0 \qquad (10.27)$$

In this equation, ψ is a column vector on which the matrices t^a in A_μ can act by matrix multiplication. Introduce a $U(x)$ defined by

$$(\partial_\mu + A_\mu)U(x) = 0 \qquad (10.28)$$

Such a U does not exist in general as a well-defined function on $\mathcal{M}$, but one can integrate the above equation along curves C from, say, y^μ to x^μ to get a path-dependent $U(x, y, C)$. In other words, we are solving the equation $C^\mu(\partial_\mu + A_\mu)U(x) = 0$, where C^μ is tangent to the curve C.

$$U(x, y, C) = [1 - A_{\mu_1}(x - \epsilon)\epsilon^{\mu_1}][1 - A_{\mu_2}(x - 2\epsilon)\epsilon^{\mu_2}] \cdots$$
$$\cdots [1 - A_{\mu_N}(y)\epsilon^{\mu_N}]$$
$$= P \, \exp\left(-\int_y^x A_\mu dx^\mu\right)$$
$$= P \, \exp\left(i\int_y^x t^a A^a_\mu dx^\mu\right) \qquad (10.29)$$

where the symbol P denotes the ordering of the matrices A_μ along the path as indicated by the first of the equalities above. We divide the path into N

intervals of small displacements ϵ^μ and write an ordered product of factors like $[1 - A_\mu(x - \epsilon)\epsilon^\mu]$ and eventually take $\epsilon \to 0$, $N \to \infty$, keeping the total displacement finite and the order of multiplications unchanged. It is then easy to check, from the first of the above equations, that $U(x, y, C)$ does indeed solve (10.28). Given U, $D_\mu \psi = 0$ is solved by

$$\psi = U \; \psi_0 \tag{10.30}$$

with $\psi_0 = contant$. Thus $U(x, y, C)$ tells us how frames rotate as we transport along C from y to x.

The set of equations $D_\mu U = 0$ (corresponding to different values of μ) have, as the integrability condition, $F_{\mu\nu} = 0$. Thus $U(x, y, C)$ would be independent of the curve C and give a well-defined function on M if and only if $F_{\mu\nu} = 0$. (If there are noncontractible loops on M, it is possible to get more general solutions. $U(x, y)$ defined by integration along an open path C need not be the same as what is obtained by integration along C with an added circuit around a noncontractible loop, since the two paths cannot be deformed into each other. For simply connected spacetimes, $F_{\mu\nu} = 0$ will give a path-independent function $U(x)$ as the result of the integration.) If $F_{\mu\nu} = 0$, we can write, for simply connected spaces,

$$A_\mu = -(\partial_\mu U) \; U^{-1} \tag{10.31}$$

If $F_{\mu\nu} \neq 0$, we see that, for a small closed loop, we have

$$U = P \; \exp\left(- \oint_C A_\mu dx^\mu\right) \approx 1 - (F_{\mu\nu}\sigma^{\mu\nu}) \tag{10.32}$$

where $\sigma^{\mu\nu}$ is the area element of the infinitesimal surface whose boundary is the curve C. (The above result is essentially Stokes' theorem. This theorem does not work, as it is, for larger loops, because of the matrix nature of the A_μ's which means that A's at different points do not necessarily commute with each other. For small loops, the above result can, however, be seen by direct expansion of the path-ordered exponential.) The above result (10.32), along with (10.30), shows that $F_{\mu\nu}$ measures the "angular deficit" for parallel transport around small loops.

Using the equation $D_\mu U = 0$, one can check that, under a gauge transformation of the potentials A_μ, we have

$$U(x, y, C, A_\mu^g) = g(x) \; U(x, y, C, A_\mu) \; g^{-1}(y) \tag{10.33}$$

For a closed loop, where x and y coincide, we have a similarity transformation of U by $g(x)$ and hence the trace is invariant. The holonomy operator or the Wilson loop operator $W(C)$ is defined by

$$W(C) = \text{Tr}\left[P \; \exp\left(- \oint_C A_\mu dx^\mu\right)\right]$$

$$= \text{Tr}\left[P \; \exp\left(i \oint_C t^a A_\mu^a dx^\mu\right)\right] \tag{10.34}$$

10.3 Charges and gauge transformations

The Lagrangian for a gauge theory is invariant under the gauge transformations

$$A_\mu \to A_\mu^g = gA_\mu g^{-1} - (\partial_\mu g)g^{-1}, \qquad \psi \to \psi^g = g\psi \qquad (10.35)$$

Consider first the constant transformations for which $g(x)$ is independent of
x^μ. For these cases, the gauge symmetry is just like any global symmetry.

$$A_\mu \to A_\mu^g = gA_\mu g^{-1}, \qquad \psi \to \psi^g = g\psi \qquad (10.36)$$

Thus we must expect a Noether current and associated charges. We can find
the current by considering the variation of the Lagrangian as discussed in
Chapter 3. As an example, we shall take the matter field to be fermions
coupled to A_μ. We then find

$$
\begin{aligned}
\delta\mathcal{L} &= -\frac{1}{2e^2}F^{a\ \mu\nu}\delta F_{\mu\nu}^a + i\bar{\psi}\gamma^\mu\partial_\mu\delta\psi + \cdots \\
&= \partial_\mu\left[-\frac{1}{e^2}(F^{a\ \mu\nu}\delta A_\nu^a) + i\bar{\psi}\gamma^\mu\delta\psi\right] + \text{terms proportional to}
\end{aligned}
$$

$$\text{equations of motion}$$

$$(10.37)$$

Thus the Noether current $J^{a\ \mu}$ is given by

$$J^{a\mu}\theta^a = -\frac{1}{e^2}(F^{a\mu\nu}\delta A_\nu^a) + i\bar{\psi}\gamma^\mu\delta\psi \qquad (10.38)$$

where θ^a are the (infinitesimal) parameters of the transformation. From
(10.36), $\delta A_\mu^a = -f^{abc}A_\mu^b\theta^c$, $\delta\psi = -it^c\theta^c\psi$. The current is thus identified
as

$$J^{a\mu} = \frac{1}{e^2}f^{abc}F^{b\mu\nu}A_\nu^c + \bar{\psi}\gamma^\mu t^a\psi \qquad (10.39)$$

This current is easily checked to be conserved by the equations of motion.
The equations of motion for the gauge field are

$$\frac{1}{e^2}(D_\mu F^{\mu\nu})^a = -\bar{\psi}\gamma^\nu t^a\psi \qquad (10.40)$$

Using this equation and the definition of the covariant derivative, we find

$$J^{a\mu} = \partial_\nu\left[\frac{F^{a\mu\nu}}{e^2}\right] \qquad (10.41)$$

The corresponding charges can thus be written as

$$Q^a = \int d^3x\, J^{a0} = \frac{1}{e^2}\oint_{|\boldsymbol{x}|\to\infty} F^{a0i}dS_i \qquad (10.42)$$

The charges are given by two-surface integrals. This is a general feature of all gauge theories. (For example, in the general theory of relativity, for asymptotically flat spacetimes, the charges corresponding to coordinate transformations, viz., momentum and angular momentum, are defined by surface integrals at spatial infinity. Generically, this has to do with the fact that charge densities cannot be defined in a gauge-invariant manner; compare with equation (10.39), for example. Also notice that equation (10.39) shows that there can be nonzero charge even when the matter fields are zero; the gauge fields themselves are charged in the nonabelian theory.)

Since the constant transformations (10.36) act like a Noether symmetry, we expect that the wave functions or states transform as representations of the corresponding symmetry group, i.e.,

$$|\Psi^g\rangle = e^{iQ^a\theta^a}\,|\Psi\rangle \tag{10.43}$$

This result can be shown as follows. The general variation of the Lagrangian is given by (10.37). From the surface term in the action, we can identify the canonical one-form as

$$
\begin{aligned}
\Theta &= \int d^3x \, \left[-\frac{1}{e^2}F^{a0i}\delta A_i^a \, + \, i\bar{\psi}\gamma^0\delta\psi \right] \\
&= \int d^3x \, \left[\frac{1}{e^2}F_{0i}^a\delta A_i^a \, + \, i\bar{\psi}\gamma^0\delta\psi \right]
\end{aligned}
\tag{10.44}
$$

This canonical one-form shows one of the difficulties of quantizing the gauge theory; there is no canonical momentum for A_0^a. However, as we have seen in the case of electrodynamics, we can set $A_0^a = 0$ by a choice of gauge. For the remaining components, the canonical one-form (10.44) leads to the commutation rules

$$
\begin{aligned}
&[A_i^a(\boldsymbol{x},t), A_j^b(\boldsymbol{y},t)] = 0 \\
&[A_i^a(\boldsymbol{x},t), \tfrac{1}{e^2}F_{0j}^b(\boldsymbol{y},t)] = i\delta^{ab}\delta_{ij}\delta^{(3)}(x-y) \\
&[F_{0i}^a(\boldsymbol{x},t), F_{0j}^b(\boldsymbol{y},t)] = 0
\end{aligned}
\tag{10.45}
$$

The fermion fields have the usual anticommutation rules.

As in the case of electrodynamics, the Gauss law requires special treatment. It is part of the Lagrangian equations of motion. Since it does not involve any time-derivatives, in a Hamiltonian formalism, it cannot be obtained as an equation of motion, but must be imposed as a condition selecting allowed initial data in the classical analysis. In our quantum theoretic treatment of electrodynamics, we solved the Gauss law, eliminating the longitudinal component of A_i which led to a set of unconstrained fields and then used the commutation rules for these unconstrained fields. One can carry through an analogous reduction here, but an alternate approach is to impose the Gauss law as a condition on the physical states. Recall that if we eliminate the longitudinal part of A_i, it is given in terms of the charge density which involves ψ and $\psi^\dagger$; as a result, those commutation rules in which the longitudinal part of A_i appears will involve the fermion fields. The alternate approach of imposing the Gauss law on states has the advantage that we can have the commutation rules (10.45) and so the operator structure is simpler. However, there are states in the Hilbert space which are generated by the action of longitudinal components of A_i^a on the vaccum. These are clearly unphysical states since they are absent if we solve the Gauss law to eliminate the longitudinal components. Physical states of the theory in this approach are then defined as those states which obey the condition

$$G^a \, |\Psi\rangle = 0 \qquad (10.46)$$

In other words, we do not require $G^a = 0$ on all states (or as an operator equality). It is a condition selecting physical states in the Hilbert space. This condition, (10.46), will still ensure that the Gauss law is obtained for matrix elements with physical states, in particular, for expectation values of gauge-invariant operators for physical states. This is sufficient for observable results of the theory to be consistent with the Gauss law.

For the nonabelian theory, the Gauss law is identified from the time-component of the equations of motion as

$$G(\theta) = \int d^3x \; \theta^a \left[\frac{1}{e^2}(D_i F^{i0})^a + \psi^\dagger t^a \psi \right] \qquad (10.47)$$

Consider now the transformation

$$\delta A_i^a \quad = \quad -(D_i \theta)^a$$
$$\delta\psi = -it^a \theta^a \psi, \qquad \delta\psi^\dagger = \psi^\dagger it^a \theta^a \qquad (10.48)$$

which is an infinitesimal gauge transformation, but we do not necessarily require that the parameters vanish at spatial infinity; instead, they could be nonzero but constant (independent of angular directions) at spatial infinity. The canonical commutation rules show that the generator of this transformation is

$$\tilde{G}(\theta) = \int d^3x \; \left[\frac{1}{e^2} F^{a0i}(D_i\theta)^a \; + \; \theta^a \psi^\dagger t^a \psi \right] \qquad (10.49)$$

(This gives $i\delta A_i^a = [A_i^a, \tilde{G}(\theta)]$, etc.) Notice that the operator (10.49) differs from the Gauss law (10.47) by a surface term. The variation of a state when the transformation (10.48) is carried out is given by

$$\delta|\Psi\rangle = i\tilde{G}(\theta) \, |\Psi\rangle \qquad (10.50)$$

Consider now states which obey the Gauss law condition (10.46), i.e., physical states. On these, for transformations for which the θ^a go to constants not equal to zero at spatial infinity, we can carry out a partial integration to obtain

$$\delta|\Psi\rangle = \frac{i}{e^2} \oint_{|\boldsymbol{x}|\to\infty} \theta^a F^{a0i} dS_i \, |\Psi\rangle$$

$$= iQ^a \theta^a \, |\Psi\rangle \tag{10.51}$$

This shows that the constant transformations do act as a Noether symmetry on the physical states with the charge operator as given by (10.42). For transformations for which θ^a go to zero at spatial infinity, there is no surface term and $G(\theta)$ and $\tilde{G}(\theta)$ coincide. The requirement (10.46) shows that physical states are invariant under such transformations.

We can now summarize the basic result of the analysis given above.

1. For transformations $g(x)$ which go to the identity 1 at spatial infinity, i.e., for $\theta^a \to 0$, the physical states $|\Psi\rangle$ are *invariant*.
2. For transformations $g(x)$ which go to a constant element g_∞ which is not the identity, or for constant g over all space,

$$\delta|\Psi\rangle = iQ^a \theta^a \, |\Psi\rangle \tag{10.52}$$

This result can also be restated as follows. Define

$$\mathcal{G}_* = \left\{ \text{set of all } g(x) \text{ such that } g(x) \to 1 \text{ as } |\boldsymbol{x}| \to \infty \right\}$$

$$\mathcal{G} = \Big\{ \text{set of all } g(x) \text{ such that } g(x) \to \text{constant element of } G,$$

$$\text{not necessarily 1, as } |\boldsymbol{x}| \to \infty \Big\}$$

The results we have obtained above then amount to saying that the physical states are *invariant* (and not just covariant) under $\mathcal{G}_*$. Thus $\mathcal{G}_*$ is the "true gauge symmetry" of the theory in the sense that its elements represent unphysical, and hence redundant, variables in the theory. Since the elements of $\mathcal{G}$ go to a constant, not necessarily the identity, at spatial infinity, we have

$$\mathcal{G}/\mathcal{G}_* \sim \text{set of constant } g\text{'s} \sim G$$

$\mathcal{G}/\mathcal{G}_* \sim G$ is the Noether symmetry of the theory defined by the charges.

10.4 Functional quantization of gauge theories

In Chapter 8, we showed that for scalar fields and fermions, we had a simple prescription to obtain the generating functional as a functional integral, viz., given the classical Euclidean action $\mathcal{S}_E$ one could obtain $Z[J, \bar{\eta}, \eta]$ as

$$Z_E[J, \bar{\eta}, \eta] = \mathcal{N} \int d\mu \, (\varphi, \psi, \bar{\psi}) \exp\left[-\mathcal{S}_E(\varphi, \bar{\psi}, \psi) + \int d^4x \, (J\varphi + \bar{\eta}\psi + \bar{\psi}\eta) \right] \tag{10.53}$$

where $d\mu$ is an integration measure on the space of configurations, that is, on

$$\mathcal{C} = \left\{ \phi(x), \psi(x), \bar{\psi}(x) \;\middle|\; \phi(x) : \mathcal{M} \to \mathbf{R}, \, \psi(x), \bar{\psi}(x) : \mathcal{M} \to \mathbf{N} \right\}$$

where $\mathbf{R}$ denotes real numbers and $\mathbf{N}$ denotes Grassmann numbers. (The specific measure we used for defining $d\mu$ was the Euclidean metric on $\mathcal{C}$.)

The functional integral for QED was also obtained in Chapter 8, but it did not use the classical Maxwell action, but rather a modified version of it, which had to do with gauge-fixing. We shall now show that the prescription used for the scalar and fermion fields will apply to gauge fields as well, with the appropriate definition of the configuration space and its volume measure.

We start by showing that the naive definition of the integration over all A_μ does not work. Let us start with the Euclidean Maxwell action

$$S = \frac{1}{4} \int F_{\mu\nu} F^{\mu\nu} = \frac{1}{2} \int A^\mu \left(-\Box \, \delta_{\mu\nu} + \partial_\mu \partial_\nu \right) A^\nu$$
$$= \frac{1}{2} \int_{x,y} A^\mu(x) M_{\mu\nu}(x,y) A^\nu(y) \tag{10.54}$$

where

$$M_{\mu\nu}(x,y) = \left(-\Box \, \delta_{\mu\nu} + \partial_\mu \partial_\nu \right) \delta^{(4)}(x-y) \tag{10.55}$$

If we had a scalar field, we would write

$$Z = \int [d\varphi] \exp\left(-\frac{1}{2} \int \varphi M \varphi + \int J\varphi \right)$$
$$= \left(\det \frac{M}{2\pi} \right)^{-\frac{1}{2}} \exp\left(\frac{1}{2} \int J M^{-1} J \right) \tag{10.56}$$

M^{-1} would then be the propagator. In the case of the Maxwell action though, we cannot do this since $M_{\mu\nu}(x,y)$ has zero modes, i.e., $\det M = 0$ and M^{-1} does not exist. The existence of zero modes is due to gauge invariance. Under a gauge transformation, $A_\mu \to A_\mu + \partial_\mu \theta$. For a mode of the form $\phi_\mu = \partial_\mu \theta$ we find

$$\int_y M_{\mu\nu}(x,y) \phi^\nu(y) = \left(-\Box \, \phi_\mu + \partial_\mu \partial \cdot \phi \right) = 0 \tag{10.57}$$

In other words, $\phi_\mu = \partial_\mu \theta$ is an eigenvector of $M_{\mu\nu}(x,y)$ with eigenvalue zero. We must remove such gauge or unphysical degrees of freedom from the A_μ's to define a proper functional integral.

In order to understand this better we must define the configuration space more carefully. We shall discuss gauge theories in general from now on, since the formalism is essentially the same for QED and other gauge theories. First we define the function space $\mathcal{A}$, which is the space of all gauge potentials. This will be the set of all four-vector-valued functions $A_\mu(x)$ which are also elements of the Lie algebra. Existence of certain integrals like $\int F^2$ will be assumed. When we write $[dA]$ in a functional integral, we are writing a measure on $\mathcal{A}$ defined by $\prod_{x,\mu,a} dA_\mu^a(x)$. This corresponds to the volume defined by a Euclidean distance function or metric on $\mathcal{A}$ given by

$$\| \, \delta A \, \|^2 = \int d^4x \, \delta A_\mu^a(x) \delta A_\mu^a(x) \tag{10.58}$$

The space $\mathcal{A}$ has the property that any two points $A_{1\mu}(x)$ and $A_{2\mu}(x)$ can be connected by a straight line, which is in fact given by

$$A_\mu(\tau, x) = \tau \, A_{1\mu}(x) + (1 - \tau) \, A_{2\mu}(x) \tag{10.59}$$

for $0 \leq \tau \leq 1$. Notice that $A_\mu(\tau, x)$ transforms, for all τ, as a gauge potential should, viz., $A_\mu^g(\tau, x) = \tau \, A_{1\mu}^g(x) + (1 - \tau) \, A_{2\mu}^g(x)$ for $g = e^{-it^a\theta^a}$. $\mathcal{A}$ is thus an affine space; i.e., we can write any configuration A_μ as $A_\mu(x) = A_\mu^{(0)} + \xi_\mu$, where $A_\mu^{(0)}$ is a fixed potential and $\xi_\mu(x)$ is a Lie-algebra-valued vector field.

In the previous section we have defined the set of gauge transformations $\mathcal{G}_*$ and showed that not only the action, but the wave functions are invariant under $\mathcal{G}_*$. The space $\mathcal{G}_*$ was defined in terms of gauge transformations at a fixed time as is appropriate for the discussion of states and operators acting on them. For the functional integral, we can generalize these ideas to a four-dimensional setting. We will use the same notation, $\mathcal{G}_*$, to denote the set of gauge transformations which go to the identity at large values of the Euclidean radius, i.e., as $\sqrt{x^\mu x^\mu} \to \infty$. Physical configurations over which we must integrate are defined on $\mathcal{A}/\mathcal{G}_*$. Any two potentials which differ by a gauge transformation, that is, by an element of $\mathcal{G}_*$, are the same physically and correspond to the same point in $\mathcal{A}/\mathcal{G}_*$. In other words, the true space of physical configurations on $\mathbf{R}^4$ is not $\mathcal{A}$ but

$$\mathcal{C} = \mathcal{A}/\mathcal{G}_* \tag{10.60}$$

Given the configuration space $\mathcal{C}$ of (10.60), we can say that the correct functional integral for a gauge theory is given by

$$Z = \int d\mu \, (\mathcal{A}/\mathcal{G}_*) \ e^{-\mathcal{S}_E + \int J^{a\mu} A_\mu^a} \tag{10.61}$$

where $\mathcal{S}_E$ is the classical Euclidean action and $d\mu(\mathcal{A}/\mathcal{G}_*)$ is to be obtained from the metric (10.58) by factoring out the action of gauge transformations.

While (10.61) is indeed the correct prescription, rarely can one evaluate the measure $d\mu(\mathcal{A}/\mathcal{G}_*)$ exactly. (This can be done for gauge fields in two dimensions.) One way to obtain the measure is the following. (There are many caveats which must be stated regarding the discussion which follows, having to do with global properties of various spaces inolved. We shall ignore them for the moment, they will be discussed separately.) Consider $A_\mu(x)$, which is a point in $\mathcal{A}$. Under a gauge transformation $A_\mu \to A_\mu^g = gA_\mu g^{-1} - \partial_\mu gg^{-1}$, which corresponds to a different point in $\mathcal{A}$. If we consider a sequence of transformations starting from $g = 1$, we see that gauge transformations generate a flow in $\mathcal{A}$. In order to get rid of gauge degrees of freedom, we can choose a surface Σ which cuts these flow lines transversally. (A "good" surface Σ should cut each flow line once and only once.) We can then pick the points on Σ as representatives of $\mathcal{A}/\mathcal{G}_*$. The choice of Σ is specified by a condition on the potentials called a gauge-fixing condition. For example, all A's which obey $\partial_\mu A^{a\mu} = 0$ lie on some surface $\Sigma_{\partial \cdot A}$, all A's which obey $f^a(A) = 0$ lie on Σ_f, etc. If the integrand is gauge-invariant, as is the case for the classical action, then the choice of Σ does not matter, since one can get from one Σ to another by a gauge transformation.

We now want to write $d\mu(\mathcal{A}/\mathcal{G}_*)$. By definition of $\mathcal{A}/\mathcal{G}_*$, we have, at least locally in this space, $\mathcal{A} \sim (\mathcal{A}/\mathcal{G}_*) \times \mathcal{G}_*$. Thus

$$[dA] = d\mu\,(\mathcal{A}/\mathcal{G}_*)\ d\mu(\mathcal{G}_*) \tag{10.62}$$

For gauge transformations which are close to the identity, namely, $g(x) \approx 1 - it^a\theta^a(x)$, we can write $d\mu(\mathcal{G}_*) \approx [d\theta^a(x)]$. Let the gauge-fixing condition be

$$f^a(A) - h^a(x) = 0 \tag{10.63}$$

where $h^a(x)$ has no A-dependence. Multiplying both sides by a common factor, we can then write (10.62) as

$$[dA]\,\left|\det\left[\frac{\delta f^a(x)}{\delta\theta^b(y)}\right]\right| = d\mu(\mathcal{A}/\mathcal{G}_*)\,[d\theta]\,\left|\det\left[\frac{\delta f^a(x)}{\delta\theta^b(y)}\right]\right| \tag{10.64}$$

The determinant of

$$\Delta_{FP}^{ab}(x,y) = \delta f^a(x)/\delta\theta^b(y) \tag{10.65}$$

is known as the Faddev-Popov determinant. We now multiply (10.64) by $\exp\left(-\mathcal{S}_E(A) + \int J^{a\mu}A_\mu^a\right)\delta[f(A) - h]$, where the δ-function is a functional δ-function, and integrate. We obtain

$$\int [dA]\,\delta[f(A) - h]\,|\det(\Delta_{FP})|\exp\left(-\mathcal{S}_E(A) + \int J^{a\mu}A_\mu^a\right)$$

$$= \int d\mu(\mathcal{A}/\mathcal{G}_*)\exp\left(-\mathcal{S}_E(A) + \int J^{a\mu}A_\mu^a\right)$$

$$\times \{[d\theta]\,|\det(\Delta_{FP})|\,\delta[f - h]\}$$

$$= \int d\mu(\mathcal{A}/\mathcal{G}_*)\exp\left(-\mathcal{S}_E(A) + \int J^{a\mu}A_\mu^a\right) \tag{10.66}$$

where we have used the identity

$$\int [d\theta]\,|\det(\Delta_{FP})|\,\delta[f - h] = \int [df]\,\delta[f - h] = 1 \tag{10.67}$$

for integration along the gauge directions. The right-hand side of the (10.66) is our definition (10.61) of the functional integral over the proper configuration space. We thus have

$$Z_h[J] = \int [dA] \; \delta[f(A) - h] \; |\det(\Delta_{FP})| \exp\left(-S_E(A) + \int J^{a\mu} A_\mu^a\right) \quad (10.68)$$

This gives a formula for $Z_h[J]$ written conveniently in terms of the Euclidean measure $[dA]$; the redundancy is factored out by use of the δ-function. Strictly speaking, $Z_h[J]$ depends on the gauge-fixing condition, as indicated by the subscript. Writing the measure $d\mu(\mathcal{A}/\mathcal{G}_*)$ in terms of A's on the $f - h = 0$ surface is correct only if the integrand is gauge-invariant. The $J^{a\mu} A_\mu^a$-term is not and thus $Z_h[J]$ will depend on the gauge-fixing condition. Of course, $Z[0]$ is gauge-invariant. The implication is that the Green's functions defined by $Z_h[J]$ will be gauge-dependent in general. However, the S-matrix elements are not and so we can simplify (10.68) even further as follows. Since h is arbitrary, we can define the S-matrix using Z_h , $Z_{h'}$ or $\int[dh]F(h)Z_h$; these are all equivalent. We can thus define, for the purposes of computing the S-matrix, another Z by choosing $F(h) = \exp(-\frac{1}{2e^2\alpha} \int h^2)$, where α is just a real number.

$$Z[J] = \int [dA] \; \delta[f(A) - h] \; |\det(\Delta_{FP})| \exp\left(-S_E(A) + \int J^{a\mu} A_\mu^a\right)$$
$$\times [dh] \exp\left(-\frac{1}{2e^2\alpha} \int_x h^2(x)\right)$$
$$= \int [dA] \; |\det(\Delta_{FP})| \exp\left(-S_E(A) - \frac{1}{2e^2\alpha} \int f^2(A) + \int J^{a\mu} A_\mu^a\right)$$
$$(10.69)$$

The δ-function has been removed in favor of a term f^2 which can be considered as an extra term in the action. We can now make one more improvement which is helpful in perturbative calculations. From our discussions of Grassmann integration

$$\int [dQ d\bar{Q}] \; e^{\int \bar{Q} M Q} = \det M \quad (10.70)$$

for independent Grassmann variables $Q(x)$, $\bar{Q}(x)$. We may thus write

$$|\det(\Delta_{FP})| = \int [dc d\bar{c}] \exp\left(-\int_{x,y} \bar{c}^a(x)\Delta_{FP}^{ab}(x,y)c^b(y)\right) \quad (10.71)$$

We have taken the determinant to be positive, which it is in the region around the classical vacuum $A_\mu = 0$. (Eventually the formula we obtain will not be applicable as it is for nonperturbative effects anyway, so that consideration of the region around $A_\mu = 0$ is not too restrictive.) This provides a convenient way of writing the determinant as part of the action with additional

(Lie algebra-valued) fields c^a and $\bar{c}^a$, which is useful in writing down the rules for Feynman diagrams in perturbation theory. The fields c^a and $\bar{c}^a$ are Grassmann-valued, yet they are not spinors, but scalars. Thus they have the wrong spin-statistics connection. They are often referred to as the Faddeev-Popov ghosts. The wrong spin-statistics relation for the ghosts is not a problem, because they do not appear in the external lines or as asymptotic states in the theory; they are merely a device to write the determinant in a convenient way. Using (10.71) in (10.69), we finally get

$$Z[J] = \int [dAdcd\bar{c}] \ \exp\left(-\mathcal{S}_q + \int J^{a\mu} A^a_\mu\right) \qquad (10.72)$$

where

$$\mathcal{S}_q = \mathcal{S}_E(A) + \int d^4x \left[\frac{1}{2e^2\alpha}f^a(A)f^a(A) + \bar{c}^a(x)\left(\frac{\delta f^a(x)}{\delta\theta^b(y)}\right)c^b(y)\right] \qquad (10.73)$$

The final prescription for the functional integral in a gauge theory is then quite simple. We have to choose a gauge-fixing function $f^a(A)$ (which, by definition, cannot be gauge-invariant) and then construct $\mathcal{S}_q$ by adding the f^2-term to the classical action and also adding the Faddeev-Popov ghost term. Integration of $e^{-\mathcal{S}_q}$ with the standard Euclidean measure for A_μ, c and $\bar{c}$ will then give the functional integral for $Z[J]$.

We must now ask the question: what is a "good" gauge-fixing? As we remarked earlier, a good gauge-fixing must produce a surface which intersects each gauge flow line transversally once and only once. The existence of such a surface depends on the global properties of $\mathcal{A}/\mathcal{G}_*$. If we choose a surface which is not transversal, then there is a direction of gauge variation which would not change f^a. This means that the matrix $\Delta^{ab}_{FP}(x, y)$ has zero modes and that the choice of f^a does not fix the gauge completely. This can be taken care of by choosing a different function, at least locally in $\mathcal{A}$ around the point $A_\mu = 0$, such that $\det \Delta_{FP} \neq 0$.

A more involved problem has to do with the fact that the gauge-fixing surface may intersect flow lines more than once. In this case, there are gauge-equivalent configurations on the gauge-fixing surface or there are nontrivial solutions $g(x)$ to the condition

$$f^a(A^g) - h = 0 \qquad (10.74)$$

(In this case, A and A^g lie on the same gauge-fixing surface.) In other words, there does not exist, globally over the configuration space, a surface which intersects the gauge flow lines once and only once. Integrating over all points on the chosen surface leads to overcounting of the degrees of freedom. This problem is known as the Gribov ambiguity. Since the gauge-flow line has to, roughly speaking, turn around to come back and intersect the gauge-fixing surface, we see that $\delta f/\delta\theta$ must vanish for some A along the flow line; the

Faddeev-Popov operator Δ_{FP} once again has a zero mode at that point. The Gribov ambiguity is easily avoided for an Abelian gauge theory such as QED. However, it is unavoidable for all smooth gauge-fixings in a nonabelian gauge theory, due to the global properties of $\mathcal{A}$ and $\mathcal{A}/\mathcal{G}_*$. (This will be discussed in some more detail in Chapter 14.) If we consider a gauge-fixing condition like $\partial_\mu A^{a\mu} = 0$, we see that Δ_{FP} is positive at $A_\mu = 0$ and so continues to be positive for a range of A's which is perturbatively accessible around $A_\mu = 0$. The Gribov problem does not affect perturbative calculations with our functional integral (10.72, 10.73).

10.5 Examples

1. *Electrodynamics with $f(A) = \partial_\mu A^\mu$*

If we choose $f(A) = \partial_\mu A^\mu$, $\Delta_{FP} = -\Box_x \delta^{(4)}(x - y)$. We then find

$$S_q = \int \left[\frac{1}{4e^2} F_{\mu\nu} F^{\mu\nu} + \frac{1}{2e^2\alpha} (\partial \cdot A)^2 + \bar{c}(-\Box)c \right] \tag{10.75}$$

For the choice $\alpha = 1$ (and scaling $A \to eA$ and adding on the fermion terms) we get our earlier expression for the QED functional integral (8.57, 8.58). The $c, \bar{c}$-dependent term just gives a constant multiplicative factor $\det(-\Box)$ which can be absorbed into the normalization constant. The limit $\alpha \to 0$ is known as the Landau gauge.

This derivation of the functional integral for QED justifies the use of the covariant propagator for the photon in calculating the S-matrix.

2. *Electrodynamics with $f(A) = \partial_\mu A^\mu + A_\mu A^\mu$*

A simple example of a nonlinear gauge in electrodynamics is given by $f(A) = \partial_\mu A^\mu + A_\mu A^\mu$. In this case, we find $\Delta_{FP} = -(\Box_x + 2A \cdot \partial)\delta^{(4)}(x - y)$, giving

$$S_q = \int \left[\frac{1}{4e^2} F_{\mu\nu} F^{\mu\nu} + \frac{1}{2e^2\alpha} (\partial \cdot A + A^2)^2 + \bar{c}(-\Box - 2A \cdot \partial)c \right] \tag{10.76}$$

The ghosts now interact with the photons and cannot be ignored.

3. *Nonabelian gauge theory with $f^a = \partial_\mu A^{a\mu}$*

In this case, with $f^a = \partial \cdot A^a$, we find $\Delta^{ab}_{FP}(x, y) = -(\partial_\mu D^\mu)^{ab} \delta^{(4)}(x - y)$ where D_μ is the covariant derivative (in the adjoint representation). S_q becomes

$$S_q = \int \left[\frac{1}{4e^2} F^a_{\mu\nu} F^{a\mu\nu} + \frac{1}{2e^2\alpha} (\partial \cdot A^a)^2 + \partial^\mu \bar{c}^a (\partial_\mu c^a + f^{abc} A^b_\mu c^c) \right] \tag{10.77}$$

Choosing $\alpha = 1$ and rescaling $A \to eA$,

$$S_q = S_0 + S_{int}$$

$$S_0 = \int \left[\frac{1}{2} \partial_\mu A_\nu^a \partial^\mu A^{a\nu} + \partial^\mu \bar{c}^a \partial_\mu c^a \right]$$

$$S_{int} = \int \left[e f^{abc} \partial_\mu A_\nu^a A^{b\mu} A^{c\nu} + \frac{e^2}{4} f^{abc} f^{ars} A_\mu^b A_\nu^c A^{r\mu} A^{s\nu} + e f^{abc} \partial_\mu \bar{c}^a A_\mu^b c^c \right]$$

$$(10.78)$$

In this case, even though the gauge-fixing condition is linear in the gauge fields, the ghosts are unavoidable since they interact with the gauge fields. The rules for Feynman diagrams can be read off from (10.78). In particular, the (Euclidean) propagators for the gauge field and the ghost field are given by

$$\langle A_\mu^a(x) A_\nu^b(y) \rangle = \delta_{\mu\nu} \delta^{ab} \int \frac{d^4 k}{(2\pi)^4} \frac{1}{k^2} e^{ik(x-y)}$$

$$\langle c^a(x) \bar{c}^b(y) \rangle = \delta^{ab} \int \frac{d^4 k}{(2\pi)^4} \frac{1}{k^2} e^{ik(x-y)} \qquad (10.79)$$

10.6 BRST symmetry and physical states

In this section, we shall discuss some aspects of the operator formulation of the quantum theory of gauge fields. The functional integral involves a modified action as given in (10.73). A canonical operator quantization of this modified action may be carried out to obtain the states of the theory. An elegant way to understand the physical states and the elimination of gauge degrees of freedom in this approach is in terms of the so-called BRST symmetry, named after Becchi, Rouet, Stora, and Tyutin. In order to display this, we go back a step and write the action (10.73) as

$$S_q = S_E(A) + \int \left[iB^a f^a + \frac{\alpha e^2}{2} B^a B^a + \bar{c}^a(x) \left(\frac{\delta f^a(x)}{\delta \theta^b(y)} \right) c^b(y) \right] \quad (10.80)$$

where $B^a(x)$ is an auxiliary field. If we eliminate it by its equation of motion, namely,

$$\alpha e^2 B^a = -i f^a \qquad (10.81)$$

or if we integrate over B^a in a functional integral with a standard Euclidean measure $[dB]$, we get back the action (10.73).

We now introduce the BRST transformation by

$$\mathcal{Q} A_\mu^a = (D_\mu c)^a$$

$$\mathcal{Q} c^a = -\frac{1}{2} f^{abc} c^b c^c$$

$$\mathcal{Q} \bar{c}^a = iB^a$$

$$\mathcal{Q} B^a = 0 \qquad (10.82)$$

The change in A_μ^a involves the ghost field; Q is Grassmann-valued and the parameters of the transformation must thus be considered as Grassmann variables. The transformations (10.82) are easy to write down. For the gauge field A_μ^a, we do an infinitesimal gauge transformation $A_\mu^a \rightarrow A_\mu^a - (D_\mu\theta)^a$ and replace the gauge parameter $-\theta^a$ by the ghost field c^a. This also means that gauge-invariant quantities are BRST-invariant. Since all observables in a gauge theory must be gauge-invariant, we see that all observables are given by BRST-invariant expressions. Once the action of Q on A_μ^a is specified, the action of Q on c^a, as given in (10.82), is determined by requiring $Q^2 = 0$. This can be easily checked as follows.

$$Q^2 A_\mu^a = \partial_\mu(Qc^a) + f^{abc}(\partial_\mu c^b)c^c + f^{abc}A_\mu^b(Qc^c)$$
$$+ f^{abc}f^{bkl}A_\mu^k c^l c^c \tag{10.83}$$

Choosing $Qc^a + \frac{1}{2}f^{abc}c^b c^c = 0$ and writing

$$f^{abc}(\partial_\mu c^b)c^c = \tfrac{1}{2}\partial_\mu(f^{abc}c^b c^c), \tag{10.84}$$

which follows from the Grassmann nature of the ghost fields, we get

$$Q^2 A_\mu^a = -\tfrac{1}{2}f^{abc}A_\mu^b f^{ckl}c^k c^l + f^{abc}f^{bkl}A_\mu^k c^l c^c$$
$$= -\tfrac{1}{2}\left[f^{abc}f^{ckl} - f^{cak}f^{cbl} - f^{cal}f^{ckb}\right]A_\mu^b c^k c^l$$
$$= 0 \tag{10.85}$$

where we have used the Jacobi identity on the structure constants, viz.,

$$f^{cab}f^{ckl} + f^{cal}f^{cbk} + f^{cak}f^{clb} = 0 \tag{10.86}$$

The Jacobi identity follows from the Lie algebra identity or matrix identity

$$[t^b, [t^k, t^l]] + [t^k, [t^l, t^b]] + [t^l, [t^b, t^k]] = 0 \tag{10.87}$$

upon using the commutation rules (10.10). One can also check, by using the Jacobi identity, that Q^2 is zero on the ghost field c^a as well, with the choice of the transformation rule $Qc^a + \frac{1}{2}f^{abc}c^b c^c = 0$. For the antighost field $\bar{c}^a$, the action of Q should produce a bosonic field; we can simply define this as iB^a. $Q B^a$ is then taken to be zero, so that $Q^2 = 0$ on $\bar{c}^a$. Q^2 is trivially zero on B^a itself.

With the BRST transformations (10.82) we can write

$$Q\left[\bar{c}^a f^a(A) - i\frac{\alpha e^2}{2}\bar{c}^a B^a\right] = iB^a f^a + \frac{\alpha e^2}{2}B^a B^a + \bar{c}^a\left(\frac{\delta f^a}{\delta\theta^b}\right)c^b \tag{10.88}$$

The modified action (10.80) to be used in the functional integral can thus be written as

$$S_q = S_E(A) + \int \left[iB^a f^a + \frac{\alpha e^2}{2} B^a B^a + \bar{c}^a(x) \left(\frac{\delta f^a(x)}{\delta \theta^b(y)} \right) c^b(y) \right]$$

$$= S_E(A) + Q \int \left[\bar{c}^a \left(f^a - i\frac{\alpha}{2} B^a \right) \right] \tag{10.89}$$

Since the action of Q on A_μ^a is an infinitesimal gauge transformation with parameter $-c^a$ and S_E is gauge-invariant, we have

$$Q \, S_q = 0 \tag{10.90}$$

using $Q^2 = 0$. This BRST-invariance expresses, for the gauge-fixed action S_q, the effect of the gauge-invariance of the theory.

The BRST-invariance leads to a conserved current and charge, which we now derive. For simplicity, consider the gauge choice $f^a = \partial \cdot A^a$. The Lagrangian for S_q can be written, in Minkowski space, as

$$\mathcal{L} = -\frac{1}{4e^2} F^2 + iB^a \partial^\mu A_\mu^a + \partial^\mu \bar{c}^a (D_\mu c)^a + \frac{\alpha e^2}{2} B^a B^a \tag{10.91}$$

The variation of this Lagrangian is

$$\delta \mathcal{L} = \partial^\mu \left[-\frac{1}{e^2} F_\mu^{a\nu} \delta A_\nu^a + iB^a \delta A_\mu^a + \delta \bar{c}^a (D_\mu c)^a + \partial_\mu \bar{c}^a \delta c^a \right]$$

$$+ \text{ equations of motion} \tag{10.92}$$

Under a BRST transformation, $\mathcal{L}$ in (10.91) changes as $\delta \mathcal{L} = \partial_\mu K^\mu = \partial_\mu [iB^a(D^\mu c)^a]$. using the general formula for Noether currents from Chapter 3 and the transformations (10.82), we can identify the current as

$$J_\mu = -\frac{1}{e^2} F_\mu^{a\nu} (D_\nu c)^a + iB^a (D_\mu c)^a + \tfrac{1}{2} f^{abc} \partial_\mu \bar{c}^a c^b c^c \tag{10.93}$$

The Grassmann nature of the variations is important for this; it is useful to introduce a Grassmann-valued parameter for the variations due to Q and then identify the current after moving this parameter to the left end of all terms in $\delta \mathcal{L}$. The charge corresponding to (10.93) is

$$Q = \int d^3x \left[\frac{1}{e^2} F_{0i}^a (D_i c)^a + iB^a (D_0 c)^a + \tfrac{1}{2} f^{abc} \partial_0 \bar{c}^a c^b c^c \right] \tag{10.94}$$

(We use Q for the canonical charge, Q for the BRST variation in the functional language.) From the Lagrangian (10.91), the canonical one-form is given by

$$\Theta = \int d^3x \left[\frac{1}{e^2} F_{0i}^a \delta A_i^a + iB^a \delta A_0^a + \delta \bar{c}^a (D_0 c)^a + \partial_0 \bar{c}^a \delta c^a \right] \tag{10.95}$$

In addition to the commutation rules for the A_i^a, F_{0j}^a given in (10.45), Θ leads to the following nontrivial commutation rules.

$$[A_0^a(\boldsymbol{x}, t), iB^a(\boldsymbol{y}, t)] = i\delta^{ab}\delta^{(3)}(x - y)$$
$$\{c^a(\boldsymbol{x}, t), \partial_0 \bar{c}^b(\boldsymbol{y}, t)\} = i\delta^{ab}\delta^{(3)}(x - y)$$
$$\{\bar{c}^a(\boldsymbol{x}, t), (D_0 c)^b(\boldsymbol{y}, t)\} = -i\delta^{ab}\delta^{(3)}(x - y) \tag{10.96}$$

One can then see immediately that Q defined by (10.94) is the canonical generator of the BRST transformations, i.e.,

$$\varphi\, Q \pm Q\, \varphi = -i\, \delta\varphi \tag{10.97}$$

where the plus sign or anticommutator applies to the fermionic fields $\varphi = \bar{c}, c$ and the minus sign or commutator applies to the bosonic fields $\varphi = A_\mu, B$. One can check that $Q^2 = 0$ in the canonical version as well.

In quantizing the theory, we must split the various fields into the creation and annihilation pieces (or negative- and positive-frequency pieces). The states can then be built up by the acting on the vacuum state many times by the creation parts, or the negative-frequency parts, of the fields. States containing ghosts would be obtained by applying $c^{a(-)}$ many times. In this context, the ghost number is a very useful concept. The Lagrangian is invariant under $c \to e^{i\chi}c$, $\bar{c} \to e^{-i\chi}\bar{c}$ for some constant χ. The corresponding charge is

$$Q_{gh} = \int d^3x \;\left[-\bar{c}^a(D_0 c)^a + \partial_0 \bar{c}^a c^a\right] \tag{10.98}$$

Consider now the case of electrodynamics again. In this case, the BRST transformation of the ghost is zero, $Qc = 0$. The ghost number simplifies to

$$Q_{gh} = \int d^3x \;\left[-\bar{c}^a(\partial_0 c)^a + \partial_0 \bar{c}^a c^a\right] \tag{10.99}$$

We now show that the physical states $|\Psi\rangle$ of electrodynamics can be specified by the conditions

$$Q\,|\Psi\rangle = 0$$
$$Q_{gh}\,|\Psi\rangle = 0 \tag{10.100}$$

We shall also see that states $|\Psi\rangle$ and $|\Psi'\rangle = |\Psi\rangle + Q\,|\lambda\rangle$ are equivalent, so that we may restrict to states which are not of the form $Q\,|\lambda\rangle$. In other words, physical states are annihilated by Q and Q_{gh} but are themselves not of the form $Q\,|\lambda\rangle$ for some state $|\lambda\rangle$. (Notice that any state of the form $Q|\lambda\rangle$ will be annihilated by Q since $Q^2 = 0$.) Introduce a field ξ defined by $Q\,\xi = c$. The field ξ is the part of A_i which is not gauge-invariant; i.e., we can write $A_i = A_i^T + \partial_i \xi$, where A_μ^T obeys $\partial_i A_i^T = 0$. We then have a "quartet" of fields ξ, B, $\bar{c}$, c, which we can call "unphysical" fields. The application of the negative-frequency parts of these operators on any state will generate a set of states with "unphysical" particles. The fields $\partial_i F_{0i}$, A_0, $\partial_0 c$ and $\partial_0 \bar{c}$ are related to the canonical conjugates of the quartet and are thus related to the positive-frequency parts of the quartet operators. Our argument will be recursive, where we start with a state with a given number of "unphysical" particles and construct a state with more "unphysical" particles and require it to obey the conditions (10.100). Let $|\Psi\rangle$ be a state with N of these "unphysical" particles. We can write

$$|\Psi\rangle = \left(B^{(-)} \mathcal{O}_1 + \xi^{(-)} \mathcal{O}_2 + \bar{c}^{(-)} \mathcal{O}_3 + c^{(-)} \mathcal{O}_4 \right) |\sigma\rangle \qquad (10.101)$$

where the $\mathcal{O}_i$ are some operators and $|\sigma\rangle$ has less number of "unphysical" particles. We can also assume that $|\sigma\rangle$ obeys the conditions (10.100). We now require $|\Psi\rangle$ to obey the same conditions (10.100), i.e.,

$$Q\,|\Psi\rangle = \Big[B^{(-)}(Q\mathcal{O}_1) + c^{(-)}\mathcal{O}_2 + \xi^{(-)}(Q\mathcal{O}_2) + iB^{(-)}\mathcal{O}_3$$

$$- \bar{c}^{(-)}(Q\mathcal{O}_3) - c^{(-)}(Q\mathcal{O}_4) \Big] |\sigma\rangle$$

$$= 0 \qquad (10.102)$$

Here $Q\mathcal{O}$ denotes the change in $\mathcal{O}$ due to the BRST transformation; it is thus the commutator or anticommutator, appropriately, of Q with $\mathcal{O}$. Equation (10.102) requires in general

$$Q\,\mathcal{O}_1 = -i\mathcal{O}_3, \qquad\qquad Q\,\mathcal{O}_3 = 0$$
$$Q\,\mathcal{O}_4 = \mathcal{O}_2, \qquad\qquad Q\,\mathcal{O}_2 = 0 \qquad (10.103)$$

The solution to these equations may be written as

$$\mathcal{O}_1 = \xi^{(-)}\mathcal{O} + \rho_1, \qquad\qquad \mathcal{O}_3 = i\,c^{(-)}\mathcal{O}$$
$$\mathcal{O}_4 = \bar{c}^{(-)}\tilde{\mathcal{O}} + \rho_4, \qquad\qquad \mathcal{O}_2 = i\,B^{(-)}\tilde{\mathcal{O}} \qquad (10.104)$$

This is the choice consistent with zero ghost number. Here $\mathcal{O}$, $\tilde{\mathcal{O}}$ are operators of zero ghost number and are BRST-invariant; ρ_1, ρ_4 are also BRST-invariant operators. Using this solution

$$|\Psi\rangle = \left[\left(B^{(-)}\xi^{(-)} + i\bar{c}^{(-)}c^{(-)} \right) \mathcal{O} + \left(i\xi^{(-)}B^{(-)} + c^{(-)}\bar{c}^{(-)} \right) \tilde{\mathcal{O}} \right] |\sigma\rangle$$

$$+ (B^{(-1)}\rho_1 + c^{(-1)}\rho_4) |\sigma\rangle \qquad (10.105)$$

Since $B^{(-)}\xi^{(-)} + i\bar{c}^{(-)}c^{(-)} = Q(-i\bar{c}^{(-)}\xi^{(-)})$ and $(B^{(-1)}\rho_1 + c^{(-1)}\rho_4) = Q(-i\bar{c}^{(-)}\rho_1 + \xi^{(-)}\rho_4)$, we get

$$|\Psi\rangle = Q\,|\lambda\rangle \qquad (10.106)$$

for some state $|\lambda\rangle$. Thus, all states which have "unphysical particles" and obey the conditions (10.100), by recursion of the above argument, are of the form $Q\,|\lambda\rangle$. Since Q is a self-adjoint operator, these are also zero-norm states, i.e.,

$$\| \, Q \, |\lambda\rangle \, \| = \langle\lambda|Q^2|\lambda\rangle = 0 \qquad (10.107)$$

These zero-norm states do not contribute to the matrix elements between physical states of any operator which is BRST-invariant, i.e.,

$$\langle\Psi_1|\mathcal{O}Q \, |\lambda\rangle = \langle\Psi_1|Q \, \mathcal{O}|\lambda\rangle = 0 \qquad (10.108)$$

since $Q|\Psi\rangle = 0$. In particular they do not contribute to the S-matrix which is built up by time-evolution by the Hamiltonian. The Hamiltonian is BRST-invariant and hence time-evolution will preserve the BRST-invariance of the physical states. We also see that $|\Psi\rangle$ and $|\Psi'\rangle = |\Psi\rangle + Q \, |\lambda\rangle$ are equivalent. (Actually going from $|\Psi\rangle$ to $|\Psi'\rangle$ is the Hilbert space version of a gauge or BRST transformation.) We may thus restrict our analysis to states which obey the conditions (10.100), but which are not of the form $Q \, |\lambda\rangle$ for some $|\lambda\rangle$. By the recursion argument above, the only states of this kind must be states with no "unphysical" particles. They are built up using BRST-invariant operators which include gauge-invariant operators as well. Since A_μ^T are Q-invariant, the general solution to (10.100) is of the form

$$|\Psi\rangle = |\Psi_T\rangle + Q \, |\lambda\rangle \qquad (10.109)$$

We have thus shown that states of transverse photons obey the conditions (10.100) and are not themselves of the form $Q \, |\lambda\rangle$. In other words, each physical state belongs to a class of $|\Psi\rangle$'s which obey the conditions (10.100) with the equivalence relation $|\Psi\rangle \sim |\Psi\rangle + Q \, |\lambda\rangle$. (One may say that the physical states are cohomology classes of the BRST operator.)

This description generalizes to the nonabelian case also, although the construction of the recursion argument is algebraically more complicated.

10.7 Ward-Takahashi identities for $\mathcal{Q}$-symmetry

We have shown that the physical states obey the condition $Q \, |\Psi\rangle = 0$. The Green's functions of interest in the theory are the N-point functions for BRST-invariant operators. The S-matrix for physical states may be obtained from such Green's functions. The symmetry translates into a set of identities for the Green's functions, the Ward-Takahashi (WT) identities, which we now derive. These identities are crucial in showing that matrix elements of physical observables are independent of the specific gauge-fixing condition.

The action to be used in the functional integral, namely (10.80), satisfies $\mathcal{Q} \, S_q = 0$. A representation of $\mathcal{Q}$ as an operator on functionals of the fields in four-dimensional space is given by

$$\mathcal{Q} = \int d^4x \, \left[(D_\mu c)^a \frac{\delta}{\delta A_\mu^a} - \frac{1}{2} f^{abc} c^b c^c \frac{\delta}{\delta c^a} + iB^a \frac{\delta}{\delta \bar{c}^a} \right] \qquad (10.110)$$

Consider now the expectation value of a function of fields and derivatives, say, $\mathcal{O}$, which is not necessarily $\mathcal{Q}$-invariant.

$$\langle \mathcal{O} \rangle = \int [dAdcd\bar{c}] \; \mathcal{O}(A, c, \bar{c}) \; \exp\left(-\mathcal{S}_q(A, c, \bar{c}\right)$$

$$= \int [dA'dc'd\bar{c}'] \; \mathcal{O}(A', c', \bar{c}') \; \exp\left(-\mathcal{S}_q(A', c', \bar{c}')\right)$$

$$(10.111)$$

where we just renamed the variables of integration in the second expression. We now take the primed variables to be the unprimed ones plus $\mathcal{Q}$-variations, i.e.,

$$A_\mu^{a'} = A_\mu^a + \omega \; \mathcal{Q} \; A_\mu^a$$
$$c^{a'} = c^a + \omega \; \mathcal{Q} \; c^a$$
$$\bar{c}^{a'} = \bar{c}^a + \omega \; \mathcal{Q} \; \bar{c}^a \qquad (10.112)$$

where ω is an arbitrary Grassmann number. If the measure of integration is invariant, which it should be for consistency of the theory, we get

$$\langle \mathcal{O} \rangle = \int [dAdcd\bar{c}] \; (\mathcal{O} + \delta\mathcal{O}) \; \exp\left(-\mathcal{S}_q - \delta\mathcal{S}_q\right)$$

$$= \langle \mathcal{O} \rangle + \langle \delta\mathcal{O} \rangle \qquad (10.113)$$

since $\delta\mathcal{S}_q = 0$. $\delta\mathcal{O} = (\mathcal{Q} \; \mathcal{O})$ is the BRST variation of the operator $\mathcal{O}$. (On the question of the invariance of the measure, see the chapters on anomalies.) We can thus write the above equation as

$$\langle \; (\mathcal{Q} \; \mathcal{O}) \; \rangle = 0 \qquad (10.114)$$

We now consider the Green's functions

$$\tilde{G}(x, x_1, x_2, ..., x_N) = \int [dAdcd\bar{c}] \; \mathcal{O}(x) \prod_i \mathcal{O}_i(x_i) \exp\left(-\mathcal{S}_q\right) \qquad (10.115)$$

where $\mathcal{O}_i$ are $\mathcal{Q}$-invariant. Following similar arguments to what was given above, from (10.111) to (10.114), we find

$$\langle (\mathcal{Q} \; \mathcal{O}) \prod_i \mathcal{O}_i(x_i) \rangle = 0 \qquad (10.116)$$

This is the basic WT identity for the $\mathcal{Q}$-symmetry. An immediate consequence is that the S-matrix and matrix elements of gauge-invariant operators are independent of the gauge-fixing condition. For instance, let f_1 and f_2 be two different choices for the gauge-fixing in $\mathcal{S}_q$. We then have

$$\mathcal{S}_{q1} - \mathcal{S}_{q2} = \int d^4x \; \mathcal{Q} \; [\bar{c}(f_1 - f_2)] \equiv \mathcal{Q} \; V_{12} \qquad (10.117)$$

The Green's function for a number of BRST-invariant operators $\mathcal{O}_i(x_i)$ calculated with the gauge-fixing f_1 can be written as

$$G_1(x_1, x_2, ..., x_N) = \int e^{-S_{q1}} \prod_i \mathcal{O}_i(x_i)$$

$$= \int e^{-S_{q2} - \mathcal{Q} \, V_{12}} \prod_i \mathcal{O}_i(x_i)$$

$$= \int e^{-S_{q2}} \prod_i \mathcal{O}_i(x_i) \; + \int e^{-S_{q2}} (\mathcal{Q} \, R_{12}) \prod_i \mathcal{O}_i(x_i)$$

$$= G_2(x_1, x_2, ..., x_N) \tag{10.118}$$

where we have used the identity (10.116) in the last step and the fact that $\exp(-\mathcal{Q}V_{12}) = 1 + \mathcal{Q} \, R_{12}$, $R_{12} = -V_{12} + \frac{1}{2} V_{12} \mathcal{Q} V_{12} + \cdots$, which follows from $\mathcal{Q}^2 = 0$. Eq.(10.118) shows that the N-point functions for $\mathcal{Q}$-invariant operators, and in particular gauge-invariant operators, are independent of the gauge choice. The S-matrix elements, which are constructed from such Green's functions, are also independent of the gauge choice. The identities for the change of the Green's functions under a change of the gauge-fixing condition are known as the Slavnov-Taylor identities. They were originally expressed in a slightly different form.

The WT identities and the result that Green's functions of Q-invariant operators are independent of the gauge choice are somewhat formal, as we have derived them. In actual calculations, there may be divergences of integrals which require regularization and suitable subtractions in order to make the theory well-defined. If the regulator does not preserve $\mathcal{Q}$-symmetry, we may lose some of these results. There are three possibilities.

1) There exists a regulator which respects $\mathcal{Q}$-symmetry and this is the one used in the calculations.

In this case, the WT identities hold and the above arguments for the gauge-independence of the S-matrix are true in the regularized, and ultimately, the renormalized theory. (Symmetries other than the $\mathcal{Q}$-symmetry may be lost due to this regularization.)

2) There exists a regulator which respects $\mathcal{Q}$-symmetry, but a different regulator which does not have manifest $\mathcal{Q}$-symmetry is used.

In this case, the regularized theory does not have $\mathcal{Q}$-symmetry; in particular, there are terms which are not $\mathcal{Q}$-invariant with potentially divergent coefficients. Counterterms to be subtracted do not have $\mathcal{Q}$-invariance, but it is possible to choose them in such a way that the renormalized theory has $\mathcal{Q}$-invariance. The S-matrix is again independent of the gauge choice. (It may not be possible to choose counterterms so as to preserve other symmetries as well, they may be lost due to regularization.)

3) There exists no regulator which preserves the $\mathcal{Q}$-symmetry.

In this case, $\mathcal{Q}$-symmetry is definitely lost in the renormalized theory and we say that the $\mathcal{Q}$-symmetry is anomalous. "Unphysical" states can contribute to matrix elements of observables and the theory loses unitarity in general. In order to have consistent theories, we must therefore eliminate possible sources of anomalies in the $\mathcal{Q}$-symmetry. This can lead to constraints on the allowed matter content of a gauge theory.

The WT identity for BRST-invariance can also be expressed in terms of the generating functional Γ for one-particle irreducible vertices. In this case, it is convenient to consider the functional

$$
e^W = \int [dAdcd\bar{c}dB] \, \exp\left(-\mathcal{S}_q + \int J^{a\mu} A_\mu^a + \bar{\eta}^a c^a + \bar{c}^a \eta \right.
$$

$$
\left. +(D_\mu c)^a K^{a\mu} - \frac{1}{2} f^{abc} c^b c^c L^a + J^a B^a \right) \qquad (10.119)
$$

W is a functional of the source functions J^μ, η, $\bar{\eta}$, K, L, and J. Γ is now defined as

$$
\Gamma[A, c, \bar{c}, K, L, J] = \int J^{a\mu} A_\mu^a + \bar{\eta}^a c^a + \bar{c}^a \eta^a \; - \; W \qquad (10.120)
$$

Here we do a Legendre transformation for $J^{a\mu}$, η^a and $\bar{\eta}^a$, but keep the other sources. The WT identity is given by

$$
0 = \int [dAdcd\bar{c}dB] \, \mathcal{Q} \, \exp\left(-\mathcal{S}_q + \int J^{a\mu} A_\mu^a + \bar{\eta}^a c^a + \bar{c}^a \eta^a \right.
$$

$$
\left. +(D_\mu c)^a K^{a\mu} - \frac{1}{2} f^{abc} c^b c^c L^a + J^a B^a \right) \qquad (10.121)
$$

This can be written as

$$
\langle \int J^{a\mu} (D_\mu c)^a + \bar{\eta}^a \frac{1}{2} f^{abc} c^b c^c + iB^a \eta^a \rangle = 0 \qquad (10.122)
$$

While we can write $J^{a\mu}$, η^a, $\bar{\eta}^a$ in terms of derivatives of Γ with respect to the fields, we are still left with the expectation values of the operators B, $D_\mu c$, and $f^{abc} c^b c^c$. The extra sources help to write these in terms of derivatives. This was why they were introduced into the functional integral. We get

$$
\int d^4 x \left[\frac{\delta\Gamma}{\delta A_\mu^a} \frac{\delta\Gamma}{\delta K^{a\mu}} - \frac{\delta\Gamma}{\delta L^a} \frac{\delta\Gamma}{\delta c^a} - i \frac{\delta\Gamma}{\delta J^a} \frac{\delta\Gamma}{\delta \bar{c}^a} \right] = 0 \qquad (10.123)
$$

This is the basic identity; by functionally differentiating with respect to various fields and sources and setting them to zero, we get an infinite set of relations among the vertex functions.

10.8 Renormalization of nonabelian theories

We now consider some aspects of the renormalization of nonabelian gauge theories. For the choice $f^a = \partial \cdot A^a$ and with $\alpha = 1$, the action (10.77) can be written as

$$\mathcal{S}_q = \int d^4x \left[\frac{1}{4}(\partial_\mu A_\nu^a - \partial_\nu A_\mu^a)(\partial^\mu A^{a\nu} - \partial^\nu A^{a\mu}) \right.$$

$$+ \frac{e}{2} f^{abc}(\partial_\mu A_\nu^a - \partial_\nu A_\mu^a) A^{b\mu} A^{c\nu} + \frac{e^2}{4} f^{abc} f^{amn} A^{b\mu} A^{c\nu} A_\mu^m A_\nu^n$$

$$\left. + \partial_\mu \bar{c}^a \partial^\mu c^a + e f^{abc} \partial_\mu \bar{c}^a A^{b\mu} c^c + \frac{1}{2}(\partial \cdot A^a)^2 \right] \qquad (10.124)$$

We have scaled the potential $A_\mu^a \to e A_\mu^a$ which is convenient for perturbation theory. All the operators involved in the action are of dimension 4. The propagators behave like k^{-2} as expected and power-counting arguments would show that potential divergences can only generate local terms of dimension 4. In order to establish renormalizability of the theory, one needs the following.

1) A subtraction procedure, essentially Bogolyubov's recursion formula, to show that naive power-counting arguments do work.

2) The WT identities for gauge- or BRST-invariance which would show that of all possible dimension 4 terms that one can write down, only those with the tensor and group structures indicated in (10.124) do arise with divergent coefficients, so that the divergences can be absorbed by redefining the coefficients of the various composite operator terms in (10.124).

The WT identities can be maintained by using a gauge-invariant regularization such as dimensional regularization. One must establish rules for power counting, separating overlapping divergences, etc., in such a regularization so that it is evident that only dimension 4 terms are generated with potentially divergent coefficients. If these conditions are satisfied, we can do calculations in a nonabelian theory with the action

$$\mathcal{S}_q = \int d^4x \left[\frac{Z_3}{4}(\partial_\mu A_\nu^a - \partial_\nu A_\mu^a)(\partial^\mu A^{a\nu} - \partial^\nu A^{a\mu}) \right.$$

$$+ \frac{e Z_1}{2} f^{abc}(\partial_\mu A_\nu^a - \partial_\nu A_\mu^a) A^{b\mu} A^{c\nu}$$

$$+ \frac{e^2 Z_4}{4} f^{abc} f^{amn} A^{b\mu} A^{c\nu} A_\mu^m A_\nu^n + \tilde{Z}_3 \partial_\mu \bar{c}^a \partial^\mu c^a$$

$$\left. + e \tilde{Z}_1 f^{abc} \partial_\mu \bar{c}^a A^{b\mu} c^c + \frac{\lambda}{2}(\partial \cdot A^a)^2 \right] \qquad (10.125)$$

where we naively put arbitrary renormalization factors for all composite operator terms. This can be split into a "classical part" $\mathcal{S}$ given by $\mathcal{S}_q$ of (10.125) with $Z_1 = Z_3 = Z_4 = \tilde{Z}_1 = \tilde{Z}_3 = \lambda = 1$, and a counterterm action $\mathcal{S}_q - \mathcal{S}$. The counterterm action is defined as what is needed to cancel potential divergences. If we have a regulator which respects the BRST-invariance, the possible divergences and hence the counterterm action will have this property. This can be seen recursively starting from the "classical action" $\mathcal{S}$ and fixing counterms loop by loop. Thus if we are using a regulator which respects the BRST-invariance, one can restrict the form of the counterterm action to one which has BRST-invariance. The WT identities or BRST invariance will then imply that, of the six renormalization factors Z_1, Z_3, Z_4, $\tilde{Z}_1$, $\tilde{Z}_3$, λ, only four are independent; i.e., one has the relation

$$\frac{Z_4}{Z_1} = \frac{Z_1}{Z_3} = \frac{\tilde{Z}_1}{\tilde{Z}_3} \tag{10.126}$$

A simple way to see how these relations arise is to define the BRST transformations

$$Q\, A_\mu^a = \partial_\mu c^a + a f^{abc} A_\mu^b c^c$$

$$Q\, c^a = -\frac{b}{2} f^{abc} c^b c^c$$

$$Q\, \bar{c}^a = \gamma\, \partial \cdot A^a \tag{10.127}$$

and require Q-invariance of the action (10.125). This will give $aZ_3 = eZ_1$, $aZ_1 = eZ_4$ and $a\tilde{Z}_3 = e\tilde{Z}_1$, leading to (10.126). Notice that $f^{abc}A_\mu^b c^c$ and $f^{abc}c^b c^c$ are composite operator terms in the transformations and as such, they do, in principle, need renormalization factors, which we have denoted by a, b. (They should not be confused with the Lie algebra indices.) From $Q^2 = 0$ on A_μ^a, b will turn out to be equal to a. We cannot impose $Q^2 = 0$ on the antighost field $\bar{c}^a$ since we have eliminated the B-field by its equation of motion. Thus Q^2 will be zero on $\bar{c}^a$ only upon using the ghost equation of motion. The Q-invariance of the action (10.125) will further give $\gamma\, \tilde{Z}_3 = \lambda$.

Gauge theories in four dimensions, with or without spontaneous symmetry breaking, are renormalizable. If one breaks gauge invariance explicitly by adding a mass term $\frac{1}{2}m^2 A^2$ to the Lagrangian, then we can lose renormalizability or unitarity. For example, for a massive Abelian vector particle we have

$$\mathcal{L} = \frac{1}{4}(\partial_\mu A_\nu - \partial_\nu A_\mu)(\partial^\mu A^\nu - \partial^\nu A^\mu) + \frac{m^2}{2}A_\mu A^\mu + \text{matter terms} \tag{10.128}$$

The propagator $D_{\mu\nu}(k)$ for A_μ obeys the equation

$$\left[(k^2 + m^2)\delta_{\mu\alpha} - k_\mu k_\alpha\right] D_{\alpha\nu} = \delta_{\mu\nu} \tag{10.129}$$

which has the solution

$$D_{\mu\nu}(k) = \frac{\delta_{\mu\nu} + (k_\mu k_\nu/m^2)}{k^2 + m^2} \tag{10.130}$$

At large $|k|$, this behaves like $k_\mu k_\nu/k^2$ and so the ordinary rules of power counting break down. One gets divergences corresponding to operators of arbitrarily high dimension. The equations of motion for the Lagrangian (10.128) have a constraint $\partial \cdot A = 0$, which can help to get rid of negative norm states. Alternatively, if one considers $\mathcal{L} = \frac{1}{2}A_\mu(-\Box)A^\mu + \frac{1}{2}m^2 A_\mu A^\mu$, the propagator is well-behaved at large $|k|$, but the A_0-component does not decouple and leads to problems with unitarity. The combination of the good features of these two cases above, viz., having enough gauge freedom to eliminate possible negative norm states and the high-energy behavior of the propagator being k^{-2}, occurs for spontaneously broken gauge theories.

10.9 The fermionic action and QED again

The BRST argument leading to relations among the renormalization constants can be extended to the case when fermions are coupled to the gauge field as well. The transformations on the fermion fields are of the form

$$Q\psi = iac^a(t^a\psi)$$
$$Q\bar{\psi} = ia(\bar{\psi}t^a)c^a \qquad (10.131)$$

(A priori, we could have a different constant, say, δ, in these equations in place of a; but $Q^2 = 0$ will show that δ should be a.) The fermionic part of the action can be written as

$$\mathcal{S} = \int d^4x\, Z_2\bar{\psi}(\gamma \cdot \partial + m)\psi \; - \; ieZ_1\bar{\psi}\gamma^\mu A_\mu^a t^a\psi \qquad (10.132)$$

With a regulator which respects BRST-invariance, we can again use a restricted action where the renormalization constants are related. Imposing Q-invariance on the fermion action (10.132), we get $Z_1 = Z_2$. We can specialize these results to the case of quantum electrodynamics which is a $U(1)$ gauge theory. We have already seen the relation $Z_1 = Z_2$ by explicit calculation at the one-loop level. Here we see that it will hold in general.

10.10 The propagator and the effective charge

As an example of one-loop calculations in a nonabelian gauge theory, we shall now calculate the gauge boson propagator to one-loop order and relate it to an effective charge.

The basic one-loop diagrams we need are

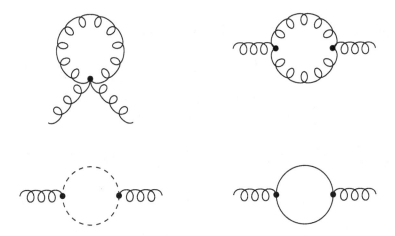

Fig 10.1. One-loop corrections to gauge boson propagator

(The dashed lines represent ghost propagators.) These diagrams will lead to divergences and so they have to be regulated. Although we used a straightforward momentum cut-off previously, the present calculations are rather invloved and can be simplified significantly by choosing a gauge-invariant regulator such as dimensional regularization. The basic idea here is to do the calculations in an arbitrary dimension $n = 4 - \epsilon$, where ϵ is taken to be very small. This is done by analytic continuation of various expressions to an arbitrary dimension. Integrals involved in Feynman digrams are well defined as a function of n, with poles at $n = 4$ corresponding to the divergences in four dimensions. The idea therefore is to do a Laurent expansion in ϵ, identify the pole terms, which are then canceled by choice of Z_i-factors, and then take the limit $\epsilon \to 0$. Since the concept of gauge-invariance does not depend on the dimension, this will be an invariant regularization. We will use such a procedure here.

In using dimensional regularization, we first do the algebraic simplification of the integrands of Feynman integrals, keeping in mind that we are in n dimensions when contracting indices and taking traces. For the basic type of integrals involved, we can carry out the integration in n dimensions to obtain the following formulae.

$$\int \frac{d^n p}{(2\pi)^n} \frac{1}{(p^2 + M^2)^\alpha} = \frac{1}{(4\pi)^{n/2}} \frac{\Gamma(\alpha - n/2)}{\Gamma(\alpha)} \frac{1}{(M^2)^{\alpha - n/2}}$$

$$\int \frac{d^n p}{(2\pi)^n} \frac{p^2}{(p^2 + M^2)^\alpha} = \frac{n}{2} \frac{1}{(4\pi)^{n/2}} \frac{\Gamma(\alpha - n/2 - 1)}{\Gamma(\alpha)} \frac{1}{(M^2)^{\alpha - n/2 - 1}}$$

$$\int \frac{d^n p}{(2\pi)^n} \frac{p_\mu p_\nu}{(p^2 + M^2)^\alpha} = \frac{\delta_{\mu\nu}}{2} \frac{1}{(4\pi)^{n/2}} \frac{\Gamma(\alpha - n/2 - 1)}{\Gamma(\alpha)} \frac{1}{(M^2)^{\alpha - n/2 - 1}}$$

$$(10.133)$$

where Γ denotes the Eulerian gamma function. The relevant Laurent expansion can be obtained by using the property $z\Gamma(z) = \Gamma(z+1)$ and

$$\Gamma(\epsilon/2) = \frac{2}{\epsilon} - \gamma + \mathcal{O}(\epsilon) \qquad (10.134)$$

γ is the Euler-Mascheroni constant, approximately equal to 0.5772.

Contribution of diagram 1

By writing out $\langle e^{-\mathcal{S}_{int}} \rangle = e^{-\Gamma}$ and carrying out the necessary Wick contractions, we obtain the contribution due to the first diagram as

$$\Gamma_{diag1} = \frac{n-1}{2} C_2 \int G(x,x) A^a_\mu(x) A^a_\mu(x) \qquad (10.135)$$

where C_2 is the quadratic Casimir invariant for the adjoint representation of the group. It is defined by $C_2 \delta^{ab} = f^{amn} f^{bmn}$ and for the group $SU(N)$ it is equal to N.

Contribution of diagram 2

The second of the Feynman diagrams shown arises from

$$\langle e^{-\mathcal{S}_{int}} \rangle = \frac{1}{2!} \left(\int G(x,y) \frac{\delta}{\delta A^m_\alpha(x)} \frac{\delta}{\delta A^m_\alpha(y)} \right)^2 \frac{1}{2!} \left(\int f^{abc} \partial_\mu A^a_\nu A^b_\mu A^c_\nu \right)^2 \qquad (10.136)$$

Carrying out the Wick contractions, we find

$$\Gamma_{diag2} = C_2 \int \left[-\frac{1}{4} f^a_{\mu\nu}(x) f^a_{\mu\nu}(y) G(x,y) G(x,y) + f^a_{\mu\nu}(x) A^a_\nu(y) \frac{\partial G(x,y)}{\partial y^\mu} G(x,y) \right.$$
$$- \frac{1}{2} A^a_\mu(x) A^a_\mu(x) G(x,x)$$
$$\left. - \frac{1}{2} A^a_\mu(x) A^a_\nu(y) \left\{ (n-1) \frac{\partial G(x,y)}{\partial x^\mu} \frac{\partial G(x,y)}{\partial x^\nu} + (2-n) \frac{\partial^2 G(x,y)}{\partial x^\mu \partial x^\nu} G(x,y) \right\} \right] \qquad (10.137)$$

where $f^a_{\mu\nu} - \partial_\mu A^a_\nu - \partial_\nu A^a_\mu$.

Contribution of diagram 3, the ghost loop

The ghost loop contribution is easily seen to be

$$\Gamma_{diag3} = \frac{C_2}{2} \int A^a_\mu(x) A^a_\nu(y) \frac{\partial G(x,y)}{\partial x^\mu} \frac{\partial G(x,y)}{\partial x^\nu} \qquad (10.138)$$

Contribution of the first three diagrams

Combining (10.135, 10.137, 10.138), we obtain

$$\Gamma = e^2 \mu^\epsilon C_2 \int \left[\frac{n-2}{2} A_\mu^a(x) A_\mu^a(x) G(x,x) - \frac{1}{4} f_{\mu\nu}^a(x) f_{\mu\nu}^a(y) G(x,y) G(x,y) \right.$$

$$+ f_{\mu\nu}^a(x) A_\nu^a(y) \frac{\partial G(x,y)}{\partial y^\mu} G(x,y)$$

$$\left. - \frac{n-2}{2} A_\mu^a(x) A_\nu^a(y) \left\{ \frac{\partial G(x,y)}{\partial x^\mu} \frac{\partial G(x,y)}{\partial x^\nu} - \frac{\partial^2 G(x,y)}{\partial x^\mu \partial x^\nu} G(x,y) \right\} \right]$$

$$(10.139)$$

where $f_{\mu\nu}^a = \partial_\mu A_\nu^a - \partial_\nu A_\mu^a$. Each of these terms carries a factor e^2, where e is the coupling constant. This is seen by the rescaling $A \to eA$. However, in dimensions other than four, the coupling constant is not dimensionless, e_n^2 in n dimensions has the mass dimension $4 - n = \epsilon$. We introduce a parameter μ with the dimensions of mass and write $e_n^2 = e^2 \mu^\epsilon$, where e will be the coupling constant in four dimensions. The factor of μ^ϵ and the four-dimensional coupling constant e^2 have been explicitly indicated in (10.139).

For the first of the integrals we find

$$\frac{n-2}{2} G(x,x) = \frac{n-2}{2} \int \frac{d^n p}{(2\pi)^n} \frac{1}{p^2 + M^2} \Bigg]_{M=0}$$

$$= \frac{1}{(4\pi)^{n/2}} \frac{n-2}{2} \Gamma\left(\frac{2-n}{2} \right) (M^2)^{1-\epsilon/2} \Bigg]_{M=0}$$

$$= -\frac{1}{(4\pi)^{n/2}} \Gamma\left(\frac{\epsilon}{2} \right) (M^2)^{1-\epsilon/2} \Bigg]_{M=0}$$

$$= 0 \qquad\qquad (10.140)$$

The integral vanishes at finite ϵ and hence can be taken to be zero in the regularized theory. Naively this integral is quadratically divergent; it is a peculiarity of dimensional regularization that quadratic divergences do not appear and such integrals can be set to zero. (This does not mean that all consequences of quadratic divergences disappear from a dimensionally regulated theory; there are examples of dimensionally regulated theories where the quadratic divergences can reappear if some summation of the perturbation series, such as Borel summation, is attempted.)

The second term in (10.139) is

$$\text{Term 2} = \frac{e^2}{4} \int f_{\mu\nu}^a(x) f_{\mu\nu}^a(y) \int \frac{d^n k}{(2\pi)^n} e^{-ik(x-y)} \Pi^{(2)}(k)$$

$$\Pi^{(2)}(k) = -(\mu^2)^{\epsilon/2} \int \frac{d^n p}{(2\pi)^n} \frac{1}{p^2 (p-k)^2}$$

$$= -(\mu^2)^{\epsilon/2} \int_0^1 du \int \frac{d^n p}{(2\pi)^n} \frac{1}{[p^2 + k^2 u(1 - u)]^2}$$

$$= -\frac{1}{16\pi^2} \int_0^1 du \, \Gamma(\epsilon/2) \exp\left[\frac{\epsilon}{2} \log\left(\frac{4\pi\mu^2}{k^2 u(1 - u)}\right)\right]$$

$$= -\frac{1}{16\pi^2} \left[\frac{2}{\epsilon} - \gamma + 2 + \log 4\pi - \log(k^2/\mu^2)\right] \tag{10.141}$$

The third term can be evaluated similarly and is equal to the second term given above. The last two terms become

$$\text{Terms } (4 + 5) = e^2 \int A_\mu^a(x) A_\nu^a(y) \int \frac{d^n k}{(2\pi)^n} e^{-ik(x-y)} \Pi^{(4+5)}(k)$$

$$\Pi^{(4+5)}(k) = (\mu^2)^{\epsilon/2}(1 - \epsilon/2) \int_0^1 du \int \frac{d^n p}{(2\pi)^n} \frac{-2p_\mu p_\nu + k_\mu k_\nu u(1 - 2u)}{[p^2 + k^2 u(1 - u)]^2}$$

$$= (k^2 \delta_{\mu\nu} - k_\mu k_\nu) \frac{1}{96\pi^2} \left[\frac{2}{\epsilon} - \gamma + \log 4\pi - \log(k^2/\mu^2) + \frac{5}{3}\right] \tag{10.142}$$

Combining terms, the one-loop contribution to the quadratic term in the effective action becomes

$$\Gamma = \frac{1}{4} \int_{x,y,k} f_{\mu\nu}^a(x) f_{\mu\nu}^a(y) e^{-ik(x-y)} \Pi(k)$$

$$\Pi(k) = -\frac{e^2 C_2}{16\pi^2} \left[\frac{5}{3}\left(\frac{2}{\epsilon} - \gamma + \log 4\pi - \log(k^2/\mu^2)\right) + \frac{31}{9}\right] \tag{10.143}$$

Contribution of fermion loops

We also consider N_f species of fermions in the fundamental representation of the group coupled to the gauge fields. The fermion action is given by $S = \bar{q}[\gamma\cdot(\partial + A) + m]q$, where $A_\mu = -it^a A_\mu^a$. t^a are matrices in the fundamental representation and are a basis of the Lie algebra. For $SU(N)$, they can be taken as traceless hermitian $(N \times N)$-matrices. We have $\text{Tr}(t^a t^b) = \frac{1}{2}\delta^{ab}$. The γ-matrices obey $\text{Tr}\gamma^\mu\gamma^\nu = 2^{n/2}\delta^{\mu\nu}$. The one-loop contribution to the term in Γ which is quadratic in the A's is

$$\Gamma = \frac{1}{2} \int_{x,y,k} A_\mu^a(x) A_\nu^a(y) \, e^{-ik(x-y)} \Pi_{\mu\nu}(k)$$

$$\Pi_{\mu\nu}(k) = e^2 N_f \, 2^{n/2-1}(\mu^2)^{\epsilon/2} \int_0^1 du$$

$$\times \int \frac{d^n p}{(2\pi)^n} \frac{2p_\mu p_\nu + (\delta_{\mu\nu}k^2 - 2k_\mu k_\nu)u(1 - u) - \delta_{\mu\nu}(p^2 + m^2)}{[p^2 + k^2 u(1 - u)]^2}$$

$$= e^2 N_f \, \frac{(k^2 \delta_{\mu\nu} - k_\mu k_\nu)}{4\pi^2} \int_0^1 du \, u(1-u)$$

$$\times \left[\frac{2}{\epsilon} - \gamma + \log 2\pi \, - \, \log \left(\frac{k^2 u(1-u) + m^2}{\mu^2} \right) \right] \qquad (10.144)$$

With this explicit formula for the integral, the fermionic contribution to Γ can then be simplified as

$$\Gamma = \frac{1}{4} \int_{x,y,k} f^a_{\mu\nu}(x) f^a_{\mu\nu}(y) e^{-ik(x-y)} \, \Pi_f(k)$$

$$\Pi_f(k) = \frac{e^2 N_f}{4\pi^2} \int_0^1 du \, u(1-u) \left[\frac{2}{\epsilon} - \gamma + \log 2\pi \, - \, \log \left(\frac{k^2 u(1-u) + m^2}{\mu^2} \right) \right]$$

$$(10.145)$$

The effective action to the first order in $\hbar$ also has a contribution from the Z_3-factor in (10.125). The quadratic term in Γ, to first order in $\hbar$, is thus

$$\Gamma = \frac{1}{4} \int_{x,y,k} f^a_{\mu\nu}(x) f^a_{\mu\nu}(y) e^{-ik(x-y)} \, V(k)$$

$$V(k) = Z_3 + \Pi_f(k) + \Pi(k) \qquad (10.146)$$

It is clear that one can choose Z_3 so as to cancel the $1/\epsilon$-term, so that the limit $n \to 4$ can be taken without divergences. This shows the renormalization of the gauge-particle propagator to one loop order. It is possible to absorb some of the constants, namely, k-independent terms, in Z_3 as well. Exactly how Z_3 is defined is a choice of the renormalization scheme. Notice that the differences can be taken to be different definitions of μ. One scheme, known as minimal subtraction, eliminates only the $1/\epsilon$-terms by choice of Z_3. For our purpose it is easier to get rid of all the constant factors by a suitable choice of Z_3. This corresponds to

$$Z_3 = 1 + e^2 \frac{5C_2 - 2N_f}{48\pi^2} \left(\frac{2}{\epsilon} - \gamma - \log 4\pi \right) + \frac{31 e^2 C_2}{144\pi^2} + \frac{e^2 N_f \log 2}{24\pi^2} + \cdots$$

$$V(k) = 1 + \frac{5 e^2 C_2}{48\pi^2} \log(k^2/\mu^2) - \frac{e^2 N_f}{4\pi^2} \int_0^1 du \, u(1-u) \log \left(\frac{k^2 u(1-u) + m^2}{\mu^2} \right)$$

$$(10.147)$$

For zero-mass fermions, which is what we shall consider for the rest of our discussion, we can take Z_3 as the expression given above minus $5 e^2 N_f / 72\pi^2$; we absorb the integral of $u(1-u) \log u(1-u)$ also into Z_3. Then

$$V(k) = 1 + \frac{e^2}{48\pi^2} (5C_2 - 2N_f) \log(k^2/\mu^2) \qquad (10.148)$$

From the calculation given above, the corrected gauge-particle propagator is of the form

$$\tilde{G}(k) = \frac{1}{k^2} \frac{1}{V(k)} \tag{10.149}$$

Among other things, this shows that the quantity μ which was introduced for dimensional reasons can now be interpreted in a more physical way. It is the value of momentum transfer k at which $V(k) = 1$ and the propagator is just $1/k^2$.

The interaction between fermions due to the gauge-particle exchange is given by

$$\Gamma = \frac{e^2}{2} \int \bar{q} t^a \gamma_\mu q(x) \ \bar{q} t^a \gamma_\mu q(y) \ \tilde{G}(x, y) \tag{10.150}$$

In the case of QED, we interpreted the modified interaction as an effective increase of the strength of interaction with k^2, or as a k-dependent effective charge. In the present case, such an interpretation is possible, but the situation is not quite so simple. The reason is that there are other diagrams which contribute to the interaction between fermions, such as vertex corrections. For QED, these other corrections are of the form shown in figure 10.2.

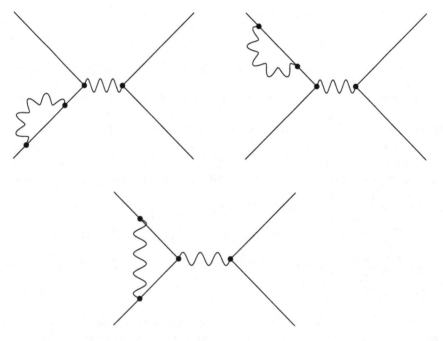

Fig 10.2. Corrections to fermion-fermion interaction by gauge boson exchange

The WT identity connects the vertex correction and fermion self-energy and in effect their contributions cancel out; they do not contribute to the effective charge. This is true in the nonabelian case as well. But in the nonabelian theory there is another contribution given by the diagram 10.3, shown below. There is a part of the mathematical expression for this diagram which is similar to the propagator contribution and it must be included in the definition of the effective charge. The expression for this diagram is given by

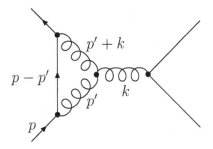

Fig 10.3. The diagram for the pinching contribution

$$\Gamma = e^4 f^{abc} \int \bar{q}(x) t^a \gamma_\mu q(x) \; \bar{q}(y) t^b t^c F_\mu(y,z) q(z)$$

$$F_\mu(y,z) = -\int_k e^{ik(x-y)+ip'(z-y)} \frac{1}{k^2} \int \frac{d^n p'}{(2\pi)^n} \frac{1}{p'^2(p'+k)^2} \Big[(\gamma \cdot k - \gamma \cdot p') S(y,z) \gamma_\mu$$

$$- \gamma_\mu S(y,z)(2\gamma \cdot k + \gamma \cdot p') + (2p'_\mu + k_\mu) \gamma_\alpha S(y,z) \gamma_\alpha \Big] \quad (10.151)$$

The integral has an interesting feature. By writing $-\gamma \cdot p' S(y,z) e^{-ip'y} = -i\gamma \cdot \partial(e^{-ip'y}) S(y,z)$, we can do a partial integration over y. This will produce a term $i\gamma \cdot \partial S(y,z)$ which is i times a delta function, $i\delta^{(n)}(y-z)$. Such terms are called pinching terms, since the points y, z become identical, which can be visualized as the fermion propagator line being pinched off the diagram. This pinching contribution is what we are interested in. The other terms, including derivatives with respect to y from the partial integration, involve either powers of p or k and do not lead to a current-current interaction structure. Therefore, even though they can be important, they are not relevant for the effective charge. After some algebraic simplifications, the pinching contribution due to (10.151) is seen to be

$$\Gamma = \frac{e^2}{2} \int \bar{q}(x) t^a \gamma_\mu q(x) \; \bar{q}(y) t^a \gamma_\mu q(y) \int \frac{d^n k}{(2\pi)^n} e^{ik(x-y)} F(k)$$

$$F(k) = e^2 (\mu^2)^{\epsilon/2} \frac{2C_2}{k^2} \int \frac{d^n p'}{(2\pi)^n} \frac{1}{p'^2(p'+k)^2}$$

$$= \frac{e^2 C_2}{8\pi^2 k^2} \left[\frac{2}{\epsilon} - \gamma + 2 + \log 4\pi - \log(k^2/\mu^2) \right] \quad (10.152)$$

The $1/\epsilon$-terms and the constant terms can be canceled by choice of the vertex renormalization in this case. By combining this result with the propagator contribution in (10.149, 10.150), we get the effective current-current interaction as

$$\Gamma_{eff} = \frac{1}{2} \int \bar{q}t^a\gamma_\mu q(x) \ \bar{q}t^a\gamma_\mu q(y) \ V_{eff}(x,y)$$

$$V_{eff}(k) = \frac{e^2}{k^2} \left[\left(1 + \frac{e^2}{48\pi^2}(5C_2 - 2N_f)\log(k^2/\mu^2)\right)^{-1} - \frac{e^2C_2}{8\pi^2}\log(k^2/\mu^2) \right]$$

$$\approx \frac{1}{k^2} \left[1 + \frac{e^2}{48\pi^2}(11C_2 - 2N_f)\log(k^2/\mu^2) \right]^{-1} \tag{10.153}$$

We now define the nonabelian analog of the fine structure constant as $\alpha = e^2/4\pi$. We can then write

$$V_{eff}(k) = 4\pi\frac{\alpha(k)}{k^2} \tag{10.154}$$

where the effective coupling constant which is a function of k^2 is identified as

$$\alpha(k) = \frac{\alpha(\mu)}{[1 + \alpha(\mu) \ b \ \log(k^2/\mu^2)]}$$

$$b = \frac{11C_2 - 2N_f}{12\pi} \tag{10.155}$$

The effective charge in QED was seen to increase with the momentum transfer k. In the present case, we see that α actually decreases with increasing k so long as the number of species of fermions is not too large, so that one has $11C_2 > 2N_f$. This means that at high energies the theory asymptotically approaches a free theory; this property is called *asymptotic freedom* in the terminology of the renormlization group introduced in Chapter 9. The discovery of asymptotic freedom in 1973 was the crucial step in using nonabelian gauge theories to construct a theory of strong interactions. Quantum chromodynamics (QCD), which describes strong interactions, is an $SU(3)$ gauge theory coupled to quarks which are in the fundamental representation of $SU(3)$, namely, triplets. In this case, $C_2 = 3$, and asymptotic freedom is possible if we have fewer than 16 species of quarks. There are six species known to date and there are reasons to believe that there are no more than six species. Asymptotic freedom is observed to hold for energies above a few GeV. At such energies, the quark mass terms in the above formulae cannot be neglected, the masses suppress their contributions in loop integrals. As a result very high mass quarks do not contribute to effective charge and so, in any case, the possible existence of very high mass quarks, beyond those we know, will not affect asymptotic freedom at presently observed energies.

The effective charge may be regarded as the running constant obtained by solving the renormalization group equation. Since we have done a calculation of the effective charge directly, we can use this to obtain the β-function for a nonabelian gauge theory, to one-loop order, as

$$\beta(e) = -\frac{e^3}{48\pi^2}(11C_2 - 2N_f) \qquad (10.156)$$

The β-function is negative, as expected for asymptotic freedom. Here we have obtained the β-function after calculating the effective charge. Going back to the action (10.125) and writing it in terms of bare fields, we see that the bare charge e_0 is related to the renormalized charge e as $e_0 = Z_1 Z_3^{-3/2}$. This shows that we can calculate the β-function, by evaluating Z_1, and using this formula and the definition (9.121); we have already obtained Z_3. This is the more conventional way of calculating the β-function in a nonabelian theory.

The behavior of QCD at high energies may be understood by solving the RG equation for the Green's functions, using the β-function (10.156). One can use perturbation theory to compute the required anomalous dimensions by virtue of asymptotic freedom. Applications of perturbative QCD to high energy scattering processes, which is a vast subject in its own right, are based on this property.

The formula for the effective charge shows that we can define a finite, dimensionful parameter Λ_{QCD} by

$$\frac{1}{\alpha(\mu)} + b \log\left(\frac{\Lambda_{QCD}^2}{\mu^2}\right) = 0 \qquad (10.157)$$

In terms of Λ_{QCD}, the effective coupling may be written as

$$\begin{aligned}
\alpha(k) &= \frac{1}{b \log(k^2/\Lambda_{QCD}^2)} \\
&= \frac{12\pi}{(11C_2 - 2N_f) \log(k^2/\Lambda_{QCD}^2)} \qquad (10.158)
\end{aligned}$$

We see that the coupling constant is entirely determined by Λ_{QCD}. The dimensionless coupling $\alpha(\mu)$ had been traded for a dimensionful parameter Λ_{QCD}. This is known as dimensional transmutation. Λ_{QCD} determines the relevance of various kinematic regimes and how they have to be analyzed. Modes of the field with values of k much larger than Λ_{QCD} are weakly coupled and one can use perturbative analysis; modes with k comparable to, or less than, Λ_{QCD}, have to be treated nonperturbatively. Λ_{QCD} is thus the basic scale parameter of the theory. The theory does not choose any value for Λ_{QCD}; it is an input parameter for the theory. The fact that we have this freedom is a residue of the classical scale invariance of the theory.

References

1. A nonabelian gauge theory was obtained by O. Klein, Lecture reported in *New Theories in Physics*, Int. Inst. of Intellectual Cooperation, Paris

(1939); gauge theories, as we know them today, are due to C.N. Yang and R.L. Mills, Phys. Rev. **96**, 191 (1954); R. Shaw, Ph.D. Thesis, Cambridge University, UK (1954); R. Utiyama, Phys. Rev. **101**, 1597 (1956).

2. The Chern-Simons term is due to S.S. Chern and J. Simons, Ann. Math. **99**, 48 (1974). Its introduction into physics literature is by R.Jackiw and S.Templeton, Phys. Rev. **D23**, 2291 (1981); J.Schonfeld, Nucl. Phys. **B185**, 157 (1981); S.Deser, R.Jackiw and S.Templeton, Phys. Rev. Lett. **48**, 975 (1982); Ann. Phys. **140**, 372 (1982).

3. The notion of parallel transport is from the general theory of relativity; see, for example, S. Weinberg, *Gravitation and Cosmology* Wiley Text Books (1972). Its importance for electrodynamics was recognized by F. London, *Superfluids*, John Wiley and Sons, Inc. (1950); Y. Aharonov and D. Bohm, Phys. Rev. **115**, 485 (1957). For a formulation of electrodynamics using the parallel transport operator $U(x, y, C)$, see S. Mandelstam, Ann. Phys. **19**, 1 (1962).

4. The Wilson loop operator is defined by K.G. Wilson, Phys. Rev. **D14**, 2455 (1974).

5. The proper mathematical definitions of the spaces of potentials and gauge transformations can be found in I.M. Singer, Phys. Scripta **T24**, 817 (1981); Commun. Math. Phys. **60**, 7 (1978).

6. The Faddeev-Popov method is due to B. de Witt, Phys. Rev. **162**, 1195 (1967); L.D. Faddeev and V.N. Popov, Phys. Lett. **B25**, 29 (1967). An earlier version of such diagrammatic rules, in the context of gravity, was given by R.P. Feynman, Acta Phys. Polonica **24**, 697 (1963).

7. For the Gribov ambiguity, see V.N. Gribov, Nucl. Phys. **B139**, 1 (1978); I.M. Singer, Commun. Math. Phys. article cited in reference 5. A clear and general discussion is by T. Killingback and E.J. Rees, Class. Quant. Grav. **4**, 357 (1987).

8. The BRST symmetry was discovered by C. Becchi, A. Rouet and R. Stora, Commun. Math. Phys. **42**, 127 (1975); I. Tyutin, Lebedev Institutre preprint N39 (1975). There is a beautiful geometric interpretation of this symmetry in terms of transformations of the fiber in a fiber-bundle approach to gauge theories; this is given in J. Thierry-Mieg, J. Math. Phys. **21**, 2834 (1980); R. Stora, in *Progress in Gauge Theories: Proceedings of a Symposium at Cargèse, 1983*, G. 't Hooft *et al* (eds.), Plenum Press (1984).

9. The analysis of physical states as BRST invariant states is given by T. Kugo and I. Ojima, Prog. Theor. Phys. **60**, 1869 (1978); Prog. Theor. Phys. Suppl. **66**, 1 (1979).

10. The WT identities for Γ can be found in J. Zinn-Justin, in *Trends in Elementary Particle Theory-International Summer Institute on Theoretical Physics in Bonn, 1974*, Springer-Verlag (1975). The Slavnov-Taylor identities, which are very similar and which were very useful in the original proofs of renormalizability of nonabelian gauge theories, were given

by A.A. Slavnov, Theor. Math. Phys. **10**, 152 (1972); J.C. Taylor, Nucl. Phys. **B33**, 436 (1971).

11. The first proofs of renormalizability of nonabelian gauge theories were in G. 't Hooft and M. Veltman, Nucl. Phys. **B50**, 318 (1972); B.W. Lee and J. Zinn-Justin, Phys. Rev. **D5**, 3121 (1972); *ibid.* **D7**, 1049 (1972). Later proofs used the BRST symmetry.

12. Dimensional regularization is due to G. 't Hooft and M. Veltman, Nucl. Phys. **B44**, 189 (1973).

13. The computation of the effective charge was also the discovery of asymptotic freedom in D.J. Gross and F. Wilczek, Phys. Rev. Lett. **30**, 1343 (1973); H.D. Politzer, Phys. Rev. Lett. **30**, 1346 (1973).

14. The pinching technique of computation is due to J.M. Cornwall, Phys. Rev. **D26**, 1453 (1982); J.M. Cornwall and J. Papavassiliou, Phys. Rev. **D40**, 3474 (1989).

11 Symmetry

11.1 Realizations of symmetry

We will begin with the discussion of continuous symmetries. In Chapter 3, we saw that, if a Lagrangian has a continuous global symmetry, then there is a conserved current J_μ^A and a conserved charge Q^A associated with it. Here A takes the values $1, 2, \cdots N$, where N is the number of continuous symmetries. The charge Q^A is conserved, which implies that it commutes with the Hamiltonian; i.e., $[Q^A, H] = 0$. This also shows that $[Q^A, Q^B]$ commutes with H and so leads to new conserved charges. If we have already included all the conserved charges, the commutator $[Q^A, Q^B]$ must be a function of the Q's themselves.

We begin with the case of the Q's generating an internal symmetry. Let $|\alpha\rangle$ be an eigenstate of H with eigenvalue E_α, $H|\alpha\rangle = E_\alpha|\alpha\rangle$. Since $[H, Q^A] = 0$, $H\left(Q^A|\alpha\rangle\right) = E_\alpha\left(Q^A|\alpha\rangle\right)$. Thus the action of Q^A on a state gives another state which is degenerate with it. By applying Q's on $|\alpha\rangle$ many times we get a sequence of states which are degenerate with it. This process can be continued until no new states are generated and further applications of Q's only produce linear combinations of states already included. This leads to a set of states generated in this way from $|\alpha\rangle$ and which is closed under the action of Q's. Let $\mathcal{H}_\alpha$ denote the subspace of the Hilbert space corresponding to this, and let $\{|\alpha_i\rangle\}$ be an orthonormal basis for this subspace. Then $Q^A|\alpha_i\rangle \in \mathcal{H}_\alpha$ for $|\alpha_i\rangle \in \mathcal{H}_\alpha$. Thus we get a matrix representation of Q^A on the subspace $\mathcal{H}_\alpha$ by $(Q^A)_{ij} = \langle\alpha_i|Q^A|\alpha_j\rangle$. This forms a representation of the algebra of the Q's under matrix multiplication since the Q's do not connect any state in $\mathcal{H}_\alpha$ to any state in the complement of $\mathcal{H}_\alpha$. This representation is also irreducible by construction since any state in $\mathcal{H}_\alpha$ is connected to the others by suitable Q-actions. Since the Q's are also realized as hermitian operators, this is a unitary irreducible representation (UIR). By carrying out a similar procedure with the other eigenstates of H, we see that the states of the system can be grouped into UIR's of the algebra of the symmetry operators.

This result holds for discrete symmetries, such as parity, as well. When we include continuous spacetime symmetries such as the Poincaré transformations, the Hamiltonian H does not commute with them, but becomes part of the symmetry algebra. In this case, the states again form UIR's of the symmetry algebra, including spacetime symmetries. The states within each UIR are degenerate, not for eigenvalues of the Hamiltonian but for eigenvalues of some invariant operators. (The representations are also infinite dimensional when spacetime symmetry, which is a noncompact symmetry, is included.) Our discussion of relativistic invariance in the appendix is an example of this, where the representations are labeled by mass and spin.

This nice result can be vitiated if $Q^A|\alpha\rangle$ is not normalizable and hence does not belong to the Hilbert space $\mathcal{H}$. In this case, we do not have a unitary representation of the symmetry and it is said to be spontaneously broken. To understand this structure better, we start with the fact that, in field theory, the states are generated by the application of various local operators on the ground state or the vacuum state $|0\rangle$. We can therefore write $|\alpha\rangle = A_\alpha|0\rangle$. From the commutation rules, we can evaluate $[Q^A, A_\alpha] \equiv B_\alpha^A$. $B_\alpha^A|0\rangle$ will be some other local operator in the theory. Generally, the normalization of the states $A_\alpha|0\rangle$ and $B_\alpha^A|0\rangle$ will have to defined with some regularization; with this understood, we may take them to be normalizable. We now write

$$Q^A|\alpha\rangle = B_\alpha^A|0\rangle \; + \; A_\alpha \, Q^A|0\rangle \qquad (11.1)$$

If the ground state has the symmetry, viz., if $Q^A|0\rangle = 0$, $Q^A|\alpha\rangle$ is normalizable in the sense defined above, and it is clear that our argument of the previous paragraphs will go through. States can be grouped into UIR's of the algebra of symmetries. The contrapositive of this statement is that if $Q^A|\alpha\rangle$ is not normalizable, the vacuum cannot be invariant. Thus, spontaneous symmetry breaking corresponds to the situation when we have symmetry at the level of operator algebra, but the vacuum (or ground state) is not invariant. There is no unitary realization of the symmetry algebra.

The discussion so far indicates that there can be two realizations of symmetry in quantum field theory, which can be summarized as follows. If a Lagrangian has a continuous global symmetry given by a set of operators $\{Q^A\}$ forming an algebra $\mathcal{A}$, we have two possibilities.

1. The ground state is symmetric, $Q^A|0\rangle = 0$ and the states can be grouped into unitary irreducible representations of $\mathcal{A}$.
2. $Q^A|0\rangle \neq 0$; in this case the symmetry is spontaneously broken and there is no unitary representation of the symmetry.

The first case is often referred to as the Wigner realization of symmetry and the second as the Goldstone realization of symmetry. (Spontaneous breaking of discrete global symmetries is also possible. A field-theoretic way of implementing the spontaneous breaking of discrete symmetries will be clear from what follows, although we do not discuss them in detail.)

In the real world, ignoring gravitational and cosmological effects, we have Poincaré symmetry. This is realized in the Wigner mode since the particles we observe fall into UIR's of the Poincaré group. (This has been used to construct relativistic wave functions in the appendix.) For this to be possible, the vacuum state must be Lorentz-invariant and this justifies the use of

Lorentz symmetry to eliminate the zero-point energy in Chapter 3. We discuss Wigner realization of symmetries in this chapter; spontaneous symmetry breaking is taken up in the next chapter.

11.2 Ward-Takahashi identities

In quantum field theory, the quantities of interest are the Green's functions. In the Euclidean formulation, these are generated by

$$Z[J] = \mathcal{N} \int [d\varphi] \, \exp\left(-\mathcal{S}[\varphi] + \int J_A \varphi^A\right) \tag{11.2}$$

where $[d\varphi]$ is a suitable measure for functional integration over the fields φ^A. If the theory has a certain symmetry, it manifests itself as relations among various Green's functions. These relations, known as Ward-Takahashi (WT) identities, can be derived as follows. Since φ is a variable of integration in (11.2) we can write

$$\int [d\varphi'] \exp\left(-\mathcal{S}[\varphi'] + \int J_A \varphi'^A\right) = \int [d\varphi] \, \exp\left(-\mathcal{S}[\varphi] + \int J_A \varphi^A\right) \tag{11.3}$$

We shall first consider continuous transformations. Let $\varphi(x) \to \varphi(x) + \xi(x,\varphi)$ under the symmetry transformtion of interest, where $\xi(x,\varphi)$ is considered infinitesimal. The strategy is to choose the new variable of integration φ' in (11.3) as $\varphi + \xi$. Under this transformation, the measure of integration transforms as

$$[d\varphi'] = [d\varphi] \, \det M \approx [d\varphi](1 + \delta\mathcal{J}) \tag{11.4}$$

where

$$M^{AB}(x,y) = \delta^{AB} \delta^{(4)}(x - y) + \frac{\delta \xi^A(x,\varphi)}{\delta \varphi^B(y)}$$

$$\delta\mathcal{J} = \mathrm{Tr}\left(\frac{\delta \xi^A(x,\varphi)}{\delta \varphi^B(y)}\right) \tag{11.5}$$

(Tr denotes the trace over the indices A, B as well as a functional trace; in other words, we take a trace considering (A, x) and (B, y) as matrix indices.)

We also have $\mathcal{S}(\varphi') = \mathcal{S}(\varphi + \xi) \approx \mathcal{S}(\varphi) + \delta_\xi \mathcal{S}$. Using these in (11.3) and collecting together terms which are of the first order in ξ, we get

$$\langle -\delta_\xi \mathcal{S} + \int J_A \xi^A + \delta\mathcal{J} \rangle = 0 \tag{11.6}$$

where the angular brackets denote the functional average

$$\langle \mathcal{O} \rangle = \mathcal{N} \int [d\varphi] \exp\left(-\mathcal{S}[\varphi] + \int J_A \varphi^A\right) \mathcal{O} \tag{11.7}$$

Equation (11.6) is the basic Ward-Takahashi identity. This can be transformed into an infinite set of relations among all the Green's functions by
expansion in powers of J_A. We shall illustrate this by examples.

Consider a theory of N scalar fields φ^a, $a = 1, 2, \cdots, N$, with the Euclidean action

$$S = \int d^4x \left[\frac{1}{2} \partial \varphi^a \partial \varphi^a + \frac{1}{2} m^2 \varphi^a \varphi^a + \lambda (\varphi^a \varphi^a)^2 \right] \tag{11.8}$$

This theory has invariance under the transformation

$$\varphi^a \to \varphi'^a \approx \varphi^a + \omega^{ab} \varphi^b \tag{11.9}$$

where ω^{ab} is antisymmetric in a, b and independent of x^μ. We may write
$\omega^{ab} = \omega^A (T^A)^{ab}$ where $(T^A)^{ab}$, $A = 1, 2, \cdots, \frac{1}{2} N(N-1)$, are generators of
$O(N)$, the set of orthogonal transformations in N variables. Thus (11.8) has a
global $O(N)$ symmetry. The symmetry is global, as opposed to local, because
the parameters are independent of x^μ.

We now consider the change of variables $\varphi^a(x) \to \varphi'^a(x) + \omega^{ab}(x) \varphi^b(x)$
in the functional integral. (For the symmetry (11.9), the parameters ω^{ab}
are independent of spacetime. To obtain the Ward-Takahashi identities, we
consider a change of variables with ω^{ab} which are functions of x^μ.) In this
case

$$\delta S = - \int \partial_\mu \omega^{ab} (\varphi^a \partial^\mu \varphi^b)$$
$$= \int \omega^{ab} \, \partial_\mu (\varphi^a \partial^\mu \varphi^b) \tag{11.10}$$

For the change of variables $\varphi^a(x) \to \varphi'^a(x) + \omega^{ab}(x) \varphi^b(x)$, the matrix
$M^{ab}(x, y) = (\delta^{ab} + \omega^{ab}(x)) \delta^{(4)}(x - y)$. The Jacobian is thus independent
of φ and since the determinant of an $O(N)$-transformation is 1, it is possible
to regulate in such a way that $\delta J = 0$. The general identity (11.6) becomes

$$\langle -\partial_\mu (\varphi^a \partial^\mu \varphi^b - \varphi^b \partial^\mu \varphi^a) + J^a \varphi^b - J^b \varphi^a \rangle = 0 \tag{11.11}$$

Writing $Z[J] = e^{W[J]}$, we have

$$-\frac{\partial}{\partial x^\mu} \left[\frac{\delta W}{\delta J^a(x)} \partial^\mu \frac{\delta W}{\delta J^b(x)} + \frac{\partial}{\partial y^\mu} \left[\frac{\delta^2 W}{\delta J^a(x) \delta J^b(y)} \right]_{y \to x} \right] - (a \leftrightarrow b)$$
$$+ J_a(x) \frac{\delta W}{\delta J^b(x)} - (a \leftrightarrow b) = 0 \tag{11.12}$$

By expanding $W[J]$ in powers of J_a, this becomes a set of relations among
the connected Green's functions of the theory.

If we now take ω^{ab} to be independent of x^μ, then $\delta S = 0$, and we get

$$\int d^4x \left(J_a(x) \frac{\delta W}{\delta J^b(x)} - J_b(x) \frac{\delta W}{\delta J^a(x)} \right) = 0 \qquad (11.13)$$

Written in terms of $\Gamma[\varphi]$, using $J_a = \delta\Gamma/\delta\varphi^a$, this equation reads

$$\int d^4x \left(\Phi^a \frac{\delta\Gamma}{\delta\Phi^b} - \Phi^b \frac{\delta\Gamma}{\delta\Phi^a} \right) = 0 \qquad (11.14)$$

In other words, the quantum effective action $\Gamma[\Phi]$ has the same $O(N)$-symmetry as the classical action.

11.3 Ward-Takahashi identities for electrodynamics

We now discuss the WT identities related to gauge invariance in quantum electrodynamics. Calculations in QED are done by functional integration over the Euclidean action

$$S_E(A, \bar\psi, \psi) = \int d^4x \left[\frac{Z_3}{4} F_{\mu\nu}F^{\mu\nu} + \frac{\lambda}{2}(\partial \cdot A)^2 \right.$$

$$\left. + Z_2 \left[\bar\psi \left(\gamma \cdot \partial + m - \delta m \right) \psi - ie\bar\psi\gamma^\mu\psi A_\mu \right] \right] \qquad (11.15)$$

In general, we have Z_1 and Z_2 in the fermionic part of the action, but we have already seen in the last chapter that when we have a regulator which preserves the BRST invariance, we can set $Z_1 = Z_2$. We have assumed this in the above expression.

This action does not have gauge-invariance due to the gauge-fixing term $(\partial \cdot A)^2$. Thus one can only expect BRST-invariance and the associated WT identities. Nevertheless, it is possible to obtain WT identities for gauge transformations using the action (11.15). Their derivation will be slightly more involved than in the case of BRST identities, but they have the simplicity that we do not need the full formalism of the BRST-transformations.

We start with the functional integral

$$Z[J, \eta, \bar\eta] = \int d\mu[A, \psi, \bar\psi] \exp\left(-S_E(A, \psi, \bar\psi) + \int J \cdot A + \bar\eta\psi + \bar\psi\eta \right)$$

$$= \int d\mu[A', \psi', \bar\psi'] \exp\left(-S_E(A', \psi', \bar\psi') + \int J \cdot A' + \bar\eta\psi' + \bar\psi'\eta \right) \qquad (11.16)$$

where in the second step we have made a semantic change of variables, just renaming the variables of integration. We now take the new variables as

$$A' = A + \partial_\mu \theta$$
$$\psi' = e^{ie\theta} \psi$$
$$\bar{\psi}' = \bar{\psi} \, e^{-ie\theta} \tag{11.17}$$

This is now substituted into the second expression for $Z[J, \psi, \bar{\psi}]$ in (11.16) and taking θ to be small, we collect the terms which are of the first order in θ. If the measure of integration is invariant under this set of transformations, we get

$$\langle -\lambda \int \partial \cdot A \partial^2 \theta + \int J \cdot \partial\theta + ie \int (\bar{\eta}\psi - \bar{\psi}\eta)\theta \rangle = 0 \tag{11.18}$$

Introducing $\log Z = W$ and the generating functional for the 1PI vertices, namely, $\Gamma[A, \psi, \bar{\psi}]$ by

$$W[J, \eta, \bar{\eta}] = -\Gamma[A, \psi, \bar{\psi}] + \int \bar{\eta}\psi + \bar{\psi}\eta + J \cdot A$$

$$J^\mu = \frac{\delta\Gamma}{\delta A_\mu}$$

$$\eta = \frac{\delta\Gamma}{\delta\bar{\psi}}, \qquad \bar{\eta} = -\frac{\delta\Gamma}{\delta\psi} \tag{11.19}$$

we can rewrite equation (11.18) as

$$-\lambda \int \theta \partial^2 \partial \cdot A + \int \partial\theta \cdot \frac{\delta\Gamma}{\delta A} - ie\theta \left[\bar{\psi}\frac{\delta\Gamma}{\delta\bar{\psi}} + \frac{\delta\Gamma}{\delta\psi}\psi \right] = 0 \tag{11.20}$$

(Here $A_\mu, \psi, \bar{\psi}$ are independent fields and not variables of integration as in (11.18).) Since Γ is a functional of the fields $A, \psi, \bar{\psi}$, its change under an infinitesimal gauge transformation (11.17) is given by

$$\delta_\theta \Gamma = \int \partial\theta \cdot \frac{\delta\Gamma}{\delta A} + \delta\bar{\psi} \frac{\delta\Gamma}{\delta\bar{\psi}} + \delta\psi \frac{\delta\Gamma}{\delta\psi}$$

$$= \int \partial\theta \cdot \frac{\delta\Gamma}{\delta A} - ie\,\theta \left[\bar{\psi} \frac{\delta\Gamma}{\delta\bar{\psi}} - \psi \frac{\delta\Gamma}{\delta\psi} \right]$$

$$= \int \partial\theta \cdot \frac{\delta\Gamma}{\delta A} - ie\,\theta \left[\bar{\psi} \frac{\delta\Gamma}{\delta\bar{\psi}} + \frac{\delta\Gamma}{\delta\psi}\psi \right] \tag{11.21}$$

Equation (11.20) can thus be written as

$$\delta_\theta \left[\Gamma - \frac{\lambda}{2} \int (\partial \cdot A)^2 \right] = 0 \tag{11.22}$$

If we define Γ^* by

$$\Gamma = \Gamma^* + \frac{\lambda}{2} \int (\partial \cdot A)^2 \tag{11.23}$$

equation (11.22) shows that Γ^* is gauge invariant. If we expand Γ or Γ^* in terms of vertex functions, this translates into an infinite set of identities connecting various vertex functions.

As an example, consider the two-point term in Γ^* given as

$$\Gamma^{*(2)} = \int A_\mu(x) V^{\mu\nu}(x, y) A_\nu(y) \tag{11.24}$$

The invariance of this expression under infinitesimal gauge transformations, namely, equation (11.22), gives us immediately

$$\partial_\mu V^{\mu\nu}(x, y) = 0 \tag{11.25}$$

Writing $V_{\mu\nu}(x, y) = (-\Box\, \delta_{\mu\nu} + \partial_\mu \partial_\nu) \delta^{(4)}(x - y) + \Pi_{\mu\nu}(x - y)$ and taking Fourier transforms, this is equivalent to

$$k_\mu \Pi^{\mu\nu}(k) = 0 \tag{11.26}$$

Lorentz invariance requires that $\Pi_{\mu\nu}$ should only be a function of k^2; along with the transversality condition (11.26), this gives

$$\Pi_{\mu\nu} = (k^2 \delta_{\mu\nu} - k_\mu k_\nu)\, \Pi(k^2) \tag{11.27}$$

As another example, consider the terms involving $\bar{\psi}\psi$ and $\bar{\psi}\psi A$; these can mix under gauge transformations, so we must consider them together. We write

$$\Gamma^{*(2,3)} = \int \bar{\psi}(x) F(x, z)\psi(z) - ie \int \bar{\psi}(x) V_\mu(x, y, z)\psi(z) A^\mu(y) + \cdots \tag{11.28}$$

The invariance of this under an infinitesimal gauge transformation gives

$$ie \left[\int (\theta(z) - \theta(x))\, \bar{\psi}(x) F(x, z)\psi(z) + \int \theta(y)\bar{\psi}(x) \frac{\partial}{\partial y_\mu} V_\mu(x, y, z)\psi(z) \right]$$
$$- e^2 \int (\theta(x) - \theta(z))\, \bar{\psi}(x) V_\mu(x, y, z)\psi(z) A^\mu(y) + \cdots = 0 \tag{11.29}$$

Setting the coefficient of the $\bar{\psi}\psi$ term to zero, we get

$$F(x, z) \left(\delta^{(4)}(z - y) - \delta^{(4)}(x - y) \right) + \frac{\partial}{\partial y_\mu} V_\mu(x, y, z) = 0 \tag{11.30}$$

For the terms involving A^μ in (11.29), we need to consider the variation of terms involving two A's; so we cannot set the coefficient of this term, as it is written in (11.29), to zero. In (11.30) we introduce the Fourier transforms

$$V_\mu(x, y, z) = \int_{p,q} V_\mu(p, q) e^{ip(x-y)} e^{-iq(z-y)}$$

$$F(x, z)\delta^{(4)}(z - y) = \int_{p,q} F(p) e^{ip(x-y)} e^{-iq(z-y)}$$

$$F(x, z)\delta^{(4)}(x - y) = \int_{p,q} F(q) e^{ip(x-y)} e^{-iq(z-y)} \tag{11.31}$$

In the last two equations, we recover the usual forms if we make the replacements $q \to q + p$, $p \to p + q$, respectively. Using (11.31), we can simplify (11.30) as

$$F(p) - F(q) + i(q - p)^\mu V_\mu(p, q) = 0 \tag{11.32}$$

Expanding this equation for small values of $q - p$, we get the equation

$$-\frac{\partial F}{\partial p^\mu} + iV_\mu(p, p) = 0 \tag{11.33}$$

Equations (11.32) and (11.33) are the forms in which the WT identities were originally obtained. Since $F(p) = i\gamma \cdot p + m + \Sigma(p)$, and $V_\mu(p, q) = \gamma_\mu + \Gamma_\mu(p, q)$, we see that these equations give an identity relating the electron self-energy $\Sigma(p)$ to the vertex correction $\Gamma_\mu(p, q)$.

11.4 Discrete symmetries

We now discuss some of the discrete symmetries of a gauge theory. These will include parity, charge conjugation, and time-reversal. When used in conjunction with the WT identities expressing the gauge-invariance of the effective action, these can be very powerful in restricting the form of the action (and the effective action) and can lead to low-energy theorems which are of general validity, not necessarily restricted to perturbation theory.

In the terms of functional integral representations, discrete symmetries can be understood as follows. Let $\tilde\varphi$ denote the transform of the field variable φ under the discrete symmetry like parity, charge conjugation or time-reversal. This means that the Green's function

$$\tilde{G} = \mathcal{N} \int [d\varphi]\, e^{iS(\varphi)}\ \tilde\varphi(x_1)\tilde\varphi(x_2) \cdots \tilde\varphi(x_N) \tag{11.34}$$

will describe the time-evolution of the transformed particle states. If the action has the property that $S(\varphi) = S(\tilde\varphi)$, and the measure has thus property as well, then we can write

$$\tilde{G} = \mathcal{N} \int [d\varphi]\, e^{iS(\varphi)}\ \tilde\varphi(x_1)\tilde\varphi(x_2) \cdots \tilde\varphi(x_N)$$

$$= \tilde{G} = \mathcal{N} \int [d\tilde\varphi]\, e^{iS(\tilde\varphi)}\ \tilde\varphi(x_1)\tilde\varphi(x_2) \cdots \tilde\varphi(x_N)$$

$$= G \tag{11.35}$$

since the change $\varphi \to \tilde{\varphi}$ is reduced to a semantic change of variables in the integral. We will now work out these transformations for parity, charge conjugation, and time-reversal. Some of these symmetries require the Minkowski metric and so in this section we shall use the functional integral in Minkowski space.

Parity

Parity corresponds to $\boldsymbol{x} \to -\boldsymbol{x}$. On Dirac spinors this is implemented by

$$\psi(\boldsymbol{x}, t) = \gamma^0 \chi(-\boldsymbol{x}, t) \tag{11.36}$$

where $\chi(\boldsymbol{x}, t)$ is the parity image or parity transform of $\psi(\boldsymbol{x}, t)$. On the gauge field, parity is given by

$$\begin{aligned} A_0(\boldsymbol{x}, t) &= \tilde{A}_0(-\boldsymbol{x}, t) \\ A_i(\boldsymbol{x}, t) &= -\tilde{A}_i(-\boldsymbol{x}, t) \end{aligned} \tag{11.37}$$

where the tildes denote parity transforms.

We can check the parity invariance of QED by checking the various terms involved. As an example, consider $\int \bar{\psi}\gamma \cdot \partial\psi$. Writing $\tilde{x} = (-\boldsymbol{x}, t)$, we find

$$\begin{aligned} \int d^4x \ \bar{\psi}i\gamma \cdot \partial\psi &= \int d^4x \ \chi^\dagger(\tilde{x})i\gamma^\mu\partial_\mu\gamma^0\chi(\tilde{x}) \\ &= \int d^4x \ \chi^\dagger(\tilde{x})i(\partial_0 - \gamma^0\gamma^i\partial_i)\chi(\tilde{x}) \\ &= \int d^4x \ \bar{\chi}(\tilde{x})i\gamma^\mu\frac{\partial}{\partial\tilde{x}^\mu}\chi(\tilde{x}) \\ &= \int d^4\tilde{x} \ (\bar{\chi}i\gamma^\mu\partial_\mu\chi)_{x\to\tilde{x}} \end{aligned} \tag{11.38}$$

(The change of limits compensates for the replacement of x by $\tilde{x}$ in the measure of integration.) Equation (11.38) shows the parity invariance of the kinetic term of the Dirac Lagrangian.

For the standard bilinear covariants of the Dirac theory, we have the parity property

$$\begin{aligned} \bar{\psi}\psi(x) &= \bar{\chi}\chi(\tilde{x}) \\ \bar{\psi}\gamma^5\psi(x) &= -\bar{\chi}\gamma^5\chi(\tilde{x}) \end{aligned} \tag{11.39}$$

$$\begin{aligned} \bar{\psi}\gamma^0\psi(x) &= \bar{\chi}\gamma^0\chi(\tilde{x}) \\ \bar{\psi}\gamma^i\psi(x) &= -\bar{\chi}\gamma^i\chi(\tilde{x}) \end{aligned} \tag{11.40}$$

$$\begin{aligned} \bar{\psi}\gamma^0\gamma^5\psi(x) &= -\bar{\chi}\gamma^0\gamma^5\chi(\tilde{x}) \\ \bar{\psi}\gamma^i\gamma^5\psi(x) &= \bar{\chi}\gamma^i\gamma^5\chi(\tilde{x}) \end{aligned} \tag{11.41}$$

$$\bar{\psi}\,[\gamma^0, \gamma^i]\,\psi(x) = -\bar{\chi}\,[\gamma^0, \gamma^i]\,\chi(\tilde{x})$$
$$\bar{\psi}\,[\gamma^i, \gamma^j]\,\psi(x) = \bar{\chi}\,[\gamma^i, \gamma^j]\,\chi(\tilde{x}) \tag{11.42}$$

Thus $\bar{\psi}\psi$ is a scalar, $\bar{\psi}\gamma^5\psi$ is a pseudoscalar, $\bar{\psi}\gamma^\mu\psi$ is a vector, $\bar{\psi}\gamma^\mu\gamma^5\psi$ is an axial vector, and $\bar{\psi}[\gamma^\mu, \gamma^\nu]\psi$ is a rank 2 tensor.

The definition of parity (11.36) shows also that

$$\tfrac{1}{2}(1 \pm \gamma^5)\psi(\boldsymbol{x}, t) = \gamma^0\,\tfrac{1}{2}(1 \mp \gamma^5)\chi(-\boldsymbol{x}, t) \tag{11.43}$$

Thus the chiral projections $\psi_L = \tfrac{1}{2}(1 + \gamma^5)\psi$ and $\psi_R = \tfrac{1}{2}(1 - \gamma^5)\psi$ are transformed into each other under parity. A theory which has only one type of chiral fermions or a theory in which there are gauge fields which couple only to one type of chiral projection (e.g., $A_\mu\bar{\psi}\gamma^\mu(1+\gamma^5)\psi$) will necessarily violate parity invariance. The standard model of weak interactions is an example of such a theory.

Charge conjugation

This corresponds to the exchange of charges, for example, $e^+ \leftrightarrow e^-$, in any process. It is given by

$$\psi(x) = C\bar{\chi}^T(x)$$
$$A_\mu(x) = -\tilde{A}_\mu(x) \tag{11.44}$$

where C is the charge conjugation matrix defined by

$$C^{-1}\gamma^\mu C = -\gamma^{\mu T} \tag{11.45}$$

Recall that the γ-matrices are defined by the algebra

$$\gamma^\mu\gamma^\nu + \gamma^\nu\gamma^\mu = 2\eta^{\mu\nu}\mathbf{1} \tag{11.46}$$

By taking the transpose of this equation, we see that $\gamma^{\mu T}$ should also obey the same algebra. Therefore, by the general theorem on the Clifford algebra mentioned in Chapter 1, we see that there should exist a matrix C with the property (11.45). The specific form of the matrix C will depend on the representation chosen for the γ^μ. For the choice

$$\gamma^0 = \begin{pmatrix} 1 & 0 \\ 0 & 1 \end{pmatrix}, \qquad\qquad \gamma^i = \begin{pmatrix} 0 & \sigma^i \\ -\sigma^i & 0 \end{pmatrix} \tag{11.47}$$

we have

$$C = i\gamma^0\gamma^2, \qquad\qquad C^{-1} = -C = C^\dagger \tag{11.48}$$

In this case, we can also write $\psi(x) = C\gamma^0\chi^*(x) = -i\gamma^2\chi^*(x)$ as the transformation of the fermion field.

Based on the rules given above, the C-invariance of any term in a Lagrangian can be easily checked. For example, consider the electron kinetic term in QED.

$$\bar{\psi}\gamma\cdot(\partial - ieA)\psi = \chi^T\gamma^0 C^\dagger\gamma^0\gamma^\mu(\partial_\mu + ie\tilde{A}_\mu)C\gamma^0\chi^*$$
$$= \chi^T\gamma^0\gamma^{0T}\gamma^{\mu T}\gamma^0(\partial_\mu + ie\tilde{A}_\mu)\chi^*$$
$$= -\partial_\mu\chi^\dagger\gamma^0\gamma^\mu\chi - ie\tilde{A}_\mu\chi^\dagger\gamma^0\gamma^\mu\chi$$
$$= -\partial_\mu\bar{\chi}\gamma^\mu\chi - ie\tilde{A}_\mu\bar{\chi}\gamma^\mu\chi \qquad (11.49)$$

where we have used (11.45). Notice also that in the third step we get an extra minus sign in rearranging χ^T and χ^* because they are Grassmann-valued. In the action, we can do a partial integration to get

$$\mathcal{S}(A,\psi,\bar{\psi}) = \int d^4x\ \bar{\psi}\ i\gamma\cdot(\partial - ieA)\psi$$
$$= \int d^4x\ i\left[-\partial_\mu\bar{\chi}\gamma^\mu\chi - ie\tilde{A}_\mu\bar{\chi}\gamma^\mu\chi\right]$$
$$= \int d^4x\ i\left[\bar{\chi}\gamma^\mu\partial_\mu\chi - ie\tilde{A}_\mu\bar{\chi}\gamma^\mu\chi\right]$$
$$= \mathcal{S}(\tilde{A},\chi,\bar{\chi}) \qquad (11.50)$$

The C-invariance of the other terms in the Lagrangian can be checked in a similar way. For the standard bilinear covariants we have

$$\bar{\psi}\psi = \bar{\chi}\chi$$
$$\bar{\psi}\gamma^5\psi = \bar{\chi}\gamma^5\chi$$
$$\bar{\psi}\gamma^\mu\psi = -\bar{\chi}\gamma^\mu\chi$$
$$\bar{\psi}\gamma^\mu\gamma^5\psi = \bar{\chi}\gamma^\mu\gamma^5\chi$$
$$\bar{\psi}[\gamma^\mu,\gamma^\nu]\psi = \bar{\chi}[\gamma^\mu,\gamma^\nu]\chi \qquad (11.51)$$

Notice that $C^{-1}\gamma^5 C = \gamma^{5T}$ using $\gamma^5 = i\gamma^0\gamma^1\gamma^2\gamma^3$.

When there are nonabelian gauge fields, the notion of charge conjugation involves conjugation in the Lie algebra as well. Consider, for example, the term $\bar{\psi}\gamma^\mu A_\mu\psi$, where ψ transforms as some representation of the gauge group. ψ is a column vector on which the Lie algebra matrices t^a can act as a linear transformation. $A_\mu = -it^a A_\mu^a$ and t^a are the generators of the Lie algebra in the representation corresponding to ψ. Using the charge conjugation property (11.44), we have

$$-iA_\mu^a\bar{\psi}\gamma^\mu t^a\psi = -iA_\mu^a\ \chi^T\gamma^0 C^\dagger\gamma^0\gamma^\mu t^a C\gamma^0\chi^*$$
$$= -iA_\mu^a\chi^T\gamma^0\gamma^{0T}\gamma^{\mu T}\gamma^0 t^a\chi^*$$
$$= iA_\mu^a\chi^\dagger\gamma^0\gamma^\mu t^{aT}\chi \qquad (11.52)$$

This term will have invariance if we define the charge conjugation property of the gauge field as $t^a A_\mu^a = (-t^{aT})\tilde{A}_\mu^a$; in this case, $-iA_\mu^a\bar{\psi}\gamma^\mu t^a\psi = -i\tilde{A}_\mu^a\bar{\chi}\gamma^\mu t^a\chi$. This definition of charge conjugation corresponds exactly to conjugation in the Lie algebra. Writing the Lie algebra commutation rules as

$[t^a, t^b] = i f^{abc} t^c$, we see that if t^a form a matrix representation, then so do $-t^{aT}$. The transformation $t^a \rightarrow -t^{aT}$ is an automorphism of the Lie algebra. In terms of representations, this corresponds to replacing a given representation by its conjugate. Thus $A^a_\mu t^a = \tilde{A}^a_\mu(-t^{aT})$ is indeed the correct definition of charge conjugation.

Time − reversal

Time-reversal transformation is defined by $(\boldsymbol{x}, t) \rightarrow (\boldsymbol{x}, -t) \equiv \tilde{x}$. The time-reversal transformation of fields is given by

$$
\begin{aligned}
\psi(\boldsymbol{x}, t) &= \gamma^5 C \chi^*(\boldsymbol{x}, -t) \\
A_0(\boldsymbol{x}, t) &= \tilde{A}_0(\boldsymbol{x}, -t) \\
A_i(\boldsymbol{x}, t) &= -\tilde{A}_i(\boldsymbol{x}, -t)
\end{aligned}
\tag{11.53}
$$

Time-reversal is an antiunitary transformation. In terms of operators, this means that if $\tilde{A}$, $\tilde{B}$ denote the transforms of operators A and B, then $\widetilde{AB} = \tilde{B}\tilde{A}$.

Consider now the change of various terms in the action in the functional integral for QED under (11.53) viewed as a change of variables. It is easily seen that the bosonic terms are invariant, i.e., $\mathcal{S}(A) = \mathcal{S}(\tilde{A})$. For the kinetic term of the Dirac action, by rearrangement of fields and a partial integration, we find

$$
\int d^4x \, i\bar{\psi}\gamma \cdot (\partial - ieA)\psi = i \int d^4x \left[\chi^T C^\dagger \gamma^5 \gamma^0 \gamma^\mu \partial_\mu \gamma^5 C \chi^* \right.
$$
$$
\left. -ie\chi^T C^\dagger \gamma^5 \gamma^0 \gamma^\mu \gamma^5 C \chi^* A_\mu \right]
$$
$$
= i \int d^4x \left[\chi^\dagger \gamma^\mu \gamma^0 \partial_\mu \chi + ie\chi^\dagger \gamma^\mu \gamma^0 \chi A_\mu \right]
$$
$$
= i \int d^4x \left[\bar{\chi}(\gamma^0 \partial_0 - \gamma^i \partial_i)\chi + ie\bar{\chi}(\gamma^0 A_0 - \gamma^i A_i)\chi \right]
$$
$$
= i \int d^4x \left[-\left(\bar{\chi}\gamma^\mu \frac{\partial}{\partial \tilde{x}^\mu} \chi \right) + ie\bar{\chi}\gamma^\mu \tilde{A}_\mu(\tilde{x})\chi \right]
$$
$$
= -\int d^4\tilde{x} \left[\bar{\chi}(\tilde{x}) \, i\gamma^\mu \left(\frac{\partial}{\partial \tilde{x}^\mu} - ie\tilde{A}_\mu(\tilde{x}) \right) \chi(\tilde{x}) \right]
$$
$$
= -\int d^4x \, \bar{\chi} \, i\gamma^\mu(\partial_\mu - ie\tilde{A}_\mu)\chi
\tag{11.54}
$$

In the last step, we renamed the variable of integration $\tilde{x}$ as x. We get the same term with the time-reversed fields substituted in, but there is an extra minus sign. A similar result holds for the mass term in QED, so that for the whole fermionic part of the action we get

$$
\mathcal{S}(A, \psi, \bar{\psi}) = -\mathcal{S}(\tilde{A}, \chi, \bar{\chi})
\tag{11.55}
$$

This change of sign is related to the antiunitary nature of the time-reversal transformation. At the level of the S-matrix, this corresponds, as should be expected, to the replacement of processes by time-reversed processes. At the level of the generating functional Z, invariance of QED follows from the fact that the determinant obtained by integration over fermions, namely, $\det(i\gamma \cdot (\partial - ieA) - m)$, is also equal to $\det[-(i\gamma \cdot (\partial - ieA) - m)]$.

Furry's theorem

The invariance of QED under charge conjugation leads to the result that all Feynman diagrams with only an odd number of external photon lines vanish. This is known as Furry's theorem. It is easily obtained as follows. We carry out the change of variables $\psi \to C\gamma^0 \chi^*$ in the functional integral. We do not make any change for the electromagnetic potential A_μ. From (11.50), we have $\mathcal{S}(A, \psi, \bar{\psi}) = \mathcal{S}(\tilde{A}, \chi, \bar{\chi}) = \mathcal{S}(-A, \chi, \bar{\chi})$ for the fermionic action. The bosonic part is even in A and so we have indeed

$$\mathcal{S}(A, \psi, \bar{\psi}) = \mathcal{S}(-A, \chi, \bar{\chi}) \tag{11.56}$$

for the whole theory. We then find

$$\begin{aligned}
Z[J] &= \int [dA d\psi d\bar{\psi}]\; e^{i\mathcal{S}(A,\psi,\bar{\psi})} e^{\int J^\mu A_\mu} \\
&= \int [dA d\chi d\bar{\chi}]\; e^{i\mathcal{S}(-A,\chi,\bar{\chi})} e^{\int J^\mu A_\mu} \\
&= \int [dA d\chi d\bar{\chi}]\; e^{i\mathcal{S}(A,\chi,\bar{\chi})} e^{-\int J^\mu A_\mu} \\
&= Z[-J]
\end{aligned} \tag{11.57}$$

This shows immediately that all diagrams with only an odd number of photon external lines will vanish. (If there are external fermion lines, the result is different since there will be changes for the fermion sources.)

CPT theorem

While QED has the discrete symmetries of C, P and T, the standard model of particle interactions violates parity invariance and, to a small extent, T invariance. However, if we have a Lorentz-invariant theory with an interaction Hamiltonian which is the integral of a local hermitian density and if the fields are quantized with the proper spin-statistics connection, then, it can be proved that CPT, the combined operation of all three, is always a symmetry. This is the celebrated CPT theorem. The product can be taken in any order.

11.5 Low-energy theorem for Compton scattering

The WT idenitity for gauge-invariance (11.22) tells us that Γ^* for QED is gauge-invariant. By combining this with the discrete symmetries, it is possible to derive general low-energy theorems for certain scattering processes. Here we illustrate how this can be done for Compton scattering. We take the incoming fermion momentum to be p, the outgoing fermion momentum to be p'; k, k' will denote the incoming and outgoing photon momenta, respectively. We want to consider the forward scattering amplitude for which $p = p'$ and $k = k'$. For the forward scattering amplitude, the only kinematic invariant is $\omega = p \cdot k/m$, and we want to calculate the scattering amplitude to the linear order in ω.

As we discussed in Chapter 8, Γ is the quantum effective action, quantum effects are already included in it, and so, we can do classical scattering theory using Γ as the action to obtain the full quantum S-matrix of the theory. In other words, the S-matrix can be constructed from Γ by considering only the tree diagrams. Consider an expansion of the Γ^* in powers of the photon field A_μ. For Compton scattering, with one incoming and one outgoing photon, we need the terms with one and two powers of A_μ only. (Higher powers of A can be eliminated since we should not form photon loops when using Γ.) We are also interested in low energy scattering, so we can consider various vertex functions, as in (11.28), expanded in powers of the momenta or derivatives of fields. For example, the two-point function for the fermions is of the form

$$\Gamma^*_{\bar\psi\psi} = \int \bar\psi(x)\ K(x,y)\psi(y) \tag{11.58}$$

The equation $K\psi = 0$ must have as solutions the usual plane wave solutions of the Dirac equation. This tells us that, in momentum space, K must have the form $K(p) = (\gamma \cdot p - m)h(s)$ where $h(s)$ is some function which can be expanded in powers of $s = (p^2 - m^2)/m^2$ and $h(0) = 1$ to ensure that the fields are normalized properly. (If we consider g as a function of p^2 rather than s, then we need $h(p^2 = m^2) = 1$. One can rearrange the series to write it in terms of s.) In the interacting theory, WT identities connect this two-point function to the photon vertices. In coordinate-space, momenta are derivatives and terms with derivatives are not gauge invariant. The photon vertices are such that they combine with the two-point function appropriately to ensure gauge-invariance. Since we know that derivatives must become covariant derivatives, $\partial \to D = \partial - ieA$, for reasons of gauge invariance, it is easy to write down the general form of the function. We need

$$\Gamma^*_{\bar\psi\psi} = \frac{1}{2} \int \bar\psi \ [h(S)\,(i\gamma \cdot D - m) + (i\gamma \cdot D - m)\,h(S)]\,\psi \tag{11.59}$$

Here $m^2 S = -D^2 - m^2$. Notice that S does not commute with $\gamma \cdot D$, so the ordering of h and $\gamma \cdot D$ is important. The symmetric ordering is consistent with

C-invariance and the fact that we have free photons in Compton scattering. Notice that the relevant commutator is

$$\bar{\psi}[i\gamma \cdot D, -D^2]\psi = -2e\bar{\psi}\gamma^\mu F_{\mu\nu}D_\nu\psi - e\bar{\psi}\gamma^\mu\psi(D_\nu F_{\mu\nu}) \tag{11.60}$$

The second term is irrelevant for Compton scattering which does not need photon propagators and so the photon field can be taken to obey the equation of motion. The first term is ruled out by C-invariance. This justifies the symmetric ordering in (11.59).

In addition to the two-point function given above, we can have terms like

$$\Gamma^* = \int \frac{ie}{16m}\bar{\psi}[\gamma^\mu, \gamma^\nu]f(S)\psi F_{\mu\nu} + aF_{\mu\nu}F^{\mu\nu}\bar{\psi}\psi + \cdots \tag{11.61}$$

The term with the two F's is also nonlocal in general. In terms of photon momenta k, it is already of order k^2, so higher derivatives are irrelevant at low energies; even this term can be neglected to the order we are calculating. Similarly, for the first term in (11.61) we only need $f(0)$, the higher terms will be negligible for k small compared to the mass of the fermion. $f(0)$ is related to the anomalous magnetic moment of the particle and, in fact, by comparing this term with the nonrelativistic limit of the Dirac theory, we can identify $f(0) = g - 2$.

For low-energy Compton scattering, we only need $h(S) \approx 1 + h_1 S = 1 + h_1(m^2 s - eA \cdot p - ep \cdot A + e^2 A^2)/m^2$. The relevant terms are then obtained as

$$\Gamma^* = \Gamma^{(0)} + \Gamma^{(1)} + \Gamma^{(1')} + \Gamma^{(1'')} + \Gamma^{(2)}$$

$$\Gamma^{(0)} = \bar{\psi}(i\gamma \cdot \partial - m)h(s)\psi$$

$$\Gamma^{(1)} = e\bar{\psi}\frac{1}{2}\left[\gamma \cdot Ah(s) + h(s)\gamma \cdot A\right]\psi$$

$$\Gamma^{(1')} = \frac{ie}{16m}(g - 2)\bar{\psi}[\gamma^\mu, \gamma^\nu]\psi F_{\mu\nu}$$

$$\Gamma^{(1'')} = \frac{eh_1}{2m^2}\bar{\psi}\Big[(i\gamma \cdot \partial - m)(A \cdot P + P \cdot A)$$

$$+ (A \cdot P + P \cdot A)(i\gamma \cdot \partial - m)\Big]\psi$$

$$\Gamma^{(2)} = \frac{e^2 h_1}{2m^2}\bar{\psi}\left[(i\gamma \cdot \partial - m)A^2 + A^2(i\gamma \cdot \partial - m)\right]\psi$$

$$- \frac{e^2 h_1}{2m^2}\bar{\psi}\left[\gamma \cdot A(A \cdot P + P \cdot A) + (A \cdot P + P \cdot A)\gamma \cdot A\right]\psi \tag{11.62}$$

In this expression $P = i\partial$, and we have already dropped the F^2-term from (11.61) for reasons cited.

Consider the term $\Gamma^{(2)}$ with two A's first. Here ψ and $\bar{\psi}$ will be replaced by free-particle wave functions which obey the equations of motion. So the first term in $\Gamma^{(2)}$ does not contribute to Compton scattering. The second term will involve $(2p + k) \cdot e$ or $(2p + k) \cdot e'$. It is possible to choose a gauge where $p \cdot e = 0$ and at low energies we get $p \cdot e' \approx 0$. (This is evident if we go to the rest frame of the initial fermion.) All of $\Gamma^{(2)}$ is not relevant to the order in photon momentum that we are calculating. For similar reasons, the term $\Gamma^{(1'')}$ (with $(A \cdot P + P \cdot A)$) can be seen to be negligible. The only terms which are important to the order we are interested in are $\Gamma^{(0)}$, $\Gamma^{(1)}$ and $\Gamma^{(1')}$. In computing the amplitude using these terms, we encounter $h(s)$ for $s = ((p+k)^2 - m^2)/m^2 = 2p \cdot k/m^2 = 2\omega/m$ and for $s = ((p-k)^2 - m^2)/m^2 = -2\omega/m$.

The probability amplitude due to $\Gamma^{(1)}$, taken to second order, is given by

$$
\begin{aligned}
\mathcal{M}^{(1)} &= -e^2 \bar{u}_p \left[\gamma \cdot e' \left(1 + h_1 \frac{\omega}{m} \right)^2 \frac{i}{(\gamma \cdot (p+k) - m)(1 + 2h_1\omega/m)} \gamma \cdot e \right. \\
&\quad \left. + \gamma \cdot e \left(1 - h_1 \frac{\omega}{m} \right)^2 \frac{i}{(\gamma \cdot (p-k) - m)(1 - 2h_1\omega/m)} \gamma \cdot e' \right] u_p \\
&\approx -ie^2 \frac{1}{2p \cdot k} \bar{u}_p \left[\gamma \cdot e' \gamma \cdot k \gamma \cdot e + \gamma \cdot e \gamma \cdot k \gamma \cdot e' \right] u_p \\
&\approx -i \frac{e^2}{m} e \cdot e'
\end{aligned}
\tag{11.63}
$$

where we have used the fact that, because we chose $p \cdot e = 0$,

$$
(\gamma \cdot p + m) \, \gamma \cdot e u_p = -\gamma \cdot e (\gamma \cdot p + m) u_p = 0
\tag{11.64}
$$

and the similar result for e'.

The amplitude which involves the product of $\Gamma^{(1)}$ and $\Gamma^{(1')}$ is zero to this order. The contribution due to $\Gamma^{(1')}$ taken to second order gives

$$
\begin{aligned}
\mathcal{M}^{(1')} &= \frac{e^2(g-2)^2}{64m^2} \bar{u}_p \left[[\gamma^\mu, \gamma^\nu] k_\mu e'_\nu \frac{i}{\gamma \cdot (p+k) - m} [\gamma^\alpha, \gamma^\beta] k_\alpha e_\beta \right. \\
&\quad \left. + [\gamma^\mu, \gamma^\nu] k_\mu e_\nu \frac{i}{\gamma \cdot (p-k) - m} [\gamma^\alpha, \gamma^\beta] k_\alpha e'_\beta \right] u_p
\end{aligned}
\tag{11.65}
$$

By rationalizing the propagator and rearranging the terms in the numerator and using $(\gamma \cdot k)^2 = 0$, we can simplify this to

$$
\mathcal{M}^{(1')} = \frac{ie^2}{16m^2} (g-2)^2 \bar{u}_p \gamma^\mu [\gamma^\nu, \gamma^\alpha] u_p \, k_\mu e'_\nu e_\alpha
\tag{11.66}
$$

We have also dropped some more terms which are negligible for small k. Since $\gamma \cdot p u_p = m u_p$ we may write

$$\bar{u}_p \gamma^\mu [\gamma^\nu, \gamma^\alpha] u_p = \frac{1}{2m} \bar{u}_p \left[\gamma \cdot p \gamma^\mu [\gamma^\nu, \gamma^\alpha] + \gamma^\mu [\gamma^\nu, \gamma^\alpha] \gamma \cdot p \right] u_p \qquad (11.67)$$

Simplifying the γ-matrix algebra, we find

$$\bar{u}_p \gamma^\mu [\gamma^\nu, \gamma^\alpha] u_p = \bar{u}_p \left[\frac{p^\mu}{m} [\gamma^\nu, \gamma^\alpha] + \frac{2p^\alpha}{m} \gamma^\mu \gamma^\nu - \frac{2p^\nu}{m} \gamma^\mu \gamma^\alpha \right] u_p \qquad (11.68)$$

Substituting this in (11.66), we get

$$\mathcal{M}^{(1')} = \frac{e^2}{8m^2} (g-2)^2 \, \omega \, \bar{u}_p \sigma^i u_p (e' \times e)_i \qquad (11.69)$$

The forward scattering amplitude can be parametrized in general as

$$\mathcal{M} = i\bar{u}_p \left[f_1(\omega) \, e \cdot e' + i\omega f_2(\omega) \, (e' \times e)_i \, \sigma^i \right] u_p \qquad (11.70)$$

where $f_1(\omega)$ and $f_2(\omega)$ are the (photon) helicity-preserving and helicity-flipping amplitudes. Combining our results (11.63) and (11.69), we get the total scattering amplitude in the forward direction at low energies as

$$\mathcal{M} \approx i\bar{u}_p \left[-\frac{e^2}{m} e \cdot e' - i\omega \frac{e^2}{8m^2} (g-2)^2 (e' \times e)_i \sigma^i \right] u_p \qquad (11.71)$$

In other words,

$$f_1(0) = -\frac{e^2}{m}$$
$$f_2(0) = -\frac{e^2}{8m^2} (g-2)^2 \qquad (11.72)$$

The low-energy forward Compton scattering is entirely determined by the charge and magnetic moment of the fermion on which the photon scatters. This result is very general. We have not made any assumptions of perturbation theory, nor have we assumed that there are only electromagnetic interactions. We have only used the invariance properties of QED coupled with a low-energy expansion. This result can hold for even composite particles like the proton.

Direct verification of (11.72) for the proton, for example, is difficult; usually it is converted to a sum rule on cross sections for scattering of polarized photons on the target fermion by using dispersion relations. This sum rule is known as the Drell-Hearn sum rule. It is in good agreement with experimental data.

References

1. The WT identities for QED, although not in the functional form, were given by J.C. Ward, Phys. Rev. **78**, 182 (1950); Y. Takahashi, Nuovo Cim. **6**, 371 (1957).

2. Furry's theorem is due to W.H. Furry, Phys. Rev. **51**, 125 (1937).

3. For a proof of the CPT theorem, see, for example, R.F. Streater and A.S. Wightman, *PCT, Spin and Statistics and All That*, W.A. Benjamin, Inc. (1964).

4. The original references on low energy theorems are F. Low, Phys. Rev. **96**, 1428 (1954); **110**, 1178 (1958); M. Gell-Mann and M.L. Goldberger, Phys. Rev. **96**, 1433 (1954); S.D. Drell and A.C. Hearn, Phys. Rev. Lett. **16**, 908 (1966).

12 Spontaneous symmetry breaking

12.1 Spontaneous breaking of a continuous global symmetry

In this chapter we give a more detailed description of spontaneous symmetry breaking or the Goldstone realization of symmetry, focusing on continuous global symmetries at first. The characterization of the ground state or vacuum state by $Q^A|0\rangle \neq 0$, for a spontaneously broken symmetry generated by Q^A, is not very convenient for calculational purposes. Therefore we shall rewrite this in somewhat different ways. Let A_α be a set of operators which transform nontrivially under the continuous symmetry Lie group G. For an infinitesimal transformation, we have $\delta A_\alpha = (t^A_{\alpha\beta}\theta^A)\, A_\beta$ for some group parameters θ^A and $t^A_{\alpha\beta}$ are the generators of the Lie algebra of G in the matrix representation to which A_α belong. Since Q^A generate group transformations in the quantum theory, this means that we can write

$$i\delta A_\alpha \equiv [A_\alpha, Q \cdot \theta]$$
$$= i(t^A_{\alpha\beta}\theta^A)\, A_\beta \tag{12.1}$$

The finite version of this relation is

$$e^{iQ\cdot\theta}\, A_\alpha\, e^{-iQ\cdot\theta} = \mathcal{D}_{\alpha\beta}(\theta)\, A_\beta \tag{12.2}$$

where $\mathcal{D}_{\alpha\beta}(\theta)$ is the matrix representing the transformation corresponding to the parameters θ^A in the representation to which A_α belong. Taking the vacuum (ground state) expectation value of this equation, we obtain

$$\langle 0|e^{iQ\cdot\theta}\, A_\alpha\, e^{-iQ\cdot\theta}|0\rangle = \mathcal{D}_{\alpha\beta}(\theta)\, \langle 0|A_\beta|0\rangle \tag{12.3}$$

From this equation we see that, if $Q^A|0\rangle = 0$, $\langle 0|A_\alpha|0\rangle = \mathcal{D}_{\alpha\beta}(\theta)\, \langle 0|A_\beta|0\rangle$. Since $\mathcal{D}_{\alpha\beta}(\theta)$ is not the identity matrix for all θ, because A_α is taken to transform nontrivially, this means that $\langle 0|A_\alpha|0\rangle$ must be zero, giving us the statement

$$Q^A|0\rangle = 0 \quad \Rightarrow \quad \langle 0|A_\alpha|0\rangle = 0 \tag{12.4}$$

for all A_α which are not invariant under G. The contrapositive of this statement is that if an operator which is not invariant under G develops a nonzero vacuum (ground state) expectation value , then the symmetry G is spontaneously broken. This gives us a way of implementing spontaneous symmetry breaking by assigning vacuum expectation values to operators which are not invariant under G. Of course, we cannot assign expectation values arbitrarily. The ground state is obtained by minimizing the Hamiltonian. So what we mean is that, in a theory with spontaneous symmetry breaking, the Hamiltonian is such that its minimization leads to nonzero expectation values for certain operators which transform nontrivially under G.

Semiclassical construction of the ground state

The construction of a ground state with the property of spontaneous symmetry breaking can be explicitly carried out if a semiclassical approach is valid for the theory under consideration. We illustrate this by considering the example of spontaneous breaking of a $U(1)$-symmetry. Consider the theory of a complex scalar field φ with the action

$$S = \int d^4x \left[\partial_\mu \varphi^* \partial^\mu \varphi - a\, \varphi^* \varphi - b\, (\varphi^* \varphi)^2 \right] \tag{12.5}$$

This has an obvious $U(1)$ symmetry, the transformations on the field being $\varphi \to \varphi' = e^{i\theta}\varphi$. The Hamiltonian corresponding to (12.5) is given by

$$H = \int d^3x \left[\partial_0 \varphi^* \partial_0 \varphi + \nabla \varphi^* \nabla \varphi + a\, \varphi^* \varphi + b\, (\varphi^* \varphi)^2 \right] \tag{12.6}$$

Classically it is easy to choose the ground state. We must minimize the Hamiltonian. The field configuration $(\varphi, \partial_0 \varphi)$ which minimizes H is the ground state configuration. This configuration will depend on the values of the parameters a, b. If b is negative, H is minimized by taking $\partial_0 \varphi = 0$, $\varphi \to \infty$, in which case $H \to -\infty$. There is no ground state; this is an unphysical case. We must therefore require that $b > 0$. Then there are two cases of interest.
1) $a > 0$. In this case every term in H is positive and H is minimized by the configuration $\partial_0 \varphi = 0$, $\varphi = 0$.
2) $a < 0$. In this case, we rewrite the Hamiltonian as

$$H = \int d^3x \left[\partial_0 \varphi^* \partial_0 \varphi + \nabla \varphi^* \nabla \varphi + b\, (\varphi^* \varphi - v^2)^2 \right] - \int d^3x \, \frac{|a|^2}{4b} \tag{12.7}$$

where $v^2 = |a|/2b$. The configuration which minimizes H is evidently given by

$$\partial_0 \varphi = 0, \qquad \nabla \varphi = 0$$
$$\varphi^* \varphi = v^2 \quad \text{or} \quad \varphi = v e^{i\alpha} \tag{12.8}$$

(α is independent of x since $\partial_\mu \varphi = 0$.)

The classical values can be realized as expectation values in the quantum theory. Denoting the operator corresponding to φ as ϕ, notice that the expectation value of a product like $\phi^* \phi$ is given by

$$\langle 0|\phi^*\phi|0\rangle = \langle 0|\phi^*|0\rangle \langle 0|\phi|0\rangle + \mathcal{O}(\hbar) \tag{12.9}$$

Thus, at least semiclassically, we can specify the properties of the ground state by writing $\langle 0|\phi|0\rangle = v$ where v is obtained by minimizing the classical Hamiltonian. The two possibilities are thus

1. $a > 0$ $\langle 0|\phi|0\rangle = 0$
2. $a < 0$ $\langle 0|\phi|0\rangle = \sqrt{|a|/2b}\, e^{i\alpha} + \mathcal{O}(\hbar).$

Since $\phi \to \phi' = e^{i\theta}\phi$ under the $U(1)$ transformation, the second case where ϕ has an expectation value in the ground state corresponds to spontaneous breaking of $U(1)$ symmetry.

One can construct the wave function of the ground state (vacuum state) explicitly as follows. Consider the first case where $\langle 0|\phi|0\rangle = 0$. We first write $\phi = (\phi_1 + i\phi_2)/\sqrt{2}$, where ϕ_1, ϕ_2 are real fields. We then introduce the mode expansions

$$\phi_i(x) = \sum_k q_{ik} u_k(\boldsymbol{x}) \tag{12.10}$$

for $i = 1, 2$ and where $u_k(x)$ is a complete set of real functions for $\boldsymbol{x}$ within a box of volume $V = L^3$. We can take $u_k(\boldsymbol{x})$ to be eigenfunctions of $-\nabla^2$ with eigenvalues $\boldsymbol{k}^2$. Explicitly, if we impose the boundary condition $\phi = 0$ on ∂V, we may take

$$u_k(\boldsymbol{x}) = \left(\frac{2}{L}\right)^{\frac{3}{2}} \sin\frac{n_1\pi x_1}{L} \sin\frac{n_2\pi x_2}{L} \sin\frac{n_3\pi x_3}{L} \tag{12.11}$$

with

$$\int d^3x\, u_k(\boldsymbol{x})u_l(\boldsymbol{x}) = \delta_{kl} \tag{12.12}$$

In a similar way we can write

$$\partial_0\phi_i = \sum_k p_{ik} u_k(\boldsymbol{x}) \tag{12.13}$$

The canonical commutation rules give

$$[q_{ik}, p_{jl}] = i\delta_{ij}\, \delta_{kl} \tag{12.14}$$

and the Hamiltonian can be written as

$$H = \frac{1}{2}\sum_{k,i} (p_{ik}^2 + \omega_k^2\, q_{ik}^2) + H_{int} \tag{12.15}$$

$\omega_k^2 = \boldsymbol{k}^2 + a$. If we ignore the interaction Hamiltonian H_{int} with the idea of including it perturbatively, then the Hamiltonian is that of the harmonic oscillator form and the ground state wave function can be immediately written down. It is given by

$$\Psi_0 = \langle q|0\rangle = \mathcal{N} \prod_{i,k} \exp\left(-\frac{1}{2}\omega_k q_{ik}^2\right) = \mathcal{N} \exp\left(-\frac{1}{2}\sum_{i,k}\omega_k q_{ik}^2\right) \quad (12.16)$$

($\mathcal{N}$ is a normalization factor.) In this case, evidently,

$$\langle 0|\phi_i|0\rangle = \sum_k \langle 0|q_{ik}|0\rangle \, u_k(\boldsymbol{x}) \, = 0 \quad (12.17)$$

The semiclassical construction of the vacuum wave function can be done along similar lines for the case of spontaneous symmetry breaking. In order to achieve $\langle 0|\phi|0\rangle = v \, e^{i\alpha}$, we write

$$\phi = e^{i\alpha}\left[v + \frac{1}{\sqrt{2}}(\eta_1 + i\eta_2)\right] \quad (12.18)$$

and treat η_1, η_2 as ordinary quantum fields with $\langle 0|\eta_i|0\rangle = 0$. Substituting this into the action (12.5), we find

$$\mathcal{S} = \int \frac{1}{2}\left[(\partial\eta_1)^2 + (\partial\eta_2)^2 - 2|a|\eta_1^2\right] - \sqrt{|a|b}\,\eta_1(\eta_1^2 + \eta_2^2) - \frac{b}{4}(\eta_1^2 + \eta_2^2)^2 \quad (12.19)$$

up to the additive constant $|a|^2/4b$. The term linear in η is zero since v was chosen to be the classical minimum of the action, $v^2|a|/2b$. Treating the cubic and higher terms perturbatively, we can construct the vacuum state in terms of η_1 and η_2. η_1 is a massive field, of mass $\sqrt{2|a|}$. η_2 is massless and it is referred to as the Goldstone boson. The appearance of a massless field is a very general feature of spontaneous breaking of continuous symmetries known as Goldstone's theorem. Introducing a mode expansion

$$\eta_1(x) = \sum_k \eta_{1k} u_k(\boldsymbol{x})$$

$$\eta_2(x) = \sum_k \eta_{2k} u_k(\boldsymbol{x}) \quad (12.20)$$

the ground state wave function can be obtained as

$$\Psi_0 = \mathcal{N} \exp\left(-\frac{1}{2}\sum_k \Omega_k \eta_{1k}^2 - \frac{1}{2}\sum_k \omega_k \eta_{2k}^2\right) \quad (12.21)$$

where $\Omega_k = \sqrt{\boldsymbol{k}\cdot\boldsymbol{k} + 2|a|}$, $\omega_k = \sqrt{\boldsymbol{k}\cdot\boldsymbol{k}}$. With this wave function, it is clear that

$$\langle 0|\phi|0\rangle = v \, e^{i\alpha} \quad (12.22)$$

since $\langle 0|\eta_1|0\rangle = 0$, $\langle 0|\eta_2|0\rangle = 0$, at least in the approximation of neglecting the interactions. (The interactions do involve the Goldstone mode and may vitiate this result in some cases.) From the point of view of separating out the Goldstone mode, a better parametrization for ϕ is

$$\phi = \left(v + \frac{\rho(x)}{\sqrt{2}}\right) \exp\left(i\frac{\lambda(x)}{v\sqrt{2}} + i\alpha\right) \tag{12.23}$$

where $\rho(x)$ and $\lambda(x)$ are the dynamical fields. Substituting this into the action (12.5) we get

$$\mathcal{S} = \int \frac{1}{2}\left[(\partial\rho)^2 + \left(1 + \frac{\rho}{v\sqrt{2}}\right)^2 (\partial\lambda)^2 - 2|a|\rho^2\right] - \sqrt{2}\, vb\, \rho^3 - \frac{b}{4}\, \rho^4 + \text{constant}$$

$$= \int \frac{1}{2}\left[(\partial\rho)^2 + (\partial\lambda)^2 - 2|a|\rho^2\right] + \mathcal{O}(\rho^3) + + \text{constant} \tag{12.24}$$

ρ is a massive field, of mass $\sqrt{2|a|}$ and λ is the massless Goldstone boson. Notice that the potential term does not depend on the Goldstone mode λ. Again, treating the cubic and higher terms perturbatively, we can construct the vacuum state in terms of ρ and λ as

$$\Psi_0 = \mathcal{N}\exp\left(-\frac{1}{2}\sum_k \Omega_k \rho_k^2 - \frac{1}{2}\sum_k \omega_k \lambda_k^2\right) \tag{12.25}$$

where the fields have the mode expansions

$$\rho(x) = \sum_k \rho_k u_k(x)$$

$$\lambda(x) = \sum_k \lambda_k u_k(x) \tag{12.26}$$

The calculation of the expectation value is now more involved because of the composite nature of the operator $\exp\left(i\lambda/v\sqrt{2}\right)$. Introduce a renormalized operator by

$$\mathcal{O} = \exp\left(i\frac{\lambda}{v\sqrt{2}} + A\right) \tag{12.27}$$

where A is a renormalization constant to be fixed by canceling the possible ultraviolet divergence. We can evaluate the expectation value using the wave function (12.25). Since it involves $\langle\lambda(x)\lambda(x)\rangle$, we can also evaluate it in a simpler way from the equal-time value of the propagator, since $\lambda(x)$ is a free field. (There is some difference between these methods, since we used periodic boundary conditions for the propagator whereas, here, we have fields vanishing on the boundary. Since there is arbitrariness due tot he cut-off anyway, this difference is immaterial; the qualitative results are the same.) With a high momentum cut-off Λ, we then find

$$\langle 0|\mathcal{O}|0\rangle = \exp\left(A - \frac{\Lambda^2}{16\pi^2 v^2}\right)$$

$$= 1 \tag{12.28}$$

where we choose the renormalization constant A to be $\Lambda^2/16\pi^2 v^2$; this gives $\langle 0|\phi|0\rangle = ve^{i\alpha}$, as expected from the minimization of the Hamiltonian, perturbatively in the interactions of ρ.

This calculation also shows that there would be exceptions in a theory in one spatial dimension. In this case the k-integration in the evaluation of $\langle 0|\mathcal{O}|0\rangle$ has infrared divergences as well and gives

$$\langle 0|\mathcal{O}|0\rangle = \exp\left(A - \frac{1}{16\pi}\log(\Lambda/k_{min})\right) \qquad (12.29)$$

where k_{min} is a low-momentum or infrared cut-off. v is dimensionless in one spatial dimension and we have set it to 1 without loss of generality. We can renormalize $\mathcal{O}$ at some scale μ, choosing $A = \frac{1}{16\pi}\log\Lambda/\mu$. This leads to

$$\langle 0|\mathcal{O}|0\rangle = \left[\frac{k_{min}}{\mu}\right]^{16\pi}$$
$$\to 0 \qquad (12.30)$$

Thus the expectation value $\langle 0|\phi|0\rangle$ vanishes as the infrared cut-off goes to zero. This is an infrared effect and has nothing to do with our renormalization prescription. Large-scale (small k) fluctuations of the potential Goldstone field destabilize the ground-state expectation value. This problem exists for any massless field; in general, one cannot have spontaneous breaking of continuous symmetries in one spatial dimension. This result is known as the Mermin-Wagner-Coleman theorem.

The construction of the ground state is semiclassical in the sense that we choose the classical ground-state configuration $ve^{i\alpha}$ and build a wave function that peaks around this value. This will be a good starting point if one can do an $\hbar$-expansion of the theory where the higher-order terms are less significant. In an $\hbar$-expansion, we can include corrections to Ψ_0 and $\langle 0|\phi|0\rangle$. However, if an $\hbar$-expansion is not suitable, there is no general rule for the construction of the ground-state wave function, even though it can still be indirectly characterized by the expectation values of various operators.

12.2 Orthogonality of different ground states

The ground state wave function we have constructed gives $\langle 0|\phi|0\rangle = ve^{i\alpha}$. Different values of α correspond to different ground states, so for this section, we will write $|\alpha\rangle$ for the ground state which gives the expectation value $ve^{i\alpha}$. Classically we see that they all have the same energy and so we have a set of degenerate classical ground states. What is the value of the phase α that we should use for the quantum ground state? Does the true ground state correspond to a specific value of α or is it a superposition of $|\alpha\rangle$ over different α's? This is the natural question in using the states $|\alpha\rangle$. We now show the following results.

1. $\langle \alpha | 0 \rangle = 0$. More generally $\langle \alpha | \alpha' \rangle = 0$ for $\alpha \neq \alpha'$.
2. Physical results are independent of the value of α.

These results show that one can build up the quantum theory by choosing one value of α; one must make a choice but it does not matter which value is chosen.

The second statement is easy to see; the substitution of the field ϕ as in (12.23) led to the action (12.24) which is manifestly independent of α. Therefore all results obtained from this action will be independent of α. As for the first statement, notice that a change in α is equivalent to a shift of the Goldstone field λ. The operator which generates this is $\partial_0 \lambda$. Thus, formally we may write

$$|\alpha\rangle = \exp\left(-iv\alpha\sqrt{2}\int d^3x\,\partial_0\lambda\right)|0\rangle \qquad (12.31)$$

The expectation value of $\exp(i\lambda/v\sqrt{2})$ for this state is $e^{i\alpha}$. We will use a cut-off on the spatial volume $V = L^3$ and define the integral of $\partial_0\lambda$ as

$$\int d^3x\,\partial_0\lambda = \left[\int d^3x\,e^{-x^2/L^2}\,\partial_0\lambda\right]_{L\to\infty} \qquad (12.32)$$

The overlap of two states $|0\rangle$ and $|\alpha\rangle$ is given by

$$\langle 0|\alpha\rangle = \langle 0|\exp\left(-iv\alpha\sqrt{2}\int\partial_0\lambda\right)|0\rangle$$

$$= \exp\left(-v^2\alpha^2\int\langle\partial_0\lambda(x)\partial_0\lambda(y)\rangle\right) \qquad (12.33)$$

Evaluating the expectation value in the exponent we find

$$\int\langle\partial_0\lambda(x)\partial_0\lambda(y)\rangle = \frac{1}{2}\int\frac{d^3k}{(2\pi)^3}d^3x\,d^3y\,\omega_k\,e^{-i\boldsymbol{k}\cdot(\boldsymbol{x}-\boldsymbol{y})}e^{-x^2/L^2-y^2/L^2}$$

$$= \frac{\pi}{2}L^2 \qquad (12.34)$$

so that

$$\langle 0|\alpha\rangle = \exp\left(-\frac{\pi}{2}\alpha^2v^2L^2\right) \qquad (12.35)$$

The overlap vanishes as $\exp(-V^{\frac{2}{3}})$. This applies also for general α, α' and further for $\langle 0|\mathcal{O}|\alpha\rangle$ for any *local* operator $\mathcal{O}$. The ground states are thus orthogonal in the limit of large volumes and there are no transitions between them induced by local operators. (Notice that the result is obtained only for $V \to \infty$ which will also correspond to infinite number of degrees of freedom; for systems with finite number of degrees of freedom, truly there is no spontaneous symmetry breaking. In practice, large systems, e.g., macroscopic crystals, can be approximated by the infinite system.) Notice also that the vanishing of $\langle 0|\alpha\rangle$ for all $\alpha \neq 0$ shows that the symmetry cannot be unitarily implemented in the quantum theory.

In summary, we have found that the $U(1)$ theory (12.5) can exist in two phases. For $a > 0$, one has the symmetric phase where the $U(1)$ symmetry is manifest, realized in the Wigner mode. For $a < 0$, the $U(1)$ symmetry is spontaneously broken and we have the Goldstone realization.

12.3 Goldstone's theorem

We shall first consider Goldstone's theorem classically using an $O(N)$-symmetric theory as an example. We consider the action

$$S = \int d^4x \left[\frac{1}{2} \partial_\mu \varphi_a \partial^\mu \varphi_a - V(\varphi) \right] \tag{12.36}$$

The $O(N)$-transformations are given by $\delta\varphi_a = \theta^A (T^A)_{ab}\varphi_b$, where $(T^A)_{ab}$ are generators of the symmetry group $G = O(N)$ given as $N \times N$ antisymmetric matrices. The kinetic term is obviously invariant under these. The invariance of the potential term gives

$$\frac{\partial V}{\partial \varphi_a} (T^A)_{ab}\varphi_b = 0 \tag{12.37}$$

Differentiating this once more, we find

$$\frac{\partial^2 V}{\partial \varphi_a \partial \varphi_b}(T^A)_{ac}\varphi_c + \frac{\partial V}{\partial \varphi_a} (T^A)_{ab} = 0 \tag{12.38}$$

We evaluate this at the vacuum expectation value $\varphi_a = v_a$, which is the solution of $(\partial V / \partial \varphi_a) = 0$. Equation (12.38) then becomes

$$\left(\frac{\partial^2 V}{\partial \varphi_a \partial \varphi_b} \right)_{\varphi=v} (T^A)_{ac} v_c = 0 \tag{12.39}$$

The mass matrix which appears upon expanding S around the vacuum expectation value is

$$M_{ab} = \left(\frac{\partial^2 V}{\partial \varphi_a \partial \varphi_b} \right)_{\varphi=v} \tag{12.40}$$

Equation (12.39) can thus be written as

$$M_{ab}\xi_a^A = 0 \tag{12.41}$$

where $\xi_a^A = (T^A)_{ac} v_c$.

The generators T^A can be divided into two classes. The generators which annihilate v^c form the isotropy subgroup of v_c (or little group of v_c) denoted by $H \in G$; these generators of H will be denoted by t^α. The generators of G which correspond to the broken symmetries, i.e., those which are not in the algebra of H, will be denoted by S^i. For the H-generators $\xi^A = 0$ and we have no information about their masses from (12.41). For the generators which correspond to the broken symmetries, $\xi^A = \xi^i$ are in general not zero. We then see that they are eigenstates of the mass matrix with zero eigenvalue; the zero eigenvalues of the mass matrix correspond to massless particles. If all ξ^i are linearly independent, we have shown that for every broken generator there is a massless particle. This is the classical version of Goldstone's theorem. The linear independence of the ξ^i, which is needed to complete the proof of the theorem, can be seen as follows. If ξ^i are linearly dependent, then there is a nonzero solution for c_i with $\xi^i c_i = 0$. This means that the quantities $S_{ij} = \xi_{ia}\xi_{ja}$, considered as the elements of a matrix, correspond to a matrix of zero determinant. Since S_{ij} is symmetric, we can diagonalize it by an orthogonal transformation R_{ij}. (This is not in $O(N)$ but is an orthogonal transformation on the directions corresponding to the generators, in a subgroup of $O[N(N-1)/2]$.) The diagonal version of S_{ij} is $S_{ij} = \zeta_{ia}\zeta_{ja}$ where $\zeta_{ia} = R_{ji}\xi_{ja}$. The diagonal elements are manifestly positive and nonzero. If ζ_{ia} is zero, then we have a larger isotropy subgroup for the vacuum expectation value v_a. Thus ξ^i are linearly independent and Goldstone's theorem follows.

Transformations which leave the vacuum invariant form a subgroup $H \in G$, which is called the *invariance group* of the vacuum. This is also the little group or isotropy group of the vacuum expectation value. Equation (12.3) tells us that $\langle 0|e^{iQ\cdot\theta} \phi_a e^{-iQ\cdot\theta}|0\rangle = \mathcal{D}_{ab}(\theta) \langle 0|\phi_b|0\rangle$, so that

$$e^{-iQ\cdot\theta}|0\rangle = |0\rangle \quad \Longrightarrow \quad \mathcal{D}_{ab}(\theta) \langle 0|\phi_b|0\rangle = \langle 0|\phi_a|0\rangle \tag{12.42}$$

The invariance group of the vacuum, H, is called the "unbroken" subgroup of the symmetry group G; it is the largest unitarily realized symmetry group of the theory. If there are several fields with corresponding vacuum expectation values, H is the largest intersection of the respective little groups.

We now turn to Goldstone's theorem in the quantum theory. Starting with a continuous global symmetry of the Lagrangian, by Noether's theorem, there is a conserved current J^μ, i.e.,

$$\partial_\mu J^\mu = 0 \tag{12.43}$$

Consider now an operator $A(x)$ which transforms nontrivially under the symmetry, or does not commute with the charge Q.

$$[Q(t), A(0)] = B(0) \tag{12.44}$$

The strategy of the proof is to consider $\langle 0|B|0\rangle$ which has to be nonzero if we have spontaneous symmetry breaking. From the conservation law (12.43), we get

$$0 = \int d^3x \, [\partial_\mu J^\mu(x), A(0)]$$

$$= \oint dS_i [J^i(x), A(0)] + \frac{d}{dt} [Q(t), A(0)]$$

$$= \frac{d}{dt} [Q(t), A(0)] \qquad (12.45)$$

where we used the fact that $\oint dS_i J^i$ vanishes at spatial infinity if the fields vanish sufficiently fast or if we have periodic boundary conditions. We can thus conclude that

$$\langle 0|B|0 \rangle = \langle 0| [Q(t), A(0)] |0 \rangle \qquad (12.46)$$

is independent of t for any t, i.e., time-independent. Using

$$J^0(\boldsymbol{x}, t) = e^{-i\boldsymbol{p}\cdot\boldsymbol{x}} e^{iHt} J^0(0,0) e^{-iHt} e^{i\boldsymbol{p}\cdot\boldsymbol{x}} \qquad (12.47)$$

and inserting a complete set of states, equation (12.45) becomes

$$(2\pi)^3 \sum_n \delta^{(3)}(k_n) \Bigg[\langle 0|J^0(0)|n \rangle \, \langle n|A(0)|0 \rangle e^{-iE_n t}$$

$$- \langle 0|A(0)|n \rangle \, \langle n|J^0(0)|0 \rangle e^{iE_n t} \Bigg] = \langle 0|B|0 \rangle$$

$$(12.48)$$

where E_n, k_n are the energy and momentum of the state $|n\rangle$. The right-hand side of this equation is independent of time because of (12.46), and it is nonvanishing because we have the premise of spontaneous symmetry breaking. Since positive and negative frequency contributions cannot cancel out mutually, this equation can be satisfied only if we have a state $|G\rangle$ with $E_G = 0$ contributing on the left hand side. Because of the $\delta^{(3)}(k_n)$ only states of zero momentum contribute on the left-hand side and so we have the following result.

> For every continuous global symmetry of the Lagrangian which is spontaneously broken, there exists a state with energy $E \to 0$ as the momentum $k \to 0$ (*Goldstone's theorem*).

This state corresponds to a massless particle if we have the relativistic relation between energy and momentum. The theorem is valid even for nonrelativistic situations such as those that occur frequently in solid state physics. The field excitation corresponding to this state is called the Goldstone mode. For the Goldstone state $|G\rangle$, it also follows that we must have

$$\langle 0|A(0)|G \rangle \neq 0, \qquad\qquad \langle 0|J^0(0)|G \rangle \neq 0 \qquad (12.49)$$

We may thus consider the current J^0 or the operator A as generating the Goldstone state. The field corresponding to the Goldstone particle is a Lorentz scalar. This follows from the fact that it is created by the current and the parameters of internal symmetry transformations are scalars. If a Lorentz tensor gets a nonzero vacuum expectation value, we would break Lorentz invariance; therefore, in a relativistic theory, B and A are Lorentz scalars. The Goldstone state, created by A, is also a scalar and hence a spin-zero boson. B and A need not be fundamental scalar fields; they can be composite operators. For example, for the spontaneous chiral symmetry breaking in quantum chromodynamics, they are of the form $\bar{q}\gamma^5 q$ or $\bar{q}q$, where q are the quark fields.

12.4 Coset manifolds

Consider a compact Lie group G and a subgroup $H \subset G$. We define a quotient or coset G/H as given by all elements g of G with the identification $g \sim g'$ for all g, $g' \in G$ if $g = g'\, h$ for some element $h \in H$. In other words, elements of G differing by an element of H are considered equivalent and correspond to the same element of G/H. Just as a Lie group G is a differentiable Riemannian space, so is G/H. The Lie algebra of the generators T^A of G can be split into the subalgebra of generators t^α of H and an orthogonal set of generators S^i. By considering infinitesimal elements of G of the form $g \approx 1 - iT^A\theta^A = 1 - it^\alpha\theta^\alpha - iS^i\theta^i \approx (1 - iS^i\theta^i)(1 - it^\alpha\theta^\alpha) \approx (1 - iS^i\theta^i)h$, we see that S^i define the coset directions near the identity element; in particular, $dim(G/H) = dim G - dim H$.

For many applications we are interested in functions on a coset space. These can be obtained from functions on G. Consider the set of all functions on the Lie group G. We denote the matrix representation of an arbitrary group element g by $\mathcal{D}^R_{mn}$ where R denotes the irreducible representation and m, n are matrix labels taking values $1, 2, \cdots, dim R$. The functions $\mathcal{D}^R_{mn}$ are called the Wigner $\mathcal{D}$-functions. For the angular momentum algebra, these are of the form $\mathcal{D}^j_{mn}(\theta) = \langle j, m| \exp(iJ^a\theta^a)|j, n\rangle$, where $|j, m\rangle$ are the standard angular momentum states and J^a is the angular momentum operator. The standard completeness theorem for groups says that, for a compact Lie group G, the functions $\{\mathcal{D}^R_{mn}\}$ form a complete set, where we include all unitary irreducible representations (which are also finite dimensional). In other words, we can expand an arbitrary function on the group as

$$f(g) = \sum_{R,m,n} C^R_{mn}\, \mathcal{D}^R_{mn}(g) \tag{12.50}$$

Functions on the coset space G/H, by definition, must be invariant under $g \to g\, h$, for elements h of the subgroup H. We can obtain a basis for functions on G/H by restricting to the subset of $\mathcal{D}$'s which obey

$$\mathcal{D}^R_{mn}(gh) = \mathcal{D}^R_{mn}(g) \tag{12.51}$$

This means that the possible choices for the right index n in $\mathcal{D}^R_{mn}$ must correspond to states which are singlets of $H \in G$.

On a Lie group there is a natural Riemannian metric known as the Cartan-Killing metric. With g denoting a general element of G, $g^{-1}dg$ can be expanded as $g^{-1}dg = -iT^A E_I^A(\varphi)d\varphi^I$, where φ^I, $I = 1, 2, \cdots, dimG$, are the parameters of the group. E_I^A are called the frame fields for the space G. We may also write this as $E^A = E_I^A d\varphi^I = 2i\, \text{Tr}(T^A g^{-1}dg)$, where we have used the normalization $\text{Tr}(T^A T^B) = \frac{1}{2}\delta^{AB}$. The Cartan-Killing metric can be given as

$$
\begin{aligned}
ds^2 &= -2\, \text{Tr}(g^{-1}dg\, g^{-1}dg) \\
&= E^A E^A \\
&= E_I^A E_J^A\, d\varphi^I d\varphi^J
\end{aligned}
\tag{12.52}
$$

(It should be kept in mind that any parametrization of G is only a choice of local coordinates on G; in general one cannot find a single global coordinate system.)

A set of frame fields for the coset space can be defined by $E^i = 2i\, \text{Tr}(S^i g^{-1}dg)$; the metric for the coset is then given by

$$
\begin{aligned}
ds^2 &= E^i E^i \\
&= G_{ij}(\varphi)d\varphi^i d\varphi^j
\end{aligned}
\tag{12.53}
$$

For $h \in H$, we find $E^i(gh) = h_j^i E^j(g)$ where h_j^i is H-transformation matrix in the representation to which the S^i belong, $hS^i h^{-1} = h_j^i S^j$. The metric ds^2 for the coset is invariant under this, $ds^2(gh) = ds^2(g)$, so it can be taken to be independent of the parameters corresponding to $H \subset G$. The remaining parameters, the φ's indicated in (12.53), are local coordinates for G/H.

We now give some examples of coset spaces.

1. $SU(2)/U(1)$

The space $SU(2)/U(1)$ is the usual two-sphere S^2. A general element of $SU(2)$ can be parametrized as $g = a + ib^i\sigma_i$ where σ_i are the Pauli matrices. The condition of unitarity and the condition of unit determinant give $a^2 + b^i b^i = 1$ with a, b^i real. Thus $SU(2)$ is topologically S^3. Define a $U(1)$ subgroup by the σ_3-direction by a general element $h = a + ib'\sigma_3$ with $a^2 + b'^2 = 1$. We write $b^i = b'\xi^i$, which gives the remaining parameters as ξ^i obeying $\xi^i\xi^i = 1$ corresponding to a two-sphere S^2. Another way to parametrize g is

$$
g = \frac{1}{\sqrt{1 + \bar{z}z}}
\begin{pmatrix} 1 & z \\ -\bar{z} & 1 \end{pmatrix}
\begin{pmatrix} e^{i\chi/2} & 0 \\ 0 & e^{-i\chi/2} \end{pmatrix}
\tag{12.54}
$$

By computing $g^{-1}dg$ and identifying E^i, we get the coset metric as

$$
ds^2 = 4\frac{dz\, d\bar{z}}{(1 + \bar{z}z)^2}
\tag{12.55}
$$

This is the standard metric for the two-sphere; the usual parametrization in terms of θ, φ is obtained by $z = \tan(\theta/2)e^{i\varphi}$.

2. *Other coset spaces*

There are many other examples of coset spaces. Two simple sets which are spheres of various dimensions are $SU(N)/SU(N-1) = S^{2N-1}$ and $SO(N)/SO(N-1) = S^{N-1}$. Another interesting example is the complex projective space $\mathbf{CP}^{n-1}$, which is given as $SU(n)/U(n-1)$. We will not discuss how the identification of the spaces corresponding to these cosets can be made; some of it will become clearer later in this chapter.

12.5 Nonlinear sigma models

Consider a set of fields $\varphi^A(x)$ which take values on some Riemannian manifold $\mathcal{M}$. In other words, φ^A are coordinates on $\mathcal{M}$. $\varphi^A(x)$ then provides a mapping from spacetime to $\mathcal{M}$,

$$\varphi^A(x) \ : \ \mathbf{R}^4 \longrightarrow \mathcal{M}$$

The space $\mathcal{M}$, into which the mapping is done, is called the target space.

Let $G_{AB}(\varphi)$ denote the Riemannian metric tensor on $\mathcal{M}$. The action

$$S = \int d^4x \, \frac{1}{2} G_{AB}(\varphi) \partial_\mu \varphi^A \partial^\mu \varphi^B \tag{12.56}$$

defines a nonlinear sigma model on $\mathcal{M}$. In Chapter 8, we have already discussed how to set up a functional integral for sigma models.

In our present context we are interested in target manifolds which are cosets of groups. In this case, using the form (12.53) of the metric, we can write

$$S = -2f^2 \int d^4x \, \mathrm{Tr}(S^i g^{-1} \partial_\mu g) \, \mathrm{Tr}(S^i g^{-1} \partial^\mu g) \tag{12.57}$$

f is a quantity with the dimensions of mass; it will play the role of the coupling constant. An important particular case is when we have G itself, or $H = 1$. This is called a principal chiral model and the action is

$$S = -f^2 \int d^4x \, \mathrm{Tr}\left(g^{-1}\partial_\mu g \, g^{-1}\partial^\mu g\right) \tag{12.58}$$

12.6 The dynamics of Goldstone bosons

The dynamics of Goldstone bosons is given by a nonlinear sigma model with G/H as the target space. We will illustrate this first by an $O(N)$ theory. As noted before, the action is of the form

$$S = \int d^4x \left[\frac{1}{2} \partial \varphi_a \partial \varphi_a - V(\varphi) \right] \tag{12.59}$$

The $O(N)$-transformations are given by $\varphi_a \rightarrow \varphi'_a = R_{ab}\varphi_b$, where R_{ab} are the matrix elements of an $(N \times N)$- rotation matrix. The potential $V(\varphi)$ is taken to be invariant under these. Further, $V(\varphi)$ will be taken to have a minimum at $\varphi_a = \langle \phi_a \rangle$ with $\langle \phi_a \rangle \langle \phi_a \rangle \neq 0$. This will be the vacuum expectation value in the quantum theory. As a particular choice we may take

$$\langle \phi_a \rangle = v_a \equiv \begin{pmatrix} 0 \\ 0 \\ \cdot \\ \cdot \\ \cdot \\ 0 \\ v \end{pmatrix} \tag{12.60}$$

where we have represented v_a as the components of a column vector. Rotations on the first $N - 1$ entries which are all zeroes do not change this vector; so the little group of this vector is $O(N-1)$. Thus the vacuum expectation value (12.60) will spontaneously break the $O(N)$ symmetry down to $O(N - 1)$. The $O(N - 1)$-transformations are rotations matrices of the form

$$h = \begin{pmatrix} r & 0 \\ 0 & 1 \end{pmatrix} \tag{12.61}$$

where r is an arbitrary $(N-1) \times (N-1)$- rotation matrix. Clearly, $h_{ab}v_b = v_a$.

One can take the vacuum value $\langle \phi_a \rangle$ to have some other components nonzero, rather than $\langle \phi_N \rangle$ as we have done in (12.60), a different orientation of $\langle \phi \rangle$ in N-dimensional space, but this is equivalent to an overall $O(N)$-rotation of (12.60). The minimum of the potential fixes $\langle \phi_a \rangle \langle \phi_a \rangle$, so that any other choice must be an $O(N)$-rotation of (12.60). For such a choice $\langle \phi' \rangle = R_0 \langle \phi \rangle$ for a specific matrix R_0, the little group elements are of the form $R_0 h R_0^{-1}$, where h is given in (12.61). Thus the little group is still isomorphic to $O(N - 1)$, but its particular form is a similarity transform of the previous one. As emphasized in our discussion of the $U(1)$ case, we have to make a choice of orientation for the vacuum value, but physics is independent of the particular choice.

The fields φ_a, as fluctuations around the vacuum value can be parametrized as

$$\varphi_a = R_{ab}(\varphi)\tilde{v}_b \tag{12.62}$$

where

$$\tilde{v}_a = \begin{pmatrix} 0 \\ 0 \\ \cdot \\ \cdot \\ \cdot \\ 0 \\ v + \rho(x) \end{pmatrix} \tag{12.63}$$

R_{ab} is a rotation matrix which is spacetime-dependent; it can be thought of as made up of the components φ_a. Since $h_{ab}v_b = v_a$ (and this is true even if the $O(N-1)$ matrix r depends on the spacetime coordinates), we find that R_{ab} and $(Rh)_{ab}$ lead to the same fields in (12.62). Thus only the parameters corresponding to $O(N)/O(N-1)$ will appear in the expression for φ_a. For any x, $\varphi^2 = \varphi_a\varphi_a$ is the same for any R; thus the degrees of freedom contained in R correspond to the field degrees of freedom where φ^2 is fixed. Since there are N φ's, this means that we have the degrees of freedom corresponding to an $(N-1)$-dimensional sphere S^{N-1}, which shows that $O(N)/O(N-1) = S^{N-1}$.

We now substitute the parametrization (12.62) in the action and obtain

$$S = \int d^4x \; \left[\frac{1}{2}(\partial\rho)^2 - V(v+\rho) - \frac{1}{2}\tilde{v}^T(R^{-1}\partial_\mu R \, R^{-1}\partial^\mu R)\tilde{v}\right] \quad (12.64)$$

The potential energy V depends only on ρ and generally gives a mass term for this field. The term involving the field R gives the kinetic term for the $(N-1)$ Goldstone bosons; they are massless as expected. There is also an interaction term between the Goldstone particles and ρ since we have $\tilde{v}$ in the term involving R. At energies which are low compared to the mass of the ρ-particle, only the Goldstone particles can be excited and the action describing their dynamics is

$$S = -\frac{1}{2}\int d^4x \; v^T(R^{-1}\partial_\mu R \, R^{-1}\partial^\mu R)v \quad (12.65)$$

Expanding $R^{-1}\partial_\mu R$ as $-iT^A E_B^A \partial_\mu\varphi^B$ we are led to the expression $v^T T^A T^B v$. $T^A v = 0$ for all generators in the little group $O(N-1)$, so only the generators in the orthogonal complement can contribute. These are matrices of the form $(T^i)_{ab} = (S^i)_{ab} = \frac{i}{2}(\delta_{ia}\delta_{bN} - \delta_{ib}\delta_{aN})$, so that $v^T S^i S^j v = \frac{1}{4}\delta^{ij}$. The action (12.65) can be written as

$$S = \frac{1}{2}f^2 \int d^4x \; E_a^i E_b^i \partial_\mu\varphi^a \partial^\mu\varphi^b$$
$$= \frac{1}{2}f^2 \int d^4x \; G_{ab}(\varphi)\partial_\mu\varphi^a \partial^\mu\varphi^b \quad (12.66)$$

where $f = v/2$. The dynamics of the Goldstone bosons is thus described by a nonlinear sigma model with the target space G/H.

The action (12.65) obviously has the full global G-symmetry, where $R \to MR$ and where M is a constant $O(N)$-rotation matrix. This is to be expected since the breaking is only spontaneous, i.e., by choice of vacuum state. If we make an infinitesimal variation $R \to (1 - iT^A\theta^A(x))R$, the variation of the action is

$$\delta S = \int d^4x \; i \, v^T(R^{-1}T^A\partial_\mu R)v \; \partial^\mu\theta^A \quad (12.67)$$

(We have used the antisymmetry of T^A.) This shows that the action has the invariance for global G-symmetry when the parameters θ^A are constant. Further it identifies the current associated with this symmetry as

$$J_\mu^A = i \, v^T (R^{-1} T^A \partial_\mu R) v \tag{12.68}$$

Also we see that the equation of motion for the fields in R is just the conservation of this current.

Our discussion so far has been classical. In quantum theory, even at low energies for the external particles in a scattering process, we cannot neglect the massive particles; they can contribute as internal lines in loop diagrams. Secondly, there are interactions among the Goldstone bosons contained in the action (12.66) because there are many nonlinear terms in general. Thus there are loop contributions due to the Goldstone bosons as well. There are many types of corrections to the quantum action from the loop diagrams. First of all, there will be many corrections to the terms involving the ρ-field. Such terms in the effective action Γ will not be important for low-energy processes which do not involve ρ. Secondly, the addition of loop corrections will replace the parameters of the action (in particular the potential V and hence the vacuum value) by renormalized parameters. Following through the simplifications above, we see that the effect is to replace f by some renormalized value. Finally there will also be many terms involving higher-order derivatives of the Goldstone fields. To quartic order in the derivatives, we can have, for example, terms like $(v^T R^{-1} \partial_\mu R R^{-1} \partial^\mu R v)^2$, $(v^T [R^{-1} \partial_\mu R, R^{-1} \partial_\nu R] v)^2$. Such terms do occur and must be included in Γ, but their contribution to scattering processes at low energies is small compared to the leading quadratic term, since the derivatives become momenta of the particles involved. Thus at low energies, they can be neglected too. (f is the scale parameter in the action, so this means that we are looking at $k/f \ll 1$ for typical momenta k.) Summarizing, we can conclude that the low-energy effective action for the Goldstone bosons for spontaneous breaking of a continuous symmetry G down to H is given by the G/H-sigma model. The parameter f is, in principle, calculable from the full theory, but one can discuss the dynamics of Goldstone bosons using the sigma model, taking the value of the one parameter f from experiment.

Even though we have used a Lagrangian model with scalar fields, the result regarding the dynamics of Goldstone bosons is quite general. Thus we may have a fermionic theory with symmetry breaking via a composite operator developing a vacuum expectation value, for example, $\langle 0|\bar{q}q|0\rangle \neq 0$. The Goldstone bosons themselves are then bound states of the fermions. Nevertheless, the low-energy effective action for the Goldstone particles is given by a G/H-sigma model. This follows from the fact that the Goldstone particles are created by the current, namely, $\langle 0|J^0(0)|G\rangle \neq 0$. The only expression for $\langle 0|J_\mu(x)|G\rangle$ consistent with this, as well as Lorentz invariance and current conservation, is

$$\langle 0|J_\mu^i(x)|G,k\rangle = \text{(constant)} \; k_\mu \qquad (12.69)$$

(This is consistent with conservation since $k^2 = 0$.) This equation tells us that the components of the current corresponding to the broken generators produce φ's from the vacuum, irrespective of whether they are bound states or not. For the current itself, we must, therefore, have a relation of the form

$$J_\mu^i = f^2 \partial_\mu \varphi^i + \mathcal{O}(\varphi^2) \qquad (12.70)$$

for some constant f, and φ represents the Goldstone field. The action which gives a current of this form is

$$\mathcal{S} = \frac{1}{2}f^2 \int d^4x \; \partial_\mu \varphi^i \partial^\mu \varphi^i + \mathcal{O}(\varphi^3) \qquad (12.71)$$

The symmetry transformation for the broken symmetries is of the form $\varphi^i \to \varphi^i + \theta^i$, for small θ^i. Since we only have spontaneous breaking of the symmetry, i.e., symmetry breaking only in the choice of the vacuum state, this action must have the full G-symmetry. The terms which are higher order in φ's can be thus fixed by G-invariance. This leads immediately to the sigma model. In other words, our result for the low energy dynamics of Goldstone bosons is quite general.

Finally, we note that, if we parametrize R as $R = \exp(-iT^A\varphi^A) = 1 - iT^A\varphi^A + \mathcal{O}(\varphi^2)$, we can simplify the expression for the current (12.68) of the sigma model as

$$J_\mu^i = f^2 \partial_\mu \varphi^i + \mathcal{O}(\varphi^2) \qquad (12.72)$$

This is for the generators corresponding to the broken ones. This result is in agreement with the general formula (12.70). For the $O(N-1)$ directions, we do not get a term linear in the fields since the unbroken generators t^α vanish acting on the vacuum expectation value v. The corresponding currents are of the form

$$J_\mu^\alpha = f^2 f^{\alpha ij} \; \varphi^i \partial_\mu \varphi^j + \mathcal{O}(\varphi^3) \qquad (12.73)$$

where $[t^\alpha, S^i] = if^{\alpha ij}S^j$ define the structure constants $f^{\alpha ij}$.

12.7 Summary of results on spontaneous symmetry breaking

1. G is the symmetry group of the Lagrangian. It is also the symmetry group of the Hamiltonian.
2. $G|0\rangle \neq |0\rangle$, $H|0\rangle = |0\rangle$ implies that G is spontaneously broken down to $H \subset G$.
3. This can be realized if minimization of energy levels leads to nonzero vacuum expectation values $\langle 0|A|0\rangle \neq 0$ where A transforms nontrivially under G and the isotropy group of the vacuum expectation value is H.

4. Transitions among different degenerate vacua do not occur in the limit of infinite number of degrees of freedom, $\langle \alpha | \alpha' \rangle \to 0$.

5. For every broken generator of a continuous global symmetry of the Lagrangian, there exists a massless spin-zero particle (the Goldstone boson).

6. The low-energy dynamics of the Goldstone bosons, including their mutual interactions, is given by a nonlinear sigma model with target space G/H.

7. The current for the broken symmetries has the form $J_\mu^i = f^2 \partial_\mu \varphi^i + \mathcal{O}(\varphi^2)$, where φ^i are the Goldstone fields and f is a constant.

12.8 Spin waves

The simplest example of spontaneous breaking of continuous symmetry is the ferromagnet. This is a spin system where we have a spin variable at each lattice site of a three-dimensional lattice. Neglecting other degrees of freedom, a good approximation for the Hamiltonian of the system is the Heisenberg Hamiltonian

$$H = -\sum_{ij} J_{ij} \boldsymbol{S}_i \cdot \boldsymbol{S}_j \tag{12.74}$$

where i, j refer to lattice sites and $\boldsymbol{S}_i$ denotes the spin vector at site i. J_{ij} is the so-called exchange integral, and this falls off for large separations of the sites; very often it is a good approximation to keep only the nearest-neighbor interaction in (12.74). (We use the letter H for the Hamiltonian and for the unbroken subgroup since this is conventional; it should be clear from the context what is meant.) The Hamiltonian (12.74) has symmetry under rotations, so we have an $O(3)$ symmetric theory. The ground state is ferromagnetic with neighboring spins aligned along the same direction with a net overall magnetization if $J_{i\,i+1}$ is positive and antiferromagnetic with neighboring spins having opposite orientation if $J_{i\,i+1}$ is negative. Either way, the ground state breaks the $O(3)$ symmetry. For a ferromagnet, with the spins aligned, the ground state has net magnetization. There is a residual symmetry for rotations around the axis of net magnetization. Therefore the $O(3)$ symmetry is broken down to $O(2)$. In the limit of large lattices and for long-wavelength excitations, we can approximate this by a continuum field theory. The symmetry breaking $O(3) \to O(2)$ will give two Goldstone modes. These modes are the spin waves. Since $O(3)/O(2)$ is S^2, the dynamics of these modes is given by a sigma model with S^2 as the target space. There is no Lorentz-invariance here, so it does not make sense to write a relativistic effective Lagrangian. However, we can say that the effective Hamiltonian is of the form of a sigma model, i.e.,

$$H = \frac{f^2}{2} \int d^3x \left[4 \frac{\nabla Z \, \nabla \bar{Z}}{(1 + \bar{Z}Z)^2} \right] \tag{12.75}$$

The field Z describes the spin waves. By expanding this powers of the field $Z, \bar{Z}$, this Hamiltonian can describe the interactions among spin waves.

12.9 Chiral symmetry breaking in QCD

Quantum chromodynamics is the theory of strong interactions, the theory of quarks and gluons. It is an $SU(3)$ gauge theory, with the gluons being the gauge particles and the matter fields being the quarks transforming in the fundamental representation of $SU(3)$. The degrees of freedom associated with this $SU(3)$ symmetry are called the color degrees of freedom. Thus each quark field has a color index which takes values $1, 2, 3$, the different components transforming into each other under $SU(3)$ gauge transformations. There are six species of quarks known, denoted by u, d, s, c, b, and t, called the up, down, strange, charm, bottom, and top, respectively. We will denote these by Q_α^i, $\alpha = 1, 2, \cdots, 6$, respectively. i is the color index taking values $1, 2, 3$. The up and down quarks have masses of the order of a few MeV, the strange quark has a mass around $150\ MeV$; these are the light quarks. Masses for charm, bottom, and top are, approximately, $1.2\ GeV$, $4.2\ GeV$, and $174\ GeV$, respectively; these are the heavy quarks. The quarks get masses from the spontaneous breakdown of the gauge symmetry of electroweak interactions; the scale for this is approximately $246\ GeV$. Thus for the dynamics of quarks at energies well below this scale, we can take a Lagrangian where the masses are included in by hand. The QCD Lagrangian then has the form

$$\mathcal{L} = -\frac{1}{4} F_{\mu\nu}^a F^{a\mu\nu} + \sum_\alpha \bar{Q}_\alpha^i (i\gamma \cdot D_{ij} - m_\alpha \delta_{ij}) Q_\alpha^j \qquad (12.76)$$

where the covariant derivative is given by

$$\begin{aligned} (D_\mu)_{ij} &= (\partial_\mu + A_\mu)_{ij} \\ &= \partial_\mu \delta_{ij} - i e_s A_\mu^A (T^A)_{ij} \end{aligned} \qquad (12.77)$$

Here A_μ^A, $A = 1, 2, \cdots, 8$ are the gauge potentials for the $SU(3)$ gauge symmetry. T^A are hermitian, traceless (3×3)-matrices which form the generators of $SU(3)$ in the fundamental representation. e_s is the strong interaction coupling constant.

Consider the quark Lagrangian without the masses, namely,

$$\begin{aligned} \mathcal{L} &= \sum_\alpha \bar{Q}_\alpha^i\, i\gamma \cdot D_{ij}\, Q_\alpha^j \\ &= \sum_\alpha \bar{Q}_{L\alpha}^i\, i\gamma \cdot D_{ij}\, Q_{L\alpha}^j + \sum_\alpha \bar{Q}_{R\alpha}^i\, i\gamma \cdot D_{ij}\, Q_{R\alpha}^j \end{aligned} \qquad (12.78)$$

where in the second step we have split the quark fields into the left and right chiral components $Q_L = \frac{1}{2}(1 + \gamma^5)Q$ and $Q_R = \frac{1}{2}(1 - \gamma^5)Q$. This Lagrangian has a global $U_L(6) \times U_R(6)$ symmetry given by

$$\begin{aligned} Q_{L\alpha}^i &\to Q_{L\alpha}^{'i} = U_{\alpha\beta} Q_{L\beta}^i \\ Q_{R\alpha}^i &\to Q_{R\alpha}^{'i} = V_{\alpha\beta} Q_{R\beta}^i \end{aligned} \qquad (12.79)$$

where U, V are unitary (6×6)-matrices. This is the chiral symmetry of strong interactions. The mass terms do not have this symmetry; so it would seem that this symmetry is not very useful. However, for the light quarks, this chiral symmetry is very useful as an approximate symmetry, since the explicit breaking of chiral symmetry due to the masses is small compared to the relevant scale of the gluon interactions. So we write QCD for the three species (or flavors as they are called) of quarks by restricting α, β to the values $1, 2, 3$ corresponding to the up, down and strange quarks. Taking the α, β indices as understood, the quark part of the Lagrangian is

$$\mathcal{L} = \bar{Q}_L i\gamma \cdot D Q_L + \bar{Q}_R i\gamma \cdot D Q_R - \left(\bar{Q}_L \mathcal{M}^q Q_R + \bar{Q}_R \mathcal{M}^q Q_L \right) \quad (12.80)$$

where $\mathcal{M}^q$ is the quark mass matrix

$$\mathcal{M}^q = \begin{pmatrix} m_u & 0 & 0 \\ 0 & m_d & 0 \\ 0 & 0 & m_s \end{pmatrix} \quad (12.81)$$

The symmetries of this Lagrangian are as follows.

1. The color gauge symmetry corresponding to

$$Q^i_\alpha \rightarrow Q'^i_\alpha = g^i_j(x) \, Q^j_\alpha$$
$$A_\mu \rightarrow A'_\mu = g(x) \, A_\mu \, g^{-1}(x) - \partial_\mu g \, g^{-1}$$

$$(12.82)$$

where $g(x)$ is an $SU(3)$ matrix. This symmetry is not broken but leads to confinement of quarks and gluons. The spectrum of the theory has only bound states of quarks and gluons which are singlets of this symmetry.

2. Discrete symmetries like parity, charge conjugation, and time-reversal are respected by this theory. Quantum effects related to the topology of the gauge field configuration space and instantons lead to a possible PT-violating term, the so-called θ-term. This is discussed in Chapter 16. Experimentally, $\theta < 10^{-9}$. The theoretical reason for the smallness of this value is not currently clear; nevertheless, based on experiment, the PT-violation due to this effect can be taken to be zero.

3. There is the $U(3)_L \times U(3)_R$ chiral symmetry given as in (12.79). At the algebraic level of infinitesimal generators, $U(N) \sim SU(N) \times U(1)$; so we will discuss the $U(1)$'s separately. We have $SU_L(3) \times SU_R(3)$ symmetry. The $SU_V(3)$ subgroup of this symmetry group is defined by the transformations

$$Q'_{L\alpha} = U_{\alpha\beta} Q_{L\beta}$$
$$Q'_{R\alpha} = U_{\alpha\beta} Q_{R\beta} \quad (12.83)$$

In other words, $SU_V(3)$ corresponds to $U = V$ subset of (12.79).

4. The $U(1)$ transformations are given by

$$U(1)_L: \qquad Q'_L = e^{i\theta}Q_L, \qquad\qquad Q'_R = Q_R$$
$$U(1)_R: \qquad Q'_L = Q_L, \qquad\qquad Q'_R = e^{i\alpha}Q_R$$

$$\text{(12.84)}$$

We can form combinations $U_V(1)$, $U_A(1)$ defined by

$$U_V(1): \qquad Q'_L = e^{i\varphi/3}Q_L, \qquad\qquad Q'_R = e^{i\varphi/3}Q_R$$
$$U_A(1): \qquad Q'_L = e^{i\lambda}Q_L, \qquad\qquad Q'_R = e^{-i\lambda}Q_R$$

$$\text{(12.85)}$$

$U_V(1)$ is even under parity and its conserved current is a vector. It can be identified as the baryon number. $U_A(1)$ is odd under parity and its current is an axial vector.

The chiral symmetries are broken in many ways.

1. The $U(3)_L \times U(3)_R$ chiral symmetry is spontaneously broken down to $U(3)_V \sim SU_V(3) \times U_V(1)$ by the effects of the gluonic interactions (the gauge field A^a_μ). The energy scale at which this happens is approximately $140\ MeV$. The effective Lagrangian for the Goldstone bosons due to this symmetry breaking is valid up to approximately $1\ GeV$ since there are some additional numerical factors in the scattering amplitudes. It is a theorem, due to Vafa and Witten, that vector-like symmetries cannot be spontaneously broken in a theory with vector-like coupling between gauge particles and the fermions. So $SU_V(3) \times U_V(1)$ is the smallest group to which the chiral symmetry can be spontaneously broken.

2. The mass terms break the chiral symmetry explicitly. If $\mathcal{M}_q = m_0 \mathbf{1}$, i.e., all the quarks have the same mass, then $SU_V(3) \times U_V(1)$ is the unbroken subgroup. If the quark masses are different, then $SU_V(3)$ is further broken. This breaking due to masses is explicit and ultimately comes from the electroweak interactions since the quark masses are generated by coupling of quarks to the electroweak Higgs field.

3. There is further explicit breaking due to electroweak interactions coming from the coupling of electroweak gauge bosons and the quarks. Among other effects, this leads to the mass difference between the $\pi^\pm$ and π^0 mesons.

4. The axial $U_A(1)$ symmetry is explicitly broken down to the cyclic group Z_6 (Z_{2N_f} for N_f species or flavors of light quarks) by anomalies which are due to quantum corrections.

The baryon number symmetry $U_V(1)$ is not spontaneously or explicitly broken by the strong interactions; it is actually broken by anomalies in the electroweak sector and $B - L$, baryon number minus lepton number, is the only global symmetry which is obtained in the standard model of particle interactions.

The hierarchy of these breakings is also important. The strongest breaking occurs for the spontaneous chiral symmetry breaking and the anomalous breaking of $U_A(1)$. The energy scales for these are comparable and of the order of 1 GeV. The explicit breaking of chiral symmetry due to masses is next in order, followed by the breaking of $SU_V(3)$ due to mass differences. Electroweak interactions come next and the weakest breaking is that of baryon number $U_V(1)$ due to electroweak anomalies.

The broken generators for the first breaking $SU_L(3) \times SU_R(3) \to SU_V(3)$ are axial in nature and this leads to eight Goldstone bosons which are pseudoscalar. These correspond to the $\pi^\pm$, π^0, $K^\pm$, K^0, $\bar{K}^0$, and η mesons. Because of the explicit chiral symmetry breaking due to masses, we are starting from an imperfect symmetry, and as a result, these mesons are not true Goldstone bosons; they are actually massive and are called pseudo-Goldstone bosons. Their masses should go to zero as the electroweak couplings and the quark masses are taken to zero. Further, since the explicit breaking scale is smaller than the spontaneous breaking scale, it is still useful to think of them as Goldstone bosons, with masses added to the effective action. The masses are indeed significantly smaller than the scale of the spontaneous breaking of approximately 1 GeV, so this is consistent. There are also mass splittings among the Goldstone bosons because $SU_V(3)$ is not a perfect symmetry either.

$SU_V(3)$ with breaking effects due to mass differences is an approximate symmetry of the theory, realized in the Wigner mode, and the states (mesons and baryons) can be classified by this. The hierarchy of breakings also tells us why it is not useful to include the heavy quarks in this analysis. Their masses, which are higher than the 1 GeV scale of spontaneous breaking, explicitly break the chiral symmetry so badly that it is not useful to think of a chiral symmetry for them, even as a first approximation.

12.10 The effective action for chiral symmetry breaking in QCD

We can now construct an effective action for the pseudoscalar mesons and for the baryons. The dynamics of the mesons should be governed by a nonlinear sigma model with G/H as the target space, where $G = SU_L(3) \times SU_R(3)$ and $H = SU_V(3)$. We can represent an element of $SU_L(3) \times SU_R(3)$ by $G = (G_L, G_R)$ and the vector subgroup corresponding to the transformations (12.83) by $h = (g, g)$, where G_L, G_R, g are (3×3)-matrices which are elements of $SU(3)$. Since g is arbitrary, we can write $g = G_R^\dagger V$ for some unitary matrix V of unit determinant. We then regard V as the variable in $h = (G_R^\dagger V, G_R^\dagger V)$, so that by considering all $V \in SU(3)$ we get $SU_V(3)$. In the coset, we identify the elements G and $G\,h$.

$$G\,h = (G_L G_R^\dagger V,\ V) = (G_L G_R^\dagger, 1)(V, V) \qquad (12.86)$$

This shows that the coset can be represented by the $SU(3)$-matrix $U = G_L G_R^\dagger$. In other words,

$$\frac{SU_L(3) \times SU_R(3)}{SU_V(3)} \sim SU(3) \tag{12.87}$$

The Goldstone particles, the mesons in this case, can be represented by an $SU(3)$ element U. An element (g_L, g_R) of the chiral symmetry group $SU_L(3) \times SU_R(3)$ acts on G_L and G_R by translating them to new elements of the group as $(g_L G_L, g_R G_R)$. Their action on U is thus of the form

$$U \rightarrow U' = g_L \, U \, g_R^\dagger \tag{12.88}$$

Choosing $U = 1$ to represent the vacuum, namely $\langle 0|U|0\rangle = 1$, shows that the isotropy group is $g_L = g_R$, which is the vector subgroup of transformations. (We can choose U to be any constant matrix in the vacuum, the orientation of the vector subgroup in G is slightly different compared to $\langle 0|U|0\rangle = 1$, that is all.) Following our general discussion, the effective action for the pseudoscalar mesons is given by

$$\mathcal{S} = \frac{f_\pi^2}{4} \int d^4x \, \text{Tr}(\partial_\mu U^\dagger \partial^\mu U) \tag{12.89}$$

We can parametrize U as

$$U = \left(1 + i\frac{M}{\sqrt{2}\,f_\pi}\right) \left(1 - i\frac{M}{\sqrt{2}\,f_\pi}\right)^{-1} \tag{12.90}$$

where M is a hermitian traceless matrix. The effective action becomes

$$\begin{aligned}
\mathcal{S} &= \int d^4x \, \frac{1}{2}\text{Tr}\left[\left(1 + \frac{M^2}{2f_\pi^2}\right)^{-1} \partial_\mu M \left(1 + \frac{M^2}{2f_\pi^2}\right)^{-1} \partial^\mu M\right] \\
&= \int d^4x \, \left[\frac{1}{2}\text{Tr}(\partial_\mu M \partial^\mu M) - \frac{1}{2f_\pi^2}\text{Tr}(M^2 \partial_\mu M \partial^\mu M) + \cdots\right]
\end{aligned} \tag{12.91}$$

The identification of the physical meson fields is given by $M = t^a \phi^a/\sqrt{2}$, where ϕ^a are the meson fields and t^a for a basis for the $SU(3)$ algebra; they are traceless hermitian matrices normalized as $\text{Tr}(t^a t^b) = \frac{1}{2}\delta^{ab}$. Written out as a matrix,

$$M = \begin{pmatrix} \frac{1}{\sqrt{2}}\pi^0 + \frac{1}{\sqrt{6}}\eta & \pi^+ & K^+ \\ \pi^- & -\frac{1}{\sqrt{2}}\pi^0 + \frac{1}{\sqrt{6}}\eta & K^0 \\ K^- & \bar{K}^0 & -\sqrt{\frac{2}{3}}\,\eta \end{pmatrix} \tag{12.92}$$

where we have used the particle names for the fields representing them. (Actually the η in this equation is not quite the physical η because of mixing with the $SU(3)$ singlet pseudoscalar meson; we will not discuss these details here.) The eight fields here form the octet representation of $SU_V(3)$.

The quartic terms in (12.91) can describe the scattering of the mesons. The mesons interact with the baryons as well. This can be included as follows. The baryons also transform as representations of $SU_V(3)$; they are made of three quarks each. In terms of $SU_V(3)$ indices, they are of the form $Q_\alpha^i Q_\beta^j Q_\gamma^k \epsilon_{ijk}$. This is antisymmetric in the color indices; so we can have the fully symmetric representation for the flavor indices α, β, γ, which is also symmetric in spin, in accordance with the exclusion principle. This gives a (ten-dimensional) decuplet of spin-$\frac{3}{2}$ baryons. Another possibility is to antisymmetrize a pair of flavor indices and corresponding spin indices, giving an octet of spin-$\frac{1}{2}$ baryons. The lowest mass baryons are the octet; they are of the form $B_{\alpha\beta\gamma} = Q_{\alpha n}^i (Q_{\beta r}^j Q_{\gamma s}^k - Q_{\gamma r}^j Q_{\beta s}^k)\epsilon_{ijk}$, where n, r, s are spin indices; the antisymmetrization gives spin-$\frac{1}{2}$. The particle identification is given in terms of $B_\alpha^\lambda = \frac{1}{2} B_{\alpha\beta\gamma} \epsilon^{\lambda\beta\gamma}$,

$$
B_\alpha^\lambda = \begin{pmatrix} \frac{1}{\sqrt{2}}\Sigma^0 + \frac{1}{\sqrt{6}}\Lambda & \Sigma^+ & p \\ \Sigma^- & -\frac{1}{\sqrt{2}}\Sigma^0 + \frac{1}{\sqrt{6}}\Lambda & n \\ \Xi^- & \Xi^0 & -\sqrt{\frac{2}{3}}\,\Lambda \end{pmatrix} \tag{12.93}
$$

with a similar structure for the antibaryons. The baryon kinetic energy term should be $\bar{B}_\lambda^\alpha i\gamma \cdot \partial B_\alpha^\lambda$.

As for the meson-baryon coupling and baryon mass term, they have to be determined so that the Lagrangian has the full chiral symmetry and symmetry breaking should be apparent only when we expand U around the vacuum value 1, consistent with the fact that the chiral symmetry is broken only spontaneously. The left-chirality spinor is a two-component spinor χ_r, $r = 1, 2$, which transforms by the (2×2)-matrix representation of Lorentz transformations. Since this matrix has unit determinant, we can form invariants by contracting the spinor indices with ϵ^{rs}. Thus a left-chirality baryon can operator can be of the form $Q_L Q_L Q_L$ with two Q_L's forming a Lorentz-invariant combination or $Q_L Q_R Q_R$ with the two Q_R's forming an invariant. To illustrate how we can form interaction terms, we will use one combination, say, $Q_L Q_L Q_L$, in what follows. Since $\epsilon_{\alpha\beta\gamma} g_{\alpha\alpha'} g_{\beta\beta'} = \epsilon_{\alpha'\beta'\gamma'} g_{\gamma'\gamma}^\dagger$, for the baryon matrix, the transformation rules are of the form

$$
B_L \to g_L B g_L^\dagger, \qquad B_R \to g_R B g_R^\dagger \tag{12.94}
$$

It is easily checked that the baryon mass term $\mathrm{Tr}(\bar{B}_L B_R + \bar{B}_R B_L)$ is not chirally invariant. By use of the transformation property of U, an invariant term can be constructed, leading to the effective baryon action

$$
\mathcal{S}_B = \int d^4x \left[\mathrm{Tr}(\bar{B}\, i\gamma \cdot \partial\, B) + g f_\pi\, \mathrm{Tr}\left(\bar{B}_L U B_R U^\dagger + \bar{B}_R U^\dagger B_L U \right) \right]
$$

$$
\tag{12.95}
$$

Here g is some constant. By writing this out in terms of the fields, and using the expansion for U, we then find the baryon mass to be $m_B = g f_\pi$ and the pion-nucleon coupling to be $g_{\pi \bar N N} = g$. The pion-nucleon interaction $i g \bar N \gamma^5 \tau^a N \pi^a$ is the original Yukawa interaction, slightly modified to take account of the pseudoscalar nature of the pion. The relation $m_B = g_{\pi \bar N N} f_\pi$ is known as the Goldberger-Treiman relation. (This is to the lowest order; there are corrections to this relation, related to form factors for the axial vector current.)

So far we have not addressed the explicit breaking of chiral symmetry. The most important consequence is the mass term for the mesons. This is of the form

$$S_{mass} = \frac{f_\pi^2}{4} \int d^4x \ \mathrm{Tr} \mathcal{M} \left(U + U^\dagger - 2 \right)$$

$$\approx \int d^4x \left[-\frac{1}{2} \mathrm{Tr}(\mathcal{M} M^2) + \frac{1}{4 f_\pi^2} \mathrm{Tr}(\mathcal{M} M^4) + \cdots \right] \quad (12.96)$$

The matrix $\mathcal{M}$ should have a structure similar to the quark mass matrix to incorporate the pattern symmetry breaking due to the quark masses and mass differences. We can take it to be of the form

$$\mathcal{M} = \begin{pmatrix} a & 0 & 0 \\ 0 & a' & 0 \\ 0 & 0 & b \end{pmatrix} \quad (12.97)$$

with $a \approx a'$ because the up and down masses are approximately the same, and $b > a, a'$. The as yet unknown coefficents a, a', b may be related to the experimental meson masses. Expanding out (12.96) and identifying the masses in terms of a, a', b, we get one of the Gell-Mann-Okubo mass formulae,

$$3 \ m_\eta^2 \approx 4 \ m_K^2 - m_\pi^2 \quad (12.98)$$

which is in reasonable agreement with the experimental values. There are other sources of symmetry breaking as discussed earlier, and some of these can also be incorporated nicely in the effective action; we will not discuss them further.

As another example of the use of the effective action, we consider the $\pi - \pi$ scattering at low energies. For this purpose we can neglect all the K, η terms in (12.92) and write $M = \tau \cdot \pi / \sqrt{2}$, where the τ^i are the Pauli matrices and $\pi^\pm = (\pi^1 \pm i\pi^2)/\sqrt{2}$ and $\pi^3 = \pi^0$. There is an obvious $SU(2)_V$ subgroup which acts on these fields; it is called the isospin symmetry. We use τ for the Pauli matrices because here they are applied to the flavor isospin symmetry. The terms in the effective action (12.91) which are relevant for $\pi - \pi$ scattering are

$$S_{int} = -\frac{1}{4 f_\pi^2} \int d^4x \left[\pi \cdot \pi \ \partial_\mu \pi \cdot \partial^\mu \pi - \frac{1}{2} m_\pi^2 \ \pi \cdot \pi \ \pi \cdot \pi \right] \quad (12.99)$$

We consider pion-pion scattering with the isospin labels and momenta as $(k_1a)(k_2b) \rightarrow (k_3c)(k_4d)$. The scattering amplitude for the process is given by

$$\langle k_3c, k_4d|\hat{S}|k_1a, k_2b\rangle = \prod_i \frac{1}{\sqrt{2\omega_{k_i}V}} \, \delta^{(4)}(k_1 + k_2 - k_3 - k_4) \, \mathcal{A}$$

$$\mathcal{A} = i\left[\delta_{ab}\delta_{cd}\left(\frac{s - m_\pi^2}{f_\pi^2}\right) + \delta_{ac}\delta_{bd}\left(\frac{t - m_\pi^2}{f_\pi^2}\right)\right.$$

$$\left. + \delta_{ad}\delta_{bc}\left(\frac{u - m_\pi^2}{f_\pi^2}\right)\right] \tag{12.100}$$

where we have introduced the so-called Mandelstam kinematic variables, $s = (k_1 + k_2)^2$, $t = (k_1 - k_3)^2$, $u = (k_1 - k_4)^2$, which also obey $s + t + u = 4m_\pi^2$. By taking $a = b = c = d = 3$, we get the amplitude for $\pi^0\pi^0 \rightarrow \pi^0\pi^0$ as $\mathcal{A} = im_\pi^2/f_\pi^2$. For $\pi^+\pi^- \rightarrow \pi^0\pi^0$, we need $a = b = 1$, $a = b = 2$ and $a = 1, b = 2$, $a = 2, b = 1$ and $c = d = 3$. This gives $\mathcal{A} = i(s - m_\pi^2)/f_\pi^2$.

It is also interesting to work out the currents for the chiral symmetries. We find

$$J_{L\mu}^a = -i\frac{f_\pi^2}{2}\text{Tr}(t^a\partial_\mu U \, U^{-1})$$

$$J_{R\mu}^a = i\frac{f_\pi^2}{2}\text{Tr}(t^a U^{-1}\partial_\mu U)$$

$$J_{V\mu}^a = J_{L\mu}^a + J_{R\mu}^a = i\frac{f_\pi^2}{2}\text{Tr}t^a\left(U^{-1}\partial_\mu U - \partial_\mu U \, U^{-1}\right)$$

$$J_{A\mu}^a = J_{L\mu}^a - J_{R\mu}^a = -i\frac{f_\pi^2}{2}\text{Tr}t^a\left(U^{-1}\partial_\mu U + \partial_\mu U \, U^{-1}\right) \tag{12.101}$$

When these are expanded in powers of M, the terms involving just the pion fields are

$$J_{L\mu}^a \approx \frac{1}{2}f_\pi\partial_\mu\pi^a - \frac{1}{2}\epsilon^{abc}\pi^b\partial_\mu\pi^c + \cdots$$

$$J_{R\mu}^a \approx -\frac{1}{2}f_\pi\partial_\mu\pi^a - \frac{1}{2}\epsilon^{abc}\pi^b\partial_\mu\pi^c + \cdots$$

$$J_{V\mu}^a \approx -\epsilon^{abc}\pi^b\partial_\mu\pi^c + \cdots$$

$$J_{A\mu}^a \approx f_\pi\partial_\mu\pi^a + \cdots \tag{12.102}$$

Notice that the formula for the axial vector current, which corresponds to the spontaneously broken symmetries, is in agreement with the general result (12.70). With the one-pion state $|kb\rangle$ we get

$$\langle 0|J_{A\mu}^a|kb\rangle = -if_\pi k_\mu \delta^{ab}\frac{e^{-ikx}}{\sqrt{2\omega_k V}} \tag{12.103}$$

If $J^a_{A\mu}$ is conserved, the above relation gives $k^2 = 0$ as is appropriate for massless particles. In the present case, the current is not conserved because of the explicit breaking due to the quark masses. From the above relation

$$\langle 0|\partial_\mu J^{a\mu}_A|kb\rangle = -m_\pi^2 f_\pi \delta^{ab} \frac{e^{-ikx}}{\sqrt{2\omega_k V}} \tag{12.104}$$

This is known as the PCAC (partial conservation of axial vector current) relation.

12.11 The range of validity of effective Lagrangians and unitarity of the S-matrix

The effective action for chiral symmetry breaking in QCD captures the low-energy features of the theory. This action was written down based on symmetries, although, in principle, such an action can be derived starting from the QCD action. There are many other situations where such an effective low-energy action can be constructed and used for many calculations. In some sense, all actions we use are of this nature, since the high-energy behavior of the standard model of particle interactions has not been tested beyond a certain point, approximately a TeV. Clearly there is a certain range of validity for the use of any effective Lagrangian, since it is based on a low momentum approximation. For scatterings and other processes beyond a certain cut-off value of momenta, the effective Lagrangian will have to be modified. Also, generically, an effective Lagrangian will involve terms which are not renormalizable. Calculations have to be done with a cut-off. One can then ask whether this cut-off value can be estimated starting from the effective Lagrangian itself. This is especially important in situations where we do not know the high-energy version of the theory, having obtained only a phenomenological theory based on low-energy experiments. We will now show how the unitarity of the S-matrix provides one way to make an estimate of the cut-off.

As shown in Chapter 5, the scattering operator is given by

$$\hat{S} = U(\infty, -\infty)$$
$$U(x^0, y^0) = T \exp\left[\int_{y^0}^{x^0} d^4x \, \mathcal{L}_{int}(\phi_{in}, \chi_{in}, ...)\right] \tag{12.105}$$

U is a unitary operator and this leads to the unitarity condition $\hat{S}^\dagger \hat{S} = 1$. This condition is a statement of the conservation of probability in the theory. The fields in $\mathcal{L}_{int}$ are the free in-fields.

The unitarity of the S-matrix gives certain general constraints on cross sections, decay rates, etc. For example, consider the scattering of two particles of momenta k_1, k_2 into any number of particles. The initial state is given by

$|i\rangle = |k_1, k_2\rangle$. For simplicity assume that incoming particles are not identical particles. By writing $\hat{S} = 1 + iT$, the unitarity condition becomes

$$i(T - T^\dagger) + T^\dagger T = 0 \tag{12.106}$$

Taking the matrix elements of this relation for the state $|i\rangle$, and using the completeness relation for the states, $\sum_f |f\rangle\langle f| = 1$, we get

$$\sum_f |\langle f|T|i\rangle|^2 = -i \langle i|(T - T^\dagger)|i\rangle \tag{12.107}$$

(Here we will assume that the forces are of short range, so that there is no divergence in the forward scattering amplitude.) $\langle f|T|i\rangle$ is the transition amplitude for the initial state $|i\rangle$ to be scattered to the state $|f\rangle$. It is of the form

$$\langle f|T|i\rangle = \frac{1}{\sqrt{2\omega_{k_1}V}} \frac{1}{\sqrt{2\omega_{k_2}V}} (2\pi)^4 \delta^{(4)}\left(\sum p - k_1 - k_2\right) \mathcal{M}_{fi} \prod_j \frac{1}{\sqrt{2\omega_{p_j}V}} \tag{12.108}$$

where $\mathcal{M}_{fi}$ is the invariant amplitude. Calculating the total cross section σ from here, we find

$$\sum_f |\langle f|T|i\rangle|^2 = \sigma\tau \frac{\sqrt{(k_1 \cdot k_2)^2 - m_1^2 m_2^2}}{\omega_{k_1}\omega_{k_2}V} \tag{12.109}$$

where τ is the total interaction time and we have used the formula for the flux (5.52).

In $\langle i|T|i\rangle$, we have the same initial and final states, so that the momentum conserving δ-function should be replaced by $V\tau$ as explained in Chapter 5. We can then write

$$\langle i|T|i\rangle = \frac{1}{2\omega_{k_1}V} \frac{1}{2\omega_{k_2}V} V\tau \, \mathcal{M}_{ii} \tag{12.110}$$

$\mathcal{M}_{ii}$ is the invariant amplitude for the (elastic) forward scattering since the final state is the same as the initial state. Using results (12.109) and (12.110) in the unitarity conditon (12.107), we get

$$\sigma = -i \frac{(\mathcal{M}_{ii} - \mathcal{M}_{ii}^*)}{4\sqrt{(k_1 \cdot k_2)^2 - m_1^2 m_2^2}}$$

$$= \frac{\mathrm{Im}\mathcal{M}_{ii}}{[(s - m_1^2 - m_2^2)^2 - 4m_1^2 m_2^2]^{\frac{1}{2}}} \tag{12.111}$$

where $s = (k_1 + k_2)^2$. Equation (12.111) between the total cross section and the imaginary part of the forward scattering amplitude is known as the optical theorem.

Unitarity also imposes restrictions on how fast cross sections can grow with energy. The general way to demonstrate this explicitly requires finding a basis which partially diagonalizes the scattering matrix, such as an angular momentum basis which corresponds to the partial wave analysis of the amplitude. Here we will do a simpler analysis just to illustrate the point about unitarity constraints. Consider the elastic scattering of two spinless particles with the momenta k_1, $k_2 \to p_1$, p_2. We will use the center-of-momentum frame, in which case $\boldsymbol{k}_1 + \boldsymbol{k}_2 = 0$. The basic kinematic variables are thus $s = (k_1 + k_2)^2$ and the scattering angle $\boldsymbol{k}_1 \cdot \boldsymbol{p}_1 = |\boldsymbol{k}_1|\,|\boldsymbol{p}_1|\cos\theta$. The invariant matrix element $\mathcal{M}$ is a function of s and $\cos\theta$. Using the formula (12.108), we can calculate the total cross section as

$$
\begin{aligned}
d\sigma_{el} &= \frac{1}{4\sqrt{(k_1 \cdot k_2)^2 - m_1^2 m_2^2}} \int \frac{d^3 p_1}{(2\pi)^3} \frac{1}{2\omega_{p_1}} \frac{d^3 p_2}{(2\pi)^3} \frac{1}{2\omega_{p_2}} \\
&\quad \times (2\pi)^4 \delta^{(4)}(p_1 + p_2 - k_1 - k_2)|\mathcal{M}|^2 \\
&= \frac{1}{64\pi^2 s}|\mathcal{M}|^2 d\Omega_p
\end{aligned}
\tag{12.112}
$$

where we have used the kinematic result

$$
|\boldsymbol{p}_1|^2 = |\boldsymbol{p}_2|^2 = \frac{[(s - m_1^2 - m_2^2)^2 - 4m_1^2 m_2^2]}{4s}
\tag{12.113}
$$

The quantity $\mathrm{Im}\mathcal{M}_{ii}$ which occurs in (12.111) is the imaginary part of the elastic amplitude evaluated at $\theta = 0$. We thus have an obvious inequality $(\mathrm{Im}\mathcal{M}_{ii})^2 \le |\mathcal{M}|^2(\theta = 0)$, which leads to

$$
\sigma^2 \le \frac{64\pi^2 s}{[(s - m_1^2 - m_2^2)^2 - 4m_1^2 m_2^2]} \left(\frac{d\sigma_{el}}{d\Omega_p}\right)_{\theta=0}
\tag{12.114}
$$

We can now apply this line of reasoning to terms in the effective Lagrangian which are of dimension higher than zero and hence non-renormalizable. As an example, consider a term like

$$
\mathcal{L}_{int} = g \int d^4x \, (\partial\varphi)^2 \chi^2
\tag{12.115}
$$

where we have two scalar fields φ, χ of masses m_1 and m_2, respectively. g is a coupling constant of dimension $(\text{mass})^{-2}$, analogous to the parameter $1/f_\pi^2$ in the effective Lagrangian for mesons. To lowest order in g, the elastic scattering amplitude for $\varphi \chi \to \varphi \chi$ is given by

$$
\mathcal{M} = 4g \, k_1 \cdot p_1
\tag{12.116}
$$

where k_1, p_1 are the momenta of the φ-particle before and after collision, respectively. For the cross sections we get

$$\left(\frac{d\sigma_{el}}{d\Omega_p} \right)_{\theta=0} = \frac{g^2}{4\pi^2 s} m_1^4$$

$$\sigma_{el} = \frac{g^2}{\pi s} \left(\omega_{k_1}^4 + \frac{1}{3} p_1^4 \right) \tag{12.117}$$

Notice that σ_{el} grows like s for large s. Since $\sigma_{el} \leq \sigma$, the inequality (12.114) becomes, for large s,

$$s^2 \leq \frac{48\pi m_1^2}{g} \tag{12.118}$$

The unitarity bound is not respected by the lowest-order calculation for $s \geq \sqrt{48\pi m_1^2/g}$. The effective theory has to be replaced by a more fundamental theory, or at least a theory with more fields included, before we get to this energy. (This is a very approximate bound; more stringent bounds can be obtained by a more detailed analysis of unitarity.) In essence, this argument is valid even if loop calculations are taken into account. One can impose a cut-off on the loop momenta of the effective theory, construct an effective Γ which will be a series of terms with higher and higher dimensions. S-matrix elements can be calculated by solving the equations of motion for Γ and evaluating Γ on the solutions, following (8.115). The calculation is essentially what we have done here. Unitarity bounds can be viewed as giving constraints on the parameters of Γ.

12.12 Gauge symmetry and the Higgs mechanism

So far we have discussed the spontaneous breaking of a global symmetry where the transformation matrices are constant as a function of the spacetime coordinates. When we have spontaneous symmetry breaking for a symmetry which is local, in other words, for a gauge symmetry, Goldstone's theorem does not apply. In this case, we get massive gauge bosons. This can be illustrated by considering the $U(1)$ gauge theory.

The action for a $U(1)$ gauge theory with a complex scalar field is

$$\mathcal{S} = \int d^4x \left[-\frac{1}{4} F_{\mu\nu} F^{\mu\nu} + (D_\mu\phi)^*(D^\mu\phi) - \lambda \left(\phi^*\phi - \frac{v^2}{2} \right)^2 \right] \tag{12.119}$$

where the gauge transformations are given by

$$A_\mu \rightarrow A'_\mu = A_\mu + \partial_\mu\theta(x)$$
$$\phi \rightarrow \phi' = \exp\left(ie\theta(x) \right) \phi \tag{12.120}$$

This model is often referred to as the Abelian Higgs model. (In this section, we will use ϕ for both the classical and quantum fields, since we need to go back and forth in our discussion.) The energy corresponding to (12.119) is minimized when ϕ has a value $v/\sqrt{2}$ which is realized in the quantum theory, as before, as the vacuum expectation value. For fluctuations of the field around this value, we parametrize ϕ as

$$\phi(x) = \exp\left(\, ie\xi(x)\,\right)\,\frac{1}{\sqrt{2}}(v+\rho) \tag{12.121}$$

Substituting this into the action (12.119), we get

$$S = \int d^4x \left[-\frac{1}{4}F_{\mu\nu}F^{\mu\nu} + \frac{1}{2}(\partial_\mu\rho\partial^\mu\rho) + \frac{1}{2}e^2W_\mu W^\mu(v+\rho)^2 \right.$$

$$\left. -\lambda v^2\rho^2 - \lambda v\rho^3 - \frac{\lambda}{4}\rho^4 \right] \tag{12.122}$$

where $W_\mu = A_\mu - \partial_\mu\xi$. We can rewrite the theory in terms of W rather than A. Since the relation between the two is in the form of a gauge transformation, $F_{\mu\nu}(A) = F_{\mu\nu}(W)$. Thus,

$$S = S_W + S_\rho + S_{int}$$

$$S_W = \int d^4x \left[-\frac{1}{4}F_{\mu\nu}(W)F^{\mu\nu}(W) + \frac{1}{2}e^2v^2W_\mu W^\mu \right]$$

$$S_\rho = \int d^4x \left[\frac{1}{2}(\partial\rho)^2 - \frac{1}{2}(2\lambda v^2)\rho^2 \right]$$

$$S_{int} = \int d^4x \left[e^2v\rho W_\mu W^\mu + \frac{e^2}{2}\rho^2 W_\mu W^\mu - \lambda v\rho^3 - \frac{\lambda}{4}\rho^4 \right] \tag{12.123}$$

The particle content can be read off this action. We have a massive gauge-particle, of mass ev. The kinetic term for ϕ has led to this mass term for W_μ. There is a massive scalar particle ρ with $m^2 = 2\lambda v^2$. Then there are a number of interaction terms. Notice that there is no massless Goldstone particle. The would-be Goldstone field is $\xi(x)$ in the parametrization (12.121). But, from the gauge-transformation property (12.120), we see that this field is like a gauge parameter. In fact, since ϕ and $e^{ie\theta}\phi$ are physically equivalent, we can even write the vacuum value of ϕ as $e^{ie\theta}v/\sqrt{2}$ for some function θ and choose θ appropriately to get rid of ξ from the field ϕ. This is to say that it can be moved into the gauge potential by a gauge transformation, which is what we have done when it is absorbed into the vector field W_μ. The action (12.123) is thus displayed in a particular gauge; it is called the unitary gauge. We can do further gauge transformations to write this in other gauges, when the W^2-term will have the form $(W + \partial\theta)^2$. This method wherein the would-be Goldstone field is absorbed by the gauge field, which then becomes a massive field, is known as the Higgs mechanism.

It is interesting to see how the gauge symmetry is important in removing the massless modes using a Hamiltonian framework somewhat along the lines of our discussion of the electromagnetic field in Chapter 6. We have already seen that we can choose a gauge where $A_0 = 0$ and then split the spatial components as $A_i = A_i^T + \partial_i f$ where A_i^T is transverse, i.e., $\partial_i A_i^T = 0$. Using the parametrization (12.121), the ϕ-part of the action can be written as

$$S_\phi = \int d^4x \, \frac{e^2v^2}{2} \left[\dot{\xi}^2 - \left(\partial_i(\xi - f) - A_i^T \right)^2 \right]$$

$$= \int d^4x \, \frac{e^2v^2}{2} \left[\dot{\xi}^2 - (\partial_i(\xi - f))^2 - A_i^T A_i^T \right] \qquad (12.124)$$

where we have ignored the ρ-field since it is not important for this discussion. It is clear that the transverse part of A_i has a mass ev. The equation of motion for ξ is

$$\ddot{\xi} - \partial_i \partial_i \tilde{\xi} = 0 \qquad (12.125)$$

where $\tilde{\xi} = \xi - f$. The equation of motion for the A_0-component, or the Gauss law for the theory, is

$$\partial_i(\partial_i \dot{f}) - e^2v^2 \dot{\xi} = 0 \qquad (12.126)$$

This can be solved for $\dot{\xi}$, in terms of Fourier modes, as

$$\dot{\xi}_k = \left[\frac{\mathbf{k} \cdot \mathbf{k}}{\mathbf{k} \cdot \mathbf{k} + e^2v^2} \right] \dot{\tilde{\xi}}_k \qquad (12.127)$$

Using this in equation (12.125), we find

$$\mathbf{k} \cdot \mathbf{k} \left[\frac{\partial^2}{\partial t^2} + \mathbf{k} \cdot \mathbf{k} + e^2v^2 \right] \tilde{\xi}_k = 0 \qquad (12.128)$$

For $\mathbf{k} \neq 0$ we see that $\omega^2 = \mathbf{k} \cdot \mathbf{k} + e^2v^2$, showing that this mode is also massive, with the same mass ev. For this mode the result can be extended to $\mathbf{k} = 0$ by continuity. (It is possible to have a mode which is spatially constant which is related to the orientation of the vacuum as before.)

We have already seen that the true gauge group of the theory is the set of all gauge transformations which go to the identity at spatial infinity, namely, $\mathcal{G}_*$ in the notation of Chapter 10. It should be emphasized that there is no breakdown of this true gauge symmetry here. In fact gauge symmetry is crucial in removing the massless modes. What is broken is the global part of the symmetry, corresponding to $\mathcal{G}/\mathcal{G}_*$. To show how this works out in some detail, we will use a gauge-invariant field χ. In terms of the Coulomb Green's function $G_C(x, y)$ we can construct a gauge-invariant combination of ϕ and A_i as

$$\chi = \exp\left(-ie \int d^3y \, G_C(x, y) \partial_i A_i(y) \right) \phi \qquad (12.129)$$

This is invariant under transformations which go to the identity at spatial infinity. Under transformations which go to a constant at spatial infinity,

χ transforms as $\chi \rightarrow e^{ie\theta(\infty)}\chi$. Thus χ is gauge-invariant but has charge since it transforms under $\mathcal{G}/\mathcal{G}_*$. The potential $V(\phi) = V(\chi)$, so the vacuum expectation value of χ is $v/\sqrt{2}$. This nonzero value for χ is consistent with the invariance of the vacuum under $\mathcal{G}_*$. In terms of ϕ we can write

$$\langle 0|\phi|0\rangle = \exp\left(-ie\int d^3y\, G_C(x,y)\partial_i A_i(y)\right)\frac{v}{\sqrt{2}} \tag{12.130}$$

The parametrization of ϕ in (12.121) becomes

$$\phi = \exp(ie\tilde{\xi})\,\exp\left(-ie\int d^3y\, G_C(x,y)\partial_i A_i(y)\right)\frac{1}{\sqrt{2}}(v+\rho) \tag{12.131}$$

Combining this with the Gauss law, we get

$$S_\phi = \int d^4x\,\frac{1}{2}\left[\dot{\varphi}^2 - (\nabla\varphi)^2 - e^2 v^2 \varphi^2\right] \tag{12.132}$$

where

$$\varphi = ev\sqrt{\frac{\mathbf{k}\cdot\mathbf{k}}{\mathbf{k}\cdot\mathbf{k}+e^2 v^2}}\,\tilde{\xi} \tag{12.133}$$

The fact that the vacuum is defined by $\langle 0|\chi|0\rangle = v/\sqrt{2}$, or equivalently (12.130), shows that the $\mathcal{G}_*$-symmetry is not broken, even though we refer to this situation as the spontaneous breaking of gauge symmetry. The vacuum expectation value for χ, however, does break the global part of the $U(1)$ symmetry.

The Higgs mechanism does not change the number of degrees of freedom. Before symmetry breaking, we have the modulus and phase of ϕ and the two transverse polarizations of the gauge particle for a total of four physical fields. After symmetry breaking, the gauge boson is massive and has three polarizations. Combined with ρ, this still leaves four independent fields. The would-be Goldstone boson becomes the third polarization of the massive vector particle.

We will now consider how this extends to nonabelian symmetries. The mass term for the gauge fields arises from the kinetic term for the scalar field which is of the form

$$S_\phi = \int d^4x\,\left[(D_\mu\phi)_a^*(D^\mu\phi)_a - V(\phi_a^*\phi_a)\right] \tag{12.134}$$

where ϕ_a transforms as some representation of the symmetry group G. The covariant derivatives are $(D_\mu\phi)_a = \partial_\mu\phi_a - ieA_\mu^A(T^A)_{ab}\phi_b$. When the vacuum is such that ϕ has a nonzero expectation value, we can parametrize it as

$$\phi_a = g_{ab}(x)\,\frac{1}{\sqrt{2}}(v_b + \rho_b) \tag{12.135}$$

The symmetry G is broken down to $H \subset G$. Since an H-transformation is identity acting on $v_b + \rho_b$, only the G/H elements really appear in (12.135). Substituting this in (12.134) and going to the unitary gauge, we get

$$S_\phi = \int d^4x \left[\frac{1}{2} g^2 \, v^T (T^A T^B) v \, W^A_\mu A_{B\mu} + \rho\text{-terms} \right] \tag{12.136}$$

where W^A_μ is the unitary gauge version of A^A_μ. The generators of the Lie algebra of H annihilate the vacuum value v^a, so the mass matrix is nonzero for the coset directions. The gauge bosons corresponding to the broken generators get masses, with a mass matrix given by $(M^2)^{AB} = g^2 \, v^T (T^A T^B) v$. One can also easily check that the counting of degrees of freedom works out as well.

Finally, even though we have used a scalar field to describe the symmetry breaking, it is possible, just as in the case of chiral symmetry breaking in QCD, for composite operators to get vacuum expectation values and so break the symmetry. The effective action for the Goldstone bosons then becomes the mass term for the gauge particles.

12.13 The standard model

All the particles and interactions known to date, except for gravity, are described by the standard model. This is based on the gauge group $SU(3) \times SU(2) \times U(1)$, where $SU(3)$ corresponds to QCD and $SU(2) \times U(1)$ corresponds to the electroweak interactions. There are gauge bosons corresponding to these groups. The matter content falls into three types, the leptons, the quarks, and the Higgs scalar field. The electroweak $SU(2) \times U(1)$ symmetry is spontaneously broken down to a $U(1)$ subgroup which is identified as the gauge group for electromagnetic interactions. The electroweak gauge fields couple differently to the left and right chiral components $\psi_{L,R} = \frac{1}{2}(1 \pm \gamma^5)\psi$ thereby breaking parity symmetry explicitly. It is therefore easier to specify the transformation properties of various fields after separating the chiral components. The covariant derivatives in general can be written as

$$D_\mu \chi = \left(\partial_\mu - i e_s A^A_\mu T^A - i g b^a_\mu t^a - i g' \frac{Y}{2} c_\mu \right) \chi \tag{12.137}$$

for any field χ. A^A_μ, b^a_μ, c_μ are the gauge fields corresponding to $SU(3)$, $SU(2)$, and $U(1)$, respectively. e_s is the strong interaction (QCD) coupling constant, g, g' are the coupling constants for the $SU(2)$ and $U(1)$ groups. T^A, t^a are matrices corresponding to the generators of $SU(3)$ and $SU(2)$, respectively, which are in the representations to which the field χ belongs. Y is the $U(1)$ charge of the field χ; it is called the weak hypercharge. The matter field content of the theory is given in the table shown below. The fields l denote the lepton fields, Q's the quarks, and ϕ is the Higgs scalar field. Notice that the same pattern of fermionic fields repeats itself three times; these are referred to as the three generations, usually called the electron generation, the muon generation, and the tau generation. We could combine the notation by writing l^i, Q^i, etc., where $i = 1, 2, 3$, label the three generations.

In the table, we have denoted the representations by their dimensions. For the singlet representations, T^A and t^a are zero. For the quarks which are triplets under the QCD $SU(3)$ group, we may take $T^A = \frac{1}{2}\lambda^A$, $A = 1, 2, \cdots, 8$ with

$$\lambda_1 = \begin{pmatrix} 0 & 1 & 0 \\ 1 & 0 & 0 \\ 0 & 0 & 0 \end{pmatrix}, \quad \lambda_2 = \begin{pmatrix} 0 & -i & 0 \\ i & 0 & 0 \\ 0 & 0 & 0 \end{pmatrix},$$

$$\lambda_4 = \begin{pmatrix} 0 & 0 & 1 \\ 0 & 0 & 0 \\ 1 & 0 & 0 \end{pmatrix}, \quad \lambda_5 = \begin{pmatrix} 0 & 0 & -i \\ 0 & 0 & 0 \\ i & 0 & 0 \end{pmatrix},$$

$$\lambda_6 = \begin{pmatrix} 0 & 0 & 0 \\ 0 & 0 & 1 \\ 0 & 1 & 0 \end{pmatrix}, \quad \lambda_7 = \begin{pmatrix} 0 & 0 & 0 \\ 0 & 0 & -i \\ 0 & i & 0 \end{pmatrix}, \quad (12.138)$$

$$\lambda_3 = \begin{pmatrix} 1 & 0 & 0 \\ 0 & -1 & 0 \\ 0 & 0 & 0 \end{pmatrix}, \quad \lambda_8 = \frac{1}{\sqrt{3}}\begin{pmatrix} 1 & 0 & 0 \\ 0 & 1 & 0 \\ 0 & 0 & -2 \end{pmatrix}$$

Matter fields in the standard model

Field χ	$SU(3)$ representation	$SU(2)$ representation	$U(1)$ charge
$l_L = \begin{pmatrix} \nu_e \\ e \end{pmatrix}_L, \begin{pmatrix} \nu_\mu \\ \mu \end{pmatrix}_L, \begin{pmatrix} \nu_\tau \\ \tau \end{pmatrix}_L$	1	2	-1
$l_R = e_R, \mu_R, \tau_R$	1	1	-2
$Q_L = \begin{pmatrix} u \\ d \end{pmatrix}_L, \begin{pmatrix} c \\ s \end{pmatrix}_L, \begin{pmatrix} t \\ b \end{pmatrix}_L$	3	2	$\frac{1}{3}$
$U_R = u_R, c_R, t_R$	3	1	$\frac{4}{3}$
$D_R = d_R, s_R, b_R$	3	1	$-\frac{2}{3}$
$\phi = \begin{pmatrix} \phi^+ \\ \phi^0 \end{pmatrix}$	1	2	1

The T^A obey the commutation rules $[T^A, T^B] = i f^{ABC} T^C$, where the structure constants f^{ABC} may be worked out from the explicit choice of matrices given above. For the doublets of $SU(2)$, we can write $t^a = \frac{1}{2}\tau^a$, where τ^a, $a = 1, 2, 3$, are the Pauli matrices and we have $[t^a, t^b] = i\epsilon^{abc} t^c$. The electric charge of the field is given by $Q = t^3 + \frac{1}{2}Y$. Given these matter representations, we notice that the transformation

$$\hat{g} = \exp\left(i\frac{2\pi}{3}T^8 2\sqrt{3}\right) \, \exp\left(i2\pi t^3\right) \, \exp\left(i\pi Y\right) \qquad (12.139)$$

acts as identity. All the fields are invariant under this. $\hat{g}$ and its multiples form a discrete subgroup C of $SU(3) \times SU(2) \times U(1)$; and thus, strictly speaking, the gauge group of the standard model is

$$G = \frac{SU(3) \times SU(2) \times U(1)}{C} \qquad (12.140)$$

The relevant field-strength tensors for the gauge fields are

$$F^A_{\mu\nu} = \partial_\mu A^A_\nu - \partial_\nu A^A_\mu + e_s f^{ABC} A^B_\mu A^C_\nu$$
$$G^a_{\mu\nu} = \partial_\mu b^a_\nu - \partial_\nu b^a_\mu + g\epsilon^{abc} b^b_\mu b^c_\nu$$
$$f_{\mu\nu} = \partial_\mu c_\nu - \partial_\nu c_\mu \qquad (12.141)$$

The action for the theory can now be written as

$$S = \int d^4x \, [\mathcal{L}_g + \mathcal{L}_f + \mathcal{L}_{Yuk} + \mathcal{L}_\phi]$$

$$\mathcal{L}_g = -\frac{1}{4}F^A_{\mu\nu}F^{A\mu\nu} - \frac{1}{4}G^a_{\mu\nu}G^{a\mu\nu} - \frac{1}{4}f_{\mu\nu}f^{\mu\nu}$$

$$\mathcal{L}_f = \sum_i \bar{Q}^i_L i\gamma^\mu \left(\partial_\mu - ie_s T^A A^A_\mu - igt^a b^a_\mu - i\frac{g'}{6}c_\mu\right) Q^i_L$$

$$+ \bar{U}^i_R i\gamma^\mu \left(\partial_\mu - ie_s T^A A^A_\mu - i\frac{2g'}{3}c_\mu\right) U^i_R$$

$$+ \bar{D}^i_R i\gamma^\mu \left(\partial_\mu - ie_s T^A A^A_\mu + i\frac{g'}{3}c_\mu\right) D^i_R$$

$$+ \bar{l}^i_L i\gamma^\mu \left(\partial_\mu - igt^a b^a_\mu + i\frac{g'}{2}c_\mu\right) l^i_L + \bar{l}^i_R i\gamma^\mu(\partial_\mu + ig'c_\mu)l^i_R$$

$$\mathcal{L}_{Yuk} = -f^l_{ij}\bar{l}^i_L \phi \, l^j_R - f^u_{ij}\bar{Q}^i_L \tilde{\phi} \, U^j_R - f^d_{ij}\bar{Q}^i_L \phi \, D^j_R + \text{h.c.}$$

$$\mathcal{L}_\phi = \left|\left(\partial_\mu - igt^a b^a_\mu - i\frac{g'}{2}c_\mu\right)\phi\right|^2 - \lambda\left(\phi^\dagger \phi - \frac{v^2}{2}\right) \qquad (12.142)$$

Here $\tilde{\phi} = i\tau_2 \phi^*$ and h.c. stands for the hermitian conjugate. $\tilde{\phi}$ transforms under $SU(2)$ exactly as ϕ does, but has the opposite $U(1)$ charge. l^i_R stands for the three charged leptons fields e_R, μ_R, τ_R. We have neglected neutrino masses, which are very small, and the right-handed neutrinos. The right chirality neutrinos and neutrino masses can be included by use of the so-called Majorana-type coupling and Majorana masses. We will not discuss these.

We now consider the simplification of the $SU(2) \times U(1)$-part of the theory. The scalar field action has a potential energy term which gives a nonzero vacuum expectation value to ϕ^0. In other words,

$$\langle 0|\phi|0 \rangle = \frac{1}{\sqrt{2}} \begin{pmatrix} 0 \\ v \end{pmatrix} \tag{12.143}$$

This vacuum expectation value breaks the $SU(2) \times U(1)$ symmetry spontaneously; the isotropy group of the expectation value (12.141) is the $U(1)_{em}$ corresponding to electromagnetism which is generated by the electric charge operator Q. Thus we have the breakdown $SU(2) \times U(1) \rightarrow U(1)_{em}$. The combination

$$A_\mu = \frac{g' \, b_\mu^3 + g \, c_\mu}{\sqrt{g^2 + g'^2}} = \sin\theta_W b_\mu^3 + \cos\theta_W c_\mu \tag{12.144}$$

is the electromagnetic gauge field and remains massless. Here $\sin\theta_W = g'/\sqrt{g^2 + g'^2}$; θ_W is known as the Weinberg angle. The orthogonal combination

$$Z_\mu = \frac{g \, b_\mu^3 - g' \, c_\mu}{\sqrt{g^2 + g'^2}} = \cos\theta_W b_\mu^3 - \sin\theta_W c_\mu \tag{12.145}$$

and the two fields $W_\mu^\pm = (b_\mu^1 \mp ib_\mu^2)/\sqrt{2}$ become massive.

One can understand the particle content and the nature of the interactions by exapnding around the vacuum expectation value given in (12.143). For the Higgs field, a general parametrization is given by

$$\phi = U(\zeta) \frac{1}{\sqrt{2}} \begin{pmatrix} 0 \\ v + \eta \end{pmatrix} \tag{12.146}$$

where $U(\zeta) = \exp(i\tau^a \zeta^a/2)$ is an element of $SU(2)$, $\zeta^a(x)$ are the would-be Goldstone bosons. When this is substituted into the Lagrangian, we see that, by virtue of the gauge invariance, we can absorb the fields ζ^a into the gauge potentials; this is the transformation to the unitary gauge. In this gauge, the gauge and Higgs terms of the electroweak part of the Lagrangian simplify as follows:

$$\mathcal{L} = -\frac{1}{4}(\partial_\mu A_\nu - \partial_\nu A_\mu)^2 - \frac{1}{4}(\partial_\mu Z_\nu - \partial_\nu Z_\mu)^2 - \frac{1}{2}|\partial_\mu W_\nu - \partial_\nu W_\mu|^2$$

$$+ M_W^2 W_\mu^+ W^{-\mu} + \frac{1}{2}M_Z^2 Z_\mu Z^\mu + \frac{1}{2}(\partial\eta)^2 - \frac{1}{2}M_H^2 \eta^2 + \mathcal{L}_{int} \tag{12.147}$$

$$\mathcal{L}_{int} = -\frac{e^2}{2}|A_\mu W_\nu - A_\nu W_\mu|^2 - \frac{g^2}{2}\cos^2\theta_W |Z_\mu W_\nu - Z_\nu W_\mu|^2$$
$$-ieA^\mu \left[W^{-\nu}(\partial_\mu W_\nu^+ - \partial_\nu W_\mu^+) - W^{+\nu}(\partial_\mu W_\nu^- - \partial_\nu W_\mu^-) \right]$$
$$-ig\cos\theta_W Z^\mu \left[W^{-\nu}(\partial_\mu W_\nu^+ - \partial_\nu W_\mu^+) - W^{+\nu}(\partial_\mu W_\nu^- - \partial_\nu W_\mu^-) \right]$$
$$+ieW^{+\mu}W^{-\nu}(\partial_\mu A_\nu - \partial_\nu A_\mu) + ig\cos\theta_W W^{+\mu}W^{-\nu}(\partial_\mu Z_\nu - \partial_\nu Z_\mu)$$
$$-eg\cos\theta_W Z^\mu \left[(A_\mu W_\nu^- - A_\nu W_\mu^-)W^{+\nu} + (A_\mu W_\nu^+ - A_\nu W_\mu^+)W^{-\nu} \right]$$
$$+\frac{g^2}{4}(W_\mu^+ W_\nu^- - W_\nu^+ W_\mu^-)^2 + 2\frac{M_W^2}{v}W^{+\mu}W_\mu^-\eta + \frac{M_W^2}{v^2}W^{+\mu}W_\mu^-\eta^2$$
$$+\frac{M_Z^2}{v}Z^\mu Z_\mu\eta + \frac{M_Z^2}{2v^2}Z^\mu Z_\mu\eta^2 - \frac{M_H^2}{2v}\eta^3 - \frac{M_H^2}{8v^2}\eta^4 \tag{12.148}$$

Here $e = g\sin\theta_W$ is the value of the electric charge of the electron. The masses of the gauge bosons, to this lowest tree-level order, are $M_Z = \frac{1}{2}\sqrt{g^2 + g'^2}\, v$ and $M_W = \frac{1}{2}g\, v$. Experimentally, $M_Z \approx 91$ GeV and $M_W \approx 80$ GeV and $\sin^2\theta_W = 0.223$. The electroweak scale $v \approx 246$ GeV. These values are obtained from experiments. The mass of the Higgs scalar particle represented by the field η is $M_H = \sqrt{2\lambda}\, v$. There is no numerical prediction for this mass yet, since λ has to be experimentally determined.

The Yukawa interactions $\mathcal{L}_{Yuk}$ give fermions masses because the vacuum value of ϕ is nonzero. These are generally mass matrices because the coupling constants f_{ij} are not necessarily diagonal. Consider how this happens in the quark sector. The mass term which arises may be written as

$$\mathcal{L}_{mass} = -M_{ij}^u \bar{U}_L^i U_R^j - M_{ij}^d \bar{D}_L^i D_R^j + \text{h.c.}$$
$$M_{ij}^u = f_{ij}^u \frac{v}{\sqrt{2}}, \qquad M_{ij}^d = f_{ij}^d \frac{v}{\sqrt{2}} \tag{12.149}$$

U_L, D_L are the chiral components of up and down quark fields, respectively. The physical particles are mass eigenstates, so we have to diagonalize the mass matrices M^u, M^d. We will take these matrices to be of nonzero determinant. If the determinant is zero, that means there is a zero mass field; we can then separate it and apply the following argument to the rest of the matrix. Now an arbitrary complex matrix M of nonzero determinant can be decomposed as $M = H\, U$, where H is hermitian and U is unitary. We diagonalize H by a unitary matrix S as $H = S^\dagger M_{diag} S$. This shows that we can write M as $M = S^\dagger M_{diag}(SU) = S^\dagger M_{diag} T$, where S and T are unitary matrices. Thus M can be diagonalized by a biunitary transformation. We can use this diagonalization to write the mass term as

$$\mathcal{L}_{mass} = \bar{U}_L S^{u\dagger} \mathcal{M}^u T^u U_R + \bar{D}_L S^{d\dagger} \mathcal{M}^d T^d D_R + \text{h.c.} \tag{12.150}$$

where $\mathcal{M}^u$ and $\mathcal{M}^d$ are diagonal matrices. Now we redefine the quark fields as

$$U_R \to T^{u\dagger} U_R, \qquad\qquad D_R \to T^{d\dagger} D_R$$
$$U_L \to S^{u\dagger} U_L, \qquad\qquad D_L \to S^{d\dagger} D_L \tag{12.151}$$

so that

$$\mathcal{L}_{mass} = \bar{U}_L \mathcal{M}^u U_R + \bar{D}_L \mathcal{M}^d D_R + \text{h.c.} \tag{12.152}$$

The redefinition of the quark fields makes the mass term diagonal, but changes the interaction terms with the W_μ.

$$\begin{aligned}
\bar{D}_L \gamma_\mu U_L &\to \bar{D}_L S^d S^{u\dagger} \gamma_\mu U_L \\
\bar{U}_L \gamma_\mu D_L &\to \bar{U}_L S^u S^{d\dagger} \gamma_\mu D_L
\end{aligned} \tag{12.153}$$

Thus the interaction terms with the W-bosons is

$$\begin{aligned}
\mathcal{L} &= \frac{g}{\sqrt{2}} (W^{+\mu} J_\mu^- + W^{-\mu} J_\mu^+) \\
J_\mu^- &= V_{ij} \bar{U}_L^i \gamma_\mu D_L^j \\
J_\mu^+ &= V_{ji}^* \bar{D}_L^i \gamma_\mu U_L^j
\end{aligned} \tag{12.154}$$

where $V_{ij} = (S^u S^{d\dagger})_{ij}$ is called the Cabibbo-Kobayashi-Maskawa (CKM) matrix. It gives the mixing angles for the weak interactions of quarks.

The electromagnetic current is given by

$$J_\mu^{em} = \frac{2}{3} \left(\bar{U}_L \gamma_\mu U_L + \bar{U}_R \gamma_\mu U_R \right) - \frac{1}{3} \left(\bar{D}_L \gamma_\mu D_L + \bar{D}_R \gamma_\mu D_R \right) \tag{12.155}$$

This is still diagonal after the redefinition of fields as in (12.151). Likewise the current

$$J_\mu^3 = \frac{1}{2} \left(\bar{U}_L \gamma_\mu U_L - \bar{D}_L \gamma_\mu D_L \right) \tag{12.156}$$

is not affected. Further the Higgs-fermion interaction terms are proportional to the mass matrices and get diagonalized when the matrices are diagonalized, ensuring that the Higgs will not mediate flavor changing processes, in a one-Higgs model such as the one we are considering here.

Collecting these results together, the quark part of the theory becomes

$$\begin{aligned}
\mathcal{L}_{quark} &= \bar{U}^i (i\gamma \cdot \partial - \mathcal{M}_i^u) U^i + \bar{D}^i (i\gamma \cdot \partial - \mathcal{M}_i^d) D^i \\
&\quad - \frac{1}{v} \bar{U}^i \mathcal{M}_i^u U^i \eta - \frac{1}{v} \bar{D}^i \mathcal{M}_i^d D^i \eta + e A^\mu J_\mu^{em} \\
&\quad + \frac{g}{\cos\theta_W} Z^\mu J_\mu^Z + \frac{g}{\sqrt{2}} (W^{+\mu} J_\mu^- + W^{-\mu} J_\mu^+)
\end{aligned} \tag{12.157}$$

The neutral current J_μ^Z is given by

$$J_\mu^Z = J_\mu^3 - \sin^2\theta_W \, J_\mu^{em} \tag{12.158}$$

For the currents in the lepton sector, we have similar expressions. Neutrino masses are taken into account by including right-handed neutrinos; the left- and right-handed neutrinos do not have the same mass, so that one needs to use so-called Majorana masses. If we neglect the neutrino masses, the lepton terms become

$$\mathcal{L}_{lepton} = \bar{N}_L^i i\gamma \cdot \partial N_L^i + \bar{E}^i (i\gamma \cdot \partial - \mathcal{M}_i^e) E^i + \mathcal{L}_{int} \qquad (12.159)$$

where N^i, for $i = 1, 2, 3$ stands for the electron-neutrino, the muon-neutrino, and the tau-neutrino; likewise, E^i denotes e, μ and τ for $i = 1, 2, 3$. The interactions of the leptons are given by

$$\mathcal{L}_{int} = eA^\mu J_\mu^{em} + \frac{g}{\cos\theta_W} Z^\mu J_\mu^Z + \frac{g}{\sqrt{2}} (W^{+\mu} J_\mu^- + W^{-\mu} J_\mu^+)$$
$$-\frac{1}{v} \bar{E}^i \mathcal{M}_i^e E^i \eta$$
$$J_\mu^- = \bar{N}_L^i \gamma_\mu E_L^i$$
$$J_\mu^+ = \bar{E}_L^i \gamma_\mu N_L^i \qquad (12.160)$$
$$J_\mu^{em} = -\bar{E}^i \gamma_\mu E^i$$
$$J_\mu^Z = \frac{1}{2} \left(\bar{N}_L^i \gamma_\mu N_L^i - \bar{E}_L^i \gamma_\mu E_L^i \right) - \sin^2\theta_W \, J_\mu^{em}$$

The CKM matrix V is an $n \times n$ unitary matrix for n generations. But it does not have n^2 physical parameters in it, since some of the angles can still be absorbed into the definition of the fields. For example, we can write $V = U_{diag} \tilde{V} U'_{diag}$, where U, U' are diagonal unitary matrices; they are just phases of the form $e^{i\varphi}$ along the diagonal. There are n such phases in U, $n-1$ phases in U', avoiding double counting of an overall common phase. U and U' can be absorbed into the definition of $\bar{U}_L$ and D_L; the currents J_μ^3 and J_μ^{em} are not affected by this, since U, U' are diagonal. The mass terms change, but we can simultaneously redefine U_R and D_R, so that the mass terms are also not affected. This shows that we can get rid of $2n - 1$ parameters leaving $n^2 - (2n - 1) = (n - 1)^2$ physical parameters in the CKM matrix. For three generations, we thus have four parameters, corresponding to three real rotation angles and one remaining phase. V is not real in this case. The fact that V is not real leads to CP-violation due to weak interactions. The CP-transformation of a fermion field is defined as

$$\psi(x) = C \, \tilde{\psi}^*(-\boldsymbol{x}, x^0) \qquad (12.161)$$

$\tilde{\psi}$ denotes the CP image of ψ. From the invariance of the Dirac Lagrangian $\bar{\psi} i\gamma \cdot (\partial - igb^a t^a)\psi$ we see that the transformation of the gauge fields is

$$-b_0^T(-\boldsymbol{x}, x^0) = \tilde{b}_0, \qquad b_i^T(-\boldsymbol{x}, x^0) = \tilde{b}_i \qquad (12.162)$$

where T denotes the transpose of b as a matrix. In terms of the components, this becomes

$$\begin{aligned} b_0^3 &= -\tilde{b}_0^3(-\boldsymbol{x}) & b_i^3 &= \tilde{b}_i^3(-\boldsymbol{x}) \\ W_0^+ &= -\tilde{W}_0^-(-\boldsymbol{x}) & W_i^+ &= \tilde{W}_i^-(-\boldsymbol{x}) \\ W_0^- &= -\tilde{W}_0^-(-\boldsymbol{x}) & W_i^- &= \tilde{W}_i^+(-\boldsymbol{x}) \end{aligned} \qquad (12.163)$$

For the charged current interactions we then find

$$\int W^{+\mu} J_\mu^- = \int \tilde{W}^{-\mu} \left[\bar{\tilde{E}}_L^i \gamma_\mu \tilde{N}_L^i + \bar{\tilde{D}}_L^i \gamma_\mu \tilde{U}^j V_{ji} \right]$$

$$\int W^{-\mu} J_\mu^+ = \int \tilde{W}^{+\mu} \left[\bar{\tilde{N}}_L^i \gamma_\mu \tilde{E}_L^i + \bar{\tilde{U}}_L^i \gamma_\mu \tilde{D}^j V_{ij}^* \right] \qquad (12.164)$$

It is then easy to show that the interaction part of the action obeys the transformation rule

$$S_{int}(N, E, U, D, V_{ij}) = S_{int}(\tilde{N}, \tilde{E}, \tilde{U}, \tilde{D}, V_{ij}^*) \qquad (12.165)$$

We have CP invariance if $V = V^*$. For three generations of quarks, not all matrix elements V_{ij} are real and we have CP violation.

Finally, notice that the presence of the gauge interactions and the Yukawa couplings (which lead to quark masses as well) show clearly the explicit breaking of chiral symmetry for the quark sector discussed earlier. From the point of view of the electroweak gauge symmetry, the spontaneous breakdown of chiral symmetry in QCD will break the electroweak gauge symmetry and this gives additional masses to the W, Z particles. But the energy scale of chiral symmetry breaking is so small compared to the electroweak scale that this effect is negligible compared to the masses due to the Higgs vacuum expectation value.

References

1. Goldstone's theorem is due to J. Goldstone, Nuovo Cim. **9**, 154 (1961); Y. Nambu, Phys. Rev. Lett. **4**, 380 (1960); J. Goldstone, A. Salam and S. Weinberg, Phys. Rev. **127**, 965 (1962).
2. The Mermin-Wagner-Coleman theorem was given by N.D. Mermin and H. Wagner, Phys. Rev. Lett. **17**, 1133 (1966); S. Coleman, Commun. Math. Phys. **31**, 259 (1973).
3. A good reference on coset spaces is R. Gilmore, *Lie Groups, Lie Algebras and Some of Their Applications*, John Wiley and Sons, Inc. (1974).
4. Historically the nonlinear sigma models were obtained from the linear sigma model of M. Gell-Mann and M. Lévy, Nuovo Cim. **16**, 705 (1960).
5. The general construction of effective Lagrangians for describing the dynamics of Goldstone bosons via coset spaces is given in S. Coleman, J. Wess and B. Zumino, Phys. Rev. **177**, 2239 (1969); C.G. Callan, S. Coleman, J. Wess and B. Zumino, Phys. Rev. **177**, 2247 (1969).
6. For more on spin waves, see, for example, R.P. Feynman, *Statistical Mechanics*, Addison-Wesley Publishing Co. (1972).
7. The idea that the pion is a Goldstone boson is essentially due to Y. Nambu, Phys. Rev. Lett. **4**, 380 (1960); Y. Nambu and G. Lona Lasinio, Phys. Rev. **122**, 345 (1961).
8. For estimate of light quark masses, see S. Weinberg, in *A Festschrift for I.I. Rabi*, Trans. N.Y. Acad. Sci. **38**, 185 (1977).

9. The Vafa-Witten theorem is in C. Vafa and E. Witten, Phys. Rev. Lett. **53**, 535 (1984); Nucl. Phys. **B234**, 173 (1984); Commun. Math. Phys. **95**, 257 (1984).

10. The effective Lagrangian approach to calculations for pions started with S. Weinberg, Phys. Rev. Lett. **18**, 188 (1967). There have been a large number of papers extending and improving the effective Lagrangian approach; see J.F. Donoghue, E. Golowich and B.R. Holstein, *The Dynamics of the Standard Model*, Cambridge University Press (1992).

11. Our formula for the Goldberger-Treiman relation sets the axial vector coupling g_A to 1, because our effective Lagrangian treats baryons as point particles, not taking account of form factor effects (effects due to the bound state wave functions of the quarks). The full relation is $m_B g_A = g_{\pi \bar{N} N} f_\pi$ as given by M.L. Goldberger and S. Treiman, Phys. Rev. Lett. **110**, 1178 (1958). The value of $g_A \approx 1.26$ can be calculated from integrated cross sections for pion-nucleon scattering by the Adler-Weisberger sum rule, S. Adler, Phys. Rev. Lett. **14**, 1051 (1965); W.I. Weisberger, Phys. Rev. Lett. **14**, 1047 (1965).

12. The Gell-Mann-Okubo formula for mesons and baryons is given by M. Gell-Mann, CalTech Synchrotron Laboratory Report CTSL-20 (1961), reprinted in M. Gell-Mann and Y. Ne'eman, *The Eightfold Way*, W.A. Benjamin (1964); S. Okubo, Prog. Theor. Phys. **27**, 949 (1962).

13. PCAC relation is due to J. Bernstein, S. Fubini, M. Gell-Mann and W. Thirring, Nuovo Cim. **17**, 757 (1960). In this connection, see also Gell-Mann and Lévy in reference 4 and Nambu's papers in reference 7.

14. The optical theorem is due to N. Bohr, R. Peierls and G. Placzek, Nature **144**, 200 (1939). For general discussions of unitarity bounds, see Steven Weinberg, *The Quantum Theory of Fields: Volume I Foundations*, Cambridge University Press, (1995). A number of different bounds on cross sections are derived and listed in R.J. Eden, *High Energy Collisions of Elementary Particles*, Cambridge University Press (1967).

15. The original references on the Higgs mechanism are: P.W. Higgs, Phys. Lett. **12**, 132 (1964); Phys. Rev. Lett. **13**, 508 (1964); Phys. Rev. **145**, 1156 (1966); F. Englert and R. Brout, Phys. Rev. Lett. **13**, 321 (1964); G.S. Guralnik, C.R. Hagen and T.W.B. Kibble, Phys. Rev. Lett. **13**, 585 (1964); T.W.B. Kibble, Phys. Rev. **155**, 1554 (1967).

16. The $SU(2) \times U(1)$ symmetry in the standard model is due to S.L. Glashow, Nucl. Phys. **22**, 579 (1961); A. Salam and J. Ward, Phys. Lett. **13**, 168 (1964). The $SU(2) \times U(1)$ model, incorporating the Higgs mechanism, was made by S. Weinberg, Phys. Rev. Lett. **19**, 1264 (1967); A. Salam, in *Elementary Particle Physics*, N. Svartholm (ed.), Almqvist and Wiksells, Stockholm (1968). The importance of the charm quark was recognized by S.L. Glashow, J. Iliopoulos and L. Maiani, Phys. Rev. **D2**, 1285 (1970).

17. $SU(3)$ gauge symmetry for strong interactions was suggested by H. Fritzsch and M. Gell-Mann, 16th International Conference on High Energy Physics, Chicago-Batavia, 1972; H. Fritzsch, M. Gell-Mann and H. Leutwyler, Phys. Lett. **B47** 365 (1973). But this model really became viable with the discovery of asymptotic freedom by D.J. Gross and F. Wilczek, Phys. Rev. Lett. **30**, 1343 (1973); H.D. Politzer, Phys. Rev. Lett. **30**, 1346 (1973).

18. Mixing angles for weak interactions were introduced by N. Cabibbo, Phys. Rev. Lett. **10**, 531 (1963); M. Kobayashi and M. Maskawa, Prog. Theor. Phys. **49**, 652 (1973).

13 Anomalies I

13.1 Introduction

Consider a classical field theory with a symmetry group G which may be partially a gauge symmetry and partially a global symmetry. In defining the quantum theory, one has to evaluate loop diagrams, some of which may be divergent, and therefore the quantum theory has to be defined with the help of a regulator. There are situations for which there exists no regulator preserving all the symmetries. Symmetries of the classical theory which are broken by quantum corrections (by choice of regulators) are said to be anomalous. The corresponding currents have nonzero divergences which are called anomalies.

This point about regulators and symmetries can be illustrated by some examples. Dimensional regularization has the great virtue that it preserves Lorentz and vector gauge-invariance (or the corresponding BRST-invariance). But $\gamma^5 = (i/4!)\epsilon_{\mu\nu\alpha\beta}\gamma^\mu\gamma^\nu\gamma^\alpha\gamma^\beta$ has no natural extension to arbitrary dimensions ($\neq 4$) since it uses the ϵ-tensor. Thus chiral symmetries are potentially anomalous if we use dimensional regularization. Another common regulator is the Pauli-Villars regulator, where we add very massive unphysical particles with negative Hilbert space norm, the mass M serving as the regularization parameter. But a mass term like $M\bar{\psi}\psi$ is not invariant under chiral transformations and so chiral symmetries are potentially anomalous.

Anomalies are of two types, anomalous global symmetries and anomalous gauge symmetries. In the case of anomalous global symmetries, the symmetry is not realized in the quantum theory, but otherwise the theory is consistent. Examples are the axial $U_A(1)$ anomalies in QED and QCD. If such a symmetry is spontaneously broken, we have a Goldstone boson classically. Quantum theoretically, since there is no symmetry, there is no Goldstone boson. In fact the quantum corrections generate a mass for the potential Goldstone boson.

Anomalies for a gauge symmetry can lead to unphysical results. The gauge-invariance (or the related BRST-invariance) of a theory is crucial for the proof of unitarity of the S-matrix or unitarity of time-evolution in general. Gauge-invariance removes the unphysical polarizations of the gauge particles; if gauge-invariance is lost, they can become propagating modes and ruin unitarity. Thus in a consistent physical theory there should be no anomaly for the gauge symmetries. There can be an anomaly for gauge symmetry in a subsector of the theory, but all the individual contributions to the anomaly must cancel out in the end.

13.2 Computation of anomalies

We now turn to the computation of the anomalies, using the Euclidean field theory for most of the calculations. In four dimensions, the anomalies are due to fermion one-loop diagrams only. Consider the Euclidean action

$$
\begin{aligned}
\mathcal{S} &= \int \left[\frac{1}{4e_L^2} F_L^2 + \frac{1}{4e_R^2} F_R^2 + \bar{\psi}_L \gamma \cdot (\partial + L)\psi_L + \bar{\psi}_R \gamma \cdot (\partial + R)\psi_R \right] \\
&= \int \left[\frac{1}{4e_L^2} F_L^2 + \frac{1}{4e_R^2} F_R^2 + \bar{\psi}\gamma \cdot (\partial + V + \gamma^5 A)\psi \right]
\end{aligned}
\tag{13.1}
$$

where $\psi_L = \frac{1}{2}(1+\gamma^5)\psi$, $\psi_R = \frac{1}{2}(1-\gamma^5)\psi$, and $V = \frac{1}{2}(L+R)$, $A = \frac{1}{2}(L-R)$. We consider N species of fermions with the chiral symmetry $U(N)_L \times U(N)_R$. We have taken all the chiral symmetries of the fermion to be gauge symmetry so as to be very general. This means that we can write $L_\mu = -iT^A L_\mu^A$, $R_\mu = -iT^A R_\mu^A$, T^A being generators of $U(N)$. One can specialize the results to any subgroup of this maximal symmetry by setting some of the gauge fields to be zero.

Notice that we do not write a fermion mass term because we are gauging the axial symmetries as well; there is no gauge-invariant mass term if we including the axial symmetries as well as vector symmetries. When we restrict to a subgroup, we may have the possibility of gauge-invariant mass terms, but the calculation of the anomalies is not affected by the mass term, so we do not need to consider it. This will become clear later. The field-strength tensors are defined by

$$
\begin{aligned}
F_{L\mu\nu}^A &= \partial_\mu L_\nu^A - \partial_\nu L_\mu^A + f^{ABC} L_\mu^B L_\nu^C \\
F_{R\mu\nu}^A &= \partial_\mu R_\nu^A - \partial_\nu R_\mu^A + f^{ABC} R_\mu^B R_\nu^C \\
F_{V\mu\nu}^A &= \partial_\mu V_\nu^A - \partial_\nu V_\mu^A + f^{ABC} V_\mu^B V_\nu^C + f^{ABC} A_\mu^B A_\nu^C \\
F_{A\mu\nu}^A &= \partial_\mu A_\nu^A - \partial_\nu A_\mu^A + f^{ABC}(V_\mu^B A_\nu^C - V_\nu^B A_\mu^C) \\
&= (D_\mu A_\nu - D_\nu A_\mu)^A
\end{aligned}
\tag{13.2}
$$

where D is the covariant derivative with respect to the vector gauge fields only.

One can add to the action (13.1) a term of the form

$$
\mathcal{S}_{Reg} = -\sum_{L,R} \int \text{Tr}\left(F_{\mu\nu} \frac{(-D^2)}{\Lambda^2} F^{\mu\nu} \right)
\tag{13.3}
$$

for both the vector and axial vector gauge fields. This term, because of the higher derivatives, makes the gauge boson propagator take the form

$$G \sim \frac{\Lambda^2}{p^4 + \Lambda^2 p^2} \sim \frac{\Lambda^2}{p^4} \qquad (13.4)$$

for large-momenta p. This can regulate the Feynman diagrams which contain gauge boson propagators, and since (13.3) is gauge-invariant, this provides a gauge-invariant regulator for all potentially divergent diagrams except the one-loop diagrams. Λ plays the role of a gauge-invariant cutoff. One-loop diagrams can involve the fermion loop, which is not regulated by (13.3). Also (13.3) introduces new vertices which can lead to new divergent one-loop graphs which need separate regularization. So we cannot make conclusions about the one-loop graphs. Since the gauge symmetry covers all the possible continuous symmetries, because we have gauged all the symmetries, this shows that the anomalies are confined to one-loop diagrams. By power counting, one can then see that the anomalies are in one-loop fermion diagrams. (Higher loops can renormalize the parameters appearing in the expression for the anomaly.)

The potentially divergent graphs which can contribute to the anomaly are shown below.

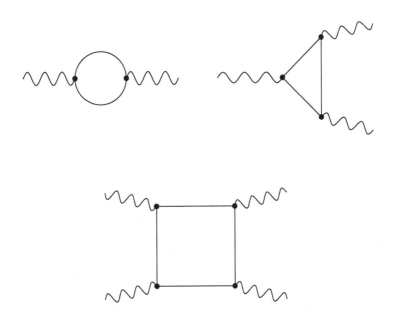

Fig 13.1. Potentially divergent diagrams for the anomaly calculation

In these, the external lines can be either vector (V) or axial vector (A). The two-point function is the vacuum polarization graph, which we have evaluated in a gauge-invariant fashion and leads to no anomaly. For the triangle diagram, the VVV-graph is zero by charge conjugation invariance (Furry's theorem), since

$$\int d^4x \; \bar{\psi}\gamma \cdot (\partial + V + \gamma^5 A)\psi = \int d^4x \; \bar{\psi}^c \gamma \cdot (\partial - V + \gamma^5 A)\psi^c \qquad (13.5)$$

where ψ^c denote the charge-conjugate fields. Thus the possibly anomalous graphs are the AVV and AAA graphs. The contribution to the effective action from the AVV type of graphs is

$$\Gamma^{(3)} = \int d^4x d^4y d^4z \; \text{Tr} \left[S(x,y)\gamma^5\gamma \cdot A(y)S(y,z)\gamma \cdot V(z)S(z,x)\gamma \cdot V(x) \right]$$

$$= \int d\mu(q,p_1,p_2) \; \text{Str}[A_\mu(2q)V_\alpha(p_1)V_\beta(p_2)] \; \mathcal{I}^{\mu\alpha\beta}(q,p_1,p_2) \qquad (13.6)$$

where

$$d\mu(q,p_1,p_2) = \frac{d^4(2q)}{(2\pi)^4} \frac{d^4p_1}{(2\pi)^4} \frac{d^4p_2}{(2\pi)^4} \delta^{(4)}(2q - p_1 - p_2)$$

$$\mathcal{I}^{\mu\alpha\beta}(q,p_1,p_2) = \int \frac{d^4k}{(2\pi)^4} \text{Tr} \left[\frac{1}{i\gamma \cdot (k-q)} \gamma^5\gamma^\mu \frac{1}{i\gamma \cdot (k+q)} \gamma^\alpha \frac{1}{i\gamma \cdot (k-r)} \gamma^\beta \right]$$

$$(13.7)$$

Here $r = \frac{1}{2}(p_1 - p_2)$, and in $\text{Str}[A_\mu(2q)V_\alpha(p_1)V_\beta(p_2)]$ we are taking the symmetrized trace over the generators of the group, i.e.,

$$\text{Str}[AVV] = \frac{1}{3}\text{Tr}[AVV + VAV + VVA] \qquad (13.8)$$

The integral $\mathcal{I}^{\mu\alpha\beta}(q,p_1,p_2)$ in (13.7) has superficial linear divergence. The infinitesimal gauge transformations are

$$V_\alpha \rightarrow V_\alpha + D_\alpha\theta$$
$$A_\alpha \rightarrow A_\alpha + D_\alpha\varphi \qquad (13.9)$$

which reads in momentum space as

$$V_\alpha(p) \rightarrow V_\alpha(p) + ip_\alpha\theta(p) + \cdots$$
$$A_\alpha(p) \rightarrow A_\alpha(p) + ip_\alpha\varphi(p) + \cdots \qquad (13.10)$$

The variation of the term corresponding to the triangle graph in the effective action under the vector gauge transformation is then given as

$$\delta_\theta \Gamma^{(3)} = -\int d\mu(q,p_1,p_2) \; \text{Str}[A_\mu(2q)\theta(p_1)V_\alpha(p_2)] \; \mathcal{F}^{\mu\alpha}(q,p_1,p_2)$$

$$\mathcal{F}^{\mu\alpha}(q,p_1,p_2) = \int \frac{d^4k}{(2\pi)^4} \text{Tr} \left[\frac{1}{\gamma \cdot (k-q)} \gamma^5\gamma^\mu \frac{1}{\gamma \cdot (k-r)} \gamma^\alpha \right.$$

$$\left. - \frac{1}{\gamma \cdot (k+r)} \gamma^5\gamma^\mu \frac{1}{\gamma \cdot (k+q)} \gamma^\alpha \right] \qquad (13.11)$$

where we have used $p_1 = (k+q) - (k-r)$ and $p_2 = (k-r) - (k-q)$. Carrying out the traces and writing $k' = k + p_1$, we find

$$\mathcal{F}^{\mu\alpha}(q, p_1, p_2) = -4\epsilon^{\mu\nu\alpha\beta} \int \frac{d^4k}{(2\pi)^4} \left[\frac{(k-q)_\nu (k-r)_\beta}{(k-q)^2(k-r)^2} - (k \leftrightarrow k') \right] \quad (13.12)$$

Thus, if we could shift the variable of integration to k' in the second term, $\mathcal{F}^{\mu\alpha}(q, p_1, p_2)$ would be zero, giving gauge-invariance of $\Gamma^{(3)}$ for the vector gauge transformations. But the integral in (13.12) is linearly divergent and the shift can produce a nonzero surface term since

$$\int \frac{d^4k}{(2\pi)^4} [f(k+a) - f(k)] = a_\nu \int \frac{d^4k}{(2\pi)^4} \frac{\partial f}{\partial k^\nu} + \cdots$$

$$= \frac{1}{(2\pi)^4} \left[\int f a_\nu \hat{k}^\nu k^3 \, d\Omega^{(3)} \right]_{k \to \infty} \quad (13.13)$$

(The higher terms are zero as $k \to \infty$.) Using this result,

$$\mathcal{F}^{\mu\alpha}(q, p_1, p_2) = 4\epsilon^{\mu\nu\alpha\beta} \int \frac{(k-q)_\nu (k-r)_\beta}{(k-q)^2(k-r)^2} k^3 \, p_{1\lambda} \hat{k}^\lambda \left. \frac{d\Omega^{(3)}}{(2\pi)^4} \right|_{k \to \infty}$$

$$= 4\epsilon^{\mu\nu\alpha\beta} \int \frac{d\Omega^{(3)}}{(2\pi)^4} \hat{k}_\nu (q-r)_\beta \, p_{1\lambda} \hat{k}^\lambda$$

$$= 4\epsilon^{\mu\nu\alpha\beta} \left(\frac{\delta_{\nu\lambda}}{4} \right) p_{1\lambda} \, p_{2\beta} \frac{\text{Vol}(S^3)}{16\pi^4}$$

$$= \frac{1}{8\pi^2} \epsilon^{\mu\nu\alpha\beta} p_{1\nu} p_{2\beta} \quad (13.14)$$

The result (13.11) can now be simplified as

$$\delta_\theta \Gamma^{(3)} = -\frac{1}{8\pi^2} \int d^4x \, \epsilon^{\mu\nu\alpha\beta} \, \text{Tr} \, (A_\mu \partial_\nu \theta \, \partial_\alpha V_\beta) \quad (13.15)$$

where we have transformed back to coordinate space.

Under an axial transformation we find

$$\delta_\varphi \Gamma^{(3)} = -\int d\mu(q, p_1, p_2) \, \text{Str}[\varphi(2q) V_\alpha(p_1) V_\beta(p_2)] \mathcal{H}^{\alpha\beta}(q, p_1, p_2)$$

$$\mathcal{H}^{\alpha\beta}(q, p_1, p_2) = \int \frac{d^4k}{(2\pi)^4} \text{Tr} \left[\frac{1}{\gamma \cdot (k-q)} \gamma^5 \, 2\gamma \cdot q \frac{1}{\gamma \cdot (k+q)} \right.$$

$$\left. \gamma^\alpha \frac{1}{\gamma \cdot (k-r)} \gamma^\beta \right]$$

$$= \int \frac{d^4k}{(2\pi)^4} \text{Tr} \left[\gamma^5 \frac{1}{\gamma \cdot (k+q)} \gamma^\alpha \frac{1}{\gamma \cdot (k-r)} \gamma^\beta \right.$$

$$\left. + \frac{1}{\gamma \cdot (k-q)} \gamma^5 \gamma^\alpha \frac{1}{\gamma \cdot (k-r)} \gamma^\beta \right] \quad (13.16)$$

where in the last step we have used $\gamma^5 2\gamma \cdot q = \gamma \cdot (k-q)\gamma^5 + \gamma^5\gamma \cdot (k+q)$. Evaluating the trace over the γ-matrices and using the symmetry $V_\alpha(p_1) \leftrightarrow V_\beta(p_2)$, we can simplify this as

$$
\begin{aligned}
\mathcal{H}^{\alpha\beta} &= -2\epsilon^{\mu\nu\alpha\beta}\left[\frac{(k+q)_\mu(k-r)_\nu}{(k+q)^2(k-r)^2} - \frac{(k+q)_\mu(k+r)_\nu}{(k+q)^2(k+r)^2}\right.\\
&\qquad\qquad\left. + \frac{(k-q)_\nu(k-r)_\mu}{(k-q)^2(k-r)^2} - \frac{(k-q)_\nu(k+r)_\mu}{(k-q)^2(k+r)^2}\right]\\
&= -2\epsilon^{\mu\nu\alpha\beta}\left[\left\{\frac{(k-q)_\nu(k-r)_\mu}{(k-q)^2(k-r)^2} - (k \to k+p_1)\right\}\right.\\
&\qquad\qquad\left. - \left\{\frac{(k-q)_\nu(k+r)_\mu}{(k-q)^2(k+r)^2} - (k \to k+p_2)\right\}\right]\\
&= \frac{1}{8\pi^4}\epsilon^{\mu\nu\alpha\beta}\int d\Omega^{(3)}\left[\hat{k}_\nu(q-r)_\mu p_{1\sigma}\hat{k}^\sigma - \hat{k}_\nu(q+r)_\mu p_{2\sigma}\hat{k}^\sigma\right]\\
&= -\frac{1}{8\pi^2}\epsilon^{\mu\nu\alpha\beta}p_{1\mu}p_{2\nu}
\end{aligned} \tag{13.17}
$$

For $\delta_\varphi\Gamma^{(3)}$ we then find

$$
\delta_\varphi\Gamma^{(3)} = -\frac{1}{8\pi^2}\int d^4x\, \epsilon^{\mu\nu\alpha\beta}\mathrm{Tr}\left(\varphi\partial_\mu V_\nu \partial_\alpha V_\beta\right) \tag{13.18}
$$

Combining (13.15) and (13.18), we have

$$
\begin{aligned}
\delta_\theta\Gamma^{(3)} &= \frac{1}{8\pi^2}\int d^4x\, \epsilon^{\mu\nu\alpha\beta}\,\mathrm{Tr}\left(A_\mu\partial_\nu\theta\, \partial_\alpha V_\beta\right)\\
\delta_\varphi\Gamma^{(3)} &= \frac{1}{8\pi^2}\int d^4x\, \epsilon^{\mu\nu\alpha\beta}\mathrm{Tr}\left(\varphi\partial_\mu V_\nu \partial_\alpha V_\beta\right)
\end{aligned} \tag{13.19}
$$

Both variations are nonzero. However, notice that

$$
\delta_\theta\left[\frac{1}{8\pi^2}\int d^4x\, \epsilon^{\mu\nu\alpha\beta}\mathrm{Tr}\left(A_\mu V_\nu \partial_\alpha V_\beta\right)\right] = \frac{1}{8\pi^2}\int d^4x\, \epsilon^{\mu\nu\alpha\beta}\,\mathrm{Tr}\left(A_\mu\partial_\nu\theta\, \partial_\alpha V_\beta\right) \tag{13.20}
$$

Therefore, we define

$$
\tilde{\Gamma}^{(3)} = \Gamma^{(3)} + \frac{1}{8\pi^2}\int d^4x\, \epsilon^{\mu\nu\alpha\beta}\mathrm{Tr}\left(A_\mu V_\nu \partial_\alpha V_\beta\right) \tag{13.21}
$$

so that

$$
\begin{aligned}
\delta_\theta\tilde{\Gamma}^{(3)} &= 0\\
\delta_\varphi\tilde{\Gamma}^{(3)} &= -\frac{1}{4\pi^2}\int d^4x\, \epsilon^{\mu\nu\alpha\beta}\mathrm{Tr}\left(\varphi\partial_\mu V_\nu \partial_\alpha V_\beta\right)
\end{aligned}
$$

$$= -\frac{1}{16\pi^2} \int d^4x \; \epsilon^{\mu\nu\alpha\beta} \mathrm{Tr}\, (\varphi f_{\mu\nu} f_{\alpha\beta})$$

$$= -\frac{1}{8\pi^2} \int d^4x \; \mathrm{Tr}\, \left(\varphi f_{\mu\nu} \tilde{f}^{\mu\nu}\right) \tag{13.22}$$

where $f_{\mu\nu} = \partial_\mu V_\nu - \partial_\nu V_\mu$ and $\tilde{f}^{\mu\nu} = \frac{1}{2}\epsilon^{\mu\nu\alpha\beta} f_{\alpha\beta}$. (This is essentially the result of Adler and Bell and Jackiw.) The local term (of dimension zero) which we added in (13.21) is a counterterm. Redefining the quantum action by adding this is equivalent to a change of regularization. Thus we see that, for the redefined action $\tilde{\Gamma}^{(3)}$, the vector gauge-invariance can be preserved by a suitable choice of regulator, but not simultaneously the axial gauge-invariance. For the $U(1)$ symmetry, one could also choose to preserve the axial symmetry, breaking the vector symmetry. For the nonabelian case, this is not so, since the axial transformations, by themselves, do not close into a group.

The variation of the action can also be expressed in terms of the conservation of the current. From the definition $e^{-\Gamma} = \int e^{-S}$,

$$\delta\Gamma = \int \bar{\psi}\gamma^\mu(\partial_\mu\theta + \gamma^5\partial_\mu\varphi)\psi \tag{13.23}$$

Expressed as conservation laws, equations (13.22) read

$$\partial_\mu(\bar{\psi}\gamma^\mu T^A\psi) = 0$$

$$\partial_\mu(\bar{\psi}\gamma^\mu\gamma^5 T^A\psi) = \frac{1}{8\pi^2}\mathrm{Tr}\left(T^A f_{\mu\nu}\tilde{f}^{\mu\nu}\right) \tag{13.24}$$

The result we have derived is for the $AVV + VAV + VVA$ graphs only. There is also the AAA-graph. We now specialize to the purely left-handed gauge fields and transformations, for which $V_\mu = A_\mu = \frac{1}{2}L_\mu$. In this case, the AAA-graph is equal to the AVV-graph, so that we can get the full result by taking $4/3$ times the value we have calculated. For the left-handed gauge transformation by parameter ξ, we find, from the formula for V_μ, A_μ, that $\theta = \frac{1}{2}\xi$ and $\varphi = \frac{1}{2}\xi$, so that

$$\delta_\xi\Gamma^{(3)} = \frac{4}{3}\left\{\left[\delta_\theta\Gamma^{(3)}\right]_{\theta=\frac{1}{2}\xi} + \left[\delta_\varphi\Gamma^{(3)}\right]_{\varphi=\frac{1}{2}\xi}\right\}$$

$$= \frac{1}{24\pi^2}\int d^4x \; \epsilon^{\mu\nu\alpha\beta}\mathrm{Tr}\,(\xi\;\partial_\mu L_\nu\;\partial_\alpha L_\beta)$$

$$= -\frac{1}{24\pi^2}\int d^4x \; \epsilon^{\mu\nu\alpha\beta}\mathrm{Tr}\,(\partial_\mu\xi\;L_\nu\;\partial_\alpha L_\beta) \tag{13.25}$$

Results (13.19, 13.22, 13.25) are from the triangle diagram only. Correspondingly we have only analyzed the behavior of Γ under the linearized gauge transformations (13.10). For example, for the left-handed gauge fields and transformations, there can be L^3-terms which are required for consistency under the full transformations, including the $f^{ABC}L^B\xi^C$-terms. These corrections come from the 4-point fermion loop and from a pentagon diagram for the case of vector and axial vector transformations. The pentagon diagram is actually convergent, but nevertheless a piece of it is needed to ensure proper transformation of the anomaly. Rather than compute these diagrams in turn, we can identify what is required from the consistency conditions. Consider again the left-handed fields. For the full gauge transformation $L_\mu^A \to L_\mu^A + (D_\mu\xi)^A = L_\mu^A + \partial_\mu\xi^A + f^{ABC}L_\mu^B\xi^C$ we define the anomaly $G(\xi)$ by

$$\delta_\xi\Gamma \equiv G(\xi) \tag{13.26}$$

The infinitesimal gauge transformation δ_ξ can be explicitly written in terms of the gauge fields as

$$\delta_\xi = \int d^4x \, (D^\mu\xi)^A(x)\frac{\delta}{\delta L_\mu^A(x)} \tag{13.27}$$

These are translation operators along the gauge directions of the potentials and so they must obey an algebra determined by the group commutation rules. In fact, from the definition (13.27) we find

$$\delta_\xi \, \delta_{\xi'} \, - \, \delta_{\xi'} \, \delta_\xi \, - \, \delta_{\xi\times\xi'} = 0 \tag{13.28}$$

where $(\xi \times \xi')^A = f^{ABC}\xi^B\xi'^C$. (If we include vertex functions with matter fields in our consideration of terms in Γ, there are additional terms in the definition of δ_ξ which carry out the gauge transformation of the matter fields. These can be easily included; they do not change the algebra or the consistency conditions derived below.) Applying the algebra (13.28) to the definition of the anomaly we get the consistency conditions for the anomaly, first obtained by Wess and Zumino,

$$\delta_\xi G(\xi') - \delta_{\xi'}G(\xi) - G(\xi \times \xi') = 0 \tag{13.29}$$

Any expression for the anomaly must obey this condition. Having obtained the first set of terms by explicit calculation, we can now postulate that $G(\xi)$ has the general form

$$G(\xi) = -\frac{1}{24\pi^2} \int d^4x \, \epsilon^{\mu\nu\alpha\beta}\text{Str}\,[\partial_\mu\xi\,(L_\nu\partial_\alpha L_\beta + cL_\nu L_\alpha L_\beta)] \tag{13.30}$$

where c is a constant which can be determined by imposing the Wess-Zumino consistency conditions. Carrying this out we find

$$\delta_\xi\Gamma = G(\xi) = -\frac{1}{24\pi^2} \int d^4x \, \epsilon^{\mu\nu\alpha\beta}\text{Str}\,\left[\partial_\mu\xi\left(L_\nu\partial_\alpha L_\beta + \frac{1}{2}L_\nu L_\alpha L_\beta\right)\right] \tag{13.31}$$

This gives the complete form of the anomaly for left-handed gauge fields.

13.3 Anomaly structure: why it cannot be removed

The WZ consistency conditions give a precise way of stating what we mean by an anomaly in a regulator-independent way. Notice that any quantity of the form $\delta_\xi W$ for any functional W of the gauge fields will obey the consistency conditions simply by virtue of the general algebra (13.28). This does not constitute an anomaly. In the first place we cannot add arbitrary functionals W to Γ, which is basically determined by the theory. However, we do have some freedom; if W is the integral of a local monomial of fields and derivatives and is of total dimension less than or equal to zero, then the addition of such a term is equivalent to a change of regularization. W is then a "counterterm". Thus if $G(\xi)$ is a solution of (13.29), then $G(\xi) + \delta_\xi W$ is also an acceptable solution for any W of dimension ≤ 0. This would simply give the form of the anomaly in a different regularization scheme. If every solution to (13.29) is of the form $\delta_\xi W$, then all the anomaly can be removed by a choice of regulator, which amounts to defining the effective action as $\tilde{\Gamma} = \Gamma - W$. So a true anomaly exists in the theory only if there is a solution to (13.29) which cannot be written as $\delta_\xi W$ for some W which is the integral of a local term, the total dimension being ≤ 0, i.e.,

$$\delta_\xi G(\xi') - \delta_{\xi'} G(\xi) - G(\xi \times \xi') = 0$$
$$G(\xi) \neq \delta_\xi W \qquad (13.32)$$

By writing out all possible monomials of fields and derivatives of dimension ≤ 4 (which correspond to terms in W with dimension ≤ 0), we can check that the solution (13.31) satisfies this requirement. This shows that the anomaly (13.31) cannot be removed by any choice of regulator. Later, in Chapter 17, we shall see a topological rephrasing of this statement.

Another important propery of the anomaly is expressed by the Adler-Bardeen theorem. This states that the expression for the anomaly is not renormalized by higher loop corrections except to the extent of replacing the fields and coupling constants by their renormalized values. We will not prove this theorem; the discussion of anomalies in terms of the index theorem and topological properties will show why such a theorem would hold true. It is a useful result, showing that the necessary cancellation of the anomalies for gauge transformations only need to be checked at the one-loop level.

Since the anomaly shifts as $G(\xi) \to G(\xi) + \delta_\xi W$ under a change of regulator, the form of the anomaly can be changed to some extent by a choice of regulators. We have already seen that the vector anomaly can be made zero, or either the vector or axial one can be made zero for the $U(1)$ case, but not both, by the use of counterterms as in (13.21). This can be very useful. For example, if we have a left-handed gauge theory, we would need to cancel the anomalies in the gauge transformations, but for the remainder, for which we have set the gauge fields to zero, and which are, therefore, global symmetries, we can live with the anomaly. There would be no inconsistency. However, we need to have the anomaly in a form which is invariant under the left-handed gauge transformations. This is achieved by use of counterterms. In a similar way, the anomaly for axial transformations can be expressed in a form which is vector gauge-invariant. This form, as mentioned before, involves the square and pentagon diagrams, but the final form can be determined by symmetry and WZ consistency arguments as

$$
G(\varphi) = -\frac{1}{4\pi^2} \int d^4x \, \epsilon^{\mu\nu\alpha\beta} \, \text{Str}\left[\varphi \left(\frac{1}{4} F_{V\mu\nu} F_{V\alpha\beta} + \frac{1}{12} F_{A\mu\nu} F_{A\alpha\beta} \right. \right.
$$
$$
\left. -\frac{2}{3}(A_\mu A_\nu F_{V\alpha\beta} + A_\mu F_{V\nu\alpha} A_\beta + F_{V\mu\nu} A_\alpha A_\beta) \right.
$$
$$
\left. \left. +\frac{8}{3} A_\mu A_\nu A_\alpha A_\beta \right) \right] \tag{13.33}
$$

This form was first derived by Bardeen.

We now consider the group structure of the anomaly. Consider first the left-handed anomaly (13.31). Since $L_\mu = -iT^A L_\mu^A$, $\xi = -iT^A \xi^A$, we get

$$
\delta_\xi \Gamma = -\frac{i}{24\pi^2} d^{ABC} \int d^4x \, \epsilon^{\mu\nu\alpha\beta} \left[\partial_\mu \xi^A \left(L_\nu^B \partial_\alpha L_\beta^C + \frac{1}{4} f^{CRS} L_\nu^B L_\alpha^R L_\beta^S \right) \right] \tag{13.34}
$$

where $d^{ABC} = \text{Str}(T^A T^B T^C)$. d^{ABC} is the symmetric rank 3 invariant of the algebra of the generators of the transformation. For the case of gauge transformations, we can calculate d^{ABC} for various fermion representations and check the cancellation of the anomalies.

For right-handed fermions and the corresponding gauge fields, in addition to the replacement $L_\mu \to R_\mu$, the anomaly has a minus sign relative to the anomaly for the left-handed fermions. This is because we have $V_\mu = -A_\mu = \frac{1}{2} R_\mu$ as opposed to $V_\mu = A_\mu = \frac{1}{2} L_\mu$ for the left handed case.

Finally, notice that $\delta_\xi \Gamma$ is imaginary. With our Euclidean calculation, Γ is usually real. This shows that the lack of gauge-invariance is in the imaginary part of the effective action.

13.4 Anomalies in the standard model

In the standard model, the gluon fields are vectorial in their coupling to fermions. As a result, we can choose a regulator for which there are no gauge anomalies. Alternatively, the gluons couple equally to the left and right chiralities; so the anomaly from the left and right sectors cancel out.

For the electroweak transformations, the couplings are not left-right symmetric; as a result we can have possible anomalies for the gauge transformations. These have to cancel out and give zero gauge anomaly for the consistency of the theory. The relevant anomalies arise from the b^3, $b^2 c$ and c^3

triangle diagrams. Thus we need to calculate d^{ABC} for each of these; we do this here for one generation of quarks and leptons.

b^3 graph

In this case, fields are either singlets for which there is no coupling to gauge fields or doublets for which the generators are $t^a = \frac{1}{2}\tau^a$.

$$d^{abc} = \text{Str}\left(\frac{\tau^a}{2}\frac{\tau^b}{2}\frac{\tau^c}{2}\right) = \frac{1}{8}\text{Tr}(\tau^a\delta^{bc}) = 0 \tag{13.35}$$

Thus all the anomalies of this type are zero.

$b^2\,c$ graph

In this case we need

$$d^{Yab} = \frac{1}{4}\text{Str}(Y\tau^a\tau^b) = \frac{1}{4}\delta^{ab}\text{Tr}(Y) \tag{13.36}$$

where Y is the weak hypercharge matrix for the left-handed fermions. (Since the other vertices of the triangle diagram involve the left-handed gauge field b_μ, right-handed fermions do not circulate in the loop; so we only need the trace of Y over the left-handed fermions.) For one generation of quarks and leptons, we have

$$\text{Tr}(Y) = \underbrace{(-1)}_{\nu_L} + \underbrace{(-1)}_{e_L} + \left(\underbrace{\frac{1}{3}}_{u_L} + \underbrace{\frac{1}{3}}_{d_L}\right) \times 3$$
$$= 0 \tag{13.37}$$

The extra multiplicity of three for the quarks is due to the number of colors being three, i.e., because they are triplets under the QCD group $SU(3)$. This anomaly cancels between quarks and leptons.

c^3 graph

The c^3 anomaly is given by $\text{Tr}(Y^3)$ and for this, both left and right fermions can contribute. We get

$$\text{Tr}(Y^3) = \left[\underbrace{(-1)}_{\nu_L} + \underbrace{(-1)}_{e_L} + \left(\underbrace{\frac{1}{27}}_{u_L} + \underbrace{\frac{1}{27}}_{d_L}\right) \times 3\right]$$
$$-\left[\underbrace{-8}_{e_R} + \left(\underbrace{\frac{64}{27}}_{u_R} - \underbrace{\frac{8}{27}}_{d_R}\right) \times 3\right]$$
$$= \left(-\frac{16}{9}\right)_L - \left(-\frac{16}{9}\right)_R$$
$$= 0 \tag{13.38}$$

In this case, the cancellation involves quarks and leptons and both chiralities.

So far we have considered anomalies in gauge invariance. But there are also anomalies in the standard model for continuous global symmetries. Two global symmetries of interest correspond to baryon number B and lepton number L and these are both anomalous. Only $(B - L)$ is anomaly-free. These anomalies have no bearing on the consistency of the theory unless we try to make these into gauge symmetries, in which case only the combination $(B - L)$ can be consistently gauged.

Consider again one generation of quarks and leptons, say the electron generation. Lepton number corresponds to the transformation

$$\begin{pmatrix} \nu \\ e \end{pmatrix}_L \quad \rightarrow \quad e^{i\alpha} \begin{pmatrix} \nu \\ e \end{pmatrix}_L, \qquad\qquad e_R \rightarrow e^{i\alpha} e_R$$

$$Q \rightarrow Q \tag{13.39}$$

The leptons ν, e have lepton number equal to 1, while the quarks have no lepton number. Baryon number corresponds to the transformation

$$\nu \rightarrow \nu, \qquad\qquad e \rightarrow e$$

$$\begin{pmatrix} u \\ d \end{pmatrix}_L \rightarrow e^{i\beta/3} \begin{pmatrix} u \\ d \end{pmatrix}_L$$

$$u_R \rightarrow e^{i\beta/3} u_R, \qquad\qquad d_R \rightarrow e^{i\beta/3} d_R \tag{13.40}$$

The leptons have zero baryon number, while the quarks carry a baryon number of $\frac{1}{3}$.

The anomalies arise from triangle diagrams with either baryon or lepton number at one vertex and b^2 or c^2 at the other vertices. (Other diagrams are important for the covariant completion of the expression for the anomaly.) The variation of the effective quantum action is then worked out from our general formulae as

$$\delta_{\alpha,\beta}\Gamma = -i \int d^4x \; \alpha(x) \left[\underbrace{c_2[b]}_{\nu_L, e_L} + \underbrace{c_2[c]}_{\nu_L} + \underbrace{c_2[c]}_{e_L} - \underbrace{4c_2[c]}_{e_R} \right]$$

$$-i \int d^4x \; \beta(x) \left[\underbrace{c_2[b]}_{u_L, d_L} + \underbrace{\frac{1}{9}c_2[c]}_{u_L} + \underbrace{\frac{1}{9}c_2[c]}_{d_L} - \underbrace{\frac{16}{9}c_2[c]}_{u_R} - \underbrace{\frac{4}{9}c_2[c]}_{d_R} \right]$$

$$= -i \int d^4x \; (\alpha(x) + \beta(x)) \left(c_2[b] - 2\, c_2[c] \right) \tag{13.41}$$

where

$$c_2[b] = \frac{1}{32\pi^2} \epsilon^{\mu\nu\alpha\beta} \, \text{Tr} \, F_{\mu\nu}(b) F_{\alpha\beta}(b) = -\frac{1}{64\pi^2} \epsilon^{\mu\nu\alpha\beta} G^a_{\mu\nu} G^a_{\alpha\beta}$$

$$c_2[c] = -\frac{1}{32\pi^2}\epsilon^{\mu\nu\alpha\beta}f_{\mu\nu}f_{\alpha\beta}$$
$$G^a_{\mu\nu} = \partial_\mu b^a_\nu - \partial_\nu b^a_\mu + \epsilon^{abc}b^b_\mu b^c_\nu$$
$$f_{\mu\nu} = \partial_\mu c_\nu - \partial_\nu c_\mu \tag{13.42}$$

in the notation for the field strength tensors used in (12.141). (c_2 is the second Chern class and is related to the instanton number as we shall see in Chapter 16.) In equation (13.41), for the quarks, there is degeneracy factor of 3 due to the QCD group $SU(3)$, which cancels the factor of $\frac{1}{3}$ due to the baryon number. From (13.41) we see that both baryon number and lepton number are anomalous, but the combination $(B - L)$ corresponding to $\beta = -\alpha$ is anomaly-free.

Baryon and lepton numbers are vectorial symmetries. It is thus possible to choose the regulator such that they are not anomalous. However, for the electroweak theory, we have to choose a regulator which makes the (nonvectorial) gauge symmetries anomaly-free; for this regulator baryon and lepton numbers must be anomalous.

We can define a lepton number current and a baryon number current by

$$J^{L\mu} = \bar{l}_L\gamma^\mu l_L + \bar{e}_R\gamma^\mu e_R$$
$$J^{B\mu} = \bar{u}\gamma^\mu u + \bar{d}\gamma^\mu d \tag{13.43}$$

The fact that the effective action is not invariant under the transformations corresponding to these symmetries can be expressed as the (lack of) conservation equations

$$\partial_\mu J^{L\mu} = \partial_\mu J^{B\mu} = c_2[b] - 2\,c_2[c] \tag{13.44}$$

(It is to facilitate this identification that we kept α, β in (13.41) as arbitrary x-dependent functions, even though the symmetry is global.) We see directly from (13.44) that there is baryon number violation (and lepton number violation) in the standard model, if there are field configurations for which the integrals of $c_2[b]$ or $c_2[c]$ are nonzero. For the $SU(2)$ gauge field b_μ such configurations exist; examples are the weak instantons.

The gluon field does not contribute to these anomalies, but it does contribute to the axial $U(1)$ anomaly in QCD. The axial $U(1)$ transformation is given by

$$Q'_L = e^{i\lambda}Q_L, \qquad\qquad Q'_R = e^{-i\lambda}Q_R$$
$$Q' = e^{i\lambda\gamma^5}Q \tag{13.45}$$

This is not a perfect symmetry of the theory, even without anomalies, due to quark masses and other weak interaction effects. However, the strongest breaking of this symmetry is due to anomalies and so it is interesting to write this down as well. In this case, the computation is a straightforward generalization of (13.22) and gives

$$\delta_\lambda \Gamma = -i \, 2N_f \int d^4x \, \lambda \, \frac{1}{32\pi^2} \epsilon^{\mu\nu\alpha\beta} \, \text{Tr} \left(F_{\mu\nu} F_{\alpha\beta} \right)$$

$$= i \, 2N_f \int d^4x \, \lambda \, \rho[A] \tag{13.46}$$

$$\rho[A] = -\frac{1}{32\pi^2} \epsilon^{\mu\nu\alpha\beta} \, \text{Tr} \left(F_{\mu\nu} F_{\alpha\beta} \right)$$

where A is the $SU(3)$-gauge field corresponding to the QCD gauge group. N_f is the number of light fermion flavors. (For the heavier fermion flavors, the breaking of the axial symmetry due to the fermion mass terms is so strong that separating out the anomaly contribution is not particularly useful.) The quantity $\rho[A]$ is the density for the instanton number; the integral of $\rho[A]$ is the instanton number which is an integer. The change in the action for $\lambda = 2\pi n/(2N_f)$, where n is an integer, is given by $\Delta\Gamma = 2\pi i n$, so that $\exp(-\Gamma)$ is invariant. We then see that, although the $U_A(1)$ symmetry is broken by the anomaly, we still have a Z_{2N_f} symmetry for all the physical results.

13.5 The Lagrangian for π^0 decay

We have calculated the lifetime of the π^0-meson in Chapter 7 using the effective Lagrangian term

$$\mathcal{L}_{int} = \frac{\alpha}{\pi f_\pi} \boldsymbol{E} \cdot \boldsymbol{B} \, \pi^0 \tag{13.47}$$

(In Chapter 7, we used the letter ϕ for the pion field, here we use π^0 in agreement with Chapter 12.) The field π^0 can be understood as a Goldstone boson arising from the spontaneous chiral symmetry breaking in the standard model. We can combine this piece of information with our calculation of the axial anomaly to derive the Lagrangian (13.47).

The pion is made up of the up and down quarks only and so for this calculation we need just those two flavors of quarks. Consider a chiral symmetry transformation

$$u \to \exp(i\gamma_5\varphi) \, u, \qquad\qquad d \to \exp(-i\gamma_5\varphi) \, d \tag{13.48}$$

This may be combined as

$$q = \begin{pmatrix} u \\ d \end{pmatrix} \to \exp(i\gamma_5\tau_3\varphi) \, q \tag{13.49}$$

The field U representing the pseudoscalar mesons behaves like $q_L \bar{q}_R$ under chiral transformations, transforming as

$$U \to U' = g_L \, U \, g_R^\dagger \tag{13.50}$$

Comparing these with the expansion of U in terms of the pion field given in (12.90) and (12.92), we see that, when the quark fields transform as in (13.48), the pion field changes as

$$\pi^0 \to \pi^0 + 2f_\pi \varphi \tag{13.51}$$

The up and down quarks have electrical charges $\frac{2}{3}e$ and $-\frac{1}{3}e$, respectively, and there are three colors of each which contribute to the term in the anomaly with the electromagnetic field. Keeping these in mind, we find, from (13.22),

$$\delta_\varphi \Gamma = -i\frac{e^2}{8\pi^2} \int d^4x \; \varphi F_{\mu\nu} \tilde{F}_{\mu\nu} \left[\left(\frac{2}{3}\right)^2 - \left(\frac{1}{3}\right)^2 \right] \times 3$$

$$= -i\frac{\alpha}{2\pi} \int d^4x \; \frac{\delta\pi^0}{2f_\pi} F_{\mu\nu} \tilde{F}_{\mu\nu} \tag{13.52}$$

This integrates to give the effective interaction term

$$\Gamma = -i\frac{\alpha}{4\pi f_\pi} \int d^4x \; \pi^0 F_{\mu\nu} \tilde{F}_{\mu\nu}$$

$$= -i\frac{\alpha}{\pi f_\pi} \int d^4x \; \boldsymbol{E} \cdot \boldsymbol{B} \; \pi^0 \tag{13.53}$$

This calculation has been done in Euclidean space; $\exp(-\Gamma)$ continues to Minkowski space as $\exp\left(i \int \mathcal{L}_{int}\right)$, with $\mathcal{L}_{int}$ given by (13.47).

13.6 The axial $U(1)$ problem

We have seen that the approximate chiral symmetry $SU(3)_L \times SU(3)_R$ for the three light flavors of quarks is spontaneously broken due to the nonperturbative effects of the gluon fields. If the chiral symmetry were a perfect symmetry of the Lagrangian, this spontaneous breaking would lead to massless Goldstone bosons. Since the starting Lagrangian does not have the full chiral symmetry due to mass effects, weak interactions, etc., the potential Goldstons bosons become massive, but remain light compared to the scale of the strong interactions. They are pseudo-Goldstone bosons. The effective Lagrangian we wrote down in equations (12.91, 12.96) capture these features of the mesons. There is, however, one puzzle in this regard, whose solution requires the axial $U(1)$ anomaly; this is the so-called $U_A(1)$ problem. Since $U(3)_L \times U(3)_R$ is broken down to $U(3)_V$, we expect nine Goldstone bosons. The candidate meson corresponding to the $U_A(1)$ generator is the η' meson which has a mass of approximately 958 MeV, which is significantly higher than the masses of the octet of pseudoscalar mesons. This is puzzling, but not a contradiction yet, since all these mesons get masses due to the fact that we have an imperfect starting symmetry. However, one can do a calculation of the mass of the η' *allowing for quark masses* and this leads to the result $m_{\eta'} \leq \sqrt{3}\, m_\pi$. This inequality is badly violated in nature, this is the $U_A(1)$ problem. Clearly something is wrong about assuming that the violation of axial $U(1)$ symmetry is not stronger than the scale of quark masses.

The anomaly of the $U_A(1)$ symmetry provides the answer. Even though the classical Lagrangian does not show such a lack of symmetry, it is violated at the scale of strong interactions due to the gluon fields. Thus one cannot really interpret the η' as a psuedo-Goldstone boson. Under an axial $U(1)$ transformation, the meson field U transforms as $U \rightarrow \exp(2i\varphi)U$. The variation of the action given in (13.46) then shows that the effective action at the level of the mesons must contain a term

$$
\begin{aligned}
S_{eff} &= \frac{i}{2}\left(\log \det U - \log \det U^\dagger\right)\left(\frac{1}{32\pi^2}\epsilon^{\mu\nu\alpha\beta}\mathrm{Tr}F_{\mu\nu}F_{\alpha\beta}\right) \\
&= \frac{\sqrt{2N_f}}{f_\pi}\,\eta'\,\rho[A,x]
\end{aligned}
\tag{13.54}
$$

(This is now written in Minkowski space.) If the two-point function for the instanton density $\rho[A]$ has the form

$$
\langle \rho(x)\rho(y)\rangle = m_0^4\delta^{(4)}(x-y) + \mathcal{O}(\partial)
\tag{13.55}
$$

then the η' gets a mass term

$$
S_{\eta'\ mass} = \frac{1}{2}\left[\frac{2N_f m_0^4}{f_\pi^2}\right]\eta'^2
\tag{13.56}
$$

This would be the extra mass for the η' due to the anomaly and can provide a solution to the $U_A(1)$ problem. This formula for the η'-mass is known as the Veneziano-Witten formula. The relevant two-point function can be estimated in some nonperturbative approaches to QCD, such as lattice QCD. (For simplicity, we have neglected mixing among the mesons.)

There is one other interesting point about this analysis. The anomalous conservation law for the axial vector current is

$$
\partial_\mu J_A^\mu = 2N_f\frac{1}{32\pi^2}\epsilon^{\mu\nu\alpha\beta}\mathrm{Tr}F_{\mu\nu}F_{\alpha\beta}
\tag{13.57}
$$

The right-hand side is actually a total derivative and can be written as $-2N_f\partial_\mu K^\mu$, where

$$
K^\mu = -\frac{1}{8\pi^2}\epsilon^{\mu\nu\alpha\beta}\mathrm{Tr}\left(A_\nu\partial_\alpha A_\beta + \frac{2}{3}A_\mu A_\alpha A_\beta\right)
\tag{13.58}
$$

Thus, we do have a conserved current $J_A^\mu + 2N_f K^\mu$. This might seem to lead back to the argument that we should have a Goldstone boson. However, K^μ is not gauge-invariant. Of course, what is relevant for this argument is really the question of whether the charge $\int d^3x\, K^0$ is gauge-invariant. (Notice that K^0 is the Chern-Simons term mentioned in Chapter 10.) Under a gauge transformation K^0 transforms as

$$K^0(A^g) - K^0(A) = -\frac{1}{8\pi^2}\epsilon^{ijk}\partial_i \mathrm{Tr}(g^{-1}\partial_j g\, A_k)$$

$$-\frac{1}{24\pi^2}\epsilon^{ijk}\mathrm{Tr}(g^{-1}\partial_i g\, g^{-1}\partial_j g\, g^{-1}\partial_k g) \quad (13.59)$$

The first term on the right-hand side is a total derivative and so will not contribute to the charge. Thus $\int K^0$ will be gauge-invariant and lead to a Goldstone boson, if the second term does not contribute. For an Abelian gauge theory, the second term is zero and hence an anomaly in $U_A(1)$ due to Abelian gauge fields will not resolve the $U_A(1)$ problem. For a nonabelian gauge theory, there are gauge transformations for which the integral of the second term is not zero; it is the winding number of the transformation, as discussed later. Further one cannot restrict attention to only those transformations for which the integral of the second term is zero, because there are field configurations, specifically the instantons, for which such transformations are needed to represent them in a nonsingular way. Thus with a nonabelian symmetry, and the associated instantons, we do have a resolution of the $U_A(1)$ problem.

References

1. A general reference is R.A. Bertlmann, *Anomalies in Quantum Field Theory*, Clarendon Press (1996). There is collection of reprints which is very useful, S.B. Treiman *et al*, *Current Algebra and Anomalies*, Princeton University Press (1986).
2. The anomaly appears in the π^0-decay calculation of J. Steinberger, Phys. Rev. **76**, 1180 (1949); H. Fukuda and Y. Miyamoto, Prog. Theor. Phys. **4**, 347 (1949). It also appears in Schwinger's beautiful paper on "Gauge Invariance and Vacuum Polarization", J. Schwinger, Phys. Rev. **82**, 664 (1951). But the real understanding of the anomaly, with its origin as the quantum breaking of classical symmetries, starts with S. Adler, Phys. Rev. **177**, 2426 (1969); J. Bell and R. Jackiw, Nuov. Cim. **60A**, 47 (1969).
3. The WZ consistency conditions were given by J. Wess and B. Zumino, Phys. Lett. **B37**, 95 (1971).
4. The nonrenormalization theorem for anomalies is due to S. Adler and W.A. Bardeen, Phys. Rev. **182**, 1517 (1969). A simple proof using the renormalization group is given by S.S. Shei and A. Zee, Phys. Rev. **D8**, 597 (1973).
5. The Bardeen form of the anomaly is in W.A. Bardeen, Phys. Rev. **184**, 1848 (1969). Counterterms to go from the WZ-consistent form to the covariant Bardeen form are discussed by W.A. Bardeen and B. Zumino, Nucl. Phys. **B244**, 421 (1984); M. Paranjape, Phys. Lett. **B 156**, 376 (1985).

6. The $U_A(1)$ problem was first noted by S.L. Glashow, in *Hadrons and Their Interactions*, A. Zichichi (ed.) Academic Press (1967). The bound $m_{\eta'} \leq \sqrt{3} m_\pi$ is due to S. Weinberg, Phys. Rev. **D11**, 3583 (1975). The role of instantons in solving the problem was pointed out by G. 't Hooft, Phys. Rev. Lett. **37**, 8 (1976). The mass formula for η' is due to E. Witten, Nucl. Phys. **B156**, 269 (1979); G. Veneziano, Nucl. Phys. **B159**, 213 (1979); Phys. Lett. **B95**, 90 (1980). We follow the effective Lagrangian approach of C. Rosenzweig, J. Schechter and C.G. Trahern, Phys. Rev. **D21**, 3388 (1980).

14 Elements of differential geometry

14.1 Manifolds, vector fields, and forms

This chapter is meant as an elementary introduction to the ideas of differential geometry. The emphasis will be on ideas and computations. Precise definitions and rigorous formulation of these ideas can be found in many books on mathematics.

We will assume some basics of open sets, topology, topological spaces, continuity, differentiability, homeomorphism as well as some understanding of analysis on $\mathbf{R}^n$.

Topological manifold

An n-dimensional topological manifold $\mathcal{M}$ is a topological space such that every point $p \in \mathcal{M}$ has a neighborhood which is homeomorphic to an open disc in $\mathbf{R}^n$.

Recall that a homeomorphism is a continuous, one-to-one (and hence invertible) map. The above definition means that we can consider open sets U_i such that $\mathcal{M} = \bigcup_i U_i$, with the collection $\{U_i\}$ forming an open cover of M. Then

$$\varphi_U \ : \ U \to D^n$$

where D^n is an open disc in $\mathbf{R}^n$, assigns local coordinates to $\mathcal{M}$. The real numbers corresponding to the points of D^n as assigned by the map φ_U are the local coordinates. φ^{-1} exists and is continuous. Each such map φ is called a chart.

Consider U and V, which are open sets of $\mathcal{M}$, with $U \cap V \neq \emptyset$. We then have $\varphi_U : U \to D^n$, $\varphi_V : V \to D^n$. On the intersection, we have two possible choices of coordinates. One can make a transformation from one set of coordinates, say, as assigned by φ_V to the other, those assigned by φ_U by using the function $\varphi_U \varphi_V^{-1}$. This is called a transition function. The transition functions are continuous.

A manifold is a differentiable manifold if all the transition functions are C^∞; i.e., they have well defined derivatives to arbitrarily high order. Most of the manifolds that we normally encounter, such as $\mathbf{R}^n$, the n-sphere S^n, etc., are examples of differentiable manifolds. One can choose different charts for coordinatizing a manifold. A mapping of one set of coordinates to another is a coordinate transformation. A C^∞ transformation on $\mathcal{M}$ is referred to as a diffeomorphism.

The one-dimensional sphere S^1 (a circle) is one of the simplest examples of a differentiable manifold which will illustrate some of the ideas presented here. In this case, we can use the angle θ, which is a real number, as the local coordinate. However, we see immediately that we cannot cover S^1 by a single chart. $\theta = 0$ and $\theta = 2\pi$ correspond to the same point on the manifold S^1. Since we want one-to-one map to a neighborhood in $\mathbf{R}^1$, this is not acceptable. We have to use at least two charts to provide a proper coordinatization of S^1.

Vector fields on $\mathcal{M}$

We now consider curves on $\mathcal{M}$. These are maps $U_t : t \in [0, 1] \to \mathcal{M}$. The image of the interval $[0, 1]$ is a curve on $\mathcal{M}$ with the initial point $p \in \mathcal{M}$ corresponding to the value $t = 0$. We will denote the points on the curve as $U_t(p)$ with $U_0(p) = p$.

Let $\mathcal{F}(\mathcal{M})$ denote the set of all differentiable functions on $\mathcal{M}$. Thus if $f \in \mathcal{F}(\mathcal{M})$, $f : \mathcal{M} \to \mathbf{R}^1$, and f is C^∞. We can define a vector X associated to the curve $U_t(p)$ at the point p as

$$(X \cdot f)(p) \equiv \frac{d}{dt} f(U_t(p)) \bigg]_{t=0} \tag{14.1}$$

X is the tangent to the curve $U_t(p)$ at the point p. By considering all curves with p as the starting point, we get a vector space at p. This is the tangent space $T_p\mathcal{M}$ and X is an element of this tangent space to $\mathcal{M}$ at the point p. By considering a set of curves with different points on the manifold as the initial points, we get a vector field, defined for all points on $\mathcal{M}$. This is equivalent to putting all tangent spaces at all points together to get $T\mathcal{M} = \bigcup_p T_p\mathcal{M}$, which is called the tangent bundle of $\mathcal{M}$. The vector field is thus obtained by choosing an element of $T_p\mathcal{M}$ for each p.

From (14.1), we also see that a vector field X can be thought of as a mapping $\mathcal{F}(\mathcal{M}) \to \mathcal{F}(\mathcal{M})$, given by $f \to X \cdot f$. In local coordinates, we can write $U_t(p) = x^i + t\xi^i(x) + \cdots$, where x^i are the coordinates of the point p. (We are actually writing $\varphi(U_t(p))$ here since we are using local coordinates. When there is no possibility of confusion, we use the local coordinates without explicitly specifying that a chart is involved.) From the formula given above

$$X \cdot f(x) = \frac{d}{dt} f(x^i + t\xi^i + \cdots) \bigg]_{t=0} = \xi^i \frac{\partial}{\partial x^i} f \tag{14.2}$$

Thus the vector field is given by

$$X = \xi^i(x) \frac{\partial}{\partial x^i} \tag{14.3}$$

in the local coordinates we have used. $\partial/\partial x^i$ define a basis for vector fields and ξ^i are the components of X in this basis. This basis is usually referred to as the coordinate basis. ξ^i are what we usually call the contravariant vectors. The definition of a vector given here is independent of the coordinates used, being defined entirely geometrically.

Given a curve $U_t(p)$, we can define the vector field at the starting point as we have done. A question which may naturally arise is whether we can obtain $U_t(p)$ given ξ^i. Locally, the answer is yes; this is the theorem of Frobenius. The differential equation

$$\frac{dx^i}{dt} = \xi^i(x) \tag{14.4}$$

has a solution at least locally, given the initial condition that $x^i(t)]_{t=0}$ corresponds to the point p. The trajectories so defined are the integral curves of the vector field X.

Differential one−forms on $\mathcal{M}$

Since $T_p\mathcal{M}$ is a vector space, we can define a dual vector space $T_p^*\mathcal{M}$. We denote the basis of the dual space as dx^i in the local coordinate basis and define it by the rule

$$\left(\frac{\partial}{\partial x^i}, dx^j\right) = \delta_i^j \tag{14.5}$$

(The bracket specifies the rule of associating basis vectors of the two vector spaces.) This is often called the interior contraction of $\partial/\partial x^i$ on dx^j. $T_p^*\mathcal{M}$ is called the cotangent space at the point p. The union of all such spaces gives us the cotangent bundle $T^*\mathcal{M}$. A differential one-form is then defined by giving an element of this dual vector space at each point of the manifold. With dx^i as a basis, a dual vector is of the form $\omega = \omega_i dx^i$, where $\omega_i(x)$ are differentiable functions on each chart U. The interior contraction of a vector field $X = \xi^i\partial/\partial x^i$ on this is given by

$$i_X\omega = \left(\xi^i \frac{\partial}{\partial x^i}, \omega_j dx^j\right) = \xi^i\omega_i \tag{14.6}$$

By requiring this to be a globally defined function on the manifold $\mathcal{M}$ (i.e., $i_X\omega \in \mathcal{F}(\mathcal{M})$) for all vector fields X, we are able to define an intrinsic notion of an element of the cotangent space.

The components of a differential one-form, namely, ω_i in writing $\omega = \omega_i dx^i$, are covariant vectors.

Coordinate invariance

Consider a coordinate transformation from a set x^i to another set y^i, which are specified as functions of the x's. For a function $f \in \mathcal{F}(\mathcal{M})$, the function in terms of x is given by $f(y)$ where we substitute for y as a function of x, i.e., $f(x) = f(y)\Big]_{y=y(x)}$. This gives

$$\frac{\partial}{\partial x^i} f(y) = \frac{\partial f}{\partial y^j} \left(\frac{\partial y^j}{\partial x^i} \right) \tag{14.7}$$

Vector fields and one-forms are defined in a coordinate-invariant way. Thus

$$X = \xi^i(x) \frac{\partial}{\partial x^i} = \tilde{\xi}^j(y) \frac{\partial}{\partial y^j}$$

$$\omega = \omega_i(x) dx^i = \tilde{\omega}_j(y) dy^j \tag{14.8}$$

This tells us how the components transform. Using (14.7), we find, for a vector field,

$$X = \xi^i(x) \frac{\partial}{\partial x^i} = \xi^i(x) \left(\frac{\partial y^j}{\partial x^i} \right) \frac{\partial}{\partial y^j}$$

$$\equiv \tilde{\xi}^j(y) \frac{\partial}{\partial y^j} \tag{14.9}$$

which gives

$$\tilde{\xi}^j(y) = \xi^i(x) \left(\frac{\partial y^j}{\partial x^i} \right) \tag{14.10}$$

In a similar way

$$\tilde{\omega}_j(y) = \omega_i(x) \left(\frac{\partial x^i}{\partial y^j} \right) \tag{14.11}$$

It should be emphasized that the vector fields and one-forms are evaluated at the same geometrical point in all this; only the coordinates used to describe the point have changed.

Differential k−forms

Starting from $T_p\mathcal{M}$ and taking the k-fold tensor product, we get $(T_p\mathcal{M})^k$. An element of this space will define a tensor of rank k at the point p. The collection of such tensors over the entire manifold will give a tensor field of rank k. In a similar fashion, we may take the tensor products of the cotangent spaces and define dual tensors.

There is a class of dual tensors which play a special role in differential geometry. These are obtained by taking the antisymmetrized tensor products of $T^*\mathcal{M}$. A differential k-form is then given by an element of the antisymmetrized product $\wedge_k(T_p^*\mathcal{M})$ for all points of the manifold. (The wedge symbol denotes the antisymmetrized tensor product.) In local coordinates, a k-form ω will be given by

$$\omega = \frac{1}{k!} \omega_{i_1 i_2 \cdots i_k} \, dx^{i_1} \wedge dx^{i_2} \wedge \cdots \wedge dx^{i_k} \tag{14.12}$$

where the components $\omega_{i_1 i_2 \cdots i_k}$ are antisymmetric in the k-indices.

We can also define an exterior (or outer) product of differential forms by taking a wedge or antisymmetrized product. If α is a k-form and β is a p-form, the product $\alpha \wedge \beta$ is a $(k+p)$-form, which is given in local coordinates by

$$\alpha = \frac{1}{k!} \alpha_{i_1 \cdots i_k} \, dx^{i_1} \wedge \cdots dx^{i_k}$$

$$\beta = \frac{1}{p!} \beta_{i_1 \cdots i_p} \, dx^{i_1} \wedge \cdots dx^{i_p}$$

$$\alpha \wedge \beta = \frac{1}{k!} \frac{1}{p!} \alpha_{[i_1 \cdots i_k} \beta_{i_{k+1} \cdots i_{k+p}]} \, dx^{i_1} \wedge \cdots dx^{i_{k+p}} \qquad (14.13)$$

where the square brackets indicate antisymmetrization of the indices enclosed.

Because of the antisymmetrization, all k-forms on a manifold of dimension n vanish for $k > n$. Thus the differential form of the highest rank is an n-form on an n-dimensional manifold.

Pullback of a k−form

Consider a map from a manifold $\mathcal{M}$ to $\mathcal{N}$, $\varphi : \mathcal{M} \to \mathcal{N}$. If x denote the coordinates on $\mathcal{M}$, then $\varphi(x)$ are the coordinates of the image of $\mathcal{M}$ under the map φ. Let ω be a differential k-form defined on N, which can be written in local coordinates y on $\mathcal{N}$ as

$$\omega = \frac{1}{k!} \omega_{i_1 i_2 \cdots i_k}(y) \, dy^{i_1} \wedge dy^{i_2} \wedge \cdots \wedge dy^{i_k} \qquad (14.14)$$

We now consider ω on the image of the map φ. This can be considered as a k-form on $\mathcal{M}$ as follows. For the y's, we substitute the map $\varphi(x)$ to obtain

$$\varphi^* \omega = \frac{1}{k!} \omega_{i_1 i_2 \cdots i_k}(\varphi(x)) \, d\varphi(x)^{i_1} \wedge d\varphi(x)^{i_2} \wedge \cdots \wedge d\varphi(x)^{i_k}$$

$$= \frac{1}{k!} \omega_{i_1 i_2 \cdots i_k}(\varphi(x)) \left(\frac{\partial \varphi^{i_1}}{\partial x^{j_1}} \right) \cdots \left(\frac{\partial \varphi^{i_k}}{\partial x^{j_k}} \right) \, dx^{j_1} \wedge \cdots \wedge dx^{j_k}$$

$$(14.15)$$

$\varphi^* \omega$ is a differential k-form on $\mathcal{M}$. It is referred to as the pullback of the k-form ω on $\mathcal{N}$ by the map φ.

The exterior derivative

The exterior derivative, denoted by d, is a differentiation operation on differential forms, which takes k-forms to $(k+1)$-forms. In local coordinates

$$d\omega = \frac{1}{k!} \frac{\partial \omega_{i_1 i_2 \cdots i_k}}{\partial x^j} dx^j \wedge dx^{i_1} \wedge dx^{i_2} \wedge \cdots \wedge dx^{i_k}$$

$$= \frac{1}{(k+1)!} \omega_{i_1 i_2 \cdots i_k i_{k+1}} \, dx^{i_1} \wedge dx^{i_2} \wedge \cdots \wedge dx^{i_k} \wedge dx^{i_{k+1}}$$

$$\omega_{i_1 i_2 \cdots i_k i_{k+1}} = \frac{\partial}{\partial x^{i_1}} \omega_{i_2 i_3 \cdots i_{k+1}} - \frac{\partial}{\partial x^{i_2}} \omega_{i_1 i_3 \cdots i_{k+1}} + \cdots + (-1)^k \frac{\partial}{\partial x^{i_{k+1}}} \omega_{i_1 i_2 \cdots i_k}$$

$$(14.16)$$

Applied to a zero-form (which is a function) and a one-form, this becomes

$$df = \frac{\partial f}{\partial x^i} dx^i$$

$$dw = \frac{\partial \omega_i}{\partial x^j} dx^j \wedge dx^i$$

$$= \frac{1}{2} \left(\partial_i \omega_j - \partial_j \omega_i \right) dx^i \wedge dx^j \tag{14.17}$$

From this, we see that the exterior derivative is the generalization of the familiar curl of a vector.

Two of the most important properties of the exterior derivative are

$$d^2 = 0$$

$$d(\alpha \wedge \beta) = d\alpha \wedge \beta + (-1)^k \alpha \wedge d\beta \tag{14.18}$$

where α is a k-form. The first of these is obvious, since $\partial_i \partial_j - \partial_j \partial_i = 0$ on any well-defined quantity. The second is also easily checked from the anti-symmetric property of the wedge products.

It is interesting to see how the antisymmetrization in the definition of the exterior derivative is important for the coordinate invariance or invariance under diffeomorphisms of $\mathcal{M}$. Consider a one-form ω which can be written in local coordinates as $\omega = \omega_i dx^i = \tilde{\omega}_j dy^j$. For the exterior derivative, we can write

$$dw = \frac{\partial \omega_i}{\partial x^j} dx^j \wedge dx^i$$

$$= \frac{\partial}{\partial x^j} \left(\tilde{\omega}_i \frac{\partial y^i}{\partial x^k} \right) dx^j \wedge dx^k$$

$$= \left[\frac{\partial \tilde{\omega}_i}{\partial y^m} \frac{\partial y^m}{\partial x^j} \frac{\partial y^i}{\partial x^k} + \tilde{\omega}_i \frac{\partial^2 y^i}{\partial x^j \partial x^k} \right] dx^j \wedge dx^k$$

$$= \frac{\partial \tilde{\omega}_i}{\partial y^k} dy^k \wedge dy^i \tag{14.19}$$

We see that the form of the derivative is the same in terms of the y-coordinates or x-coordinates. This invariance is possible only because the term involving the second derivative of y^i cancels out in this expression due to the antisymmetrization in j, k. This shows the importance of the antisymmetrization in defining the exterior derivative.

Integration of forms

One of the reasons that differential forms are important is that they can be integrated over submanifolds of suitable dimension. Consider an n-form on an n-dimensional manifold. For the product of the dx's we can write

$$dx^{i_1} \wedge \cdots \wedge dx^{i_n} = \epsilon^{i_1 i_2 \cdots i_n} d^n x \tag{14.20}$$

where $\epsilon^{i_1 i_2 \cdots i_n}$ is the purely numerical Levi-Civita symbol which is antisymmetric under exchange of any two indices. It has the value zero if any two indices have the same value and $\epsilon^{12\cdots n} = 1$. If $i_1 i_2 \cdots i_n$ is a permutation P of $12 \cdots n$, $\epsilon^{i_1 i_2 \cdots i_n}$ is 1 if P is an even permutation and -1 if P is an odd permutation. In (14.20), $d^n x$ is the standard integration measure for the n real variables, $x^1, x^2, \cdots, x^n$.

If ω is an n-form, we can write, locally,

$$\omega = \frac{1}{n!} \omega_{i_1 i_2 \cdots i_n} \, dx^{i_1} \wedge \cdots \wedge dx^{i_n}$$

$$= \frac{1}{n!} \omega_{i_1 i_2 \cdots i_n} \, \epsilon^{i_1 i_2 \cdots i_n} \, d^n x$$

$$= \rho(x) \, d^n x \tag{14.21}$$

This is of the form of a function ρ with the integration measure $d^n x$ for the region in $\mathbf{R}^n$ which corresponds to the neighborhood on $\mathcal{M}$ that we are considering. We can thus integrate this by the standard techniques of integration over n real variables. We can do this for each neighbourhood of $\mathcal{M}$ to get the integral of the n-form ω over $\mathcal{M}$. In putting together the values for the various neighborhoods, we have to use the transition functions, which is equivalent to a coordinate transformation. Suppose we use the coordinates x on one patch and the coordinates y on a neighboring patch. On the intersection region

$$dx^{i_1} \wedge \cdots \wedge dx^{i_n} = \frac{\partial x^{i_1}}{\partial y^{j_1}} \cdots \frac{\partial x^{i_n}}{\partial y^{j_n}} dy^{j_1} \wedge \cdots \wedge dy^{j_n}$$

$$= \frac{\partial x^{i_1}}{\partial y^{j_1}} \cdots \frac{\partial x^{i_n}}{\partial y^{j_n}} \epsilon^{j_1 \cdots j_n} \, d^n y$$

$$= \det \left(\frac{\partial x^{i_1}}{\partial y^{j_1}} \right) dy^{i_1} \wedge \cdots \wedge dy^{i_n}$$

$$= J \, dy^{i_1} \wedge \cdots \wedge dy^{i_n} \tag{14.22}$$

J is the Jacobian of the transformation. Since $\omega_{i_1 i_2 \cdots i_n}$ transforms by the inverse Jacobian, the value of ω is the same irrespective of which coordinates are used, and so we can extend the integration over all of $\mathcal{M}$.

There is one complication that could arise in this integration. Consider the computation of the integral of an n-form which has $\rho(x) = 1$ on the intersection region of two coordinate patches. In integrating over this region, we can use either the x-coordinates or the y-coordinates. We then get the coordinate volume as measured by the x's or the integral of $J d^n y$ over this region. If the Jacobian is not positive, we can get inconsistent results for the volume of the region of integration. A manifold for which it is possible to choose the transition Jacobians to be positive is said to be orientable. We shall be interested in orientable manifolds in what follows.

If we have a k-form, we can integrate it over a k-dimensional subspace or submanifold. Neighborhoods on the submanifold can be mapped to regions of $\mathbf{R}^k$, by an invertible map. The restriction of the k-form to the submanifold can then be pulled back to the region of $\mathbf{R}^k$ and integrated as an integral over real variables.

The exterior algebra of forms

Differential forms of the same degree can be added to get another form of the same degree; we can also take arbitrary linear combinations with real coefficients. (Generally, the coefficients can be in some field of numbers, but we shall use just real numbers in our discussions.)

The wedge product of two differential forms is again a differential form. The exterior derivative of a k-form gives a $(k + 1)$-form. Under these operations, the differential forms on a manifold form a closed algebra, which is known as the exterior algebra of $\mathcal{M}$.

Differential forms, their algebra of exterior (or wedge) products and the exterior derivative are somewhat special in terms of analysis on a manifold $\mathcal{M}$. We have seen that k-forms can naturally be integrated over k-dimensional subspaces. This is one of the main reasons that differential forms are important. We have also seen that the antisymmetrization in the derivative makes possible a coordinate independent definition of a derivative. The important point is that up to now we have not introduced the notion of a metric or any other additional structures on the manifold $\mathcal{M}$. Differential forms and the exterior derivative are independent of any such structure. They are thus very useful in expressing some of the topological properties of the manifold.

Exact and closed forms

If $d\omega = 0$, then ω is said to be closed. If $\omega = d\alpha$, then ω is said to be exact. Obviously, exactness implies closure, since $\omega = d\alpha \Rightarrow d\omega = d^2\alpha = 0$ by virtue of $d^2 = 0$. The converse is in general not true. The Poincaré lemma is the statement that every closed form is locally exact. In other words, if $d\omega = 0$, then $\omega = d\alpha$ for some α in a local region. In general, this will not hold globally.

For example, on $M = S^1$ with the usual angular coordinate θ, the one-form $d\theta$ is evidently closed, but it is not exact, because the zero-form whose derivative it is, namely, θ, is not globally defined because of the problem of single-valuedness discussed earlier. θ is well defined on a local neighborhood and so Poincaré lemma is also illustrated by this example. By contrast, the one-form $\sin\theta \, d\theta$ is also closed and we can write $\sin\theta \, d\theta = d(-\cos\theta)$; $\cos\theta$ is a well defined function on S^1, and so $\sin\theta \, d\theta$ is exact as well.

As another example, consider the 2-form which is given in the usual angular coordinates as $\sin\theta \, d\theta \wedge d\varphi$ on $M = S^2$. This is closed but not exact.

Stokes' theorem

If ω is a k-form and C is a $(k + 1)$-dimensional subspace of $\mathcal{M}$

$$\int_C d\omega = \int_{\partial C} \omega \tag{14.23}$$

where ∂C is the boundary of C. This result is a generalization of the usual Stokes' theorem of three-dimensional vector analysis. It can be proved in the same way as the usual Stokes' theorem is proved. If C is a closed manifold, i.e., its boundary is null, then $\int_C d\omega = 0$.

Stokes' theorem is very useful in operations involving differential forms. One can even use it to prove the nonexactness of a differential form in some cases. For example, consider $\omega = \sin\theta \, d\theta \wedge d\varphi$ on $\mathcal{M} = S^2$. By direct evaluation of the integral

$$\int_{S^2} \omega = 4\pi \tag{14.24}$$

ω is evidently closed, since $d\theta \wedge d\theta = 0$. If it were exact, say, $\omega = d\alpha$, then one could write

$$\int_{S^2} \omega = \int_{\partial S^2} \alpha = 0 \tag{14.25}$$

since the boundary of S^2 is null. Therefore, the previous result (14.24) shows that ω cannot be exact. Thus $\omega = \sin\theta \, d\theta \wedge d\varphi$ is a closed, but not exact, 2-form on S^2.

The volume form and the Hodge dual

Consider a Riemannian manifold $\mathcal{M}$ with a metric or distance function given by $ds^2 = g_{ij}dx^i dx^j$. The volume element on this space is given in these coordinates by $\sqrt{\det g} \, d^n x$. We can write this as an n-form V on $\mathcal{M}$ which is called the volume form. Explicitly

$$V = \frac{1}{n!}\sqrt{\det g} \, \epsilon_{i_1 \cdots i_n} \, dx^{i_1} \wedge \cdots \wedge dx^{i_n} \tag{14.26}$$

where $\epsilon_{i_1 \cdots i_n}$ is again the purely numerical Levi-Civita symbol.

If ω is a k-form on an n-dimensional manifold M, then we can define an associated $(n-k)$-form, called the Hodge dual of ω and denoted by $^*\omega$, by using the volume form. Given ω and its local components $\omega_{j_1 \cdots j_k}$, we can define a contravariant tensor associated with it, using the metric, by

$$
\begin{aligned}
T_\omega &= \frac{1}{k!}\omega^{i_1 \cdots i_k}\frac{\partial}{\partial x^{i_k}}\cdots\frac{\partial}{\partial x^{i_1}} \\
&= \frac{1}{k!}g^{i_1 j_1}\cdots g^{i_k j_k}\omega_{j_1 \cdots j_k}\frac{\partial}{\partial x^{i_k}}\cdots\frac{\partial}{\partial x^{i_1}}
\end{aligned}
\tag{14.27}
$$

g^{ij} is the inverse to the metric defined by $g^{ij}g_{jk} = \delta^i_k$. Notice that we have reversed the order of indices for the $\partial/\partial x$'s; this will avoid some unpleasant minus signs. The contraction of this tensor with the volume form is the $(n-k)$-form $^*\omega$. In local coordinates

$$
\begin{aligned}
^*\omega &\equiv (T_\omega, V) \\
&= \frac{1}{(n-k)!}\left[\frac{1}{k!}\sqrt{\det g}\,\epsilon_{i_1 \cdots i_k i_{k+1}\cdots i_n}g^{j_1 i_1}\cdots g^{j_k i_k}\omega_{j_1 \cdots j_k}\right] \\
&\quad \times dx^{i_{k+1}} \wedge \cdots \wedge dx^{i_n}
\end{aligned}
$$

$$\tag{14.28}$$

From this definition we can check immediately that $*(*\omega) = (-1)^{k(n-k)} \omega$ with Euclidean signature for the metric. (With Minkowski signature when $\det g$ is negative, we have $*(*\omega) = -(-1)^{k(n-k)} \omega$.)

The definition of the Hodge dual depends on the metric; it is not a metric-independent quantity.

The Hodge dual also gives a notion of an inner product of forms on a manifold. Let α and β be two differential k-forms on $\mathcal{M}$, which we will take to have Euclidean signature for the metric. We can then define an inner product of these two forms as

$$
\begin{aligned}
(\alpha, \beta) &= \int_{\mathcal{M}} \alpha \wedge {}^* \beta \\
&= \frac{1}{k!} \int_{\mathcal{M}} \sqrt{\det g} \; \alpha_{i_1 \cdots i_k} \beta_{j_1 \cdots j_k} \; g^{i_1 j_1} \cdots g^{i_k j_k} \qquad (14.29) \\
&= (\beta, \alpha)
\end{aligned}
$$

Using this inner product, we can define an adjoint δ for the exterior derivative d by

$$
(\alpha, d\gamma) = (\delta \alpha, \gamma) \qquad (14.30)
$$

where α is a k-form and γ is a $(k-1)$-form. The operator δ is defined only on forms for which the inner product (14.29) is finite. This is the case for compact manifolds of finite volume and also for suitably restricted class of forms on manifolds of infinite volume. While d is a mapping from k-forms to $(k+1)$-forms, δ is a mapping from k-forms to $(k-1)$-forms, as is clear from (14.30). By using the definition of the Hodge dual we find $\delta = (-1)^{nk+n+1} *d*$. In local coordinates,

$$
\delta \alpha = -\frac{1}{(k-1)!} \; \partial_j(g^{j i_1} \alpha_{i_1 \cdots i_k}) \; dx^{i_2} \wedge \cdots \wedge dx^{i_k} \qquad (14.31)
$$

We see that δ is a generalization of the notion of the divergence of a tensor.

Another operator of interest related to these is the Laplace operator Δ defined by

$$
\Delta = d\, \delta + \delta\, d \qquad (14.32)
$$

The Laplace operator is a mapping from k-forms to k-forms. A differential form ω for which $\Delta \omega = 0$ is said to be a harmonic form.

Lie derivative

Another differential geometric concept which is very useful is the Lie derivative. A transformation $x^i \rightarrow x^i + \xi^i(x)$ on a manifold $\mathcal{M}$ can be considered as being generated by the vector field $\xi = \xi^i \frac{\partial}{\partial x^i}$. In particular, for a function f on $\mathcal{M}$, the change under this transformation is given by

$$
\delta_\xi f = \xi^i \frac{\partial f}{\partial x^i} \equiv L_\xi f \qquad (14.33)
$$

This is called the Lie derivative of f with respect to the vector field ξ. In the case of a differential one-form $A = A_i dx^i$, there will be two sources of change, due to the change of $A_i(x)$ and due to the change of dx^i or the change in the basis with respect to which the components A_i are defined. We thus find

$$
\begin{aligned}
L_\xi A &= \xi^j \frac{\partial A_i}{\partial x^j} dx^i + A_i \frac{\partial \xi^i}{\partial x^j} dx^j \\
&= \xi^j (\partial_j A_i - \partial_i A_j) dx^i + \partial_i (A_j \xi^j) dx^i \\
&= i_\xi dA + d(i_\xi A) = (i_\xi d + d i_\xi)\, A
\end{aligned}
\tag{14.34}
$$

where i_ξ denotes the interior contraction of the vector field ξ with A, which was introduced in equation (14.6). Notice that the formula (14.33) for the Lie derivative of a function may also be written as $(i_\xi d + d i_\xi)f$, since $i_\xi f = 0$ anyway. For a differential k-form, by virtue of the antisymmetry of indices, i_ξ is defined by

$$
\begin{aligned}
i_\xi \omega &= i_\xi \left(\frac{1}{k!} \omega_{i_1 i_2 \cdots i_k} dx^{i_1} \wedge dx^{i_2} \cdots \wedge dx^{i_k} \right) \\
&= \frac{1}{k!} \left(\xi^{i_1} \omega_{i_1 i_2 \cdots i_k} dx^{i_2} \cdots \wedge dx^{i_k} - \xi^{i_2} \omega_{i_1 i_2 \cdots i_k} dx^{i_1} \wedge dx^{i_3} \cdots \wedge dx^{i_k} + \cdots \right)
\end{aligned}
\tag{14.35}
$$

With this definition, it is easy to check that for any differential form

$$
L_\xi = i_\xi d + d i_\xi
\tag{14.36}
$$

Consider now the Lie derivative of a vector $\eta = \eta^i \frac{\partial}{\partial x^i}$ with respect to ξ. $\frac{\partial}{\partial x^i}$ being dual to dx^i, we find

$$
\begin{aligned}
L_\xi \eta &= \xi^j \frac{\partial \eta^i}{\partial x^j} \frac{\partial}{\partial x^i} - \eta^i \frac{\partial \xi^j}{\partial x^i} \frac{\partial}{\partial x^j} \\
&= \left[\xi^i \frac{\partial}{\partial x^i}, \eta^j \frac{\partial}{\partial x^j} \right]
\end{aligned}
\tag{14.37}
$$

The Lie derivative is given by the commutator of the two vector fields.

In summary, we see that the Lie derivative gives the change in any quantity under the action of a vector field, taking account of the change in the frames used to define the components and the change in the components themselves. Thus it is a notion of the derivative with a direct physical meaning. It gives the change of any quantity as we move from a point to an infinitesimally nearby point by the action of the vector field.

14.2 Geometrical structures on manifolds and gravity

14.2.1 Riemannian structures and gravity

We have used the notion of a Riemannian metric on a manifold in a few places in our discussion. In this section, metrics and other geometrical structures on a manifold will be considered in some more detail.

A metric is a measure of distance between nearby points on a manifold. If the coordinate separation is dx^μ, we define the square of the distance between the points as

$$ds^2 = g_{\mu\nu}dx^\mu dx^\nu \qquad (14.38)$$

$g_{\mu\nu}$ is a symmetric covariant tensor; it characterizes the geometry of the space. We also require that the metric be nonsingular, i.e., $g_{\mu\nu}$, considered as matrix, should have $\det g \neq 0$.

Since $g_{\mu\nu}$ is a symmetric matrix, it can be brought to the form $(E^T E)_{\mu\nu}$; explicitly

$$g_{\mu\nu} = E^a_\mu \, E^a_\nu \qquad (14.39)$$

where a is an index taking values 1 to n on an n-dimensional manifold. E^a_μ is invertible since $g_{\mu\nu}$ is nonsingular. The quantities $E^a = E^a_\mu dx^\mu$ are a basis of one-forms on the manifold and are called the frame fields. Here we are taking a metric of Euclidean signature so that $g_{\mu\nu}$ is a positive matrix. For a space of Lorentzian signature, the frame fields are defined by $g_{\mu\nu} = E^a_\mu \eta^{ab} E^b_\nu$ where $\eta^{ab} = diag(1, -1, -1, -1)$; we will continue with Euclidean signature, most of the results are easily continued to Lorentzian signature. (Rather than starting with the definition of the metric, as we have done here, one could also take the frame fields as the fundamental objects and introduce the metric as a derived notion.)

There is an ambiguity in the identification of the frame fields given the metric tensor $g_{\mu\nu}$. Notice that E^a and $R^{ab}E^b$, where R^{ab} is a local rotation matrix, give the same metric. The rotation group $SO(n)$ should thus be a gauge symmetry in all of our considerations on the geometry of the space. Therefore we introduce a gauge potential or connection $\omega^{ab} = \omega^{ab}_\mu dx^\mu$ for the rotation group. This is a one-form taking values in the Lie algebra of $SO(n)$. The torsion T^a and the Riemann curvature $\mathcal{R}^{ab}$ are then defined by

$$dE^a + \omega^{ab} \wedge E^b \equiv T^a$$
$$d\omega^{ab} + \omega^{ac} \wedge \omega^{cb} \equiv \mathcal{R}^{ab} \qquad (14.40)$$

Here $dE^a + \omega^{ab} \wedge E^b$ is the gauge-covariant exterior derivative of E^a. The connection ω, usually known as the spin connection, transforms under rotations as $\omega^{ab} \to (\omega')^{ab} = R^{ac}\omega^{cd}(R^{-1})^{db} - dR^{ac}\,(R^{-1})^{cb}$. The curvature and torsion are gauge-covariant two-forms; the torsion transforms under rotations as the vector representation of $SO(n)$, while $\mathcal{R}^{ab}$ takes values in the Lie algebra of $SO(n)$.

A manifold $\mathcal{M}$ endowed with metric and connection as described above is said to be a Riemannian manifold if we have the further condition of zero torsion, $T^a = 0$. In such cases, the spin connection is determined, up to the usual freedom of gauge choice, by the metric. For, if $T^a = 0$, in local coordinates we may write

$$\left[\partial_\mu E_\nu^a + \omega_\mu^{ab} E_\nu^b\right] - \left[\partial_\nu E_\mu^a + \omega_\nu^{ab} E_\mu^b\right] = 0 \tag{14.41}$$

We then introduce $\Gamma_{\mu\nu}^\lambda$ by

$$\partial_\mu E_\nu^a + \omega_\mu^{ab} E_\nu^b - \Gamma_{\mu\nu}^\lambda E_\lambda^a = 0 \tag{14.42}$$

$\Gamma_{\mu\nu}^\lambda$ is symmetric in μ, ν by (14.41). From the definition of the metric, keeping in mind the antisymmetry of ω^{ab} in a, b, we get

$$\partial_\alpha g_{\mu\nu} - \Gamma_{\alpha\mu}^\lambda g_{\lambda\nu} - \Gamma_{\alpha\nu}^\lambda g_{\mu\lambda} = 0 \tag{14.43}$$

Writing out this equation with permutations of indices and adding terms suitably, we find

$$\Gamma_{\mu\nu}^\lambda = \frac{1}{2} g^{\lambda\alpha} \left(\partial_\mu g_{\alpha\nu} + \partial_\nu g_{\alpha\mu} - \partial_\alpha g_{\mu\nu}\right) \tag{14.44}$$

We can now go back and write equation (14.42) as

$$\omega_\mu^{ab} = (E^{-1})^{b\nu} \Gamma_{\mu\nu}^\lambda E_\lambda^a - (E^{-1})^{b\nu} \partial_\mu E_\nu^a \tag{14.45}$$

We see that, for a Riemannian manifold, the metric $g_{\mu\nu}$ determines $\Gamma_{\mu\nu}^\lambda$; since it also determines E_μ^a up to gauge rotations, ω_μ^{ab} can be constructed from the metric. $\Gamma_{\mu\nu}^\lambda$ are known as the Christoffel symbols.

The Riemann curvature of a Riemannian manifold may be simplified, using the expression (14.45), as

$$\mathcal{R}_{\mu\nu}^{ab} = E_\lambda^a (E^{-1})^{b\alpha} \mathcal{R}_{\mu\nu\alpha}^\lambda$$
$$\mathcal{R}_{\mu\nu\alpha}^\lambda = \partial_\mu \Gamma_{\nu\alpha}^\lambda - \partial_\nu \Gamma_{\mu\alpha}^\lambda + \Gamma_{\mu\beta}^\lambda \Gamma_{\nu\alpha}^\beta - \Gamma_{\nu\beta}^\lambda \Gamma_{\mu\alpha}^\beta \tag{14.46}$$

As mentioned earlier, the frames E^a define a basis for one-forms or the local cotangent space. One can regard E_μ^a and $(E^{-1})^{a\alpha}$ as transformation matrices connecting local coordinate frames and local tangent and cotangent frames. One can then use these to transform tensors from a coordinate basis to the local tangent frame basis and vice versa. $\mathcal{R}_{\mu\nu\alpha}^\lambda$ can thus be regarded as the curvature tensor in a local coordinate basis. The covariant derivative of a vector V^a is given by

$$(D_\mu V)^a = \partial_\mu V^a + \omega^{ab} V^b \tag{14.47}$$

Defining the components in the coordinate basis as $V_\mu = E_\mu^a V^a$, we find

$$D_\mu V_\nu = \partial_\mu V_\nu - \Gamma^\lambda_{\mu\nu} V_\lambda \tag{14.48}$$

The left-hand side of equation (14.43) can thus be identified as the covariant derivative of the metric tensor; we see that the metric is covariantly constant on a Riemannian manifold.

The Riemann tensor measures the curvature of space. Consider the parallel transport of a vector V^a around a closed curve C. The change in the orientation of the vector for a complete circuit of the curve is given by

$$\delta V^a = U^{ab} V^b$$

$$U^{ab} = P \exp \left(\oint_C \omega^{ab} S^{ab} \right) \tag{14.49}$$

$$\approx 1 + \int_\sigma \mathcal{R}^{ab} S^{ab}$$

where S^{ab} are the generators of the rotation group; in the vector representation, which is relevant here, they are given by $(S^{ab})_{cd} = \delta^a_c \delta^b_d - \delta^a_d \delta^b_c$. U^{ab} is the gravitational analog of the parallel transport operator in a gauge theory given in (10.32). As in that case, in the last line of equation (14.49), we have simplified the integral for a closed curve of small area; σ is the area of the surface with C as the boundary. Equation (14.49) shows that $\mathcal{R}^{ab}$ is indeed the curvature of space.

The Ricci tensor $\mathcal{R}_{\nu\alpha}$ and the Ricci scalar $\mathcal{R}$ are related to the Riemann tensor $\mathcal{R}^\lambda_{\mu\nu\alpha}$ and are obtained from it by contraction of indices.

$$\mathcal{R}_{\nu\alpha} \equiv \mathcal{R}^\lambda_{\lambda\nu\alpha}$$

$$= \partial_\lambda \Gamma^\lambda_{\nu\alpha} - \partial_\nu \Gamma^\lambda_{\lambda\alpha} + \Gamma^\lambda_{\lambda\beta} \Gamma^\beta_{\nu\alpha} - \Gamma^\lambda_{\nu\beta} \Gamma^\beta_{\lambda\alpha}$$

$$\mathcal{R} \equiv g^{\mu\nu} \mathcal{R}_{\mu\nu} = g^{\mu\nu} \mathcal{R}^\lambda_{\lambda\mu\nu} \tag{14.50}$$

The Ricci tensor $\mathcal{R}_{\mu\nu}$ is a symmetric tensor, as can be checked from the explicit expression for it.

According to Einstein's theory of gravity, spacetime is a Riemannian manifold with a Lorentzian metric of the form $g_{\mu\nu} = E^a_\mu \eta^{ab} E^b_\nu$. The metric is dynamically determined by matter; the equation which gives the metric of spacetime for a given distribution of matter is the Einstein equation, which is

$$\mathcal{R}_{\mu\nu} - \frac{1}{2} g_{\mu\nu} \mathcal{R} = 8\pi G \, T_{\mu\nu} \tag{14.51}$$

Here G is Newton's constant of gravity and $T_{\mu\nu}$ is the energy-momentum tensor of matter. We have continued the definition of the curvature tensors to Lorentzian signature in this equation. The Einstein equation may be obtained as the variational equation for the action

$$\mathcal{S} = -\frac{1}{16\pi G} \int d^4x \, \sqrt{-g} \, \mathcal{R} \; + \; \mathcal{S}_{matter}$$

$$= -\frac{1}{64\pi G} \int \epsilon_{abcd} \, \mathcal{R}^{ab} \wedge E^c \wedge E^d \; + \; \mathcal{S}_{matter} \tag{14.52}$$

In the second line, we have written the integral as the integral of a four-form involving the curvature two-form. The variation of the gravitational part of the action gives

$$\delta S = -\frac{1}{16\pi G} \int \sqrt{-g}\, d^4x \left[\mathcal{R}_{\mu\nu}\delta g^{\mu\nu} + \mathcal{R}\frac{1}{\sqrt{-g}}\delta\sqrt{-g} + \delta\mathcal{R}_{\mu\nu}g^{\mu\nu} \right] \quad (14.53)$$

The variation of $\mathcal{R}_{\mu\nu}$ involves covariant derivatives and integrates to surface terms since $D_\alpha g_{\mu\nu} = 0$. Using $\delta\sqrt{-g} = -\frac{1}{2}g_{\mu\nu}\delta g^{\mu\nu}$ and the result (3.107) for the variation of the matter part, we can check equation (14.51).

The coupling of matter to gravity is achieved by the gauge principle of replacing ordinary derivatives by covariant derivatives. Thus for a scalar field φ coupled to gravity the action is

$$S = \int \sqrt{-g}\, d^4x \left[\frac{1}{2}g^{\mu\nu}\partial_\mu\varphi\partial_\nu\varphi - \frac{1}{2}m^2\varphi^2 + \mathcal{L}_{int} \right] \quad (14.54)$$

For the Dirac theory, the corresponding action would be

$$S_{Dirac} = \int \sqrt{-g}\, d^4x\, \bar\psi \left[i\gamma^a(E^{-1})^{a\mu}\left(\partial_\mu - i\omega^{ab}S^{ab} \right) - m \right] \psi \quad (14.55)$$

Here S^{ab} is the spin matrix for the Dirac spinors, $S^{ab} = -[\gamma^a, \gamma^b]/4i$. For the gauge theory, the action is

$$S_{YM} = -\frac{1}{4} \int \sqrt{-g}\, d^4x\, g^{\mu\alpha}g^{\nu\beta}F^a_{\mu\nu}F^a_{\alpha\beta} \quad (14.56)$$

$F = dA + A \wedge A$ does not involve the metric, and so one need not use derivatives which are covariant with respect to coordinate transformations or local rotations. If one uses the explicit formula and covariant derivatives in the coordinate basis, the Christoffel symbols cancel out.

14.2.2 Complex manifolds

Given an even-dimensional differential manifold, one can ask the question whether it is possible to combine the coordinates into complex conjugate pairs. This question is motivated by the fact that in two dimensions, it is often very useful to combine the coordinates into complex coordinates $z = x_1 - ix_2$ and $\bar z = x_1 + ix_2$. If T_x denotes the tangent space at a point (with coordinates x) on a general $2n$-dimensional manifold, an almost complex structure J is defined as a tensor field which gives a map $T_x \to T_x$, for each x, such that $J^2 = -1$. In local coordinates, this means that we have a set of quantities J^μ_ν such that $J^\mu_\alpha J^\alpha_\nu = -\delta^\mu_\nu$. If we have such a structure, then we can define the holomorphic coordinate differentials $(J^\mu_\nu + i\,\delta^\mu_\nu)dx^\nu$. (Only n of these are independent.) Thus J is basically a rule to combine the coordinate differentials into complex components. (Even though an almost complex manifold can be shown to be orientable, it is not automatic that such a structure can be defined on any even-dimensional manifold, S^4 is a counterexample.)

An almost complex structure gives a rule for combining differentials pointwise, but under certain conditions, the structure J can be integrated to define holomorphic coordinates in a neighborhood and eventually to define a complex structure for the manifold. The idea is that we can then define the notion of holomorphic functions and also achieve a separation of differential forms into different types. (The different types of differential forms may be labeled as (p, q), which indicates forms which have p differentials of the holomorphic type and q differentials of the antiholomorphic type.) They will remain of the same type under holomorphic change of coordinates. The integrability condition can be obtained as follows. Starting with the holomorphic differentials $\sigma^\mu = (J^\mu_\nu + i\delta^\mu_\nu)dx^\nu$, we can write

$$d\sigma^\mu = A^\mu_{\alpha\beta}\sigma^\alpha\sigma^\beta + B^\mu_{\alpha\beta}\sigma^\alpha\bar{\sigma}^\beta + C^\mu_{\alpha\beta}\bar{\sigma}^\alpha\bar{\sigma}^\beta \tag{14.57}$$

If there is a consistent separation into holomorphic and antiholomorphic forms, the derivative of $(1, 0)$-form can generate a $2, 0)$-form and a $(1, 1)$-form but not a $(0, 2)$-form, since we have a holomorphic differential to begin with. This means that the derivative in (14.57) cannot have a term proportional to $\bar{\sigma}^\alpha\bar{\sigma}^\beta$; i.e., we must require $C^\mu_{\alpha\beta}$ to vanish as the necessary condition for defining a complex structure using J. This can be reduced to the vanishing of the Nijenhuis tensor

$$N^\alpha_{\mu\nu} = J^\gamma_\mu(\partial_\gamma J^\alpha_\nu - \partial_\nu J^\alpha_\gamma) - J^\gamma_\nu(\partial_\gamma J^\alpha_\mu - \partial_\mu J^\alpha_\gamma) \tag{14.58}$$

The Newlander-Nirenberg theorem says that a manifold with an almost complex structure J satifying $N^\alpha_{\mu\nu} = 0$ can be given a complex structure, if J obeys certain smoothness conditions such as being C^∞. Again, integrability is far from automatic; there are many examples of manifolds with almost-complex structures which are not complex manifolds. S^6 is a famous example of a manifold with a nonintegrable almost-complex structure. In fact any almost-complex manifold of $dim \geq 4$ can always have a nonintegrable almost-complex structure, for even if one is given an integrable one, it can be perturbed to obtain a nonintegrable one.

Of special interest to us are Kähler manifolds, which are defined as follows. Let $\mathcal{M}$ be a $2n$-dimensional manifold with a complex structure J (so that J satisfies $N^\alpha_{\mu\nu} = 0$) and a Riemannian metric $g_{\mu\nu}$. The metric is said to be hermitian if

$$g_{\mu\nu} J^\mu_\alpha J^\nu_\beta = g_{\alpha\beta} \tag{14.59}$$

Given a hermitian metric, the fundamental two-form Ω is defined as

$$\Omega = \frac{1}{2} J^\alpha_\mu g_{\alpha\nu} \, dx^\mu \wedge dx^\nu \tag{14.60}$$

(The definitions (14.58) and (14.60) are in terms of local coordinates, but can be easily written in more invariant ways.) Now given $g_{\mu\nu}$, J^μ_ν and Ω, $\mathcal{M}$ is said to be Kähler if $d\Omega = 0$. In this case, Ω is the Kähler form, $g_{\mu\nu}$ is the Kähler metric. For a Kähler manifold, one can choose complex coordinates z^a, $\bar{z}^{\bar{a}}$ such that

$$ds^2 = g_{a\bar{b}}dz^a d\bar{z}^{\bar{b}}$$

$$\Omega = \frac{i}{2}g_{a\bar{b}}dz^a \wedge d\bar{z}^{\bar{b}} \qquad (14.61)$$

Further, the metric can be obtained from a potential K as

$$g_{a\bar{b}} = \frac{\partial}{\partial z^a}\frac{\partial}{\partial \bar{z}^{\bar{b}}}K \qquad (14.62)$$

K is known as the Kähler potential. Examples of Kähler manifolds are Riemann surfaces and complex projective spaces $\mathbf{CP}^k$. There are many manifolds which are complex but not Kähler, $S^3 \times S^1$ is a simple example.

14.3 Cohomology groups

We have seen that there can be closed forms which are not exact. Since Poincaré lemma shows that locally every closed form is exact, the existence of closed but not exact forms is related to nontrivial global topological properties of the manifold. The cohomology classes capture this feature of a manifold.

We define an equivalence relation that two forms are equivalent if they differ by an exact form; i.e., $\omega_1 \sim \omega_2$ if there is a form α such that $\omega_1 = \omega_2 + d\alpha$. The equivalence classes defined by this relation are called the cohomology classes. More specifically, let $\mathcal{C}^k(\mathcal{M}, \mathbf{R})$ denote the set of closed k-forms on a manifold $\mathcal{M}$, and let $\mathcal{Z}^k(\mathcal{M}, \mathbf{R})$ be the set of exact k-forms on $\mathcal{M}$, where linear combinations of forms are taken with real numbers as coefficients; this is indicated by $\mathbf{R}$ as one of the arguments for these sets. We then define the k-th cohomology group of $\mathcal{M}$ over the real numbers as

$$\mathcal{H}^k(\mathcal{M}, \mathbf{R}) = \mathcal{C}^k(\mathcal{M}, \mathbf{R})/\mathcal{Z}^k(\mathcal{M}, \mathbf{R}) \qquad (14.63)$$

(Cohomology classes can be defined with complex coefficients, $\mathbf{Z}_2$-valued coefficients, etc., but we do not consider them here.)

The cohomology classes by themselves do not give a complete characterization of the topology of the manifold; there are many other quantities needed to do that.

The two cohomology groups $\mathcal{H}^0(\mathcal{M}, \mathbf{R})$ and $\mathcal{H}^n(\mathcal{M}, \mathbf{R})$ for an n-dimensional manifold $\mathcal{M}$ are of some special significance. Let $\mathcal{M}$ be compact and orientable without boundary and let V be the volume form. Since $\oint_{\mathcal{M}} V \neq 0$, V cannot be exact. V is evidently closed for dimensional reasons. Thus V is an element of $\mathcal{H}^n(\mathcal{M}, \mathbf{R})$. In fact, up to multiplication by a constant, it is the only element of $\mathcal{H}^n(\mathcal{M}, \mathbf{R})$, for if ω is another element of $\mathcal{H}^n(\mathcal{M}, \mathbf{R})$, we can write $\omega = c\, V$; then c has to be constant due to $d\omega = 0$, $dV = 0$. Therefore we may write $\mathcal{H}^n(\mathcal{M}, \mathbf{R}) = \mathbf{R}$, with V as its representative element.

On a compact manifold with no boundary, we can also define an inner product between elements $\alpha \in \mathcal{H}^k(\mathcal{M}, \mathbf{R})$ and $\beta \in \mathcal{H}^{n-k}(\mathcal{M}, \mathbf{R})$ by

$$(\alpha, \beta) = \oint_{\mathcal{M}} \alpha \wedge \beta \tag{14.64}$$

We see that the changes $\alpha \to \alpha + d\varphi$, $\beta \to \beta + d\xi$ leave this invariant by partial integration, so that this is really an inner product on the cohomologies. (The restriction to compact M with no boundary is to ensure that the inner product exists and the partial integration can be done. If we consider forms with sufficiently fast fall-off behavior at infinity or at any boundary M may have, these conditions can be relaxed.) Poincaré duality on the cohomology groups is the statement that $\mathcal{H}^k(\mathcal{M}, \mathbf{R})$ and $\mathcal{H}^{n-k}(\mathcal{M}, \mathbf{R})$ are duals to each other as vector spaces under this inner product. For a compact manifold with only one connected component, we define $\mathcal{H}^0(\mathcal{M}, \mathbf{R}) = \mathbf{R}$ so that Poincaré duality holds for $k = 0$ as well.

The dimension of $\mathcal{H}^k(\mathcal{M}, \mathbf{R})$ is known as the k-th Betti number b_k of $\mathcal{M}$. The Euler number of a manifold is given by $\chi(\mathcal{M}) = \sum_k (-1)^k \dim \mathcal{H}^k(\mathcal{M}, \mathbf{R}) = \sum_k (-1)^k b_k$. For a two-dimensional sphere for which $\mathcal{H}^1(\mathcal{M}, \mathbf{R}) = 0$, $\chi(S^2) = 2$; for Riemann surface with genus g, $\chi = 2 - 2g$. (It is easy to see that $\mathcal{H}^1(\mathcal{M}, \mathbf{R}) = 0$; if it were not so, there would be a one-form α, for which $d\alpha = 0$. The integral of α over any closed loop on S^2 is then an invariant under deformations of the loop. Since every loop on S^2 can be contracted to a point, the value of the integral is zero. This implies that α must be exact.)

Cohomology of Lie groups

The cohomologies of Lie groups are important for many questions in physics. These were worked out many years ago by Borel, Hirzebruch, and others. We will not give the general analysis here but will go through a simple construction of the corresponding differential forms.

Consider a Lie group G, a typical element of which may be denoted by g. There are many parametrizations possible for g. For example, we may think of it as given in the form $g = \exp(it^a \theta^a)$, where θ^a are the continuous group parameters and t^a are matrices which give infinitesimal generators of g in some matrix representation. For most of this discussion, t^a may be taken to be in the fundamental representation. Thus, for the group $SU(n)$, t^a may be considered as a basis of hermitian, traceless $(n \times n)$-matrices, the index taking values $1, 2, \cdots, (n^2 - 1)$. Given g, we can construct the Lie-algebra-valued one-form

$$\omega = g^{-1} dg = -it^a E_i^a d\theta^i = -it^a E^a \tag{14.65}$$

where the one-form frame fields $E^a = E_i^a(\theta) d\theta^i$ were introduced in our discussion of spontaneous symmetry breaking in Chapter 12. Starting with ω, by taking exterior products and traces, we can construct $\text{Tr}(\omega^k)$. When k is even, these vanish by cyclicity of the trace, e.g.,

$$
\begin{aligned}
\text{Tr}(\omega^2) &= -\text{Tr}(t^a t^b) E^a \wedge E^b \\
&= -\text{Tr}(t^b t^a) E^a \wedge E^b \\
&= \text{Tr}(t^b t^a) E^b \wedge E^a \\
&= -\text{Tr}(\omega^2)
\end{aligned} \tag{14.66}
$$

For odd values of k, $\Omega^{(k)} = \text{Tr}\omega^k$ are nonzero in general. Since $d\omega = d(g^{-1}dg) = dg^{-1}\, dg = -g^{-1}dg\, g^{-1}dg = -\omega^2$, we find

$$
\begin{aligned}
d\Omega^{(k)} &= -\text{Tr}\left(\omega^2 \omega^{k-1} - \omega\, \omega^2\, \omega^{k-2} \cdots\right) \\
&= 0
\end{aligned} \tag{14.67}
$$

since $\text{Tr}\omega^{even} = 0$. Thus $\Omega^{(k)}$ are closed differential forms. Further, these are not exact. The simplest way to show this is by a case by case analysis; we will consider the $SU(n)$ groups. For $k = 1$, $\Omega^{(1)} = \text{Tr}\omega$. This is zero by tracelessness of t^a unless we consider $U(1)$ for which $g = \exp(i\theta)$. For $U(1)$, $\Omega^{(1)} = id\theta$. Integrating this over the circle $0 \leq \theta \leq 2\pi$,

$$
-i \oint \Omega^{(1)} = 2\pi \tag{14.68}
$$

If $\Omega^{(1)}$ were exact, then we would have $\Omega^{(1)} = d\alpha$ for some α which is a periodic function of θ (so that it is a proper function on $U(1)$), and so $\oint \Omega^{(1)} = 0$ by Stokes' theorem. The result (14.68) shows that $\Omega^{(1)}$ cannot be exact. Thus it is an element of $\mathcal{H}^1(U(1), \mathbf{R})$. For $k = 3$, we have $\Omega^{(3)} = \text{Tr}(g^{-1}dg)^3$. For the group $SU(2)$ we take the parametrization $g = a + ib_i\sigma_i$ with $a^2 + b_1^2 + b_2^2 + b_3^2 = 1$, a, b_i real. This shows that the group $SU(2)$ is a three-dimensional sphere S^3. We integrate $\Omega^{(3)}$ over this S^3. For this purpose, we may write

$$
a = \frac{u^2 - 1}{u^2 + 1}, \qquad\qquad b_i = \frac{2u_i}{u^2 + 1} \tag{14.69}
$$

where u_i are three unrestricted real variables, $-\infty \leq u_i \leq \infty$, $u^2 = u_i u_i$. We then find

$$
\oint_{S^3} \Omega^{(3)} = -24\,\pi^2 \tag{14.70}
$$

showing that $\Omega^{(3)}$ is not exact. Thus it is an element of $\mathcal{H}^3(SU(2), \mathbf{R})$. $\Omega^{(3)}$ is proportional to the volume of $SU(2)$ for dimensional reasons and so $\mathcal{H}^3(SU(2), \mathbf{R}) = \mathbf{R}$. This result holds for any compact Lie group G, $\mathcal{H}^3(G, \mathbf{R}) \neq 0$ for any compact Lie group, $\Omega^{(3)}$ being a representative element.

A similar argument can be made for $\Omega^{(5)} = \text{Tr}\omega^5$, where we can expect that demonstrating the nontriviality of $\Omega^{(5)}$ involves integration over a five-dimensional sphere S^5. Such an integration requires a parametrization of group elements in a way similar to (14.69), but there is a simpler argument one can use. First of all, notice that since the group $SU(2)$ is S^3, for this group $\text{Tr}\omega^5$ must be zero. Generally $\Omega^{(5)}$ is of the form

$$
\begin{aligned}
\Omega^{(5)} &= -i\,\text{Tr}(t^a t^b t^c t^d t^e)\,E^a \wedge \cdots \wedge E^e \\
&= -\frac{i}{4}\text{Tr}(t^a[t^b,t^c][t^d,t^e])\,E^a \wedge \cdots \wedge E^e \\
&= \frac{i}{8}f^{bcm}f^{den}\text{Tr}\left(t^a(t^m t^n + t^n t^m)\right)E^a \wedge \cdots \wedge E^e \\
&= \frac{i}{8}d^{amn}f^{bcm}f^{den}E^a \wedge \cdots \wedge E^e
\end{aligned}
\tag{14.71}
$$

(For $SU(2)$, $d^{amn} = 0$; this is another way to see that $\Omega^{(5)}$ should be zero for this group.) $\Omega^{(5)}$ can be nonzero only for $SU(n)$, $n \geq 3$. Consider $SU(3)$ first, which is eight-dimensional. In this case, $\Omega^{(5)} \wedge \Omega^{(3)}$ should be proportional to the volume element of $SU(3)$. In fact, we have

$$
\begin{aligned}
\Omega^{(5)} \wedge \Omega^{(3)} &= \text{Tr}(t^a t^b t^c t^d t^e)\text{Tr}(t^f t^g t^h)E^a \wedge \cdots \wedge E^h \\
&= i\frac{45}{2\sqrt{3}}V_{SU(3)}
\end{aligned}
\tag{14.72}
$$

where we have used

$$
\epsilon_{ab\cdots h}\text{Tr}(t^a t^b t^c t^d t^e)\text{Tr}(t^f t^g t^h) = i\frac{45}{2\sqrt{3}}
\tag{14.73}
$$

and $V_{SU(3)}$ is the volume element for $SU(3)$ given by $E^a \wedge \cdots \wedge E^h = \epsilon^{a\cdots h}V_{SU(3)}$. We see that the integral of $\Omega^{(5)} \wedge \Omega^{(3)}$ over $SU(3)$ is nonzero, since the volume integrates to a nonzero value. Since we have already shown that $\Omega^{(3)}$ is cohomologically nontrivial, this result shows that $\Omega^{(5)}$ is closed but not exact. In other words, $\mathcal{H}^5(SU(3),\mathbf{R})$ is nonzero. Notice that this result for $SU(3)$ also follows from Poincaré duality, since we have already shown that $\mathcal{H}^3(SU(3),\mathbf{R})$ is nonzero. As in the case of $\mathcal{H}^3(SU(3),\mathbf{R})$, the dimension of $\mathcal{H}^5(SU(3),\mathbf{R})$ is 1, and the nontrivial element is generated by $\Omega^{(5)}$ up to constant factors. This result can be shown to be true for all $SU(n)$, $n \geq 3$, $\mathcal{H}^5(SU(n),\mathbf{R}) = \mathbf{R}$, $n \geq 3$.

From the definition of $\Omega^{(5)}$, we also find

$$
\begin{aligned}
\Omega^{(5)}(gh) = \Omega^{(5)}(g) + \Omega^{(5)}(h) + d\,\text{Tr}\bigg[&-5dh\,h^{-1}\omega d\omega - 5\omega(dh\,h^{-1})^3 \\
&+\frac{5}{2}dh\,h^{-1}\omega dh\,h^{-1}\omega \bigg]
\end{aligned}
\tag{14.74}
$$

If h is an element of an $SU(2)$ subgroup of G, $\Omega^{(5)}(h) = 0$, $\Omega^{(5)}(gh)$ and $\Omega^{(5)}(g)$ differ only by an exact form. For $G = SU(3)$, the coset space

$SU(3)/SU(2)$ is S^5. So in integrating $\Omega^{(5)}$ over S^5, there are two possibilities; one may consider a sphere in $SU(3)$ or one may consider a disc in $SU(3)$ with the boundary lying in an $SU(2)$ subgroup, this disc being a sphere in $SU(3)/SU(2)$. In both cases, the integral of $\Omega^{(5)}$ gives a number characteristic of the cohomology involved, although the two results are not the same.

Similar arguments can be made for $\Omega^{(k)}$, for $k = 7, 9, \cdots$. One finds in general that $\mathcal{H}^k(SU(n), \mathbf{R})$ has one generating element $\Omega^{(k)}$ for odd values of k, $k = 2r + 1$, $r = 1, 2, ..., (n-1)$. For $SU(n)$, this gives $(n-1)$ nonzero cohomology groups. The rank of $SU(n)$ is also $(n-1)$ and it has $(n-1)$ invariant tensors or Casimir invariants. This is not an accident, there is a correspondence between invariant tensors and cohomology elements. Evaluating $\Omega^{(k)}$ in an arbitrary representation of G we find

$$\Omega^{(k)} \sim d_R^{a_1 a_2 \cdots a_{r+1}} f^{b_2 b_3 a_2} f^{b_4 b_5 a_3} \cdots E^{a_1} \wedge E^{b_2} \wedge E^{b_3} \wedge E^{b_4} \wedge E^{b_5} \cdots \quad (14.75)$$

where $d_R^{a_1 a_2 \cdots a_{r+1}}$ is given by the trace of $t^{a_1} t^{a_2} \cdots t^{a_{r+1}}$ with all indices symmetrized, the t^a's being in the representation R. $d_R^{a_1 a_2 \cdots a_{r+1}}$ is an invariant tensor of the Lie algebra of G. Thus the $\Omega^{(k)}$ are, in fact, given in terms of the invariant tensors and suitable products of the frame field one-forms. There is a correspondence, known as Weil homomorphism, between Chern classes (to be discussed later) and Casimir invariants; the present connection between invariant tensors and cohomology classes for Lie groups is clearly related to this.

14.4 Homotopy

Let $f_1(x) : \mathcal{M} \to \mathcal{N}$ and $f_2(x) : \mathcal{M} \to \mathcal{N}$ be two continuous mappings from a manifold $\mathcal{M}$ to another manifold $\mathcal{N}$. Suppose we can find a continuous mapping $f(x, \tau)$ depending on a real parameter $0 \leq \tau \leq 1$ such that $f(x, 0) = f_1(x)$ and $f(x, 1) = f_2(x)$. We then say that $f_1(x)$ and $f_2(x)$ are homotopic to each other. Basically this means that $f_1(x)$ can be continuously deformed to $f_2(x)$, τ being the parameter characterizing the deformation.

Given a number of mappings from $\mathcal{M}$ to $\mathcal{N}$, we can define an equivalence relation $f_1 \sim f_2$, if f_1 and f_2 are homotopic to each other. Mappings from $\mathcal{M}$ to $\mathcal{N}$ can then be grouped into equivalence classes where the members of each class are homotopic to each other. These homotopy classes of maps from $\mathcal{M}$ to $\mathcal{N}$ are important because there are many situations where certain physical quantities are invariant under smooth deformations of the mappings (or functions) involved. The notion of homotopy is then important in giving a precise characterization of their properties.

While homotopy classes can be defined for arbitrary manifolds $\mathcal{M}$ and $\mathcal{N}$, maps from spheres to a manifold $\mathcal{N}$ are particularly important and useful. The homotopy classes of maps from the k-sphere S^k to a manifold $\mathcal{N}$ are denoted by $\Pi_k(\mathcal{N})$. We will consider these in some detail.

We start with $\Pi_1(\mathcal{N})$, which is also very often referred to as the fundamental group of $\mathcal{N}$. It refers to the homotopy classes of maps $S^1 \to \mathcal{N}$. Such a map describes a closed loop in $\mathcal{N}$. Homotopically equivalent maps are thus loops in $\mathcal{N}$ which are continuously deformable to each other. $\Pi_1(\mathcal{N})$ thus characterizes the loop-connectivity of $\mathcal{N}$.

The simplest example of a nontrivial fundamental group, which is also, at least partially, a paradigm for more complicated cases, is that of the 2-plane with one point, say the origin, removed, namely, $\mathcal{N} = \mathbf{R}^2 - \{0\}$. Clearly there are paths which loop around the origin which are not contractible. We can characterize the "nontriviality" of paths on this space as follows. Consider first loops which start at the point x and return there without going around the origin. There is evidently a composition law for these paths. If C_0 is a closed loop and C_0' is another loop with the same starting point, the composite path $C_0 + C_0'$ is defined as the path which traces over C_0 first and then continues over C_0'. We can also define a path C_1 which loops around the origin once. The composition $C_1' = C_1 + C_0$ then denotes a path which starts from x, goes around the origin once along C_1 and then goes along C_0. Paths are equivalent if they can be smoothly deformed into each other, so $C_0 \sim C_0'$, $C_1 \sim C_1'$, but evidently $C_0 \not\sim C_1$. We can also define $-C$ as the loop C traversed the opposite way. Under these conditions we see that loops on the space have the following structure. $C_0 + C_0 \sim C_0$, $C_1 + C_0 \sim C_1$, $C_1 + C_1 \sim C_2$. The loops thus behave like the group of integers $\mathbf{Z}$ with C_0 acting as zero. In other words, $\Pi_1(\mathbf{R}^2 - \{0\}) = \mathbf{Z}$. It has the structure of a group, with the geometrical loop composition represented by addition in $\mathbf{Z}$. More generally, $\Pi_k(\mathcal{N})$ are groups with the composition of maps represented in a suitable way. For this reason, $\Pi_k(\mathcal{N})$ are referred to as the homotopy groups.

Continuing with $\mathbf{R}^2 - \{0\}$, the relation between loop compositions and the group of integers can be made more precise by defining an invariant for loops as follows. For any loop C from x to x introduce

$$\nu(C) = \int_{x,C}^{x} \alpha_i dx^i \tag{14.76}$$

where $\alpha_i dx^i$ is a one-form. We want ν to be invariant under small deformations of the loop C. This requires

$$\partial_i \alpha_j - \partial_j \alpha_i = 0 \tag{14.77}$$

Further for a path which winds once around the origin, we need $\oint \alpha_i dx^i = 1$. This condition, along with (14.77), determines α_i for the two-dimensional problem as

$$\alpha_i = -\frac{1}{2\pi} \frac{\epsilon_{ij} x^j}{x^2} \tag{14.78}$$

In polar coordinates (r, φ), $\alpha_i dx^i = d\varphi$. Evidently, $\nu(C_0) = 0$, $\nu(C_1) = 1$, $\nu(C_2) = 2$, $\nu(-C_1) = -1$, etc. $\nu(C)$ is a topological invariant for a closed curve C, i.e., it is unchanged under small deformations of the curve and is an integer for any closed curve C. It gives the element of $\Pi_1(\mathbf{R}^2 - \{0\}) = \mathbf{Z}$. This integer measures the number of times the path winds around the origin, taking account of the orientation of the path. $\nu(C)$ is called the winding number of C.

The statements given above show an interesting relation with cohomology groups. If we consider $\alpha = \alpha_i dx^i$ as a one-form on $\mathcal{N} = \mathbf{R}^2 - \{0\}$, the condition (14.77) says that $d\alpha = 0$. Thus α is a closed one-form. On the other hand, if it is exact, $\alpha = dh$ for some (single-valued) function on $\mathcal{N}$. In this case, $\nu(C)$ would be identically zero. ($\alpha = d\varphi$ does not say that it is exact, since φ is not single-valued.) Thus, we are using a closed but not exact one-form, or an element of the cohomology group $\mathcal{H}^1(\mathcal{N}, \mathbf{R})$ to represent the element of $\Pi_1(\mathcal{N})$. The composition rule shows that it is enough to consider elements of $\mathcal{H}^1(\mathcal{N}, \mathbf{R})$ with the coefficients for taking linear combinations of forms restricted to being integers, i.e., $\mathcal{H}^k(\mathcal{N}, \mathbf{Z})$. The generating elements of this group are the same as $\mathcal{H}^1(\mathcal{N}, \mathbf{R})$. Because it is a restriction to integer coefficients of $\mathcal{H}^1(\mathcal{N}, \mathbf{R})$, the cohomology group $\mathcal{H}^1(\mathcal{N}, \mathbf{Z})$ is the group of integers $\mathbf{Z}$ and α is its representative element. We are integrating this one-form over the loop to obtain $\nu(C)$; more precisely, we use the map $x(\tau) : S^1 \to \mathcal{N}$ to pull back the differential form on $\mathcal{N}$ to a form on S^1 (which has coordinate τ) and then integrate over S^1. In many situations where the homotopy group is $\mathbf{Z}$ we are able to represent elements of the homotopy groups $\Pi_k(\mathcal{N})$ in terms of integrals over S^k of a nontrivial element of $\mathcal{H}^k(\mathcal{N}, \mathbf{R})$ pulled back via the map $f : S^k \to N$.

This relation with cohomology groups is, however, not generic. Homotopy and cohomology are different concepts and do not agree in general. For example, one has $\Pi_3(S^2) = \mathbf{Z}$ but clearly $\mathcal{H}^3(S^2, \mathbf{R}) = 0$ since we cannot have a differential three-form on a two-dimensional manifold. A few more general remarks about homotopy groups are the following. Π_1 can be a nonabelian group, while $\mathcal{H}^1(\mathcal{N}, \mathbf{R})$ is Abelian; $\Pi_k(\mathcal{N})$ for $k > 1$ are Abelian. In many cases where the homotopy group is the group of integers, it is possible to use elements of the cohomology group to define winding numbers which characterize the elements of the homotopy group. The homotopy group $\Pi_0(\mathcal{N})$ is defined as the set of connected components of $\mathcal{N}$. If there is only one connected component of N, $\Pi_0(\mathcal{N})$ has only one element.

Homotopy groups of spheres

A class of examples where the homotopy groups can be obtained in terms of the cohomology groups is the case of spheres for which $\Pi_n(S^n) = \mathbf{Z}$ and $\Pi_k(S^n) = 0$, $k \leq n-1$. We use cohomology elements to write down a winding number which represents $\Pi_n(S^n)$. In this case, we are considering the maps $n^a(x) : S_M^n \to S_N^n$, where n^a, $a = 1, 2, \cdots, (n+1)$, with $n^a n^a = 1$, represent the coordinates of the target sphere S_N^n. x's represent the coordinates of the base sphere S_M^n. The volume form on the target space S_N^n is given by

$$dV = \frac{1}{n!}\epsilon_{a_1 a_2 \cdots a_{n+1}} \, n^{a_1} \, dn^{a_2} \wedge dn^{a_3} \wedge \cdots \wedge dn^{a_{n+1}} \tag{14.79}$$

Pulling this back by the map $n^a(x)$, which means that we simply substitute $n^a(x)$ for the n^a's in the above formula, we get a differential n-form on the starting n-sphere S_M^n. By integrating this over S_M^n we can define a winding number for the map $n^a(x)$ by

$$Q[n] = \frac{1}{n! \, vol(S^n)} \int_{S_M^n} \epsilon_{a_1 a_2 \cdots a_{n+1}} \, n^{a_1} \, dn^{a_2} \wedge dn^{a_3} \wedge \cdots \wedge dn^{a_{n+1}} \tag{14.80}$$

where $vol(S^n)$ stands for the standard volume of an n-sphere, given in terms of the Eulerian gamma function by

$$vol(S^n) = \frac{2 \, \pi^{\frac{n+1}{2}}}{\Gamma\left(\frac{n+1}{2}\right)} \tag{14.81}$$

It is easily verified that $Q[n]$ is indeed the same for homotopically equivalent maps. Consider an infinitesimal deformation of the map $n^a(x)$ given by $n^a(x) + \delta n^a(x)$. By direct computation we find

$$Q[n + \delta n] - Q[n] = \frac{1}{n! vol(S^n)} \int_{S_M^n} (n + 1) \, \delta n^{a_1} \omega_{a_1} - n \, d\omega'$$

$$\omega_{a_1} = \epsilon_{a_1 a_2 \cdots a_{n+1}} \, dn^{a_2} \wedge \cdots \wedge dn^{a_{n+1}} \tag{14.82}$$

$$\omega' = \epsilon_{a_1 a_2 \cdots a_{n+1}} \left(\delta n^{a_1} \, n^{a_2} \, dn^{a_3} \wedge \cdots \wedge dn^{a_{n+1}}\right)$$

Since $n^a n^a = 1$ we have $n^a dn^a = 0$. The dn's being orthogonal to n^a, the n-fold antisymmetric product $\epsilon_{a_1 a_2 \cdots a_{n+1}} (\, dn^{a_2} \wedge \cdots \wedge dn^{a_{n+1}})$ is proportional to n^{a_1}. For the first term in the variation of $Q[n]$, we then have $\delta n^a \, n^a$ which is zero since the variations $\delta n^a(x)$ must also obey the requirement $n^a \delta n^a = 0$ to preserve the condition $n^a n^a = 1$. The second term is a total derivative and integrates to zero on the closed space S_M^n. Thus $Q[n + \delta n] = Q[n]$, showing that $Q[n]$ is invariant under continuous deformations of the map $n^a(x)$. It is thus a homotopic invariant. It is also independent of metrical or other geometrical structures on the two spaces S_M^n and S_N^n involved, since we have only used differential forms.

The particular map $n^1 = 1$, all other n's being zero, maps the sphere S_M^n to a point on the target space S_N^n. Evidently, Q is zero for this case. The map $n^a = x^a$ where x^a are the coordinates of the sphere S_M^n, with $x^a x^a = 1$, identifies the target sphere and the base sphere. For this map, $Q = (1/vol(S^n)) \int dV = 1$. Thus any smooth deformation of the map $n^a = x^a$ belongs to the equivalence class with winding number 1.

Another general result of interest is $\Pi_k(S^1) = 0$ for all $k \geq 2$. There are also many intriguing results for the homotopy groups of spheres of the form $\Pi_{n+k}(S^n)$, for example, $\Pi_3(S^2) = \mathbf{Z}$, $\Pi_4(S^2) = \mathbf{Z}_2$, $\Pi_{11}(S^6) = \mathbf{Z}$, to name a few. There is no simple way to obtain these in terms of cohomology groups, so we will not discuss them further. The exact sequence of homotopy groups, which we give later in this chapter, can be used to relate some of them to winding numbers as constructed above.

Homotopy groups for compact connected Lie groups

The homotopy groups for compact connected Lie groups G (where $G = SU(n)$, $SO(n)$, $Sp(n)$, G_2, F_4, E_6, E_7, or E_8) have all been calculated and are listed in standard mathematical tables. Some of the general results are

$$\Pi_1(G) = \begin{cases} \mathbf{Z}, & G = SO(2) \\ \mathbf{Z}_2, & G = SO(n), \ n \geq 3 \\ 0, & all \ other \ G \end{cases}$$

$$\Pi_2(G) = 0 \ for \ all \ G$$

$$\Pi_3(G) = \begin{cases} \mathbf{Z} \times \mathbf{Z}, & G = SO(4) \\ \mathbf{Z}, & all \ other \ G \end{cases}$$

$$\Pi_{2r+1}(SU(n)) = \mathbf{Z}, \ n - 1 \geq r$$

(14.83)

The result that the fundamental group of $SO(3)$ is $\mathbf{Z}_2$ is the statement that $SO(3)$ is doubly connected. This is, of course, well known and is related to the double-valued spinor representations of the rotation group.

There are further general results beyond what is given in (14.83), but here we will just consider some cases for which we can write winding numbers in terms of the cohomology elements. We have seen in our discussion of the cohomology groups of Lie groups that the $SU(n)$ groups have one nontrivial generating element for the cohomologies $\mathcal{H}^{2r+1}(SU(n), \mathbf{R})$ which is of the form $\mathrm{Tr}(g^{-1}dg)^{2r+1}$, $n - 1 \geq r$. We can use this to write a winding number for the corresponding homotopy groups which is of the form $Q = \lambda \int_{S^{2r+1}} \mathrm{Tr}(g^{-1}dg)^{2r+1}$ for a map $g(x) : S^{2r+1} \to SU(n)$, $n - 1 \geq r$, λ being a normalization constant. This is a realization of the last result in (14.83). Specifically for the case of Π_3 we find

$$Q[g] = -\frac{1}{24\pi^2} \int_{S^3} \mathrm{Tr}(g^{-1}dg)^3 \tag{14.84}$$

Intuitively, we can understand this result by considering $SU(2)$, which is S^3, so that this result reduces to the previous case, $\Pi_3(S^3) = \mathbf{Z}$. For higher groups, there is always an $SU(2)$ subgroup into which the 3-sphere S^3 can be mapped. From the definition (14.84), we can easily check that

$$Q[gh] = Q[g] + Q[h] \tag{14.85}$$

The extra crossterms which arise combine to total derivatives and integrate to zero. This result shows that if h is a small deformation, of zero winding number, $Q[gh] = Q[g]$ which is the homotopy invariance of Q. Secondly, if g_1 has winding number 1, a mapping with winding number 2 is given by $g_1 g_1$, and an element with winding number -1 is given by $g_1^\dagger$. It is then easily seen that we generate $\mathbf{Z}$ from using Q to represent the homotopy group $\Pi_3(SU(n))$.

14.5 Gauge fields

In this section, we will see that the formalism of differential forms give a very natural setting for discussing gauge theories. We start with the electromagnetic field.

14.5.1 Electrodynamics

The electromagnetic field is described by the vector potential A_μ which enters the theory via the covariant derivative $D_\mu = \partial_\mu - ieA_\mu$. (Here we will display the coupling constant e explicitly in the covariant derivative, since this is more conventional for the Maxwell theory.) A_μ is thus naturally a covariant vector, and so we define the one-form

$$A = A_\mu dx^\mu \tag{14.86}$$

Applying the exterior derivative on this we find

$$dA = \partial_\mu A_\nu dx^\mu \wedge dx^\nu = \frac{1}{2}\left(\partial_\mu A_\nu - \partial_\nu A_\mu\right) dx^\mu \wedge dx^\nu$$

$$= \frac{1}{2}F_{\mu\nu} dx^\mu \wedge dx^\nu$$

$$\equiv F \tag{14.87}$$

The field strength tensor is thus the exterior derivative of the one-form potential A. The identification of the usual electric and magnetic components is given by

$$F = F_{0i}\, dx^0 \wedge dx^i + \frac{1}{2}F_{ij}\, dx^i \wedge dx^j$$

$$= E_i\, dx^0 \wedge dx^i + \frac{1}{2}\epsilon_{ijk}B_k\, dx^i \wedge dx^j \tag{14.88}$$

Because $d^2 = 0$, we have immediately $dF = 0$. This is written out as

$$dF = \frac{1}{3!}\left(\partial_\mu F_{\nu\alpha} + \partial_\alpha F_{\mu\nu} + \partial_\nu F_{\alpha\mu}\right) dx^\mu \wedge dx^\nu \wedge dx^\alpha = 0 \tag{14.89}$$

This is the so-called Bianchi identity and is identical to the sourceless Maxwell equations.

The Hodge dual of F is also a two-form in four dimensions. The contravariant tensor obtained from F of (14.88) is

$$T_F = -E_i \frac{\partial}{\partial x^i} \frac{\partial}{\partial x^0} + \frac{1}{2}\epsilon_{ijk}B_k \frac{\partial}{\partial x^j} \frac{\partial}{\partial x^i} \qquad (14.90)$$

Taking the contraction of this with the volume form $dx^0 \wedge dx^1 \wedge dx^2 \wedge dx^3$, we find

$$^*F = B_i \, dx^0 \wedge dx^i - \frac{1}{2}\epsilon_{ijk}E_k \, dx^i \wedge dx^j \qquad (14.91)$$

(In equations (14.88) to (14.91) we do not distinguish between upper and lower indices for the spatial components , all minus signs due to the Minkowski metric are explicitly written out and raising and lowering of spatial indices are done by using the Kronecker delta.)

The current for the Maxwell theory, being a vector, can also be thought of as a one-form. Its dual is a three-form given by

$$^*J = J^0 d^3x - \frac{1}{2}\epsilon_{ijk}J_k \, dx^0 \wedge dx^i \wedge dx^j \qquad (14.92)$$

It is now easily checked that the equations $d^*F =^* J$ are identical to the Maxwell equations with the sources. In summary, the Maxwell equations are then

$$F = dA$$
$$d^*F = {}^*J \qquad (14.93)$$

Again from $d^2 = 0$, the second equation leads to $d^*J = 0$. This is the conservation of the current.

The action for the Maxwell theory can be written as

$$S = -\frac{1}{2}\int_{\mathcal{M}} F \wedge^* F + \int_{\mathcal{M}} A \wedge^* J \qquad (14.94)$$

For the classical electromagnetic theory, instead of starting with the potential, we could take the two-form F as the basic variable and write the equations as

$$d\, F = 0$$
$$d^*F = {}^*J \qquad (14.95)$$

Since F is closed, the Poincaré lemma tells us that we can write $F = dA$ at least locally. This takes us back to the potential. The existence of the potential is only guaranteed locally, and one could have situations where we do not have a globally defined potential. From our definition of the cohomology group, we see that this can happen if the manifold has nontrivial second cohomology, namely, $\mathcal{H}^2 \neq 0$. An example of this is the magnetic monopole, which we shall now discuss briefly.

14.5.2 The Dirac monopole: A first look

Consider a point magnetic monopole with magnetic charge g. Taking the origin of coordinates as the location of the magnetic monopole, the magnetic field is

$$B_i = g \, \frac{x_i}{|x|^3} \tag{14.96}$$

This is singular at the origin $|x| = 0$, so, in order to have a nonsingular field, we must consider this as being defined on $\mathbf{R}^3 - \{0\}$, i.e., $\mathbf{R}^3$ with the origin removed. For $\mathcal{M} = \mathbf{R}^3 - \{0\}$, there are noncontractible spheres we can draw around the origin. Thus the space is topologically nontrivial. The two-form F corresponding to (14.96) is

$$F = \frac{1}{2}\epsilon_{ijk} \, g \, \frac{x_k}{|x|^3} dx^i \wedge dx^j \tag{14.97}$$

By direct computation, $d \, F = 0$. (Actually, we find a delta function $\delta^{(3)}(x)$, but this is zero on $\mathcal{M}$ since it does not include the origin.) Integrating F on a sphere around the origin

$$\int_{S^2} F = \int_{S^2} \frac{1}{2}\epsilon_{ijk} \, g \, \frac{x_k}{|x|^3} dx^i \wedge dx^j = g \int_{S^2} \frac{r^2 \sin\theta d\theta d\varphi}{r^2}$$
$$= 4\pi \, g \tag{14.98}$$

Thus F is closed but cannot be exact on $\mathcal{M}$. This also shows that $\mathcal{H}^2(\mathbf{R}^3 - \{0\}) \neq 0$.

The sphere surrounding the origin can be described by two coordinate patches, one covering the northern hemisphere ($x^3 > 0 - \epsilon$) and the other covering the southern hemisphere ($x^3 < 0 + \epsilon$), with a small overlap region of width 2ϵ around the equator. On each of these patches, by Poincaré lemma, we should be able to write F as dA for some A. By direct calculation we find

$$A_N = \; g \, \frac{\epsilon_{abc} n_a x_b}{r(r + n \cdot x)} \, dx^c$$
$$A_S = -g \, \frac{\epsilon_{abc} n_a x_b}{r(r - n \cdot x)} \, dx^c \tag{14.99}$$

where $n_1 = n_2 = 0$, $n_3 = 1$ and $r^2 = x \cdot x$. A_N can be used for the northern hemisphere, it has a singularity at the south pole where $n \cdot x = -r$. Likewise A_S can be used for the southern hemisphere and has a singularity at the north pole. The singularities of these potentials is a reflection of the fact that F is closed but not exact and so cannot give a nonsingular potential. The singularity is at a point (either the south pole or the north pole) and so in the full space $\mathcal{M}$ we have a line of singularities, stretching to infinity, for A_N and A_S considered as separate functions. This line of singularity is called a Dirac string. By using the two potentials, each in its own patch, we can avoid the singularity over all of $\mathcal{M}$.

The two potentials overlap at the equator where $n \cdot x = 0$. In this region we find

$$A_N - A_S = 2g \, \frac{x_1 dx_2 - x_2 dx_1}{(x_1^2 + x_2^2)} = 2g \, d\varphi \tag{14.100}$$

where φ is the azimuthal angle. The difference in the overlap region is a gauge transformation with gauge parameter $\Lambda = 2g \, \varphi$. Thus the physical field $\boldsymbol{B}$ is the same in the overlap region, whether computed using A_N or A_S.

14.5.3 Nonabelian gauge fields

We now turn to nonabelian gauge fields. In this case, the potential is also an element of the Lie algebra, and so we introduce a Lie-algebra-valued one-form A written in local coordinates as

$$A = (-it^a)A_\mu^a dx^\mu \tag{14.101}$$

where t^a are hermitian matrices which form a basis of the Lie algebra of the group. As we have written it, A is antihermitian; this is convenient for many calculations.

The covariant exterior derivative of a function ϕ is given by $D\phi = (d + A)\phi$. Therefore, $D^2\phi = (d + A)(d + A)\phi = (dA + A^2)\phi$ using $d^2 = 0$ and $d(A\phi) = dA\phi - Ad\phi$. (We will not write the wedge sign anymore whenever it is clear from the context.) This calculation shows that we must define the field strength as

$$F = dA + A^2 \tag{14.102}$$

Using the expression for the potential, we may simplify this as

$$\begin{aligned}
F &= \left[\frac{\partial}{\partial x^\mu} (-it^a)A_\nu^a + (-it^b)(-it^c)A_\mu^b A_\nu^c \right] dx^\mu \wedge dx^\nu \\
&= \frac{1}{2} \left[(-it^a)(\partial_\mu A_\nu^a - \partial_\nu A_\mu^a) + [(-it^b),(-it^c)]A_\mu^b A_\nu^c \right] dx^\mu \wedge dx^\nu \\
&= \frac{1}{2}(-it^a) \left[\partial_\mu A_\nu^a - \partial_\nu A_\mu^a + f^{abc} A_\mu^b A_\nu^c \right] dx^\mu \wedge dx^\nu
\end{aligned} \tag{14.103}$$

In the second step we used the antisymmetry of the wedge product. We see that the field strength is indeed what we expect from our previous discussions about gauge fields.

The gauge transformation of the potential by the group element g may be written as

$$A \to A^g = gAg^{-1} - dgg^{-1} \tag{14.104}$$

We can directly check that $F^g = gFg^{-1}$ using (14.102).

The Bianchi identity is now more involved. We find

$$dF = dA\ A - A\ dA$$
$$= (F - A^2)\ A - A\ (F - A^2)$$
$$= F\ A - A\ F \tag{14.105}$$

In components, this works out to be the usual Bianchi identity

$$D_\mu F_{\nu\alpha} + D_\alpha F_{\mu\nu} + D_\nu F_{\alpha\mu} = 0 \tag{14.106}$$

The Yang-Mills action can be written as

$$S = -\frac{1}{2e^2} \int_M F^a \wedge^* F^a$$
$$= \frac{1}{e^2} \int_M \text{Tr}(F \wedge^* F) \tag{14.107}$$

where we have chosen the normalization of the matrices t^a as $\text{Tr}(t^a t^b) = \frac{1}{2}\delta^{ab}$.

As an example of a calculation with differential forms for the nonabelian gauge theory, we consider the quantity

$$\nu[A] = -\frac{1}{8\pi^2} \int_M \text{Tr}(F\ F)$$
$$\equiv \int_M \Omega \tag{14.108}$$

This is known as the instanton number and its relevance to physics will be discussed later. For now, we notice that

$$d\,\Omega = -\frac{1}{8\pi^2} \text{Tr}\,(dF\ F + F\ dF)$$
$$= -\frac{1}{8\pi^2} \text{Tr}\,((FA - AF)\ F + F\ (FA - AF))$$
$$= 0 \tag{14.109}$$

where we have used the cyclicity of trace. (For a matrix-valued k-form α and a p-form β, $\text{Tr}(\alpha\ \beta) = (-1)^{pk}\text{Tr}(\beta\ \alpha)$ as can be checked easily using the local coordinate expressions.) The four-form Ω is thus closed and so, by Poincaré lemma, we should be able to find a three-form K with

$$\Omega = -\frac{1}{8\pi^2}\text{Tr}(F\ F) = dK \tag{14.110}$$

Since K is a three-form built out of d's and A's, it has to be of the form

$$K = a\ \text{Tr}(AdA + b\ A^3) \tag{14.111}$$

where a, b are constants. We then find

$$dK = a \operatorname{Tr} \left(dAdA + b \, dA \, A^2 - b \, AdA \, A + b \, A^2dA \right)$$
$$= a \operatorname{Tr} \left(dAdA + 3b \, dA \, A^2 \right)$$
$$= a \operatorname{Tr} \left((F - A^2)(F - A^2) + 3b \, (F - A^2)A^2 \right)$$
$$= a \operatorname{Tr} \left(F \, F + (3b - 2) \, FA^2 \right) \tag{14.112}$$

(We used $\operatorname{Tr} A^4 = 0$.) Choosing $3b = 2$ and $a = -1/8\pi^2$, we get the result

$$K = -\frac{1}{8\pi^2} \operatorname{Tr} \left(A \, dA + \frac{2}{3} \, A^3 \right)$$
$$= -\frac{1}{8\pi^2} \operatorname{Tr} \left(A \, F - \frac{1}{3} \, A^3 \right) \tag{14.113}$$

K is known as the Chern-Simons three-form.

Notice that, even though Ω is gauge-invariant, K is not. From the gauge transformation (14.104), we find

$$K[A^g] = -\frac{1}{8\pi^2} \operatorname{Tr} \left((A + v)F - \frac{1}{3}(A + v)^3 \right)$$
$$= K[A] - \frac{1}{8\pi^2} \operatorname{Tr} \left(vdA - v^2A - \frac{1}{3}v^3 \right)$$
$$= K[A] + d \left[\frac{1}{8\pi^2} \operatorname{Tr}(vA) \right] + \frac{1}{24\pi^2} \operatorname{Tr}(dg \, g^{-1})^3 \tag{14.114}$$

where $v = dg \, g^{-1}$ and we have used the result $dv = v^2$. The last term is at least cubic in the parameters of g and so, for infinitesimal transformations, K changes by an exact form. The last term can give nontrivial contributions upon integration over three-spaces of nontrivial topology. This will be discussed in more detail in Chapter 16. Notice that, because K is not gauge-invariant, we cannot conclude that the integral of Ω over a closed four-manifold, namely, ν, is zero. Just as in the case of the monopole, if ν is nonzero, K will have singularities and we have to define a separate K for each coordinate patch, with K in different patches related by a gauge transformation on the overlap regions.

14.6 Fiber bundles

Fiber bundles are the proper mathematical notion for introducing internal symmetries in a field theory. Consider a set of fields ϕ^i which take values in some space $\mathcal{F}$ on the spacetime manifold $\mathcal{M}$. Consider a neighborhood U of $\mathcal{M}$. On this the information we need for the theory is given by (U, ϕ^i). We thus have a direct product structure $U \times \mathcal{F}$ with $\phi^i(x) : U \to \mathcal{F}$ as a map from U to $\mathcal{F}$. A fiber bundle is the formalization of this idea.

A fiber bundle over a manifold $\mathcal{M}$ is a manifold $\mathcal{E}$ which is, locally on $\mathcal{M}$, a direct product of $\mathcal{M}$ and $\mathcal{F}$, where $\mathcal{F}$ is also a manifold. Thus U_i are open sets of $\mathcal{M}$, $\mathcal{E} = U_i \times \mathcal{F}$ for each U_i. We denote a fiber bundle by the triplet $(\mathcal{F}, \mathcal{E}, \mathcal{M})$, $\mathcal{F}$ is called the fiber, $\mathcal{E}$ is the bundle, and $\mathcal{M}$ is called the base manifold. Just as the coordinates of U_i are related to the coordinates over U_j by transition functions on the overlap region $U_i \cap U_j$, the fiber over U_i will be related to the fiber over U_j by a set of transition functions Φ_{ij} on the overlap region. This is a reflection of the freedom of overall coordinate changes of $\mathcal{E}$. It may be worth emphasizing that the choice of the space $\mathcal{F}$ is the same over all of $\mathcal{M}$; we cannot change $\mathcal{F}$ from one neighborhood to another.

If $\mathcal{F}$ is a vector space, we say that $\mathcal{E}$ is a vector bundle. The transition functions are then general linear transformations of $\mathcal{F}$ (or a subset of them). If $\mathcal{F}$ is a Lie group G, the bundle is called a principal bundle.

A local section $s(U_i)$ of the bundle is a rule which assigns a point of $\mathcal{F}$ to each point of the neighborhood U_i; in other words, $s(U_i) : U_i \to \mathcal{F}$. This is precisely the notion of the field $\phi^i(x)$.

A bundle is said to be trivial if $\mathcal{E} = \mathcal{M} \times \mathcal{F}$, so that the direct product structure holds globally on $\mathcal{M}$. In this case, sections are globally defined. Effectively the transition functions Φ_{ij} for the fibers can be chosen to be identity. A general result is that a bundle over a contractible space is trivial.

Examples

1. The tensor bundles and bundles of forms

We have earlier introduced the space of tangent vectors at each point of a manifold. Thus we do indeed have the local structure of a direct product of an n-dimensional vector space and U_i for each neighborhood U_i of an n-dimensional manifold $\mathcal{M}$. This defines the tangent bundle $T\mathcal{M}$, a term that we have already introduced. Sections of the tangent bundle are vector fields.

In a similar way, we can introduce the cotangent bundle $T^*\mathcal{M}$, sections of which are covariant vector fields or one-forms. More generally, we can talk of tensor bundles of different ranks and also of bundles of differential forms.

2. Scalar fields on $\mathcal{M}$

Consider complex-valued fields as an example. In this case, $\mathcal{E}$ is locally $U_i \times$ $\mathbf{C}$, sections are complex-valued fields $\phi(x)$, which are in general only locally defined. If we consider two neighborhoods U and V, we need a transition function Φ_{UV} which relates ϕ_U and ϕ_V. Such a transition function could be a phase transformation, for example. This corresponds to the rule

$$\phi_U(x) = e^{i\theta_{UV}(x)} \phi_V(x) \tag{14.115}$$

on the overlap region $U \cap V$. If the bundle is trivial, $\mathcal{E} = \mathcal{M} \times \mathbf{C}$, $\phi(x)$ are globally defined functions on $\mathcal{M}$.

3. The Möbius strip

This is one of the simplest examples of a nontrivial bundle. In this case, $\mathcal{M} = S^1$ and $\mathcal{F} = [-1, 1]$. We describe S^1 using two coordinate patches U and V, corresponding to $-\epsilon < \theta < \pi + \epsilon$ and $\pi - \epsilon < \theta < 2\pi + \epsilon$, respectively, in terms of the usual angular coordinate. There are two overlap regions $R_I = (\pi - \epsilon, \pi + \epsilon)$ and $R_{II} = (-\epsilon, \epsilon)$. $\mathcal{E}$ is given by the two neighborhoods $U \times t_U$ and $V \times t_V$ where t's denote neighborhoods of $\mathcal{F}$. The transition rules are taken as

$$
\begin{aligned}
t_U &= t_V & \text{on } I \\
t_U &= -t_V & \text{on } II
\end{aligned}
\tag{14.116}
$$

There is a change of orientation for $\mathcal{F}$ from U to V on one transition region. A general section s, if we try to extend over all of $\mathcal{M}$, will have a singularity at one point, because the value of s must have opposite signs approaching this point from the two sides. Thus this bundle is nontrivial. (It is also nonorientable.)

Connection on a fiber bundle

Consider a vector bundle over $\mathcal{M}$ where the fiber is some N-dimensional vector space. Sections of this bundle are of the form $\phi^i(x)$ and can be used as an N-component scalar field. Strictly speaking, ϕ^i are the components of the ϕ along some chosen basis vectors, say, e_i in $\mathcal{F}$, with $\phi = \phi^i e_i$. The basis $\{e_i\}$ defines a frame for $\mathcal{F}$. In the case of tangent vectors, we have seen that vectors are properly written as $X = \xi^i(x)e_i$, where

$$
e_i = \frac{\partial}{\partial x^i}
\tag{14.117}
$$

in the coordinate basis. Likewise, one-forms are $\omega_i e^i$, $e^i = dx^i$ in the coordinate basis. The components ξ^i, ω_i transform nontrivially under coordinate changes, but the vectors and one-forms are invariant. In a similar way, the invariant way to write the sections of the vector bundle we are considering is $\phi(x) = \phi^i(x)e_i(x)$. The frames for $\mathcal{F}$ can in general be different at different points on $\mathcal{M}$. The natural question that arises is then how the frames change over the manifold $\mathcal{M}$. Let δe_i be the difference of frames at nearby points with coordinates x^μ and $x^\mu + \delta x^\mu$. δe_i can be expanded in terms of $\{e_i\}$ themselves and it should also be proportional to δx^μ. Therefore

$$
\delta e_i = e_j A^j_{\mu i} \delta x^\mu
\tag{14.118}
$$

This defines a one-form $A^j_{\mu i} dx^\mu$ which is called the connection one-form of the bundle. The derivative of a section ϕ is now given as

$$
\begin{aligned}
d\phi &= de_i \phi^i + e_i d\phi^i \\
&= e_i(\partial_\mu \phi^i + A^i_{\mu j}\, \phi^j)\, dx^\mu \\
&= e_i(D_\mu \phi)^i\, dx^\mu
\end{aligned}
\tag{14.119}
$$

We see the emergence of the covariant derivative; the connection one-form is seen to be the gauge potential. In this way, fiber bundles naturally lead to the idea of gauge theories. (More generally, the frames could also depend on the coordinates of the fiber space $\mathcal{F}$; then we would have $A^i_{Aj}dx^A$, where x^A denote all the coordinates, on the base and on the fiber. This would be the general form of a connection on a bundle.) The field strength F for the connection is referred to as the curvature of the bundle.

There is also arbitrariness in the choice of the frame. We could do an invertible linear transformation $e_i \to e'_i = e_k(g^{-1})_{ki}$; this can be compensated for by a change in the components $\phi^i \to \phi'^i = g^{ij}\phi^j$. The transformation matrix g does not have to be constant over $\mathcal{M}$. Mathematically it only arises from splitting ϕ into components ϕ^i and the basis vectors e_i; the intrinsic quantity ϕ is insensitive to the transformation effected by g. Physically, it is sensible to expect that we can do a local basis transformation; it would be strange if we could not and if we had to align the frames in our local region with frames in a region light-years away before we could properly set up the physics. Physical results must be independent of the local basis transformation. This is the essence of gauge-invariance. The transformation $\phi \to \phi' = g\phi$ is a gauge transformation.

The present discussion highlights the fact that gauge theories have a very deep geometrical origin. We have already discussed the physics of gauge theories in some detail. The concept of fiber bundles clarifies some of our discussion and will also help in understanding the geometrical and topological properties of gauge fields.

Fiber bundles provide the natural setting for all physical fields. Matter fields are sections of various vector bundles over the spacetime manifold, the fiber being complex numbers or spinors of the Lorentz group. Gauge fields are connections on these vector bundles. As we have already discussed earlier in this chapter, connections on the tangent bundle to spacetime lead to Christoffel symbols (or, equivalently, spin connections) and the theory of gravity.

The exact homotopy sequence for fiber bundles

There is an interesting technique for relating the homotopy groups for the fiber, bundle, and base of a fiber bundle which is very useful. This is the exact sequence of homotopy groups which gives a series of maps between the homotopy groups involved; it can be written as

$$\cdots \to \Pi_{k+1}(\mathcal{M}) \to \Pi_k(\mathcal{F}) \to \Pi_k(\mathcal{E}) \to \Pi_k(\mathcal{M}) \to \Pi_{k-1}(\mathcal{F}) \to \cdots$$
$$(14.120)$$

This sequence is exact, which means that the kernel of a particular map, namely, those elements which are mapped to zero or the trivial element, is the image of the previous map. For any given k, the order of the spaces involved is fiber, bundle, base. The sequence is in principle infinite, but by focusing on subsequences, one can obtain useful results on homotopies.

As a simple application of this idea, recall from Chapter 12 on spontaneous symmetry breaking that we can write the two-sphere as a coset space $SU(2)/U(1) = S^3/S^1$. This is equivalent to saying that $SU(2)$ or S^3 is a bundle over S^2 with $U(1) = S^1$ as the fiber. The sequence (14.120) around $k = 3$ becomes

$$\rightarrow \Pi_3(S^1) \rightarrow \Pi_3(S^3) \rightarrow \Pi_3(S^2) \rightarrow \Pi_2(S^1) \rightarrow \qquad (14.121)$$

From what we have said before, $\Pi_2(S^1) = 0$. Thus all elements of $\Pi_3(S^2)$ are mapped to zero. The exactness of the sequence tells us that all the elements of $\Pi_3(S^2)$ must therefore be images of elements of $\Pi_3(S^3)$. Further, since $\Pi_3(S^1)$ is zero, the only element of $\Pi_3(S^3)$ which can come from a previous map is the trivial element. The other elements of $\Pi_3(S^3)$, which are not images of the previous map, cannot be mapped to zero in $\Pi_3(S^2)$. The mapping $\Pi_3(S^3) \rightarrow \Pi_3(S^2)$ must therefore be one-to-one. This shows that $\Pi_3(S^2) = \mathbf{Z}$, since we already know that $\Pi_3(S^3) = \mathbf{Z}$.

Another example of interest is the coset space G/H, where G is simply connected, so that $\Pi_1(G) = 0$. Since $\Pi_2(G) = 0$ as well, we get the result $\Pi_2(G/H) = \Pi_1(H)$. For example, choosing $G = SU(n)$ and $H = U(n-1)$, we get the complex projective space $\mathbf{CP}^{n-1} = SU(n)/U(n-1)$. Since $\Pi_1(U(n-1)) = \mathbf{Z}$, we find $\Pi_2(\mathbf{CP}^{n-1}) = \mathbf{Z}$. There are many examples of the application of the exact sequence, some of which are of interest in physics. We will take up some more cases as the occasion arises.

14.7 Applications of the idea of fiber bundles

14.7.1 Scalar fields around a magnetic monopole

Consider scalar charged fields in the background of a magnetic monopole. The radial dependence of the fields is not important for this discussion, so we will take $\mathcal{M} = S^2$. The fields are sections of a bundle with the fiber being $\mathbf{C}$. We then have the field ϕ_N on the northern hemisphere and ϕ_S on the southern hemisphere. The intersection of these two hemispheres is the equator. The restriction of the fields ϕ_N and ϕ_S to the equator need not be identical, but can differ by a gauge transformation. The gauge parameter of the transformation was already worked out in the last section as $\Lambda(\varphi) = 2g\varphi$, where φ is the azimuthal angle. Since the gauge potentials undergo transformation by Λ, the charged fields must be related by $\phi_N(\varphi) = e^{ie\Lambda(\varphi)} \phi_S(\varphi)$. This means that the transition function $e^{ie\Lambda(\varphi)}$ is an element of $U(1)$, the gauge group. If we take the relation $\phi_N(\varphi) = e^{ie\Lambda(\varphi)} \phi_S(\varphi)$ from $\varphi = 0$ to $\varphi = 2\pi$, thus coming back to the same point on the equator, single-valuedness of the wave function will require

$$\exp\left[ie\left(\Lambda(2\pi) - \Lambda(0)\right)\right] = 1 \qquad (14.122)$$

Using the formula for Λ given after (14.100), we find that this is equivalent to $4\pi e g = 2\pi n$, where n is an integer or

$$e\,g = \frac{n}{2} \qquad (14.123)$$

The product of magnetic and electric charges must be quantized. This is the famous Dirac quantization condition. It follows from the singlevaluedness of the wave function, but its origin is purely topological and not related to the dynamics.

The strength of the monopole g is an integer multiple of $1/2e$. We can therefore consider a monopole of charge $1/2e$ as the fundamental monopole and n as the number of such monopoles which go into making the configuration (14.96).

If $n \neq 0$, the bundle is nontrivial since we cannot take the transition function to be 1. For example, try to define ϕ_N over all of the northern hemisphere, not just the equator, via the relation $\phi_N(\varphi) = e^{ie\Lambda(\varphi)}\,\phi_S(\varphi)$. As we take this to the north pole, we see that the function $e^{ie\Lambda(\varphi)}$ becomes singular when $n \neq 0$. Thus we will not be able to make the transition function equal to 1. This nontriviality of the bundle is measured by

$$n = \int_{S^2} \frac{eF}{2\pi} \qquad (14.124)$$

14.7.2 Gribov ambiguity

In our discussion of gauge fixing and the Fadeev-Popov procedure for functional quantization of gauge theories, we have seen that there could be the problem of Gribov ambiguity. This refers to the fact that there could be different field configurations which obey the same gauge-fixing condition, but which are related by a gauge transformation. This means that we are unable to find a "good gauge fixing" where the gauge-fixing condition chooses one and only one representative configuration for all potentials which are gauge equivalent. This Gribov ambiguity can be given a precise mathematical characterization in the language of fiber bundles.

We start by recalling some concepts defined in Chapter 10 on gauge theories. Let $\mathcal{A}$ denote the set of gauge potentials, i.e., the set of all Lie-algebra-valued one-forms on $\mathbf{R}^4$ which may be taken to obey the condition of finiteness for the Yang-Mills action. A point on this space is a particular gauge potential on $\mathbf{R}^4$. Further let $\mathcal{G}_*$ denote the set of gauge transformations on Euclidean four-dimensional space,

$$\mathcal{G}_* = \left\{ \text{set of all } g(x) \text{ such that } g(x) \to 1 \text{ as } \sqrt{x^\mu x_\mu} \to \infty \right\}$$

Because of the boundary condition, these transformations are topologically equivalent to transformations on S^4. The gauge-invariant set of configurations is given by $\mathcal{C} = \mathcal{A}/\mathcal{G}_*$. This is the space we are interested in. A neighborhood of $\mathcal{C}$ is a set of field configurations which are physically distinct. A set of configurations in $\mathcal{A}$ may be thought of as being given by a set of representative configurations which are points on $\mathcal{C}$ and the set of gauge transformations one can perform on them. Thus we have a natural splitting

$$\mathcal{A} = \mathcal{C} \times \mathcal{G}_* \qquad (14.125)$$

for any neighborhood in $\mathcal{C}$. This leads to the structure that $\mathcal{A}$ is a fiber bundle over $\mathcal{C}$ with $\mathcal{G}_*$ as the fiber. This bundle need not be trivial, so the splitting (14.125) is not globally true.

We can think of gauge fixing as follows. When we do gauge fixing, we choose a representative potential A (obeying some gauge fixing condition) for each physical configuration. Thus we are specifying the physical configuration C and a gauge transformation g_C associated to it which takes it into A. We have an assignment of a point on the fiber, namely, g_C, for each point C in $\mathcal{C}$. In other words, gauge fixing is the choice of a section for the bundle $\mathcal{A}$. If we can choose a section globally, then we have the splitting $\mathcal{A} = \mathcal{C} \times \mathcal{G}_*$ globally. There is no problem with gauge fixing and no Gribov ambiguity. The existence of the Gribov problem is thus equivalent to the statement that the bundle $(\mathcal{G}_*, \mathcal{A}, \mathcal{C})$ is nontrivial and does not have a global section. This gives a precise characterization of the Gribov problem as a topological property of the bundle of gauge potentials.

Our argument so far does not prove that there is a Gribov problem; to do that, we must prove that the bundle $(\mathcal{G}_*, \mathcal{A}, \mathcal{C})$ is nontrivial. For this we need some information about the topology of the spaces involved. As we have argued in Chapter 10, the space $\mathcal{A}$ has the property that any two points $A_{1\mu}(x)$ and $A_{2\mu}(x)$ can be connected by a straight line. This line in $\mathcal{A}$ is given by

$$A_\mu(\tau, x) = \tau \, A_{1\mu}(x) + (1 - \tau) \, A_{2\mu}(x) \qquad (14.126)$$

for $0 \le \tau \le 1$. Notice that the intermediate configurations $A_\mu(\tau, x)$ transform, for all τ, as a gauge potential is expected to transform; viz., we have the transformation rule $A_\mu^g(\tau, x) = \tau \, A_{1\mu}^g(x) + (1 - \tau) \, A_{2\mu}^g(x)$. $\mathcal{A}$ is thus an affine space; i.e., we can write any configuration $A_\mu(x) = A_\mu^{(0)} + \xi_\mu$ where $\xi_\mu(x)$ is a Lie-algebra-valued vector field. This shows that $\mathcal{A}$ is a contractible space. In other words, its homotopy groups are trivial.

If $\mathcal{A}$ is of the form $\mathcal{C} \times \mathcal{G}_*$ globally, then the factor spaces $\mathcal{C}$ and $\mathcal{G}_*$ must also be contractible. The strategy of proving the Gribov ambiguity is then to show that the factor spaces are not contractible, so that by *reductio ad absurdum*, we can conclude that the bundle $(\mathcal{G}_*, \mathcal{A}, \mathcal{C})$ is nontrivial.

Consider a one-parameter sequence of gauge transformations given by $g(s, x)$, where s is a real variable, with $g(s, x) \to 1$ as $s \to \pm\infty$. This is a closed curve in $\mathcal{G}_*$ starting at a point corresponding to the identity transformation (at $s = -\infty$) and ending back there. Thus homotopy classes of such closed curves in $\mathcal{G}_*$ will give $\Pi_1(\mathcal{G}_*)$. We can also think of the map $g(s, x)$ as a map from $\mathbf{R}^5$ to the group G with the boundary condition that g goes to 1 at infinity in any direction. Such maps are equivalent to maps from S^5 to G. The homotopy classes of such maps are then given by $\Pi_5(G)$. We therefore have the result that $\Pi_1(\mathcal{G}_*) = \Pi_5(G)$. Since $\Pi_5(SU(n)) = \mathbf{Z}$ for all $n \geq 3$, this immediately shows that $\mathcal{G}_*$ is not contractible for $SU(n)$, $n \geq 3$, proving the existence of the Gribov problem for such gauge groups. If the gauge group is $SU(2)$, there is the result that $\Pi_4(SU(2)) = \mathbf{Z}_2$; this leads, via a similar argument, to $\Pi_0(\mathcal{G}_*) = \mathbf{Z}_2$, proving the Gribov problem for this group as well.

The exact sequence of homotopy groups can be used to make some more interesting statements about the bundle $(\mathcal{G}_*, \mathcal{A}, \mathcal{C})$. If we choose $k = 2$, the sequence gives

$$\rightarrow \Pi_2(\mathcal{A}) \rightarrow \Pi_2(\mathcal{C}) \rightarrow \Pi_1(\mathcal{G}_*) \rightarrow \Pi_1(\mathcal{A}) \rightarrow \qquad (14.127)$$

Since $\mathcal{A}$ is homotopically trivial, this shows that $\Pi_2(\mathcal{C}) = \mathbf{Z}$ for gauge groups $SU(n)$, $n \geq 3$. For $SU(2)$ one can similarly obtain $\Pi_1(\mathcal{C}) = \mathbf{Z}_2$.

If we do a similar analysis in three dimensions, then, by arguments similar to those given above, we can relate $\Pi_0(\mathcal{G}_*)$ to $\Pi_3(G)$. Since the latter is $\mathbf{Z}$ for compact Lie groups G, (except for $SO(4)$ for which one has $\mathbf{Z} \times \mathbf{Z}$), we get the result that $\Pi_1(\mathcal{C}) = \mathbf{Z}$ ($\mathbf{Z} \times \mathbf{Z}$ for $SO(4)$), where $\mathcal{C}$ is the gauge-invariant configuration space of three-dimensional gauge potentials. This space $\mathcal{C}$ also corresponds to the configuration space at fixed time in the $A_0 = 0$ gauge of the four-dimensional theory, which is appropriate for a Hamiltonian analysis of the four-dimensional theory. The result $\Pi_1(\mathcal{C}) = \mathbf{Z}$ is then related to the existence of instantons.

14.8 Characteristic classes

The Dirac magnetic monopole is an example of a nontrivial fiber bundle. The nontriviality of the bundle resides in the fact that F is closed but not exact. We have also seen from the discussion of scalar fields on a monopole background that the nontriviality of the bundle is measured by $\int F/2\pi$. Characteristic classes generalize this idea. We measure or characterize the nontriviality of a bundle in terms of a set of differential forms constructed from the curvature F (or the Riemann curvature $\mathcal{R}$ in the case of gravity). There are a number of characteristic classes which are relevant for different questions of topological nature. We shall list some of them here.

Chern class
    This is relevant for a complex vector bundle on a manifold M. Let F be the curvature of the bundle. The Chern classes are then defined by

$$c(F) = \det\left(1 + i\frac{F}{2\pi}\right)$$

$$= 1 + c_1(F) + c_2(F) + \cdots \qquad (14.128)$$

c_1 is callled the first Chern class; c_2 is the second Chern class and so on. Using

$$\det(1 + iX) = \exp[\mathrm{Tr}\log(1 + iX)]$$
$$= 1 + \mathrm{Tr}(iX) + \frac{1}{2}\left[\mathrm{Tr}iX\ \mathrm{Tr}iX - \mathrm{Tr}(iX)^2\right] + \cdots,$$

$$(14.129)$$

we can work out the c_i's. F is a Lie algebra valued two-form and in expanding the formula (14.128), the wedge product and matrix product are understood. We thus get

$$c_1(F) = \frac{i}{2\pi}\ \mathrm{Tr}F$$

$$c_2(F) = \frac{1}{8\pi^2}\left[\mathrm{Tr}(F \wedge F) - (\mathrm{Tr}F) \wedge (\mathrm{Tr}F)\right] \qquad (14.130)$$

For an $SU(N)$ gauge field, $\mathrm{Tr}F = 0$, so that

$$c_2(F) = \frac{1}{8\pi^2}\ \mathrm{Tr}(F \wedge F) \qquad (14.131)$$

$\det(1+iX)$, for arbitrary X in the Lie algebra of a group is the generating function for the Casimir invariants; thus, the Chern classes are all seen to be proportional to the Casimir invariants of the Lie algebra. This is the Weil homomorphism.

Chern character

This class also pertains to a complex vector bundle. The Chern character $Ch(F)$ is defined by

$$Ch(F) = \mathrm{Tr}\exp\left(i\frac{F}{2\pi}\right)$$
$$= 1 + Ch_1(F) + Ch_2(F) + \cdots \qquad (14.132)$$

$Ch_1(F)$ coincides with $c_1(F)$, while

$$Ch_2(F) = -c_2(F) + \frac{1}{2}c_1^2 \qquad (14.133)$$

$\hat{A}-$ genus

Here we consider a background gravitational field with a Riemann curvature two-form $\mathcal{R}$ which also take values in the Lie algebra of $SO(2n)$ for a $2n$-dimensional manifold. The generating function for the $\hat{A}$-genus is then given by

$$\hat{A}(\mathcal{R}) = \prod_i \frac{x_i/2}{\sinh x_i/2} \qquad (14.134)$$

where the x_i's are defined by

$$
\frac{\mathcal{R}}{2\pi} =
\begin{bmatrix}
0 & x_1 & 0 & 0 & . & . \\
-x_1 & 0 & 0 & 0 & . & . \\
0 & 0 & 0 & x_2 & . & . \\
0 & 0 & -x_2 & 0 & . & . \\
. & . & . & . & . & .
\end{bmatrix}
\tag{14.135}
$$

The idea of this representation is that invariant polynomials of $\mathcal{R}$ can be expressed in terms of sums of products of x_i's. The interpretation of (14.134) is that we expand it and re-express the sums of products of x_i's involved in terms of traces of products of $\mathcal{R}/2\pi$ using (14.135). Expanding (14.134) to the lowest nontrivial order

$$
\begin{aligned}
\hat{A}(\mathcal{R}) &= \prod_i \frac{x_i/2}{\sinh x_i/2} = \prod_i \left(1 - \frac{1}{24}x_i^2 + \cdots\right) \\
&= \left(1 - \frac{1}{24}\sum_i x_i^2 + \cdots\right) \\
&= 1 + \frac{1}{24}\frac{1}{8\pi^2}\mathrm{Tr}(\mathcal{R} \wedge \mathcal{R}) + \cdots
\end{aligned}
\tag{14.136}
$$

Euler class

The Euler class is zero for an odd-dimensional manifold. For an even-dimensional manifold it is defined as follows. Consider a real $(2n \times 2n$-matrix M. Its determinant is a perfect square, so that we can write $\det M = (e)^2$. This defines the Pfaffian $e(M)$. The Euler class is now given by substituting $\mathcal{R}/2\pi$ for M with wedge products of the $\mathcal{R}$'s taken. For a 2×2 matrix, we can bring it to the form

$$
M = \begin{pmatrix} 0 & x \\ -x & 0 \end{pmatrix}
\tag{14.137}
$$

to evaluate $\det M = \frac{1}{4}(\epsilon_{ab}M^{ab})^2$, giving

$$
e(\mathcal{M}) = \frac{1}{4\pi}\epsilon_{ab}\mathcal{R}^{ab}
\tag{14.138}
$$

In four dimensions, we start with

$$
M = \begin{pmatrix}
0 & x_1 & 0 & 0 \\
-x_1 & 0 & 0 & 0 \\
0 & 0 & 0 & x_2 \\
0 & 0 & -x_2 & 0
\end{pmatrix}
\tag{14.139}
$$

with $\det M = (\epsilon_{abcd}M^{ab}M^{cd})^2/64$. This gives

$$
e(\mathcal{M}) = \frac{1}{32\pi^2}\epsilon_{abcd}\mathcal{R}^{ab}\mathcal{R}^{cd}
\tag{14.140}
$$

The Euler number of a closed compact manifold was defined earlier as

$$\chi(\mathcal{M}) = \sum_k (-1)^k b_k \qquad (14.141)$$

where b_k are the Betti numbers, the dimensions of the cohomology groups of $\mathcal{M}$ with real coefficients. The Gauss-Bonnet theorem is the statement that this is given by the integral of $e(\mathcal{M})$ over $\mathcal{M}$. Thus for dimensions 2 and 4,

$$\chi(\mathcal{M}) = \frac{1}{4\pi} \int_{\mathcal{M}} \epsilon_{ab} \mathcal{R}^{ab}, \qquad n = 2$$

$$= \frac{1}{32\pi^2} \int_{\mathcal{M}} \epsilon_{abcd} \mathcal{R}^{ab} \mathcal{R}^{cd}, \qquad n = 4 \qquad (14.142)$$

There are many other characteristic classes, such as the Todd class, the Hirzebruch polynomial, etc., which are of interest in the context of index theorems and anomalies. We do not give them here since we do not discuss any application involving them.

The characteristic classes consist of differential forms built from products of two-forms; for a manifold of dimension $2n$, the terms in these expansions beyond rank $2n$ differential forms are obviously zero. One of the places where the topological properties embodied in the characteristic classes arise naturally is in the context of index theorems related to zero modes of physically interesting operators like the Dirac operator. This will be discussed in Chapter 17.

References

1. The topics here are mostly standard from the mathematical point of view. We give a general set of references.
   a) A.P. Balachandran, *Classical Topology and Quantum States*, World Scientific Pub. Co. (1991).
   b) The mathematical part of R.A. Bertlmann, *Anomalies in Quantum Field Theory*, Clarendon Press (1996).
   c) S.S. Chern, *Complex Manifolds without Potential Theory*, Springer-Verlag (1979).
   d) T. Eguchi, P. Gilkey and A. Hanson, Phys. Rep. **66C**, 213 (1980).
   e) S.I. Goldberg, *Curvature and Homology*, Dover Publications, Inc. (1962 & 1982).
   f) C. Nash and S. Sen, *Topology and Geometry for Physicists*, Academic Press (1988).
   g) N. Steenrod, *The Topology of Fibre Bundles*, Princeton University Press (1951 & 1995).
   h) Michael Stone, *Methods of Mathematical Physics*, a set of lecture notes, http://w3.physics.uiuc.edu/ m-stone5/

2. Any book on general relativity will give the Einstein equations; see, for example, S. Weinberg, *Gravitation and Cosmology*, Wiley Text Books (1972).

3. Lists of cohomology and homotopy groups can be found in *Encyclopedic Dictionary of Mathematics*, S. Iyanaga and Y. Kawada (eds.), MIT Press (1980).

4. The Dirac monopole comes from P.A.M. Dirac, Proc. Roy. Soc. **A133**, 60 (1931).

5. The Gribov problem was identified by V.N. Gribov, Nucl. Phys. **B139**, 1 (1978). The argument we present is from I.M. Singer, Commun. Math. Phys. **60**, 7 (1978). For a general argument for all four-dimensional spin manifolds, see T. Killingback and E.J. Rees, Class. Quant. Grav. **4**, 357 (1987).

15 Path Integrals

15.1 The evolution kernel as a path integral

In this chapter, we will consider the path integral representation of time-evolution in quantum theory.

Consider a state $|i\rangle$ at time t'. Time-evolution of this state is given by the operator $e^{-iH(t-t')}$. Denoting the time-evolved state by $|\psi\rangle$, we can write

$$|\psi, t\rangle = e^{-iH(t-t')} |i, t'\rangle \tag{15.1}$$

where H is the Hamiltonian. The scalar product or overlap integral $\langle f|\psi, t\rangle$ gives the amplitude for the time-evolution of the state $|i, t'\rangle$ by H (which may contain interactions) to generate $|f\rangle$. Thus the transition amplitude is given by $\langle f|e^{-iH(t-t')}|i\rangle$. Usually, we consider transitions to happen over an infinite interval of time and the preparation of $|i\rangle$ to be in the distant past and the measurement or detection of $|f\rangle$ to be in the distant future. This is an idealization of the usual case where the measurements are done before the interactions under study become effective and after they cease to be effective. In this idealization, we can define the S-matrix element by

$$S_{fi} = \lim_{t \to \infty} \lim_{t' \to -\infty} \langle f|e^{-iH(t-t')}|i\rangle \tag{15.2}$$

We shall derive the path integral formulae for (15.1) and (15.2), first for a quantum mechanical system and then the generalization to field theory.

Consider a three-dimensional quantum mechanical system, with q_i, p_i, $i = 1, 2, 3$, being the coordinates and momenta, respectively. The Hamiltonian is taken to be

$$H = \frac{1}{2}p^2 + V(q) \tag{15.3}$$

(We absorb the mass into the momenta.) For this system, we can choose a set of q-diagonal states $|q\rangle$ with the orthogonality and completeness relations

$$\langle q|q'\rangle = \delta^{(3)}(q - q'), \qquad \int |q\rangle \, d^3q \, \langle q| = 1 \tag{15.4}$$

By taking the scalar product of (15.1) with $\langle q|$ and also using the completeness relation, we get

$$\langle q | \psi, t \rangle = \int d^3 q' \; \langle q | e^{-iH(t-t')} | q' \rangle \langle q' | i, t' \rangle \tag{15.5}$$

We can write this as

$$\Psi(q, t) = \int d^3 q' \; K(q, t, q', t') \Psi_i(q', t') \tag{15.6}$$

We have used the definition of the wave functions, $\langle q | \psi, t \rangle = \Psi(q, t)$ and $\langle q' | i \rangle = \Psi_i(q')$. The evolution kernel $K(q, t, q', t')$ is given by

$$K(q, t, q', t') = \langle q | e^{-iH(t-t')} | q' \rangle \tag{15.7}$$

Evidently the S-matrix element is given by

$$S_{fi} = \lim_{t \to \infty} \lim_{t' \to -\infty} \int d^3 q d^3 q' \; \Psi_f^*(q) K(q, t, q', t') \Psi_i(q') \tag{15.8}$$

We now write a path integral representation for $K(q, t, q', t')$. Divide the time-interval $(t - t')$ into N steps, each of length ϵ, so that $N\epsilon = t - t'$; eventually we shall take $\epsilon \to 0$, $N \to \infty$, keeping $N\epsilon$ fixed. We label the intermediate times by t_k, with $t_0 = t'$, $t_N = t$ and denote $t_k - t_{k-1} = \epsilon_k$. We can then write

$$K(q, t, q', t') = \langle q | e^{-iH\epsilon_N} e^{-iH\epsilon_{N-1}} \cdots e^{-iH\epsilon_1} | q' \rangle \tag{15.9}$$

Inserting the completeness relation of (15.4) between every two factors of $e^{-iH\epsilon}$, we get

$$K(q, t, q', t') = \int d^3 q_{N-1} \cdots d^3 q_1 \; \langle q | e^{-iH\epsilon_N} | q_{N-1} \rangle \langle q_{N-1} | e^{-iH\epsilon_{N-1}} | q_{N-2} \rangle$$
$$\cdots \langle q_1 | e^{-iH\epsilon_1} | q' \rangle \tag{15.10}$$

In order to evaluate this, we need the matrix element $\langle q_k | e^{-iH\epsilon_k} | q_{k-1} \rangle$. The difficulty in evaluating this is primarily due to the fact that the Hamiltonian has both p- and q-dependent terms, and they are in the exponent of this matrix element. We begin by separating them. Using the formula $e^{A+B} = e^A e^B e^{-\frac{1}{2}[A,B]} \cdots$, we can write

$$e^{-i\epsilon(p^2/2 + V(q))} = e^{-i\epsilon V(q)} e^{-i\epsilon p^2/2} e^{-\frac{1}{2}\epsilon^2[p^2/2, V(q)]} \cdots \tag{15.11}$$

The ellipsis refers to terms of order ϵ^3 which are not displayed. Since ϵ goes to zero eventually, one can neglect the term of order ϵ^2. The term of order ϵ^2 goes like $i\epsilon^2(pF + Fp)/4$, where $F = -\frac{\partial V}{\partial q}$ is the force. Since $p\epsilon$ is the distance traveled by a particle in time ϵ, we see that this term is roughly ϵ times the kinetic energy transferred to the particle in time ϵ by the potential. In most situations, where the potential is not too singular, this will go to zero as $\epsilon \to 0$. This is the case we shall consider. (There could be exceptional situations where the operator $Fp + pF$ has divergent matrix elements and this argument will have to be modified.) In the case where we can neglect the terms of order ϵ^2, we can write

$$\langle q_k|e^{-iH\epsilon_k}|q_{k-1}\rangle = \langle q_k|e^{-i\epsilon_k V(q)}e^{-i\epsilon_k p^2/2}|q_{k-1}\rangle + \mathcal{O}(\epsilon^2)$$
$$= e^{-i\epsilon_k V(q_k)}\langle q_k|e^{-i\epsilon_k p^2/2}|q_{k-1}\rangle + \mathcal{O}(\epsilon_k^2) \quad (15.12)$$

The matrix element involving the momentum operator can be evaluated by using a complete set of momentum eigenstates.

$$\langle q_k|e^{-i\epsilon_k p^2/2}|q_{k-1}\rangle = \int \frac{d^3 p_k}{(2\pi)^3}\langle q_k|p_k\rangle\langle p_k|e^{-i\epsilon_k p^2/2}|q_{k-1}\rangle$$
$$= \int \frac{d^3 p_k}{(2\pi)^3} e^{-i\epsilon_k p_k^2/2}\langle q_k|p_k\rangle\langle p_k|q_{k-1}\rangle$$
$$= \int \frac{d^3 p_k}{(2\pi)^3} e^{ip_k(q_k-q_{k-1})}e^{-i\epsilon_k p_k^2/2} \quad (15.13)$$

The momentum of the eigenstates used has been specified with the subscript k, since this pertains to the matrix element between q_k, q_{k-1}. The integrand is oscillatory and the integral can be evaluated by giving ϵ a small imaginary part. Equivalently, we can define the integral by

$$\int \frac{d^3 p_k}{(2\pi)^3} e^{ip_k(q_k-q_{k-1})}e^{-i\epsilon_k p_k^2/2} = \left[\int \frac{d^3 p_k}{(2\pi)^3} e^{ip_k(q_k-q_{k-1})}e^{-\tilde{\epsilon}_k p_k^2/2}\right]_{\tilde{\epsilon}_k=i\epsilon_k}$$
$$= \left[\frac{1}{(2\pi\tilde{\epsilon}_k)^{\frac{3}{2}}}e^{-(q_k-q_{k-1})^2/2\tilde{\epsilon}_k}\right]_{\tilde{\epsilon}_k=i\epsilon_k}$$
$$= \frac{1}{(2\pi i\epsilon_k)^{\frac{3}{2}}}e^{i(q_k-q_{k-1})^2/2\epsilon_k} \quad (15.14)$$

We define a path $q(t)$ by specifying the values of q at various times as follows: $q(t') \equiv q_0 = q'$, $q(t_1) \equiv q_1, \cdots, q(t_k) \equiv q_k, \cdots, q(t) \equiv q_N = q$. This is defined by a sequence of straight line segments and is only continuous in general. The quantity $(q_k - q_{k-1})/\epsilon_k$ becomes the velocity or tangent to the path as $\epsilon_k \to 0$. Denoting this by $\dot{q}_k$, we can write

$$\langle q_k|e^{-iH\epsilon_k}|q_{k-1}\rangle = \frac{1}{(2\pi i\epsilon_k)^{\frac{3}{2}}}\exp\left[i\left(\tfrac{1}{2}\dot{q}_k^2 - V(q_k)\right)\epsilon_k\right] \quad (15.15)$$

Using this result in (15.10) and taking the limit $\epsilon \to 0$ so as to eliminate errors due to the neglect of $\mathcal{O}(\epsilon^2)$-terms,

$$K(q,t,q',t') = \lim_{N\to\infty,\epsilon\to 0} \int \prod_{k=1}^{N-1} \frac{d^3 q_k}{(2\pi i\epsilon_{k+1})^{\frac{3}{2}}}\frac{1}{(2\pi i(t_1 - t'))^{\frac{3}{2}}}$$
$$\times e^{i\sum_{k=1}^{N}\left[\frac{1}{2}\dot{q}_k^2 - V(q_{k-1})\right]\epsilon_k}$$

$$= \lim_{N \to \infty, \epsilon \to 0} \int \prod_{k=1}^{N-1} \frac{d^3 q_k}{(2\pi i \epsilon_{k+1})^{\frac{3}{2}}} \frac{1}{(2\pi i (t_1 - t'))^{\frac{3}{2}}} e^{iS[q,t,q',t']}$$

(15.16)

where we have defined

$$S[q,t,q',t'] = \sum_{k=1}^{N} \left[\frac{1}{2} \dot{q}_k^2 - V(q_k) \right] \epsilon_k$$

$$= \sum_{k=1}^{N} \left[\frac{(q_k - q_{k-1})^2}{2(t_k - t_{k-1})} - V(q_k)(t_k - t_{k-1}) \right]$$

(15.17)

This has the following interpretation. $q(t)$ as given by the values of q_k defines a path from q' to q in time $(t - t')$. As we vary $q_k = q(t_k)$, we are considering different paths. The integration over all q_k is thus equivalent to summing or integrating over all paths connecting q' and q in the time-interval $(t - t')$. The weight factor for each path (or the amplitude for each path) is $\exp(iS[q,t,q',t'])$. The quantity $S[q,t,q',t']$ is the classical action for the specified path connecting (q',t') and (q,t), since $\mathcal{L} = \frac{1}{2} \dot{q}^2 - V(q)$ is the Lagrangian corresponding to the Hamiltonian $H = \frac{1}{2} p^2 + V(q)$. The sum in (15.17) may be considered as the evaluation of the integral $\int dt \mathcal{L}$ by a discretization procedure. Define e^{iS_P} as the amplitude for the particle to go from (q',t') to (q,t) along a path P, S_P being the classical action for the path. Expression (15.16) then tells us that the total amplitude for a particle to go from q' at time t' to q at time t is given by the sum (or integral) of the individual amplitudes e^{iS_P} over all paths. The summation or integration over all paths is to be defined by a discretization procedure and involves the measure

$$[\mathcal{D}q] = \lim_{N \to \infty, \epsilon \to 0} \prod_{k=1}^{N-1} \frac{d^3 q_k}{(2\pi i \epsilon_{k+1})^{\frac{3}{2}}}$$

(15.18)

The integration over all paths, because of the oscillatory nature of the amplitude, has to be defined for imaginary time and the result is to be continued to real time. In other words

$$\int [\mathcal{D}q] e^{iS[q,t,q',t']} = \left[\int [\mathcal{D}q] e^{-S_E(q,\tau,q',\tau')} \right]_{\tau = it}$$

(15.19)

where S_E is the Euclidean or imaginary time action defined by

$$S_E = \int d\tau \left[\frac{1}{2} \left(\frac{dq}{d\tau} \right)^2 + V(q) \right]$$

(15.20)

15.2 The Schrödinger equation

For a point particle in three dimensions, the path integral gives

$$\Psi(q_N, t_N) = \int \prod_{i=0}^{N-1} \frac{d^3 q_{N-i-1}}{[2\pi i(t_{N-i} - t_{N-i-1})]^{\frac{3}{2}}} \; e^{i\mathcal{S}} \; \Psi(q_0, t_0) \qquad (15.21)$$

where

$$\mathcal{S} = \sum_{i=0}^{N-1} \left[\frac{(q_{N-i} - q_{N-i-1})^2}{2(t_{N-i} - t_{N-i-1})} - V(q_{N-i})(t_{N-i} - t_{N-i-1}) \right] \qquad (15.22)$$

Differentiation of (15.21) yields

$$i\frac{\partial \Psi(q_N, t_N)}{\partial t_N} = \int \left[\frac{1}{2}\left(\frac{q_N - q_{N-1}}{t_N - t_{N-1}} \right)^2 + V(q_N) \right] e^{i\mathcal{S}} \Psi(q_0, t_0)$$
$$- i\frac{3}{2} \frac{1}{(t_N - t_{N-1})} \Psi(q_N, t_N) \qquad (15.23)$$

From (15.21), we also get

$$\frac{\partial \Psi(q_N, t_N)}{\partial q_N^i} = \int i \frac{(q_N - q_{N-1})^i}{(t_N - t_{N-1})} e^{i\mathcal{S}} \Psi(q_0, t_0)$$
$$- i(t_N - t_{N-1}) \frac{\partial V}{\partial q_N^i} \Psi(q_N, t_N) \qquad (15.24)$$

Differentiating once again,

$$\frac{\partial^2 \Psi(q_N, t_N)}{\partial q_N^2} = i\frac{3}{(t_N - t_{N-1})} \Psi(q_N, t_N) - \int \frac{(q_N - q_{N-1})^2}{(t_N - t_{N-1})^2} e^{i\mathcal{S}} \Psi(q_0, t_0)$$
$$- i(t_N - t_{N-1}) \left[\frac{\partial^2 V}{\partial q_N^2} \Psi(q_N, t_N) + \frac{\partial V}{\partial q_N} \frac{\partial \Psi(q_N, t_N)}{\partial q_N} \right]$$
$$+ (q_N - q_{N-1})^i \frac{\partial V}{\partial q_N^i} e^{i\mathcal{S}} \Psi(q_0, t_0) \qquad (15.25)$$

Comparing (15.23, 15.25), we get, as $(t_N - t_{N-1}) \to 0$,

$$i\frac{\partial \Psi(q_N, t_N)}{\partial t_N} = \left[-\frac{1}{2}\frac{\partial^2}{\partial q_N^2} + V(q_N) \right] \Psi(q_N, t_N) \qquad (15.26)$$

Thus the wave function as defined by the path integral, not surprisingly, obeys the Schrödinger equation.

15.3 Generalization to fields

In this section, we extend the path integral to field theories. Consider a scalar field theory described by the Lagrangian

$$\mathcal{L} = \tfrac{1}{2}\left[\dot{\varphi}^2 - (\nabla\varphi)^2 - m^2\varphi^2\right] - V(\varphi) \qquad (15.27)$$

We have

$$\pi \equiv \dot{\phi} \qquad (15.28)$$

$$\mathcal{H} \equiv \int d^3x \, (\pi\dot{\phi} - \mathcal{L})$$

$$= \int d^3x \, \left[\tfrac{1}{2}\pi^2 + \tfrac{1}{2}(\nabla\phi)^2 + \tfrac{1}{2}m^2\phi^2 + V(\phi)\right] \qquad (15.29)$$

where ϕ is the operator corresponding to φ. In discussing this theory in terms of a many-particle interpretation, we have used the expansion

$$\phi(x) = \sum_k [a_k u_k(x) + a_k^\dagger u_k^*(x)] \qquad (15.30)$$

where $u_k(x) = e^{-ikx}/\sqrt{2\omega_k V}$, $kx = k_0 t - \boldsymbol{k}\cdot\boldsymbol{x} = \omega_k t - \boldsymbol{k}\cdot\boldsymbol{x}$, $\omega_k = \sqrt{\boldsymbol{k}^2 + m^2}$. The above relation can be inverted as follows:

$$a_k = i \int d^3x \, [u_k^*(x)\partial_0\phi(x) - (\partial_0 u_k^*)\phi(x)]$$

$$\equiv i \int d^3x \, \left[u_k^*(x)\overleftrightarrow{\partial}_0\phi(x)\right]$$

$$a_k^\dagger = -i \int d^3x \, [u_k(x)\partial_0\phi(x) - (\partial_0 u_k)\phi(x)]$$

$$\equiv -i \int d^3x \, \left[u_k(x)\overleftrightarrow{\partial}_0\phi(x)\right] \qquad (15.31)$$

The integrals are taken at fixed time.

We want to define the states $|\varphi\rangle$ which are eigenstates of the field operator ϕ at some chosen time. Toward this, consider the free theory first. At $t = 0$ define

$$\varphi_k = \int d^3x \, \frac{1}{\sqrt{V}} \, e^{i\boldsymbol{k}\cdot\boldsymbol{x}} \, \varphi(x)$$

$$\pi_k = \int d^3x \, \frac{1}{\sqrt{V}} \, e^{i\boldsymbol{k}\cdot\boldsymbol{x}} \, \pi(x) \qquad (15.32)$$

The canonical commutation rules can be written in terms of the Fourier components (15.32) as

$$[\varphi_k, \varphi_l] = 0$$
$$[\pi_k, \pi_l] = 0$$
$$[\varphi_k, \pi_l] = i \, \delta_{k+l,0} \qquad (15.33)$$

The situation is similar to ordinary quantum mechanics, with a countable infinity of dynamical variables $\varphi_k, \pi_k, \varphi_k$ being the analogue of the coordinates and π_k being the conjugate canonical momenta. Since the φ_k commute among themselves, we can define φ-diagonal states as states which diagonalize the φ_k's. Going back to (15.31), we see that

$$
\begin{aligned}
a_k &= \quad i(\pi_k - i\omega_k\, \varphi_k) \\
a_k^\dagger &= -i(\pi_{-k} + i\omega_k\varphi_{-k})
\end{aligned}
\tag{15.34}
$$

Diagonalizing φ_k is equivalent to diagonalizing $(a_k + a_k^\dagger)$, which is different from diagonalizing the number operator $a_k a_k^\dagger$. The many-particle states, which are also eigenstates of the free Hamiltonian, can be represented in terms of wave functions which diagonalize the φ_k's. For example, the wave function of a state with n_1 particles of momentum k_1, n_2 particles of momentum k_2, etc., can be written as

$$
\Psi_{n_1,n_2,\ldots,n_r}[\varphi] = \langle\varphi|n_1,k_1,n_2,k_2,\ldots,n_r,k_r\rangle = \langle\varphi|\frac{a_{k_1}^\dagger{}^{n_1}}{\sqrt{n_1!}}\frac{a_{k_2}^\dagger{}^{n_2}}{\sqrt{n_2!}}\cdots\frac{a_{k_r}^\dagger{}^{n_r}}{\sqrt{n_r!}}|0\rangle
\tag{15.35}
$$

On such a wave function, the π_k's act as differential operators. From the commutation rules (15.33), we see that

$$
\pi_k\Psi[\varphi] = -i\frac{\delta}{\delta\varphi_{-k}}\Psi[\varphi]
\tag{15.36}
$$

The ground-state wave function satisfies $a_k\Psi_0[\varphi] = 0$, which may be rewritten as

$$
\left[\frac{\partial}{\partial\varphi_{-k}} + \omega_k\varphi_k\right]\Psi_0[\varphi] = 0
\tag{15.37}
$$

As in the quantum mechanics of the harmonic oscillator, we can solve this to get

$$
\begin{aligned}
\Psi_0[\varphi] &= \mathcal{N}\exp\left[-\tfrac{1}{2}\sum_k \omega_k\varphi_k\varphi_{-k}\right] \\
&= \mathcal{N}\exp\left[-\tfrac{1}{2}\int d^3x\, d^3y\ \varphi(x)\Omega(x,y)\varphi(y)\right] \\
\Omega(x,y) &= \frac{1}{V}\sum_k \omega_k e^{i\boldsymbol{k}\cdot(\boldsymbol{x}-\boldsymbol{y})} \\
\varphi(x) &= \frac{1}{\sqrt{V}}\sum_k \varphi_k e^{-i\boldsymbol{k}\cdot\boldsymbol{x}}
\end{aligned}
\tag{15.38}
$$

where $\mathcal{N}$ is a normalization constant. Ψ_0 is an infinite product of the ground-state wave functions of harmonic oscillators, one for each $\boldsymbol{k}$. This result could also have been obtained by solving the Schrödinger equation, after writing the Hamiltonian in terms of φ_k, π_k.

Given the states $|\varphi\rangle$ and the fact that we have an infinite number of quantum mechanical systems put together to form the field theory, we can immediately obtain

$$\Psi[\varphi, t] = \int [d\varphi] \; e^{iS(\varphi,t,\varphi',t')} \; \Psi_i[\varphi', t'] \tag{15.39}$$

where

$$[d\varphi] = \prod_k \lim_{N \to \infty} \prod_{\alpha=0}^{N-1} \frac{d\varphi_k^\alpha}{\sqrt{(2\pi i\epsilon_{\alpha+1})}} \tag{15.40}$$

The action $S(\varphi, t, \varphi', t')$ is the classical action $\int dt d^3x \; \mathcal{L}(\varphi, \partial\varphi)$ evaluated for $\varphi(\boldsymbol{x}, t)$ with $\varphi(\boldsymbol{x}, t') = \varphi'(\boldsymbol{x})$, $\varphi(\boldsymbol{x}, t) = \varphi(\boldsymbol{x})$. It is to be interpreted by a discretization of the time-interval, splitting it into N intervals, with $N \to \infty$ eventually. The integration over time in defining the action is thus replaced by summation. As before, the integral is defined by the Euclidean integral appropriately continued. The S-matrix element can be written, up to overall normalization, as

$$S_{fi} = \lim_{t\to\infty, t'\to-\infty} \int [d\varphi]\Psi_f^*[\varphi, t]\Psi_i[\varphi', t'] \; e^{iS(\varphi,t,\varphi',t')} \tag{15.41}$$

We shall rewrite this in the form of a reduction formula. Consider the process where we have incoming particles of momenta $k_1, k_2, \cdots, k_N$ and outgoing particles of momenta $p_1, p_2, \cdots, p_M$. From (15.35)

$$\begin{aligned}
\Psi_i[\varphi] &= \langle\varphi|a_{k_1}^\dagger a_{k_2}^\dagger \cdots a_{k_N}^\dagger |0\rangle \\
&= \int d^3x_1 \cdots d^3x_N \left[-iu_{k_1}(x_1)\overleftrightarrow{\partial}_{01}\right] \cdots \left[-iu_{k_N}(x_N)\overleftrightarrow{\partial}_{0N}\right] \\
&\quad \times \langle\varphi|\phi(x_1)\phi(x_2)\cdots\phi(x_N)|0\rangle \\
&= \int \prod_1^N \left[-iu_{k_i}(x_i)\overleftrightarrow{\partial}_{0i}\varphi(x_i)\right] \Psi_0[\varphi] \tag{15.42}
\end{aligned}$$

where we have used (15.31) and the fact that $\langle\varphi|\phi = \langle\varphi|\varphi$. With a similar way of writing the wave function for the final state, we find

$$\begin{aligned}
S_{fi} &= \int \prod \left[iu_{p_j}^*(y_j)\overleftrightarrow{\partial}_{0j}\right] \left[-iu_{k_i}(x_i)\overleftrightarrow{\partial}_{0i}\right] \int [d\varphi]\Psi_0^*\Psi_0 \\
&\quad \times \varphi(y_1)\cdots\varphi(y_M)\varphi(x_1)\cdots\varphi(x_N) \; e^{iS(\varphi)} \\
&= \int \left[\prod_1^M iu_{p_j}^*(y_j)\overleftrightarrow{\partial}_{0j} \prod_1^N -iu_{k_i}(x_i)\overleftrightarrow{\partial}_{0i}\right] G(y_1, y_2, \cdots y_M, x_1, x_2, \cdots x_N)
\end{aligned} \tag{15.43}$$

In the above formula, it is understood that we should take $y_j^0 \to \infty$, $x_i^0 \to -\infty$. We have also defined the Green's function

$$G(x_1, x_2, \cdots, x_N) = \int [d\varphi]\Psi_0^*[\varphi]\Psi_0[\varphi] \; e^{iS(\varphi)} \; \varphi(x_1)\varphi(x_2)\cdots\varphi(x_N) \tag{15.44}$$

The Green's function involved in (15.43) can be written in terms of the vertex function defined as follows:

$$G(y_1, y_2, \cdots, y_M, x_1, x_2, \cdots, x_N)$$
$$= \int d^4 y_1' \cdots d^4 x_N' \ G(y_1, y_1')...G(x_N, x_N')V(y_1', ..., x_N')$$

$$(15.45)$$

where $G(y_1, y_1')$, etc., are the two-point Green's functions or propagators. The support of the vertex function is roughly over the interaction region. If we assume the interactions are negligible in the far past and far future, then we may use the fact that eventually $y_j^0 \to \infty$, $x_i^0 \to -\infty$ to write

$$G(y_1, y_1') = i \int \frac{d^4 k}{(2\pi)^4} \frac{e^{ik(y_1 - y_1')}}{k^2 - m^2 + i\epsilon}$$
$$= \int \frac{d^3 k}{(2\pi)^3} \frac{1}{2\omega_k} e^{-ik(y_1 - y_1')}$$
$$= \sum_k u_k(y_1) u_k^*(y_1'), \qquad y_1^0 > y_1'^0$$

$$G(x_1, x_1') = \int \frac{d^3 k}{(2\pi)^3} \frac{1}{2\omega_k} e^{ik(x_1 - x_1')}$$
$$= \sum_k u_k^*(x_1) u_k(x_1'), \qquad x_1^0 < x_1'^0 \qquad (15.46)$$

We shall simplify (15.43) by using the formulae (15.45, 15.46). The simplification of the integrals involved may done using

$$\int d^3 y u_p^*(y) \ i \overleftrightarrow{\partial}_0 \sum_k u_k(y) u_k^*(y') = u_p^*(y') \qquad (15.47)$$

This result finally leads to

$$S_{fi} = \int d^4 y_1' \cdots d^4 x_N' \ \prod \left[u_{p_j}^*(y_j') u_{k_i}(x_i') \right] V(y_1', ..., x_N') \qquad (15.48)$$

The propagators satisfy $i(\Box + m^2)G(x, y) = \delta^{(4)}(x - y)$, so that

$$\prod_i i(\Box_i + m^2)G(x_1, \cdots, x_N) = V(x_1, \cdots, x_N) \qquad (15.49)$$

We can thus write the formula for the S-matrix element as

$$S_{fi} = \prod_1^M iu_{p_j}^*(y_j)(\Box_j + m^2) \prod_1^N iu_{k_i}(x_i)(\Box_i + m^2) \ G(y_1, ..., y_M, x_1, ..., x_N)$$

$$(15.50)$$

where the Green's function G is given by (15.44). (Of course, G is to be evaluated in Euclidean space and then continued to Minkowski space. The overall normalization factor must be fixed by requiring $\langle 0|S|0 \rangle = 1$.)

In such a framework, as in the discussion of functional integrals and renormalization earlier, calculations have to be done by rewriting the action in terms of renormalized fields and the renormalization factors Z's. The action is then of the form

$$S = \int Z_3 \left[\tfrac{1}{2}(\partial\varphi)^2 + (m^2 - \delta m^2)\varphi^2\right] + Z_1 \lambda \varphi^4 \qquad (15.51)$$

where the renormalization constants $Z_1, Z_3, \delta m^2$ are terms to be fixed by requiring the elimination of potentially divergent terms when the short distance cut-off goes to zero. We have written out (15.51) with a specific choice of the interaction term. More generally, every monomial of the fields requires a separate Z-factor.

15.4 Interpretation of the path integral

The field configurations $\varphi(x)$ at a fixed time form the configuration space $\mathcal{C}$, i.e.,

$$\mathcal{C} = \left\{\text{Set of all } \varphi(\boldsymbol{x}) : \mathbf{R}^3 \to \mathbf{R}\right\} \qquad (15.52)$$

Thus each point of $\mathcal{C}$ is a field configuration $\varphi(\boldsymbol{x})$ for all spatial points $\boldsymbol{x}$ taken together. The wave function $\Psi[\varphi]$ gives a complex number at each point of $\mathcal{C}$. As the time label t varies, $\varphi(\boldsymbol{x}, t)$ gives a path in $\mathcal{C}$. Thus (15.39) can be interpreted as

$$\Psi[\varphi, t] = \sum_{all\ paths\ P} e^{i\mathcal{S}_P[\varphi]}\, \Psi[\varphi', t'] \qquad (15.53)$$

where $\mathcal{S}[\varphi]$ is the action for a path P in $\mathcal{C}$ connecting $\varphi'(\boldsymbol{x})$ at time t' to $\varphi(\boldsymbol{x})$ at time t. Basically this result is the Huygens' principle of wave optics; the only difference is that the paths are now in the configuration space $\mathcal{C}$, rather than spacetime.

We shall now turn to some of the topological aspects of the path integral. To recapitulate the results so far, the basic path integral can be written as

$$\Psi[Q, t] = \int [dQ]\, e^{i\mathcal{S}[Q,t,Q',t']} \Psi_i[Q', t'] \qquad (15.54)$$

where Q specifies a point in the configuration space $\mathcal{C}$ and the integration is over all paths connecting the point Q' at time t' to the point Q at time t. The time-variable t parametrizes the paths. When the configuration space $\mathcal{C}$ has nontrivial topology, there are many subtleties involved in defining the quantum theory. Two cases of particular interest are when the fundamental group of $\mathcal{C}$, viz., $\Pi_1(\mathcal{C})$ is nontrivial, and secondly when there are two-dimensional closed surfaces in $\mathcal{C}$ which are not the boundaries of any three-volume along with the existence of closed two-forms which are not exact, i.e., $\mathcal{H}^2(\mathcal{C}) \neq 0$. We shall briefly consider these two possibilities.

15.5 Nontrivial fundamental group for $\mathcal{C}$

The simplest example of a nontrivial fundamental group for $\mathcal{C}$ is that of particle motion on a 2-plane with one point, say, the origin, removed; we have discussed this case in the last chapter. $\Pi_1(\mathbf{R}^2 - \{0\}) = \mathbf{Z}$ and the nontriviality of a closed loop C is measured by the winding number

$$\nu(C) = -\frac{1}{2\pi} \oint_C \frac{\epsilon_{ij}x^j}{x^2} dx^i = \oint_C \alpha$$

$$\alpha = -\frac{1}{2\pi} \frac{\epsilon_{ij}x^j}{x^2} dx^i \tag{15.55}$$

Since α is a closed one-form,

$$\int_P \alpha = \int_{P'} \alpha \tag{15.56}$$

for two open paths P, P' which are homotopic to each other.

Given the structure of paths and the winding number, we can generalize the path integral as

$$\Psi(x,t) = \int [dx] \, e^{iS(x,t,x',t')} \, e^{i\theta \int_{x'}^x \alpha} \, \Psi(x',t') \tag{15.57}$$

θ is an arbitrary parameter. A general path from x' to x may be written as $P + C_\nu$ where P is an open path with no noncontractible loops around the origin and C_ν is a loop around the origin with winding number ν. Evidently

$$\int_{P+C_\nu} \alpha = \nu + \int_P \alpha \tag{15.58}$$

We can thus write the path integral (15.57) as

$$\Psi(x,t) = \left[\sum_{all\tilde{P}\sim P} e^{iS_{\tilde{P}}+i\theta\int_P \alpha} + \sum_{all\tilde{P}\sim P+C_1} e^{iS_{\tilde{P}}+i\theta\int_P \alpha+i\theta} + \cdots \right] \Psi(x',t')$$

$$= \sum_\nu \sum_{all\tilde{P}\sim P+C_\nu} e^{iS_{\tilde{P}}+i\theta\int_P \alpha+i\nu\theta} \, \Psi(t') \tag{15.59}$$

The summation is over all paths $\tilde{P}$ which are homotopic to $P + C_\nu$, for each value of ν. The overall phase $\exp(i\int_P \alpha)$ does not matter for matrix elements, so we have

$$\Psi(x,t) = \sum_{\nu} \sum_{all\tilde{P}\sim P+C_\nu} e^{iS_{\tilde{P}}+i\nu\theta} \, \Psi(x',t') \tag{15.60}$$

Since ν is an integer, we see that Ψ is unchanged under $\theta \to \theta + 2\pi$. The parameter θ may thus be taken to be in the interval $[0, 2\pi]$.

In the path integral $\hbar$ appears in $S/\hbar$ in the exponent. The term $\theta\nu(C)$ has no $\hbar$ in it. This is evident since θ is an arbitrary parameter between zero and 2π and $\nu(C)$ is normalized to be an integer. Thus the $\theta\nu$-term has no classical effect. We must regard θ as an extra parameter (which does not appear in the classical action) which arises in quantizing theories with $\Pi_1(C) \neq 0$. If we want to mimic the effect of $\theta\nu$ by a term in the action, we can do so by writing

$$S = \int dt \, \mathcal{L} - \hbar\frac{\theta}{2\pi} \int dt \, \epsilon_{ij} \frac{\dot{x}^i x^j}{x^2} \tag{15.61}$$

The extra "topological" term, we notice, has an explicit $\hbar$ in it. (It is also singular at $x = 0$; thus the point $x = 0$ has to be removed for this to make sense.)

In general, $\Pi_1(C)$ can be something other than $\mathbf{Z}$. The general rule is obtained as follows. In addition to e^{iS}, include a factor $K(C)$ for curves C where K has the following properties:

1) $K(C + \delta C) = K(C)$
2) $K(C_1)K(C_2) = K(C_1 + C_2)$
3) $K(-C) = K^*(C)$

The first property gives the invariance of $K(C)$ under small deformations of the curve C; or in other words, $K(C)$ should be a topological invariant of the path. The second property gives the composition law for the paths, while the third tells us that $K(C)$ is a phase which changes sign under change of orientation. All these properties follow easily from the discussion so far. In total these properties say that $K(C)$ is a one-dimensional unitary representation of the group $\Pi_1(C)$. $K(C)$ will possibly involve new parameters (like θ); such parameters are new parameters which arise in quantizing the theory with the given nontrivial $\Pi_1(C)$. They are to be treated as extra coupling constants in addition to whatever coupling constants appear in the classical Lagrangian. The values of these parameters will take in the quantum theory have to be determined experimentally.

The quantization rule can thus be restated as follows:

$$\Psi(Q,t) = \int [dQ] \, e^{iS(Q,t,Q',t')} \, K(C) \, \Psi(Q',t') \tag{15.62}$$

where Q's denote the coordinates of the configuration space. Some examples where this kind of topological feature is important are:

1. Charged particle dynamics in the presence of a magnetic vortex.

2. Particles of fractional spin and statistics in two spatial dimensions. In this case, θ can be taken to be in the interval $[-\pi, \pi]$ with $\theta = 0$ giving bosons and $\theta = \pi$ or $(-\pi)$ giving fermions. The values in between give "anyons" or particles of any statistics.

3. Quantum chromodynamics. The configuration space $\mathcal{C}$ is not simply connected and in fact $\Pi_1(\mathcal{C}) = \mathbf{Z}$. This leads to an additional term $\theta \nu[A]$ in the functional integral for chromodynamics. $\nu[A]$, which measures the nontriviality of paths in $\mathcal{C}$, is the instanton number.

15.6 The case of $\mathcal{H}^2(\mathcal{C}) \neq 0$

We now turn to the second type of quantization problem which could arise when the configuration space has noncontractible two-surfaces. The quintessential example of this is the motion of a particle on a two-sphere. The path integral gives

$$\Psi(x, t) = \int [dx] \, W(P, x, x') \, \Psi(x', t')$$
$$W(P, x, x') = e^{i\mathcal{S}_P(x, x')} \tag{15.63}$$

We can think of $e^{i\mathcal{S}_P}$ as follows. For every path P between x' and x, $e^{i\mathcal{S}_P}$ gives a complex number of unit modulus; i.e.,

$$W(P, x, x') = e^{i\mathcal{S}_P(x, x')} : \quad \text{set of paths} \to U(1) \tag{15.64}$$

The path integral gives the summation over all paths with this factor $e^{i\mathcal{S}}$ as the weight factor for each path in the summation. The new idea we explore is that we can associate weight factors for the path integral with surfaces rather than directly with paths. One way to do this is as follows. Choose a particular path, referred to as standard path from now on, between the two points x' and x; call this P_0. The path of interest is denoted by P. Now $P - P_0$ forms a closed loop and we consider the surface Σ whose boundary is $P - P_0$. (On the two-sphere such a surface always exists; more generally, the following construction holds in situations where there is such a surface. When $P - P_0$ is not the boundary of a surface, our construction has to be modified.) The coordinates $x^\mu(t)$ which give the paths can be generalized to $x^\mu(\sigma, t) = x^\mu(\xi)$ where $\xi^i = (\sigma, t)$ parametrize the surface Σ. The area element of the surface Σ can be written as $\frac{1}{2}\partial_i x^\mu \partial_j x^\nu \epsilon^{ij}$, $i, j = 1, 2$, $\xi_1 = t, \xi_2 = \sigma$. The weight factor associated with the path P is now generalized as

$$W(P, x, x') = e^{i\mathcal{S}_P(x, x')} \, e^{i\Gamma(\Sigma)}$$
$$\Gamma(\Sigma) = k \int_\Sigma d^2\xi \, \frac{1}{2} B_{\mu\nu} \partial_i x^\mu \partial_j x^\nu \epsilon^{ij} \tag{15.65}$$

where $B_{\mu\nu}$ is a function of x^μ, i.e., $B_{\mu\nu} = B_{\mu\nu}(x)$ and is antisymmetric, i.e., $B_{\mu\nu} = -B_{\nu\mu}$.

There are many consistency requirements on such a term which we shall now examine. First of all, we may think of the extra term as an addition to the action. In defining the weight factor (15.65), we need the surface Σ or $x^\mu(\sigma, t)$. $x^\mu(t)$ defines the path, but the variable σ has no meaning physically. The first property of the term (15.65) should be that the σ-dependence of x^μ does not affect the physics. Alternatively, note that there are many surfaces which have $P - P_0$ as the boundary. Physics should not depend on which surface we choose to define (15.65). The choice of different surfaces can be made to correspond to different functional dependences of x^μ on σ and thus we must require that $\Gamma(\Sigma)$ be invariant under small deformations of Σ or small variations of $x^\mu(\xi)$. We have

$$\delta\Gamma = \frac{k}{2} \int_\Sigma d^2\xi \left[\partial_\alpha B_{\mu\nu} \delta x^\alpha \partial_i x^\mu \partial_j x^\nu \epsilon^{ij} \right.$$
$$+ B_{\mu\nu} \partial_i (\delta x^\mu) \partial_j x^\nu \epsilon^{ij}$$
$$\left. + B_{\mu\nu} \partial_i x^\mu \partial_j (\delta x^\nu) \epsilon^{ij} \right]$$
$$= \frac{k}{2} \int_\Sigma d^2\xi \left[\partial_\alpha B_{\mu\nu} + \partial_\mu B_{\nu\alpha} + \partial_\nu B_{\alpha\mu} \right] \partial_i x^\mu \partial_j x^\nu \epsilon^{ij} \delta x^\alpha$$
$$+ k \oint_{\partial\Sigma} d\xi_i B_{\mu\nu} \partial_j x^\nu \epsilon^{ij} \delta x^\mu \tag{15.66}$$

The first term is sensitive to the σ-dependence of x^μ (or choice of surface Σ). We can obtain physics independent of the σ-dependence of x^μ or the choice of Σ, at least for small variations of the surface, if we require

$$\partial_\alpha B_{\mu\nu} + \partial_\mu B_{\nu\alpha} + \partial_\nu B_{\alpha\mu} = 0 \tag{15.67}$$

The second term depends only on the path $P - P_0$, which is $\partial\Sigma$. Now δx^μ is zero for the standard path P_0 since P_0 is fixed once and for all. Thus $\delta\Gamma$ depends only on δx^μ along P or on infinitesimal variations of P, provided (15.67) is satisfied. It is the only requirement for small variations of the path. We shall rewrite $\delta\Gamma$ in a more useful way shortly, but before we do so, a couple of other properties have to be discussed. First of all, consider solutions of the condition (15.67). One set of solutions is obviously given by

$$B_{\mu\nu} = \partial_\mu C_\nu - \partial_\nu C_\mu \tag{15.68}$$

where C_μ is some vector potential. In this case, we have

$$\Gamma = 2\frac{k}{2} \int_\Sigma d^2\xi \, (\partial_\mu C_\nu) \partial_i x^\mu \partial_j x^\nu \epsilon^{ij}$$
$$= k \int_\Sigma d^2\xi \, \frac{\partial}{\partial\xi^i} [C_\nu \partial_j x^\nu \epsilon^{ij}]$$
$$= k \oint_{\partial\Sigma} d\xi_i \, C_\nu \partial_j x^\nu \epsilon^{ij}$$

$$= k \left[\int_P dt \, C_\mu \frac{dx^\mu}{dt} - \int_{P_0} dt \, C_\mu \frac{dx^\mu}{dt} \right] \tag{15.69}$$

The contribution of the standard path is a common phase factor for all wave functions since it is common for all paths. Therefore, we can drop it from the path integral. Thus we get $W(P, x, x') = e^{i\tilde{S}_P(x,x')}$ for each path P, with $\tilde{S} = S + k \int dt \, C_\mu \dot{x}^\mu$. We simply have another term in S and all this discussion about Σ is not relevant. The kind of topological term we are discussing becomes relevant only for $B_{\mu\nu}$ which satisfies (15.67) but which cannot be written as a curl of a vector as in (15.68), i.e.,

$$\partial_\alpha B_{\mu\nu} + \partial_\mu B_{\nu\alpha} + \partial_\nu B_{\alpha\mu} = 0 \tag{15.70}$$
$$B_{\mu\nu} \neq \partial_\mu C_\nu - \partial_\nu C_\mu \tag{15.71}$$

Notice that $B_{\mu\nu}$ is antisymmetric in μ, ν by its definition (15.65); it is the component-version of a two-form on the configuration space, $B = \frac{1}{2} B_{\mu\nu} dx^\mu \wedge dx^\nu$. The condition (15.70) is equivalent to $dB = 0$ and (15.71) to $B \neq dC$ for some one-form C. Thus the B's we are interested in are closed two-forms on $\mathcal{C}$ which are not exact, i.e., elements of $\mathcal{H}^2(\mathcal{C}, \mathbf{R})$.

Consider as an example the monopole field

$$B_{\mu\nu} = \frac{1}{4\pi} \frac{\epsilon_{\mu\nu\alpha} x^\alpha}{|x|^3} = \frac{1}{4\pi} \frac{\epsilon_{\mu\nu\alpha} \hat{x}^\alpha}{r^2}$$
$$B = \frac{1}{8\pi} \frac{\epsilon_{\mu\nu\alpha} \hat{x}^\alpha}{r^2} dx^\mu \wedge dx^\nu \tag{15.72}$$

We then find

$$dB = \frac{1}{3!} (\partial_\alpha B_{\mu\nu} + \partial_\mu B_{\nu\alpha} + \partial_\nu B_{\alpha\mu}) dx^\alpha dx^\mu dx^\nu$$
$$= \frac{1}{2} \epsilon_{\mu\nu\alpha} \epsilon^{\mu'\nu'\alpha'} (\partial'_\alpha B_{\mu'\nu'}) \, dx^\alpha dx^\mu dx^\nu$$
$$= \frac{1}{8\pi} \epsilon_{\mu\nu\alpha} \epsilon^{\mu'\nu'\alpha'} \left[\frac{\epsilon_{\mu'\nu'\alpha'}}{r^3} - 3 \frac{\epsilon_{\mu'\nu'\beta} x^\beta x'_\alpha}{r^5} \right] \, dx^\alpha dx^\mu dx^\nu$$
$$= 0, \qquad \text{for } r \neq 0 \tag{15.73}$$

On $\mathbf{R}^3 - \{0\}$, dB is zero. Integrating B over a two-sphere surrounding the origin

$$\oint_{S^2} B = 1 \tag{15.74}$$

so that $B \neq dC$. The two-form (15.72) furnishes an example for $\mathbf{R}^3 - \{0\}$. For the path integral for the motion of a charged particle in the background of a monopole we include a term $\exp(i\Gamma(\Sigma))$.

This particular solution we have found is valid in $\mathbf{R}^3 - \{0\}$; this space has noncontractible two-surfaces and indeed, we have used such a surface in arriving at the argument for the nontriviality of (15.72).

Consider now the case of $\mathbf{R}^3$ with two points removed from it. In this case, we have different types of closed noncontractible surfaces. We can consider a surface S_1 surrounding one of the removed points, or a surface S_2 surrounding the other removed point. Evidently, these cannot be deformed into each other. We can also consider a single surface surrounding both points, but this surface can be decomposed, up to smooth deformations, into $S_1 + S_2$. In this case, we say that S_1 and S_2 are generators of noncontractible two-surfaces in the sense that any noncontractible two-surface can be decomposed into copies of S_1 and S_2. More generally, we need $\mathcal{H}_2(\mathcal{C})$ which is the set of all closed two-surfaces which are themselves not the boundaries of any three-volume. $\mathcal{H}_2(\mathcal{C})$ is called the second homology group of $\mathcal{C}$. In the example of $\mathbf{R}^3 - \{0\}$, $\mathcal{H}_2(\mathcal{C}) = \mathbf{Z}$ and there is one generator, namely, the unit sphere surrounding the origin. For the example of $\mathbf{R}^3$ with two points removed, $\mathcal{H}_2 = \mathbf{Z} + \mathbf{Z}$, and we have two generators, S_1 and S_2. In general, we can define a $B_{\mu\nu}$ for each generator as proportional to the area element of the generating surface. Such a $B_{\mu\nu}$ can be normalized by requiring $\oint_S B_{\mu\nu} dS^{\mu\nu} = 1$.

We now go back to the path integral and consider the specification of Σ again. In general, there are several surfaces which have $P - P_0$ as the boundary. Consider two such surfaces Σ and Σ', where one is a small deformation of the other. Since they have a common boundary, we have a three-volume V in between the surfaces, such that $\partial V = \Sigma - \Sigma'$. Then by Stokes' theorem, we have

$$
\begin{aligned}
\Gamma_\Sigma - \Gamma_{\Sigma'} &= k \int_{\Sigma - \Sigma'} d^2\xi \, \frac{1}{2} B_{\mu\nu} \partial_i x^\mu \partial_j x^\nu \epsilon^{ij} = k \int_{\Sigma - \Sigma'} B \\
&= k \int_V dB \\
&= 0, \qquad\qquad\qquad \text{by (15.70)}
\end{aligned}
$$

$$\tag{15.75}$$

Thus it does not matter which surface we choose. This argument works only for Σ, Σ' such that $\Sigma - \Sigma'$ is the boundary of volume V. If $\mathcal{H}_2(\mathcal{C})$ is not trivial, there are cases where $\Sigma - \Sigma'$ is not the boundary of any volume. For example, for $\mathbf{R}^3 - \{0\}$, we can use Σ and Σ' such that $\Sigma - \Sigma'$ is the two-sphere surrounding the origin. This is noncontractible (since the origin has been removed) and further there is no volume V such that $\partial V = \Sigma - \Sigma'$. In this case

$$
\begin{aligned}
\Gamma_\Sigma - \Gamma_{\Sigma'} &= k \oint_{S^2} B \\
&= k
\end{aligned}
\tag{15.76}
$$

where we have used the result (15.74). In the path integral, we thus get

$$
e^{i\Gamma_\Sigma} = e^{i\Gamma_{\Sigma'}} \, e^{ik}
\tag{15.77}
$$

We can get results independent of the surface if we impose the further condition $e^{ik} = 1$ or $k = 2\pi n$, $n \in \mathbf{Z}$. This is a topological quantization rule. In
the context of the charged particle in a monopole field, this is again the Dirac
quantization rule. With this condition, we may summarize the results as follows. If the configuration space $\mathcal{C}$ has noncontractible two-surfaces, (and if
every closed curve is the boundary of some surface), it is possible to generalize
the path integral as

$$\Psi(Q,t) = \int [dQ] e^{iS(Q,t,Q',t')+i\Gamma_\Sigma} \Psi(Q',t') \tag{15.78}$$

where Σ is a surface whose boundary is the path P and a standard path
$-P_0$. Γ_Σ is given by

$$\Gamma_\Sigma = 2\pi n \int_\Sigma B \tag{15.79}$$

where n is an integer and B is a two-form on $\mathcal{C}$ with

$$dB = 0 \tag{15.80}$$
$$B \neq dC \tag{15.81}$$

and B is normalized, i.e.,

$$\int_S B = 1 \tag{15.82}$$

where S is a generator of the group of $\mathcal{H}_2(\mathcal{C})$. A term like (15.79) with the
conditions (15.80, 15.81) is called a Wess-Zumino term. All conditions are
such that no $\hbar$ appears anywhere and this term also has no classical effect.
(Generally, nontrivial $\mathcal{H}_2(\mathcal{C})$, rather than $\mathcal{H}^2(\mathcal{C})$, is more relevant to the kind
of topological feature of the path integral we are considering. We can specify
weights for paths in terms of surfaces whose boundaries they are. We have
chosen to write these weights in terms of integrals of a differential form,
thereby bringing in the cohomology group. Such a differential form is obtained
in all the problems we are considering and so this is adequate for our purpose.)

The Wess-Zumino term occurs and makes a difference in a number of
physical situations. Examples are:

1. Motion of a charged particle in a monopole background
2. Low-energy behavior of quantum chromodynamics, in particular in interpreting solitons of meson fields as baryons. (Such solitons are called
   skyrmions.)
3. Anomalous gauge theories
4. Boson-fermion equivalence in (1+1) dimensions.

References

1. Path integrals originated with Dirac's discussion of the role of action in
   quantum mechanics. It became the basis for a quantization scheme in

R.P. Feynman, Rev. Mod. Phys. **20**, 367 (1948). General books include: R.P. Feynman and A.R. Hibbs, *Quantum Mechanics and Path Integrals*, McGraw Hill (1965); L.S. Schulman, *Techniques and Applications of Path Integration*, John Wiley and Sons, Inc. (1981); H. Kleinert, *Path Integrals in Quantum Mechanics, Statistics, Polymer Physics, and Financial Markets*, World Scientific Pub. Co., 3rd edition (2004).

2. The role of the nontrivial fundamental group is discussed by M.G.G. Laidlaw and C.M. De Witt, Phys. Rev. **D3**, 1375 (1970); J.S. Dowker, J. Phys. **A5**, 936 (1972); R.D. Sorkin, Phys. Rev. **D27**, 1787 (1983); Yong-Shi Wu, Phys. Rev. Lett. **52**, 2103 (1984).

3. Implications of topology for quantization are discussed in many books by now, see, for example, A.P. Balachandran, *Classical Topology and Quantum States*, World Scientific Pub. Co. (1991).

16 The Configuration Space in Nonabelian Gauge Theory

16.1 The configuration space

We now apply some of the general analysis of the previous chapters to the specific case of the configuration space of an unbroken nonabelian gauge theory. Quantum chromodynamics (QCD) is, of course, the most interesting theory of this type. We will see that $\Pi_1(\mathcal{C}) = \mathbf{Z}$ for QCD. As a result we get a new parameter θ, which is needed to characterize QCD.

Consider a pure gauge theory. We use $A_0 = 0$ gauge to simplify the analysis. The theory is thus described by potentials A_i, $i = 1, 2, 3$, which are Lie algebra valued. They can be taken to be antihermitian matrices, $A_i = -i A_i^a t^a$, where t^a are hermitian matrices which form a basis of the Lie algebra of the gauge group. For simplicity we shall consider an $SU(2)$ gauge theory. Gauge transformations act on A_i by

$$A_i \rightarrow A_i^g = g \, A_i \, g^{-1} - \partial_i g \, g^{-1} \tag{16.1}$$

where $g(\boldsymbol{x}) : \mathbf{R}^3 \rightarrow SU(2)$. One cannot consider $g(\boldsymbol{x})$ which depend on time, since such transformations would change the gauge choice $A_0 = 0$. Thus, once we choose $A_0 = 0$, the transformations above represent the residual gauge freedom of the theory. The spaces of relevance to us are

$$\mathcal{A} = \left\{ \text{space of gauge potentials } A_i \right\} \tag{16.2}$$

$$\mathcal{G}_* = \left\{ \text{space of gauge transformations } g(\boldsymbol{x}) : \mathbf{R}^3 \rightarrow SU(2) \right.$$

$$\left. \text{such that } g \rightarrow 1 \text{ as } |\boldsymbol{x}| \rightarrow \infty \right\} \tag{16.3}$$

(This is similar to what we discussed in Chapter 10; there we defined and used similar spaces for potentials in four-dimensional Euclidean space, for the purpose of the functional integral. Here the potentials and gauge transformations are at a fixed time and are functions of the spatial coordinates. Even though we use the same notation, the context should make clear what is meant.)

As we have discussed before, transformations $g(x)$ which go to a constant element $g_\infty \neq 1$ act as a Noether symmetry. The states fall into unitary irreducible representations of such transformations, which are isomorphic to the gauge group $SU(2)$, up to $\mathcal{G}_*$-transformations. The true gauge freedom is only $\mathcal{G}_*$. The configuration space of the theory is thus $\mathcal{C} = \mathcal{A}/\mathcal{G}_*$. In analyzing the structure of this space we need information about $\mathcal{A}$ and $\mathcal{G}_*$.

(Note: The action of $\mathcal{G}$, which is the set of gauge transformations which go to a constant $g_\infty \neq 1$, on $\mathcal{A}$ has fixed points; e.g., $A_i = 0$ is left invariant by the action of constant g's. Thus $\mathcal{A}/\mathcal{G}$ will have singularities. Mathematically, to avoid such singularities, one must use $\mathcal{A}/\mathcal{G}_*$. Our discussion of Chapter 10 shows that this is also the relevant quantity physically.)

Topology of $\mathcal{A}$

$\mathcal{A}$ is an affine space, i.e., any potential A_i can be written as $A_i^{(0)} + h_i$ where $A_i^{(0)}$ is a fixed potential and h_i is an arbitrary Lie algebra valued vector field. Equivalently, any two points in $\mathcal{A}$ can be connected by a straight line. If $A_i^{(1)}$ and $A_i^{(2)}$ are two points in $\mathcal{A}$, i.e., two gauge potentials, the sequence of configurations

$$A_i(x, \tau) = A_i^{(1)}(1 - \tau) + \tau A_i^{(2)} \tag{16.4}$$

where $0 \leq \tau \leq 1$ provides a straight-line interpolation between $A_i^{(1)}$ and $A_i^{(2)}$. $\mathcal{A}$ thus looks like an infinite-dimensional Euclidean space. Topologically it is a rather trivial space. In particular $\Pi_0(\mathcal{A}) = 0$, $\Pi_1(\mathcal{A}) = 0$.

Topology of $\mathcal{G}_*$

$\mathcal{G}_*$ is made up of mappings $g(x) : \mathbf{R}^3 \to SU(2)$ with the condition $g(x) \to 1$ as $|x| \to \infty$. Consider an arbitrary element of $SU(2)$. The group $SU(2)$ can be thought of as the group of (2×2) unitary matrices of unit determinant. We can parametrize an element of $SU(2)$ as

$$g = \phi_0 + i\phi_i \sigma_i \tag{16.5}$$

where σ_i are the Pauli matrices and the condition $g^\dagger = g^{-1}$ implies that ϕ_0, ϕ_i are real. The condition $\det g = 1$ requires

$$\phi_0^2 + \sum_i \phi_i^2 = 1 \tag{16.6}$$

The group $SU(2)$ is thus topologically a three-sphere. The $\phi_\mu(x)$, $\mu = 0, 1, 2, 3$, thus give a mapping $\mathbf{R}^3 \to S^3$. Further we have $g \to 1$ as $|x| \to \infty$. For the case of such maps, we can think of space $\mathbf{R}^3$ itself as being a three-sphere. Explicitly this can be realized as follows. We define a mapping

$$y_0 = \frac{x^2 - 1}{x^2 + 1}, \qquad\qquad y_i = \frac{2x_i}{x^2 + 1} \tag{16.7}$$

Evidently $y_0^2 + \sum_i y_i^2 = 1$; the y's define a three-sphere. (16.7) thus gives a description of $\mathbf{R}^3$ as a three-sphere, spatial infinity corresponding to $|\boldsymbol{x}| \to \infty$ being mapped to the pole $y_0 = 1$, $y_i = 0$. Now, given an $SU(2)$-valued function on S^3, i.e., given $g(y)$ where $\sum_\mu y_\mu^2 = 1$, we can use (16.7) to write it as a function on $\mathbf{R}^3$ with $g \to g_\infty$, g_∞ not dependent on angles as $|\boldsymbol{x}| \to \infty$. In particular, choosing $g_\infty = 1$ requires $g(1, 0, 0, 0) = 1$. We are thus concerned with maps $S^3 \to S^3$, where the first S^3 represents space $\mathbf{R}^3$ via the map (16.7) and the second S^3 is $SU(2)$. The classes of such maps will be given by $\Pi_3(S^3) = \mathbf{Z}$.

We shall analyze these maps by defining a winding number Q. Consider $\phi^\mu : S^3 \to S^3$. We want to determine how many times the target sphere S^3 is covered by the map $\phi^\mu(x)$ as we cover the spatial S^3 once. The volume element for the image generated by $\phi^\mu(x)$ is given by

$$dS_\mu \epsilon^{\mu\nu\alpha\beta} = \partial_i \phi^\nu \partial_j \phi^\alpha \partial_k \phi^\beta \; \epsilon^{ijk} d^3x \tag{16.8}$$

or

$$dS_\mu = \frac{1}{3!} \epsilon_{\mu\nu\alpha\beta} \epsilon^{ijk} \partial_i \phi^\nu \partial_j \phi^\alpha \partial_k \phi^\beta d^3x$$
$$= \phi^\mu d\Omega^{(3)} \tag{16.9}$$

where

$$d\Omega^{(3)} = \frac{1}{3!} \epsilon_{\mu\nu\alpha\beta} \epsilon^{ijk} \phi^\mu \partial_i \phi^\nu \partial_j \phi^\alpha \partial_k \phi^\beta d^3x \tag{16.10}$$

Since the volume of S^3 is $2\pi^2$, we get for the winding number

$$Q[g] = \frac{1}{2\pi^2} \{\text{volume traced out by } \phi^\mu(x)\}$$
$$= \frac{1}{12\pi^2} \int d^3x \; \epsilon_{\mu\nu\alpha\beta} \epsilon^{ijk} \phi^\mu \partial_i \phi^\nu \partial_j \phi^\alpha \partial_k \phi^\beta \tag{16.11}$$

This can also be written directly in terms of $g(x)$ as

$$Q[g] = -\frac{1}{24\pi^2} \int d^3x \, \text{Tr}\left(g^{-1} \partial_i g \; g^{-1} \partial_j g \; g^{-1} \partial_k g\right) \epsilon^{ijk}$$
$$= -\frac{1}{24\pi^2} \int \text{Tr}(g^{-1} dg)^3 \tag{16.12}$$

We have already seen this formula from our discussion of the cohomology and homotopy of Lie groups. The arguments given here will provide a more explicit realization.

We now show some important properties of $Q[g]$.

a) Q is invariant under smooth deformations of ϕ.

We have seen this in terms of g in Chapter 14; we will now show the same result in terms of ϕ. Consider deformations $\phi^\mu(x) \to \phi^\mu(x) + \delta\phi^\mu(x)$, which preserve the boundary condition $\phi^\mu(x) \to (1,0,0,0)$ as $|x| \to \infty$. This requires $\delta\phi^\mu \to 0$ as $|x| \to \infty$. Further $\delta\phi^\mu \phi_\mu = 0$, since we have to preserve $\phi^\mu \phi_\mu = 1$. For the change in Q to first order in $\delta\phi^\mu$ we find

$$
\begin{aligned}
Q[\phi + \delta\phi] - Q[\phi] = \frac{1}{12\pi^2} \int d^3x \Big[&\delta\phi^\mu d\phi^\nu \wedge d\phi^\alpha \wedge d\phi^\beta \\
+ &\phi^\mu d(\delta\phi^\nu) \wedge d\phi^\alpha \wedge d\phi^\beta \\
+ &\phi^\mu d\phi^\nu \wedge d(\delta\phi^\alpha) \wedge d\phi^\beta \\
+ &\phi^\mu d\phi^\nu \wedge d\phi^\alpha \wedge d(\delta\phi^\beta) \Big] \epsilon_{\mu\nu\alpha\beta} \\
= \frac{1}{3\pi^2} \int d^3x \Big[&\delta\phi^\mu d\phi^\nu \wedge d\phi^\alpha \wedge d\phi^\beta \Big] \epsilon_{\mu\nu\alpha\beta}
\end{aligned}
$$

$$(16.13)$$

where we have done partial integrations on terms involving $d(\delta\phi)$. The surface terms are zero, since $\delta\phi \to 0$ as $|x| \to \infty$. Now, since $\delta\phi \cdot \phi = 0$, we have $\delta\phi^\mu = \epsilon^{\mu\nu\alpha\beta}\phi_\nu \omega_{\alpha\beta}$ for some arbitrary antisymmetric tensor $\omega_{\alpha\beta}$. Using this in the above equation,

$$
\begin{aligned}
Q[\phi + \delta\phi] - Q[\phi] &= \frac{1}{3\pi^2} \int \delta^{\alpha\beta\gamma}_{\mu\nu\sigma} \phi_\alpha \omega_{\beta\gamma} d\phi^\mu \wedge d\phi^\nu \wedge d\phi^\sigma \\
&= 0
\end{aligned}
$$

$$(16.14)$$

since all terms, upon expanding $\delta^{\alpha\beta\gamma}_{\mu\nu\sigma}$, involve one power of $\delta\phi \cdot \phi$. Thus Q is invariant under smooth deformations of $\phi^\mu(x)$ which preserve $\phi^\mu \to (1,0,0,0)$ as $|x| \to \infty$.

b) $Q[g_1 g_2] = Q[g_1] + Q[g_2]$.

Consider $Q[g_1 g_2]$. We shall use (16.12) to evaluate this. We have

$$
\begin{aligned}
(g_1 g_2)^{-1} d(g_1 g_2) &= g_2^{-1}(g_1^{-1} dg_1)g_2 + g_2^{-1} dg_2 \\
&= g_2^{-1}(A + B)g_2
\end{aligned}
$$

$$(16.15)$$

where $A = g_1^{-1} dg_1$ and $B = dg_2 \, g_2^{-1}$. Using this decomposition

$$
\begin{aligned}
Q[g_1 g_2] &= -\frac{1}{24\pi^2} \int \mathrm{Tr}(A+B)(A+B)(A+B) \\
&= -\frac{1}{24\pi^2} \int \Big[\mathrm{Tr}A^3 + \mathrm{Tr}B^3 + (3A^2 B + 3B^2 A) \Big] \\
&= Q[g_1] + Q[g_2] - \frac{1}{8\pi^2} \int \mathrm{Tr}(-dA \, B + A dB) \\
&= Q[g_1] + Q[g_2] + \frac{1}{8\pi^2} \int d(\mathrm{Tr}AB) \\
&= Q[g_1] + Q[g_2]
\end{aligned}
$$

$$(16.16)$$

$$(16.17)$$

(The wedge products or antisymmetrized products are left understood in this and some of the following equations.) We see that the winding numbers add when we take products of the group-valued functions g_1 and g_2. The possible surface contribution in passing from (16.16) to (16.17) is zero, because the antisymmetric product of A and B falls off sufficiently fast as $|x| \to \infty$.

We now consider an example of a configuration for which $Q = 1$. This is given by

$$g_1(x) = \frac{x^2 - 1}{x^2 + 1} + i\frac{2x^i}{x^2 + 1}\sigma_i \tag{16.18}$$

which is equivalent to $\phi^\mu(x) = y^\mu$. The ϕ's form a three-sphere; so do the y's. $\phi^\mu = y^\mu$ gives one covering of the S^3, which is $SU(2)$, when y's cover the S^3 corresponding to space (or equivalently $\mathbf{R}^3$). Thus $Q = 1$ for this configuration. This may also be verified directly from the integral (16.12). $g_1(x)$ is a smooth configuration with $g_1 \to 1$ as $|x| \to \infty$. Now the configuration $g(x) = 1$ everywhere evidently has $Q = 0$. Since we have $Q = 0$ for $g = 1$ and $Q = 1$ for $g = g_1(x)$ and Q is invariant under smooth deformations, it is clear that $g_1(x)$ cannot be smoothly deformed to the identity everywhere. Further we can consider $g(x) = g_1(x)g_1(x)$, which has $Q = 2$ by (16.17). Also we have $g = g_1^\dagger(x)$, which has $Q = -1$, which follows from $Q[g^\dagger g] = Q[g^\dagger] + Q[g]$ and $Q[g^\dagger g] = Q[1] = 0$. The classes of maps $g(x) : S^3 \to S^3$ are now clear. We can write $\mathcal{G}_*$ as the sum of different components, each of which is connected and is characterized by the winding number Q; i.e.,

$$\mathcal{G}_* = \sum_{Q=-\infty}^{+\infty} \oplus \mathcal{G}_{*Q} \tag{16.19}$$

where $\mathcal{G}_{*0}$ consists of all maps smoothly deformable to 1 everywhere; $\mathcal{G}_{*1}$ consists of all maps smoothly deformable to $g_1(x)$ everywhere; and more generally, $\mathcal{G}_{*Q}$ consists of all maps smoothly deformable to the Q-fold product of $g_1(x)$'s. $\mathcal{G}_{*-Q}$ is similarly defined using $g_1^\dagger(x)$. $\mathcal{G}_{*Q}$ and $\mathcal{G}_{*Q'}$ are disconnected from each other for $Q \neq Q'$, since if they are connected, $(g_1)^Q$ and $(g_1)^{Q'}$ should be deformable to each other and this is impossible since $Q \neq Q'$. Q's add upon taking the product of two maps as in (16.17) and hence this structure is isomorphic to the additive group of integers $\mathbf{Z}$.

We can now analyze the structure of $\mathcal{A}/\mathcal{G}_*$. Consider the line in $\mathcal{A}$ given by

$$A_i(x, \tau) = A_i(x)(1 - \tau) + A_i^{g_1}\tau \tag{16.20}$$

for $0 \leq \tau \leq 1$ or more generally

$$A_i(x, \tau) \quad \text{with} \quad \begin{aligned} A_i(x, 0) &= A_i(x) \\ A_i(x, 1) &= A_i^{g_1}(x) \end{aligned} \tag{16.21}$$

where $A_i^{g_1}$ is the gauge transform of A_i by $g_1(x)$ as in (16.1). This is an open path in $\mathcal{A}$. But since $A_i^{g_1}$ is the gauge transform of A_i, both configurations

A_i and $A_i^{g_1}$ represent the same point in $\mathcal{C} = \mathcal{A}/\mathcal{G}_*$. Thus $A_i(x, \tau)$ describes a closed loop in $\mathcal{C}$. The question is whether this loop is contractible. If the loop is contractible, we can deform the trajectory to a curve purely along the gauge flow directions which would then connect $g = 1$ to $g = g_1(x)$. This would imply that $g_1(x)$ is smoothly deformable to the identity. But we know that this is impossible from our discussion of the structure of $\mathcal{G}_*$. In turn this implies that $A_i(x, \tau)$ of (16.21) is a noncontractible loop. Thus $\Pi_1(\mathcal{C}) \neq 0$. Further, we can repeat the argument with $g_2 = g_1 g_1$, or with g_3, g_4, etc., and also with $g_1^\dagger$. With the composition rule $Q[gg'] = Q[g] + Q[g']$, we see immediately that

$$\Pi_1(\mathcal{C}) = \Pi_1(\mathcal{A}/\mathcal{G}_*) = \mathbf{Z} \qquad (16.22)$$

Another way to see this is to use the exact homotopy sequence. From (16.19), $\mathcal{G}_*$ has connected components labeled by Q, so that $\Pi_0[\mathcal{G}_*] = \mathbf{Z}$. Then

$$\rightarrow \Pi_1[\mathcal{A}] \rightarrow \Pi_1[\mathcal{A}/\mathcal{G}_*] \rightarrow \Pi_0[\mathcal{G}_*] \rightarrow \Pi_0[\mathcal{A}] \rightarrow \qquad (16.23)$$

gives $\Pi_1[\mathcal{A}/\mathcal{G}_*] = \mathbf{Z}$, since $\Pi_1[\mathcal{A}] = \Pi_0[\mathcal{A}] = 0$.

16.2 The path integral in QCD

From our general discussion of the path integral and the result (16.22) above, it follows that QCD has a parameter $0 \leq \theta \leq 2\pi$ in addition to the gauge coupling constant. The path integral is given by

$$\Psi[A, t] = \int d\mu[\mathcal{A}/\mathcal{G}_*] \, e^{iS(A, A')} e^{i\theta\nu(A, A')} \, \Psi[A', t'] \qquad (16.24)$$

We still have to determine $\nu[A]$. As in our general discussion, $\nu[A]$ should be invariant under small changes of the path and sensitive only to the "boundary values" $A_i(x)$ and $A_i^g(x)$, in fact only to g. Further it must be Lorentz invariant with the parameter τ now being the time variable t, since paths in the path integral are parametrized by time. We may thus expect

$$\nu[A] = \int d^4x \, \partial_\mu K^\mu(A) \qquad (16.25)$$

Carrying out the time-integration will give

$$\nu(A, A') = \int_{t=\infty} d^3x \, K^0(A) \; - \; \int_{t=-\infty} d^3x \, K^0(A') \qquad (16.26)$$

which is clearly insensitive to the path in between. Since $Q[g]$ gives the winding number for closed curves, we should have

$$\int d^3x \, [K^0(A^g) - K^0(A)] = Q[g] = -\frac{1}{24\pi^2} \int \mathrm{Tr}(g^{-1}dg)^3 \qquad (16.27)$$

or

$$K^0(A^g) - K^0(A) = -\frac{1}{24\pi^2}\text{Tr}(g^{-1}dg)^3 + \text{(total derivative in } \boldsymbol{x}) \quad (16.28)$$

Now, A^g involves $gAg^{-1} - dg\, g^{-1}$; further dA^g has $(dg\, g^{-1})^2$. Thus in order to obtain $(g^{-1}dg)^3$, we need $K^0(A)$ to be made up of $\epsilon^{ijk}A_i\partial_j A_k \sim AdA$ and $\epsilon^{ijk}A_i A_j A_k \sim A^3$. We thus consider

$$K^0(A) = \text{Tr}\left[aAF + bA^3\right] \quad (16.29)$$

which gives

$$K^0(A^g) - K^0(A) = \text{Tr}\left[-avF - 3bvA^2 + 3bv^2A - bv^3\right] \quad (16.30)$$

where $v = g^{-1}dg$. The last term with v^3 gives the required form for $Q[g]$ and identifies b as $(1/24\pi^2)$. The terms involving A must vanish or combine as a total derivative. This identifies $a = -3b$. We thus get

$$\begin{aligned}
K^0(A) &= -\frac{1}{8\pi^2}\text{Tr}[AF - \frac{1}{3}A^3] \\
&= -\frac{1}{8\pi^2}\text{Tr}(A_i\partial_j A_k + \frac{2}{3}A_i A_j A_k)\epsilon^{ijk}
\end{aligned} \quad (16.31)$$

The covariant version is evidently

$$K^\mu(A) = -\frac{1}{8\pi^2}\epsilon^{\mu\nu\alpha\beta}\text{Tr}\left[A_\nu\partial_\alpha A_\beta + \frac{2}{3}A_\nu A_\alpha A_\beta\right] \quad (16.32)$$

$K^0(A)$ is called the (three-dimensional) Chern-Simons term. From (16.32) we find

$$\partial_\mu K^\mu(A) = -\frac{1}{32\pi^2}\epsilon^{\mu\nu\alpha\beta}\text{Tr}\left[F_{\mu\nu}F_{\alpha\beta}\right] \quad (16.33)$$

Thus the path integral for QCD becomes

$$\Psi[A,t] = \int d\mu[A/\mathcal{G}_*]\, e^{iS(A,t,A',t')+i\theta\int_{t'}^t d^4x\, \partial_\mu K^\mu}\Psi[A',t'] \quad (16.34)$$

For Green's functions, one is interested in $t \to \infty$, $t' \to -\infty$. The path integral, which becomes the generating functional for Green's functions, is then given by

$$Z[J] = \int d\mu[A/\mathcal{G}_*]\, e^{iS(A)+i\theta\nu[A]}\, e^{\int d^4x\, iA_\mu^a J^{a\mu}} \quad (16.35)$$

where we have included a source term $e^{i\int A\cdot J}$ to facilitate the calculations of arbitrary correlators. $\nu[A]$ is given by

$$\nu[A] = -\frac{1}{32\pi^2}\int d^4x\, \epsilon^{\mu\nu\alpha\beta}\text{Tr}\left[F_{\mu\nu}F_{\alpha\beta}\right] \quad (16.36)$$

$\nu[A]$ with integration over all time (and space) is known as the instanton number. The density

$$c_2 = \frac{1}{32\pi^2} \, \epsilon^{\mu\nu\alpha\beta} \mathrm{Tr}\, [F_{\mu\nu}F_{\alpha\beta}] = \frac{1}{8\pi^2} \mathrm{Tr}(F \wedge F) \qquad (16.37)$$

is the second Chern class of the gauge field A; $-\nu[A]$ is the second Chern number.

An important property of $\epsilon^{\mu\nu\alpha\beta}\mathrm{Tr}\,[F_{\mu\nu}F_{\alpha\beta}]$ is that, because of the ϵ-symbol, it has only one time index and three spatial indices. As a result, it is odd under parity P and time-reversal T. Its presence in the path integral thus leads to T-violation by strong interactions if $\theta \neq 0$. (Our discussion has been for $SU(2)$; the gauge group has to be extended to $SU(3)$ for the case of QCD, but all the above analysis goes through for all compact gauge groups since they always contain $SU(2)$-subgroups and $\Pi_0[\mathcal{G}_*] = \Pi_3(G) = \mathbf{Z}$ for all compact Lie groups, except for $SO(4)$ in which case it is $\mathbf{Z} \times \mathbf{Z}$.) One consequence of this T-violation is that the neutron can have a static electric dipole moment. From experimental measurements of the neutron electric dipole moment, we get the experimental limit on θ as

$$|\theta| \leq 10^{-9} \qquad (16.38)$$

This abnormally small value of θ has been a puzzle, referred to as the strong CP-problem. It suggests that $\theta = 0$ by virtue of some symmetry. Attempts to understand this small value of θ has led to the concept of axions and associated interesting ideas.

Another important consequence of the presence of the $\epsilon^{\mu\nu\alpha\beta}$ is that the Euclidean continuation of $\nu[A]$ does not pick up any factors of $i = \sqrt{-1}$.

$$d^4x \, \epsilon^{\mu\nu\alpha\beta}F_{\mu\nu}F_{\alpha\beta} \sim dt d^3x \, \epsilon^{ijk}F_{0i}F_{jk} = dx^4 d^3x \, \epsilon^{ijk}F_{4i}F_{jk} \qquad (16.39)$$

where $x^4 = it = ix^0$. The formula (16.35) for $Z[J]$ can be written, in terms of a Euclidean functional integral, as

$$Z[J] = Z_E[J]\Big]_{x^4=ix^0, A_4=-iA_0}$$

$$Z_E[J] = \int d\mu[A/\mathcal{G}_*] \, e^{-S_E(A)+i\theta\nu[A]} \, e^{i\int A \cdot J} \qquad (16.40)$$

16.3 Instantons

We have shown that $\Pi_1(\mathcal{C}) = \mathbf{Z}$. There are noncontractible paths in $\mathcal{A}/\mathcal{G}_*$. An example of such a path is

$$A_i(\boldsymbol{x}, x_4) = A_i(\boldsymbol{x})\frac{1}{e^{x_4}+1} + \frac{e^{x_4}}{e^{x_4}+1}A_i^{g_1}(\boldsymbol{x}) \qquad (16.41)$$

which has the property that $A_i(\boldsymbol{x}, x_4) \to A_i(\boldsymbol{x})$ as $x_4 \to -\infty$ and $A_i(\boldsymbol{x}, x_4) \to A_i^{g_1}(\boldsymbol{x})$ as $x_4 \to +\infty$. By construction $\nu[A_i(\boldsymbol{x}, x_4)] = 1$. The contribution of such a path to the path integral is given by $\exp\left(-\mathcal{S}[A(x)] + i\theta\right)$. Of course there are an infinity of paths which can be deformed to (16.41). In summing over such paths, the dominant contribution will come from paths with the least Euclidean action. We thus expect certain extremal paths with the properties

1. $\frac{\delta \mathcal{S}_E}{\delta A_\mu} = 0$, i.e., $D_\mu F^{\mu\nu} = 0$ for pure Yang-Mills theory
2. $A_0 \equiv A_4 = 0$ (our gauge choice)
3. $A_i(\boldsymbol{x}, x_4) \to 0$ as $x_4 \to -\infty$, $A_i(\boldsymbol{x}, x_4) \to -\partial_i g_1 \, g_1^{-1}$ as $x_4 \to \infty$
4. $\mathcal{S}_E[A_i(x)] < \infty$ (otherwise it is irrelevant for the path integral since $e^{-\mathcal{S}} \to 0$ for $\mathcal{S} \to \infty$)

In condition 3, we have considered, for simplicity, the starting configuration as $A_i(\boldsymbol{x}) = 0$. Apart from our gauge choice, these conditions are equivalent to

$$D_\mu F^{\mu\nu} = 0$$
$$\nu[A] = 1$$
$$\mathcal{S}_E[A] < \infty \tag{16.42}$$

Configurations $A_\mu(x)$ satisfying (16.42) are called instantons. Configurations with $\nu[A] = -1$ will give anti-instantons. $A_\mu(x)$'s with $\nu[A] =$ integer, ($\neq -1, 0, 1$) give multi-instanton or anti-instanton configurations appropriately.

The one-instanton configuration can be written as

$$A_\mu(x) = \frac{x^2}{x^2 + \lambda^2} \omega^{-1} \partial_\mu \omega$$
$$\omega = \frac{x_4 + i\boldsymbol{\tau} \cdot \boldsymbol{x}}{\sqrt{x^2}} = \frac{e_\mu x^\mu}{\sqrt{x^2}} \tag{16.43}$$

where $x^2 = x_\mu x^\mu$ and $e_\mu = (1, i\tau_i)$. λ is a scale factor giving the "size" of the instanton. Since our arguments leading to $\nu[A]$ were presented in the gauge $A_0 = A_4 = 0$, we note that the configuration (16.43) can be transformed into this gauge. It then looks like

$$A_i = U\left(\frac{x^2}{x^2 + \lambda^2} \omega^{-1} \partial_i \omega\right) U^{-1} \; - \; \partial_i U \, U^{-1} \tag{16.44}$$
$$U = \exp\left(i\boldsymbol{\tau} \cdot \hat{x}\rho\right] \tag{16.45}$$
$$\rho = -\frac{|\boldsymbol{x}|}{\sqrt{|\boldsymbol{x}|^2 + \lambda^2}} \left[\arctan\left(\frac{x_4}{\sqrt{|\boldsymbol{x}|^2 + \lambda^2}}\right) + \frac{\pi}{2}\right] \tag{16.46}$$
$$\frac{\partial \rho}{\partial x_4} = -\frac{|\boldsymbol{x}|}{x^2 + \lambda^2} \tag{16.47}$$

(It is worth noting at this stage that the formulae for the gauge potentials do have singularities in general. To get a nonsingular description, one has to define different expressions for A on different coordinate patches, with the gauge transformations as transition functions on the intersections. Gauge invariant quantities, such as $\mathrm{Tr}(F_{\mu\nu}F_{\alpha\beta})$ and its integral, are nonsingular and well defined for the instanton; however, if we choose to express the instanton number in terms of $K^0(A)$ which is not gauge invariant, a patchwise description is needed.)

Instanton configurations obey a self-duality $(F_{\mu\nu} = \tilde{F}_{\mu\nu})$ or antiself-duality $(F_{\mu\nu} = -\tilde{F}_{\mu\nu})$ condition which is best seen using the so-called Bogomol'nyi inequality. The dual of $F^a_{\mu\nu}$ is given as $\tilde{F}^a_{\mu\nu} = \frac{1}{2}\epsilon_{\mu\nu\alpha\beta}F^a_{\alpha\beta}$. We start with the inequality

$$\int d^4x \; (F^a_{\mu\nu} - \tilde{F}^a_{\mu\nu})^2 \geq 0 \qquad (16.48)$$

Expanding out and using the definition of the instanton number $\nu[A]$ from (16.36), we find for the Euclidean Yang-Mills action

$$\mathcal{S}_E = \frac{1}{4g^2}\int d^4x \; F^a_{\mu\nu}F^a_{\mu\nu}$$
$$\geq \frac{8\pi^2}{g^2}\left|\nu[A]\right| \qquad (16.49)$$

This is the Bogomol'nyi inequality. (If the instanton number is negative, we can use the inequality $(F^a_{\mu\nu} + \tilde{F}^a_{\mu\nu})^2 \geq 0$ to obtain the bound given above.) This shows that the solutions for $\nu = 1$ which minimize the action saturating this bound must be self-dual. Notice that if they are self-dual or antiself-dual, then the equations we have to solve are first order in the derivatives and so they are simpler. The solutions given in (16.43, 16.47) obey the self-duality property. In fact, the Bogomol'nyi inequality shows that the solutions for any $\nu \neq 0$ must be either self-dual or antiself-dual. Of course, there are many configurations with $\nu \neq 0$ which are neither self-dual or antiself-dual, which are not solutions of the equations of motion. Also not every self-dual or antiself-dual configuration has to minimize the action; there are many which are just extrema of the action.

The one-instanton configuration of (16.43) can also be written out as

$$A^a_\mu = 2\frac{\eta^a_{\mu\nu}x^\nu}{(x^2 + \lambda^2)} \qquad (16.50)$$

where the three $\eta^a_{\mu\nu}$ is given by

$$\eta^a_{\mu\nu} = -\delta^a_\mu\delta_{\nu4} + \delta^a_\nu\delta_{\mu4} + \epsilon^{abc}\delta_{\mu b}\delta_{\nu c} \qquad (16.51)$$

$\eta^a_{\mu\nu}$ is known as the 't Hooft tensor.

Instanton solutions can be used to carry out a semiclassical evaluation of the functional integral. This will require expanding the action around the instanton solution by writing $A = A_{inst}(x) + \xi(x)$, where $\xi(x)$ is a small fluctuation in the field; keeping terms in the action up to the quadratic order in ξ, one can then do a Gaussian integral to get $Z[J]$. This should be done, not with just one instanton; one should sum over all classical solutions, which include multi-instanton solutions. In practice, the calculation is limited to a dilute gas of far separated instantons and anti-instantons. Some features of the theory are exposed by this calculation, but this is still within the small coupling regime.

16.4 Fermions on an instanton background and an index theorem

Consider a four-dimensional space $\mathcal{M}$ with Eucliden signature for the metric. We want to analyze the eigenmodes of the Dirac operator. The spinors on the manifold can be separated into components of left and right chirality.

$$\psi = \tfrac{1}{2}(1 + \gamma_5) \, \psi \; + \; \tfrac{1}{2}(1 - \gamma_5) \, \psi$$
$$\equiv \phi_L + \phi_R \qquad (16.52)$$

We also define $\gamma \cdot D = D - D^\dagger$ where

$$D = (\gamma \cdot D) \left(\frac{1 + \gamma_5}{2} \right)$$
$$D^\dagger = \left(\frac{1 + \gamma_5}{2} \right) (\gamma \cdot D)^\dagger = (-\gamma \cdot D) \left(\frac{1 - \gamma_5}{2} \right) \qquad (16.53)$$

We denote the set of *normalizable* spinor modes of left chirality on the manifold as E_L; E_R will denote the set of normalizable modes of right chirality. Since γ_μ changes the chirality, we have the mappings

$$\begin{aligned} D &: & E_L &\longrightarrow & E_R \\ D^\dagger &: & E_R &\longrightarrow & E_L \end{aligned} \qquad (16.54)$$

For example, if ϕ belongs to E_L,

$$\gamma_5(D\phi) = \gamma_5(\gamma \cdot D)\tfrac{1}{2}(1 + \gamma_5)\phi = -(\gamma \cdot D)\tfrac{1}{2}(1 + \gamma_5)\phi$$
$$= -D\phi \qquad (16.55)$$

showing that $D\phi$ belongs to E_R with $\gamma_5 = -1$. This also means that if $D\phi \neq 0$, we have a $1 - 1$ mapping from E_L to E_R. Likewise, $D^\dagger$ provides a $1 - 1$ mapping $E_R \rightarrow E_L$. Thus we have the result that the nonzero modes are paired up. The zero modes or the kernels of the two operators need not match. We define

$$\text{index}(\gamma \cdot D) = dim(ker\ D) - dim(ker\ D^\dagger) \qquad (16.56)$$

We can express this in another way. We cannot define the eigenmodes of D or $D^\dagger$ since these connect different spaces. But $D^\dagger D : E_L \to E_L$ and $DD^\dagger : E_R \to E_R$. If ϕ is an eigenfunction of $D^\dagger D$ with eigenvalue λ, i.e., $D^\dagger D\phi = \lambda\phi$, we get $D\,D^\dagger D\phi = \lambda D\phi$, showing that $D\phi$ is an eigenfunction of $DD^\dagger$ with the same eigenvalue. All nonzero modes are exactly paired up. Further, $D^\dagger D$ and $DD^\dagger$ are elliptic operators; i.e., the term with the highest number of derivatives is elliptic. Thus we can also define

$$\text{index}(\gamma \cdot D) = \dim(\ker D^\dagger D) - \dim(\ker DD^\dagger)$$

$$= \begin{pmatrix} \text{number of zero} \\ \text{modes of } D^\dagger D \end{pmatrix} - \begin{pmatrix} \text{number of zero} \\ \text{modes of } DD^\dagger \end{pmatrix} \quad (16.57)$$

We assume that $\mathcal{M}$ is a compact orientable Riemannian manifold with $\partial\mathcal{M} = \emptyset$. In our case of flat Euclidean space, this is equivalent to requiring that $F_{\mu\nu}$ vanish as $|x| \to \infty$. In this case, the operators $D^\dagger D$ and $DD^\dagger$ are Fredholm operators. They are continuous operators and have finite multiplicity of eigenvalues. Thus the index is finite.

The index can also be written as

$$\text{index}(\gamma \cdot D) = \text{Tr}\left[e^{-D^\dagger D/M^2}\right]_{E_L} - \text{Tr}\left[e^{-DD^\dagger/M^2}\right]_{E_R}$$

$$= \text{Tr}\left[\gamma_5 \exp\left((\gamma \cdot D)^2/M^2\right)\right]_{E=E_L+E_R}$$

$$\equiv \text{Tr}(\gamma_5) \quad (16.58)$$

The trace is over the infinite-dimensional spaces of eigenmodes and over the Dirac matrices. The nonzero eigenvalues and eigenmodes are paired and so will cancel out in the sum. For the zero eigenvalues, it does not matter what M^2 is, so we can calculate this as $M^2 \to \infty$. We then find

$$\text{index}(\gamma \cdot D) = \lim_{M^2 \to \infty} \text{Tr}\left[\gamma_5 e^{(\gamma \cdot D)^2/M^2}\right]$$

$$= \int \frac{d^4p}{(2\pi)^4} \,\text{Tr}\left[\langle y|p\rangle\langle p|\gamma_5 e^{-(\gamma\cdot(p-iA))^2/M^2}|x\rangle\right]_{M^2\to\infty,\,y\to x}$$

$$= \int d^4x \frac{d^4p}{(2\pi)^4} \,\text{Tr}\left[\gamma_5 \exp\left(-(p-iA)^2/M^2 + \tfrac{1}{2}F_{\mu\nu}\gamma^\mu\gamma^\nu/M^2\right)\right]$$

$$= \int d^4x\,\text{Tr}\left[\frac{1}{2!M^4}\gamma_5\tfrac{1}{2}F_{\mu\nu}\gamma^\mu\gamma^\nu\tfrac{1}{2}F_{\alpha\beta}\gamma^\alpha\gamma^\beta\right] \int \frac{d^4p}{(2\pi)^4}e^{-p^2/M^2}$$

$$= \frac{1}{16\pi^2}\int d^4x\,\text{Tr}\left(F_{\mu\nu}\tilde{F}_{\mu\nu}\right)$$

$$= -\nu[A] \quad (16.59)$$

(Here we use the convention $\epsilon_{4123} = \epsilon_{0123} = 1$ and $\text{Tr}(\gamma^5\gamma^\mu\gamma^\nu\gamma^\alpha\gamma^\beta) = 4\epsilon^{\mu\nu\alpha\beta}$.) We have thus shown that

$$\dim(\ker D) - \dim(\ker D^\dagger) = -\nu[A] \quad (16.60)$$

This is the Atiyah-Singer index theorem applied to the special case of the spin complex in four dimensions. The general index theorem gives an integral formula for the index of any elliptic complex in any even number of dimensions.

The index is a topological invariant, i.e., it is invariant under small changes of the potential A_μ. If we write the index as a surface integral given by

$$\nu[A] = \oint d\Sigma_\mu K^\mu$$

$$K^\mu = -\frac{1}{8\pi^2} \text{Tr} \left(A_\nu \partial_\alpha A_\beta + \frac{2}{3} A_\nu A_\alpha A_\beta \right) \epsilon^{\mu\nu\alpha\beta} \qquad (16.61)$$

we can easily check that

$$\nu[A + \xi] = \nu[A] - \frac{1}{4\pi^2} \oint d\Sigma_\mu \text{Tr}(\xi_\nu F_{\alpha\beta}) \epsilon^{\mu\nu\alpha\beta}$$

$$= \nu[A] \qquad (16.62)$$

using the fact that $F_{\alpha\beta} \to 0$ as $|x| \to \infty$. $\oint d\Sigma_\mu K^\mu$ is not zero by this argument, since it is not gauge-invariant and A_μ does not have to be zero, but can be a pure gauge potential as $|x| \to \infty$. Thus the situation is as follows. For a given potential A_μ, $D^\dagger D$ and $DD^\dagger$ have, say, n_+ and n_- zero modes, respectively. What we have shown is that $n_+ - n_- = -\nu[A]$, so that, except for $|\nu[A]|$ zero modes, the remainder are paired up. The paired zero modes, if any, are generally not preserved under small changes in A_μ. Thus they disappear, become nonzero modes, when we make a small change in the potential. The unpaired zero modes are preserved under small deformations of the potential A_μ.

Consider now the functional integral over fermions. We have

$$Z = \sum_\nu \int d\mu(A)[d\psi d\bar\psi] \, e^{i\nu\theta} \exp\left(-\mathcal{S}[A] - \int \bar\psi\gamma \cdot D\psi \right)$$

$$= \sum_\nu Z_\nu \qquad (16.63)$$

For the sector with $\nu = -1$, we can write

$$Z_1 = \int d\mu(A)[d\psi d\bar\psi] \, e^{-i\theta} \exp\left(-\mathcal{S}[A] - \int \bar\psi\gamma \cdot D\psi \right) \qquad (16.64)$$

The Dirac Lagrangian can be split as $\bar\psi\gamma \cdot D\psi = \bar\phi_L D\phi_L - \bar\phi_R D^\dagger \phi_R$ where

$$\phi_L \in E_L, \qquad\qquad \bar\phi_L \in E_R^*$$
$$\phi_R \in E_R, \qquad\qquad \bar\phi_R \in E_L^* \qquad (16.65)$$

Based on this, we introduce the mode expansions

$$\phi_L = a\phi_0 + \sum_n b_n \phi_n$$

$$\bar{\phi}_L = \sum_n \bar{b}_n \chi_n^*$$

$$\phi_R = \sum_n c_n \chi_n$$

$$\bar{\phi}_R = \bar{a}\phi_0^* + \sum_n \bar{c}_n \phi_n^* \qquad (16.66)$$

where ϕ_0 is the zero mode and the coefficients $a, \bar{a}, b_n, \bar{b}_n, c_n \bar{c}_n$ are Grassman variables. The action does not involve $a, \bar{a}$ because ϕ_0 is a zero mode. In fact, for the Dirac action, we get

$$\mathcal{S}_D = \sum_{m,n} \left[\bar{b}_n b_m \int \chi_n^* D\phi_m - \bar{c}_n c_m \int \phi_n^* D^\dagger \chi_m \right] \qquad (16.67)$$

The fermionic measure of integration contains the factor $dad\bar{a}$. Thus

$$Z_1 = \int d\mu(A)[db d\bar{b} dc d\bar{c}] dad\bar{a} \ e^{-i\theta} \ \exp\left(-\mathcal{S}_D(b, c, \bar{b}, \bar{c}) - \mathcal{S}(A)\right)$$
$$= 0 \qquad (16.68)$$

since $\int dad\bar{a} = 0$ for Grassman variables a and $\bar{a}$. The one-instanton (and more generally the many-instanton) contribution to the functional integral is zero if there are massless fermions. The nonzero amplitude must involve $\bar{\phi}_R \phi_L$, which can produce factors of a and $\bar{a}$. We find

$$\bar{\phi}_R \phi_L = \bar{a}a \int \phi_0^* \phi_0 + \cdots$$

$$\langle \bar{\phi}_R \phi_L \rangle = \int dad\bar{a} \ \bar{a}a \left(\int d^4 x \phi_0^* \phi_0 \right) \times$$

$$\int d\mu(A)[db d\bar{b} dc d\bar{c}] \ e^{-i\theta} \ \exp\left[-\mathcal{S}_D(b, c, \bar{b}, \bar{c}) - \mathcal{S}(A)\right] \qquad (16.69)$$

More generally, there are $-\nu$ zero modes and we have to have a factor like $(\bar{\phi}_R \phi_L)^{-\nu}$ so as to get a sufficient number of Grassman variables for a nonzero amplitude. If there are several species of fermions, we have zero modes corresponding to each species and so we must have a factor like $\bar{\phi}_R \phi_L$ for each species to have a nonzero value for the integral. By the antisymmetry properties of the Grassman variables, we see that the nozero amplitude involves $\det(\bar{\phi}_R \phi_L)$. The result of the integration over fermions can then be represented as an effective interaction

$$\mathcal{S}_{eff} = -i \ \nu \int \log \det \bar{\phi}_R \phi_L \qquad (16.70)$$

This is the 't Hooft effective interaction. The effective action which we obtained for the axial anomaly is closely related to this. It provides one way to understand the $U(1)_A$ problem in QCD; the extra mass of the η' particle is essentially due to this interaction induced by the instantons of QCD. A similar term generated by the instantons of $SU(2) \times U(1)$ gauge theory of electroweak interactions leads to baryon-number-violating effects in the standard model.

Under the axial $U(1)$ transformation $\phi_L \to e^{-i\alpha}\phi_L$, $\phi_R \to e^{i\alpha}\phi_R$. In the fermionic measure of integration, the exponentials from the transformation of the b's and c's cancel out. The measure thus transforms as

$$[dad\bar{a}]\,[dbd\bar{b}dcd\bar{c}] \to e^{2i\alpha}\,[dad\bar{a}]\,[dbd\bar{b}dcd\bar{c}] \tag{16.71}$$

In general, for instanton number ν, we get

$$\begin{aligned}[d\psi d\bar{\psi}] \quad &\to e^{-2i\alpha\nu}\,[d\psi d\bar{\psi}]\\ &= \exp\left(2i\alpha(\mathrm{Tr}\gamma_5)\right)\,[d\psi d\bar{\psi}]\end{aligned} \tag{16.72}$$

where we have used the definition of the functional trace over γ_5 in (16.58). The exponential factor can be interpreted as the Jacobian of the axial $U(1)$ transformation $\psi \to \exp(-i\gamma_5\alpha)\psi$. (The Jacobian for the transformation of the fermionic measure is the inverse of the Jacobian for bosonic fields, which explains the sign in the exponent.) Thus, although the action is invariant under axial $U(1)$ transformations, the measure of integration is not and this leads to the axial anomaly. The integrated version of the axial anomaly is the index theorem (16.60). This gives a new understanding of anomalies as arising from the lack of invariance of the functional measure; this will be discussed in more detail in the next chapter.

16.5 Baryon number violation in the standard model

We have seen in our discussion of anomalies that baryon number (B) and lepton number (L) are not good symmetries in the standard model, being violated by anomalies. Only $B - L$ is a good anomaly-free symmetry. Thus the standard model does contain baryon-number-violating interactions. We can use the index theorem to show how this arises in some detail.

The violation of baryon and lepton numbers was obtained in Chapter 13 as

$$\partial_\mu J^{B\mu} = \partial_\mu J^{L\mu} = c_2[b] - 2\,c_2[c] \tag{16.73}$$

The expression $-c_2[b]$ is the instanton density for the $SU(2)$ gauge field of weak interactions. Consider, therefore, a gauge-field configuration for which this $SU(2)$ instanton number is not zero, say, $\nu = -1$, as an example. In this case, there are zero modes for all the massless fermion fields which couple to the $SU(2)$ gauge field. In the standard model with one generation of quarks and leptons, the relevant fermions are the left-handed quark doublet of u and d quarks and the e, ν_e doublet. The quarks come in three colors, which is just a degeneracy factor for the coupling to $SU(2) \times U(1)$ gauge fields. Thus we expect three quark zero modes, one for each color, and one lepton zero mode. Each field can be expanded as

$$\phi_L = a\phi_0 + \sum_n b_n\phi_n$$

$$\bar{\phi}_L = \sum_n \bar{b}_n\chi_n^* \tag{16.74}$$

Since the Grassman coefficients anticommute with each other, we need antisymmetrization of the fields to get a nonzero value. Keeping this in mind, the nonvanishing amplitude in the one-instanton sector is of the form

$$\epsilon_{abc}\epsilon_{\alpha\beta}\epsilon_{\gamma\delta}\epsilon^{AB}\epsilon^{CD}\langle Q_A^{a\alpha}(x)Q_B^{b\beta}(x)Q_C^{c\gamma}(x)l_D^{\delta}(x)\rangle$$

$$= \int d\mu[A, \psi, \bar{\psi}] \; e^{-S_E}\epsilon_{abc}\epsilon_{\alpha\beta}\epsilon_{\gamma\delta}\epsilon^{AB}\epsilon^{CD}Q_A^{a\alpha}(x)Q_B^{b\beta}(x)Q_C^{c\gamma}(x)l_D^{\delta}(x)$$

$$\tag{16.75}$$

Here a, b, c refer to color indices; $\alpha, ..., \delta$ are $SU(2)$ indices; and $A, ..., D$ are Lorentz indices for the two-component left-handed fields. Notice that the quantity on the left-hand side is invariant under color $SU(3)$, weak $SU(2)$, and the hypercharge $U(1)_Y$ as well as under Lorentz transformations. This combination will give the right number of Grassman coefficients for the zero modes to make the integral nonzero. The actual value of the integral is small since the action for an instanton is at least $8\pi^2/g^2$, where g is the $SU(2)$ coupling constant. Thus we get a suppression factor $\exp(-8\pi^2/g^2)$ from the exponential, if we expand around the one-instanton configuration. Since g is small, this factor is very small. If we consider an amplitude similar to (16.75), but with a number of Higgs fields also at the same point x, effectively it corresponds to a local baryon number violating interaction with a number of accompanying Higgs particles. The total phase space integral for the cross section can give an enhancement which can compensate for the exponential suppression to some extent. Nevertheless, the rate remains small in general. The question of whether it may be sufficient to explain the baryon asymmetry of the universe is interesting; it seems it is inadequate unless there is further enhancement due to the characteristics of the electroweak symmetry breaking phase transition.

References

1. For general information on the nature of the configuration space, see I.M. Singer, Phys. Scripta **T24**, 817 (1981); Commun. Math. Phys. **60**, 7 (1978).

2. The instanton solution was discovered by A.A. Belavin, A.M. Polyakov, A.S. Schwarz and Yu.S. Tyupkin, Phys. Lett. **B59**, 85 (1975). For the instanton and other classical solutions, see A. Actor, Rev. Mod. Phys. **51**, 461 (1979); R. Rajaraman, *Solitons and Instantons*, Elsevier Science Ltd. (1982).

3. The existence of self-dual and antiself-dual solutions which are extrema of the Yang-Mills action, and not minima, was shown in L.M. Sibner, R.J. Sibner and K. Uhlenbeck, Proc. Nat. Acad. Sci. USA **86**, 8610 (1989); L. Sadun and J. Segert, Commun. Math. Phys. **145**, 391 (1992); J. Schiff, Phys. Rev. **D44**, 528 (1991).

4. The semiclassical calculation around the instanton was done by G. 't Hooft, Phys. Rev. **D14**, 3432 (1976), Erratum *ibid.* **D18**, 2199 (1978). A nice review of instanton calculations is in S. Coleman, *Aspects of Symmetry*, Cambridge University Press (1988). In this context, see also M.A. Shifman, *Instantons in Gauge Theories*, World Scientific Pub. Co. (1994).

5. For the physical implications of the instanton and the θ-parameter, see C.G. Callan, R. Dashen and D. Gross, Phys. Lett. **B63**, 334 (1976); R. Jackiw and C. Rebbi, Phys. Rev. Lett. **37**, 172 (1976); R. Jackiw, Rev. Mod. Phys. **49**, 681 (1977).

6. For instantons in relation to hadron structure, see C.G. Callan, R. Dashen and D. Gross, Phys. Rev. **D17**, 2717 (1978); *ibid.* **D18** 4684 (1978); *ibid.* **D19**, 1826 (1979).

7. The CP-violation effect of the θ-parameter is investigated in V. Baluni, Phys. Rev. **D19**, 2227 (1978); R.J. Crewther, P. Di Vecchia, G. Veneziano and E. Witten, Phys. Lett. **B88**, 123 (1979).

8. One way to explain the smallness of θ is by making it dynamically determined. This leads to the Peccei-Quinn symmetry introduced in R.D. Peccei and H. Quinn, Phys. Rev. Lett. **38**, 1440 (1977); Phys. Rev. **D16**, 1791 (1977). One consequence of this is the axion, as shown by S. Weinberg, Phys. Rev. Lett. **40**, 223 (1978); F. Wilczek, Phys. Rev. Lett. **40**, 279 (1978). Variants of this idea, which makes the axion consistent with experimental bounds, are in J.E. Kim, Phys. Rev. Lett. **43**, 103 (1979); M.A. Shifman, A. Vainshtein and V. Zakharov, Nucl. Phys. **B166**, 493 (1980); A. Zhitnitsky, Sov. J. Nucl. Phys. **31**, 260 (1980); M. Dine, W. Fischler and M. Srednicki, Phys. Lett. **B104**, 199 (1981).

9. The general index theorem is due to M. Atiyah and I.M. Singer, Ann. Math. **87**, 485, 546 (1968); *ibid.* **93**, 119, 139 (1971); M. Atiyah and G.B. Segal, Ann. Math. **87**, 531 (1968). They can be derived by heat equation methods which are very similar to the calculation we do; see P.B. Gilkey, *Invariance Theory, the Heat Equation and the Atiyah-Singer Index Theorem*, CRC Press (1995). The index theorem can also be derived by using supersymmetric quantum mechanics, see L. Alvarez-Gaumé, Commun. Math. Phys. **90**, 161 (1983); D. Friedan and P. Windey, Nucl. Phys. **B235**, 395 (1984).

10. The 't Hooft effective action and baryon number violation via weak instantons were obtained by G. 't Hooft, Phys. Rev. Lett. **37**, 8 (1976).

17 Anomalies II

17.1 Anomalies and the functional integral

Anomalies arise when a classical symmetry cannot be realized in the quantum theory because there is no choice of regularization which preserves all the required or desired symmetries. We have calculated the anomalies in perturbation theory using Feynman diagrams. Fujikawa has given an elegant interpretation of anomalies in the functional integral language. As we have seen, a quantum field theory can be defined by the functional integral of the exponential of the Euclidean action. All questions of regularization and renormalization are equivalent to the problem of a proper regularized definition of the functional integral. There are two basic ingredients in the functional integral, the classical action and the functional measure of integration. In a situation with an anomalous symmetry, the classical action has the symmetry. Fujikawa's observation was that the quantum anomalies can be understood as arising from the nontrivial transformation of the measure under the symmetry transformation.

Consider the functional integral for a set of massless Dirac fermions vectorially coupled to a nonabelian gauge field V_μ.

$$Z = \int [d\psi d\bar{\psi}] \; e^{-\mathcal{S}(\psi,\bar{\psi})}$$

$$\mathcal{S}(\psi,\bar{\psi}) = \int d^4x \; \bar{\psi}\gamma \cdot (\partial + V)\psi \tag{17.1}$$

The classical action has the chiral $U(1)$ symmetry

$$\psi \to e^{-i\gamma_5\theta} \, \psi, \qquad\qquad \bar{\psi} \to \bar{\psi} \, e^{-i\gamma_5\theta} \tag{17.2}$$

where the parameter θ is independent of x^μ. We now consider a change of variables in the functional integral given by $\psi' = e^{-i\gamma_5\theta(x)} \, \psi$, $\bar{\psi}' = \bar{\psi} \, e^{-i\gamma_5\theta(x)}$, where $\theta(x)$ is taken to be an arbitrary function of the coordinates. Z is unchanged under this since it is just a change of variables. Thus

$$Z = \int [d\psi' d\bar{\psi}'] \; e^{-\mathcal{S}(\psi',\bar{\psi}')}$$

$$= \int [d\psi d\bar{\psi}] \; \det(e^{2i\gamma_5\theta}) \; e^{-\mathcal{S}(\psi,\bar{\psi})} \; \exp\left[-\int d^4x \; \theta(x)\partial_\mu J_{\mu5}\right]$$

$$= \int [d\psi d\bar{\psi}] \ e^{-\mathcal{S}(\psi,\bar{\psi})} \exp \left[2i \text{Tr}(\gamma_5 \theta) - \int d^4x \ \theta \partial_\mu J_{\mu 5} \right] \qquad (17.3)$$

where J_μ^5 is the $U(1)$ axial vector current

$$J_{\mu 5} = i \bar{\psi} \gamma_\mu \gamma_5 \psi \qquad (17.4)$$

(Since we have Grassmann variables, the Jacobian is the determinant of the inverse of the transformation.) The invariance of the classical action gives the classical conservation law $\partial_\mu J_{\mu 5} = 0$. Notice that in (17.3) one gets an additional contribution, $\det(e^{2i\gamma_5 \theta})$, which is the Jacobian of the transformation of the measure. Comparing the result (17.3) with (17.1), we get the Ward-Takahashi (WT) identity

$$\int [d\psi d\bar{\psi}] \ e^{-\mathcal{S}(\psi,\bar{\psi})} \left[\int d^4x \ \theta \partial_\mu J_{\mu 5} \ - \ 2i \text{Tr}(\gamma_5 \theta) \right] = 0 \qquad (17.5)$$

$\text{Tr}(\gamma_5 \theta)$ involves the functional trace and a trace over the Dirac matrices and must be calculated with proper regularization. This has been done in the previous chapter and gives the result

$$\text{Tr}(\gamma_5 \theta) = \frac{1}{16\pi^2} \int d^4x \ \theta \ \text{Tr}(F_{\mu\nu} \tilde{F}_{\mu\nu}) \qquad (17.6)$$

The WT identity now becomes

$$\int [d\psi d\bar{\psi}] \ e^{-\mathcal{S}(\psi,\bar{\psi})} \int d^4x \ \theta \left[\partial_\mu J_{\mu 5} \ - \ \frac{i}{8\pi^2} \text{Tr}(F_{\mu\nu} \tilde{F}_{\mu\nu}) \right] = 0 \qquad (17.7)$$

The WT identity for correlation functions can be obtained by the same procedure with sources inserted in the functional integral. The lack of conservation of the $U(1)$ axial current shows that the symmetry is not obtained at the quantum level. The variation of the quantum effective action can be read off from the above equation as

$$\delta_\theta \Gamma = \int d^4x \ \theta \ \frac{i}{8\pi^2} \ \text{Tr}(F_{\mu\nu} \tilde{F}_{\mu\nu}) \qquad (17.8)$$

This is in agreement with our calculation using Feynman diagrams in Chapter 13.

We have shown the calculation for the $U(1)$ transformtion. This approach to the anomalies can be used for nonabelian symmetries as well. In this case, one finds $\text{Tr}(\gamma_5 \theta)$ again, where $\theta = -it^a \theta^a(x)$ is a parameter of the nonabelian transformation and the trace involves a functional trace, Dirac trace and a trace over the products of Lie algebra matrices. One can also include vector and axial vector gauge fields. The result of the calculation is the Bardeen form of the anomaly which was given in equation (13.33).

$$\text{Tr}(\gamma_5\theta) = -\frac{1}{8\pi^2} \int d^4x \; \epsilon^{\mu\nu\alpha\beta} \; \text{Str}\left[\varphi\left(\frac{1}{4}F_{V\mu\nu}F_{V\alpha\beta} + \frac{1}{12}F_{A\mu\nu}F_{A\alpha\beta}\right.\right.$$

$$-\frac{2}{3}(A_\mu A_\nu F_{V\alpha\beta} + A_\mu F_{V\nu\alpha}A_\beta + F_{V\mu\nu}A_\alpha A_\beta)$$

$$\left.\left.+\frac{8}{3}A_\mu A_\nu A_\alpha A_\beta\right)\right] \tag{17.9}$$

(The axial anomaly will be $2i$ times $\text{Tr}(\gamma_5\theta)$; $F_{V\mu\nu}$ and $F_{A\mu\nu}$ are given in Chapter 13, equation (13.2).)

17.2 Anomalies and the index theorem

There is a simple and very elegant relationship between anomalies and index theorems for differential operators which occur in the kinetic terms of particle Lagrangians. We have already seen an example of this in the last chapter where the index of the Dirac operator in four dimensions was given by $\text{Tr}(\gamma_5)$, which is the integral of the anomaly for the axial $U(1)$ transformation. This result generalizes to other dimensions.

Let $\mathcal{M}$ be a $2n$-dimensional manifold with Euclidean signature. One can define a set of Dirac γ-matrices by the algebra

$$\gamma_\mu \gamma_\nu + \gamma_\nu \gamma_\mu = 2 \delta_{\mu\nu} 1 \tag{17.10}$$

One can realize them explicitly as $(2^n \times 2^n)$-matrices. The chirality matrix or the analogue of γ_5 matrix is then given by

$$\gamma_{2n+1} = i^n \gamma_1 \gamma_2 \cdots \gamma_{2n} \tag{17.11}$$

(Here we have slightly changed our definition of γ_{2n+1} compared to our definition of γ_5 in four dimensions. In four dimensions, it was convenient to have $\epsilon^{0123} = 1$; in generalizing, it is easier to take $\epsilon 12 \cdots (2n) = 1$.) For a Dirac spinor in $2n$ dimensions with 2^n components, the chiral projections are defined by

$$\psi_\pm = \frac{1}{2}(1 \pm \gamma_{2n+1}) \, \psi \tag{17.12}$$

The index of the Dirac operator $\gamma \cdot D$ gives the number of normalizable zero modes of positive chirality (n_+) minus the number of normalizable zero modes of negative chirality (n_-). This is given by $\text{Tr}(\gamma_{2n+1})$. One can show by direct calculation

$$n_+ - n_- = \int_{\mathcal{M}} \hat{A}(\mathcal{R}) \wedge Ch(F) \tag{17.13}$$

where $\hat{A}(\mathcal{R})$ is the characteristic class $\hat{A}$-genus and $Ch(F)$ is the Chern character defined in Chapter 14. This result is a special case of the very general index theorem due to Atiyah and Singer. The axial $U(1)$ transformations are given by

$$\psi \rightarrow \exp(-i\gamma_{2n+1}\theta)\,\psi \tag{17.14}$$

and the anomaly of the theory under these transformations is given by

$$2\,\mathrm{Tr}(\gamma_{2n+1}\theta) = 2\int_{\mathcal{M}} \theta\,\hat{A}(\mathcal{R}) \wedge Ch(F) \tag{17.15}$$

The index density gives the anomaly.

The anomaly for the nonabelian transformations is also related to the index density. As we shall show below, for a fermion theory in $2n$ dimensions, it is the index density in $2n+2$ dimensions, which we will denote by $\mathcal{I}_{2n+2}$, that is relevant. Thus for four dimensions we must start with the index density corresponding to six dimensions. It is given by

$$\mathcal{I}_6 = -\frac{i}{48\pi^3}\mathrm{Tr}F^3 \;+\; \frac{i}{384\pi^3}\mathrm{Tr}F\,\mathrm{Tr}\mathcal{R}^2 \tag{17.16}$$

(Wedge products are understood, we have omitted them to avoid cluttering the notation.) Concentrating on the $\mathrm{Tr}F^3$ term first, notice that we have $d\mathrm{Tr}F^3 = \mathrm{Tr}(DF\,F^2 + F\,DF\,F + F^2\,DF) = 0$ by Bianchi identity. Thus, we expect that, at least locally, we can write

$$-\frac{i}{48\pi^3}\mathrm{Tr}F^3 \;=\; d\omega_5 \tag{17.17}$$

ω_5 defined by this equation is the Chern-Simons five-form. There is a neat way to calculate this and other Chern-Simons forms which we will now describe. Let $A_s = sA$, where s is a real parameter between zero and one. The curvature or field strength corresponding to this is $F_s = sF + (s^2 - s)A^2$; also $dF_s/ds = dA + 2sA^2 = D_s A$, where D_s is the covariant derivative with respect to A_s; i.e., $D_s C = dC + sAC + CsA$ for a one-form C. The Bianchi identity is $D_s \wedge F_s = 0$. Evidently we can then write

$$\begin{aligned}
\frac{i^{n+1}}{(2\pi)^{n+1}(n+1)!}\mathrm{Tr}F^{n+1} &= \int_0^1 ds\,\frac{d}{ds}\frac{i^{n+1}}{(2\pi)^{n+1}(n+1)!}\mathrm{Tr}F_s^{n+1} \\
&= \int_0^1 ds\,\frac{i^{n+1}}{(2\pi)^{n+1}(n+1)!}\,(n+1)\mathrm{Tr}D_s A F_s^n \\
&= \int_0^1 ds\,\frac{i^{n+1}}{(2\pi)^{n+1}(n+1)!}\,(n+1)d\,\mathrm{Tr}A F_s^n \\
&= d\omega_{2n+1}
\end{aligned} \tag{17.18}$$

where

$$\omega_{2n+1} = \int_0^1 ds\,\frac{i^{n+1}}{(2\pi)^{n+1}n!}\mathrm{Tr}A F_s^n \tag{17.19}$$

Using $F_s = sF + (s^2 - s)A^2$, we can evaluate the integral to obtain the Chern-Simons form ω_{2n+1}. Explicitly, for $n = 2$, we find

$$\omega_5(A) = -\frac{i}{48\pi^3} \text{Tr} \left(AdAdA + \frac{3}{2}A^3 dA + \frac{3}{5}A^5 \right)$$

$$= -\frac{i}{48\pi^3} \text{Tr} \left(AF^2 - \frac{1}{2}A^3 F + \frac{1}{10}A^5 \right) \qquad (17.20)$$

ω_5 is not gauge-invariant; however, since its derivative is proportional to $\text{Tr}F^3$ and this is gauge-invariant, we have $d[\omega_5(A^g) - \omega_5(A)] = 0$, where $A^g = gAg^{-1} - dgg^{-1}$ is the gauge transform of A. This shows that $\omega_5(A^g) - \omega_5(A)$ is a closed five-form. It can be explicitly calculated as

$$\omega_5(A^g) - \omega_5(A) = d\alpha_4 + \frac{i}{480\pi^3} \text{Tr}(dg\ g^{-1})^5$$

$$\alpha_4 = -\frac{i}{48\pi^3} \text{Tr} \left[g^{-1}dg \left(\frac{1}{2}AdA + \frac{1}{2}dAA + \frac{1}{2}A^3 \right) \right.$$

$$\left. + \frac{1}{4}(g^{-1}dg\ A\ g^{-1}dg\ A) - \frac{1}{2}(g^{-1}dg)^3 A \right]$$

$$(17.21)$$

α_4 is a four-form. For infinitesimal transformations, $g \approx 1 - \theta$, $A \to A + D\theta$, we can write

$$\alpha_4 \approx \frac{i}{48\pi^3} \text{Tr} \left[d\theta \left(\frac{1}{2}AdA + \frac{1}{2}dAA + \frac{1}{2}A^3 \right) \right] \qquad (17.22)$$

Comparing with (13.31), we see that we can write the anomaly for a fermion of left chirality as $2\pi i \int \alpha_4$.

We can regard $\omega_5(A^g) - \omega_5(A)$ as giving the integrated version of the anomaly or the change in the effective action under a finite gauge transformation g. The WZ consistency conditions are the integrability conditions for composing infinitesimal transformations to obtain the change of the effective action under a finite transformation. Since $\omega_5(A^g) - \omega_5(A)$ already furnishes an integrated version, it is clear that expression (17.22) will obey the WZ consistency conditions.

The result for finite gauge transformations may be represented neatly as follows. On the four-dimensional space, we are interested in fields obeying the condition $F \to 0$ and $g \to 1$ as $|x| \to \infty$. Thus $\mathcal{M}$ is effectively compact with no boundary ($\sim S^4$) and we can consider a five-dimensional space $\mathcal{D}$ (a unit disc in $\mathbf{R}^5$) with boundary $\partial\mathcal{D} = \mathcal{M} = S^4$. Let U denote an extension of the function g from spacetime to $\mathcal{D}$, so that U restricted to $\partial\mathcal{D} = \mathcal{M}$ is g. The change in the effective action under a finite gauge transformation can be expressed as

$$\Delta\Gamma = 2\pi i \int_{\mathcal{D}} [\omega_5(A^U) - \omega_5(A)] \qquad (17.23)$$

Equivalently, we may write for the Dirac determinant of a fermion of left chirality

$$\det(\gamma \cdot D^g) = \det(\gamma \cdot D) \, \exp\left(2\pi i \int_{\mathcal{D}} [\omega_5(A^U) - \omega_5(A)] \right) \tag{17.24}$$

In $[\omega_5(A^U) - \omega_5(A)]$, we encounter the term

$$\Omega^{(5)} = \frac{i}{480\pi^3} \mathrm{Tr}(dU \, U^{-1})^5 \tag{17.25}$$

As shown in Chapter 14, this five-form is closed but not exact; it is the pullback to $\mathcal{D}$ of an element of $\mathcal{H}^5(G)$. Therefore (17.23) cannot be integrated to get something defined on the four-dimensional spacetime. This shows that the anomaly cannot be eliminated by a local counterterm, thus completing the identification of the four-dimensional anomaly in terms of the the six-dimensional index density. The expression $2\pi i \int_{\mathcal{D}} [\omega_5(A^U) - \omega_5(A)]$ was first obtained, in a different form, by Wess and Zumino by integrating the anomaly and is therefore referred to as a Wess-Zumino term. The form in which we have given it is due to Witten. Since the differential forms do not involve metrical factors, equations (17.23- 17.25) only need $\mathcal{M}$ to be topologically S^4. Thus the formulae apply for $\mathcal{M} = \mathbf{R}^4$ with the field strengths vanishing as $|x| \to \infty$.

The specification of the finite anomaly (17.23) requires the extension U of g from the four dimensions to the five-dimensional space $\mathcal{D}$. (The gauge fields only occur in α_4 and hence need no extension.) There are many ways to extend g to the disc $\mathcal{D}$. Consider two different extensions U_1, U_2 We can think of spacetime as the equator of a five-sphere and the two extensions as U's defined on the upper and lower hemispheres. In other words, we take $\mathcal{D}$ to be the upper hemisphere for the first extension and the lower hemisphere for the other extension. We then see that the difference in the finite anomaly is given by

$$\Delta\Gamma(U_1) - \Delta\Gamma(U_2) = \oint_{S^5} \Omega^{(5)}(U) \tag{17.26}$$

where $U = U_1$ for the upper hemisphere and $U = U_2$ for the lower hemisphere. On the equator $U_1 = U_2 = g$, so there is no difficulty of continuity of the functions on S^5. The integral in (17.26) gives the winding number of the map $U : \; S^5 \to G$ considered as an element of $\Pi_5(G)$. This is an integer and so the ambiguity of different extensions will not affect equation (17.24) or the exponential of the effective action.

The reason we cannot eliminate the anomaly by a counterterm has to do with the term $\Omega^{(5)}$, which is an element of $\mathcal{H}^5(G)$ for the gauge group G; it cannot be written as an exact form and so it does not become the integral of a local function on spacetime. $\mathcal{H}^5(G)$ is zero for all compact Lie groups, except for $SU(n)$, $n \geq 3$. In four dimensions anomalies can possibly occur only for these gauge groups.

17.3 The mixed anomaly in the standard model

The relationship between the higher-dimensional index density and the anomaly for nonabelian transformations holds in general. We can in fact generalize the formula (17.24) as

$$\det(\gamma \cdot D^g) = \det(\gamma \cdot D) \ \exp\Big(2\pi i [\omega_{2n+1}(A^U) - \omega_{2n+1}(A)] \Big) \qquad (17.27)$$

In general, there can also be gravitational anomalies corresponding to the quantum breaking of the symmetry of coordinate transformations or diffeomorphisms. This can also be obtained from the index density in higher dimensions. Since the index density $\mathcal{I}_{2n+2}$ corresponding to $2n+2$ dimensions is a closed $2n+2$ form, we define

$$\mathcal{I}_{2n+2} = d\omega_{2n+1} \qquad (17.28)$$

where the Chern-Simons form ω_{2n+1} can depend on the gauge and gravitational fields. Let $\tilde{\omega}_{2n+1}$ denote the Chern-Simons form with the transformed gauge and gravitational fields. For the left chiral Dirac determinant on a $2n$-dimensional space $\mathcal{M}^{2n}$, we then have

$$\det (\gamma \cdot \tilde{D}) = \det(\gamma \cdot D) \ \exp\Big(2\pi i \int_{\mathcal{D}} [\tilde{\omega}_{2n+1} - \omega_{2n+1}] \Big) \qquad (17.29)$$

where $\partial \mathcal{D} = \mathcal{M}^{2n}$. The index density $\mathcal{I}_{2n+2}$ is referred to as the anomaly polynomial for $2n$ dimensions. It is very useful in checking for the absence of anomalies; if this polynomial vanishes when the traces are evaluated over the assigned representations of the fields, then the anomaly is zero. This is much easier in practice than using the explicit formula for the anomalies. This is the procedure used for verifying anomaly cancellation in string theories.

The full index density in four dimensions is given by

$$\mathcal{I}_4 = -\frac{1}{8\pi^2} \text{Tr} F^2 + \frac{1}{192\pi^2} \text{Tr} \mathcal{R}^2 \qquad (17.30)$$

This shows that chiral fermions in two dimensions can have gravitational anomalies as well as anomalies under nonabelian gauge transformations.

The index density corresponding to six dimensions given in (17.16) shows that, in a four-dimensional theory, we can have gauge anomalies and anomalies which mix gauge and gravitational fields. This means that the expression for the change in the effective action under a coordinate transformation is not zero unless the gauge field is zero.) There are no purely gravitational anomalies in four dimensions.

We have already checked the absence of gauge anomalies for the standard model. However, since the $U(1)_Y$ charge is nonzero for each fermion, the mixed anomaly can potentially occur for the standard model, due to the

TrF term in the formula (17.16). Actually, with the assigned charges of the fermions, Tr$Y = 0$ if taken over the quarks in each generation or over the leptons in each generation. Thus the mixed anomaly is also zero for the standard model; one can consistently couple it to gravity.

17.4 Effective action for flavor anomalies of QCD

The theory of strong interactions, namely, QCD, involves quarks and gluons. These are eventually bound into states which are color singlets which are in turn mesons, baryons, and glueballs. They are observable as asymptotic states. This picture of confinement of colored states leads to certain requirements on the effective theory of mesons and baryons. For example, some of the electroweak anomalies cancel between quarks and leptons. In the low-energy regime of confined quarks, the leptonic anomaly is unchanged. There must therefore be some terms in the effective action for mesons and baryons which represent the flavor anomalies of QCD and which ensure the cancellation of the leptonic anomaly. Otherwise, the picture of confinement cannot be consistently implemented. One could turn this around and argue that for any theory, not necessarily QCD, there must be matching of anomalies between different phases of the theory. This anomaly-matching condition was first proposed by 't Hooft as a way of understanding phases of a gauge theory. There is one potential caveat to this argument as we have outlined it. The anomaly arose as a high-energy effect, having to do with regularization of short-distance singularities. So one might argue that there is no reason for it to be the same in the low-energy effective theory. The topological nature of the anomalies shows that they are invariant under change of scales and so should be the same for the low-energy theory as well. In other words, if anomalies cancel between two sectors of the theory at high energy, they must also cancel at low energies, even if one sector is in a different phase. There is another elegant argument leading to the same conclusion. In the triangle diagram which leads to the anomaly, one can check that the anomaly is due to the imaginary part of the diagram. The imaginary part of a one-loop diagram, by the unitarity result (12.111), is related to the tree-level cross section, which can be calculated in terms of asymptotic states in the effective theory. Therefore, the anomaly can be viewed as a high-energy effect or as a low-energy effect. For this reason, we must demand a matching of anomalies between the high- and low-energy regimes of a theory.

Coming back to the specific case of QCD, we may thus ask how the flavor anomalies are represented in terms of mesons and baryons. The effective theory arises from the chiral symmetry breaking $U_L(3) \times U_R(3) \rightarrow U_V(3)$ and involves the meson field U, as discussed in Chapter 12. The Goldstone bosons behave in a way similar to the parameters of gauge transformations. (This is why they can be absorbed into the gauge field in a unitary gauge for the Higgs mechanism.) Therefore, we can expect the effective action for

QCD to have a term of the form $\omega_5(A^U) - \omega_5(A)$, where U is the Goldstone field for chiral symmetry breaking and A's are flavor gauge fields. This is almost but not quite right. The reason is that this expression would give the anomaly for a set of left-handed fermions and so would be appropriate for a symmetry breaking pattern $U_L(3) \to 1$. The chiral transformation rule for U is $U \to U' = g_L \, U \, g_R^\dagger$. From (17.21), we see that the relevant term involving U should thus be of the form

$$\Gamma_{WZ} = -i\frac{N}{240\pi^2} \int_D (\mathrm{Tr}(dU \, U^{-1})^5 + 2\pi N[\alpha_4(U^{-1}, A_L) - \alpha_4(U, A_R)] \; + \Gamma_{count}$$

$$(17.31)$$

where we have a factor N corresponding to the number of colors which are just the degeneracy for quarks for the computation of flavor anomalies. For QCD, $N = 3$, but we display the more general result. Γ_{count} are extra counterterms which can be added, if necessary, to ensure that the flavor anomalies for the nongauge directions are expressed in a gauge-invariant way, analogous to the Bardeen form of the anomaly. This term can be worked out by considering transformations of Γ_{WZ}, and the full result is

$$\begin{aligned}
\Gamma_{WZ} = &-\frac{iN}{240\pi^2} \int_D (\mathrm{Tr}(dU \, U^{-1})^5 \\
&+\frac{iN}{48\pi^2} \int_{\mathcal{M}} \mathrm{Tr}[(A_L dA_L + dA_L A_L + A_L^3)dUU^{-1}] \\
&+\frac{iN}{48\pi^2} \int_{\mathcal{M}} \mathrm{Tr}[(A_R dA_R + dA_R A_R + A_R^3)U^{-1}dU] \\
&-\frac{iN}{96\pi^2} \int_{\mathcal{M}} \mathrm{Tr}[A_L dUU^{-1} A_L dUU^{-1} - A_R U^{-1}dU A_R U^{-1}dU] \\
&-\frac{iN}{48\pi^2} \int_{\mathcal{M}} \mathrm{Tr}[A_L (dUU^{-1})^3 - A_R (U^{-1}dU)^3] \\
&-\frac{iN}{48\pi^2} \int_{\mathcal{M}} \mathrm{Tr}[dA_L dU A_R U^{-1} - dA_R d(U^{-1})A_L U] \qquad (17.32) \\
&-\frac{iN}{48\pi^2} \int_{\mathcal{M}} \mathrm{Tr}[A_R U^{-1} A_L U(U^{-1}dU)^2 - A_L U A_R U^{-1}(dUU^{-1})^2] \\
&+\frac{iN}{48\pi^2} \int_{\mathcal{M}} \mathrm{Tr}[(dA_R A_R + A_R dA_R)U^{-1}A_L U \\
&\qquad\qquad -(dA_L A_L + A_L dA_L)U A_R U^{-1}] \\
&+\frac{iN}{48\pi^2} \int_{\mathcal{M}} \mathrm{Tr}[A_L U A_R U^{-1} A_L dUU^{-1} + A_R U^{-1} A_L U A_R U^{-1}dU] \\
&-\frac{iN}{48\pi^2} \int_{\mathcal{M}} \mathrm{Tr}[A_R^3 U^{-1} A_L U - A_L^3 U A_R U^{-1} \\
&\qquad\qquad +\tfrac{1}{2} U A_R U^{-1} A_L U A_R U^{-1} A_L]
\end{aligned}$$

This effective action, first given by Witten, will describe all processes which are mediated by flavor anomalies. It is odd under $U \leftrightarrow U^{-1}$, $A_L \leftrightarrow A_R$ as it should be. The combination $U \leftrightarrow U^{-1}$, $\boldsymbol{x} \leftrightarrow -\boldsymbol{x}$ is a symmetry, except for the A_L, A_R gauging. This corresponds to parity, which is preserved by the strong interactions. The gauge fields of weak interactions break parity symmetry, so there is no reason to expect this symmetry with A_L, A_R. The term proportional to $\text{Tr}(dUU^{-1})^5$ can describe certain purely mesonic processes due to the anomaly; an example is $K^+K^- \to \pi^+\pi^-\pi^0$.

The gauge fields in (17.32) are to be evenually restricted to the electroweak gauge fields. If we consider only the electromagnetic field, the terms which depend on the gauge field are

$$\Gamma_{WZ} = \frac{e}{16\pi^2} \int_{\mathcal{M}} \text{Tr}[A(dUU^{-1})^3 + A(U^{-1}dU)^3]$$
$$+i\frac{e}{16\pi^2} \int_{\mathcal{M}} \text{Tr}[dAA(dUU^{-1} + U^{-1}dU)]$$
$$-i\frac{e}{32\pi^2} \int_{\mathcal{M}} \text{Tr}[dAdU\,AU^{-1} + dAU\,AdU^{-1}] \qquad (17.33)$$

By expanding U, it is easily checked that this leads to the effective term for $\pi^0 \to 2\gamma$ decay, which was separately analyzed in Chapter 13.

17.5 The global or nonperturbative anomaly

The anomalies in symmetry transformations discussed so far can be expressed in terms of the change in the effective action or the fermion determinant under an infinitesimal transformation. This change is computable in terms of Feynman diagrams or the Jacobian of the fermion measure for infinitesimal transformations. The change for finite transformations is obtained by integrating the anomaly. The consistency conditions for integrability are the WZ conditions given in Chapter 13 and the integrated form is obtained as in equation(17.23) or (17.29). But there can be transformations which cannot be continuously connected to the identity and so cannot be obtained by integrating infinitesimal transformations. It is possible for the fermion measure or the fermion determinant to change in a nontrivial way under such a transformation. This was shown by Witten in the classic case of an $SU(2)$ gauge theory in four dimensions. This is generally referred to as a global anomaly or nonperturbative anomaly.

For an $SU(2)$ gauge field in four dimensions, there is no perturbative anomaly, since $d^{abc} \sim \text{Tr}(t^a t^b t^c + t^a t^c t^b) = 0$. Thus one might expect an $SU(2)$ gauge theory with one doublet of chiral fermions to be a consistent theory. Witten showed that such a theory would be inconsistent due to the global anomaly.

For the gauge theory under consideration, the gauge transformation is an $SU(2)$ matrix $g(x)$: $\mathbf{R}^4 \to SU(2)$. As usual, we will consider transformations with $g \to 1$ as $|x| \to \infty$. Functions $g(x)$ with this condition are equivalent to functions from S^4 to $SU(2)$. The possible homotopy classes of these are given by $\Pi_4[SU(2)]$, which is $\mathbf{Z}_2$. This means that there are two types of gauge transformations for an $SU(2)$ gauge theory in four dimensions. The first type of transformation corresponds to the trivial element of $\Pi_4[SU(2)] = \mathbf{Z}_2$ and hence such transformations are all smoothly deformable to the identity everywhere on spacetime. They are thus connected to the identity and any anomaly they can lead to can be computed from infinitesimal transformations. (For $SU(2)$, there is no such anomaly.) The second type of transformation belongs to the nontrivial element of $\mathbf{Z}_2$; and hence, by definition, these transformations are not smoothly deformable to the identity everywhere on spacetime. Let $\tilde{g}(x)$ be a transformation which belongs to the homotopically nontrivial class of transformations. Because the homotopy group is $\mathbf{Z}_2$, two successive transformations by $\tilde{g}(x)$, which is equivalent to a transformation by $\tilde{g}^2(x)$, will correspond to the trivial element of $\mathbf{Z}_2$ and can be deformed to the identity.

The change in the fermion measure under $\tilde{g}$ cannot be computed from infinitesimal transformations. The gauge potentials A and $A^{\tilde{g}}$ correspond to the same physical configuration. The configuration space of potentials is connected and one can therefore construct a path $A(x, \tau)$ from $A(x, 0) = A(x)$ to $A(x, 1) = A^{\tilde{g}}(x)$. By using an index theorem, Witten then showed that the fermion measure changes sign as we go from A to $A^{\tilde{g}}$,

$$\det(\gamma \cdot D^{\tilde{g}}) = -\det(\gamma \cdot D) \tag{17.34}$$

This renders the theory inconsistent since one cannot restrict the integration over the gauge fields to regions where one sign is obtained for the determiant.

We will now show this result by a different argument which relates the global anomaly to the usual perturbative anomaly. We think of the gauge group $SU(2)$ as being embedded in $SU(3)$. Since $\Pi_4[SU(3)] = 0$, there is no transformation which is not deformable to the identity and hence there is no global anomaly in this enlarged theory. However, while $SU(2)$ is perturbatively anomaly-free, $SU(3)$ is not. Thus the global anomaly of the $SU(2)$ theory can be recovered from the usual result (17.24). When we go to $SU(3)$, we must consider a triplet of fermions, since the original fermions form a doublet under $SU(2)$. The extra added fermion field is a singlet under $SU(2)$ and so it will not affect arguments regarding the consistency of the theory for $SU(2)$ transformations. Under a general $SU(3)$ transformation $A \to A^g$, we have

$$\det(\gamma \cdot D^g) = \det(\gamma \cdot D) \, \exp\left[i \, \Gamma(U, A)\right]$$

$$\Gamma(U, A) = 2\pi \int_{\mathcal{D}} [\omega_5(A^U) - \omega_5(A)]$$

$$= 2\pi \left[\int_{\mathcal{M}} \alpha_4(g, A) + \int_{\mathcal{D}} \Omega^{(5)}(U) \right] \tag{17.35}$$

where U has the property that $U = g$ on $\partial \mathcal{D} = \mathcal{M}$. The gauge fields occur only in α_4 and we can keep them as $SU(2)$ gauge fields. α_4 is then zero, since the perturbative anomaly is zero for $SU(2)$; and so, for the rest of our argument, we can take

$$\Gamma(U, A) = \Gamma(U) = 2\pi \int_{\mathcal{D}} \Omega^{(5)}(U) \tag{17.36}$$

If we consider two extensions of g into the disc $\mathcal{D}$, say, U and U' with the same boundary value g, then the diference $\Gamma(U) - \Gamma(U')$ is 2π times an integer by the arguments given after (17.26). This integer is the winding number of the map $S^5 \to SU(3)$, which labels the elements of $\Pi_5[SU(3)] = \mathbf{Z}$. Alternatively, we can write $U' = UV$, where $V : \mathcal{D} \to SU(3)$ with the condition $V = 1$ on $\partial \mathcal{D} = \mathcal{M}$. Such a function is equivalent to a map from S^5 to $SU(3)$ and so we get the winding number for $S^5 \to SU(3)$ again. Let V_1 denote the basic map generating all the nontrivial maps $\mathcal{D} \to SU(3)$, with $V_1 = 1$ on $\partial \mathcal{D} = \mathcal{M}$. V_1 has winding number 1 and so

$$\Gamma(V_1) = 2\pi \tag{17.37}$$

Once we have shown the consistency of different extensions of g into $\mathcal{D}$, it is sufficient to consider only one extension U in the formula (17.35). $U : \mathcal{D} \to SU(3)$ gives a map of the disc $\mathcal{D}$ into $SU(3)$. Instead of pulling back the five-form to $\mathcal{D}$ via this map and integrating, we can think of (17.36) as the integral of $\Omega^{(5)}$ over the image disc in $SU(3)$.

We now argue that this can be regarded as an integral over a sphere in $SU(3)/SU(2)$. We easily check that $\Gamma(Uh) = \Gamma(U)$ for any $h \in SU(2) \subset SU(3)$, since the perturbative anomaly is zero for $SU(2)$. Thus Γ is defined on maps from $\mathcal{D}$ to $SU(3)/SU(2)$. Since the boundary value of the functions U is in $SU(2)$, the image of the disc $\mathcal{D}$, while it is a disc in $SU(3)$, is a sphere S^5 in $SU(3)/SU(2)$. Further, the value of the integral $\Gamma(U)$ is invariant under small deformations of U, since the integrand is a closed two-form. Thus it is a topological invariant for the maps $S^5 \to SU(3)/SU(2)$. It gives a real number for every element of $\Pi_5[SU(3)/SU(2)] = \mathbf{Z}$. (The space $SU(3)/SU(2)$ is the five-sphere S^5.) We also see from expanding out the five-form

$$\Gamma(UU') = \Gamma(U) + \Gamma(U') \tag{17.38}$$

for maps U, U' with boundary values in $SU(2)$.

Consider now the nontrivial $SU(2)$ transformation $\tilde{g}(x)$. There is no way to extend this over the entire disc $\mathcal{D}$ staying within $SU(2)$, but we can find $\tilde{U} : \mathcal{D} \to SU(3)$ such that $\tilde{U} = \tilde{g}(x)$ on $\partial \mathcal{D} = \mathcal{M}$. For the transformation of the fermion determinant we need $\Gamma(\tilde{U})$. This number can be evaluated using the exact homotopy sequence discussed in Chapter 14; the part of sequence which is relevant for this case reads

$$0 \to \, \to \Pi_5[SU(3)] \to \Pi_5\left[\frac{SU(3)}{SU(2)}\right] \to \Pi_4[SU(2)] \to 0$$

$$\begin{array}{cccc} \mathbf{Z} \to & \mathbf{Z} \to & \mathbf{Z}_2 \to & 0 \\ & 2k+1 \to & 1 \to & 0 \\ n \to & 2k \to & 0 \to & 0 \end{array} \qquad (17.39)$$

The exactness of the sequence tells us that what maps to zero must come as the image of the previous map. Therfore the nontrivial element 1 of $\mathbf{Z}_2$ must be the image of some elements of $\Pi_5[SU(3)/SU(2)]$. None of those elements can come as the image of the previous step, so all of $\Pi_5[SU(3)]$ must map to some other elements of $\Pi_5[SU(3)/SU(2)]$. Given the composition rules for the homotopy elements, we then see that all even winding numbers in $\Pi_5[SU(3)/SU(2)]$ are images of $\Pi_5[SU(3)]$. The odd winding numbers of $\Pi_5[SU(3)/SU(2)]$ are not images of elements in $\Pi_5[SU(3)]$ and map onto the nontrivial element of $\Pi_4[SU(2)]$.

There are many $\tilde{U}$'s giving the same boundary value $\tilde{g}$, corresponding to different elements of $\Pi_5[SU(3)/SU(2)]$. Let $\tilde{U}_1$ denote the extension of $\tilde{g}$ which is the basic nontrivial map of winding number 1 in $\Pi_5[SU(3)/SU(2)]$. The function $\tilde{U}_1^2$ corresponds to the trivial element of $\Pi_4[SU(2)]$ and can be deformed to the identity on the boundary ∂D. Since the boundary is the identity and not any $SU(2)$ element, the corresponding image disc is a sphere in $SU(3)$ and hence, using (17.37) and (17.38), we have the result

$$\begin{aligned} \Gamma(\tilde{U}_1) &= \frac{1}{2}\left[\Gamma(\tilde{U}_1) + \Gamma(\tilde{U}_1)\right] \\ &= \frac{1}{2}\,\Gamma(\tilde{U}_1^2) \\ &= \frac{1}{2}\,\Gamma(V_1) \\ &= \pi \end{aligned} \qquad (17.40)$$

This gives the result (17.34).

In the standard model, there are an even number of doublets of fermions for the $SU(2)$ group of weak interactions. Thus the standard model is free of the global anomaly. In seeking extensions of the standard model, by changing gauge groups and particle contents, one must ensure the absence of global anomalies as well as the perturbative anomalies for gauge transformations. This can impose some additional constraints on the model.

It is easy to see that these arguments for the global anomaly can be generalized to other dimensions as well as other representations of fermions. For a gauge group G in n dimensions, there are gauge transformations which are not connected to the identity if $\Pi_n(G)$ is nonzero. From the homotopy groups of Lie groups, some of which were listed in Chapter 14, this can be seen to occur in many dimensions. The investigation of whether there is a global anomaly for the given fermion representations can be carried out in a way similar to what we have done here.

17.6 The Wess-Zumino-Witten (WZW) action

The Wess-Zumino-Witten (WZW) action is intimately related to anomalies in two dimensions and has many applications. It is used for nonabelian bosonization. It defines a conformal field theory in two dimensions; various rational conformal field theories can be obtained as either a WZW model or gauged versions of it.

The field variables of the WZW action are invertible matrices $M(x)$, i.e., elements of $GL(N, \mathbf{C})$ or suitable subgroups and cosets of it, so it may be regarded as a particular type of sigma model. We will denote this target space by G. We shall discuss the action in two-dimensional space with Euclidean signature; the Minkowski version is briefly discussed in the Chapter 20 in the context of geometric quantization. The action is given by

$$\mathcal{S}_{WZW} = \frac{1}{8\pi} \int_{\mathcal{M}^2} d^2x \sqrt{g} \; g^{ab} \text{Tr}(\partial_a M \partial_b M^{-1}) \; + \; \Gamma[M] \qquad (17.41)$$

$$\Gamma[M] = \frac{i}{12\pi} \int_{\mathcal{M}^3} d^3x \; \epsilon^{\mu\nu\alpha} \text{Tr}(M^{-1}\partial_\mu M M^{-1}\partial_\nu M M^{-1}\partial_\alpha M) \qquad (17.42)$$

Here $\mathcal{M}^2$ denotes the two-dimensional space on which the fields and action are defined. It can in general be a curved manifold with a metric tensor g_{ab}. (g^{ab} is the inverse metric and g denotes the determinant of g_{ab} as a matrix.) $\mathcal{M}^2$ will be taken as a closed manifold. One can also use this for fields on $\mathbf{R}^2$, by choosing the boundary condition $M \to 1$ (or some fixed value independent of directions) as $|\boldsymbol{x}| \to \infty$; topologically, such fields are equivalent to fields on a closed manifold. $\Gamma[M]$ is the Wess-Zumino term. It is defined by integration over a three-dimensional space $\mathcal{M}^3$ which has $\mathcal{M}^2$ as its boundary. The integrand is a differential three-form; it does not require metrical factors for the integration. However, it requires an extension of the fields to the three-space $\mathcal{M}^3$. There can be many spaces $\mathcal{M}^3$ with the same boundary $\mathcal{M}^2$, or equivalently, there can be many different ways to extend the fields to the three-space $\mathcal{M}^3$. If M and M' are two different extensions of the same field into the three-space, we write $M' = MN$, where $N = 1$ on $\mathcal{M}^2$, the boundary of $\mathcal{M}^3$. By direct computation, we observe that

$$\Gamma[MN] = \Gamma[M] + \Gamma[N] - \frac{i}{4\pi} \int_{\mathcal{M}^2} d^2x \; \epsilon^{ab} \text{Tr}(M^{-1}\partial_a M \; \partial_b N N^{-1}) \quad (17.43)$$

The last term vanishes for $N = 1$ on $\mathcal{M}^2 = \partial \mathcal{M}^3$.

Since $N = 1$ on the boundary of $\mathcal{M}^3$, N is equivalent to a map from a closed three-space to G. In general, there are homotopically distinct classes of such maps. For example, if we take $\mathcal{M}^2 = S^2$ (or $\mathbf{R}^2$ with the boundary condition indicated), $\mathcal{M}^3$ is a ball in three dimensions. With the prescribed behavior for N, it is equivalent to a map from the three-sphere S^3 to G. As we have seen many times before, the homotopy classes of such maps are given by $\Pi_3(G)$. If G contains any compact nonabelian Lie group, then this is nonzero. In particular, $\Pi_3(G) = \mathbf{Z}$ for all simple nonabelian Lie groups, except for $SO(4)$, in which case it is $\mathbf{Z} \times \mathbf{Z}$. The winding number of the map $N(x) : S^3 \to G$ is given by

$$Q[N] = -\frac{1}{24\pi^2} \oint_{S^3} d^3x \; \epsilon^{\mu\nu\alpha} \mathrm{Tr}(N^{-1}\partial_\mu N N^{-1}\partial_\nu N N^{-1}\partial_\alpha N) \qquad (17.44)$$

$Q[N]$ is an integer for any $N(x)$. Thus, for $\Gamma[N]$, we have two cases to discuss. $\Gamma[N]$ is zero for N close to identity, to linear order in $\partial N N^{-1}$; hence, by successive transformations, $\Gamma[M]$ is independent of the extension to $\mathcal{M}^3$ for all N connected to identity, i.e., for N belonging to a homotopically trivial element. On the other hand, if N is homotopically nontrivial, the integral $\Gamma[N]$ gives $2\pi i$ times the winding number of the map $N(x) : S^3 \to G$. Since $Q[N]$ is an integer, $\exp(-k\,\Gamma[M])$ is independent of how the extension into the three-space is made if k is an integer. Thus, by using the action

$$S = k\,\mathcal{S}_{WZW} \qquad (17.45)$$

where k is an integer, we can construct a field theory on the two-space $\mathcal{M}^2$. Since the theory can be defined by using $\exp(-S) = \exp(-k\mathcal{S}_{WZW})$ to construct the functional integral, this will be well-defined, not requiring more than the specification of field configurations on $\mathcal{M}^2$ itself. The action (17.45) defines the WZW theory; k is referred to as the level number of this theory. Even though we presented the arguments for quantization of the coefficient of the action for $\mathcal{M}^2 = S^2$, similar arguments and results hold more generally.

For the simplest case of $\mathcal{M}^2 = \mathbf{R}^2$ with the appropriate boundary conditions on the field $M(x)$, we can write the WZW action using complex coordinates as

$$\mathcal{S}_{WZW} = \frac{1}{2\pi} \int_{\mathcal{M}^2} \mathrm{Tr}(\partial_z M \partial_{\bar{z}} M^{-1}) + \Gamma[M] \qquad (17.46)$$

For the first term in this expression, we have used complex coordinates $z = x_1 - ix_2$, $\bar{z} = x_1 + ix_2$. This action obeys a very useful identity known as the Polyakov-Wiegmann identity. Using (17.43), one can easily verify that

$$\mathcal{S}_{WZW}[M\,h] = \mathcal{S}_{WZW}[M] + \mathcal{S}_{WZW}[h] - \frac{1}{\pi} \int_{\mathcal{M}^2} \mathrm{Tr}(M^{-1}\partial_{\bar{z}}M \; \partial_z h \; h^{-1}) \qquad (17.47)$$

Notice the chiral splitting; we have only the antiholomorphic derivative of M and the holomorphic derivative of h. This shows that the equations of motion are given by

$$\partial_z(M^{-1}\partial_{\bar{z}}M) = M^{-1}\partial_{\bar{z}}(\partial_z M \; M^{-1})M = 0 \qquad (17.48)$$

The WZW action has invariance under infinitesimal left translations of M by a holomorphic function, $M \to (1 + \theta(z))M$, and right translations of

M by an antiholomorphic function, $M \to M(1+\chi(\bar{z}))$. This is easily checked using (17.47). These transformations have the associated currents

$$J_z = -\frac{k}{\pi} \, \partial_z M \, M^{-1}$$

$$J_{\bar{z}} = \frac{k}{\pi} \, M^{-1}\partial_{\bar{z}}M \qquad (17.49)$$

(These currents are for a level k WZW model.) By the equations of motion (17.48), these currents obey

$$\partial_{\bar{z}}J_z = 0, \qquad \partial_z J_{\bar{z}} = 0 \qquad (17.50)$$

The Polyakov-Wiegmann property also gives another result for the WZW action which is very useful. Consider a small variation of the field M given by $M + \delta M = (1 + \theta)M$, where $\theta = \delta M \, M^{-1}$ is infinitesimal. Using the Polyakov-Wiegmann property, we then get

$$\delta S_{WZW} = -\frac{1}{\pi} \int \text{Tr} \left(\partial_{\bar{z}}(\delta M M^{-1})\partial_z M M^{-1} \right)$$

$$= -\frac{1}{\pi} \int \text{Tr}(\delta M M^{-1}\partial_{\bar{z}}A_z) = -\frac{1}{\pi} \int \text{Tr}(\delta M M^{-1}D_z \bar{A})$$

$$= -\frac{1}{\pi} \int \text{Tr}(\bar{A}\delta A_z) \qquad (17.51)$$

where $A_z = -\partial_z M M^{-1}$ and D_z is the covariant derivative in the adjoint representation, $D_z\bar{A} = \partial_z\bar{A} + [A_z, \bar{A}]$. $\bar{A}$ is defined by

$$\bar{A} = -\partial_{\bar{z}}M \, M^{-1} \qquad (17.52)$$

Notice that this obeys the equation

$$\partial_{\bar{z}}A_z - \partial_z\bar{A} + [\bar{A}, A_z] = 0 \qquad (17.53)$$

17.7 The Dirac determinant in two dimensions

The functional integrals over the fermion fields in two dimensions lead to the determinant of the two-dimensional Dirac operator. For massless fermions, this determinant can be exactly evaluated using the WZW action.

The Dirac matrices relevant for two dimensions can be taken as σ_i, $i = 1, 2$, since they obey the relation $\sigma_i\sigma_j + \sigma_j\sigma_i = 2\delta_{ij}$. Consider a set of massless fermion fields in two dimensions which belong to an irreducible representation R of $U(N)$ and which are coupled to a $U(N)$ gauge field. (Other groups can be treated similarly.) The Lagrangian for these fermions can be written as

$$\mathcal{L} = \bar{\psi}(D_1 + iD_2)\psi + \bar{\chi}(D_1 - iD_2)\chi \qquad (17.54)$$

where ψ, χ may be taken as the two chiral components of a two-spinor. $D_i = \partial_i + A_i$ are the covariant derivatives. It is convenient to use the complex components $D_z = \frac{1}{2}(D_1 + iD_2)$ and $D_{\bar{z}} = \frac{1}{2}(D_1 - iD_2)$, since the Lagrangian is naturally split in terms of these.

A parametrization for gauge fields

The evaluation of the determinants $\det D_z$ and $\det D_{\bar{z}}$ can be done most efficiently using an elegant general parametrization of the complex components of the gauge field A_z and $A_{\bar{z}}$ given by

$$A_z = -\partial_z M \; M^{-1}$$
$$A_{\bar{z}} = M^{\dagger -1}\partial_{\bar{z}} M^{\dagger} \tag{17.55}$$

where M is a complex matrix; it has unit determinant if the gauge group is $SU(N)$. (We shall consider this case first.) The possibility of this parametrization can be seen as follows. ∂_z is invertible in two dimensions with suitable boundary conditions, since there are no nonsingular antiholomorphic functions by Liouville's theorem. The number of zero modes of its covariant version $\partial_z + A_z$ is given by an index theorem, and this is zero for an $SU(N)$ gauge field. One may still have paired zero modes for specific A_z's but generically these will become nonzero modes under a small change in A. Thus, generically we may take $\partial_z + A_z$ to be invertible and then for a given A_z, we can define a new potential $\mathcal{A}_{\bar{z}}$ by

$$\partial_{\bar{z}} A_z - \partial_z \mathcal{A}_{\bar{z}} + [\mathcal{A}_{\bar{z}}, A_z] = 0$$
$$\mathcal{A}_{\bar{z}} = (D_z^{-1})\,\partial_{\bar{z}} A_z \tag{17.56}$$

The required matrix M can then be constructed as

$$M(x, 0, C) = P \exp\left(-\int_{0\,C}^{x} A_z dz + \mathcal{A}_{\bar{z}} d\bar{z}\right) \tag{17.57}$$

Equation (17.56) shows that M is independent of the path C from the origin to x. To see this, consider more generally the parallel transport operator along a path C from the origin to a point with coordinates x^i on a simply connected $2n$-dimensional space $\mathcal{M}$ given by

$$\mathcal{P}(x, 0, C) = P \exp\left(-\int_{0\,C}^{x} A_i dx^i\right) \tag{17.58}$$

Generally $\mathcal{P}(x, 0, C)$ depends on the path of integration C; if C' is a path which differs from C by a small deformation by an area element σ^{ij} at the point y along the path, we have

$$\delta\mathcal{P}(x, 0, C) = -P \exp\left(-\int_{yC}^{x} A_i dx^i\right) \frac{1}{2}F_{ij}\sigma^{ij} \; P \exp\left(-\int_{0\,C}^{y} A_i dx^i\right) \tag{17.59}$$

The path-dependence is due to the nonzero field strength F_{ij}. In the case of M, we see that the flatness condition (17.56) ensures the path-independence of the integral. (The starting point of the integral in (17.57) may be taken as any point, not necessarily the origin.) From the construction of M, we also see that $A_z = -\partial_z M M^{-1}$ in agreement with (17.55); the second of equations (17.55) is obtained by hermitian conjugation.

If the gauge group is $U(1)$, we can use the elementary fact that A_i can be written as a gradient plus a curl, $A_i = \partial_i \theta + \epsilon_{ij} \partial_j \phi$. This leads to the parametrization (17.55) with $M = \exp(\phi + i\,\theta)$. Thus the result (17.55) holds for $U(N)$ in general.

If the space is not simply connected, one can have zero modes for ∂_z; there are in general flat potentials a which are not gauge equivalent to zero. (The transformation which would transform a to zero will not be singlevalued.) Further, there are different types of closed paths, those which are contractible and those which are not deformable to a contractible closed path. For the latter type of path, the integral $\mathcal{P}(0,0,C)$ for the flat potential around a closed noncontractible curve C has a value which is invariant under small deformations of the path; it is a topological invariant of the gauge field. (It is known as the holonomy of the field around C.) One can then generalize (17.55) to include this degree of freedom. As an example, consider the torus $S^1 \times S^1$. This can be described by two real coordinates ξ_1, ξ_2, $0 \leq \xi_i \leq 1$, with the identification of $\xi_1 = 0$ and $\xi_1 = 1$ and likewise for ξ_2. The complex coordinate can be taken as $z = \xi_1 + \tau \xi_2$, where $\tau = \mathrm{Re}\ \tau + i\ \mathrm{Im}\ \tau$ is a complex number called the modular parameter of the torus. Noncontractible closed curves are generated by two basic cycles called the α and β cycles. Holonomies around these cycles can be generated by a constant gauge field a. For the remaining degrees of freedom in A, the previous argument goes through and so altogether we can write

$$A_z = M \left[\frac{i\pi a}{\mathrm{Im}\ \tau} \right] M^{-1} - \partial_z M\ M^{-1} \tag{17.60}$$

There is an ambiguity in the parametrization (17.55). Notice that M and $MV(\bar{z})$, where $V(\bar{z})$ is purely antiholomorphic will lead to the same potential A_z. On a sphere, or on the complex plane with suitable boundary conditions, there are no nonsingular antiholomoprhic functions (Liouville's theorem), and there is no ambiguity. Any gauge-invariant function of $M, M^\dagger$ will be an observable. On other spaces where ∂_z has zero modes, if we use the parametrization analogous to (17.55), we have to ensure that only quantities which are independent of this ambiguity are considered as observables. Even on the sphere, such an ambiguity can exist when there are charges since singularities are possible at the locations of charges. Physical quantities defined in terms of the potential will be free of such ambiguities.

One of the advantages of the parametrization (17.55) or (17.60) is with regard to gauge transformations. Under a gauge transformation by a $U(N)$

group element $g(x)$, $M \rightarrow M^g = gM$. It is easy to see that this reproduces the transformation law $A \rightarrow A^g = gAg^{-1} - dg\, g^{-1}$. Since M is complex, a gauge transformation cannot be used to set M to the identity matrix. While the unitary part of M can be eliminated by a gauge choice, the hermitian part of M is gauge-invariant. Alternatively, we may use $M^\dagger M$ as the gauge-invariant quantity for two-dimensional gauge fields.

Evaluation of the determinant

The evaluation of the Dirac determinant will require the regularization of the Green's functions for D_z and $D_{\bar{z}}$. The inverse of ∂_z is the Euclidean chiral Dirac propagator and is given by

$$G(x, x') = \frac{1}{\pi(\bar{x} - \bar{x}')} \tag{17.61}$$

For $D_z = \partial_z + A_z = \partial_z - \partial_z M M^{-1}$, we then get

$$D_z^{-1}(x, x') = \frac{M(x)M^{-1}(x')}{\pi(\bar{x} - \bar{x}')} \tag{17.62}$$

We define a regularized version by a simple procedure of separating points, or point-splitting, and write

$$D_z^{-1}(x, x')_{Reg} \equiv \mathcal{G}(x, x') = \int d^2y\, \frac{M(x)M^{-1}(y)}{\pi(\bar{x} - \bar{y})}\, \sigma(x', y; \epsilon)$$

$$\sigma(x', y; \epsilon) = \frac{1}{\pi\epsilon} \exp\left(-\frac{|x' - y|^2}{\epsilon}\right) \tag{17.63}$$

Here ϵ is the regularization parameter. Notice that as $\epsilon \rightarrow 0$, $\sigma(x', y; \epsilon)$ tends to a two-dimensional δ-function and we recover $D_z^{-1}(x, x')$. Thus, $\mathcal{G}(x, x')$ is indeed a regulated form of $D_z^{-1}(x, x')$ with better short-distance properties.

Now let $S_{eff} \equiv \log \det D_z = \mathrm{Tr}\log D_z$ where the second expression involves the functional trace and the matrix trace. Taking the variation of this quantity, we find $\delta S_{eff} = \mathrm{Tr}(D_z^{-1}\delta D_z)$. Since $\delta D_z = \delta A_z$,

$$\frac{\delta S_{eff}}{\delta A_z^a(x)} = \mathrm{Tr}\left[D_z^{-1}(x, x')(-it^a)\right]_{x' \rightarrow x}$$

$$= \mathrm{Tr}\left[\mathcal{G}(x, x)(-it^a)\right]_{\epsilon \rightarrow 0} \tag{17.64}$$

where the trace is now just over the matrices. In the second line, we have used the regularized version of the variation by using $\mathcal{G}$ for D_z^{-1}. When $x' = x$, since $\sigma(x', y) = \sigma(x, y)$ is sharply peaked at $x \approx y$, we can expand $\mathcal{G}(x, x)$ as

$$\mathcal{G}(x, x) = \int d^2y\, \frac{\sigma(x, y)}{\pi} \left[\frac{1}{(\bar{x} - \bar{y})} - M\partial_z M^{-1}(x)\left(\frac{x - y}{\bar{x} - \bar{y}}\right) - M\partial_{\bar{z}}M^{-1}\right.$$

$$\left. + \cdots \right] \tag{17.65}$$

The angular integration gives zero for the first two terms. From the third term we get a finite contribution; higher terms vanish as $\epsilon \to 0$. Thus we get

$$
\begin{aligned}
\delta S_{eff} &= \int d^2x \mathrm{Tr}\left[\mathcal{G}(x,x)(-it^a)\right]_{\epsilon \to 0} \delta A_z^a(x) \\
&= \frac{1}{\pi} \int d^2x \ \mathrm{Tr}\left[\partial_{\bar{z}} M M^{-1} \delta A_z\right] \\
&= -\frac{1}{\pi} \int d^2x \ \mathrm{Tr}(\bar{A}\delta A_z)
\end{aligned}
\tag{17.66}
$$

A_z and $\bar{A} = \mathcal{A}_{\bar{z}} = -\partial_{\bar{z}} M M^{-1}$ are matrices in the Lie algebra in the representation R. For the trace, we can thus use the formula $\mathrm{Tr}(t^a t^b)_R = A_R \mathrm{Tr}(t^a t^b)_F$, where F denotes the fundamental representation and A_R is an integer known as the index of the representation R. Equation(17.66) can then be written as

$$
\begin{aligned}
\delta S_{eff} &= -\frac{A_R}{\pi} \int d^2x \ \mathrm{Tr}(\bar{A}\delta A_z)_F \\
&= A_R \ \delta \mathcal{S}_{WZW}(M)
\end{aligned}
\tag{17.67}
$$

where we have used the property (17.51) for the WZW action. This shows that, up to a constant, S_{eff} is given by the WZW action. When $A = 0$, $\det D_z = \det \partial_z$, which identifies the constant, giving the final result for the chiral Dirac determinant as

$$
\det D_z = \det(\partial_z) \ \exp\left(A_R \ \mathcal{S}_{WZW}(M)\right)
\tag{17.68}
$$

Notice that this part is not gauge-invariant. In fact, under an infinitesimal gauge transformation, we find

$$
\delta S_{eff} = -\frac{1}{\pi} \int d^2x \ \mathrm{Tr}(\partial_{\bar{z}} A_z \ \delta g \ g^{-1})
\tag{17.69}
$$

There is anomaly for the gauge transformation. Since M belongs to the complexification of the gauge group, we may regard our derivation of the determinant as the integration of the anomaly $\partial_{\bar{z}} A/\pi$ over complex parameters.

A similar result is obtained for the determinant of $D_{\bar{z}}$,

$$
\det D_{\bar{z}} = \det(\partial_{\bar{z}}) \exp\left(A_R \ \mathcal{S}_{WZW}(M^\dagger)\right)
\tag{17.70}
$$

For the full Dirac determinant $\det(D_z D_{\bar{z}})$, we can take the product of the expressions for $\det D_z$ and $\det D_{\bar{z}}$. This would lead to

$$
\det(D_z D_{\bar{z}}) = \det(\partial_z \partial_{\bar{z}}) \ \exp\left[A_R\left(\mathcal{S}_{WZW}(M) + \mathcal{S}_{WZW}(M^\dagger)\right)\right]
\tag{17.71}
$$

This expression is still not gauge-invariant. A gauge-invariant expression is given by

$$
\begin{aligned}
\det(D_z D_{\bar{z}}) &= \det(\partial_z \partial_{\bar{z}}) \ \exp\left[A_R \mathcal{S}_{WZW}(M^\dagger M)\right] \\
&= \det(\partial_z \partial_{\bar{z}}) \ \exp\left[A_R \mathcal{S}_{WZW}(H)\right]
\end{aligned}
\tag{17.72}
$$

$H = M^\dagger M$ is gauge-invariant. Expanding out $\mathcal{S}_{WZW}(H)$ using the Polyakov-Wiegmann identity, we find

$$
\begin{aligned}
\mathcal{S}_{WZW}(H) &= \mathcal{S}_{WZW}(M) + \mathcal{S}_{WZW}(M^\dagger) \\
&\quad - \frac{1}{\pi} \int d^2x \; \mathrm{Tr}(M^{\dagger-1}\partial_{\bar{z}}M^\dagger \; \partial_z M \; M^{-1}) \\
&= \mathcal{S}_{WZW}(M) + \mathcal{S}_{WZW}(M^\dagger) + \frac{1}{\pi} \int d^2x \; \mathrm{Tr}(A_{\bar{z}}A_z)
\end{aligned}
$$

$$(17.73)$$

Thus the formula (17.72) can be understood as arising from the formula (17.71) by the addition of a local counterterm $(1/\pi) \int \mathrm{Tr}(A_{\bar{z}}A_z)$. As we discussed in Chapter 13 and earlier in this chapter, one has the freedom of adding local counterterms; this is equivalent to choosing a different regularization. Effectively, in obtaining the result (17.72) we are using a gauge-invariant regularization.

The Abelian version of (17.72), including the counterterm for gauge invariance, was first obtained by Schwinger in 1961. For the Abelian case, equation(17.72) simplifies as

$$
\det(D_z D_{\bar{z}}) = \det(\partial_z \partial_{\bar{z}}) \; \exp\left[-\frac{1}{4\pi} \int_{x,y} F_{\mu\nu}(x)G(x-y)F_{\mu\nu}(y)\right]
$$

$$
G(x-y) = \int \frac{d^2p}{(2\pi)^2} \frac{1}{p^2} \; \exp[ip \cdot (x-y)] \tag{17.74}
$$

This is the fermion determinant for two-dimensional electrodynamics, which is also known as the Schwinger model; we see that the effect of fermions is to generate a gauge-invariant nonlocal mass term for the gauge field.

References

1. The following books are useful for some of the material in this chapter: A.P. Balachandran, *Classical Topology and Quantum States*, World Scientific Pub. Co. (1991); R.A. Bertlmann, *Anomalies in Quantum Field Theory*, Clarendon Press (1996).

2. The interpretation of the anomaly as arising from the functional Jacobian is due to K. Fujikawa, Phys. Rev. Lett. **42**, 1195 1979.

3. The evaluation of the nonabelian anomaly using the Jacobian is given by A.P. Balachandran, G. Marmo, V.P. Nair and C.G. Trahern, Phys. Rev. **D25**, 2713 (1982).

4. For the general relationship between nonabelian anomalies and the index theorem, see B. Zumino, Les Houches Lectures, 1983, reprinted in S.B. Treiman *et al*, *Current Algebra and Anomalies*, Princeton University Press (1986); R. Stora, Lectures at the Cargèse Summer Institute on Progress in Gauge Field Theory, 1983; B. Zumino, Y-S. Wu and A. Zee, Nucl. Phys. **B239**, 477 (1984). In this context, see also A. Niemi and G.W. Semenoff, Phys. Rev. Lett. **51**, 2077 (1983).

5. The geometrical interpretation of anomalies is analyzed in M. Atiyah and I.M. Singer, Proc. Nat. Acad. Sci. USA **81**, 2597 (1984); O. Alvarez, I.M. Singer and B. Zumino, Commun. Math. Phys. **96**, 409 (1984); P. Nelson and L. Alvarez-Gaumé, Commun. Math. Phys. **99**, 103 (1985); L. Alvarez-Gaumé and P. Ginsparg, Ann. Phys. **161**, 423 (1985); Erratum *ibid.* **171**, 233 (1986).

6. The WZ action in terms of $\Omega^{(5)}$ and its subsequent analysis were given by E. Witten, Nucl. Phys. **B223**, 422 (1983). This paper also gives the effective action for flavor anomalies in the standard model. (We give the minimally gauged effective action in text; the version given in Witten's paper has an extra term which is gauge invariant by itself. In this context, see also O. Kaymakcalan, S. Rajeev and J. Schechter, Phys. Rev. **D30**, 594 (1984).)

7. Anomaly matching conditions are given in G. 't Hooft, Cargèse Lectures, 1979, reprinted in G. 't Hooft, *Under the Spell of the Gauge Principle*, World Scientific Pub. Co. (1994). In this context, see also S. Coleman and B. Grossman, Nucl. Phys. **B203**, 205 (1982).

8. The mixed anomaly, and gravitational anomalies in general, were calculated in L. Alvarez-Gaumé and E. Witten, Nucl. Phys. **B234**, 269 (1984).

9. The global anomaly was discovered by E. Witten, Phys. Lett. **B117**, 324 (1982). The argument presented is from S. Elitzur and V.P. Nair, Nucl. Phys. **B243**, 205 (1984).

10. The WZW action is also due to E. Witten, Commun. Math. Phys. **92**, 455 (1983). Another relevant paper is S.P. Novikov, Usp. Mat. Nauk. **37**, 3 (1982).

11. The PW identity and the evaluation of the determinant are given in A.M. Polyakov and P.B. Wiegmann, Phys. Lett. **B141**, 223 (1984).

12. The Schwinger model was introduced in J. Schwinger, Phys. Rev. **128**, 2425 (1962).

18 Finite temperature and density

18.1 Density matrix and ensemble averages

In the previous chapters, we have focused mostly on processes with a small number of real particles in a background which is the vacuum state. However, there are many physical situations where one encounters processes in a medium such as, for example, the propagation of particles and scattering processes in a gas at finite temperature and density. If the system is in a pure state, we can still write the transition amplitude for a scattering process as

$$\mathcal{A}(i \rightarrow f) = \langle f|\hat{S}|i \rangle \tag{18.1}$$

where $\hat{S}$ is the scattering operator. The evaluation of the matrix element may be difficult calculationally, but this does indeed give the answer. However, if the initial state is only statistically specified, which is the case for systems at finite temperature and density, we are interested in averaging over initial states with appropriate probabilities. (This is the probability of the choice of a particular state in a statistical ensemble and is not the quantum probability of the collapse of the wave function onto a specific eigenstate of the observable being measured.) The choice of states in quantum mechanics is given by the density matrix ρ. Consider the averaged expectation value of an observable A, given by

$$\langle A \rangle = \sum_{\alpha} c_{\alpha} \langle \alpha|A|\alpha \rangle \tag{18.2}$$

where $|\alpha\rangle$'s form a complete set of states, and c_{α} is the probability of finding the state $|\alpha\rangle$ in the statistical ensemble. Because c_{α} is a probability, we need $\sum_{\alpha} c_{\alpha} = 1$. Defining

$$\rho = \sum_{\alpha} c_{\alpha} |\alpha\rangle \langle \alpha| \tag{18.3}$$

we can write the above formula as

$$\langle A \rangle = \text{Tr} \, (\rho A) \tag{18.4}$$

The normalization of the c_{α}'s translates to $\text{Tr} \, \rho = 1$. These two equations may be taken as the definition of the density matrix ρ; in other words, ρ, with $\text{Tr} \, \rho = 1$, is an operator which specifies the choice of states by giving the average as in (18.4). Notice that this gives a basis-independent definition of ρ. The choice of a pure state $|\lambda\rangle$ corresponds to $c_\lambda = 1$, $c_\alpha = 0$ for $\alpha \neq \lambda$. In this case, $\rho^2 = \rho$, and we may take this as the definition of a pure state. If $\rho^2 \neq \rho$, we have a statistical mixture of states.

The evolution of states via the Schrödinger equation leads to

$$i\frac{\partial \rho}{\partial t} = H\,\rho - \rho\,H \tag{18.5}$$

The density matrix ρ may be considered as the quantum analogue of the phase-space density of classical statistical mechanics; equation (18.5) is the quantum version of the Liouville equation.

The density matrix is an operator corresponding to the choice of states. The Hilbert space of a quantum system gives all the possible states of the system, but when we are interested in discussing a system which has been prepared in a particular state (or a particular mixture of states), we need an operator to specify this choice. This is done by the density matrix. It is somewhat special in that it evolves with time in the Schrödinger picture but is time-independent in the Heisenberg picture. If the system under study is in a pure state $|a\rangle$, $\rho_a = |a\rangle\langle a|$. The probability of finding the system in a state $|b\rangle$ is then given by $|\langle a|b\rangle|^2 = \text{Tr}(\rho_a\rho_b)$. If we start with a state described by $\rho_a(t)$ at time t, it evolves into $\rho_a(t')$ by a later time t' according to (18.5); the probability of observing a state given by ρ_b in this evolved state is given by

$$P(a \to b, t, t') = \text{Tr}\big(\rho_a(t')\rho_b(t)\big)$$
$$= \text{Tr}\big(e^{-iH(t'-t)}\rho_a e^{iH(t'-t)}\rho_b\big) \tag{18.6}$$

This may be interpreted as the scattering probability for a state $|a\rangle$ at time t to evolve into $|b\rangle$ by the time t'. As $t \to -\infty$, $t' \to +\infty$, this is the absolute square of the S-matrix element. It is possible to rephrase many of the calculations we have done in terms of the density matrix.

Since the time-evolution of ρ is given by a unitary transformation, it is possible to write an action for the density matrix which leads to the equation of motion (18.5). Suppose we start with ρ_0 at time $t = 0$. Let U be an arbitrary unitary operator on the Hilbert space of the theory. The action is then given by

$$S = \int dt \left[i\text{Tr}\left(\rho_0 U^\dagger \frac{\partial}{\partial t} U\right) - \text{Tr}(\rho_0 U^\dagger H U) \right] \tag{18.7}$$

The entire dynamical information of the system is in U, the density matrix at time t is given by $\rho = U\rho_0 U^\dagger$. The variation of the action (18.7) is given by

$$\delta S = \int dt \text{Tr}\left[U\rho_0 U^\dagger i\frac{\partial \xi}{\partial t} + U\rho_0 U^\dagger \xi H - U\rho_0 U^\dagger H\xi \right]$$

$$= \int dt \operatorname{Tr}\left[\left(-i\frac{\partial\rho}{\partial t} + H\rho - \rho H\right)\xi\right] \tag{18.8}$$

where $\xi = \delta U\, U^\dagger$ and we have done a partial integration. We see that the extremization of the action (18.7) by varying with respect to U leads to (18.5). It is possible in some circumstances to use (18.7) as an effective action in a quantum theory by writing U in terms of the subset of operators which are important in the chosen kinematic regime.

Consider now a dynamical system in thermodynamic equilibrium. At equilibrium, ρ must be independent of time and so we must have $[H, \rho] = 0$. The general solution is that ρ should be a function of operators which commute with the Hamiltonian; in other words, it should be a function of conserved quantities. If we consider a number of such quantities which are additively conserved, it is easy to fix the form of this function as follows. Consider two completely independent systems I and II. The Hilbert space of the two systems together is the tensor product of the corresponding Hilbert spaces, $\mathcal{H} = \mathcal{H}_I \otimes \mathcal{H}_{II}$, and the probability of a particular state will go as the product of individual probabilities; therefore $\rho = \rho_I \cdot \rho_{II}$. If $\mathcal{O}$ is an additively conserved quantity, we have $\mathcal{O} = \mathcal{O}_I + \mathcal{O}_{II}$. Thus ρ must be an exponential function of the additively conserved quantities. The general solution is of the form

$$\rho = Z^{-1}\,\exp\left(-\sum_i \beta_i \mathcal{O}_i\right) \tag{18.9}$$

$\mathcal{O}_i$ are additively conserved operators, such as energy, momentum, angular momentum, charge, etc. $Z = \operatorname{Tr} \exp\left(-\sum_i \beta_i \mathcal{O}_i\right)$ to ensure the correct normalization of ρ; it is the partition function. β_i are constants, to be specified by the average values for the conserved quantities for the system under consideration. Since ρ commutes with H, we can also diagonalize it, thus writing equation (18.9) as

$$\rho = V\,\rho_{diag}\,V^\dagger \tag{18.10}$$

where V is a unitary transformation which commutes with H, corresponding, for example, to rotations, charge rotations, etc., The parameters in the transformation V, as well as the parameters in ρ_{diag}, will serve to define the β_i in (18.9) and hence the average values of the conserved quantities. For a nonrelativistic theory, particle number can be a conserved quantity, leading to the familiar chemical potential term $\beta\mu N$ in the exponent of (18.9). In the relativistic case, particle creation and annihilation are possible and one can only require conservation of appropriate charges.

If we consider a fluid with local equilibrium, one can take the form (18.10) to be valid for small fluid elements. The parameters of the unitary transformation are then approximately constant for each fluid element but can vary over larger regions. An approximate description of the dynamics is obtained by restricting the action (18.7) to just the modes corresponding to these local parameters; this could provide a way of obtaining the hydrodynamic description of fluids.

18.2 Scalar field theory

For an application of these results, consider a real scalar field theory. In this case, the only conserved quantities are energy, momentum, and angular momentum. The equilibrium ρ is thus a function of these quantities. We can bring the system to rest in a particular Lorentz frame and then rotate it such that the angular momentum is in a fixed direction, say, along the third axis. Going back to the general situation is accomplished by a general Lorentz transformation (including spatial rotations). The form of ρ is then

$$\rho = V \exp(-\beta_0 H - l_3 M_{12}) V^\dagger \tag{18.11}$$

where $M_{12} = J_3$ is the third component of angular momentum. V represents the unitary transformation corresponding to a general Lorentz transformation. β_0 and l_3 are real numbers. From the transformation properties of the operators, we can write

$$\rho = \exp\left(-\beta_0 L_0^\mu P_\mu - l_3 L_1^\mu L_2^\nu M_{\mu\nu}\right) \tag{18.12}$$

where $P_\mu = (H, P_i)$ are the total energy and momentum operators and $M_{\mu\nu}$ are the generators of Lorentz transformations, including rotations. L_α^μ is the Lorentz transformation matrix for a vector; the parameters in L_α^μ arise from parameters in the unitary transformation V. L_0^μ can be identified as the four-velocity of the moving medium, $\frac{1}{2}l_3(L_1^\mu L_2^\nu - L_1^\nu L_2^\mu)$ is related to the angular momentum of the medium, with the parameters of the Lorentz boost transformations in L_α^μ determining the velocity and the rotation parameters in L_α^μ determining the angular momentum. By evaluating the average expectation value of energy, one can identify β_0 as the inverse temperature in the rest frame of the medium, $\beta_0 = T_0^{-1}$.

In the rest frame, and in the case of zero angular momentum, the partition function is given by $Z = \mathrm{Tr}\exp(-\beta_0 H) = \mathrm{Tr}\exp(-\beta H)$. We will drop the subscript on β from now on for simplicity of notation; all quantities below are in the rest frame of the medium. For a free field the partition function is easily evaluated as follows:

$$Z = \mathrm{Tr}\exp(-\beta \sum_k \omega_k a_k^\dagger a_k)$$

$$= \prod_k \mathrm{Tr}_k \exp(-\beta \omega_k a_k^\dagger a_k)$$

$$= \prod_k \left(1 + e^{-\beta\omega_k} + e^{-2\beta\omega_k} + \cdots\right)$$

$$= \prod_k \frac{1}{1 - e^{-\beta\omega_k}} \tag{18.13}$$

Here Tr_k involves the sum over states at fixed value of $\boldsymbol{k}$. Since H commutes with $a_k^\dagger a_k$, it is clear that $\text{Tr}\, a_k^\dagger a_p^\dagger \rho = \text{Tr}\, a_k a_p \rho = 0$; further $\text{Tr}\, a_k^\dagger a_p \rho = 0$ for $\boldsymbol{k} \neq \boldsymbol{p}$. The average occupation number n_k is obtained as

$$n_k = \langle a_k^\dagger a_k \rangle = -\frac{\partial}{\partial \beta \omega_k} \log Z$$
$$= \frac{1}{e^{\beta \omega_k} - 1} \tag{18.14}$$

The average value of the Hamiltonian is given as

$$\langle H \rangle = \text{Tr}\left(\rho \sum_k \omega_k a_k^\dagger a_k \right)$$
$$= V \int \frac{d^3 k}{(2\pi)^3} \frac{\omega_k}{e^{\beta \omega_k} - 1} \tag{18.15}$$

We have taken the large volume limit in the last step. This is the standard Bose-Einstein formula, as expected, since the scalar field describes bosons.

The propagator for the free scalar field can be evaluated, using $\langle a_k^\dagger a_l \rangle = n_k \delta_{kl}$, $\langle a_k^\dagger a_l^\dagger \rangle = \langle a_k a_l \rangle = 0$, as

$$G(x, y) \equiv \text{Tr}\,[\rho\, T\, \phi(x)\phi(y)]$$
$$= \theta(x^0 - y^0) \int \frac{d^3 k}{(2\pi)^3} \frac{1}{2\omega_k} \left[(1 + n_k)e^{-ik(x-y)} + n_k e^{ik(x-y)} \right]$$
$$+ \theta(y^0 - x^0) \int \frac{d^3 k}{(2\pi)^3} \frac{1}{2\omega_k} \left[n_k e^{-ik(x-y)} + (1 + n_k)e^{ik(x-y)} \right]$$
$$= \theta(x^0 - y^0) \sum_k \frac{e^{-ik(x-y)}}{2\omega_k V} + \theta(y^0 - x^0) \sum_k \frac{e^{ik(x-y)}}{2\omega_k V}$$
$$+ \sum_k \frac{n_k}{2\omega_k V} \left(e^{-ik(x-y)} + e^{ik(x-y)} \right)$$
$$= \int \frac{d^4 k}{(2\pi)^4} \left[\frac{i}{k^2 - m^2 + i\epsilon} + n_k\, 2\pi\, \delta(k^2 - m^2) \right] e^{-ik(x-y)} \tag{18.16}$$

In addition to the standard zero-temperature contribution, we have a term proportional to the occupation number. The propagator gives the amplitude for detecting a particle at the spacetime point x if a particle is introduced into the system at y. The first term describes the probability for the particle to propagate to y from x; however, there is also the possibility that the detected particle is from the background distribution, this is accounted for by the second term.

The temperature-dependent propagator can be used for calculations involving scattering, corrections to parameters, etc. For example, we have seen that the mass correction for a $\lambda\phi^4$-theory can be given as $12\lambda G(x, x)$. Using the propagator in (18.16), the temperature-dependent part can be evaluated as

$$\delta m^2 = 12\lambda \int \frac{d^3k}{(2\pi)^3 2\omega_k} \frac{2}{e^{\beta\omega_k} - 1}$$
$$= \lambda T^2 \, \mathcal{I}(\alpha) \tag{18.17}$$

where $\alpha = m/T$ and

$$\mathcal{I}(\alpha) = \frac{3}{2\pi^3} \int d^3u \frac{1}{\sqrt{u^2 + \alpha^2}} \frac{1}{e^{\sqrt{u^2 + \alpha^2}} - 1} \tag{18.18}$$

As the particle propagates in the medium, it can interact with the background distribution of particles. One consequence is that the effective mass of the particle is increased. At high temperatures, $\mathcal{I}(\alpha) \approx \mathcal{I}(0) = 1$ and $\delta m^2 = \lambda T^2$.

18.3 Fermions at finite temperature and density

The creation and annihilation operators for fermions obey anticommutation rules. We shall consider the case of relativistic fermions, with creation and annihilation operators $a^\dagger_{k,r}$, $a_{k,r}$ for the particles and $b^\dagger_{k,r}$, $b_{k,r}$ for the antiparticles. Generally, we also have the conservation of fermion number which would be proportional to the electric charge for electrically charged particles. For free fermions, we can write $H = \sum_{k,r}(E_k a^\dagger_{k,r} a_{k,r} + E_k b^\dagger_{k,r} b_{k,r})$ and the conserved fermion number is $N = \sum_{k,r}(a^\dagger_{k,r} a_{k,r} - b^\dagger_{k,r} b_{k,r})$. The partition function is

$$Z = \mathrm{Tr}\exp\left[-\beta(H - \mu N)\right]$$
$$= \mathrm{Tr}\exp\left[-\beta \sum_{k,r}(E_k - \mu)a^\dagger_{k,r} a_{k,r} - \beta \sum_{k,r}(E_k + \mu)b^\dagger_{k,r} b_{k,r}\right]$$
$$= \prod_{k,r} \mathrm{Tr}_{k,r} \exp\left[-\beta(E_k - \mu)a^\dagger_{k,r} a_{k,r}\right] \prod_{p,r} \mathrm{Tr}_{p,s} \exp\left[-\beta(E_p + \mu)b^\dagger_{p,s} b_{p,s}\right]$$
$$= \prod_{k,r}\left(1 + e^{-\beta(E_k - \mu)}\right) \prod_{p,s}\left(1 + e^{-\beta(E_p + \mu)}\right) \tag{18.19}$$

$\mathrm{Tr}_{k,r}$ denotes the trace over occupation numbers at the fixed value of momentum, k, and for fixed spin r. The average occupation numbers are given by

$$n_k \, \delta_{r,s} = \langle a^\dagger_{k,r} a_{k,s}\rangle = \delta_{r,s} \frac{1}{e^{\beta(E_k - \mu)} + 1}$$
$$\bar{n}_k \, \delta_{r,s} = \langle b^\dagger_{k,r} b_{k,s}\rangle = \delta_{r,s} \frac{1}{e^{\beta(E_k + \mu)} + 1} \tag{18.20}$$

As expected for fermions, we have the Fermi-Dirac distribution. For positive μ we have a net excess of particles over antiparticles. The average value of the Hamiltonian is

$$\langle H \rangle = \text{Tr}\,(\rho\,H)$$
$$= \sum_r V \int \frac{d^3k}{(2\pi)^3}\ E_k\,(n_k + \bar{n}_k) \tag{18.21}$$

The propagator for a Dirac particle can be evaluated as

$$S(x,y) = \text{Tr}\left[\rho\,T\psi(x)\bar\psi(y)\right]$$
$$= \left[\theta(x^0 - y^0)S_> - \theta(y^0 - x^0)S_<\right] \tag{18.22}$$

where

$$S_> = \int \frac{d^3p}{(2\pi)^3}\frac{1}{2E_p}\left[(1-n_p)(\gamma\cdot p + m)e^{-ip(x-y)} + \bar{n}_p(\gamma\cdot p - m)e^{ip(x-y)}\right]$$
$$S_< = \int \frac{d^3p}{(2\pi)^3}\frac{1}{2E_p}\left[n_p(\gamma\cdot p + m)e^{-ip(x-y)} + (1-\bar{n}_p)(\gamma\cdot p - m)e^{ip(x-y)}\right]$$
$$\tag{18.23}$$

These expressions may also be combined and written as

$$S(x,y) = i\int \frac{d^4p}{(2\pi)^4}\frac{\gamma\cdot p + m}{p^2 - m^2 + i\epsilon}e^{-ip(x-y)}$$
$$+ \int \frac{d^3p}{(2\pi)^3}\frac{1}{2E_p}\left[-n_p(\gamma\cdot p + m)e^{-ip(x-y)} + \bar{n}_p(\gamma\cdot p - m)e^{ip(x-y)}\right]$$
$$\tag{18.24}$$

We see that the propagator again splits into the standard zero-temperature part and a part which depends on the distribution functions.

18.4 A condition on thermal averages

Since the Hamiltonian is the operator for translations in time, $e^{-\beta H}$ can be interpreted as shifting the time-arguments of operators by an imaginary amount $-i\beta$. This can be very useful in certain calculations as well as for a functional integral version of the partition function. For an explicit formula, consider the average of the product of two operators $A(x)$ and $B(y)$ given by $\text{Tr}(e^{-\beta H}\,A(x)B(y))$. By the cyclicity of the trace, we may bring $B(y)$ to the left end of the expression inside the trace and write

$$\langle A(x)B(y) \rangle = \text{Tr}(e^{-\beta H}\,A(x)B(y))$$
$$= \text{Tr}(B(y)e^{-\beta H}\,A(x)) \tag{18.25}$$

From the time-translation property of H, $B(y)e^{-\beta H} = e^{-\beta H}B(y^0 - i\beta, \boldsymbol{y})$. Alternatively, we can Fourier analyze $B(y)$ in time and write

$$B(y) = \int d\omega \, B(\omega, \boldsymbol{y}) \, e^{-i\omega y^0} \tag{18.26}$$

The Heisenberg equation of motion then shows that

$$[B(\omega, \boldsymbol{y}), H] = \omega \, B(\omega, \boldsymbol{y}) \tag{18.27}$$

Using this result, we can write

$$
\begin{aligned}
B(y)e^{-\beta H} &= \int d\omega \, e^{-\beta H} \left[e^{\beta H} B(\omega, \boldsymbol{y}) e^{-\beta H} \right] \, e^{-i\omega y^0} \\
&= \int d\omega \, e^{-\beta H} \, B(\omega, \boldsymbol{y}) \, e^{-\beta\omega} e^{-i\omega y^0} \\
&= e^{-\beta H} \int d\omega \, B(\omega, \boldsymbol{y}) \, e^{-i\omega(y^0 - i\beta)} \\
&= e^{-\beta H} B(y^0 - i\beta, \boldsymbol{y})
\end{aligned}
\tag{18.28}
$$

Finally, using this condition in (18.25), we get

$$\langle A(x) \, B(y) \rangle = \langle B(y^0 - i\beta, \boldsymbol{y}) \, A(x) \rangle \tag{18.29}$$

This statement on thermal averages is known as the Kubo-Martin-Schwinger (KMS) condition.

18.5 Radiation from a heated source

As another application of the concepts we have introduced, we consider radiation from a heated object. The source is kept at a constant temperature by electric currents or some other means of supplying energy; it could also be producing energy by itself like a star. The object is in vacuum and radiates. The radiation escapes and the source is not in thermal equilibrium with the surroundings. We want to calculate the spectrum of the emitted radiation. (There are well-known arguments which lead to a Planck-type spectrum if one can assume that the source is in equilibrium with the radiation. Such a situation is clearly not obtained in most cases, for example, for a star which radiates in empty space. Our idea is to treat this problem without further assumptions.)

The general form of the interaction part of the action is

$$S_{int} = \int d^4x \, eA_\mu J^\mu \tag{18.30}$$

Consider a transition from a definite initial quantum state $|i\rangle$ to a final state $|f\rangle$. The transition amplitude for photon emission of four-momentum k and polarization $\epsilon_\mu^{(\lambda)}$ is given by

$$\mathcal{A} = -ie \int d^4x \, \frac{\epsilon_\mu^{(\lambda)}}{\sqrt{2\omega_k V}} \langle f|J^\mu(x) \, e^{ikx}|i\rangle \tag{18.31}$$

We take the square of this and sum over final states; the final states correspond to the states of the photon and the states of the source. We write the summation over the photon polarizations as

$$\sum_\lambda \epsilon_\mu^{(\lambda)} \epsilon_\nu^{(\lambda)} = P_{\mu\nu} \tag{18.32}$$

We have $\epsilon_0^{(\lambda)} = 0$, so that $P_{00} = P_{0i} = 0$ and

$$P_{ij} = \delta_{ij} - \frac{k_i k_j}{\mathbf{k} \cdot \mathbf{k}} \tag{18.33}$$

For the transition probability from the initial state $|i\rangle$ we thus get

$$\sum_f |\mathcal{A}|^2 = e^2 \frac{d^3k}{(2\pi)^3} \frac{1}{2\omega_k} \int d^4x d^4y \, e^{-ik(x-y)} P_{\mu\nu} \, \langle i|J^\mu(x)J^\nu(y)|i\rangle \tag{18.34}$$

We have used the completeness relation for the final states of the source.

Since the choice of the initial state is specified only statistically, we must average the above result over initial states with $\rho = e^{-\beta H}$ corresponding to the fact that the source is kept at a temperature β^{-1}. This gives

$$\langle \sum_f |\mathcal{A}|^2 \rangle = e^2 \int d^4x d^4y \, \frac{d^3k}{(2\pi)^3} \frac{1}{2\omega_k} \, e^{-ik(x-y)} P_{\mu\nu} \, \langle J^\mu(x)J^\nu(y)\rangle \tag{18.35}$$

Introducing $\rho^{\mu\nu}(\omega, \mathbf{x}, \mathbf{y})$ by

$$\langle J^\mu(x)J^\nu(y)\rangle = \int d\omega \, e^{i\omega(x^0-y^0)} \, \rho^{\mu\nu}(\omega, \mathbf{x}, \mathbf{y}) \tag{18.36}$$

we get

$$\langle J^\nu(y)J^\mu(x)\rangle = \int d\omega \, e^{i\omega(y^0-x^0)} \, \rho^{\nu\mu}(\omega, \mathbf{y}, \mathbf{x})$$

$$= \int d\omega \, e^{i\omega(x^0-y^0)} \, \rho^{\nu\mu}(-\omega, \mathbf{y}, \mathbf{x}) \tag{18.37}$$

Equation (18.29) applied to this average gives

$$\rho^{\nu\mu}(-\omega, \mathbf{y}, \mathbf{x}) = e^{\beta\omega} \, \rho^{\mu\nu}(\omega, \mathbf{x}, \mathbf{y}) \tag{18.38}$$

Defining

$$A^{\mu\nu}(\omega, \boldsymbol{x}, \boldsymbol{y}) = \rho^{\nu\mu}(-\omega, \boldsymbol{y}, \boldsymbol{x}) - \rho^{\mu\nu}(\omega, \boldsymbol{x}, \boldsymbol{y}) \qquad (18.39)$$

we can solve (18.38) in terms of $A^{\mu\nu}$ to get

$$\rho^{\mu\nu}(\omega, \boldsymbol{x}, \boldsymbol{y}) = \frac{A^{\mu\nu}(\omega, \boldsymbol{x}, \boldsymbol{y})}{e^{\beta\omega} - 1} \qquad (18.40)$$

The rate of radiation for photons of momenta in a small range of values around $\boldsymbol{k}$ can be obtained by substituting this in (18.35) and dividing by the total time $\int dx^0$. The result is

$$d\Gamma = \frac{\langle \sum_f |\mathcal{A}|^2 \rangle}{\int dx^0}$$

$$= \frac{\pi e^2}{2\omega_k} \int d^3x d^3y \; e^{i\boldsymbol{k}\cdot(\boldsymbol{x}-\boldsymbol{y})} \; P_{\mu\nu} \, A^{\mu\nu}(\omega_k, \boldsymbol{x}, \boldsymbol{y}) \; dN_{Planck}$$

$$dN_{Planck} = \frac{2}{e^{\beta\omega_k} - 1} \frac{d^3k}{(2\pi)^3} \qquad (18.41)$$

dN_{Planck} is the Planck distribution function. $A^{\mu\nu}(\omega_k, \boldsymbol{x}, \boldsymbol{y})$ depends on the nature of the heated source; it is evaluated at the photon energy ω_k in the above formula. If the source has a very large number of energy levels, with every possible value of ω being equally likely, we get a pure Planck distribution for the emitted photons. Otherwise, the radiation rate will have maxima characteristic of the source.

One can also express $A^{\mu\nu}$ in terms of the commutator of currents. Combining (18.36) and (18.37), we get

$$\langle \, [J^\mu(x), J^\nu(y)] \, \rangle = - \int d\omega \; e^{i\omega(x^0 - y^0)} \; A^{\mu\nu}(\omega, \boldsymbol{x}, \boldsymbol{y}) \qquad (18.42)$$

The retarded two-point function for the currents is defined as

$$\Pi_R^{\mu\nu}(x, y) = -i\theta(x^0 - y^0)\langle \, [J^\mu(x), J^\nu(y)] \, \rangle \qquad (18.43)$$

Using the Fourier transformation in time, we can write this as

$$\Pi_R^{\mu\nu}(x, y) = \int d\omega' \; e^{i\omega'(x^0 - y^0)} \; \Pi_R^{\mu\nu}(\omega', \boldsymbol{x}, \boldsymbol{y})$$

$$\Pi_R^{\mu\nu}(\omega', \boldsymbol{x}, \boldsymbol{y}) = \int \frac{d\omega}{2\pi} \frac{A^{\mu\nu}(\omega, \boldsymbol{x}, \boldsymbol{y})}{\omega' - \omega - i\epsilon} \qquad (18.44)$$

where we have used the result

$$i \, e^{i\omega(x^0 - y^0)} \; \theta(x^0 - y^0) = \int \frac{d\omega'}{2\pi} \frac{e^{i\omega'(x^0 - y^0)}}{\omega' - \omega - i\epsilon} \qquad (18.45)$$

Equation (18.44) leads to the result

$$\int_{x,y} e^{ik\cdot(x-y)} P_{\mu\nu} \Pi_R^{\mu\nu}(\omega',x,y) = \int \frac{d\omega}{2\pi} \left[\int_{x,y} e^{ik\cdot(x-y)} P_{\mu\nu} A^{\mu\nu}(\omega,x,y)\right]$$

$$\times \frac{1}{\omega'-\omega-i\epsilon}$$

$$(18.46)$$

Comparison of this with (18.41) shows that the quantity in square brackets is real, since $d\Gamma$ and dN_{Planck} are real. Taking the imaginary part of this equation, we then have

$$\text{Im}\left[\int_{x,y} e^{ik\cdot(x-y)} P_{\mu\nu} \Pi_R^{\mu\nu}(\omega',x,y)\right] = \frac{1}{2}\int_{x,y} e^{ik\cdot(x-y)} P_{\mu\nu} A^{\mu\nu}(\omega',x,y)$$

$$(18.47)$$

where we used the standard formula

$$\frac{1}{\omega'-\omega-i\epsilon} = P\frac{1}{\omega'-\omega} + i\pi\delta(\omega'-\omega) \qquad (18.48)$$

Using (18.47), we can finally write the expression for the radiation rate as

$$d\Gamma = \frac{\pi e^2}{\omega_k} \text{Im}\Pi_R^{\mu\nu}(\omega_k,k,-k) P_{\mu\nu} dN_{Planck} \qquad (18.49)$$

where

$$\Pi_R^{\mu\nu}(\omega,k,-k) = \int d^3x d^3y \, e^{ik\cdot(x-y)} \, \Pi_R^{\mu\nu}(\omega,x,y) \qquad (18.50)$$

This shows that $\text{Im}\Pi_R^{\mu\nu}(\omega_k,k,-k)$ captures all the material properties of the source which can be seen in the radiation. In systems with a large number of closed spaced energy levels with roughly equal radiation intensities, $(1/\omega_k)\text{Im}\Pi_R^{\mu\nu}(\omega,k,-k)$ is approximately constant and this becomes the usual blackbody radiation formula.

18.6 Screening of gauge fields: Abelian case

In a plasma of positive and negative charges, electrostatic potentials can be screened due to the accumulation of oppositely charged particles near any test charge. This is the Debye screening effect. A similar effect can occur at finite densities of particles. This screening effect can be described as being due to an effective mass for the gauge fields, in this case, the electromagnetic field. The usual argument for the absence of mass for a gauge field requires gauge invariance and Lorentz invariance. In a system at finite temperature and density, there is a preferred frame, namely, the frame in which the temperature is $T = \beta^{-1}$. This is the rest frame of the plasma. In this case, we cannot demand Lorentz covariance of the polarization tensor, we can only use gauge invariance. The property of gauge invariance allows for certain (nonlocal) mass terms. Such terms are generated by the interaction of the photon field with the charged particles of the medium. This effect can be obtained by calculating the polarization diagram for the photon, not in vacuum, but in the background of the medium, using the propagators appropriate to the medium. (One still has, as one must have, Lorentz invariance when the overall motion of the plasma is taken into account, as indicated after (18.12).)

The screening effect can be demonstrated by the simple case of zero-mass fermions coupled to the photon. The induced mass term is bilinear in the photon field and so the relevant quantity is the two-photon term in the effective action Γ; it is given by

$$\Gamma^{(2)} = -\frac{i}{2} \int d^4x \, d^4y \, \text{Tr}\left[\gamma \cdot A(x)S(x,y)\gamma \cdot A(y)S(y,x)\right] \qquad (18.51)$$

(We do not show the coupling constant here, it can be restored by the scaling $A_\mu \to eA_\mu$ at the end.) Since we do not have manifest Lorentz invariance, there is no particular advantage to simplifying this in a Lorentz-invariant way. We shall carry out the time-integrations first. Using equation (18.22) and carrying out the time-integrations, we get

$$\Gamma^{(2)} = \frac{1}{2} \int d\mu(k) \frac{d^3q}{(2\pi)^3} \frac{1}{2E_p} \frac{1}{2E_q}$$

$$\left[T(p,q)\left(\frac{\alpha_p\beta_q}{E_p - E_q - k^0 - i\epsilon} - \frac{\alpha_q\beta_p}{E_p - E_q - k^0 + i\epsilon}\right) + \right.$$

$$T(p,q')\left(\frac{\alpha_p\bar{\alpha}_q}{E_p + E_q - k^0 - i\epsilon} - \frac{\beta_p\bar{\beta}_q}{E_p + E_q - k^0 + i\epsilon}\right) +$$

$$T(p',q)\left(\frac{\bar{\alpha}_p\alpha_q}{E_p + E_q + k^0 - i\epsilon} - \frac{\bar{\beta}_p\beta_q}{E_p + E_q + k^0 + i\epsilon}\right) +$$

$$\left. T(p',q')\left(\frac{\bar{\alpha}_p\bar{\beta}_q}{E_p - E_q + k^0 - i\epsilon} - \frac{\bar{\beta}_p\bar{\alpha}_q}{E_p - E_q + k^0 + i\epsilon}\right)\right] \qquad (18.52)$$

where $\alpha_p = 1 - n_p$, $\beta_p = n_p$ and

$$A_\mu(x) = \int \frac{d^4k}{(2\pi)^4} e^{ikx} A_\mu(k)$$

$$d\mu(k) = (2\pi)^4 \delta^{(4)}(k + k') \frac{d^4k}{(2\pi)^4} \frac{d^4k'}{(2\pi)^4}$$

$$T(p,q) = \text{Tr}\left[\gamma \cdot A(k) \, \gamma \cdot p \, \gamma \cdot A(k') \, \gamma \cdot q\right] \qquad (18.53)$$

and $p = q + k$ in equation(18.52). Further, $p' = (p^0, -\boldsymbol{p})$, $q' = (q^0, -\boldsymbol{q})$ and $p_0 = E_p = |\boldsymbol{p}|$, $q_0 = E_q = |\boldsymbol{q}|$; n_p is given by

$$n_p = \frac{1}{e^{p_0/T} + 1} = \frac{1}{e^{|\boldsymbol{p}|/T} + 1} \tag{18.54}$$

As an example of the kind of simplification we did, consider the term in (18.51) with $x^0 > y^0$, so that we get $S_>(x,y)$ and $S_<(y,x)$. In using (18.22), we get, among others, a term like

$$\Gamma_1 = \frac{i}{2} \int_{\boldsymbol{x},\boldsymbol{y}} \int_{-\infty}^{\infty} dx^0 \int_{-\infty}^{x^0} dy^0 \; \mathrm{Tr}[\gamma \cdot A(k) \; \gamma \cdot p \; \gamma \cdot A(k') \; \gamma \cdot q]$$

$$\times \alpha_p \bar{\alpha}_p \; e^{i(k-p-q)x} e^{i(k'+p+q)y} \tag{18.55}$$

We use a convergence factor $e^{\epsilon y^0}$ for the y^0-integration which leads to a factor $-i(k'^0 + p^0 + q^0 - i\epsilon)^{-1}$. Further integrations give energy-momentum δ-functions; one of them implies $k'^0 = -k^0$. This leads to the denominator of the $\alpha_p \bar{\alpha}_p$-term in (18.52). One of the other δ-functions implies $\boldsymbol{q} = -\boldsymbol{p} + \boldsymbol{k}$. We then make a change $\boldsymbol{q} \to -\boldsymbol{q}$ so that we may write $\boldsymbol{p} = \boldsymbol{q} + \boldsymbol{k}$. This gives the coefficient $T(p,q')$, where $q' = (q^0, -\boldsymbol{q}) = (E_q, -\boldsymbol{q})$. The other terms simplify in a similar way; we get q' and p' in some terms when we bring them to a form where we have $\boldsymbol{p} = \boldsymbol{q} + \boldsymbol{k}$ for all terms.

The convergence factors $e^{\pm \epsilon(x^0 - y^0)}$ were put in appropriately to do the time-integrations; these appear as the $i\epsilon$-factors in the denominators. The $i\epsilon$'s can be taken to go to zero at this stage. They can contribute to the imaginary part. Here we are interested in screening effects which are described by the real part of the two-point function, so we shall not need them. Also, the relevant imaginary part, for most physical situations, is that of the retarded function which is not directly given by the above time-ordered functions in (18.52). The correct imaginary part can be obtained from the expression for the real part by the prescription $k_0 \to k_0 + i\epsilon$ for the external photon momentum k.

The real part of the medium-dependent part of $\Gamma^{(2)}$ is given by

$$\Gamma^{(2)} = \frac{1}{2} \int d\mu(k) \frac{d^3q}{(2\pi)^3} \frac{1}{2E_p} \frac{1}{2E_q} \left[(n_q - n_p) \frac{T(p,q)}{E_p - E_q - k^0} \right.$$

$$- (n_p + \bar{n}_q) \frac{T(p,q')}{E_p + E_q - k^0} - (\bar{n}_p + n_q) \frac{T(p',q)}{E_p + E_q + k^0}$$

$$\left. + (\bar{n}_q - \bar{n}_p) \frac{T(p',q')}{E_p - E_q + k^0} \right] \tag{18.56}$$

The screening effects pertain to the long distance behavior of the fields; equivalently, a mass term is important at low values of k. For the nonvacuum contribution, the average value of the loop momentum $\boldsymbol{q}$ in (18.52) is given by the temperature T or the chemical potential μ, depending on which is larger. In extracting the screening mass term, we may therefore simplify the expressions by taking $|\boldsymbol{p}|, |\boldsymbol{q}| \gg |\boldsymbol{k}|$, so that $E_p - E_q - k^0 \approx -\boldsymbol{k} \cdot \boldsymbol{Q}$, $E_p - E_q +$

$k^0 \approx k \cdot Q'$, $E_p + E_q \pm k^0 \approx 2E_q$, where $Q = (1, \boldsymbol{q}/E_q)$, $Q' = (1, -\boldsymbol{q}/E_q)$. (Notice that Q and Q' are lightlike vectors, $Q^2 = Q'^2 = 0$.) Further,

$$T(p,q) \simeq 8E_q^2 A_1 \cdot Q\, A_2 \cdot Q$$
$$T(p',q') \simeq 8E_q^2 A_1 \cdot Q'\, A_2 \cdot Q'$$
$$T(p',q) \simeq T(p,q') \simeq 4E_q^2 \left(A_1 \cdot Q' A_2 \cdot Q + A_1 \cdot Q A_2 \cdot Q' - 2A_1 \cdot A_2 \right)$$
$$(18.57)$$

where $A_1 = A(k)$, $A_2 = A(k')$.

The difference of the distributions can also be approximated as $n_p - n_q \simeq (dn/dq^0)\boldsymbol{Q} \cdot \boldsymbol{k}$ and the following result can be used.

$$\int d^3q \frac{dn}{dq^0} f(Q) = - \int d^3q \frac{2n}{q^0} f(Q) \tag{18.58}$$

for any function f of Q or Q'. (For us f will be terms from $T(p,q)$ and $T(p',q')$.) We can further use $2\boldsymbol{Q} \cdot \boldsymbol{k} = k \cdot Q' - k \cdot Q$. Expression (18.56) can be simplified using (18.57) and (18.58) as

$$\Gamma^{(2)} = \frac{1}{2} \int d\mu(k) \int \frac{d^3q}{(2\pi)^3} \left[\frac{n+\bar{n}}{E_q} \left(2A_1 \cdot Q A_2 \cdot Q - A_1 \cdot Q A_2 \cdot Q' \right. \right.$$
$$\left. - A_1 \cdot Q' A_2 \cdot Q + 2A_1 \cdot A_2 \right) - \frac{n}{E_q} 2A_1 \cdot Q A_2 \cdot Q \frac{k \cdot Q'}{k \cdot Q}$$
$$\left. - \frac{\bar{n}}{E_q} 2A_1 \cdot Q' A_2 \cdot Q' \frac{k \cdot Q}{k \cdot Q'} \right] \tag{18.59}$$

The angular integration in (18.59) over the directions of $\boldsymbol{q}$ (or $\boldsymbol{Q}$) help simplify it further by virtue of

$$\int d\Omega (2A_1 \cdot Q A_2 \cdot Q - A_1 \cdot Q A_2 \cdot Q' - A_1 \cdot Q' A_2 \cdot Q + 2A_1 \cdot A_2) = \int d\Omega (2A_1 \cdot Q A_2 \cdot Q') \tag{18.60}$$

Defining

$$A_+ = \frac{A \cdot Q}{2}, \qquad A_- = \frac{A \cdot Q'}{2} \tag{18.61}$$

we can write (18.59) as

$$\Gamma^{(2)} = \frac{1}{2} \int d\mu(k) \int \frac{d^3q}{(2\pi)^3} \frac{1}{2E_q} \; f(A, n, \bar{n})$$
$$f(A, n, \bar{n}) = 16 \left[A_{1+} A_{2-} (n + \bar{n}) - n \frac{k \cdot Q'}{k \cdot Q} A_{1+} A_{2+} - \bar{n} \frac{k \cdot Q}{k \cdot Q'} A_{1-} A_{2-} \right]$$
$$(18.62)$$

At high temperatures, i.e., $T \gg \mu$, we also have $n_p \approx \bar{n}_p$. The integral can be evaluated in this case to yield

$$\Gamma^{(2)} = \frac{1}{2} \int d\mu(k) \left(\frac{T^2}{12\pi} \right) \int d\Omega \left[2A_{1+}A_{2-} - \frac{k \cdot Q'}{k \cdot Q} A_{1+}A_{2+} - \frac{k \cdot Q}{k \cdot Q'} A_{1-}A_{2-} \right]$$

$$(18.63)$$

The other limit of interest is at high densities corresponding to a degenerate gas of charged fermions; in this case, $T \ll \mu$. As $T/\mu \to 0$, the antiparticle occupation numbers $\bar{n}_p \to 0$ (for positive μ). In this case, (18.62) simplifies to

$$\Gamma^{(2)} = \frac{1}{2} \int d\mu(k) \left(\frac{\mu^2}{4\pi^3} \right) \int d\Omega \left[2A_{1+}A_{2-} - \frac{k \cdot Q'}{k \cdot Q} A_{1+}A_{2+} - \frac{k \cdot Q}{k \cdot Q'} A_{1-}A_{2-} \right]$$

$$(18.64)$$

and we have taken the limit of $T \to 0$ (small compared to μ). Notice that screening at finite density occurs even with one type of charge, because charge fluctuations above the common background charge are what are relevant.

We can discuss plasma oscillations using this screening term. By adding $\Gamma^{(2)}$ to the Maxwell action, we get an effective action which will incorporate some of the effects of the matter distribution at finite temperature,

$$\mathcal{S}_{eff} = -\frac{1}{4e^2} \int F^2 + \Gamma^{(2)} \qquad (18.65)$$

This is a low energy effective action since we have made a long wavelength approximation in simplifying it. Long-wavelength and low-frequency plasma waves are the classical solutions of the effective theory (18.65). It also includes effects such as the screening of Coulomb fields. In terms of the Fourier components of A_μ, we can write

$$\mathcal{S}_{eff} = \frac{1}{2} \int A_\mu(-k) M^{\mu\nu}(k) A_\nu(k) \frac{d^4 k}{(2\pi)^4} \qquad (18.66)$$

where

$$M^{\mu\nu} = (-k^2 \eta^{\mu\nu} + k^\mu k^\nu) + \frac{e^2 T^2}{12\pi} \left[4\pi \eta^{\mu 0} \eta^{\nu 0} - \int d\Omega \frac{k_0}{k \cdot Q} Q^\mu Q^\nu \right] \qquad (18.67)$$

$k^\mu M_{\mu\nu} = 0$ in accordance with the requirement of gauge invariance. We have restored the coupling constant e by the scaling $A_\mu \to eA_\mu$. We now split A_μ into a gauge-dependent part and gauge-invariant components as

$$A_\mu = k_\mu \Lambda(k) + \zeta_\mu + \xi_\mu \qquad (18.68)$$

where Λ shifts under gauge transformations and ζ_μ, ξ_μ are gauge-invariant. ζ, ξ correspond to the three directions orthogonal to k_μ, and they can be parametrized as

$$\zeta_0 = \left(\frac{k^2}{k^2 - k_0^2} \right) \phi$$

$$\zeta_i = \frac{k_0}{\sqrt{\boldsymbol{k}^2}} e_i^{(3)} \left(\frac{\boldsymbol{k}^2}{\boldsymbol{k}^2 - k_0^2} \right) \phi$$

$$\xi_0 = 0$$

$$\xi_i = e_i^{(\lambda)} a_\lambda, \quad \lambda = 1, 2 \tag{18.69}$$

The e_i's form a triad of spatial unit vectors. If needed, a specific choice, as indicated in Chapter 6, would be

$$e_i^{(3)} = \frac{k_i}{\sqrt{\boldsymbol{k}^2}}, \quad i = 1, 2, 3$$

$$e_i^{(1)} = \left(\epsilon_{ij} \frac{k_j}{\sqrt{k_T^2}}, 0 \right)$$

$$e_i^{(2)} = \left(\frac{k_3 k_i}{\sqrt{k_T^2 \boldsymbol{k}^2}}, -\sqrt{\frac{k_T^2}{\boldsymbol{k}^2}} \right), \quad i = 1, 2, \tag{18.70}$$

where $k_T^2 = k_1^2 + k_2^2$. Notice that $k_i e_i^{(\lambda)} = 0$, $\lambda = 1, 2$. ϕ and a_λ are the gauge-invariant degrees of freedom in (18.68). When the mode decomposition (18.68) is used in (18.66), we get

$$\mathcal{S}_{eff} = \frac{1}{2} \int \frac{d^4 k}{(2\pi)^4} \left[\xi_i(-k) \left(\frac{\boldsymbol{k}^2 \delta_{ij} - k_i k_j}{\boldsymbol{k}^2} \right) M^T(k) \xi_j(k) + \phi(-k) M^L(k) \phi(k) \right]$$
$$+ \int \phi J^0 + \xi_i J^i \tag{18.71}$$

where

$$M^T(k) = k_0^2 - \boldsymbol{k}^2 - \frac{e^2 T^2}{6} \left[\frac{k_0^2}{\boldsymbol{k}^2} + \left(1 - \frac{k_0^2}{\boldsymbol{k}^2} \right) \frac{k_0}{2|\boldsymbol{k}|} L \right]$$

$$M^L(k) = \boldsymbol{k}^2 + \frac{e^2 T^2}{3} \left[1 - \frac{k_0}{2|\boldsymbol{k}|} L \right]$$

$$L = \log \left(\frac{k_0 + k}{k_0 - k} \right) \tag{18.72}$$

We have also included an interaction term with a conserved source J_μ in (18.71); i.e., we include $\int A_\mu J^\mu$ and simplify it using (18.69). From (18.71) we see that the interaction between charges in the plasma is governed by $(M^L)^{-1}$, which shows the Debye screening with a Debye mass $m_D = \sqrt{e^2 T^2 / 3}$. The action (18.71) can also give free wavelike solutions. The dispersion rules for these plasma waves would be $M^T = 0$ for the transverse waves and $(k^2 - k_0^2/\boldsymbol{k}^2) M^L = 0$ for the longitudinal waves. (The extra factor multiplying M_L is from rewriting $\phi(k)$ in terms of the potential.)

18.7 Screening of gauge fields: Nonabelian case

We have considered the screening of Abelian gauge fields in the last section. Here we will extend those considerations to the nonabelian theory; this is relevant for the quark-gluon plasma, for instance. Consider the term in the effective action $\Gamma[A]$ which represents the screening effect. The term quadratic in A's is what we have calculated. In the nonabelian case, the structure of this term remains the same, except for the trace over the Lie algebra matrices. The prefactor also changes since the gauge bosons (or gluons) themselves can propagate in the loop and contribute to the two-point function. But there are also many higher point functions which are nonzero and which are needed for reasons of gauge invariance. Screening is an infrared or low-momentum effect and so we are interested in the simplification of loop diagrams when the external momenta are small compared to the loop momentum. The average loop momentum is of the order of the temperature T, since it is controlled by the statistical distribution function. One can then establish a set of power-counting rules, applicable when the external momenta are small (of the order of eT), and isolate the relevant diagrams. They will be one-loop diagrams with the loop momentum of the order of T and external momenta of order eT; they are called hard thermal loops. Rather than calculating such diagrams directly, we will use an argument based on gauge invariance to get the result.

There are two key properties which are important to our analysis.

1. $\Gamma[A]$ is gauge-invariant with respect to gauge transformations of the gauge potential A_μ and is independent of the gauge-fixing used to define the gluon propagators.

   We can understand how the kinematics of hard thermal loops can lead to gauge invariance. The thermal propagators satisfy the same differential equations as the zero temperature propagators. This is evident from (18.16) and (18.23). The extra term involving the distribution functions obeys the homogeneous free particle equation of motion, so the propagators differ from the zero temperature ones in the choice of boundary conditions only. As a result, the generating functional of one-particle irreducible vertices, viz., $\Gamma[A, c, \bar{c}, ...]$ obeys the standard BRST Ward-Takahashi identities. Keeping only the part of Γ relevant for external gauge fields, we thus have the identity (10.123)

$$\int d^4x \left[\frac{\delta\Gamma}{\delta A_\mu^a} \frac{\delta\Gamma}{\delta K^{a\mu}} - \frac{\delta\Gamma}{\delta L^a} \frac{\delta\Gamma}{\delta c^a} - i\frac{\delta\Gamma}{\delta J^a} \frac{\delta\Gamma}{\delta \bar{c}^a} \right] = 0 \qquad (18.73)$$

In the hard thermal loop approximation, terms involving the ghosts are subdominant. Recall that, with a gauge-fixing term $\sim (\partial \cdot A)^2$, the ghost-gluon coupling involves $f^{abc} A^{a\mu}(\partial_\mu \bar{c}^b)c^c$; the derivative is on the antighost field. In a diagram with external ghosts, this is a power of external momentum and therefore such a diagram is smaller compared to a similar diagram with the ghost lines replaced by gluons where there are contributions with all derivatives on the internal lines which give powers of T. The terms $\frac{\delta\Gamma}{\delta c}$, $\frac{\delta\Gamma}{\delta\bar{c}}$ in (10.123) are thus negligible in the hard thermal loop approximation. Further, we are interested only in one-loop terms. The identity (10.123) then gives the gauge-invariance of Γ. Thus, effectively, in a high-T-expansion, the leading term $\Gamma[A]$, which is proportional to T^2, is gauge-invariant. Since the thermal contribution to the propagator is on-shell, the T-dependent part of a one-loop diagram is classical and so it is not surprising that the BRST Ward identities reduce to the statement of gauge-invariance. The fact that Γ does not depend on the gauge choice for the gluon propagators can be seen by similar arguments; identities for the variation of Γ under changes in the gauge-fixing function can be written down and simplified as in the case of (10.123).

2. $\Gamma[A]$ has the form

$$\Gamma = (N + \tfrac{1}{2}N_f)\frac{T^2}{12\pi}\left[\int d^4x\, 2\pi A_0^a A_0^a + \int d\Omega\, W(A\cdot Q)\right]. \quad (18.74)$$

This is a restatement of the result (18.63) in the form used in (18.67). The angular integration is over the directions of Q. Here we are considering an $SU(N)$ gauge theory and N_f is the number of fermion flavours. The key point about this equation is that, for each Q, $\Gamma[A]$ involves essentially only two components of the gauge potential, A_0 and $A\cdot Q$. The $d\Omega$-integration will bring in all components of the potential, but for each Q only two components are needed. In the explicit calculation of the two-point function, Q was the angular part of the loop-momentum q. The $d\Omega$-integration in (18.74) is over the orientations of Q and is the unfinished part of the loop integration, after the integration over the modulus of the loop momentum has been done. The structure of (18.74) can be seen to be general by analysis of diagrams again. For example, for diagrams with the derivative gluon coupling due to the interaction term $f^{abc}A_\mu^a A_\nu^b \partial_\mu A_\nu^c$, since $p\cdot A \approx q\cdot A$, the possible tensor structures are $q^2 A^2$ and $(q\cdot A)^2$. The former is zero since the thermal part of the propagator involves the δ-function $\delta(q^2)$. Writing $q = |q|(1, Q)$ and carrying out the $|q|$-integration, we are left with a structure like (18.74). This argument generalizes to diagrams with an arbitrary number of external gluons.

Given these two properties of Γ one can determine W and hence Γ simply by the requirement of gauge-invariance. First of all, define two derivatives $\partial_+ = \frac{1}{2}Q\cdot\partial$ and $\partial_- = \frac{1}{2}Q'\cdot\partial$ in a way analogous to our definition of $A_\pm$ in (18.61). The condition for gauge-invariance of Γ is then

$$\int d\Omega\left(\left[D_+\left(\frac{\delta W}{\delta A_+}\right)\right]^a + 4\pi\partial_0 A_0^a\right) = 0 \quad (18.75)$$

Using the identity

$$4\pi\partial_0 A_0^a = \int d\Omega \ \partial_0(Q \cdot A^a) = 2 \ (\partial_+ + \partial_-)A_+^a \tag{18.76}$$

we see that the condition of gauge-invariance (18.75) is satisfied if

$$\left[D_+\left(\frac{\delta W}{\delta A_+}\right)\right] \ + \ 2 \ \partial_+ A_+ \ + \ 2 \ \partial_- A_+ = 0 \tag{18.77}$$

Further, if we define $\mathcal{A}_-$ by

$$\frac{\delta W}{\delta A_+^a} = -2 \ \mathcal{A}_-^a \ - \ 2 \ A_+^a \tag{18.78}$$

equation (18.77) becomes

$$D_+\mathcal{A}_- \ - \ \partial_- A_+ = 0 \tag{18.79}$$

This is similar in form to equation (17.53) of the WZW theory,

$$\partial_{\bar{z}} A_z - \partial_z \bar{A} \ + \ [\bar{A}, A_z] = 0 \tag{18.80}$$

Equation (18.80) was obtained for the WZW action in Euclidean space. In the present case, it is easier to define a WZW action with Minkowski signature. If we are in two dimensions, with $\partial_\pm = \frac{1}{2}(\partial_0 \pm \partial_1)$,

$$\mathcal{S}_{WZW}(U) = \frac{1}{2\pi} \int d^2x \ \mathrm{Tr} \left(\partial_+ U \partial_- U^{-1}\right)$$
$$+ \ \frac{1}{12\pi} \int_{\mathcal{M}^3} d^3x \ \epsilon^{\mu\nu\alpha} \mathrm{Tr}(\partial_\mu U U^{-1}\partial_\nu U U^{-1}\partial_\alpha U U^{-1}) \tag{18.81}$$

The Polyakov-Wiegman identity for this WZW action is

$$\mathcal{S}_{WZW}(hU) = \mathcal{S}_{WZW}(h) \ + \ \mathcal{S}_{WZW}(U) \ - \ \frac{1}{\pi} \int d^2x \ \mathrm{Tr}(h^{-1}\partial_- h \ \partial_+ U U^{-1}) \tag{18.82}$$

Further, if we define $A_+ = -\partial_+ U U^{-1}$, we find

$$\delta\mathcal{S}_{WZW}(U) = -\frac{1}{\pi} \int d^2x \ \mathrm{Tr}(\mathcal{A}_- \delta A_+) \tag{18.83}$$

This is the analog of (17.51).

In our four-dimensional case, we note that $\partial_+ = \frac{1}{2}Q \cdot \partial$ is invertible with

$$\partial_+ G(x, y) = \delta^{(4)}(x - y)$$
$$G(x, y) = \int \frac{d^4p}{(2\pi)^4} e^{ip(x-y)} \frac{2}{ip \cdot Q} \tag{18.84}$$

Thus it is indeed possible to define a unitary matrix U such that

$$A_+ = -\partial_+ U \, U^{-1} \tag{18.85}$$

(Notice that one can, at least as a series in A, solve this equation for U.) The equations of interest to us have no derivatives with respect to the transverse coordinates x^T, $Q \cdot x^T = Q' \cdot x^T = 0$. They simply act as some parameters on which the fields can depend. We may then use the identity (18.83) to write the solution for W as

$$W = 2 \int d^4x \; \text{Tr}(A_+ A_+) - 4\pi \int d^2 x^T S_{WZW}(U) \tag{18.86}$$

The WZW action must be integrated over the transverse coordinates, since the original WZW action was defined in two dimensions and hence carry integration only over the $x^\pm$ coordinates. Equation (18.86) leads to

$$\Gamma = (N + \tfrac{1}{2}N_f)\frac{T^2}{12\pi}\left[\int d^4x \; 2\pi A_0^a A_0^a \right.$$
$$\left. + \int d\Omega \left\{ 2 \int d^4x \; \text{Tr}(A_+ A_+) - 4\pi \int d^2 x^T S_{WZW}(U) \right\} \right] \tag{18.87}$$

We can rewrite this in a more transparent way by introducing another unitary matrix V such that

$$A_- = -\partial_- V^\dagger \, V = V^\dagger \partial_- V \tag{18.88}$$

so that $U \leftrightarrow V^\dagger$ under $Q \leftrightarrow Q'$. Since we are integrating over all directions of $\boldsymbol{Q}$, $\int d\Omega \; S_{WZW}(U) = \int d\Omega \; S_{WZW}(V^\dagger)$. Further we have the identity

$$2\pi A_0^a A_0^a + 2 \int d\Omega \; \text{Tr}(A_+ A_+) = -2 \int d\Omega \; \text{Tr}(A_+ A_-) \tag{18.89}$$

which follows by straightforward $d\Omega$-integration. Using these two results, we can finally write the effective action which describes screening for the nonabelian theory as

$$\Gamma[A] = (N + \tfrac{1}{2}N_f)\frac{T^2}{12\pi}(-2\pi) \int \left[\frac{1}{\pi}\text{Tr}(A_+ A_-) + S_{WZW}(U) + S_{WZW}(V^\dagger) \right]$$
$$= -(N + \tfrac{1}{2}N_f)\frac{T^2}{6} \int d^2 x^T d\Omega \; S_{WZW}(V^\dagger U) \tag{18.90}$$

where we have used the Polyakov-Wiegmann identity to combine terms.

The final result for the screening term in the effective action is thus very simple and transparent. It is given by the gauge-invariant WZW action for fields on the lightcone defined by $x^0 \pm \boldsymbol{Q} \cdot \boldsymbol{x}$, the unit vector $\boldsymbol{Q}$ specifying the orientation of the lightcone. In general the fields depend on the transverse coordinates as well. The WZW action is then integrated over the transverse directions and over $\boldsymbol{Q}$, i.e., over all orientations of the lightcone.

One can derive a similar result for screening for a fermion gas at high densities and low temperatures. We find, in a way analogous to (18.64),

$$\Gamma[A] = -\frac{1}{2} N_f \frac{\mu^2}{2\pi^2} \int d^2 x^T d\Omega \; \mathcal{S}_{WZW}(V^\dagger U) \tag{18.91}$$

where μ is the chemical potential. (In these equations, the gauge potentials are the antihermitian matrices $A_\mu = -it^a A_\mu^a$, whereas we used the hermitian components in (18.64) since that is more conventional for Abelian fields.)

18.8 Retarded and time-ordered functions and the Kubo formula

In the last two sections, we used time-ordered products to calculate the real part of the two-point vertex function for A_μ, and more generally the screening term in the effective action, which can also describe plasma wave propagation. However, for the imaginary part, we should use the retarded functions to get physically relevant results, as we shall explain below.

In scattering theory we ask the question that if we start with an essentially free particle state $|i\rangle$ at some time, what is the probability amplitude to observe the (essentially free particle) state $|f\rangle$ at a later time. This involves time-evolving the state $|i\rangle$ and then finding the overlap with the state $|f\rangle$. Since time-evolution is a naturally time-ordered process, we get the time-ordered products in the S-matrix. Notice that the state $|f\rangle$ evolves as in the free theory. If we use the Schrödinger picture, the amplitudes are therefore of the form $\langle f | e^{iH_0 t} e^{-iHt} | i \rangle$, where H_0 is the free Hamiltonian and H is the full Hamiltonian. If we define

$$U(t, t_0) = e^{iH_0(t-t_0)} e^{-iH(t-t_0)} \tag{18.92}$$

we find

$$
\begin{aligned}
i\frac{\partial U}{\partial t} &= e^{iH_0(t-t_0)} H_{int} e^{-iH(t-t_0)} \\
&= H_{int}(\phi_{in}) \, U
\end{aligned}
\tag{18.93}
$$

The fields ϕ_{in} in the last step are defined by $\phi_{in} = e^{iH_0(t-t_0)} \phi e^{-iH_0(t-t_0)}$. Therefore they evolve as free Heisenberg fields. Integration of equation (18.93) leads to the formula (5.72) for the S-operator

$$\hat{S} = T \, e^{i\mathcal{S}_{int}(\phi_{in})} = U(\infty, -\infty)$$

$$U(x^0, y^0) = T \, \exp\left[i \int_{y^0}^{x^0} d^4 x \, \mathcal{L}_{int}(\phi_{in}(x)) \right] \tag{18.94}$$

at least as long as we do not have derivative couplings. (With derivative couplings, we do not have $H_{int} = -\mathcal{L}_{int}$ and the situation is a bit more involved.)

However, if we have to calculate the time-evolution of field configurations, then we need to calculate matrix elements in the Schrödinger picture where both states involved evolve with the Hamiltonian H; i.e., we need

$$\langle \phi \rangle = \langle x^0 = 0 | e^{iHx^0} \phi e^{-iHx^0} | x^0 = 0 \rangle \qquad (18.95)$$

or equivalently, we need to solve the Heisenberg equations of motion for the Heisenberg field $\phi(x^0) = e^{iHx^0} \phi e^{-iHx^0}$. The question we are asking is, given the initial field operator, what is the field operator at a later time? The solution involves the retarded Green's functions as in (5.81). Similarly for electrodynamics, if we apply a perturbation to the medium via the field A_μ, we can ask what the induced current is; this is the relevant quantity for the conductivity of the medium. In these cases, unlike in the case of the S-matrix, the final states are not specified.

In the case of a scalar field, the equation of motion was obtained in equation (5.77) as

$$(\Box + m^2)\phi(x) = \rho(x) \qquad (18.96)$$

where the source $\rho(x)$ was given in equation (5.83) in terms of the scattering operator by

$$\rho(x) \equiv -i\hat{S}^{-1} \frac{\delta \hat{S}}{\delta \phi_{in}(x)}$$
$$= \frac{\delta S_{int}(\phi)}{\delta \phi(y)} \qquad (18.97)$$

The analogous equations for electrodynamics are easy to set up and read

$$\partial_\nu F^{\nu\mu}(x) = i\hat{S}^{-1} \frac{\delta \hat{S}}{\delta A_\mu(x)} \qquad (18.98)$$

where the current J^μ is

$$J^\mu(x) = i\hat{S}^{-1} \frac{\delta \hat{S}}{\delta A_\mu(x)} \qquad (18.99)$$

(While no gauge choice is specified in (18.98), it can be fixed, for example, by adding $-\frac{1}{2}(\partial \cdot A)^2$ to the Lagrangian, whereupon the left side would become $\Box A^\mu$.) The scattering operator $\hat{S}$ is given by $T \exp[ie \int d^4x A_\mu j^\mu]$, where $j^\mu = \bar{\psi}_{in} \gamma^\mu \psi_{in}$ is the current with the in-fields.

We can use (18.99) to obtain J^μ as a power series in A_μ. In particular, a Taylor expansion of (18.99) to linear order in A gives

$$J^\mu(x) = ej^\mu(x) - ie^2 \int d^4y \ \theta(x^0 - y^0)[j^\mu(x), j^\nu(y)]A_\nu(y) \qquad (18.100)$$

Using this in (18.98) and taking the average with the density matrix, we can write

$$\partial_\nu F^{\nu\mu}(x) = \int d^4y \; \Pi_R^{\mu\nu}(x,y) \; A_\nu(y) \tag{18.101}$$

where

$$\Pi_R^{\mu\nu}(x,y) = -ie^2 \; \theta(x^0 - y^0) \; \langle [j^\mu(x), j^\nu(y)] \rangle \tag{18.102}$$

The angular brackets denote the averaging, with the unperturbed density matrix, so that $\langle j^\mu \rangle$ vanishes. Thus the response function (18.101), namely, the average of the retarded commutator (or equation (18.102)), is appropriate to the situation where we perturb the medium by the field and ask how the field evolves. Equation (18.102) is known as the Kubo formula and has been used extensively in condensed matter physics.

Consider now the calculation of this for the thermal plasma. After Wick contractions and using the thermal propagators (18.22,18.23) and carrying out the time-integrations with convergence factors as before, we find

$$\Pi^{\mu\nu}(x,y) = \int \frac{d^4k}{(2\pi)^4} e^{-ik(x-y)} \Pi^{\mu\nu}(k) \tag{18.103}$$

$$\Pi^{\mu\nu}(k) = \int \frac{d^3q}{(2\pi)^3} \frac{1}{2E_p} \frac{1}{2E_q} \times$$
$$\left[T^{\mu\nu}(p,q) \left(\frac{\alpha_p \beta_q}{E_p - E_q - k_0 - i\epsilon} - \frac{\alpha_q \beta_p}{E_p - E_q - k_0 - i\epsilon} \right) \right.$$
$$+ T^{\mu\nu}(p,q') \left(\frac{\alpha_p \bar\alpha_q}{E_p + E_q - k_0 - i\epsilon} - \frac{\beta_p \bar\beta_q}{E_p + E_q - k_0 - i\epsilon} \right)$$
$$+ T^{\mu\nu}(p',q) \left(\frac{\bar\alpha_p \alpha_q}{E_p + E_q + k_0 + i\epsilon} - \frac{\bar\beta_p \beta_q}{E_p + E_q + k_0 + i\epsilon} \right)$$
$$\left. + T^{\mu\nu}(p',q') \left(\frac{\bar\alpha_p \bar\beta_q}{E_p - E_q + k_0 + i\epsilon} - \frac{\bar\beta_p \bar\alpha_q}{E_p - E_q + k_0 + i\epsilon} \right) \right] \tag{18.104}$$

where $T^{\mu\nu}(p,q) \equiv \mathrm{Tr}\left(\gamma^\mu \; \gamma \cdot p \; \gamma^\nu \; \gamma \cdot q \right)$ and $p'^\mu = (E_p, -\boldsymbol{p})$, $q'^\mu = (E_q, -\boldsymbol{q})$ and $\boldsymbol{p} = \boldsymbol{q} + \boldsymbol{k}$.

Notice that equation (18.104) is exactly what we got from $\Gamma^{(2)}$ except for the changes in sign of some of the $i\epsilon$'s. The retarded function (18.104) can be obtained by continuing the real part by the rule $k_0 \to k_0 + i\epsilon$. The real part of $\Pi^{\mu\nu}(k)$ is the same for both the time-ordered and the retarded functions. Using the formula

$$\frac{1}{z - i\epsilon} = P\frac{1}{z} + i\pi\delta(z) \tag{18.105}$$

we now calculate the imaginary part of the retarded function (18.104) for the momenta of the gauge fields small compared to the temperature.

The contributions due to the $T^{\mu\nu}(p,q')$ and $T^{\mu\nu}(p',q)$ terms in (18.104) are subdominant for k small compared to T because their imaginary parts carry the δ-functions, $\delta(E_p + E_q \pm k_0)$, which have no support as $k \to 0$. The dominant contributions to the imaginary part of $\Pi^{\mu\nu}$ are due to the $T^{\mu\nu}(p,q)$, $T^{\mu\nu}(p',q')$ terms and give

$$
\mathrm{Im}\Pi_R^{\mu\nu} = \pi \int \frac{d^3q}{(2\pi)^3} \frac{1}{2E_p} \frac{1}{2E_q} \left[T^{\mu\nu}(p,q)\,(n_q - n_p)\,\delta(E_p - E_q - k_0) \right.
$$
$$
\left. - T^{\mu\nu}(p',q')\,(\bar{n}_q - \bar{n}_p)\,\delta(E_p - E_q + k_0) \right] \qquad (18.106)
$$

For $|\boldsymbol{k}|$ small compared to $|\boldsymbol{p}|$, $|\boldsymbol{q}|$, the latter being of order T, we have the same simplifications indicated after (18.56) and in (18.57). With these, equation (18.106) can be simplified as

$$
\mathrm{Im}\Pi_R^{\mu\nu} \simeq \frac{k_0 P^{\mu\nu}}{2\pi^2} \int_0^\infty dq\, q(n_q + \bar{n}_q)
$$
$$
= \frac{k_0 T^2}{12} P^{\mu\nu} \qquad (18.107)
$$

where

$$
P^{\mu\nu} = \int d\Omega\, \delta(k \cdot Q) Q^\mu Q^\nu \qquad (18.108)
$$

and we take $n = \bar{n}$. The integration in (18.108) is over the orientations of the unit vector $\boldsymbol{Q} = \hat{q}$. $P^{\mu\nu}$ is transverse and traceless, while the δ-function has support only for spacelike k. Carrying out the angular integration

$$
P^{\mu\nu} = -k^2 \theta(-k^2) \frac{6\pi}{|\boldsymbol{k}|^3} \left[\frac{1}{3} P_1^{\mu\nu} + \frac{1}{2} P_2^{\mu\nu} \right] \qquad (18.109)
$$

where

$$
P_1^{\mu\nu} = \left(g^{\mu\nu} - \frac{k^\mu k^\nu}{k^2} \right) \qquad (18.110)
$$

and

$$
P_2^{\mu 0} = P_2^{0\nu} = 0
$$
$$
P_2^{ij} = \delta^{ij} - \frac{k^i k^j}{|\boldsymbol{k}|^2} \qquad (18.111)
$$

Here $\theta(-k^2)$ is the step function, $\theta(-k^2) = 1$ for $k^2 < 0$ and $\theta(-k^2) = 0$ for $k^2 > 0$.

18.9 Physical significance of Im $\Pi_R^{\mu\nu}$

We have already seen that in the case of radiation from a heated source the retarded function naturally emerged as the important quantity. From the above discussion it is also important for the time-evolution of field configurations, the induced current, etc. In the plasma context, the imaginary part of $\Pi_R^{\mu\nu}$ describes Landau damping, which can occur for fields with spacelike momenta; this gives a direct interpretation for $\Pi_R^{\mu\nu}$. Explicitly we consider

$$A_\mu(x) = \int \frac{d^4k}{(2\pi)^4} e^{-ikx} A_\mu(k) \qquad (18.112)$$

with

$$A_\mu(k) = \delta(k_0 - \omega(\boldsymbol{k})) A_\mu(k) + \delta(k_0 + \omega(\boldsymbol{k})) A_\mu^\dagger(k) \qquad (18.113)$$

and $\omega^2 < \boldsymbol{k}^2$. The amplitudes $A_\mu(k)$, $A_\mu^\dagger(k)$, respectively, of the positive- and negative-frequency terms, correspond to absorption and emission processes. The decay of the field in the plasma arises from absorption by fermions and antifermions. The amplitude for absorption by fermions is given, to the lowest order in coupling constant, by

$$\mathcal{A} = i\bar{u}_p(\gamma \cdot A) u_q (2\pi)^4 \delta^{(4)}(p - q - k) \sqrt{n_q(1 - n_p)} \qquad (18.114)$$

where u_p, u_q are wave functions for outgoing and incoming fermions, respectively. The factor $\sqrt{n_q(1 - n_p)}$ can be interpreted physically as follows. The initial fermion is chosen from a state of occupation number n_q, which has a probability amplitude $\sqrt{n_q}$, and the final fermion is scattered into a state of occupation number n_p which carries a factor $\sqrt{1 - n_p}$ in accordance with the exclusion principle. Absorption of a single quantum, as in (18.114), is kinematically allowed for spacelike momenta; it is in fact the inverse of Čerenkov radiation discussed in Chapter 7. One can also have creation of the mode $(\omega, \boldsymbol{k})$ by (Čerenkov) radiation from fermions, given by a formula like (18.114) with $p \leftrightarrow q$ and $A \leftrightarrow A^\dagger$. There are also similar contributions from antifermions. For the net absorption probability per unit spacetime volume, denoted by γ, we then find, with summation over all fermion states,

$$\gamma = \int \frac{d^3q}{(2\pi)^3} \frac{1}{2E_p} \frac{1}{2E_q} A_\mu^\dagger(k) T^{\mu\nu}(p,q) A_\nu(k)$$
$$T^{\mu\nu}(p,q) = 2\pi\delta(E_p - E_q - k_0) T^{\mu\nu}(p,q) (n_q(1 - n_p) - n_p(1 - n_q))$$
$$+ 2\pi\delta(E_p - E_q + k_0) T^{\mu\nu}(p',q') (\bar{n}_p(1 - \bar{n}_q) - \bar{n}_q(1 - \bar{n}_p))$$
$$\qquad (18.115)$$

Comparing this with (18.106), we see that

$$\gamma = 2A_\mu^\dagger(k) [\mathrm{Im}\Pi_R^{\mu\nu}(k_0)] A_\nu(k) \qquad (18.116)$$

In terms of the parametrization of the potentials (18.68) and (18.69), we can write

$$\gamma = T^2 \frac{\pi\omega}{|\boldsymbol{k}|} \left[\frac{1}{3} \phi^\dagger \phi + \frac{1}{6} \left(1 - \frac{\omega^2}{\boldsymbol{k}^2} \right) \boldsymbol{A}_T^\dagger \cdot \boldsymbol{A}_T \right] \qquad (18.117)$$

where $\omega = |\mathbf{k}|$. γ is manifestly positive, as expected for net absorption or damping.

In summary, we see that the retarded commutator is what is relevant for the operator equations of motion, Landau damping, etc. However, because of the asymmetry in its time-arguments, the retarded commutator cannot be obtained by varying an effective action, which necessarily involves the symmetric, time-ordered products. One can use a more general formalism where the functional integral and effective actions can be defined with time-integration running over a contour that goes from $-\infty$ to ∞ and then goes back from ∞ to $-\infty$. This is applicable not only to near-equilibrium situations like the ones considered so far, but forms the general theory of nonequilibrium phenomena in quantum field theory. This approach will be discussed briefly in the next section.

18.10 Nonequilibrium phenomena and quantum kinetic theory

In classical physics, the Boltzmann equation gives the time-evolution of the one-particle distribution function. More generally the evolution of a statistical distribution is given by the Liouville equation. Since this is very hard to analyze in full generality for a many-particle system, one makes the approximation of truncating to the one-particle, or in some situations the two-particle, distributions. The description of nonequilibrium phenomena via kinetic theory is developed from these. We will now discuss briefly the quantum field theoretic approach to kinetic theory.

The statistical averages of interest are calculated with the density matrix ρ. For a scalar field ϕ, the one-particle operators are of the form $\hat{M} = \int_{x,y} \phi(x) M(x,y) \phi(y)$, where $M(x,y)$ is independent of ϕ. The average is thus given by

$$\mathrm{Tr}\hat{M}\rho = \int_{x,y} M(x,y)\mathrm{Tr}\left[\phi(y)\rho\phi(x)\right]$$

$$= \int_{x,y} M(x,y)\, F(y,x)$$

$$F(x,y) = \mathrm{Tr}\left[\phi(x)\, \rho\, \phi(y)\right] \tag{18.118}$$

$F(x,y)$ is a one-particle distribution function. Similarly, we can define a whole set of higher point expectation values, appropriate for many-particle operators. For kinetic theory, we need some method of calculating such functions, or a set of equations for them. Notice that for $y^0 > x^0$, $F(x,y) = \mathrm{Tr}[\rho T(\phi(y)\phi(x))] = \langle T\phi(y)\phi(x)\rangle = G(y,x)$. One can obtain it in terms of the propagator calculated in the medium. However, for $x^0 > y^0$ we cannot obtain $F(x,y)$ in terms of the time-ordered functions. So we define the functional

$$Z[J, \tilde{J}] = \text{Tr} \left[T e^{\int_{-\infty}^{\infty} J\phi} \rho_0 \, \bar{T} e^{\int_{\infty}^{-\infty} \tilde{J}\phi} \right] \qquad (18.119)$$

where T denotes time-ordering as usual and $\bar{T}$ denotes antitime-ordering, i.e.,

$$\bar{T} e^{\int_{\infty}^{-\infty} \tilde{J}\phi} = e^{\int_{-\tau+\epsilon}^{-\tau} \tilde{J}\phi} \, e^{\int_{-\tau+2\epsilon}^{-\tau+\epsilon} \tilde{J}\phi} \cdots e^{\int_{\tau}^{\tau-\epsilon} \tilde{J}\phi} \Bigg]_{\tau \to \infty} \qquad (18.120)$$

The expression for $Z[J, \tilde{J}]$ in (18.119) has the free density matrix ρ_0; the effect of the interactions is carried by the fields, since we are in the Heisenberg picture and the fields evolve according to the equations of motion of the interacting theory. It is now clear that we can write

$$F(x, y) = \left[\frac{1}{Z} \frac{\delta^2 Z}{\delta J(x) \delta \tilde{J}(y)} \right]_{J=\tilde{J}=0} \qquad (18.121)$$

Higher correlators of ϕ's needed for multiparticle operators can also be obtained from (18.119) by further differentiations with respect to J, $\tilde{J}$. By cyclicity of trace we may write (18.119) as

$$Z[J, \tilde{J}] = \text{Tr} \left[T_C e^{\oint J\phi} \rho_0 \right] = \langle T_C e^{\oint J\phi} \rangle \qquad (18.122)$$

where C denotes a contour for time-integration starting from $-\infty$, going to ∞ and then returning to $-\infty$. J's on the return branch are $\tilde{J}$. If there are several fields invloved, we include a source function for each field. One can define Z by a functional integral, or by functional equations, or their partially summed versions, the Schwinger-Dyson equations. These equations form a hierarchy of equations for various multiparticle functions which completely characterize the theory. For example, let ϕ obey the equation of motion

$$[\Box_x + m^2 + U(x)]\phi(x) - \rho(x) = 0 \qquad (18.123)$$

where $\rho(x)$ depends on ϕ's and other fields which may couple to ϕ. (This $\rho(x)$ is the source for the field given in (5.83) and (18.97), not the density matrix; we will not use the density matrix in what follows.) We also include a background potential $U(x)$ for generality. The equations of motion for Z are then given by

$$K_1 \frac{\delta Z}{\delta J(1)} - \hat{\rho}(1) \, Z = -i J(1) \, Z$$

$$K_1 \frac{\delta Z}{\delta \tilde{J}(1)} - \hat{\tilde{\rho}}(1) \, Z = -i \tilde{J}(1) \, Z \qquad (18.124)$$

where $K_x = \Box_x + m^2 + U(x)$. $\hat{\rho}(1)$ stands for ρ with the fields replaced by derivatives with respect to sources and $\hat{\tilde{\rho}}$ involves derivatives with respect to the tilde-sources. As an example, if the interaction term is $S_{int} = \lambda \int \phi^2 \chi$ for two scalar fields ϕ, χ, then $\rho = 2\lambda\phi\chi$ and

$$\hat{\rho}(1) = 2\lambda \frac{\delta}{\delta J(1)} \frac{\delta}{\delta \eta(1)}$$

$$\hat{\tilde{\rho}}(1) = 2\lambda \frac{\delta}{\delta \tilde{J}(1)} \frac{\delta}{\delta \tilde{\eta}(1)} \tag{18.125}$$

where η is the source function for χ.

By taking derivatives of the first of equations (18.124) with respect to J's and setting J, $\tilde{J}$ to zero, we get equations for time-ordered correlators; from the second equation in (18.124) we get equations for antitime-ordered correlators; and both lead to equations for $F(x, y)$ and its multiparticle generalizations. At the simplest, one-particle level we find

$$K_x F(x, y) = \hat{\rho}(x) \frac{\delta}{\delta \tilde{J}(y)} Z$$

$$K_y F(x, y) = \frac{\delta}{\delta J(x)} \hat{\tilde{\rho}}(y) Z \tag{18.126}$$

We will now examine the structure of the terms of the right-hand sides of (18.126) in some detail. Taking the example of $S_{int} = \lambda \int \phi^2 \chi$, and simplifying along the lines of deriving the Schwinger-Dyson equations in Chapter 8, we find that we can write

$$\hat{\rho}(x) \frac{\delta}{\delta \tilde{J}(y)} Z = i \int_z [\Sigma(x, z)F(z, y) - \mathcal{E}(x, z)G^*(z, y)]$$

$$\frac{\delta}{\delta J(x)} \hat{\tilde{\rho}}(y) Z = i \int_z [G(x, z)\mathcal{E}(z, y) - F(x, z)\Sigma^*(z, y)] \tag{18.127}$$

where

$$\mathcal{E}(x, z) = 4\lambda^2 F_\phi(x, z) \ F_\chi(x, z)$$

$$\Sigma(x, z) = 4\lambda^2 G_\phi(x, z)G_\chi(x, z) \tag{18.128}$$

From the definitions, we have the result

$$G(x, z) = \theta(x^0 - z^0)F^*(x, z) + \theta(z^0 - x^0)F(x, z) \tag{18.129}$$

This shows that we can write

$$\Sigma(x, z) = \theta(x^0 - z^0)\mathcal{E}^*(x, z) + \theta(z^0 - x^0)\mathcal{E}(x, z) \tag{18.130}$$

For more general interactions, the form of the self-energies in (18.128) will change, but the general form of equations (18.127) and (18.130) is retained. Using (18.127) in equation (18.126) and writing out the limits of z^0-integration appropriately, we get

$$K_x F(x,y) = -i \left[\int_{-\infty}^{x^0} (\mathcal{E} - \mathcal{E}^*)(x,z) F(z,y) - \int_{-\infty}^{y^0} \mathcal{E}(x,z)(F - F^*)(z,y) \right]$$

$$K_y F(x,y) = -i \left[\int_{-\infty}^{x^0} (F - F^*)(x,z)\mathcal{E}(z,y) - \int_{-\infty}^{y^0} F(x,z)(\mathcal{E} - \mathcal{E}^*)(z,y) \right]$$

$$\text{(18.131)}$$

These equations are often known as the Kadanoff-Baym equations. They are one of the quantum versions of the one-particle Boltzmann equation.

The Boltzmann equation emerges from (18.131) when further approximations are made. First we take the difference of the two equations given above to get

$$(K_x - K_y)\, F(x,y) = -i \left[\int_{-\infty}^{x^0} [(\mathcal{E} - \mathcal{E}^*)F - (F - F^*)\mathcal{E}] \right.$$

$$\left. - \int_{-\infty}^{y^0} [\mathcal{E}(F - F^*) - F(\mathcal{E} - \mathcal{E}^*)] \right]$$

$$\text{(18.132)}$$

We now write $F(x,y)$ as

$$F(x,y) = \int \frac{d^4 p}{(2\pi)^4} e^{ip(x-y)}\, F(p, X) \qquad \text{(18.133)}$$

where $X = \frac{1}{2}(x + y)$. The potential is taken to be slowly varying on the scale of the interparticle interactions, so that one can approximate $U(x)$ by expanding to first order in the relative coordinate $x - y$. In this case

$$(K_x - K_y)\, F(x,y) \approx i \int \frac{d^4 p}{(2\pi)^4} \left[2\, p^\mu \frac{\partial F}{\partial X^\mu} + \frac{\partial F}{\partial p_\mu} \frac{\partial U}{\partial X^\mu} \right] e^{ip(x-y)} \quad \text{(18.134)}$$

For the terms on the right-hand side of (18.132), we introduce the change of variables

$$\tfrac{1}{2}(x + z) = X + \tfrac{1}{2}(\xi + \zeta)$$
$$\tfrac{1}{2}(z + y) = X + \tfrac{1}{2}\zeta$$
$$x - y = \xi, \quad z - x = \zeta \qquad \text{(18.135)}$$

and expand everything to the lowest order in ξ, ζ. For example, for the first term we get

$$\int_{-\infty}^{x^0} (\mathcal{E} - \mathcal{E}^*)F \quad = \quad \int_{-\infty}^{0} d^4\zeta\, (\mathcal{E} - \mathcal{E}^*)(X + \tfrac{1}{2}\xi + \tfrac{1}{2}\zeta, -\zeta)\, F(X + \tfrac{1}{2}\zeta, \xi + \zeta)$$

$$approx \int_{-\infty}^{0} d^4\zeta \frac{d^4 p_1}{(2\pi)^4} (\mathcal{E} - \mathcal{E}^*)(X, -\zeta)\, F(X, \xi + \zeta)$$

$$\approx \int_{-\infty}^{0} d^4\zeta \frac{d^4p_1}{(2\pi)^4} \frac{d^4p_2}{(2\pi)^4} (\mathcal{E} - \mathcal{E}^*)(X, p_2) F(X, p_1)$$
$$\times \exp\left[ip_1(\xi + \zeta) - ip_2\zeta\right] \qquad (18.136)$$

In a similar way, we find

$$\int_{-\infty}^{y^0} F(\mathcal{E} - \mathcal{E}^*) \approx \int_{-\infty}^{0} d^4\zeta \frac{d^4p_1}{(2\pi)^4} \frac{d^4p_2}{(2\pi)^4} F(X, p_2)(\mathcal{E} - \mathcal{E}^*)(X, p_1)$$
$$\times \exp\left[ip_1(\xi + \zeta) - ip_2\zeta\right]$$
$$\approx \int_{0}^{\infty} d^4\zeta \frac{d^4p_1}{(2\pi)^4} \frac{d^4p_2}{(2\pi)^4} F(X, p_1)(\mathcal{E} - \mathcal{E}^*)(X, p_2)$$
$$\times \exp\left[ip_2(\xi - \zeta) + ip_1\zeta\right] \qquad (18.137)$$

The two results (18.136) and (18.137) can now be combined and the integration over ζ and one of the momenta carried out to give

$$\int_{-\infty}^{x^0} (\mathcal{E} - \mathcal{E}^*)F + \int_{-\infty}^{y^0} F(\mathcal{E} - \mathcal{E}^*) \approx \int \frac{d^4p}{(2\pi)^4}(\mathcal{E} - \mathcal{E}^*)F(X, p)e^{ip\xi}$$
$$(18.138)$$

A similar simplification for the other two terms on the right-hand side of (18.132) gives

$$\int_{-\infty}^{y^0} \mathcal{E}(F - F^*) + \int_{-\infty}^{x^0} (F - F^*)\mathcal{E} \approx \int \frac{d^4p}{(2\pi)^4}\mathcal{E}(F - F^*)(X, p)e^{ip\xi}$$
$$(18.139)$$

Using (18.134), (18.138), and (18.139), equation (18.132) becomes

$$\left[2 \, p^\mu \frac{\partial F}{\partial X^\mu} + \frac{\partial F}{\partial p_\mu} \cdot \frac{\partial U}{\partial X^\mu}\right] = \mathcal{E}^* F - \mathcal{E} F^* + \cdots \qquad (18.140)$$

where the ellipsis indicates that a number of terms, corresponding to higher orders in $x - y$, have been neglected in arriving at this result. This equation is the Boltzmann equation; the right-hand side is the collision integral.

We will now simplify the collision integral to bring it to a more familiar form. For this purpose, we will consider the case when $U(x) = 0$, and also do a perturbative calculation. For specificity, we will continue with the two-scalar-field case with $S_{int} = \lambda \int \phi^2\chi$, ϕ and χ being fields of masses m and M, respectively. In this case, $F(x, y)$ can be expanded as

$$F_\phi(x, y) = \sum_k \frac{\alpha_k}{2\omega_k V} e^{ik\cdot(x-y)} + \frac{\beta_k}{2\omega_k V} e^{-ik\cdot(x-y)}$$

$$F_\chi(x, y) = \sum_k \frac{\tilde{\alpha}_k}{2\Omega_k V} e^{ik\cdot(x-y)} + \frac{\tilde{\beta}_k}{2\Omega_k V} e^{-ik\cdot(x-y)} \qquad (18.141)$$

where

$$\alpha_k = 1 + n_k, \qquad \beta_k = n_k$$
$$\tilde{\alpha}_k = 1 + N_k, \qquad \tilde{\beta}_k = N_k \tag{18.142}$$

where $\omega_k = \sqrt{k^2 + m^2}$, $\Omega_k = \sqrt{k^2 + M^2}$. Equation (18.141) is similar to equation (18.16), but at this stage we do not assume the Bose distribution for n_k, N_k. From this form of F, we get

$$F(p) = 2\pi \left[\frac{\alpha_p}{2\omega_p} \delta(p_0 - \omega_p) + \frac{\beta_{-p}}{2\omega_p} \delta(p_0 + \omega_p) \right]$$
$$F^*(p) = 2\pi \left[\frac{\beta_p}{2\omega_p} \delta(p_0 - \omega_p) + \frac{\alpha_{-p}}{2\omega_p} \delta(p_0 + \omega_p) \right] \tag{18.143}$$

We can use these in (18.140) and separate out an equation for α_p by multiplying by $\theta(p_0)$ and integrating over all p_0. As for the right-hand side of (18.140), we have, to the lowest nontrivial order

$$\mathcal{E}(x, z) = 4\lambda^2 F_\phi(x, z) F_\chi(x, z) \tag{18.144}$$

By Fourier transformation, we get

$$\mathcal{E}(p) = 4\lambda^2 \sum_{k_1, k_2} \frac{(2\pi)^4}{2\omega_{k_1} 2\Omega_{k_2} V^2} \Big[\alpha_{k_1} \tilde{\alpha}_{k_2} \delta^{(4)}(p - k_1 - k_2)$$
$$+ \alpha_{k_1} \tilde{\beta}_{k_2} \delta^{(4)}(p - k_1 + k_2) + \beta_{k_1} \tilde{\alpha}_{k_2} \delta^{(4)}(p + k_1 - k_2)$$
$$+ \beta_{k_1} \tilde{\beta}_{k_2} \delta^{(4)}(p + k_1 + k_2) \Big] \tag{18.145}$$

Substituting for F, F^* from (18.143), using the above expression for $\mathcal{E}$, and taking the limit of large volumes, we get the Boltzmann equation

$$\frac{p^\mu}{\omega_p} \frac{\partial \alpha_p}{\partial X^\mu} = \frac{4\lambda^2}{2\omega_p} \int d\mu_{k_1, k_2} \Big[\left(\alpha_p \beta_{k_1} \tilde{\beta}_{k_2} - \beta_p \alpha_{k_1} \tilde{\alpha}_{k_2} \right) (2\pi)^4 \delta^{(4)}(p - k_1 - k_2)$$
$$+ \left(\alpha_p \beta_{k_1} \tilde{\alpha}_{k_2} - \beta_p \alpha_{k_1} \tilde{\beta}_{k_2} \right) (2\pi)^4 \delta^{(4)}(p - k_1 + k_2)$$
$$+ \left(\alpha_p \alpha_{k_1} \tilde{\beta}_{k_2} - \beta_p \beta_{k_1} \tilde{\alpha}_{k_2} \right) (2\pi)^4 \delta^{(4)}(p + k_1 - k_2) \tag{18.146}$$
$$+ \left(\alpha_p \alpha_{k_1} \tilde{\alpha}_{k_2} - \beta_p \beta_{k_1} \tilde{\beta}_{k_2} \right) (2\pi)^4 \delta^{(4)}(p + k_1 + k_2) \Big]$$

where

$$d\mu_{k_1, k_2} = \frac{d^3 k_1}{(2\pi)^3} \frac{1}{2\omega_{k_1}} \frac{d^3 k_2}{(2\pi)^3} \frac{1}{2\Omega_{k_2}} \tag{18.147}$$

The first term on the right-hand side is the difference of the transition rates for $\chi_{k_2} + \phi_{k_1} \rightarrow \phi_p$ and $\phi_p \rightarrow \chi_{k_2} + \phi_{k_1}$. The second term gives the difference of rates for $\phi_{k_1} \leftrightarrow \chi_{k_2} + \phi_p$, the third is the similar quantity for $\chi_{k_2} \leftrightarrow \phi_{k_1} + \phi_p$. The fourth term is zero for kinematic reasons. Thus we see that the right hand side of equations (18.132) does reduce to the standard collision integral of the Boltzmann equation. Equilibrium is obtained when the collision integral vanishes, so that α_p is independent of time. From the explicit form of (18.146), we see that this requires the conditions

$$\alpha_p \beta_{k_1} \tilde{\beta}_{k_2} - \beta_p \alpha_{k_1} \tilde{\alpha}_{k_2} = 0$$
$$\alpha_p \beta_{k_1} \tilde{\alpha}_{k_2} - \beta_p \alpha_{k_1} \tilde{\beta}_{k_2} = 0$$
$$\alpha_p \alpha_{k_1} \tilde{\beta}_{k_2} - \beta_p \beta_{k_1} \tilde{\alpha}_{k_2} = 0 \qquad (18.148)$$

Writing the first one of these as

$$\frac{\alpha_p}{\beta_p} = \frac{\alpha_{k_1}}{\beta_{k_1}} \frac{\tilde{\alpha}_{k_2}}{\tilde{\beta}_{k_2}} \qquad (18.149)$$

we see that this is in the form of a conservation law; $\log(\alpha/\beta)$ must be an additively conserved quantity for collisions. The other equations in (18.148) give the same conclusion. If the conserved quantity is taken as the energy,

$$\frac{\alpha_p}{\beta_p} = \exp\left(\frac{\omega_p}{T}\right) \qquad (18.150)$$

This leads to the equilibrium solution

$$n_p = \frac{1}{e^{\omega_p/T} - 1} \qquad (18.151)$$

The parameter T is now identified as the temperature. More generally, the conservation law would include momentum; this would lead to a term proportional to p in the exponent, which can be interpreted as an overall drift motion of the medium.

Our derivation of the Boltzmann equation shows its limitations. It applies to dilute systems for which the truncation to one-particle distributions is adequate. Further, the potentials have to be slowly varying. In more general situations, one has to use $Z[J, \tilde{J}]$ as in (18.122) or analyze the full set of equations (18.124). The derivation of the Boltzmann equation was primarily to show that those equations do describe kinetic theory.

18.11 The imaginary time formalism

We have constructed propagators for various fields at finite temperature and density using the expectation values for the occupation numbers given by

the appropriate density matrix. If we are interested in equilibrium properties, rather than time-dependent phenomena, there is another method which
is very useful for certain types of calculations. This is the imaginary time
method, which we shall briefly discuss now. For simplicity, we shall again
consider a scalar-field theory with the action

$$S = \int d^4x \left[\frac{1}{2}(\partial\varphi)^2 - \frac{1}{2}m^2\varphi^2 - \lambda\varphi^4 \right] \tag{18.152}$$

The equilibrium solution for the density matrix in the rest frame of the
medium is $\rho = Z^{-1}e^{-\beta H}$. The partition function is given by $Z = \mathrm{Tr}e^{-\beta H}$.
$e^{-\beta H}$ is in the form of the time-evolution operator e^{-itH} for an imaginary
interval of time $-i\beta$. Using the path integral formula developed earlier,

$$\langle\alpha|e^{-itH}|\gamma\rangle = \int [d\varphi][d\varphi']\ \Psi_\alpha^*(\varphi')\Psi_\gamma(\varphi)\ K(\varphi',t,\varphi,0)$$

$$K(\varphi',t,\varphi,0) = \langle\varphi'|e^{-itH}|\varphi\rangle$$

$$= \int [d\varphi]\ \exp\left(i \int_0^t dt\ d^3x\ \mathcal{L}(t) \right) \tag{18.153}$$

where $\Psi_\alpha(\varphi) = \langle\varphi|\alpha\rangle$ is the wave functional of the state $|\alpha\rangle$ and the functional
integral for $K(\varphi',t,\varphi,0)$ is over all fields with the boundary values φ at $t = 0$
and φ' at t. For the partition function, we introduce the variable τ, $0 \le \tau \le \beta$,
with $t = -i\tau$. This gives the result

$$\langle\alpha|e^{-\beta H}|\gamma\rangle = \int [d\varphi][d\varphi']\ \Psi_\alpha^*(\varphi')\Psi_\gamma(\varphi)\ K_\beta(\varphi',\varphi)$$

$$K_\beta(\varphi',\varphi) = \int [d\varphi]\ \exp\left(\int_0^\beta d\tau\ d^3x\ \mathcal{L}(t = -i\tau) \right)$$

$$= \int [d\varphi]\ \exp\left(-\int_0^\beta d\tau\ d^3x\ \mathcal{L}_E \right) \tag{18.154}$$

where

$$\mathcal{L}_E = -\mathcal{L}(t = -i\tau)$$

$$= \frac{1}{2}\left[(\partial_\tau\varphi)^2 + (\nabla\varphi)^2 + m^2\varphi^2 \right] + \lambda\ \varphi^4 \tag{18.155}$$

is the Euclidean Lagrangian for the theory. Since $\sum_\alpha \Psi_\alpha(\varphi)\Psi_\alpha^*(\varphi') = \delta(\varphi,\varphi')$,
we see that, upon taking the trace, we must impose periodic boundary con-
ditions for the fields φ in the τ-direction. This leads to

$$Z = \int_{periodic} [d\varphi]\ \exp\left(-\mathcal{S}_E(\beta)\right)$$

$$\mathcal{S}_E(\beta) = \int d\tau\ d^3x\ \frac{1}{2}\left[(\partial_\tau\varphi)^2 + (\nabla\varphi)^2 + m^2\varphi^2 \right] + \lambda\ \varphi^4 \tag{18.156}$$

One can also define correlators

$$\langle \varphi(\tau_1, \boldsymbol{x}_1)...\varphi(\tau_N, \boldsymbol{x}_N)\rangle = Z^{-1} \int [d\varphi]\, e^{-S_E(\beta)}\, \varphi(\tau_1, \boldsymbol{x}_1)...\varphi(\tau_N, \boldsymbol{x}_N)$$

(18.157)

We thus see that the equilibrium theory is defined by the Euclidean theory on $\mathbf{R}^3 \times S^1$, with periodicity of fields along the τ-direction (corresponding to the S^1 component).

It is instructive to evaluate the propagators which are needed for the imaginary-time formalism. The periodicity in the τ-direction implies that the frequencies are quantized as $\omega_n = 2\pi n/\beta = 2\pi nT$; these are referred to as the Matsubara frequencies. The propagator, which is the inverse of $(-\Box_E + m^2)$, is thus given by

$$G_E(x, y) = T\sum_n \int \frac{d^3k}{(2\pi)^3}\, \frac{\exp\left[i\omega_n(\tau_x - \tau_y) + i\boldsymbol{k}\cdot(\boldsymbol{x} - \boldsymbol{y})\right]}{\omega_n^2 + \boldsymbol{k}\cdot\boldsymbol{k} + m^2}$$

(18.158)

The summation over n can be carried out using complex integration techniques. The summations involved are of the form

$$F(\tau) = \sum_n f(n, \tau) = \sum_{-\infty}^{\infty} \frac{1}{n^2 b^2 + a^2} e^{inb\tau}$$

(18.159)

Notice that this function is invariant under $b\tau \to b\tau + 2\pi m$ for any integer m. For positive τ, we can thus evaluate this in the interval $0 < b\tau < 2\pi$ and define the value of the function for other positive values of $b\tau$ by periodicity. Consider now the function

$$g(z) = \frac{1}{e^{2\pi i z} - 1}$$

(18.160)

This has poles at $z = n$; expanding around the pole, we see that the residue is $1/2\pi i$. We can therefore write the summation in (18.159) as

$$F(\tau) = \oint_C \frac{1}{(b^2 z^2 + a^2)(e^{2\pi i z} - 1)} \exp(ibz\tau)$$

(18.161)

where the contour C is as shown in figure 18.1, enclosing the poles due to

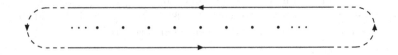

Fig 18.1. Contour C for integration in (18.161)

$g(z)$. For $\tau > 0$ and $b\tau < 2\pi$, the integrand vanishes for large positive and negative imaginary values of z. Therefore we can deform the contours as shown in figure 18.2, with the large semicircles in the upper and lower half-planes, picking up the contributions due to the poles of $1/(b^2z^2 + a^2)$. $F(\tau)$ can be evaluated as

$$F(\tau) = \frac{\pi}{ab}\left[\frac{e^{a\tau}}{e^{2\pi a/b} - 1} + e^{-a\tau}\left(1 + \frac{1}{e^{2\pi a/b} - 1}\right)\right] \tag{18.162}$$

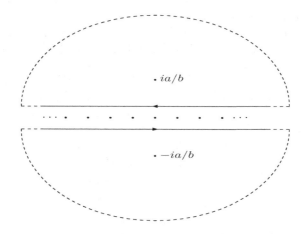

Fig 18.2. Deformed contours for the evaluation of (18.161)

This gives the value of the integral for $\tau > 0$. Using this result to carry out the summation in (18.158) and continuing to Minkowski space by $\tau = ix^0$, we get

$$
\begin{aligned}
G_E(x, y)\bigg|_{\tau_x, \tau_y \to ix^0, iy^0} &= \int \frac{d^3k}{(2\pi)^3}\frac{1}{2\omega_k}\bigg[(1 + n_k)e^{-i\omega_k(x^0 - y^0)} \\
&\qquad\qquad + n_k e^{i\omega_k(x^0 - y^0)}\bigg]e^{ik\cdot(x - y)} \\
&= \int \frac{d^3k}{(2\pi)^3}\frac{1}{2\omega_k}\bigg[(1 + n_k)e^{-ik(x-y)} + n_k e^{ik(x-y)}\bigg]
\end{aligned}
\tag{18.163}
$$

Comparing this with the propagator (18.16) we see that the Minkowski continuation of (18.158) is indeed the correct temperature-dependent propagator. Actually, we have shown this result only for $\tau_x - \tau_y > 0$. The summation can be carried out for $\tau_x - \tau_y < 0$ using the formula

$$F(\tau) = \oint_C \frac{e^{2\pi i z}}{(b^2 z^2 + a^2)(e^{2\pi i z} - 1)} \exp(ibz\tau) \tag{18.164}$$

This ensures that we have convergence when we deform the contour to the two semicircles. The integral is then obtained as

$$F(\tau) = \frac{\pi}{ab} \left[\frac{e^{-a\tau}}{e^{2\pi a/b} - 1} + e^{a\tau} \left(1 + \frac{1}{e^{2\pi a/b} - 1} \right) \right] \tag{18.165}$$

The continuation of $G_E(x, y)$ using this result shows that the Minkowski propagator (18.16) is correctly obtained by analytic continuation of the Euclidean propagator with periodicity in τ, for both $\tau_x - \tau_y < 0$ and $\tau_x - \tau_y > 0$.

The summation in (18.159) is defined in terms of the function $f(n, \tau)$ which is defined on the integers. In writing the formulae (18.161) and (18.164), we are doing an analytic continuation of this function to all values of z. Such a continuation is not unique and this leads to the possibility of using different integrands in evaluating the sum by contour integration. For $\tau > 0$ and for $\tau < 0$, the appropriate continuation is given by the one which gives convergence of integration over the semicircles when we deform the contours. This is what we have done in (18.161) and (18.164).

So far we have discussed bosonic fields. In the case of fermions, we should use antiperiodicity conditions

$$\psi(\tau + \beta) = -\psi(\tau) \tag{18.166}$$

The Matsubara frequencies in this case would be of the form $\omega_n = 2\pi(n + \frac{1}{2})T$. The Euclidean free fermion action is $\mathcal{S} = \int d^4 x \, \bar{\psi}(\gamma \cdot \partial + m)\psi$ and the fermion propagator is then given by

$$S(x, y) = T \sum_n \frac{e^{ip \cdot (x-y)}}{i\gamma \cdot p + m} \tag{18.167}$$

where $p_4 = \omega_n = 2\pi(n + \frac{1}{2})T$. By virtue of the summation formula

$$\sum_n f(n + \tfrac{1}{2}, \, \tau) = \oint \frac{f(z, \tau)}{e^{2\pi i z} + 1} \tag{18.168}$$

one can check that the Minkowski continuation of this propagator correctly reproduces the temperature-dependent propagator (18.22), (18.23). The denominator on the right-hand side in (18.168) has poles at $z = n + \frac{1}{2}$; the structure of this term is thus dictated by the antiperiodicity of the fields. It then leads to the correct thermal distribution for fermions with $1/(e^{\omega/T} + 1)$. This is essentially why the antiperiodicity conditions (18.166) are the correct prescription for fermions. Notice that since observables involve bilinears of fermions, periodicity of observables is compatible with the antiperiodicity of fermions fields.

Combining our discussions of fermions and bosons, the functional integral for a general field theory at finite temperature is given by

$$Z = \int_{P/AP} d\mu(\varphi, \psi, \bar{\psi}, A...) \ \exp\left[-S_E(\varphi, \psi, \bar{\psi}, A, ...) + \int J\varphi + \bar{\eta}\psi + \cdots\right]$$
(18.169)

where the subscript P/AP on the integral indicates that the integration is done over all fields with periodicity for boson fields and antiperiodicity for fermion fields in the imaginary time direction, the period being $\beta = 1/T$. By functionally differentiating Z with respect to various source functions up to the apprpopriate order and setting them to zero, we can obtain temperature-dependent correlation functions. For $J = \bar{\eta} = ... = 0$, Z is the thermal partition function. In the case of gauge fields, there is a qualification which is useful to emphasize. The ghost fields are anticommuting, but they must still have periodicity under $\tau \to \tau + \beta$. This is because the ghosts are, after all, just a way to rewrite the Faddeev-Popov determinant. The latter is for bosonic fields, specifically the gauge fields, and is computed with modes which have periodicity conditions. Since any rewriting must reproduce this result, the ghost fields should be periodic in τ.

Another interesting feature which emerges clearly in the imaginary time formalism is the dimensional reduction in the high-temperature limit. As T increases, the Matsubara frequencies become large, except for the $n = 0$ mode for bosons. In the mode sums in the propagators, the magnitudes of the various terms become small, except for $n = 0$ mode of bosons. Thus, as T becomes very large, only the mode with zero Matsubara frequency is important. This mode is independent of the imaginary time τ. Thus, apart from factors of β arising from integrations, the theory reduces to a three-dimensional theory. The high-temperature limit is given by the bosonic theory in one lower dimension. This reasoning can be useful, but applies only when there are no infrared divergences. If there are such divergences, one has to take account of them by resummations or other techniques, before this kind of dimensional reduction can be used.

From the real-time point of view, one may understand the dimensional reduction as follows. If a perturbation is introduced into the medium, thereby creating a nonequilibrium situation, it returns to equilibrium by collisions and other processes in a characteristic equilibration time. As one increases the temperature, the equilibration time decreases. Thus asymptotically for large temperatures, equilibration is so fast that all time-dependent processes can be ignored. Effectively, this leads to the dimensional reduction.

18.12 Symmetry restoration at high temperatures

We have already seen that the mass of a scalar particle gets temperature-dependent corrections. In a theory with spontaneous symmetry breaking, the temperature-dependent corrections can make it energetically favorable to have a symmetric ground state at high temperatures. Thus we can get the restoration of symmetry at high temperatures, with spontaneous symmetry breaking at low temperatures; the theory is realized in a symmmetric phase at high temperatures and in a broken symmetry phase at low temperatures with a phase transition at some critical temperature T_c. A simple model which can illustrate this is the $O(N)$ model with N scalar fields ϕ^i, $i = 1, 2, ..., N$, with a Euclidean action

$$
\begin{aligned}
\mathcal{S}_E &= \int d^4x \left[\frac{1}{2} \partial \phi^i \partial \phi^i + \lambda(\phi^i \phi^i - v^2)^2 \right] \\
&= \int d^4x \left[\frac{1}{2} \partial \phi^i \partial \phi^i - \frac{1}{2}(4\lambda v^2)\phi^i \phi^i + \lambda(\phi^i \phi^i)^2 + \text{constant} \right]
\end{aligned}
$$

$$(18.170)$$

To first order in λ the mass correction is given by

$$
\begin{aligned}
\delta m^2 &= 4\lambda(N+2)G_0(x,x) \\
&\approx \lambda \frac{(N+2)}{3} T^2
\end{aligned}
$$

$$(18.171)$$

where $G_0(x,y)$ is the thermal propagator in the free theory, and in the second line, we have shown only the temperature-dependent correction. (The T-independent part is absorbed into the definition of mass by the renormalization procedure as usual.) Including this correction, the term in the effective action which is quadratic in ϕ is $\Gamma^{(2)} = \frac{1}{2}M^2\phi^i \phi^i$, where

$$
\begin{aligned}
M^2 &= -4\lambda v^2 + 4\lambda(N+2)G_0(x,x) \\
&= -4\lambda \left(v^2 - \frac{(N+2)}{12} T^2 \right)
\end{aligned}
$$

$$(18.172)$$

This shows that there is a phase transition at

$$
T_c = \sqrt{\frac{12}{N+2}}\, v
$$

$$(18.173)$$

with the symmetry broken phase with $\langle \phi^i \rangle \neq 0$ being energetically favored at $T < T_c$ and the symmetric phase with $\langle \phi^i \rangle = 0$ being favored at $T > T_c$.

This simple calculation indicates that there is a phase transition as we go from low temperatures to high temperatures, with $\langle \phi \rangle \neq 0$ for $T < T_c$ and $\langle \phi \rangle = 0$ for $T > T_c$. However, the first-order perturbation theory is inadequate concerning the details of the transition. The higher-order corrections are important near the transition point. This is because the effective mass M goes to zero at the transition point and various diagrammatic contributions develop infrared divergences; they become singular as $M \to 0$. As a result, we cannot justify neglecting them. One needs to do a resummation of terms in the perturbative expansion. The Schwinger-Dyson equations provide one way to address this problem.

In the imaginary time formalism, the equations of motion and the functional integrals retain the same form as they have at zero temperature; the only difference is that they are defined on $\mathbf{R}^3 \times S^1$, rather than $\mathbf{R}^4$, reflecting the periodicity of fields in the imaginary time direction. The Schwinger-Dyson equations may then be written as

$$\int d^4x_2 \ [K(1,2) + \Sigma(1,2)] \, G(2,3) = \delta^{(4)}(x_1 - x_3) \tag{18.174}$$

where $K = (-\Box - 4\lambda v^2)$ and

$$\Sigma(1,2) = 4\lambda(N+2)G(1,1)\delta^{(4)}(x_1 - x_2) + \cdots \tag{18.175}$$

$G(1,2)$ is now the full propagator. As indicated after (8.108), the higher terms in Σ start off as $\sim \lambda^2 G(1,2)^3 + \ldots$. For a constant effective mass, equations (18.174), (18.175) simplify as

$$M^2 = -4\lambda v^2 + 4\lambda(N+2) \int \frac{d^3k}{(2\pi)^3} \frac{1}{\omega_k} \frac{1}{e^{\omega_k/T} - 1} \tag{18.176}$$

with $\omega_k^2 = k^2 + M^2$. This is in the form of a gap equation (for the energy gap M). The solution of this equation for M as a function of T will give a better approximation to the behavior near T_c. Let $\mu = M/T$ and define

$$\mathcal{I}(\mu) = \frac{6}{\pi^2} \int_0^\infty dx \frac{x^2}{\sqrt{x^2 + \mu^2}} \frac{1}{\exp(\sqrt{x^2 + \mu^2}) - 1} \tag{18.177}$$

This is the same integral as (18.18), except for μ replacing $\alpha = m/T$; the prefactor in (18.177) is such that $\mathcal{I}(0) = 1$. The gap equation can now be written as

$$\mu^2 = \lambda \frac{(N+2)}{3} \left(\mathcal{I}(\mu) - \frac{T_c^2}{T^2} \right) \tag{18.178}$$

Notice that it is consistent for μ to go to zero as we approach T_c from above. Thus the transition temperature is still given by equation (18.173), up to this order of resummation. Our calculation is valid for approach from above the transition point; below T_c, we have symmetry breaking and the calculation has to be done in terms of the Goldstone bosons and the one massive scalar field.

The terms which are resummed by the Schwinger-Dyson equations when Σ is approximated by $4\lambda(N+2)G(1,1)\delta^{(4)}(x_1 - x_2)$ are shown in figure 18.3. The diagrams in the first line are often referred to as the "daisy" diagrams and the second as the "superdaisy" diagrams.

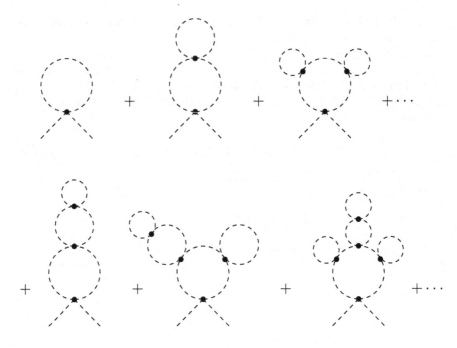

Fig 18.3. Daisy and superdaisy diagrams

When further corrections such as the $\lambda^2 G^3(1,2)$-term are added to Σ, they generate new sequences of diagrams in addition to those shown here. The various types of higher-order contributions in perturbation theory can be classified by the powers of N they generate in the large N-limit. The term we have considered so far in Σ, namely, (18.175), is of order λN. The next term in Σ is given by

$$\Sigma^{(2)}(1,2) = -\frac{32(3N+6)}{3}\lambda^2\, G^3(1,2) \tag{18.179}$$

This term is of order $\lambda^2 N$. We define a partial resummation of the perturbation theory by reorganizing the terms in the perturbative expansion in inverse powers of N, by taking λ small, with λN fixed, as N becomes large. The formula (18.175) for Σ, and the "daisy" and "superdaisy" diagrams it generates, are of the leading order in this expansion. The next correction (18.179) is of lower order, of order $1/N$. In the large N-expansion, the resummation with (18.175) is exact and the gap equation shows that the effective mass vanishes at the transition point. Indeed, by taking the derivative of $\mathcal{I}(\mu)$ and evaluating it for small μ and integrating, we find

$$\mathcal{I}(\mu) \approx 1 - \frac{3}{\pi}\,\mu + \cdots \tag{18.180}$$

The gap equation (18.178) can be solved near T_c as

$$M \approx \frac{2\pi}{3}(T - T_c) \tag{18.181}$$

showing that M vanishes linearly as the transition temperature is approached from above. Because the effective mass is zero, one gets massless excitations (long-wavelength excitations) at the transition; therefore, the transition is at least of second order from the thermodynamic point of view.

The characteristics of the transition for finite N are not clear from the calculation we have done, since the contribution (18.179) to Σ, and the further higher corrections which are possible, cannot be taken as being of higher order; they can have a nonnegligible effect. A more involved nonperturbative technique will be needed.

18.13 Symmetry restoration in the standard model

In the standard model, we have the spontaneous breaking of the symmetry via the Higgs mechanism which gives masses to the $W^\pm$ and Z particles. At high temperatures, this symmetry can be restored in a manner analogous to what happens for the $O(N)$ model. In this case there are gauge particles and fermions in addition to scalars and the details of the transition are somewhat different. We can discuss this effect by an effective potential calculation, as we did for the zero temperature case. For an arbitrary (but constant) background value of the Higgs field ϕ, the gauge particles have masses, with $M_Z^2 = \frac{1}{2}(g^2 + g'^2)\phi^\dagger\phi$, $M_W^2 = \frac{1}{2}g^2\phi^\dagger\phi$. Therefore their contribution to the free energy changes with ϕ and this energy is the effective potential. In the imaginary time description, for a single scalar particle, we have

$$\Gamma(s) = \frac{1}{2}\operatorname{Tr}\log(k^2 + s)$$

$$= \frac{1}{2}\int d^4x \int \frac{d^3k}{(2\pi)^3}\, T\sum_n \log(\omega_n^2 + \boldsymbol{k}\cdot\boldsymbol{k} + s) \tag{18.182}$$

Here s is eventually the square of the mass for the field with the Higgs background. Taking the derivative with respect to s, we get

$$\frac{\partial\Gamma}{\partial s} = \frac{1}{2}\int d^4x \int \frac{d^3k}{(2\pi)^3}T\sum_n \frac{1}{\omega_n^2 + \boldsymbol{k}\cdot\boldsymbol{k} + s}$$

$$= \frac{1}{2}\int d^4x \int \frac{d^3k}{(2\pi)^3}\frac{1}{\sqrt{\boldsymbol{k}\cdot\boldsymbol{k} + s}}\left(n_k + \frac{1}{2}\right) \tag{18.183}$$

$$n_k = \frac{1}{\exp\left(\beta\sqrt{\boldsymbol{k}\cdot\boldsymbol{k} + s}\right) - 1}$$

Integrating with respect to s, we get

$$\Gamma(m^2) = \int d^4x \int \frac{d^3k}{(2\pi)^3} \left[\frac{1}{2}\omega_k + T \log\left(1 - e^{-\beta\omega_k}\right) \right] + \text{constant} \quad (18.184)$$

where the "constant" denotes terms which are independent of m^2 and $\omega_k = \sqrt{k \cdot k + m^2}$. The first term on the right-hand side is the zero point energy and leads to the one-loop effective potential as in Chapter 8. It has to be evaluated with a cut-off and will require renormalization of the parameters in the Lagrangian. Here we are interested in the temperature-dependent second term, which we denote by Γ_T. From (18.183) we can write

$$\begin{aligned} \frac{\partial \Gamma_T}{\partial s} &= \frac{1}{2} \int d^4x \int \frac{d^3k}{(2\pi)^3} \frac{1}{\sqrt{k \cdot k + s}} \, n_k \\ &= \frac{T^2}{24} \int d^4x \, \mathcal{I}(\beta\sqrt{s}) \end{aligned} \quad (18.185)$$

Approximating $\mathcal{I}(\beta\sqrt{s})$ for small values of $\beta\sqrt{s}$ as in (18.180), we get

$$\Gamma_T = \text{constant} + \frac{T^2 m^2}{24} - \frac{T m^3}{12\pi} + \cdots \quad (18.186)$$

This is the contribution for each polarization state of each boson. For the contribution of the W and Z bosons, we thus find

$$\Gamma_T = \text{constant} + \frac{T^2}{24}(3M_Z^2 + 6M_W^2) - \frac{T}{12\pi}(3M_Z^3 + 6M_W^3) + \cdots \quad (18.187)$$

With the values of M_Z and M_W given earlier, the full effective potential, to this order of calculation, is

$$\begin{aligned} \Gamma &= \lambda(\phi^\dagger\phi)^2 - \alpha(T)\phi^\dagger\phi - \gamma(T)(\phi^\dagger\phi)^{\frac{3}{2}} + \cdots \\ \alpha(T) &= \lambda v^2 - \frac{T^2}{24}\left(\frac{9}{2}g^2 + \frac{3}{2}g'^2\right) \\ \gamma(T) &= \frac{T}{12\pi}\left[3\left(\frac{g^2 + g'^2}{2}\right)^{\frac{3}{2}} + 6\left(\frac{g^2}{2}\right)^{\frac{3}{2}}\right] \end{aligned} \quad (18.188)$$

$\alpha(T)$ changes sign at some temperature T_c defined by $\alpha(T_c) = 0$, with $\alpha > 0$ for $T < T_c$ and $\alpha < 0$ for $T > T_c$. The cubic term $\gamma(\phi^\dagger\phi)^{\frac{3}{2}}$ is very important in determining the nature of the transition. Ignoring the constant part which is irrelevant for this discussion, we see that $\Gamma = 0$ at $\phi = 0$. As ϕ increases, for small ϕ, $\alpha\phi^\dagger\phi$ is the leading term and we see that Γ decreases. As ϕ increases further, the $\gamma(\phi^\dagger\phi)^{\frac{3}{2}}$-term decreases Γ further. Eventually the $\lambda(\phi^\dagger\phi)^2$-term dominates, making Γ increase. Thus for $\alpha > 0$, the minimum is at a nonzero value of ϕ given by

$$\langle\phi\rangle = \frac{\sqrt{32\lambda\alpha + 9\gamma^2} + 3\gamma}{8\lambda} \quad (18.189)$$

This gives symmetry breaking at low temperatures. When $\alpha = 0$, Γ still decreases as we increase ϕ starting from zero. There is still a nonzero minimum and we still have symmetry breaking. When $\alpha < 0$, Γ initially increases with ϕ, then comes down due to the $\gamma(\phi^\dagger\phi)^{\frac{3}{2}}$-term and then increases again due to the $\lambda(\phi^\dagger\phi)^2$-term. There are thus two minima, $\phi = 0$ and $\phi = \langle\phi\rangle$, given by (18.189). For small $|\alpha|$, $\langle\phi\rangle$ remains the global minimum. For large $|\alpha|$, $\phi = 0$ is the global minimum. There is a potential barrier between the two minima, as shown in the figure below. Any transition from $\phi = \langle\phi\rangle$ to the minimum at $\phi = 0$ has to proceed by quantum tunneling because of this barrier.

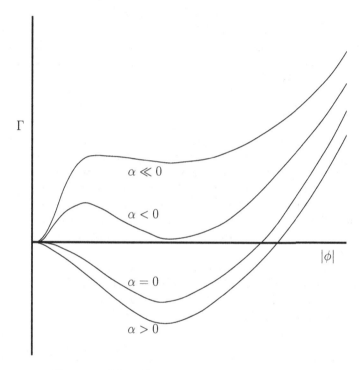

Fig 18.4. The effective potential for different values of α

Suppose we start from low temperatures in the phase with the broken symmetry. As the system heats up, eventually $\phi = 0$ becomes the global minimum. But the system remains in the local minimum at $\langle\phi\rangle$, with the broken symmetry. This is a metastable state. Eventually there is a tunneling transition to $\phi = 0$ and the symmetric phase. The barrier becomes smaller and the tunneling is easier if the temperature is further raised. What we have described is typical of a first-order transition. The energy difference between the local metastable minimum at $\langle\phi\rangle$ and the global minimum at $\phi = 0$ is the latent heat of the transition. Thus in the one-loop approximation we have used, we find that the standard model undergoes a first-order transition to the symmetric phase if the temperature is raised to a sufficiently high value.

We have included the contribution of the gauge bosons to the effective potential. A more comprehensive analysis, even within the one-loop approximation, will require the contribution of the Higgs particle and the fermions. Further, close to the transition point, one has to include higher order corrections as in the $O(N)$ model to obtain a satisfactory description of the dynamics of the transition.

References

1. For some of the material in this chapter, the following books may be useful: M. Le Bellac, *Thermal Field Theory*, Cambridge University Press (2000); A. Das, *Finite Temperature Field Theory*, World Scientific Pub. Co. (1997); J.I. Kapusta, *Finite Temperature Field Theory*, Cambridge University Press (1994).

2. For a general reference on the use of the density matrix, see R.P. Feynman, *Statistical Mechanics*, Addison-Wesley Pub. Co. (1972).

3. The KMS condition is due to R. Kubo, J. Phys. Soc. Japan **12**, 570 (1957); P.C. Martin and J. Schwinger, Phys. Rev. **115**, 1342 (1959).

4. The screening of electrostatic fields goes back to P. Debye's work on electrolytes in the 1920s. The calculation using thermal photon propagator is due to V.P. Silin, Sov. J. Phys. JETP **11**, 1136 (1960); V.N. Tsytovich, Sov. Phys. JETP, **13**, 1249(1961); O.K. Kalashnikov and V.V. Klimov, Sov. J. Nucl. Phys., **31**, 699 (1980); V.V. Klimov, Sov. J. Nucl. Phys. **33**, 934 (1981); Sov. Phys. JETP **55**, 199 (1982); H.A. Weldon, Phys. Rev. **D26**, 1394 (1982).

5. Screening of nonabelian gauge fields, by isolating hard thermal loops, was done by R. Pisarski, Physica **A158**, 246 (1989); Phys. Rev. Lett. **63**, 1129 (1989); E. Braaten and R. Pisarski, Phys. Rev. **D42**, 2156 (1990); Nucl. Phys. **B337**, 569 (1990); *ibid.* **B339**, 310 (1990); Phys. Rev. **D45**, 1827 (1992).

6. Calculations of the effective action for screening are also given in J. Frenkel and J.C. Taylor, Nucl. Phys. **B334**, 199 (1990); J.C. Taylor and S.M.H. Wong, Nucl. Phys. **B346**, 115 (1990); R. Efraty and V.P. Nair, Phys. Rev. Lett. **68**, 2891 (1992); Phys. Rev. **D47**, 5601 (1993). We have followed the last reference in our presentation.

7. A kinetic equation approach has been used by P.F. Kelly *et al*, Phys. Rev. Lett. **72**, 3461 (1994); Phys. Rev. **D50**, 4209 (1994).

8. Kubo's paper in reference 3 gives the Abelian Kubo formula.

9. For the calculation of the imaginary part of Γ we follow R. Jackiw and V.P. Nair, Phys.Rev. **D48**, 4991 (1993); Related work is also given by J.P. Blaizot and E. Iancu, Nucl. Phys. **B390**, 589 (1993); Phys. Rev. Lett. **70**, 3376 (1993).

10. For screening at high densities, see C. Manuel, Phys. Rev. **D53**, 5866 (1996); G. Alexanian and V.P. Nair, Phys. Lett. **B390**, 370 (1997).

11. A recent review on screening effects for gauge fields is U. Kraemmer and A. Rebhan, Rep. Prog. Phys. **67**, 351 (2004).

12. The method of time-contour is due to J. Schwinger, J. Math. Phys. **2**, 407 (1961); P.M. Bakshi and K. Mahanthappa, J. Math. Phys. **4**, 1 (1963); *ibid.* **4**, 12 (1963); L.V. Keldysh, Sov. Phys. JETP **20**, 1018 (1964).

13. The Kadanoff-Baym equations are in L.P. Kadanoff and G. Baym, *Quantum Statistical Mechanics*, W.A. Benjamin, Inc. (1962).

14. A general reference on quantum kinetic theory is S.R. De Groot, W.A. van Leeuwen and Ch. G. van Weert, *Relativistic Kinetic Theory*, North-Holland/ Elsevier (1980).

15. Imaginary time Green's functions were introduced by T. Matsubara, Prog. Theor. Phys. **14**, 351 (1955).

16. For relativistic field theories, this formalism was used by D.A. Kirzhnits and A.D. Linde, Phys. Lett. **B42**, 471 (1972); Ann. Phys. **101**, 195 (1976); L. Dolan and R. Jackiw, Phys. Rev. **D9**, 3320 (1974); S. Weinberg, Phys. Rev. **D9**, 3357 (1974). These papers also discuss symmetry restoration at high temperatures.

17. The observation that the ghosts must obey periodic boundary condition is due to C. Bernard, Phys. Rev. **D9**, 3312 (1974).

18. The situation regarding the order of the phase transition in the standard model, as well as the $O(N)$ model at small N, is not completely satisfactory yet; for a recent review, see P. Arnold, Lecture at *Quarks '94*, Russia, 1994, http://arXiv.org/hep-ph/9410294.

19 Gauge theory: Nonperturbative questions

19.1 Confinement and dual superconductivity

19.1.1 The general picture of confinement

We have seen in Chapter 10 some of the perturbative aspects of Yang-Mills theories. Since the physical realization of gauge theories is the standard model, most of the perturbative calculations which have been done are in this context. Various processes, and radiative corrections to them, have been calculated and checked against experimental data for the electroweak theory. The $SU(3)$ gauge theory of strong interactions has an unbroken gauge symmetry and our discussion of the effective charge and asympotic freedom shows that perturbative calculations are reliable at high energies. A large number of processes have been analyzed at high energies and compared with experimental data. The basic conclusion of all these calculations is that the standard model is in very good agreement with experiments for all processes in the kinematic regimes where reliable perturbative calculations can be done. However, there are many situations for which the perturbative analysis is not adequate; the most notable and physically relevant case is the low-energy limit of quantum chromodynamics (QCD). By now there is some understanding of the qualitative features of the theory in the nonperturbative regime, although most of the quantitative questions have not been answered. In this chapter we will discuss some of the approaches to nonperturbative issues.

In an unbroken Yang-Mills theory with a small number of fermion or scalar matter fields, we still have the property of asymptotic freedom. This shows that the effective coupling constant of the theory increases as we go to processes of lower and lower momenta. The Λ-parameter defined by the formula for the running coupling constant defines the basic scale of the theory; fields of momenta much larger than Λ can be treated perturbatively, but all modes of momenta comparable or less than Λ have to treated nonperturbatively. The perturbative formula for asymptotic freedom is no longer valid near Λ, so we can only conclude that the tendency of the theory is to move to stronger effective couplings at lower momenta. The general expectation is that this will lead to interparticle potential energies which increase with distance and hence to the confinement of all particles belonging to all representations of the group except the identity representation which has zero charge. This confinement hypothesis may be stated as follows:

> The asymptotic states of a nonabelian Yang-Mills theory with group G are all singlets (identity representation) of G. In particular, in QCD, quarks do not appear as asymptotic states; they can only be constituents of color singlet combinations and cannot be separated arbitrarily far off from the rest of the constituents.

Recall that the transformation of the gauge potential is given by $A \rightarrow A^g = gAg^{-1} - dgg^{-1}$, where $g(\boldsymbol{x}) : \mathbf{R}^3 \rightarrow G$, with g equal to a constant at spatial infinity. Among these transformations, those which go to the identity at spatial infinity are the true gauge transformations; the set of such transformations were denoted by $\mathcal{G}_*$ and they act as identity on the physical states. The transformations which go to a constant g, not necessarily equal to 1, form the set $\mathcal{G}$. We argued earlier that $\mathcal{G}/\mathcal{G}_* = G$, the Lie group in which g takes values. The transformations which go to a constant, not necessarily equal to the identity, are equivalent to a G-transformation; they act as a Noether symmetry and the states can be classified by representations of G. Nontrivial representations correspond to the charged states of the theory. Thus, at this kinematic level, there is nothing against charged states. Confinement is thus of a dynamical origin.

If quarks are confined, the obvious next question is about the interquark potential. Studies of bound states of heavy quarks such as charmonium and the upsilon show that the interquark potential, at least for these heavy quarks, is linearly increasing with the separation, $V(r) = \sigma r$. Such a linearly increasing potential shows that the chromoelectric flux between the quarks is drawn out into a thin flux tube of some constant nonzero tension, so that the energy of the tube, which is the energy of the separated quark-antiquark pair, is proportional to the length of the tube, which is r. A chromoelectric string connects the quark-antiquark pair. One may now ask how such a flux configuration can be physically realized. In the theory of superconductivity, one has the Meissner effect, which leads to exclusion of magnetic fields from superconducting regions. In type II superconductors, where the coherence length is smaller than the London penetration depth for magnetic fields, it is possible for magnetic fields to penetrate into a superconductor without destroying the superconductivity completely. In this case, the magnetic field is squeezed by the superconductor into as narrow a cross section as possible (within the limits imposed by energy minimization), and the result is a long tube of magnetic flux. These are the Abrikosov vortices, which have a constant energy per unit length. The total flux in such a vortex has to be quantized and is of the appropriate value as to connect a monopole-antimonopole pair. The interaction energy of a monopole-antimonopole pair, if such a pair were physically possible, immersed in a type II superconductor is thus proportional to the separation of the two particles. The situation is very similar to what happens with quark-antiquark pairs except that in the latter case the charge is of the electric kind rather than the magnetic kind. If we consider a magnetic superconductor due to a vacuum condensate of magnetic monopoles, we would have squeezing of electric flux lines and the formation of electric flux tubes between elctrical charges. Thus quark confinement with a linear potential for heavy charges can be understood if we take the vacuum state of QCD (or any unbroken Yang-Mills theory with a small number of matter fields) as a chromomagnetic superconductor formed due to the condensation of chromomagnetic charges.

In an ordinary superconductor, one can define a field which behaves as an order parameter. This is the effective field ϕ, representing the Cooper pairs, or the paired electrons. ϕ has a charge $2e$, couples to the electromagnetic field, and has the ground-state expectation values $\langle \phi \rangle \neq 0$ for $T < T_c$, corresponding to the superconducting state, and $\langle \phi \rangle = 0$ for $T > T_c$, corresponding to the normal state. Thus, the phenomenon of superconductivity is the Higgs phenomenon of spontaneous symmetry breaking in the Landau-Ginzburg model which is the (nonrelativistic) Abelian Higgs model. In the context of Yang-Mills theory, we have to address the question of how to go to the dual picture of chromomagnetic, rather than chromoelectric, superconductivity, whether we can find an order parameter field, how we can get vortices, how the vacuum state can be characterized, etc. We will begin by relating the interquark potential for heavy particles to the Wilson loop operator.

19.1.2 The area law for the Wilson loop

We start with the definition of the parallel transport operator

$$U^{ab}(x, y, A) = \left[P \exp \left(- \int_{yC}^{x} A_\mu dx^\mu \right) \right]^{ab} \tag{19.1}$$

Under a gauge transformation $A \to A^g$, this transforms as

$$U^{ab}(x, y, A^g) = \left[g(x) \, P \exp \left(- \int_{yC}^{x} A_\mu dx^\mu \right) g^{-1}(y) \right]^{ab}$$

$$= \left[g(x) U(x, y, A) g^{-1}(y) \right]^{ab} \tag{19.2}$$

As discussed in Chapter 10, the Wilson loop operator is given by

$$W_R(C) = \text{Tr} \left[P \exp \left(- \oint_{xC}^{x} A_\mu dx^\mu \right) \right] = \text{Tr} \, U(x, x, A, C) \tag{19.3}$$

where the integral is over a closed curve C starting and ending at x. The transformation law (19.2) shows that $W_R(C)$ is gauge-invariant. The subscript R refers to the fact that $A_\mu = -it^a A_\mu^a$ are taken to be in a representation R of the group G.

In order to relate this to the energy of a particle-antiparticle pair, we will start by considering the Euclidean spacetime process where a heavy static particle-antiparticle pair is created at a certain time x^0, propagates to time $x^0 + T$, and then is annihilated. The gauge-invariant operator which creates a particle-antiparticle pair separated by a spatial distance r is given by

$$F^\dagger(x^0, x^1, x^1 + r) = \phi^{\dagger a} (x)U^{ab}(x,y) \chi^{\dagger b}(y) \tag{19.4}$$

Here $x = (x^0, x^1)$, $y = (x^0, x^1 + r)$. For simplicity we take the separation of the pair to be along the x^1-direction. $\phi^\dagger$ and $\chi^\dagger$ are creation operators for heavy, static particle and antiparticle, respectively. Under a gauge transformation, $\phi \to g\phi$, $\chi \to g^*\chi$, , so that $F^\dagger$ is gauge-invariant. For heavy particles with the mass tending to infinity, the action is given by

$$S(\phi, \chi) = \int d^4x \; [i\phi^\dagger D_0\phi + i\chi^\dagger D_0\chi] \tag{19.5}$$

where $D_0\phi = \partial_t\phi + A_0\phi$ and $D_0\chi = \partial_t + A_0^*\chi$ are the covariant derivatives of ϕ and χ, respectively.

Let H be the Hamiltonian corresponding to the Yang-Mills theory with these matter fields ϕ, χ in the $A_0 = 0$ gauge. The A_0-dependent terms in (19.5) are then zero but will contribute to H via the Gauss law. We can then write

$$\langle 0|F \; e^{-HT} F^\dagger|0\rangle \approx \mathcal{N} \; e^{-E(r)T} \tag{19.6}$$

where $\mathcal{N}$ is some prefactor related to the normalization of F, and $E(r)$ is the energy of the pair. We will consider T to be large. When T becomes large, we get the energy of the lowest energy state which can be created by $F^\dagger$. Since the particles are heavy and static, $E(r)$ is just the interaction energy of the pair due to the gauge field. We may rewrite (19.6) as a Euclidean functional integral

$$\langle 0|F \; e^{-HT} F^\dagger|0\rangle = \int d\mu(A, \phi, \chi) \exp{[-S_E(A, \phi, \chi)]}$$
$$\chi^c(y')U^{cd}(y', x')\phi^d(x') \; \phi^{\dagger a}(x)U^{ab}(x,y)\chi^{\dagger b}(y) \tag{19.7}$$

where $x' = (x^0 + T, x^1)$, $y' = (x^0 + T, x^1 + r)$. The Euclidean action for ϕ, χ is given by

$$S_E(\phi, \chi) = \int d^4x \; \left[\phi^\dagger \frac{\partial\phi}{\partial\tau} + \chi^\dagger \frac{\partial\chi}{\partial\tau}\right] \tag{19.8}$$

with the propagators

$$\langle\phi^a(x)\phi^{\dagger b}(x')\rangle = \delta^{ab}\theta(\tau - \tau')\delta^{(3)}(x - y)$$
$$\langle\chi^a(x)\chi^{\dagger b}(x')\rangle = \delta^{ab}\theta(\tau - \tau')\delta^{(3)}(x - y) \tag{19.9}$$

The functional integral (19.7) then reduces to

$$\langle 0|F \ e^{-HT}F^\dagger|0\rangle = \int d\mu \ e^{-\mathcal{S}_E} \ U^{ba}(y',x') \ U^{ab}(x,y)$$

$$= \int d\mu \ e^{-\mathcal{S}_E} \ W_R(C) \tag{19.10}$$

where C is the rectangle with vertices x, y, x', y'. Combining (19.6) and (19.10), we see that

$$\langle W_R(C)\rangle \approx \mathcal{N}e^{-E(r)T} \tag{19.11}$$

Thus, evaluating the Euclidean expectation value of a large Wilson loop, we get the interaction energy of a heavy static particle-antiparticle pair. Even though we showed this in $A_0 = 0$ gauge, $W_R(C)$ is gauge-invariant and so are energies of gauge-invariant states, so that the argument holds true in general. For a heavy quark-antiquark pair, $E(r) = \sigma r$, and hence we may encode this piece of information as the statement

$$\langle W_R(C)\rangle \approx \mathcal{N}\exp(-\sigma \ r \ T)$$
$$\approx \mathcal{N}\exp(-\sigma \ \mathcal{A}(C) \ ) \tag{19.12}$$

where $\mathcal{A}(C)$ is the area of the minimal surface whose boundary is C.

In order to relate this "area law" for the Wilson loop to the picture of dual superconductivity, we need to discuss vortices in an ordinary superconductor, show that this leads to a linear potential between monopoles, and then go to a dual picture. So we now turn to vortices.

19.1.3 Topological vortices

The nonrelativistic version of the Abelian Higgs model is the Landau-Ginzburg theory of superconductivity. Vortex excitations in this model were first analyzed by Abrikosov. We will consider the relativistic model where the pioneering work on vortices is due to Nielsen and Olesen.

The action for the Abelian Higgs model, given in (12.119), is

$$\mathcal{S} = \int d^4x \ \left[-\frac{1}{4}F_{\mu\nu}F^{\mu\nu} + (D_\mu\phi)^*(D^\mu\phi) - V(\phi^*\phi)\right]$$
$$V(\phi^*\phi) = \lambda(\phi^*\phi - v^2/2)^2 \tag{19.13}$$

The Hamiltonian corresponding to (19.13) is

$$H = \int d^3x \ \left[\frac{1}{2}(E^2 + B^2) + \dot\phi^*\dot\phi + (D_i\phi)^*(D_i\phi) + V(\phi^*\phi)\right] \tag{19.14}$$

The minimization of H as a functional of the fields and derivatives is obtained by

$$\dot\phi = 0, \qquad\qquad D_i\phi = 0$$
$$V(\phi^*\phi) = 0, \qquad\qquad E = B = 0 \tag{19.15}$$

$\phi = v\sqrt{2}$ is a solution to these conditions. The classical vacuum is thus given by $\phi = v/\sqrt{2}$. Building up a vacuum state with this expectation value, we get the standard Higgs mechanism as discussed in Chapter 12.

We now consider static classical solutions of finite nonzero energy. For finite energy, the integrand in (19.14) has to vanish at spatial infinity so that we can have convergence of the integral. This requires that the fields go to a solution of the conditions (19.15) at spatial infinity. Since V has to be zero, the field ϕ at spatial infinity is of the form $\phi = (v/\sqrt{2})\exp(i\theta)$. This gives a mapping $e^{i\theta} : S^2 \to U(1)$, from S^2 which is spatial infinity to $U(1)$. The homotopy classes of such mappings are given by $\Pi_2(U(1)) = 0$, so that the field space has only one connected component and minimization leads to the vacuum again. However, we will get nontrivial topological sectors if we consider vortices which have finite energy per unit length. The simplest case is to take a straightline vortex, which we take to be along the third axis. The energy per unit length of static configurations is given by

$$\mathcal{E} = \int d^2x \left[\frac{1}{2}B^2 + (D_i\phi)^*(D_i\phi) + V(\phi^*\phi)\right] \tag{19.16}$$

For this to be finite we need $\phi \to (v/\sqrt{2})e^{i\theta}$ as $r^2 = x_1^2 + x_2^2 \to \infty$, giving

$$e^{i\theta} : S^1 \to U(1) \tag{19.17}$$

The homotopy classes of such maps are given by $\Pi_1(U(1)) = \mathbf{Z}$. There are winding numbers for these maps given by

$$\begin{aligned} Q &= -\frac{i}{2\pi} \int d^2x \; \epsilon^{\mu\nu} \frac{\partial_\mu \phi^* \partial_\nu \phi}{\phi^*\phi} \\ &= -\frac{i}{2\pi} \oint_{|x|\to\infty} \frac{\phi^* d\phi}{\phi^*\phi} \end{aligned} \tag{19.18}$$

The field space for configurations of finite energy per unit length has an infinity of connected components labeled by Q. In the simplest nontrivial case of $Q = 1$, we need $\phi \to (x_1 + ix_2)/r$ as $r \to \infty$. In a more general context, if we have a gauge symmetry G which is spontaneously broken to H, then the relevant homotopy group for vortices is $\Pi_1(G/H)$.

For finiteness of energy, we also need the term $(D_i\phi)^*(D_i\phi)$ to vanish as $x_1^2 + x_2^2 \to \infty$. This can be achieved by choosing an ansatz for A_i such that $\partial_i\phi$ is canceled by $-ieA_i\phi$. In this case, we can write

$$Q = \frac{e}{2\pi} \oint_{|x|\to\infty} A$$
$$\int d^2x \; F = \frac{2\pi}{e}Q \tag{19.19}$$

Thus a configuration with winding number Q may be interpreted as Q flux tubes, each carrying magnetic flux $2\pi/e$. Notice that this elementary flux is exactly what is needed to connect a Dirac monopole, which has magnetic charge $1/2e$, to a Dirac antimonopole. The quantization of the magnetic flux carried by the vortex is a consequence of the topological origin of the vortex as being due to nontrivial $\Pi_1[U(1)]$.

For the case of $Q = 1$, a suitable ansatz is given by

$$\phi = \frac{v}{\sqrt{2}}\, h(r)\, e^{i\theta}, \qquad\qquad eA_i = -\frac{\epsilon_{ij}x^j}{r^2}f(r) \qquad (19.20)$$

The equations of motion for static fields then become

$$\frac{d}{d\xi}\left(\frac{1}{\xi}\frac{df}{d\xi}\right) - \frac{h^2(f-1)}{\xi^2} = 0$$

$$\frac{1}{\xi}\frac{d}{d\xi}\left(\xi\frac{dh}{d\xi}\right) - \frac{h(f-1)^2}{\xi^2} - \frac{M_H^2}{2M_V^2}h(h^2 - 1) = 0 \qquad (19.21)$$

where $\xi = evr$. There are two relevant scales for the analysis of vortices; these are given by the mass of the vector particles $M_V = ev$ and the mass of the Higgs scalars $M_H = \sqrt{2\lambda}\, v$. The asymptotic behavior of the functions f, h, which leads to finite energy per unit length, is

$$f \to 1 + \alpha\, \exp(-\xi)$$
$$h \to 1 - \beta\, \exp(-M_H r) \qquad (19.22)$$

Type I and type II superconductors correspond to

$$\frac{M_H}{M_V} < 1 \qquad\qquad \text{Type I}$$

$$> 1 \qquad\qquad \text{Type II} \qquad (19.23)$$

For type I, the Higgs field falls off more slowly than the gauge field; as a result the flux is contained in a region where the Higgs field is close to zero. For type II, the gauge field falls off more slowly, giving a flux tube in the superconductor. While an exact analytical solution for the equations (19.21) is not known, many of the essential features follow from approximate analysis. Equations (19.21) are the variational equations for the energy functional

$$\mathcal{E} = 2\pi v^2 \int \xi d\xi\, \left[\frac{1}{2}\left(\frac{f'^2}{\xi^2} + h'^2 + \frac{h^2(f-1)^2}{\xi^2}\right) + \frac{\lambda}{4e^2}(h^2 - 1)^2\right] \qquad (19.24)$$

Using inequalities on this functional (or the more general form (19.16)) one can show that vortices in a type I superconductor attract each other, making it favorable for them to coalesce, while in a type II superconductor they repel each other, leading to well-defined multivortex configurations.

For simplicity, we have so far considered a vortex along the x^3-axis. One can consider fluctuations of this vortex corresponding to bending of the vortex. One can in general get a vortex along some curve C'. As noted before, vortices have to end in monopoles since the flux has to terminate in magnetic charges. But it is also possible for vortices to form closed loops, since there is no difficulty with flux conservation in such a situation.

We will now consider very thin vortices which can be obtained when M_H and M_V become very large, with $M_H \gg M_V$. We want to introduce an operator which can create such a vortex. Let C' denote the vortex as a curve in three-dimensional space with coordinates $z^i(\tau)$. The magnetic flux for this vortex is given by

$$F = 4\pi g \int_{C'} d\tau \, \frac{dz^k}{d\tau} \delta^{(3)}(x - z(\tau)) \, \frac{1}{2}\epsilon_{ijk} dx^i \wedge dx^j$$
$$\equiv F_v \tag{19.25}$$

where $g = \frac{1}{2e}$ is the strength of the vortex. It is easily checked that the integral of F over a surface intersecting the vortex transversally is $2\pi/e$ as required by (19.19). Let α be a one-form such that $d\alpha$ gives the right-hand side of this equation. In other words, it is a specific potential which gives the magnetic field of a vortex. Notice that F is zero everywhere except on a line C', which is the vortex. Thus α is a closed one-form on all of $\mathbf{R}^3 - \{C'\}$.

We will now construct an operator which creates a vortex. In the $A_0 = 0$ gauge, the action for a $U(1)$ gauge theory is

$$S = \int d^4x \, \left[\frac{1}{2}(\partial_0 A_i \partial_0 A_i) - \frac{1}{2}B^2\right] \tag{19.26}$$

The equal-time commutation rules are thus

$$[A_i(x), E_j(y)] = i \, \delta^{(3)}(x - y) \tag{19.27}$$

In the full Hamiltonian anlysis, we have to impose the Gauss law as well. In our discussion of the electromagnetic field, we solved the Gauss law explicitly. An alternative is to require that the physical states are invariant under gauge transformations. This is mathematically the statement

$$(\nabla \cdot E - J_0) \, |\psi\rangle = 0 \tag{19.28}$$

for all physical states $|\psi\rangle$. It is convenient to use this approach in constructing the vortex creation operator.

The operator which can create the vortex line in (19.25) is then given by

$$T(C') = \exp\left(-i \int d^3x \, \alpha_i E_i\right) \tag{19.29}$$

This gives

$$T^{-1}(C') \, A_i(x) \, T(C') = A_i(x) \, + \, \alpha_i(x) \tag{19.30}$$

where we have used the canonical commutation rules. Magnetic flux across a surface Σ can be measured by $\oint_C A$, where $\partial\Sigma = C$. Using (19.30)

$$T^{-1}(C') \left[\oint_C A\right] T(C') = \oint_C (A + \alpha)$$

$$= \oint_C A + \int_\Sigma F_v \qquad (19.31)$$

We see that the magnetic flux is shifted by the flux of a vortex by the operator $T(C')$ confirming its interpretation as a vortex creation operator. If the vortex line does not intersect with the surface Σ, then the integral $\int_\Sigma F_v$ is zero; if the vortex line intersects with the surface more than once, we get an identical contribution for each intersection. Thus,

$$\int_\Sigma F_v = 4\pi g \, L(C, C') \qquad (19.32)$$

where $L(C, C')$ is the number of times the curve C links the curve C'. It is called the Gauss linking number. This can have a topologically invariant meaning only if C' is infinitely long or if it is closed. We will consider C''s which are closed. The Gauss linking number of two closed curves C and C' may be written as

$$L(C, C') = -\frac{1}{4\pi} \oint_C \oint_{C'} \epsilon_{ijk} \frac{(x - y)^i}{|x - y|^3} dy^j \, dx^k \qquad (19.33)$$

Equation (19.31) may now be rewritten as

$$\left[\oint_C A\right] T(C') = T(C') \left[\oint_C A\right] + T(C') \, 4\pi g \, L(C, C') \qquad (19.34)$$

In the Abelian case, the Wilson loop operator becomes

$$W(C) = \exp\left[ie \oint_C A\right] \qquad (19.35)$$

Consider now the identity

$$\exp\left[-ie \oint_C A\right] E_i(x) \exp\left[ie \oint_C A\right] = E_i(x) + e \oint_C \delta^{(3)}(x - z(\tau)) \frac{dz_i}{d\tau} d\tau \qquad (19.36)$$

This shows that we may also think of the Wilson loop operator as creating a thin electric flux tube (or electric string) along the curve C. A dual interpretation for the two operators $T(C')$ and $W(C)$ emerges from these considerations. One may think of $T(C')$ as creating a magnetic flux tube, with $\oint_C A$ measuring the magnetic flux through C. $W(C)$ can be thought of as creating an electric flux tube, with the electric flux being measured by $T(C')$. The commutation rules may be written, in exponentiated form, as

$$W(C) \, T(C') = \exp(i4\pi eg) \, T(C') \, W(C)$$
$$= T(C') \, W(C) \qquad (19.37)$$

The additional phase factor gives 1 because of the quantization of the flux carried by the vortex, $eg = n/2$ for some integer n. Therefore, this commutation rule may be taken as another way of stating the Dirac quantization rule.

The operators $T(C')$ and $W(C)$ provide a way of specifying the phase of the gauge theory without relying on the details of the mechanism which produces the superconducting phases. If we introduce a dual potential $\tilde{A}$ for the electric field so that $E_i = \epsilon_{ijk}\partial_j\tilde{A}_k$, then we can write

$$T(C') = \exp\left(-i4\pi g \oint \tilde{A}\right) \tag{19.38}$$

Even though there are some restrictions on when such a dual potential can be introduced, this equation shows that we may indeed think of $T(C')$ as the dual of $W(C)$ under an electric-magnetic duality transformation. We have seen that confinement of electrically charged particles can be stated as the area law for the Wilson loop. Confinement of magnetically charged particles occurs in the usual electric superconductor with a vortex connecting monopole-antimonopole pairs. The energy of such a configuration increases linearly with the separation of the particles since the energy of a vortex is proportional to its length. Since $T(C')$ creates a magnetic flux tube, by arguments similar to what we used for the Wilson loop, the linear potential for the monopole-antimonopole pair will give an area law for the expectation value of $T(C')$. Going back to $W(C)$, we then see that the area law for its expectation value can be interpreted as being due to magnetic superconductivity. To summarize

$$\langle T(C')\rangle \sim \exp(-\sigma\mathcal{A}(C')\,), \qquad \text{electric superconductivity,}$$
$$\text{magnetic confinement} \tag{19.39}$$
$$\langle W(C)\rangle \sim \exp(-\sigma\mathcal{A}(C)\,), \qquad \text{magnetic superconductivity,}$$
$$\text{electric confinement}$$

The advantage of defining the phases in this way is that it is a gauge-invariant specification purely in terms of operators constructed from gauge fields; it does not involve other fields such as the Cooper pair field, whose nonabelian analog is not easy to construct. Further, duality is very transparent. To extend this characterization to the nonabelian case, we need to consider the nonabelian analogs of the operators $T(C')$ and $W(C)$.

The operator $T(C')$ was introduced in the context of the characterization of phases of a gauge theory by 't Hooft and it will therefore be referred to as the 't Hooft loop operator.

19.1.4 The nonabelian dual superconductivity

As a first step toward the generalization of the ideas of the last subsection to the nonabelian Yang-Mills theory, we will consider the construction of operators $T(C')$ in a $U(1)$ subgroup of an $SU(N)$ gauge theory. The $SU(N)$ group has nontrivial central elements, namely, there are some elements of $SU(N)$ which commute with all the elements of $SU(N)$. The central elements of $SU(N)$ form the cyclic group $\mathbf{Z}_N$ whose elements can be represented as the set of $(N \times N)$-matrices $\{1, \omega, \omega^2, \cdots, \omega^{N-1}\}$ with $\omega = \exp(2\pi i/N)$. As a matrix, all these are proportional to the identity and evidently commute with all of $SU(N)$. An explicit construction is obtained as follows. Let t_a, $a = 1, 2, \cdots, (N^2 - 1)$, be a basis of the Lie algebra of $SU(N)$, realized as $(N \times N)$-matrices. As usual the normalization condition is $\mathrm{Tr}\, t_a t_b = \frac{1}{2}\delta_{ab}$. Consider

$$t_{N^2-1} = \sqrt{\frac{N}{2(N-1)}} \, diag \, \left(\frac{1}{N}, \frac{1}{N}, \cdots, \frac{1}{N}, -1 + \frac{1}{N}\right)$$

$$\equiv \sqrt{\frac{N}{2(N-1)}} \, Y \tag{19.40}$$

The group element

$$\exp\left(2\pi i \sqrt{\frac{2(N-1)}{N}} t_{N^2-1}\right) = \exp(2\pi i Y) = e^{2\pi i/N} = \omega \tag{19.41}$$

This corresponds to choosing the group parameter corresponding to t_{N^2-1} as $\theta_{N^2-1} = 2\pi\sqrt{2(N-1)/N}$. Fields in the adjoint representation transform as $\phi \to \phi' = g\phi g^{-1}$, $\phi = \phi^a t_a$, $g \in SU(N)$. These are invariant under $\mathbf{Z}_N$. All representations obtained from reduction of tensor products of the adjoint representation are also invariant. For all such fields, including the gauge field, which is in the adjoint representation, the group is actually $SU(N)/\mathbf{Z}_N$. In particular, all components of all fields which are $\mathbf{Z}_N$ invariant will have integer eigenvalues for Y. Matter fields in the fundamental representation are not $\mathbf{Z}_N$ invariant since they transform as $\phi' = g\,\phi$; for them, Y has fractional eigenvalues.

The group $SU(N)$ is simply connected, but if all fields are $\mathbf{Z}_N$ invariant, the relevant group is $SU(N)/\mathbf{Z}_N$, and this is not simply connected. In fact, $\Pi_1[SU(N)/\mathbf{Z}_N] = \mathbf{Z}_N$ and hence if we consider a theory where all fields are $\mathbf{Z}_N$ invariant, the spontaneous breaking of the symmetry completely will lead to $\mathbf{Z}_N$ vortices. We want to analyze these vortices in some more detail.

We will start with just the $U(1)$ subgroup of $SU(N)$ defined by the generator t_{N^2-1} or Y. The covariant derivatives are of the form

$$D_\mu \phi = \partial_\mu \phi - ieY A_\mu \phi \tag{19.42}$$

(We use Y, absorbing the normalization factor $\sqrt{N/2(N-1)}$ into the field A_μ.) One can break this $U(1)$ group by Higgs fields which are in some representation of $SU(N)/\mathbf{Z}_N$. For example, there are many components of a Higgs field in the adjoint representation which carry nonzero eigenvalues of

Y and so transform nontrivially under this $U(1)$; they can be used to break the symmetry. If this $U(1)$ is broken, we get vortices as discussed before. Since the eigenvalues of Y are integers, the magnetic flux of the vortices will be quantized in units of $2\pi/e$. One can then construct the operator $T(C')$ as before. Consider now the Wilson loop operator defined in (19.3). Unlike the Abelian case, we cannot use $\oint_C A$ as a gauge-invariant operator. Only the exponentiated form, with the path ordering of the Lie algebra matrices and the trace, is invariant. Therefore we must ask whether the flux of a vortex can be detected by measuring it with $W(C)$. For Wilson loops in the fundamental representation, the values of Y are $1/N$ or $-1 + (1/N)$. In this case, we see immediately that

$$W_F(C) \, T(C') \;=\; \exp\left(\frac{2\pi i L(C, C')}{N}\right) \, T(C') \, W_F(C) \qquad (19.43)$$

where $T(C')$ creates the basic vortex of flux $2\pi/e$. The basic magnetic vortices can be detected by use of $W(C)$ in the fundamental representation. There are no dynamical charges in the fundamental representation, so what is meant is that by use of heavy external (nondynamical) charges which are in the fundamental representation, we can detect the basic vortices, of fluxes $k \, 2\pi/e$, $k = 1, 2, ..., (N - 1)$, or more generally fluxes modulo N, in units of $2\pi/e$.

At the level of just the $U(1)$ field, there are also vortices with fluxes which are N times the basic unit of $2\pi/e$, or multiples thereof. An operator $T(C')$ for them will commute with $W_F(C)$, and hence with $W(C)$'s in any representation. They cannot be detected. In fact they have no meaning, once we embed the chosen $U(1)$ subgroup in the full $SU(N)$; the only meaningful flux values are modulo N since $\Pi_1[SU(N)/\mathbf{Z}_N] = \mathbf{Z}_N$.

In the nonabelian theory with $\mathbf{Z}_N$ invariance, since the only relevant fluxes are $\mathbf{Z}_N$-valued, we can *define* the operator $T(C')$ by the commutation rules (19.43). Notice that if $T(C')$ is defined by (19.43), then effectively the operators $T(C')$ and $W(C)$ are expressed entirely in terms of gauge fields and do not depend on ideas of Higgs fields and spontaneous symmetry breaking which we used to arrive at it. The commutation rule (19.43) as the general algebra for the electric and magnetic vortex creation operators was first obtained by 't Hooft and hence is often called the 't Hooft algebra. The interpretation of $W_F(C)$ as an operator creating a chromoelectric string (with chromoelectric flux value appropriate for connecting a heavy particle pair in the fundamental representation) and measuring chromomagnetic flux is still valid. $T(C')$ creates a chromomagnetic string and measures chromoelectric flux. We can now use these operators to characterize a notion of duality and phases for the gauge theory. We have already seen that confinement of heavy quarks via a linear potential is characterized by an area law for the expectation value of the Wilson loop. This may be taken as the definition of the phase of chromomagnetic superconductivity. The dual phase would correspond to an area law for $T(C')$ which may be taken as a phase of chromoelectric superconductivity.

The Wilson loop $W_F(C)$ thus serves as an order parameter if all the dynamical fields are $\mathbf{Z}_N$ invariant. This analysis does not in any way prove or even provide a logical argument for confinement. All that we have done is to give a qualitative picture which singles out area law for the vacuum expectation value of $W_F(C)$ as the result to be proved in order to demonstrate mathematically the confinement of heavy charged states which are not $\mathbf{Z}_N$ invariant. It would show that the vacuum state is a state of chromomagnetic superconductivity.

We now turn to what happens in a theory with matter fields in the fundamental representation or other representations which are not $\mathbf{Z}_N$ invariant. This is clearly the case of physical interest since chromodynamics has quarks in the fundamental representation. In this case, the relevant group becomes $SU(N)$, which is simply connected, and there are no vortices. We can no longer characterize phases of this theory in terms of order parameters $W(C)$ and $T(C')$. This is not to say that there is no notion of confinement; it can be defined in terms of all the asymptotic states being singlets, but we can no longer use $\langle W_F(C) \rangle$ as a probe or signal of confinement. One can also see more explicitly why $\langle W_F(C) \rangle$ will not work when there are $\mathbf{Z}_N$ noninvariant fields such as quarks. In this case, there is some probability of creating a quark-antiquark pair if sufficient energy is available. Thus it is possible for the electric string connecting the heavy external charges to break, the flux at the broken ends getting attached to the quark and antiquark. Such a process is consistent with Gauss law. Effectively, the external charges are screened by the newly created particles which are attached to them by short strings. The operator $F^\dagger$ of equation (19.4) creates a very unstable state if the separation is large; the state of external charges connected by a long string loses identity as it acquires a large decay width.

19.2 't Hooft-Polyakov magnetic monopoles

We have argued that the condensation of monopoles forming a state of dual superconductivity can lead to the confinement of charged particles for a non-abelian gauge theory. The monopoles themselves were not directly used in this argument. Nonabelian theories which are not spontaneously broken admit configurations of magnetic charge and this is crucial for the dual super-conductor picture. However these monopoles do not appear as stable config-urations in the theory. This is why they were not explicitly used in all the arguments.

In a spontaneously broken gauge theory monopoles can appear as stable classical solutions of finite energy and these provide a way to understand many of their properties in general. These solutions, known as 't Hooft-Polyakov monopoles, retain their identity under interactions due to a topo-logically conserved charge, the monopole number. In other words, they are solitons. In the quantum theory, they will then survive as particle-like excitations. Monopoles can also appear in any grand unified theory where the electromagnetic $U(1)$ group is embedded in a semisimple nonabelian group. Therefore they can be of interest in their own right as possible particle-like excitations in the theory. We will discuss monopole solutions of a spontaneously broken theory briefly here.

A question that might arise at this stage would be about why we need to consider spontaneously broken nonabelian theories. If we are interested in a theory of monopoles, why not just build a theory of Dirac monopoles in a $U(1)$ theory? In fact we have already discussed some of the properties of a Dirac monopole and obtained the quantization condition for the charge of a monopole. However, the field has a singularity at the origin or the position of the magnetic monopole and the classical energy is infinite. This is similar to what happens for an electron and one could attempt to treat a monopole as a fundamental particle with an associated field operator and absorb the infinity of energy as a mass renormalization. However, until now, a completely satisfactory theory of charges and monopoles along these lines has not been constructed.

Consider a nonabelian gauge theory with symmetry G which is spontaneously broken down to a subgroup H. The potential energy is invariant under G-transformations of the Higgs field, and hence the finiteness of energy for static configurations requires that the asymptotic value of the Higgs field should be the vacuum value, upto a G-transformation. Since H leaves the Higgs expectation value invariant, we see that, asymptotically, Higgs field configurations are in G/H. We thus have, asymptotically, a map from S^2 (corresponding to the large radius sphere in $\mathbf{R}^3$) to G/H. The homotopy classes of such maps are given by $\Pi_2(G/H)$. If this is nonzero, we can have magnetic monopoles. The exact homotopy sequence

$$\cdots \to \Pi_2(G) \ \to \ \Pi_2(G/H) \ \to \ \Pi_1(H) \ \to \ \Pi_1(G) \ \to \cdots$$

$$\cdots \to \quad 0 \ \to \quad \Pi_2(G/H) \ \to \ \Pi_1(H) \ \to \ \Pi_1(G) \ \to \cdots$$

(19.44)

shows that, if G is simply connected, $\Pi_2(G/H) = \Pi_1(H)$.

A simple model which realizes this structure is the spontaneous breaking of an $SU(2)$ gauge theory down to a $U(1)$ subgroup. The Higgs field may be taken as a triplet under $SU(2)$ and the action is given by

$$S = \int d^4x \left[-\frac{1}{4} F_{\mu\nu}^a F^{a\mu\nu} + \frac{1}{2}(D_\mu\phi)^a (D^\mu\phi)^a - V(\phi^a\phi^a) \right] \quad (19.45)$$

where

$$(D_\mu\phi)^a = \partial_\mu\phi^a + e\epsilon^{abc} A_\mu^b \phi^c$$

$$V(\phi^a\phi^a) = \frac{\lambda}{4} \left(\phi^a\phi^a - v^2 \right)^2 \quad (19.46)$$

$a, b, c = 1, 2, 3$. Since the Higgs field is a triplet, the symmetry group is actually $SO(3)$ rather than $SU(2)$ and the symmetry breaking is $SO(3) \rightarrow SO(2)$.

The energy for static configurations is given by

$$\mathcal{E} = \int d^3x \left[\frac{1}{2} B_i^a B_i^a + \frac{1}{2} (D_i\phi)^a (D_i\phi)^a + V(\phi^a\phi^a) \right] \quad (19.47)$$

The classical vacuum configuration is at the minimum of V; we may take this as $\phi^a = \delta^{a3}v$. The isotropy subgroup is $SO(2)$ corresponding to group rotations around the $a = 3$ direction. We have a Higgs breaking of the gauge group to $SO(2) \sim U(1)$. The gauge field components $W_i^\pm = (A_i^1 \mp iA_i^2)/\sqrt{2}$ get mass $M_V = ev$. The unbroken $SO(2)$ direction will be taken as the electromagnetic $U(1)$ gauge group of this model. (Since ϕ^a are real, it is convenient to denote the vacuum expectation value as v rather than $v/\sqrt{2}$.)

For finite energy configurations, we need

$$\begin{aligned} B_i^a &\rightarrow 0, & (D_i\phi)^a &\rightarrow 0 \\ \phi^a\phi^a &\rightarrow v^2, & \text{as } r \equiv |\boldsymbol{x}| &\rightarrow \infty \end{aligned} \quad (19.48)$$

The last condition shows that as $r \rightarrow \infty$, we have a mapping $\phi^a : S^2 \rightarrow S^2$. There are homotopy classes corresponding to $\Pi_2[SO(3)/SO(2)] = \Pi_2[S^2] = \mathbf{Z}$, characterized by the winding number

$$Q = \frac{1}{8\pi} \oint_{r \rightarrow \infty} \epsilon^{abc} \, \hat{\phi}^a \, d\hat{\phi}^b \wedge d\hat{\phi}^c \quad (19.49)$$

where $\hat{\phi}^a = \phi^a/|\phi|$. (This was discussed in Chapter 14.) Q is identically conserved; classical and quantum transitions must preserve Q. For the vacuum configuration $\phi^a = \delta^{a3}v$, evidently $Q = 0$. With the asymptotic requirement of $D\phi$ vanishing, we can write

$$\oint \epsilon^{abc} \hat{\phi}^a (D\hat{\phi})^b \wedge (D\hat{\phi})^c = 0 \quad (19.50)$$

Simplifying this equation, we find

$$\oint F^a \hat{\phi}^a = -\frac{4\pi}{e} Q \quad (19.51)$$

where $F^a = dA^a + \frac{1}{2}\epsilon^{abc} A^b \wedge A^c$ is the field-strength two-form. The unbroken direction is asymptotically given, in the sector with nonzero Q, by $\hat{\phi}^a$; thus $F^a\hat{\phi}^a$ is the field strength of the unbroken group. Equation (19.51) tells us that the total magnetic flux is proportional to Q. Thus the configurations we are discussing are indeed magnetic monopoles, the elementary monopole having magnetic charge $g = \frac{1}{e}$.

For the elementary monopole configuration ($Q = -1$), the asymptotic behavior of ϕ^a becomes

$$\phi^a \rightarrow -v \, \frac{x^a}{r} \tag{19.52}$$

Finiteness of energy requires the vanishing of $(D_i\phi)$ as well. With the behavior of ϕ^a given above, this leads to the requirement

$$A_i^a \rightarrow \epsilon_{aik} \, \frac{x^k}{er^2} \tag{19.53}$$

A suitable ansatz for the fields for all r is then given by

$$\phi^a = -v \, \frac{x^a}{r} \left[\frac{H(\xi)}{\xi} \right]$$

$$A_i^a = \epsilon_{aik} \frac{x^k}{er^2} \, [1 - K(\xi)] \tag{19.54}$$

where $\xi = evr$.

As we move in from spatial infinity, if ϕ^a changes smoothly, with $\phi^2 \neq 0$, we can use r itself as a homotopy parameter. If ϕ does not vanish anywhere, this will lead to a contradiction, mapping $Q \neq 0$ configuration to $Q = 0$, the latter being what we obtain as the sphere shrinks to zero size. This means that any configuration of nonzero Q must have a zero for the Higgs field ϕ^a somewhere. Further, K must go to 1 as $r \rightarrow 0$ to avoid any singularity at $r = 0$ due to the $1/r^2$ factor. Thus we have

$$\frac{H}{\xi} \rightarrow 1, \qquad K \rightarrow 0, \qquad \text{as} \quad \xi \rightarrow \infty$$

$$\frac{H}{\xi} \rightarrow 0, \qquad K \rightarrow 1, \qquad \text{as} \quad \xi \rightarrow 0 \tag{19.55}$$

The static energy for this ansatz is given by

$$\mathcal{E} = \frac{4\pi v}{e} \int_0^\infty \frac{d\xi}{\xi^2} \left[\xi^2 (\partial_\xi K)^2 + \frac{1}{2} \left(\xi \frac{dH}{d\xi} - H \right)^2 + \frac{1}{2} (K^2 - 1)^2 \right.$$

$$\left. + K^2 H^2 + \frac{\lambda}{4e^2} (H^2 - \xi^2)^2 \right] \tag{19.56}$$

The static equations of motion are

$$\xi^2 \frac{d^2 K}{d\xi^2} = KH^2 + K(K^2 - 1)$$

$$\xi^2 \frac{d^2 H}{d\xi^2} = 2K^2 H + \frac{\lambda}{e^2} \, H(H^2 - \xi^2) \tag{19.57}$$

Exact analytic solutions of these equations are difficult, but one can find, numerically, solutions to these equations, with the conditions (19.55). The energy of the solution is generally of the form

$$\mathcal{E} = \frac{4\pi v}{e} f(\lambda/e^2) \tag{19.58}$$

with $f(0) = 1$. The asymptotic behavior is of the form

$$H - \xi \sim \exp(-M_H r), \qquad K \sim \exp(-M_V r) \tag{19.59}$$

where $M_V = ev$ and $M_H = \sqrt{2\lambda}\, v$ are the vector and scalar field masses due to symmetry breaking.

Although we cannot solve equations (19.57) analytically, there is a particular case, called the Bogomol'nyi-Prasad-Sommerfield (BPS) limit, in which this is possible. Notice that the energy functional (19.47) obeys an inequality

$$\mathcal{E} \geq \int d^3 x \left[\frac{1}{2} B_i^a B_i^a + \frac{1}{2} (D_i \phi)^a (D_i \phi)^a \right]$$

$$\geq \int d^3 x \, B_i^a (D_i \phi)^a$$

$$\geq \frac{4\pi v}{e} \tag{19.60}$$

which follows from $(B - D\phi)^2 \geq 0$. In equation (19.60) we have used the Bianchi identity $(D_i B_i)^a = 0$ and equation (19.51). In the BPS limit, we consider $\lambda \to 0$, still retaining the boundary condition $\phi^a \phi^a \to v^2$ as $r \to \infty$. The contribution of the potential term V to the energy is then zero and it is possible to saturate the inequality (19.60) by taking

$$B_i^a = (D_i \phi)^a \tag{19.61}$$

The solution to this equation is

$$H(\xi) = \xi \coth \xi - 1, \qquad K(\xi) = \frac{\xi}{\sinh \xi} \tag{19.62}$$

For this case, $\mathcal{E} = 4\pi v/e$, in agreement with $f(0) = 1$ in (19.58).

We see from equation (19.51) that the magnetic and electric charges obey the quantization condition $eg = Q$. The Dirac quantization condition was $eg = n/2$. The fact that eg is twice what is expected for the Dirac case is due to the fact that the starting group was $SO(3)$ rather than $SU(2)$. The theory can admit charges $e/2$ corresponding to the spinor representation of $SO(3)$, i.e., fundamental representation of $SU(2)$. The Dirac argument in this case shows that we again have $eg = $ integer. Another way is to notice that the exactness of the homotopy sequence

$$0 \to \Pi_2 \left[SO(3)/SO(2) \right] \to \Pi_1[SO(2)] \to \Pi_1[SO(3)] \to 0 \tag{19.63}$$

$$0 \to \qquad \mathbf{Z} \qquad \to \qquad \mathbf{Z} \qquad \to \quad \mathbf{Z}_2 \quad \to \quad 0$$

shows that elements of $\Pi_2[SO(3)/SO(2)]$ should map onto even elements of $\Pi_1[SO(2)]$.

19.3 The $1/N$-expansion

Another approach to the nonperturbative analysis of gauge theories is the $1/N$-expansion. The number of fields has been used as parameter to control the regrouping and resummation of the perturbation series in many theories. The symmetry restoring phase transition in a scalar field theory in Chapter 18 is one example. There have also been successful nonperturbative $1/N$-analyses in many lower-dimensional theories. For an $SU(N)$ gauge theory in four dimensions with no matter fields, the property of asymptotic freedom shows that there is dimensional transmutation. The coupling constant can be eliminated in favor of a single dimensionful parameter Λ which defines the scale of the theory. Asymptotic freedom implies that modes of momenta much greater than Λ can be treated perturbatively; for modes of momenta comparable to or less than Λ perturbation theory is inadequate. There is no obvious expansion parameter that could be used for all modes. 't Hooft, who introduced the $1/N$-expansion in the context of gauge theories, observed that N (of $SU(N)$ or the number of colors) may be used as an expansion parameter with the large N limit as the zeroth order starting point. To see how this is possible, we need a rule to work out the N-dependence of any Feynman diagram. For this purpose, it is best to consider the gauge potential as an $(N \times N)$-matrix, $(A_\mu)_{ij} = -i(t^a)_{ij}A_\mu^a$, $i, j = 1, 2, ..., N$; each component of the field is represented by the color indices i, j, each color index taking values $1, 2, ..., N$. We are interested in large N, so we can use the gauge group $U(N)$ rather than $SU(N)$; the extra degree of freedom will have only a small effect as N becomes large. This will simplify our analysis. Thus t^a are the generators of $U(N)$ and obey the completeness relation $(t^a)_{ij}(t^a)_{kl} = \frac{1}{2}\delta_{jk}\delta_{il}$. Since the ghosts are also in the adjoint representation, one can write $c_{ij} = c^a(t^a)_{ij}$, $\bar{c}_{ij} = \bar{c}^a(t^a)_{ij}$, etc. The gluon propagator is of the form

$$\langle (A_\mu)_{ij}(x)(A_\nu)_{kl}(y)\rangle \sim \delta^{ab}\eta_{\mu\nu}G(x,y) \sim \frac{1}{2}\eta_{\mu\nu}\delta_{jk}\delta_{il} \qquad (19.64)$$

This shows that there is a flow of color along the propagator lines in every Feynman diagram and for the purpose of obtaining the leading power of $1/N$ for a diagram, we may represent each external gluon line and each internal propagator line by a double line as shown here.

$$\langle A_{\mu ij}(x)A_{\nu kl}(y)\rangle \quad = \qquad$$

Fig 19.1. The double line representation of the propagator

One can then trace the flow of color via the double lines throughout the diagram. Color loops represent N as coming from trace over indices like i, j. For example, the following one-loop diagram for gluon self-energy is seen to be of order e^2N, where e is the coupling constant. The coupling constant

Fig 19.2. A diagram which is of the leading order in $1/N$

will be taken to be of order $1/\sqrt{N}$, so that this diagram is of order zero in the large N-limit, which is the leading order in $1/N$. Similarly, one can check

Fig 19.3. Another diagram which is of the leading order in $1/N$

that the diagram given above is of the same order in $1/N$ because there is an additional color loop giving a factor of N from the trace and a factor of e^2 due to the additional vertices. By contrast, the diagram 19.4 given below, which has one propagator overpassing another, is of order $1/N^2$.

Fig 19.4. A diagram which is of order $1/N^2$

Notice that the first two diagrams, which are of order N^0, can be drawn on a plane with no overpassing or underpassing of propagators; they are planar diagrams. The last diagram, namely figure 19.4, which is of order N^{-2}, requires a surface with a handle for us to be able to draw it with no overpassing or underpassing of propagators. For diagrams with two external lines, we see that the leading N-behavior is of the form N^{-2h}, where h is the minimal number of handles of the surface on which the diagram can be drawn. One can easily check by the double-line rule that this behavior persists more generally. We can classify all Feynman diagrams by the minimal number of handles of the surface on which they can be drawn with no overpassing or underpassing. If there are no external lines the N dependence is given by N^{2-2h}. For correlators of operators which are color singlets, the color indices i, j, etc., do not appear in the external lines. The same formula N^{2-2h} holds in those cases as well. For diagrams with external lines, the general result is of the form $a_N N^{2-2h}$ where a_N is an N-dependent prefactor which is common for all diagrams with the same number of external lines. Thus generally a correlator is of the form

$$F = a_N \sum_h f_h N^{2-2h} \qquad (19.65)$$

The $1/N$ expansion is an expansion in terms of the surfaces on which the diagrams can be drawn. In string theory, the perturbation expansion is in terms of the number of handles of the Riemann surface which corresponds to the world sheet swept out by the string as it propagates in time. The striking similarity between that expansion and $1/N$ expansion suggests that the large N limit of the Yang-Mills theory may be a string theory. This is also consistent with the dual superconductor picture, where there is a thin chromoelectric flux tube connecting the heavy quarks, essentially a string. However, to date, no one has succeeded in evaluating the planar limit of Yang-Mills theory which would be the lowest-order result for a string interpretation. Any result along such lines will shed light on confinement and other nonperturbative questions. In simpler models, and also for theories in lower dimensions, the $1/N$ expansion has been used to obtain nonperturbative results.

Another interesting property of the $1/N$ expansion is that the vacuum expectation value of gauge-invariant operators factorizes to the leading order. Let α_1, α_2 be two gauge-invariant operators. The leading terms of $\langle \alpha_1 \rangle$ and $\langle \alpha_2 \rangle$ can be represented by some set of closed planar double line loops. Consider now a term where we have a propagator connecting α_1 and α_2, which is a term that contributes to the connected part of $\langle \alpha_1 \alpha_2 \rangle$. Insertion of this propagator will bring in a factor of e^2. In addition we have to connect the two color lines of the propagator to color lines in α_1, α_2. This means that one color line loop will also be lost. Thus the diagram where α_1, α_2 are connected will be suppressed by $e^2/N \sim N^{-2}$ compared to $\langle \alpha_1 \rangle \langle \alpha_2 \rangle$. Addition of further propagators connecting α_1, α_2 will not change this behavior. This leads to the basic result that correlators of color singlet operators will factorize into a product of the expectation values each color singlet operator.

In a pure Yang-Mills theory, the spectrum of the theory consists of color singlet states which are called glueballs. Color singlet operators can generate glueballs from the vacuum, and so, glueballs can be analyzed by studying correlators of color singlet operators. Having no color quantum number, the basic quantum numbers of glueballs are spin (J), parity (P), and charge conjugation (C); there may also be additional quantum numbers characterizing higher-mass states of the same J^{PC} value. Let $\alpha(x)$ be a color singlet operator and denote the leading N-dependence of $\langle \alpha(x) \rangle$ as N^{α}, using the same symbol for the power of N as the operator itself. The connected diagram for $\langle \alpha(x)\alpha(y) \rangle$ will be of order $N^{2\alpha-2}$; this is the main contribution to the propagator of the glueball. We normalize and define a glueball operator $\mathcal{O}_{\alpha}(x) = N^{1-\alpha}\alpha(x)$. Consider now the decay amplitude for a glueball $\mathcal{O}_{\alpha}(x)$ to decay into two glueballs corresponding to two other operators, $\mathcal{O}_{\beta}(y)$ and $\mathcal{O}_{\gamma}(z)$. The connected diagram for $\langle \mathcal{O}_{\alpha}(x)\mathcal{O}_{\beta}(y)\mathcal{O}_{\gamma}(z) \rangle$ must have at least one gluon propagator connecting $\mathcal{O}_{\alpha}(x)$ to $\mathcal{O}_{\beta}(y)$ and at least another propagator connecting $\mathcal{O}_{\alpha}(x)$ to $\mathcal{O}_{\gamma}(z)$. This gives a factor N^{-4} compared to $\langle \alpha(x)\beta(y)\gamma(z) \rangle$, so that $\langle \mathcal{O}_{\alpha}(x)\mathcal{O}_{\beta}(y)\mathcal{O}_{\gamma}(z) \rangle$ goes like N^{-1}. Thus the decay amplitude for a glueball is of order $1/N$ at large N. This means that a glueball cannot decay into other glueballs at large N, we expect stable glueballs at large N. By a similar argument, the decay of a glueball into three other glueballs will be of order N^{-2}; since the $1 \to 3$ decay amplitude and the $2 \to 2$ scattering amplitude arise from the same vertex, of some hermitian effective Lagrangian, we can conclude that the glueball-glueball scattering amplitude is also of order N^{-2}. (The existence of a hermitian effective Lagrangian leads to the symmetry of the scattering amplitude under exchanges of incoming particles with outgoing particles, along with suitable changes of momenta; it is known as crossing symmetry.)

This result can be very useful on two counts. Imagine that we are able to solve the Schrödinger equation for the Yang-Mills theory and arrive at its spectrum. Then for each value of J^{PC}, we will just find one eigenstate, since the higher states can decay in general and are not eigenstates. But if we do the analysis at large N, the higher states are also stable and we can identify them as eigenstates of the Hamiltonian at large N and then put in the decays and other corrections as a second step. Thus the $1/N$ expansion can help to sort out the states of interest. Secondly, the emerging picture is that at large N, we expect a sequence of states for each value of J^{PC} with some quantum number characterizing the higher-mass states. The higher-mass states are narrow resonances with width $\sim N^{-2}$. This is again very similar to what happens in a string theory, further giving support to the idea that large N Yang-Mills theory is some kind of string theory.

19.4 Mesons and baryons in the $1/N$ expansion

In accordance with the idea of confinement, we expect the states in QCD to be color singlet combinations like mesons and baryons and glueballs. What are the implications of large N analysis for the mesons and baryons? This is the question we will consider now.

19.4.1 Chiral symmetry breaking and mesons

We consider N_f flavors of massless quarks in $U(N)$ Yang-Mills theory. First of all, since quarks transform as the N-dimensional fundamental representation of $U(N)$, quark loops have only one color line if we use the double line representation for gauge bosons. As a result, for any correlator, each quark loop gives a factor of N^{-1} in any diagram compared to similar diagrams which have no quark loops. Further, the axial $U_A(1)$ anomaly of QCD is not relevant for the following argument about chiral symmetry breaking; its effects will be seen to be at least of order $1/N$. The chiral symmetry group is then $U_L(N_f) \times U_R(N_f)$, neglecting quark masses and other weak sources of explicit chiral symmetry breaking.

Coleman and Witten have given an argument to show that large N leads to spontaneous breaking of this chiral symmetry in QCD to the diagonal subgroup $U_V(N_f)$. For this one assumes that chiral symmetry breaking can be described by a composite field of the form

$$\Phi^i_j \sim \bar{Q}^a_{Rj} Q^i_{La} \tag{19.66}$$

Under the chiral symmetry, this transforms as $\Phi \to g_L \Phi g_R^\dagger$, regarding Φ^i_j as the matrix elements of the $(N_f \times N_f)$-matrix Φ. The color indices a of the quark fields are summed over to form a color singlet.

The effective potential V for Φ must have the full chiral symmetry. The chiral transformations can be used to bring Φ to a diagonal form with real nonnegative eigenvalues. (The argument is similar to what we used for our discussion of the CKM matrix in Chapter 12.) Thus V may be taken as a function of the eigenvalues of Φ or equivalently eigenvalues of $\Phi^\dagger \Phi$ or $\Phi \Phi^\dagger$. In V we will have some terms of the form $\mathrm{Tr}(\Phi\Phi^\dagger)^m$. We could also have terms like $\mathrm{Tr}(\Phi\Phi^\dagger)^k \mathrm{Tr}(\Phi\Phi^\dagger)^l$; terms with products of three or more separate traces of $\Phi\Phi^\dagger$ are also possible. For the product of two separate traces, since Φ is made of quark operators, one has to have two quark loops in the diagrams which generate such terms from the underlying theory, namely, quantum chromodynamics. Connected diagrams contribute to the effective potential, and so, the two quark loops have to be connected by at least one gluon propagator. Such terms are, therefore, down by a power of N compared to terms with a single trace. Similarly, products of higher numbers of traces are also smaller by further factors of N^{-1}. Thus, in the large N limit, the form of the effective potential must be

$$V = N \sum_n c_n \mathrm{Tr}(\Phi\Phi^\dagger)^n = N f(\Phi\Phi^\dagger) \tag{19.67}$$

where f is some N-independent function. The overall factor of N is seen to be correct by considering a term like $\langle \Phi\Phi^\dagger \rangle$; the summation over the colors of the quark operators leads to N. In terms of eigenvalues λ_i of $\Phi\Phi^\dagger$, $V = N \sum_i f(\lambda_i)$. Assuming there is a single global minimum for the function f, we see that all eigenvalues must be the same, say, λ_0, at the minimum. Thus the expectation value of Φ must be of the form $diag(\lambda_0, \lambda_0, \cdots, \lambda_0)$. The unbroken subgroup is thus $U_V(N_f)$. If there is chiral symmetry breaking, large N considerations would show that breaking to the diagonal subgroup is preferred.

Notice that if all the eigenvalues of Φ are different at the minimum, it would correspond to an unbroken symmetry of $U(1)^{N_f}$. Thus, in using Φ, we are excluding breakdown to smaller subgroups. This might seem too stringent a starting point. However, there is a theorem due to Vafa and Witten that in any gauge theory with only vector-like coupling to quarks, parity cannot be spontaneously broken. Thus, if we neglect electroweak effects, the minimal group to which the chiral symmetry can be broken is $U_V(N_f)$. This considerably strengthens the large N argument.

A priori, it is possible that there be no chiral symmetry breaking in QCD; in other words, λ_0 could be zero. This possibility can be ruled out as follows. Consider any flavor current, say, a left-handed current of the form $J_\mu = \bar{Q}_L \xi \gamma_\mu Q_L$, where ξ is some combination of the generators of $U_L(N_f)$; we will take a combination for which $\text{Tr}(\xi^3) \neq 0$. This current has an anomaly which may be expressed as

$$r^\alpha \Gamma_{\mu\nu\alpha}(p,q,r) = i\frac{N}{12\pi^2}\text{Tr}(\xi^3)\epsilon_{\mu\nu\beta\gamma}p^\beta q^\gamma$$

$$\Gamma_{\mu\nu\alpha}(p,q,r) \equiv \int d^4x_1 d^4x_2 \, e^{ipx_1+iqx_2}\langle J_\mu(x_1)J_\nu(x_2)J_\alpha(0)\rangle \quad (19.68)$$

where $p^\alpha + q^\alpha + r^\alpha = 0$. (Result (19.68) may be obtained by starting with nonzero flavor gauge fields, obtaining the anomaly equations and then functionally differentiating twice with respect to the gauge fields and setting them to zero.) This result shows that $\Gamma_{\mu\nu\alpha}$ cannot be analytic at $p = q = r = 0$. For if it were analytic, it could be expanded in powers of p, q; but the only combination which has the correct power of momenta and the permutation symmetry of the currents is $\epsilon_{\mu\nu\alpha\beta}(p^\beta + q^\beta + r^\beta)$, which is zero. This shows that the result (19.68) is not compatible with analyticity of $\Gamma_{\mu\nu\alpha}$ at zero momenta. Therfore there must be zero-mass poles like p^{-2} or q^{-2} or r^{-2} in $\Gamma_{\mu\nu\alpha}$. This would correspond to a zero-mass particle whose propagator can lead to such a pole. J_μ has to create this zero-mass state acting on the vacuum. Since the current is a vector, it may create a vector or a scalar particle. However, for massless particles, the possibility of creating vector particles is ruled out. A simple way to see this is as follows. If a vector particle of momentum k and polarization λ is created by the current, we can write $\langle k, \lambda | J_\mu(k)|0\rangle = f_\mu^\lambda(k)$, where $J_\mu(k)$ is the Fourier transform of $J_\mu(x)$. The quantity $\alpha_{\mu\nu} = \sum_\lambda f_\mu^\lambda(k)f_\nu^\lambda(k)$ must be a function of k which obeys $k^\mu\alpha_{\mu\nu} = 0$, $k^\nu\alpha_{\mu\nu} = 0$. There is no such tensor we can construct since k_μ is a null vector. (For massive vectors of mass m, this could be $\eta_{\mu\nu} - k_\mu k_\nu/m^2$.) We are led to the conclusion that J_μ has to create scalars. The only way a current can create a massless scalar particle is due to spontaneous symmetry breaking, in which case $J_\mu \sim \partial_\mu \varphi$, for a Goldstone boson φ. Thus the possibility of the chiral symmetry being unbroken is ruled.

This part of the argument, based on anomalies, is a specialization of the general idea we mentioned in Chapter 17, due to 't Hooft, of finding possible symmetry-breaking patterns by matching anomalies in the confined phase with the calculation in terms of quarks.

Once we accept the idea of chiral symmetry breaking at large N, the effective Lagrangian of Chapter 12 can be used to represent interactions. The meson-meson interaction was found to be of order f_π^{-2}. Since $\bar{Q}Q$ has an expectation value which goes like N (due to the one quark loop), the meson operators are of the form $N^{-1/2}\bar{Q}Q$. For three such operators, the leading diagram has one quark loop again, and so, the three-meson vertex will be of order $N^{-1/2}$. Meson-meson scattering amplitude is of order N^{-1}. This shows that we must interpret f_π^2 to be of order N. This may also be seen directly from the equation

$$\langle 0|J_\mu|k \rangle = -i f_\pi^2 k_\mu \frac{e^{-ikx}}{\sqrt{2\omega_k V}} \tag{19.69}$$

by counting powers of N in a diagrammatic expansion. Equation (19.69) is essentially (12.103).

19.4.2 Baryons

We now turn to the question: how do we see baryons in the large N limit? Baryons are bound states of N quarks for a $U(N)$ gauge theory. The following argument, due to Witten, can be used to estimate the N-dependence of the mass of a baryon. The quarks interact via potentials generated by multigluon exchanges. A single gluon exchange between a pair of quarks in the baryon will give a factor of $e^2 \sim 1/N$. Every additional gluon propagator between the same quarks will give $e^2 N \sim 1$, the factor of N arising from trace over color indices. The interaction energy between a pair of quarks thus goes like $1/N$ for large N. In the baryon there are $\frac{1}{2}N(N-1)$ pairs of quarks, so the total interaction energy goes like $e^2 N(N-1) \sim N$ for large N. Since N is the expansion parameter, we see that the baryons have energy which goes like the inverse of the expansion parameter. This is typical of solitons in a field theory where the mass of the soliton goes like $1/\lambda$, where λ is the coupling constant. They cannot be obtained in perturbation theory; one has to look for nonperturbative field configurations which have stability and behave like particles. Thus, in the large N limit, baryons should be seen as solitons.

We have already seen in Chapter 12 that, due to chiral symmetry breaking in the large N limit, the effective theory is the theory of mesons described by

$$S = \frac{f_\pi^2}{4} \int d^4x \, \text{Tr}(\partial_\mu U^\dagger \partial^\mu U) - \frac{iN}{240\pi^2} \int_D (\text{Tr}(dU \, U^{-1})^5 \tag{19.70}$$

This is the action neglecting quark masses and electroweak gauge fields and is obtained by putting together (12.89) and (17.32). We will now show that this theory has soliton solutions; they are called skyrmions, after Skyrme, who found these solitons and showed that they have many of the properties of baryons, many years ago, before the gauge theory of strong interactions was invented. For the purpose of analyzing solitons, we will consider the modified model

$$
S = \frac{1}{4} f_\pi^2 \int d^4x \ \mathrm{Tr}(\partial_\mu U^\dagger \partial^\mu U) + \frac{1}{32\epsilon^2} \mathrm{Tr}([\partial_\mu U U^{-1}, \partial_\nu U U^{-1}]^2)
$$
$$
- \frac{iN}{240\pi^2} \int_{\mathcal{D}} (\mathrm{Tr}(dU \ U^{-1})^5 \tag{19.71}
$$

The extra term is known as the Skyrme term; ϵ is another coupling parameter in the theory. Such higher-derivative terms can in general exist since this is an effective action. A particular form which is convenient has been adopted here; its purpose is only to give a clear presentation of skyrmions. Addition of other possible terms will not change the essential features of the discussion. The energy functional for static configurations is

$$
\mathcal{E} = \int d^3x \left[\frac{1}{4} f_\pi^2 \mathrm{Tr}(\partial_i U \partial_i U^\dagger) - \frac{1}{32\epsilon^2} \mathrm{Tr}([\partial_i U U^{-1}, \partial_j U U^{-1}]^2) \right] \tag{19.72}
$$

Recall that the Wess-Zumino term, being first order in the time derivatives, does not contribute to the Hamiltonian. (Alternatively, being defined by a differential form, it does not involve the metric and so does not contribute to the energy.) For the vacuum configuration, we have $U = 1$. For configurations of finite energy, we need $U(x)$ to go to the vacuum value, namely, 1, as $|x| \to \infty$. Since U goes to the identity at spatial infinity, $U(x)$ is effectively a map from S^3 to $U(N_f)$. The homotopy classes of such maps are given by $\Pi_3[U(N_f)] = \Pi_3[SU(N_f)] = \mathbf{Z}$ and can be labeled by the winding number

$$
Q[U] = -\frac{1}{24\pi^2} \int \mathrm{Tr}(U^{-1}dU)^3
$$
$$
= -\frac{1}{24\pi^2} \int d^3x \ \epsilon^{ijk} \mathrm{Tr}(U^{-1}\partial_i U U^{-1}\partial_j U U^{-1}\partial_k U) \tag{19.73}
$$

This may be taken as the charge associated with a current

$$
J^\mu = -\frac{1}{24\pi^2} \epsilon^{\mu\nu\alpha\beta} \mathrm{Tr}(U^{-1}\partial_\nu U U^{-1}\partial_\alpha U U^{-1}\partial_\beta U) \tag{19.74}
$$

This current is identically conserved without the need of equations of motion. Field configurations with nonzero Q are the solitons or skyrmions. Time-evolution of a field configuration has to conserve Q and so, configurations with nonzero Q maintain their identity; the only way they can disappear is by pair annihilation of a soliton (of winding number Q) and an antisoliton (of winding number $-Q$).

The simplest set of maps with nonzero Q correspond to maps from S^3 to an $SU(2)$ subgroup of $SU(N_f)$. For $N_f = 2$, a spherically symmetric map is of the form

$$U(x) = U_S(x) \equiv \exp\left(i\phi(r)\tau \cdot \hat{x}\right)$$
$$= \cos\phi(r) + i\tau \cdot \hat{x} \sin\phi(r) \tag{19.75}$$

where τ_i are Pauli matrices. For this configuartion to be nonsingular, we need the boundary behavior $\sin\phi(0) = 0 = \sin\phi(\infty)$; thus $\phi(0)$ are $\phi(\infty)$ are multiples of π. The configuration (19.75) has

$$Q = \frac{1}{\pi}\left[\phi(0) - \phi(\infty)\right] \tag{19.76}$$

By taking $\phi(0) - \phi(\infty) = \pi$, we get a skyrmion of winding number 1. A configuration with $Q = 1$ is given by

$$\sin\phi = \frac{2Rr}{R^2 + r^2} \tag{19.77}$$

which corresponds to the stereographic projection of a sphere onto $\mathbf{R}^3$.

The actual profile of the soliton is not determined by the topological considerations. One could take (19.75) as an ansatz and solve static equations of motion (or slightly modified versions of them) to get a solution. Since this is an effective theory, it suffices for many purposes to treat this variationally. We can take a profile for $\phi(r)$ with the required boundary values and a scale parameter R and calculate the energy and minimize to determine R. The first term in (19.72) has dimension -1, while the second term has dimension 1; thus the general form of the energy is

$$\mathcal{E} = af_\pi^2 R + \frac{b}{\epsilon^2 R} \tag{19.78}$$

where a, b are constants. This gives the best value of R as $R = \sqrt{b/a\epsilon^2 f_\pi^2}$. Notice that without the Skyrme term, it is energetically favorable for the soliton to shrink to zero size. The Skyrme term is introduced to stabilize the soliton. For this purpose, any term of dimension ≥ 1 will do; the Skyrme term is just a simple choice.

In order to identify the Skyrmions as the baryons of the large N limit, we need to show that they carry baryon number, are fermions, and have the correct flavor properties. We will consider these questions now.

19.4.3 Baryon number of the skyrmion

A technique for identifying the baryon number of the soliton is to couple the theory to quarks so that the baryon number operator can be defined. We thus add to the effective action a term

$$\mathcal{S}_Q = \bar{Q}(i\gamma \cdot \partial)Q - m\left(\bar{Q}_L U Q_R + \bar{Q}_R U^\dagger Q_L\right) \tag{19.79}$$

We have included a mass term for the quarks. Since quark has baryon number $1/N$ for QCD with $SU(N)$ symmetry, the baryon number operator is defined by

$$B = \frac{1}{N} \int d^3x\ \bar{Q}\gamma^0 Q \tag{19.80}$$

The strategy for identifying the baryon number of the soliton is then to calculate the expectation value of B for the soliton state, which corresponds to having a background value for U given by the soliton configuration. Thus we calculate

$$\langle B \rangle_S = \frac{1}{N} \int d^3x\ \langle S|\bar{Q}\gamma^0 Q|S\rangle$$

$$= -\frac{1}{N} \int d^3x\ \text{Tr}\left[\gamma^0 S\left(x - \frac{\epsilon}{2}, x + \frac{\epsilon}{2}, U\right)\right]_{\epsilon \to 0} \tag{19.81}$$

The second line expresses the baryon number in terms of the short-distance behavior of the fermion propagator in the presence of the background U. Since only the short-distance behavior is needed, we can obtain it by an expansion in powers of $1/m$. (It was for this reason that a quark mass term was introduced.)

The propagator obeys the equation

$$(i\gamma \cdot \partial - mF)S(x, y) = i\delta^{(4)}(x - y) \tag{19.82}$$

where $F = \frac{1}{2}[U(1 - \gamma_5) + U^\dagger(1 + \gamma_5)]$. The solution to this may be written as $S(x, y) = -(i\gamma \cdot \partial + mF^\dagger)\ G(x, y)$, where G is defined by

$$\left(\Box + m^2 - im\gamma \cdot (\partial F^\dagger)\right) G(x, y) = i\delta^{(4)}(x - y) \tag{19.83}$$

We have used the properties $FF^\dagger = 1$ and $\gamma_\mu F^\dagger = F\gamma_\mu$. We will expand G in powers of $X = im\gamma \cdot \partial F^\dagger$ by iterating the integral equation

$$G(x, y) = G_0(x, y) - i \int_z G_0(x, z)X(z)G(z, y)$$

$$G_0(x, y) = -i \int \frac{d^4p}{(2\pi)^4} \frac{1}{p^2 - m^2 + i\epsilon} \exp[-ip(x - y)] \tag{19.84}$$

In the expression for the expectation value of the baryon current, the relevant term for us is

$$\langle J^\mu(x) \rangle = \frac{1}{N} \text{Tr}\left[\gamma^\mu(i\gamma \cdot \partial + mF^\dagger)G(x, y)\right]_{y \to x}$$

$$= (-i)^3 \text{Tr}\left[\gamma^\mu mF^\dagger X\, X\, X\right] \int \frac{d^4p}{(2\pi)^4} \left[\frac{1}{p^2 - m^2 + i\epsilon}\right]^4 + \cdots$$

$$= \frac{i}{96\pi^2} \text{Tr}\left[\gamma^\mu F^\dagger \gamma \cdot F^\dagger \gamma \cdot F^\dagger \gamma \cdot F^\dagger\right] + \cdots$$

$$= -\frac{1}{24\pi^2} \epsilon^{\mu\nu\alpha\beta} \text{Tr}(U^{-1}\partial_\nu U U^{-1}\partial_\alpha U U^{-1}\partial_\beta U) + \cdots \tag{19.85}$$

The factor of $1/N$ is canceled by the trace over the N colors of quarks. There are other terms which can contribute to the current. The terms which are potentially divergent give zero by Dirac and $SU(N_f)$ trace identities. We have neglected terms involving derivatives of X in passing from the first to the second line of this equation; such terms could give additional contributions. The term we have displayed leads to a baryon charge given by the winding number Q. The density for Q is a closed but not exact three-form belonging to an element of $\mathcal{H}^3[SU(N_f)]$. Since this term is unique, all the other terms in $\langle J^\mu \rangle$ have to be of the form $\partial_\nu K^{\mu\nu}$, where $K^{\mu\nu}$ is antisymmetric, so that we have conservation of the baryon current. In the expression for the charge, such terms integrate to zero. Thus, even though one can have corrections to the expression we derived for the current, such terms are not needed for the charge. (This is why we calculated only the one required term in $\langle J^\mu \rangle$. The key to our derivation is to show that a baryon number equal to Q is generated, not that there are no other terms in the current.) We have thus shown that skyrmions have baryon number which is given by the winding number.

19.4.4 Spin and flavor for skyrmions

For discussing the flavor and spin properties of the soliton, we will just consider $N_f = 3$ and $N = 3$; there is no substantial difference for higher numbers of flavors. The argument presented below is essentially due to Witten.

The soliton configuration (19.75), for $N_f = 3$, is

$$U_S(x) = \begin{pmatrix} \exp(i\phi(r)\tau \cdot \hat{x}) & 0 \\ 0 & 1 \end{pmatrix} \qquad (19.86)$$

This configuration is not invariant under rotations of $\boldsymbol{x}$, but is invariant under combined rotations and $SU(2)$ transformations generated by the τ's. Therefore, we see that spatial rotation may be taken to be given by $U_S \rightarrow GU_SG^\dagger$, where G is an $SU(2)$ transformation of the form

$$G = \begin{pmatrix} \exp(i\tau \cdot \theta) & 0 \\ 0 & 1 \end{pmatrix} \qquad (19.87)$$

To begin with, the theory has symmetry $SU_V(3) \times SU_J(2)$, where $SU_J(2)$ corresponds to rotations; but the configuration (19.86) does not have this full symmetry. In the quantum theory, quantum fluctuations of the configuration (19.86) occur and among these there are some which correspond to moving by (spatially constant) $SU_V(3) \times SU_J(2)$ transformations. Since this is a symmetry of the theory, the static energy does not change. One can then take the parameters of such transformations to be time-dependent and quantize fluctuations around (19.86) by just keeping these modes. This is the first step in a semiclassical quantization of the soliton and will suffice to show its spin and flavor properties. The ansatz for the soliton, with the constant flavor rotations, is

$$U(x,t) = A(t)\, U_S\, A^\dagger(t) \tag{19.88}$$

where $A(t)$ is an $SU(3)$ matrix. The symmetries are now realized in the following way.

$$A'(t) = V\, A(t), \qquad\qquad A'(t) = A(t)\, G \tag{19.89}$$

where $V \in SU_V(3)$ will give the flavor symmetries and G gives spin or rotations. We may define a set of differential operators which are infinitesimal generators for these transformations as

$$L_a\, A = t_a\, A, \qquad\qquad R_a\, A = A\, t_a \tag{19.90}$$

Writing $A = \exp(it_a\varphi^a)$ for some parameters φ^i, $a, i = 1, 2, \cdots, 8$, we can identify

$$L_a = -i(E^{-1})^i_a \frac{\partial}{\partial\varphi^i}, \qquad\qquad R_a = -i(\tilde{E}^{-1})^i_a \frac{\partial}{\partial\varphi^i} \tag{19.91}$$

where

$$dAA^\dagger = -it_a E^a_i d\varphi^i, \qquad\qquad A^\dagger dA = -it_a \tilde{E}^a_i d\varphi^i \tag{19.92}$$

To obtain the quantum theory, we now substitute (19.89) into the action (19.71) to obtain

$$S = \int dt \left[-\frac{\alpha}{2}\{\mathrm{Tr}(t_i A^\dagger \partial_t A)\}^2 - \frac{\beta}{2}\{\mathrm{Tr}(t_k A^\dagger \partial_t A)\}^2 - i\frac{QN}{\sqrt{3}}\mathrm{Tr}(t_8 A^\dagger \partial_t A) \right] \tag{19.93}$$

where t's are $SU(3)$ matrices and $i = 1, 2, 3$, $k = 4, 5, 6, 7$. α and β are some constants arising from the spatial integrations whose precise values are not important here. Notice that there is no quadratic term involving $\mathrm{Tr}(t_8 A^\dagger \partial_t A)$. In quantizing, the time-derivatives of $\mathrm{Tr}(t_a A^\dagger \partial_t A)$ for $a = 1, 2, \cdots, 7$ will be related to the differential operators L_a or R_a. The wave functions will be functions of the $SU(3)$ matrix A and the action of various operators will be defined in terms of L_a and R_a. But since there is no quadratic term for $\mathrm{Tr}(t_8 A^\dagger \partial_t A)$, we do not get a canonical momentum for this direction. Under $A \to A \exp(it_8\theta)$, the action changes as

$$S(Ae^{it_8\theta}) = S + \frac{NQ}{2\sqrt{3}}\dot\theta \tag{19.94}$$

This means that the wave functions must have the property

$$\Psi(Ae^{it_8\theta}) = \Psi(A)\, \exp\left(i\frac{NQ}{2\sqrt{3}}\theta \right) \tag{19.95}$$

This puts a constraint on the allowed wave functions. (The fact that the wave functions must obey such an equation if the action changes as in (19.94) is easily seen in a path-integral representation for the wave function.)

The wave functions can be generally written as

$$\Psi(A) = C_R \, \mathcal{D}^R(A)_{I,I_3,Y;I',I'_3,Y'} \qquad (19.96)$$

The $\mathcal{D}$'s are the matrices representing the group element A in some representation R, which is a priori arbitrary. ($\mathcal{D}$'s are the Wigner $\mathcal{D}$-matrices for $SU(3)$.) C_R is a normalization constant. The matrix labels are specified as follows. $L_1^2 + L_2^2 + L_3^2 = I(I + 1)$ is the quadratic Casimir for the $SU(2)$ subgroup of the left action of $SU(3)$ on A. I_3 is the eigenvalue of L_3 and Y is the eigenvalue of $(2/\sqrt{3})L_8$. (This is the hypercharge.) I', I'_3, and Y' refer to the corresponding quantities for the right action R_a on A; thus $R_1^2 + R_2^2 + R_3^2 = I'(I' + 1)$, $R_3 = I'_3$ and $(2/\sqrt{3})R_8 = Y'$ when acting on $\mathcal{D}^R(A)_{I,I_3,Y;I',I'_3,Y'}$. These are the quantum numbers needed to specify a state within an $SU(3)$ representation. Notice that R_1, R_2, R_3 define the rotation group for the soliton, so that $I' = J$, the spin of the soliton in the quantum theory.

The constraint (19.95) tells that we must choose $Y' = 1$ for $N = 3$ and $Q = 1$. The lowest-dimensional representations of $SU(3)$ containing any state with $Y' = 1$ are the (8-dimensional) octet, which is the adjoint representation, and the (10-dimensional) decuplet. For the octet, the state with $Y' = 1$ has $I' = J = \frac{1}{2}$. The left indices show that these fall into a flavor multiplet which is an octet; since Y' is fixed, there are only two values for the right indices, $I'_3 = \pm\frac{1}{2}$. Thus we get an octet of spin-$\frac{1}{2}$ soliton states. For the decuplet $I' = J = \frac{3}{2}$, leading to a flavor decuplet of solitons with spin-$\frac{3}{2}$. Thus, this theory leads to a set of solitons carrying baryon number 1, whose low-lying states are a spin-$\frac{1}{2}$ octet and a spin-$\frac{3}{2}$ decuplet. This is in complete agreement with the quark model, which was discussed in Chapter 12, and with what is observed. The soliton, which is a coherent field configuration of bosonic fields, has become a fermion due to the effect of the Wess-Zumino term.

For two flavors, there is no Wess-Zumino term, since $\mathcal{H}^5[SU(2)] = 0$. Since soliton configurations are maps from S^3 to $SU(2)$, we find that $\Pi_1[\mathcal{C}] = \Pi_4[SU(2)] = \mathbf{Z}_2$, where $\mathcal{C}$ denotes one component of the configuration space. Thus $\mathcal{C}$ can support double-valued wave functions leading to the possibility of fermions. This was already noted by Skyrme many years ago. The extension to $SU(3)$ gives a concrete realization. One may even regard the embedding in $SU(3)$ as similar to our argument for the global anomaly of $SU(2)$ and one can obtain the fermion nature of the solitons by a modification of that argument.

There are a large number of papers which show further properties of baryons from the soliton point of view and explore the Skyrme model as a complementary picture of baryons in the standard model.

In conclusion, we see that the large N theory can reproduce all the expected features of the particle spectrum, lending further support for the $1/N$

expansion as a sensible way to analyze the nonperturbative aspects of the gauge theory.

19.5 Lattice gauge theory

19.5.1 The reason for a lattice formulation

A very important approach to the study of nonperturbative features of a nonabelian gauge theory is lattice gauge theory. The basic idea here is the following. Quantum field theory can be described by the Euclidean functional integral. This integral has to be defined by first considering the theory truncated to a finite number of modes, calculating the finite dimensional ordinary integral and then taking the limit as the number of modes goes to infinity. In the lattice approach, the truncation to a finite number of modes is made by considering the theory on a finite lattice which is usually taken, for simplicity, as a simple cubic lattice of appropriate dimensions. Thus in four dimensions, one considers a four-dimensional hypercubic lattice with M lattice points along each direction with a lattice spacing a. The field variables are located at each lattice site or on the link connecting two nearby lattice sites, so that the total number of field variables is finite, proportional to M^4. Since the shortest possible distance between two points is a, there is no ultraviolet divergence, a serving as the cut off parameter. The continuum theory, which is what we are interested in, can be obtained as $M \to \infty$, $a \to 0$. Rescaling of fields and coupling constants and renormalization of parameters would be needed in extracting this limit. The four-dimensional Euclidean invariance can also be regained in this limit.

This approach is important for many reasons. On the theoretical side, the theory on the lattice gives a mathematically well-defined functional integral and correlators and so it is a good starting point for investigating questions like the existence of a field theory in a fully nonperturbative sense. This is the approach of constructive field theory and indeed this has been used to argue that the ϕ^4-theory in four dimensions, despite its great usefulness in perturbative calculations, may be trivial. (Basically this is the statement that if the bare theory is formulated on the lattice with bare parameters taken to be finite as $a \to 0$, then the continuum theory will be a free theory.)

On a more practical side, the truncation of the functional integral to a finite dimensional integral leads to the possibility of numerically evaluating the integral for a set of values of (M, a) and then extrapolating the results to the continuum limit $M \to \infty$, $a \to 0$, to obtain numerical estimates of physical quantities. The approach is intrinsically nonperturbative and with the increasing power of electronic computers, this becomes more and more practical. Since it is possible to formulate gauge theories on a lattice manifestly preserving gauge-invariance, the lattice approach can be particularly useful for gauge theories. In practice, we still have to do a lot of analytical work formulating the lattice versions of physical quantities and a procedure for estimating them as well as developing a number of specialized numerical algorithms for implementing these calculations via computers. A large body of work has been built up over the years. Many qualitatively general features of gauge theories such as confinement and the deconfinement transition at finite temperatures can now be analyzed in a reasonably reliable way. There have also been some estimates of glueball masses.

In this section we will review the most elementary features of a lattice gauge theory, considering a theory with no spontaneous symmetry breaking.

19.5.2 Plaquettes and the Wilson action

The basic gauge-field variables are defined on links connecting nearby lattice sites. They are elements of the Lie group $SU(N)$ for an $SU(N)$ gauge theory and are denoted by $U(x + \mu, x)$, where x denotes a lattice point and $x + \mu$ the nearby lattice point obtained from x by translation by an elementary lattice vector (of magnitude a) in the direction μ. We may take U's to be in the fundamental representation of $SU(N)$ and consider them as $(N \times N)$-matrices. On the cubic lattice we then consider the faces of the unit cells, one example of which would be the square with vertices x, $x + \mu$, $x + \mu + \nu$, $x + \nu$. The faces of the unit cells are called plaquettes; they provide a tiling of any surface we can form in the lattice. The action is defined in terms of the trace of the product of U's around the edges of a plaquette. We define a plaquette variable

$$
\begin{aligned}
W_P(x) &= W(x, x + \mu, x + \mu + \nu, x + \nu) \\
&= \mathrm{Tr}\Big[U(x, x + \nu)\, U(x + \nu, x + \mu + \nu) \\
&\qquad\quad U(x + \mu + \nu, x + \mu)\, U(x + \mu, x)\Big]
\end{aligned}
\tag{19.97}
$$

The simplest version of the lattice action for a gauge theory, first given by Wilson, is then

$$
\begin{aligned}
\mathcal{S} &= \beta \sum_P \left[1 - \frac{1}{N}\, \mathrm{Re} W_P(x)\right] \\
&= \beta \sum_P 1 + \mathcal{S}' \\
\mathcal{S}' &= -\frac{\beta}{N} \sum_P \mathrm{Re} W_P(x)
\end{aligned}
\tag{19.98}
$$

The sum is over all plaquettes P in the lattice and β is related to the bare coupling constant e_0 by $\beta = 2N/e_0^2$.

Gauge transformations are $SU(N)$-valued functions and are defined at the lattice sites. Thus, under a gauge transformation

$$U(x + \mu, x) \rightarrow U^g(x + \mu, x) = g(x + \mu)\, U(x + \mu, x)\, g^\dagger(x) \tag{19.99}$$

where $g \in SU(N)$. $W_P(x)$ is invariant under such a transformation and so is the action (19.98).

The relationship with the continuum formulation is as follows. $U(x+\mu, x)$ is the lattice version of the parallel transport matrix; this is clear from the transformation law (19.99). For small lattice spacing, the gauge potential can be defined by

$$U(x + \mu, x) \approx P \exp\left[i \int_x^{x+\mu} t^a A_\mu^a \, dx^\mu\right] \tag{19.100}$$

$W_P(x)$ is then the Wilson loop operator where the loop corresponds to the boundary of the plaquette P and so

$$W_P(x) \approx \mathrm{Tr}\left[e^{\frac{i}{2}t^a F_{\mu\nu}^a \sigma^{\mu\nu}}\right] \tag{19.101}$$

$\sigma^{\mu\nu}$ is the area of the plaquette and equation (19.101) is valid for small areas, or as $a \rightarrow 0$. We thus find

$$S \approx \int d^4x \, \frac{1}{4e_0^2} F_{\mu\nu}^a F^{a\mu\nu} \tag{19.102}$$

The summation over plaquettes, with a factor of a^4, becomes the integral over the Euclidean spacetime volume. Equation (19.102) agrees with the continuum action, showing that (19.98) does indeed correspond to the Yang-Mills theory. This limit is often refered to as the naive continuum limit since the actual continuum limit of the theory defined by evaluating the functional integral and then taking $M \rightarrow \infty$, $a \rightarrow 0$ may give a very different behavior.

The functional integral is defined by

$$Z = \int \prod_{links} d\mu(U) \, \exp(-S) \tag{19.103}$$

where $d\mu(U)$ for each link is the Haar measure for the group $SU(N)$. This is defined as follows. The Cartan-Killing metric on $SU(N)$, considered as a differential manifold, is given by

$$ds^2 = -2 \, \mathrm{Tr}(U^{-1}dU)^2 = E_i^a E_j^a d\varphi^i d\varphi^j \tag{19.104}$$

where $U^{-1}dU = -it^a E_i^a(\varphi)d\varphi^i$, φ^i being group parameters in terms of which we can write U; for example, we can take $U = \exp(-it^a \varphi^a)$. (In (19.104) we have the symmetric product, not a wedge product.) The volume element corresponding to (19.104) is

$$d\mu(U) = (\det E) \prod_i d\varphi^i \tag{19.105}$$

This is the Haar measure for the group.

There are two properties of $d\mu(U)$ which will be useful.

1. The total volume of the group defined by

$$vol[SU(N)] = \int d\mu(U) \tag{19.106}$$

   is finite.

2. Representation matrices are orthonormal with respect to the Haar measure, i.e.,

$$\int d\mu(U)\ \mathcal{D}_{AB}^{(R)}(U^\dagger)\mathcal{D}_{\alpha\beta}^{(R')}(U) = \frac{vol[SU(N)]}{dimR}\delta^{RR'}\delta_{B\alpha}\delta_{A\beta} \tag{19.107}$$

where $\mathcal{D}_{AB}^{(R)}(U)$ is the matrix representation of the group element U in the irreducible representation R. A, B take values $1, 2, \cdots, dimR$; α, β take values $1, 2, \cdots, dimR'$. (These are the Wigner $\mathcal{D}$-functions for $SU(N)$.)

In the functional integral (19.103), in integrating over all U's, we end up integrating over the directions in configuration space corresponding to gauge transformations as well. Since there are a finite number of points, this does not give a divergence and the volume of gauge transformations gets divided out in expectation values.

Strong coupling expansion

The total volume of integration in (19.103) is finite, so one may expand the exponential $\exp(-\mathcal{S}) = \exp(-\beta \sum_P 1) \exp(-\mathcal{S}')$ in (19.103) as a series in $\mathcal{S}'$ and evaluate integrals term by term to get a series in $1/e_0^2$. This is referred to as the strong coupling expansion and, a priori, would seem reasonable for large e_0 and fixed a. e_0 is the bare coupling constant for the lattice with lattice cut-off a and so this expansion is appropriate for distance scales of the order a. Asymptotic freedom tells us that the effective coupling increases with separation of charges, so that to get a finite coupling at a given separation as $a \to 0$, we must tune e_0 to smaller The strong coupling expansion is thus not appropriate for the continuum limit where $a \to 0$. Nevertheless, some interesting features emerge from such an expansion.

The lowest-order term in the expansion of the exponential will give a constant $\int \prod_{links} d\mu(U) = \{vol[SU(N)]\}^{n_L}$, where n_L is the total number of links. In the next term, which is of the form $\int d\mu(U)\ \mathcal{S}'$, we encounter integrals like $\int dU(x+\mu, x)\ U(x+\mu, x)$, which is zero by (19.107); thus $\int d\mu(U)\mathcal{S}' = 0$. To get a nonzero value we must have the product of $U(x + \mu, x)$ and its conjugate $U(x, x + \mu)$ in the integrand. This can arise from a product of W_P's for nearby plaquettes with one common edge $(x + \mu, x)$, for example, the plaquettes $(x, x + \mu, x + \mu + \nu, x + \nu)$ and $(x, x + \mu, x + \mu + \alpha, x + \alpha)$. Integration over $U(x+\mu, x)$ using (19.107) then replaces $U(x+\mu, x)U(x, x+\mu)$ by a constant, giving a surface made of two plaquettes glued together along the edge $(x + \mu, x)$ and a remaining Wilson loop along the boundary of this surface. In $\int d\mu(U)\mathcal{S}'^2$ we will find such a possibility, one factor of $\mathcal{S}'$ contributing $U(x + \mu, x)$ and the other giving $U(x, x + \mu)$. $\int d\mu(U)\mathcal{S}'^2$ is still zero since there are remaining integrations over link variables with no matching conjugate element. Continuing in this way, it is easy to see that the only way we can get conjugate elements for each link variable is if the plaquettes under consideration form a tiling of a closed surface Σ. We can have a factor of W_P for each plaquette on the closed surface which comes from a corresponding factor of S' in the expansion of the exponential; the edges would match, giving pairing of the link variables. In this case, the integral will be proportional to $\{1/e_0^2\}^{n(\Sigma)} \sim \exp[n(\Sigma) \times \text{constant}]$, where $n(\Sigma)$ is the number of plaquettes for the surface Σ. The functional integral thus reduces to some kind of summation over closed surfaces with a weight determined by an action proportional to the number of plaquettes or the area of the closed surface. The surfaces inolved are not necessarily simple; for example, one can have surfaces with different branches sharing common links. Nevertheless, this structure is again reminiscent of string theory where the functional integral is given as a summation over surfaces.

A curve C in the lattice can be considered as formed by a sequence of links $l_1, l_2, \cdots, l_K$ and the Wilson loop operator over C can be defined in terms of the product of U's along these link segments. Thus

$$W_C = \text{Tr}\,(U_{l_1} U_{l_2} \cdots U_{l_K}) \tag{19.108}$$

In the strong coupling expansion, we can see that the expectation value of the Wilson loop will give an area law. Consider

$$\langle W_C \rangle = \int d\mu(U)\, W_C\, e^{-S} \tag{19.109}$$

In matching the U's to get nonzero contributions, we see that we can now match the U's from the expansion of e^{-S} with the U's from the Wilson loop. The integration then reduces to summation over open surfaces with the boundary C. Since the minimal number of plaquettes for tiling this surface is proportional to the minimal area of a surface with boundary C, we see that the behavior of the Wilson loop is

$$\langle W_C \rangle \sim \exp\left(-\frac{\mathcal{A}(C)}{a^2} \log e_0^2\right) \tag{19.110}$$

We get an area law, but this is in the strong coupling expansion and does not provide any argument for confinement as $a \to 0$. The general expectation is that for dynamical reasons, this behavior will persist to the continuum limit.

19.5.3 The fermion doubling problem

Lattice versions of matter fields can also be introduced in a simple way. Matter fields are associated with lattice sites. Thus, for a scalar field, we introduce a field variable $\phi(x)$ at every lattice point x. Derivatives can then be defined by

$$\partial_\mu \phi \equiv \frac{\phi(x + \mu) - \phi(x)}{a} \tag{19.111}$$

It is then possible to construct actions of the form

$$S = \sum_{sites} a^4 \left[\frac{1}{2} \left(\frac{\phi(x + \mu) - \phi(x)}{a} \right)^2 + \frac{m^2}{2} \phi^2(x) + \lambda \phi^4(x) \right] \tag{19.112}$$

In the small a limit, this goes to the continuum action

$$S = \int d^4 x \left[\frac{1}{2} (\partial \phi)^2 + \frac{m^2}{2} \phi^2 + \lambda \phi^4 \right] \tag{19.113}$$

In the case of fermions, a similar procedure can be carried out but leads to the so-called fermion doubling problem. Any lattice action for fermions with locality of interactions, hermiticity of Hamiltonian and translational invariance on the lattice will give a continuum theory with an even number of fermion fields with pairs of fields of opposite chirality. This means that we cannot set up a lattice theory if the desired continuum theory is chiral in nature, such as the standard model, unless we give up some of the conditions mentioned above.

The doubling problem can be illustrated in a simple way by considering massless fermions in one spatial dimension; we will also use a Hamiltonian approach. First we shall consider the continuum theory for which we have the Dirac equation

$$i \frac{\partial \psi}{\partial t} - i\sigma_3 \frac{\partial \psi}{\partial x} = 0 \tag{19.114}$$

We use the γ-matrices, $\gamma^0 = \sigma_1$, $\gamma^1 = i\sigma_2$ and take the chirality operator as $-\gamma^0 \gamma^1 = \sigma_3$. For left chirality, the Dirac equation becomes

$$i \frac{\partial \psi}{\partial t} - i \frac{\partial \psi}{\partial x} = 0 \tag{19.115}$$

so that, with the substitution $\psi \sim \exp(-ip_0 t + ipx)$, we find $p_0 = -p$. Thus negative values of p lead to the positive energy solutions, $p_0 = \omega_p = |p|$; upon quantization, these will be associated with the annihilation operator for particles. Notice that these are left-moving modes on the line. Positive values of p lead to $p_0 = -\omega_p = -p$, corresponding to the creation operator part. For right chirality, we get $p_0 = p$, with positive values of p, or right movers, corresponding to particle annihilation. There is thus a correlation between chirality and left- and right-moving modes.

Consider now one chirality, say, the right one, on the lattice. Keeping time as a continuous variable still, the expected equation is of the form

$$i \frac{\partial \psi}{\partial t} + i \left(\frac{\psi(x + a) - \psi(x)}{a} \right) = 0 \tag{19.116}$$

The second term will correspond to an expression $-i \sum_x \psi^\dagger(x)[\psi(x+a) - \psi(x)]$ in the Hamiltonian operator. This is not hermitian by itself. The hermitian form which also gives terms which are first order in the spatial derivatives as $a \to 0$ is $\sum_x \frac{1}{2}\{-i\psi^\dagger(x)[\psi(x+a) - \psi(x)] + \text{h.c.}\}$. This choice leads to the equation

$$i\frac{\partial \psi}{\partial t} + i\left(\frac{\psi(x+a) - \psi(x-a)}{2a}\right) = 0 \tag{19.117}$$

We solve this by introducing a mode expansion of the form

$$\psi(x) = \sum_p e^{-ip_0 t + ipx} a_p \tag{19.118}$$

Periodic boundary conditions are the easiest to use for the lattice, so that for a lattice of length Ma we have $\psi(x + Ma) = \psi(x)$. This gives the possible values of p as $2\pi n/Ma$, where n is an integer. Since x is an integer multiple of a, $\exp(ip_{M+n}x) = \exp(ip_n x + 2\pi i x/a) = \exp(ip_n x)$. Thus there are only M independent modes, which we may take as

$$p_n = \frac{2\pi n}{Ma}, \qquad n = 0, \pm 1, \pm 2, \cdots, \pm\frac{M-1}{2}. \tag{19.119}$$

(We will take M to be odd for simplicity.) The points $\pm(M-1)/2$ are the edges of the Brillouin zone, giving the highest possible values for $|p|$ on the lattice. Using the mode expansion (19.118) in the Dirac equation (19.117) we get

$$p_0 = \frac{\sin(pa)}{a} \tag{19.120}$$

For modes near $n = 0$, p is small and as $a \to 0$, we get $p_0 = p$ corresponding to a massless mode of right chirality. However, we can also consider modes near the edge of the Brillouin zone. For modes near $(M-1)/2$, we can write $p = (2\pi/Ma)(M-1)/2 - \tilde{p}$, $\tilde{p} \geq 0$; we then find $\sin pa = \sin \tilde{p}a$ as $M \to \infty$. This leads to another mode with $p_0 = \tilde{p}$ as $M \to \infty$, $a \to 0$. In the mode expansion, this is of the form

$$\psi(x) \sim e^{i\pi x/a} \sum_{\tilde{p}} \exp(-i\tilde{p}t - i\tilde{p}x) \, a_{\tilde{p}} \tag{19.121}$$

Thus $e^{-i\pi x/a}\psi(x)$ is another field in the continuum which behaves as a mode of momentum $-\tilde{p}$ and $p_0 = \tilde{p}$; it is thus part of a field of left chirality. The other half of modes for this field comes from modes near $-(M-1)/2$, for which we may write $pa \approx -\pi + \tilde{p}a$ as $M \to \infty$. In this case, the momentum will be $\tilde{p}$ and $p_0 = -\tilde{p}$ consistent with left chirality again. Thus, even though we started from the lattice version of what appeared to be a purely right chirality field, we see that we end up getting both chiralities as $M \to \infty$, $a \to 0$, due to extra contributions from modes near the edge of the Brillouin zone. This is the essence of fermion doubling.

This result is quite general. It can be formulated in terms of the action with latticization of the time direction as well. For every direction, we get both chiralities, leading to a multiplicity of $2^4 = 16$ in four dimensions. The generality of this doubling is captured by the Nielsen-Ninomiya theorem. Let the equation for the lattice Dirac field be of the form

$$i\frac{\partial\psi}{\partial t} = \sum_y H(x - y)\psi(y) \qquad (19.122)$$

We consider the following three conditions on the lattice theory:

1. The theory is local in the sense that the Hamiltonian $H(x - y)$ goes to zero as $|x-y| \to \infty$ such that the Fourier transform of H has a continuous first derivative.
2. The action and Hamiltonian have translational invariance under translations by lattice vectors.
3. The Hamiltonian $H(x - y)$ as a matrix, with x, y as labels for matrix elements, is hermitian (or the action is real).

The Nielsen-Ninomiya theorem then states that given a lattice fermion theory obeying these three conditions, the fermion modes are in pairs of left and right chirality, as $M \to \infty$, $a \to 0$.

This theorem shows that it is rather difficult to construct lattice theories for chiral fermions. One is forced to give up one of the three premises of the theorem. A number of different approaches have been proposed for constructing chiral theories on the lattice. While they are important for the standard model and for questions of chiral symmetry breaking for QCD, they generally lead to a considerable increase in computational complexity. With presently available computers, we are still only at the threshold of fully dynamical numerical computations relevant to such questions.

References

1. The suggestion that confinement of quarks can be understood as dual superconductivity was made by Y. Nambu, in Phys. Rep. **23**, 250 (1976). This was developed by G. 't Hooft, Nucl. Phys. **B138**, 1 (1978); *ibid.* **B153**, 141 (1979); S. Mandelstam, Phys. Lett. **B53**, 476 (1975); Phys. Rev. **D19**, 2391 (1979).
2. The Wilson loop and the area law are due to K.G. Wilson, Phys. Rev. **D14**, 2455 (1974).
3. For topological vortices, see A.A. Abrikosov, Sov. Phys. JETP **5**, 1174 (1957); H.B. Nielsen and P. Olesen, Nucl. Phys. **B61**, 45 (1973).
4. The $T(C')$ operators and their algebra are in 't Hooft's papers, reference 1.
5. Magnetic monopoles in nonabelian gauge theory were obtained by G. 't Hooft, Nucl. Phys. **B79**, 276 (1974); A.M. Polyakov, JETP Lett. **20**,

194 (1974). This subject has developed enormously; some general results are in *Monopoles in Quantum Field Theory*, Proceedings of the Monopole Meeting, Trieste, P. Goddard. W. Nahm and N.S. Craigie (eds.), World Scientific Pub. Co. (1982); A. Jaffe and C. Taubes, *Vortices and Monopoles*, Birkhauser (1980).

6. For the $\mathcal{N} = 2$ supersymmetric gauge theory, it is possible to give an exact analysis of the vacuum structure; see N. Seiberg and E. Witten, Nucl. Phys. **B426**, 19 (1994). This has led to a number of new developments; for a recent review, see M. Strassler, in *Trieste 2001: Superstrings and Related Matters*, Abdus Salam ICTP, Trieste (2001).

7. The $1/N$ expansion for gauge theories was introduced by G. 't Hooft, Nucl. Phys. **B72**, 461 (1974); *ibid.* **B75**, 461 (1974).

8. The large N argument for chiral symmetry breaking is from S. Coleman and E. Witten, Phys. Rev. Lett. **45**, 100 (1980).

9. Anomaly matching conditions are given in G. 't Hooft, Cargèse Lectures, 1979, reprinted in G. 't Hooft, *Under the Spell of the Gauge Principle*, World Scientific Pub. Co. (1994). In this context, see also S. Coleman and B. Grossman, Nucl. Phys. **B203**, 205 (1982).

10. The Vafa-Witten theorem is in C. Vafa and E. Witten, Phys. Rev. Lett. **53**, 535 (1984); Nucl. Phys. **B234**, 173 (1984); Commun. Math. Phys. **95**, 257 (1984).

11. The suppression of glueball and meson decays, and baryons as solitons at large N, are shown in E. Witten, Nucl. Phys. **B160**, 57 (1979).

12. General references on older work on skyrmions are: T.H.R. Skyrme, Proc. Roy. Soc. **A260**, 127 (1961); Nucl. Phys. **31**, 556 (1962); J. Math. Phys. **12**, 1735 (1970); J.G. Williams, J. Math. Phys. **11**, 2611 (1970); N.K. Pak and H.Ch. Tze, Ann. Phys. **117**, 164 (1979).

13. The computation of the baryon number is from A.P. Balachandran, V.P. Nair, S.G. Rajeev and A. Stern, Phys. Rev. Lett. **49**, 1124 (1982); Phys. Rev. **D27**, 1153 (1983).

14. The analysis of the spin and flavor of skyrmions, as well as the proof of their fermionic nature, are due to E. Witten, Nucl. Phys. **B223**, 422, 433 (1983). See also G. Adkins, C. Nappi and E. Witten, Nucl. Phys. **B228**, 552 (1983). Our analysis follows A.P. Balachandran, F. Lizzi, V.G.J. Rodgers and A. Stern, Nucl. Phys. **B256**, 525 (1985).

15. For reviews on the soliton approach to baryons, see I. Zahed and G.E. Brown, Phys. Rep. **142**, 1 (1986); A.P. Balachandran, *Classical Topology and Quantum States*, World Scientific Pub. Co. (1991).

16. For further study of lattice gauge theory, the following books are useful: M. Creutz, *Quarks, Gluons and Lattices*, Cambridge University Press (1985); C. Itzykson and J-M. Drouffe, *Statistical Field Theory*, volumes 1 and 2, Cambridge University Press (1991); H.J. Rothe, *Lattice Gauge Theories: An Introduction*, World Scientific Pub. Co. (1998).

17. The theorem on fermion doubling is due to H.B. Nielsen and M. Ni-nomiya, Nucl. Phys. **B185**, 20 (1981); Erratum *ibid.* **B195**, 541 (1982); *ibid.* **B193**, 173 (1981); Phys. Lett. **B105**, 219 (1981). For a recent review which discusses new directions, such as the domain wall approach of D. Kaplan and the overlap formalism of H. Neuberger and R. Narayanan for chiral fermions on the lattice, see M. Creutz, Rev. Mod. Phys. **73**, 119 (2001).

20 Elements of Geometric Quantization

This is a special chapter on some aspects of geometric quantization. The first part of this discussion has overlap with Chapter 3 on canonical quantization. In Chapter 15, we considered some topological issues in quantization from a path integral point of view. The more canonical point of view of the same topological features is discussed here. Some examples from mechanics, as well as from field theory, are given. The example of the gauge theory configuration space has significant overlap with Chapter 16, but is included, and partly repeated, here to show how it fits in with the canonical framework.

20.1 General structures

1. Symplectic form, canonical transformations, and Poisson brackets

We shall first consider the formulation of theories in the symplectic language. The question of deriving this from an action formulation will be discussed later.

In the analytical formulation of classical physics, one starts with a phase space, i.e., a smooth even-dimensional manifold $\mathcal{M}$ endowed with a symplectic structure Ω. Ω is a differential two-form on $\mathcal{M}$ which is closed and nondegenerate. This means that $d\Omega = 0$, and further, for any vector field ξ on $\mathcal{M}$, if $i_\xi \Omega = 0$, then ξ must be zero. In local coordinates q^μ, on $\mathcal{M}$, we can write

$$\Omega = \frac{1}{2} \, \Omega_{\mu\nu} dq^\mu \wedge dq^\nu \tag{20.1}$$

The condition $d\Omega = 0$ becomes

$$
\begin{aligned}
d\Omega &= \frac{1}{2} \frac{\partial \Omega_{\mu\nu}}{\partial q^\alpha} dq^\alpha \wedge dq^\mu \wedge dq^\nu \\
&= \frac{1}{3!} \left[\frac{\partial \Omega_{\mu\nu}}{\partial q^\alpha} + \frac{\partial \Omega_{\alpha\mu}}{\partial q^\nu} + \frac{\partial \Omega_{\nu\alpha}}{\partial q^\mu} \right] dq^\alpha \wedge dq^\mu \wedge dq^\nu \\
&= 0
\end{aligned}
\tag{20.2}
$$

The interior contraction of Ω with a vector field $\xi = \xi^\mu(\partial/\partial q^\mu)$ is given by

$$i_\xi \Omega = \xi^\mu \Omega_{\mu\nu} dq^\nu \tag{20.3}$$

The vanishing of this is the condition $\xi^\mu \Omega_{\mu\nu} = 0$. Thus if Ω is degenerate, $\Omega_{\mu\nu}$, considered as a matrix, has a zero mode ξ^μ. Nondegeneracy of Ω is thus equivalent to the invertibility of $\Omega_{\mu\nu}$ as a matrix. When needed, we denote the inverse of $\Omega_{\mu\nu}$ by $\Omega^{\mu\nu}$, i.e.,

$$\Omega_{\mu\nu} \, \Omega^{\nu\alpha} = \delta^\alpha_\mu \tag{20.4}$$

(We will consider Ω's which are nondegenerate. If Ω has zero modes, one has to eliminate them by constraining the variables, or equivalently, one has to project Ω to a smaller space where there is no zero mode and use this smaller space for setting up the quantum theory. Such a situation occurs in gauge theories. In general, zero modes of Ω indicate the existence of gauge symmetries.)

With the structure Ω defined on it, $\mathcal{M}$ is a symplectic manifold.

Since Ω is closed, at least locally we can write

$$\Omega = d\mathcal{A} \tag{20.5}$$

The one-form so defined is called the canonical one-form or symplectic potential. There is an ambiguity in the definition of $\mathcal{A}$, since $\mathcal{A}$ and $\mathcal{A} + d\Lambda$ will give the same Ω for any function Λ on $\mathcal{M}$. As we shall see shortly, this corresponds to the freedom of canonical transformations.

There are two types of features associated with the topology of the phase space which are apparent at this stage. If the phase space $\mathcal{M}$ has nontrivial second cohomology, i.e., if $\mathcal{H}^2(\mathcal{M}) \neq 0$, then there are possible choices for Ω for which there is no globally defined potential. The action, as we shall see later, is related to the integral of $\mathcal{A}$, so if Ω belongs to a nontrivial cohomology class of $\mathcal{M}$, then the definition of the action requires auxiliary variables or dimensions. Such cases do occur in physics and correspond to the Wess-Zumino terms discussed in Chapter 17. They are intrinsically related to anomalies and also to central (and other) extensions of the algebra of observables.

Even when $\mathcal{H}^2(\mathcal{M}) = 0$, there can be topological problems in defining $\mathcal{A}$. If $\mathcal{H}^1(\mathcal{M}) \neq 0$, then there can be several choices for $\mathcal{A}$ which differ by elements of $\mathcal{H}^1(\mathcal{M})$. There are inequivalent $\mathcal{A}$'s for the same Ω. One can consider the integral of $\mathcal{A}$ around closed noncontractible curves on $\mathcal{M}$. The values of these integrals or holonomies will be important in the quantum theory as vacuum angles. The standard θ-vacuum of nonabelian gauge theories is an example. We take up these topological issues in more detail later.

Given the above-defined geometrical structure, transformations which preserve Ω are evidently special; these are called canonical transformations. In other words, a canonical transformation is a diffeomorphism of $\mathcal{M}$ which preserves Ω. Infinitesimally, canonical transformations are generated by vector fields ξ such that $L_\xi \Omega = 0$, where L_ξ denotes the Lie derivative with respect to ξ. This gives

$$L_\xi \Omega \equiv (d\, i_\xi + i_\xi\, d)\, \Omega$$
$$= d\, (i_\xi \Omega)$$
$$= 0 \tag{20.6}$$

where we have used the closure of Ω. For canonical transformations, $i_\xi \Omega$ is closed. If the first cohomology of $\mathcal{M}$ is trivial, we can write

$$i_\xi \Omega = -df \tag{20.7}$$

for some function f on $\mathcal{M}$. In other words, to every infinitesimal canonical transformation, we can associate a function on $\mathcal{M}$. If $\mathcal{H}^1(\mathcal{M}) \neq 0$, then there is the possibility that for some transformations ξ, the corresponding $i_\xi \Omega$ is a nontrivial element of $\mathcal{H}^1(\mathcal{M})$ and hence there is no globally defined function f for this transformation. As mentioned before this is related to the possibility of vacuum angles in the quantum theory. For the moment, we shall consider the case $\mathcal{H}^1(\mathcal{M}) = 0$. Notice that for every function f we can always associate a vector field by the correspondence

$$\xi^\mu = \Omega^{\mu\nu} \partial_\nu f \tag{20.8}$$

Thus when $\mathcal{H}^1(\mathcal{M}) = 0$ there is a one-to-one mapping between functions on $\mathcal{M}$ and vector fields corresponding to infinitesimal canonical transformations. A vector field corresponding to an infinitesimal canonical transformation is often referred to as a Hamiltonian vector field. The function f defined by (20.7) is called the generating function for the canonical transformation corresponding to the vector field.

Let ξ, η be two Hamiltonian vector fields, so that $L_\xi \Omega = L_\eta \Omega = 0$ and let their generating functions be f and g, respectively. Since $L_\xi L_\eta - L_\eta L_\xi = L_{[\xi,\eta]}$, the Lie bracket of ξ and η so defined is also a Hamiltonian vector field. The Lie bracket of ξ and η is given in local coordinates by

$$[\xi, \eta]^\mu = \xi^\nu \partial_\nu \eta^\mu - \eta^\nu \partial_\nu \xi^\mu \tag{20.9}$$

We must therefore have a function corresponding to $[\xi, \eta]$. This is called the Poisson bracket of g and f and is denoted by $\{g, f\}$. (There is a minus sign in this correspondence; $\xi \leftrightarrow f$, $\eta \leftrightarrow g$ and $[\xi, \eta] \leftrightarrow \{g, f\}$.) We define the Poisson bracket as

$$\{f, g\} = i_\xi i_\eta \Omega = \eta^\mu \xi^\nu \Omega_{\mu\nu}$$
$$= -i_\xi dg = i_\eta df$$
$$= \Omega^{\mu\nu} \partial_\mu f \partial_\nu g \tag{20.10}$$

Because of the antisymmetry of $\Omega_{\mu\nu}$ we have the property

$$\{f, g\} = -\{g, f\} \tag{20.11}$$

From the definition of the Poisson bracket, we can write, using local coordinates,

$$
\begin{aligned}
2\,\partial_\alpha\{f,g\} &= \partial_\alpha(\eta\cdot\partial f - \xi\cdot\partial g)\\
&= \partial_\alpha\eta^\mu\partial_\mu f + \eta^\mu(\partial_\mu\partial_\alpha f) - \partial_\alpha\xi^\mu\partial_\mu g - \xi^\mu(\partial_\mu\partial_\alpha g)\\
&= \partial_\alpha\eta^\mu\partial_\mu f - \partial_\alpha\xi^\mu\partial_\mu g + \eta\cdot\partial(\xi^\mu\Omega_{\alpha\mu}) - \xi\cdot\partial(\eta^\mu\Omega_{\alpha\mu})\\
&= \partial_\alpha\eta^\mu\partial_\mu f - \partial_\alpha\xi^\mu\partial_\mu g + (\xi\cdot\partial\eta - \eta\cdot\partial\xi)^\mu\Omega_{\mu\alpha}\\
&\quad + \eta^\mu\xi^\nu(\partial_\mu\Omega_{\alpha\nu} + \partial_\nu\Omega_{\mu\alpha})\\
&= [\xi,\eta]^\mu\Omega_{\mu\alpha} + \partial_\alpha(\eta^\mu\xi^\nu\Omega_{\mu\nu}) + \eta^\mu\xi^\nu(\partial_\mu\Omega_{\alpha\nu} + \partial_\nu\Omega_{\mu\alpha} + \partial_\alpha\Omega_{\nu\mu})\\
&= [\xi,\eta]^\mu\Omega_{\mu\alpha} + \partial_\alpha\{f,g\} + \eta^\mu\xi^\nu(\partial_\mu\Omega_{\alpha\nu} + \partial_\nu\Omega_{\mu\alpha} + \partial_\alpha\Omega_{\nu\mu})
\end{aligned}
$$
$$\tag{20.12}$$

In local coordinates, the closure of Ω is the statement $\partial_\mu\Omega_{\alpha\nu} + \partial_\nu\Omega_{\mu\alpha} + \partial_\alpha\Omega_{\nu\mu} = 0$. We then see that

$$
-d\{g,f\} = i_{[\xi,\eta]}\,\Omega \tag{20.13}
$$

which shows the correspondence stated earlier.

Consider the change in a function F due to a canonical transformation generated by a Hamiltonian vector field ξ corresponding to the function f. This is given by the Lie derivative of F with respect to ξ. Thus

$$
\delta F = \xi^\mu\partial_\mu F = \{F,f\} \tag{20.14}
$$

The transformation is given by the Poisson bracket of F with the generating function f corresponding to ξ.

An important property of the Poisson bracket is the Jacobi identity for any three functions f, g, h,

$$
\{f,\{g,h\}\} + \{h,\{f,g\}\} + \{g,\{h,f\}\} = 0 \tag{20.15}
$$

This can be verified by direct computation from the definition of the Poisson bracket. In fact, if ξ, η, ρ are the Hamiltonian vector fields corresponding to the functions f, g, h, then

$$
\{f,\{g,h\}\} + \{h,\{f,g\}\} + \{g,\{h,f\}\} = -i_\xi i_\eta i_\rho d\Omega \tag{20.16}
$$

and so the Jacobi identity follows from the closure of Ω. This result is also equivalent to (20.13, 20.14). Acting on a function F

$$
\begin{aligned}
(L_\xi L_\eta - L_\eta L_\xi)F &= \{\{F,g\},f\} - \{\{F,f\},g\}\\
&= -\{\{g,f\},F\}\\
&= L_{[\xi,\eta]}\,F
\end{aligned}
$$
$$\tag{20.17}$$

where we have used the Jacobi identity in the second step.

The change of the symplectic potential $\mathcal{A}$ under an infinitesimal canonical transformation can be worked out as

$$\delta\mathcal{A} = L_\xi\mathcal{A}$$
$$= d(i_\xi\mathcal{A} - f) \tag{20.18}$$

where we used the definition $L_\xi = (d\ i_\xi\ +\ i_\xi\ d)$ and equation (20.7). Thus under a canonical transformation $\mathcal{A} \to \mathcal{A} + d\Lambda$, $\Lambda = i_\xi\mathcal{A} - f$. Evidently $d\mathcal{A} = \Omega$ is unchanged under such a transformation. This suggests that we may think of $\mathcal{A}$ as a $U(1)$ gauge potential and Ω as the corresponding field strength. The transformation $\mathcal{A} \to \mathcal{A} + d\Lambda$ is a gauge transformation.

Finally we note that the symplectic two-form defines a volume form on the phase space $\mathcal{M}$ by

$$d\sigma(M) = c\frac{\Omega \wedge \Omega \wedge \cdots \wedge \Omega}{(2\pi)^n}$$
$$= c\sqrt{\det\left(\frac{\Omega}{2\pi}\right)}\ d^{2n}q \tag{20.19}$$

where we take the n-fold product of Ω's for a $2n$-dimensional phase space. (c is a constant which is undetermined at this stage.) If the dimension of the phase space is infinite, then a suitable regularized form of the determinant of $\Omega_{\mu\nu}$ has to be used. The volume measure defined by equation (20.19) is called the Liouville measure.

Finally, notice from the last line of equation (20.10) that for the phase space coordinates we have $\{q^\mu, q^\nu\} = \Omega^{\mu\nu}$. This is often interpreted as saying that the basic Poisson brackets are the inverse of the symplectic structure.

2. Darboux's theorem

A useful result concerning the symplectic form is Darboux's theorem which states that in the neighborhood of a point on the phase space it is possible to choose coordinates p_i, x_i, $i = 1, 2, \cdots, n$, (which are functions of the coordinates q^μ we start with) such that the symplectic two-form is

$$\Omega = dp_i \wedge dx_i \tag{20.20}$$

Evidently from the above equation, we see that the Poisson brackets in terms of this set of coordinates are

$$\{x_i, x_j\} = 0$$
$$\{x_i, p_j\} = \delta_{ij}$$
$$\{p_i, p_j\} = 0 \tag{20.21}$$

This is the standard separation of the phase space coordinates into momenta and configuration space coordinates; Darboux's theorem is clearly an important result. The standard proof of the theorem is by induction; an outline is as follows. (This is very elegantly discussed in Arnold's book.) Let B denote the point on phase space in the neighbourhood of which we want to reduce Ω to the Darboux form. As the first coordinate p_1, we can take any nonconstant function of the coordinates q such that its differential dp_1 is not zero at B. One can assume, without loss of generality, that $p_1 = 0$ at the point B. Associated to $p_1(q)$ there is a Hamiltonian vector field

$$P_1^\mu = \Omega^{\mu\nu} \frac{\partial p_1}{\partial q^\nu} \qquad (20.22)$$

We now choose a $(2n-1)$-dimensional surface Σ transverse to the vector P_1 and passing through the point B. The equation

$$\frac{dq^\mu}{d\tau} = -\Omega^{\mu\nu} \frac{\partial p_1}{\partial q^\nu} \qquad (20.23)$$

defines the flow due to the vector field P_1; these flow lines intersect the surface Σ transversally. Consider any point q near Σ but not necessarily on it. We can solve (20.23) with q as the initial point and choose the direction such that the motion is toward the surface Σ. At some value τ determined by the initial point q, this motion arrives at Σ. This particular value of τ, viewed as a function of the initial point q, we denote by x_1. Equation (20.23) shows that for this function $x_1(q)$, we have

$$\{x_1, p_1\} = 1 \qquad (20.24)$$

(Since we are considering τ as a function of the initial value q, there is an additional minus sign in differentiations. q^μ on the left-hand side of (20.23) is the moving point, the derivative $\partial \tau / \partial q^\mu_{initial}$ is what we want for the Poisson bracket. This eliminates the minus sign in (20.23).) p_1, x_1 give the first canonical pair of coordinates. Notice that $x_1 = 0$ for points on Σ.

Consider the surface Σ^* defined by $p_1 = 0$, $x_1 = 0$. The differentials dp_1, dx_1 are linearly independent since their Poisson bracket is nonzero. In fact if X_1 is the vector field corresponding to x_1, we have $i_{P_1} i_{X_1} \Omega = -1$. Let ξ be any vector field which induces a flow along (tangential to) Σ^*. Such a vector cannot change the value of p_1, x_1, so we have $i_\xi dp_1 = 0$, $i_\xi dx_1 = 0$. Equation (20.24) and this result show that we can write

$$\Omega = \Omega^* + dp_1 \wedge dx_1 \qquad (20.25)$$

Ω^* does not involve differentials dp_1 or dx_1. Evidently, $d\Omega^* = 0$. Further the contraction with Ω^* of any vector tangential to Σ^* is the same as its contraction with Ω. Thus Ω^* must be invertible for vectors tangential to Σ^*. Ω^* therefore defines a symplectic two-form on Σ^*, which is $(2n-2)$-dimensional. The problem is reduced to the question of choosing Darboux coordinates on the lower dimensional space. We can now proceed in a similar manner, starting with Σ^* and Ω^*, constructing another canonical pair p_2, x_2, obtaining a reduction to a $(2n-4)$-dimensional subspace, and so on inductively, to prove the theorem.

20.2 Classical dynamics

The time-evolution of any quantity is a particular canonical transformation generated by a function H called the Hamiltonian; this is the essence of the Hamiltonian formulation of dynamics. If F is any function on $\mathcal{M}$, we then have

$$\frac{\partial F}{\partial t} = \{F, H\} \tag{20.26}$$

Specifically for the local coordinates on $\mathcal{M}$ we can write

$$\frac{\partial q^{\mu}}{\partial t} = \{q^{\mu}, H\}$$
$$= \Omega^{\mu\nu} \frac{\partial H}{\partial q^{\nu}} \tag{20.27}$$

Since Ω is invertible, we can also write this equation as

$$\Omega_{\mu\nu} \frac{\partial q^{\nu}}{\partial t} = \frac{\partial H}{\partial q^{\mu}} \tag{20.28}$$

At this point we can relate this to an action and a variational principle. We define the action as

$$S = \int_{t_i}^{t_f} dt \left(A_{\mu} \frac{dq^{\mu}}{dt} - H \right) \tag{20.29}$$

where $q^{\mu}(t)$ gives a path on $\mathcal{M}$. Under a general variation of the path $q^{\mu}(t) \to q^{\mu}(t) + \xi^{\mu}(t)$, the action changes by

$$\delta S = \int dt \left(\frac{\partial A_{\nu}}{\partial q^{\mu}} \frac{dq^{\nu}}{dt} \xi^{\mu} + A_{\mu} \frac{d\xi^{\mu}}{dt} - \frac{\partial H}{\partial q^{\mu}} \xi^{\mu} \right)$$
$$= A_{\mu} \xi^{\mu} \Big]_{t_i}^{t_f} + \int dt \left(\Omega_{\mu\nu} \frac{dq^{\nu}}{dt} - \frac{\partial H}{\partial q^{\mu}} \right) \xi^{\mu} \tag{20.30}$$

The variational principle says that the equations of motion are given by the extremization of the action, i.e., by $\delta S = 0$, for resricted set of variations with the boundary data (initial and final end point data) fixed. From the above variation, we see that this gives the Hamiltonian equations of motion (20.28). There is a slight catch in this argument because q^{μ} are phase space coordinates and obey first-order equations of motion. So we can only specify the initial value of q^{μ}. However, the Darboux theorem tells us that one can choose coordinates on neighborhoods of $\mathcal{M}$ such that the canonical one-form $\mathcal{A}$ is of the form $p_i dx^i$ for each neighborhood. Therefore, instead of specifying initial data for all q^{μ}, we can specify initial and final data for the x^i's. The ξ^{μ} in the boundary term is thus just δx^i. Since the boundary values are kept fixed in the variational principle $\delta S = 0$, we may set $\delta x^i = 0$ at both boundaries and the equations of motion are indeed just (20.28).

We have shown how to define the action if Ω is given. However, going back to the general variations, notice that the boundary term is just the canonical one-form contracted with ξ^μ. Thus if we start from the action as the given quantity, we can identify the canonical one-form and hence Ω from the boundary term which arises in a general variation. In fact

$$\delta S = i_\xi \mathcal{A}(t_f) - i_\xi \mathcal{A}(t_i) + \int dt \left(\Omega_{\mu\nu} \frac{dq^\nu}{dt} - \frac{\partial H}{\partial q^\mu} \right) \xi^\mu \qquad (20.31)$$

We have used this equation in Chapter 3 to identify the canonical one-form and carry out the canonical quantization.

20.3 Geometric quantization

In the quantum theory, the algebra of observables is an operator algebra with a Hilbert space which provides an irreducible unitary representation of this algebra. The allowed transformations of variables are then unitary transformations. There are thus two essential points to quantization: 1) a correspondence between canonical transformations and unitary transformations and 2) ensuring that the representation of unitary transformations on the Hilbert space is irreducible. Since functions on phase space generate canonical transformations and hermitian operators generate unitary transformations, we get a correspondence between functions on phase space and operators on the Hilbert space. The algebra of Poisson brackets will be replaced by the algebra of commutation rules. The irreducibility leads to the necessity of choosing a polarization for the wave functions.

Before considering different aspects of the operator approach, we shall start with the notion of the wave function. In the geometric approach to the wave function, the first step is the so-called prequantum line bundle. This is a complex line bundle on the phase space with curvature Ω. Sections of this line bundle form the prequantum Hilbert space. In more practical terms, this means that we consider complex functions $\Psi(q)$ on open neighbourhoods in $\mathcal{M}$, with a suitably defined notion of covariant derivatives such that the commutator of two covariant derivatives gives Ω. Since Ω is closed, at least locally we can write $\Omega = d\mathcal{A}$, where $\mathcal{A}$ is the symplectic potential. Under a canonical transformation, Ω does not change, but the symplectic potential transforms as $\mathcal{A} \to \mathcal{A}' = \mathcal{A} + d\Lambda$, as we have seen in (20.18). In other words, $\mathcal{A}$ undergoes a $U(1)$ gauge transformation. The statement that Ψ's are sections of a line bundle means that locally they are complex functions which transform as

$$\Psi \to \Psi' = \exp(i\Lambda)\,\Psi \qquad (20.32)$$

As mentioned before, one may think of $\mathcal{A}$ as a $U(1)$ gauge potential; Ψ's are then like matter fields. The above equation is equivalent to the requirement of canonical transformations being implemented as unitary transformations.

The transition rules for the Ψ's from one patch on $\mathcal{M}$ to another are likewise given by exponentiating the transition function for $\mathcal{A}$. The functions Ψ's so defined, which are also square-integrable, form the prequantum Hilbert space; the inner product is given by

$$(1|2) \;=\; \int d\sigma(\mathcal{M})\, \Psi_1^* \,\Psi_2 \tag{20.33}$$

where $d\sigma(\mathcal{M})$ is the Liouville measure on the phase space, defined by Ω.

One can, at the prequantum level, introduce operators corresponding to various functions on the phase space. A function $f(q)$ on the phase space generates a canonical transformation which leads to the change $\Lambda = i_\xi \mathcal{A} - f$ in the symplectic potential. The corresponding change in Ψ is thus

$$
\begin{aligned}
\delta\Psi &= \xi^\mu \partial_\mu \Psi \;-\; i(\, i_\xi \mathcal{A} - f)\Psi \\
&= \xi^\mu \left(\partial_\mu - i\mathcal{A}_\mu\right)\Psi + if\Psi \\
&= \left(\xi^\mu \mathcal{D}_\mu + if\right)\,\Psi
\end{aligned}
\tag{20.34}
$$

where the first term gives the change in Ψ considered as a function and the second term compensates for the change of $\mathcal{A}$. In the last expression $\mathcal{D}_\mu$ are covariant derivatives $\partial_\mu \;-\; i\mathcal{A}_\mu$; these are the appropriate derivatives to consider in view of the $U(1)$ gauge symmetry on the wave functions. The gauge potential to be used is indeed $\mathcal{A}$, so that the curvature is Ω. Based on (20.34), we define the prequantum operator corresponding to $f(q)$ by

$$
\begin{aligned}
\mathcal{P}(f) &= -i(\xi \cdot \mathcal{D} \;+\; if) \\
&= -i\xi \cdot \mathcal{D} \;+\; f
\end{aligned}
\tag{20.35}
$$

We have seen that if the Hamiltonian vector fields for f, g, are ξ and η, respectively, then the vector field corresponding to the Poisson bracket $\{f, g\}$ is $-[\xi, \eta]$. From the definition of the prequantum operator above, we then find

$$
\begin{aligned}
[\mathcal{P}(f), \mathcal{P}(g)] &= [-i\xi \cdot \mathcal{D} + f, -i\eta \cdot \mathcal{D} + g] \\
&= -[\xi^\mu \mathcal{D}_\mu, \eta^\nu \mathcal{D}_\nu] - i\xi^\mu [\mathcal{D}_\mu, g] + i\eta^\mu [\mathcal{D}_\mu, f] \\
&= i\xi^\mu \eta^\nu \Omega_{\mu\nu} - (\xi^\mu \partial_\mu \eta^\nu)\mathcal{D}_\nu + (\eta^\mu \partial_\mu \xi^\nu)\mathcal{D}_\nu - i\xi^\mu \partial_\mu g + i\eta^\mu \partial_\mu f \\
&= i\left(-\xi^\mu \eta^\nu \Omega_{\mu\nu} + i[\xi, \eta] \cdot \mathcal{D}\right) \\
&= i\left(-i\left(i_{[\eta,\xi]}\mathcal{D}\right) + \{f, g\}\right) \\
&= i\mathcal{P}(\{f, g\})
\end{aligned}
\tag{20.36}
$$

In other words, the prequantum operators form a representation of the Poisson bracket algebra of functions on phase space.

The prequantum wave functions Ψ depend on all phase space variables. The representation of the Poisson bracket algebra on such wave functions, given by the prequantum operators, is reducible. A simple example is sufficient to illustrate this. Consider a point particle in one dimension, with the symplectic two-form $\Omega = dp \wedge dx$. We can choose $\mathcal{A} = pdx$. The vector fields corresponding to x and p are $\xi_x = -\partial/\partial p$ and $\xi_p = \partial/\partial x$. The corresponding prequantum operators are

$$\mathcal{P}(x) = i\frac{\partial}{\partial p} + x$$

$$\mathcal{P}(p) = -i\frac{\partial}{\partial x} \tag{20.37}$$

which obey the commutation rule

$$[\mathcal{P}(x), \mathcal{P}(p)] = i \tag{20.38}$$

We clearly have a representation of the algebra of $\mathcal{P}(x)$, $\mathcal{P}(p)$ in terms of prequantum functions $\Psi(x, p)$. But this is reducible. For if we consider the subset of functions on the phase space which are independent of p, namely, those which obey the condition

$$\frac{\partial \Psi}{\partial p} = 0, \tag{20.39}$$

then the prequantum operators become

$$\mathcal{P}(x) = x$$

$$\mathcal{P}(p) = -i\frac{\partial}{\partial x} \tag{20.40}$$

which obey the same algebra (20.38). Thus we are able to obtain a representation of the algebra of observables on the smaller space of Ψ's obeying (20.39), showing that the previous representation (20.37) is reducible.

In order to obtain an irreducible representation, one has to impose subsidiary conditions which restrict the dependence of the prequantum wave functions to half the number of phase space variables. This is the choice of polarization and generally leads to an irreducible representation of the Poisson algebra. For the implementation of this we need to choose a set of n vector fields P_i^μ, $i = 1, 2...n$, so that

$$\Omega_{\mu\nu} P_i^\mu P_j^\nu = 0 \tag{20.41}$$

and impose the condition

$$P_i^\mu \, \mathcal{D}_\mu \, \Psi = 0 \tag{20.42}$$

The vectors P_i^μ define the polarization. The wave functions so restricted are the true wave functions.

The next step is to define an inner product, and restrict to square-integrable functions, so that these wave functions form a Hilbert space. While the volume element and the notion of inner product can be defined on the phase space in terms of Ω, generally, there is no natural choice of inner product once we impose the restriction of polarization. However, there is one case where there is a natural inner product on the Hilbert space. This happens when the phase space is also Kähler and Ω is the Kähler form or some multiple thereof. In this case we can introduce local complex coordinates and write

$$\Omega = \Omega_{a\bar{a}} dx^a \wedge dx^{\bar{a}} \tag{20.43}$$

$a, \bar{a} = 1, 2...n$. The covariant derivatives are of the form

$$\mathcal{D}_a = \partial_a - i\mathcal{A}_a$$
$$\mathcal{D}_{\bar{a}} = \partial_{\bar{a}} - i\mathcal{A}_{\bar{a}} \tag{20.44}$$

For a Kähler manifold, there is a Kähler potential K defined by

$$\mathcal{A}_a = -\frac{i}{2}\partial_a K$$
$$\mathcal{A}_{\bar{a}} = \frac{i}{2}\partial_{\bar{a}} K \tag{20.45}$$

In this case, one can choose the holomorphic polarization

$$\mathcal{D}_{\bar{a}}\Psi = (\partial_{\bar{a}} + \frac{1}{2}\partial_{\bar{a}} K)\Psi = 0 \tag{20.46}$$

which gives

$$\Psi = \exp(-\tfrac{1}{2}K) \ F \tag{20.47}$$

where F is a holomorphic function on $\mathcal{M}$. The wave functions are thus holomorphic, apart from the prefactor involving the Kähler potential. In this case the inner product of the prequantum Hilbert space can then be retained, up to a constant of proportionality, as the inner product of the Hilbert space; specifically

$$\langle 1|2 \rangle = \int d\sigma(\mathcal{M}) \ e^{-K} \ F_1^* F_2 \tag{20.48}$$

Almost all the cases of interest to us are of this type.

Once the polarized wave functions are defined, the idea is to represent observables as linear operators on the wave functions as given by the prequantum differential operators. Let ξ be the Hamiltonian vector field corresponding to a function $f(q)$. If the commutator of ξ with any polarization vector field P_i is proportional to P_i itself, i.e., $[\xi, P_i] = C_i^j P_j$ for some functions C_i^j, then, evidently, ξ does not change the polarization. $\xi\Psi$ obeys the same polarization condition as Ψ. In this case the operator corresponding to $f(q)$ is given by $\mathcal{P}(f)$. For operators which do not preserve the polarization, the situation is more involved. Since operators corresponding to observables are generators of unitary transformations, we must construct directly infinitesimal unitary transformations whose classical limit is the required canonical transformation and identify the quantum operator from the result. This will require, in general, a pairing between wave functions obeying different polarization conditions.

20.4 Topological features of quantization

We now turn to some of the topological features of phase space and their effects on quantization. As we have mentioned before two of the key topological problems have to do with the first and second cohomology of the phase space.

1. The case of nontrivial $\mathcal{H}^1(\mathcal{M}, \mathbf{R})$

Consider first the case of $\mathcal{H}^1(\mathcal{M}, \mathbf{R}) \neq 0$. In this case for a given symplectic two-form Ω, we can have different symplectic potentials. $\mathcal{A}$ and $\mathcal{A} + A$ lead to the same Ω if A is closed, i.e., if $dA = 0$. If A is exact so that $A = dh$ for some globally defined function h on $\mathcal{M}$, then the function h is a canonical transformation, physical results are unchanged, and so this is equivalent to $A = 0$ upon carrying out a canonical transformation. If A is closed but not exact, which is to say that if it is a nontrivial element of the cohomology $\mathcal{H}^1(\mathcal{M}, \mathbf{R})$, then we cannot get rid of it by a canonical transformation. Locally we can write $A = df$ for some f, but f will not be globally defined on M. Classical dynamics, which is defined by Ω as in the equations of motion (20.27), will not be affected by this ambiguity in the choice of the symplectic potential. In the quantum theory such A's do make a difference. We see this immediately in terms of the action. The action for a path C, parametrized as $q^\mu(t)$ from a point a to a point b is

$$S = \int dt \left(\mathcal{A}_\mu \frac{dq^\mu}{dt} - H \right) + \int_a^b A_\mu dq^\mu \tag{20.49}$$

The action depends on the path but the contribution from A is topological. If we change the path slightly from C to C' with the end points fixed, we find, using Stokes' theorem,

$$\int_C A - \int_{C'} A = \oint_{C-C'} A$$
$$= \int_\Sigma dA$$
$$= 0 \tag{20.50}$$

where $C - C'$ is the path where we go from a to b along C and back from b to a along C'. (Since we are coming back the orientation is reversed, hence, the minus sign.) Σ is a surface in $\mathcal{M}$ with $C - C'$ as the boundary. The above result shows that the contribution from A is invariant under small changes of the path. (This is not true for the other terms in the action.) In particular, the value of the integral is zero for closed paths so long as they are contractible; for then we can make a sequence of small deformations of the path (which do not change the value) and eventually contract the path to zero. If there are noncontractible loops, then there can be nontrivial contributions. If $\mathcal{H}^1(\mathcal{M}, \mathbf{R}) \neq 0$, then there are noncontractible loops. In the quantum theory, it is $e^{i\mathcal{S}}$ which is important, so we need $e^{i \int A}$. Assume for simplicity that $\mathcal{H}^1(\mathcal{M}, \mathbf{R}) = \mathbf{Z}$ so that there is only one topologically distinct noncontractible loop apart from multiple traversals of the same. Let $A = \theta\alpha$, where θ is a constant and α is normalized to unity along the noncontractible loop for going around once. For all paths which include n traversals of the loop, we find

$$\exp\left(i \oint A\right) = \exp\left(i\theta \oint \alpha\right) = \exp\left(i\theta n\right) \tag{20.51}$$

Notice that a shift $\theta \to \theta + 2\pi$ does not change this value, so that we may restrict θ to be in the interval zero to 2π. Putting this back into the action (20.49), we see that, considered as a function on the set of all paths, the action has an extra parameter θ. Thus the ambiguity in the choice of the symplectic potential due to $\mathcal{H}^1(\mathcal{M}, \mathbf{R}) \neq 0$ leads to an extra parameter θ which is needed to characterize the quantum theory completely. If $\mathcal{H}^1(\mathcal{M}, \mathbf{R})$ is not just $\mathbf{Z}$ and there are more distinct paths possible, then there can be more such parameters.

It is now easy to see these results in terms of wave functions. The relevant covariant derivatives are of the form $\mathcal{D}_\mu \Psi = (\partial_\mu - i\mathcal{A}_\mu - iA_\mu)\Psi$. We can write

$$\Psi(q) = \exp\left(i \int_a^q A\right) \ \Phi(q) \tag{20.52}$$

where the lower limit of the integral is some fixed point a. By using this in the covariant derivative, we see that A is removed from $\mathcal{D}_\mu$ in terms of action on Φ. This is like a canonical transformation, except that the relevant transformation $\exp\left(i \int_a^q A\right)$ is not single valued. As we go around a closed noncontractible curve, it can give a phase $e^{i\theta}$. Since Ψ is single valued, this means that Φ must have a compensating phase factor; Φ is not single valued but must give a specific phase labeled by θ. Thus we can get rid of A from the covariant derivatives and hence the various operator formulae, but diagonalizing the Hamiltonian on such Φ's can give results which depend on the angle θ.

The θ-parameter in a nonabelian gauge theory is an example of this kind of topological feature. The description of anyons or particles of fractional statistics in two spatial dimensions is another example.

2. The case of nontrivial $\mathcal{H}^2(\mathcal{M}, \mathbf{R})$

We now turn to the second topological feature, the case of $\mathcal{H}^2(\mathcal{M}, \mathbf{R})$ $\neq 0$. In this case there are closed two-forms on $\mathcal{M}$ which are not exact. Correspondingly, there are closed two-surfaces which are not the boundaries of any three-dimensional region, i.e., noncontractible closed two-surfaces. In general, elements of $\mathcal{H}^2(\mathcal{M}, \mathbf{R})$ integrated over such noncontractible two-surfaces are not zero. If Ω is some nontrivial element of $\mathcal{H}^2(\mathcal{M}, \mathbf{R})$, then the symplectic potential cannot be globally defined. We see this easily as follows. Consider the integral of Ω over a noncontractible closed two-surface Σ,

$$I(\Sigma) = \int_{\Sigma} \Omega \qquad (20.53)$$

If Σ' is a small deformation of Σ, then

$$I(\Sigma) - I(\Sigma') = \int_{\Sigma - \Sigma'} \Omega = \int_{V} d\Omega = 0 \qquad (20.54)$$

where V is a three-dimensional volume with the two surfaces Σ and Σ' as the boundary. Thus the integral of Ω is a topological invariant, invariant under small deformations of the surface on which it is integrated. If we can write Ω as $d\mathcal{A}$ for some $\mathcal{A}$ globally defined on Σ then clearly $I(\Sigma)$ is zero by Stokes' theorem. Thus if $I(\Sigma)$ is nonzero, then $\mathcal{A}$ cannot be globally defined on Σ. We have to use different functions to represent $\mathcal{A}$ in different coordinate patches, then have transition functions relating the $\mathcal{A}$'s in overlap regions. Even though we may have different definitions of $\mathcal{A}$ in an overlap region corresponding to the different patches which are overlapping, Ω is the same, and so the transition functions must be canonical transformations. As an example, consider a closed noncontractible two-sphere, or any smooth deformation of it, which is a subspace of M. We can cover it with two coordinate patches corresponding to the two hemispheres, denoted N and S as usual. The symplectic potential is represented by $\mathcal{A}_N$ and $\mathcal{A}_S$, respectively. On the equatorial overlap region, they are connected by

$$\mathcal{A}_N = \mathcal{A}_S + d\Lambda \qquad (20.55)$$

where Λ is a function defined on the overlap region. It gives the canonical transformation between the two $\mathcal{A}$'s.

Since $\mathcal{A}$ is what is used in setting up the quantum theory and since, in particular, the canonical transformations are represented as unitary transformations on the wave functions, we see that we must have a Ψ_N for the patch N and a Ψ_S for the patch S. On the equator they must be related by the canonical transformation, which from (20.32), is given as

$$\Psi_N = \exp(i\Lambda) \ \Psi_S \qquad (20.56)$$

We now consider the integral of $d\Lambda$ over the equator E, which is a closed curve being the boundary of either N or S. From (20.55) this is given as

$$\Delta \Lambda = \oint_E d\Lambda = \int_E \mathcal{A}_N - \int_E \mathcal{A}_S$$
$$= \int_{\partial N} \mathcal{A}_N + \int_{\partial S} \mathcal{A}_S$$
$$= \int_N \Omega + \int_S \Omega$$
$$= \int_\Sigma \Omega \tag{20.57}$$

In the second step, we reverse the sign for the S-term because E considered as the boundary of S has the opposite orientation compared to itself considered as the boundary of N. The above equation shows that the change of Λ as we go around the equator once, namely, $\Delta \Lambda$, is nonzero if $I(\Sigma)$ is nonzero; Λ is not single valued on the equator. But the wave function must be single valued. From (20.56), we see that this can be achieved if $\exp(i\Delta \Lambda) = 1$ or if $\Delta \Lambda = 2\pi n$ for some integer n. Combining with (20.57), we can state that single-valuedness of wave functions in the quantum theory requires that

$$\int_\Sigma \Omega = 2\pi n \tag{20.58}$$

The integral of the symplectic two-form on closed noncontractible two-surfaces must be quantized as 2π times an integer. We have given the argument for surfaces which are deformations of a two-sphere, but a similar argument can be made for noncontractible two-surfaces of different topology as well. The result (20.58) is quite general.

The typical example of this kind of topological feature is the motion of a charged particle in the field of a magnetic monopole. The condition (20.58) is then the Dirac quantization condition. The Wess-Zumino terms occuring in many field theories are another example.

20.5 A brief summary of quantization

In summary, the key features of the quantization of a system using the holomorphic polarization are the following:
1) We need a phase space which is also Kähler; the symplectic two-form being a multiple of the Kähler form.
2) The polarization condition is chosen as $\mathcal{D}_{\bar{a}} \Psi = 0$.
3) The inner product of the prequantum Hilbert space, which is essentially square integrability on the phase space, is retained as the inner product on the true Hilbert space in the holomorphic polarization.
4) The operator corresponding to an observable $f(q)$ which preserves the chosen polarization is given by the prequantum operator $\mathcal{P}(f)$ acting on true (polarized) wave functions.

5) For observables which do not preserve the polarization, one has to construct infinitesimal unitary transformations whose classical limits are the required canonical transformations.

6) If the phase space $\mathcal{M}$ has noncontractible two-surfaces, then the integral of Ω over any of these surfaces must be quantized in units of 2π.

7) If $\mathcal{H}^1(\mathcal{M}, \mathbf{R})$ is not zero, then there are inequivalent $\mathcal{A}$'s for the same Ω and we need extra parameters to specify the quantum theory completely.

20.6 Examples

20.6.1 Coherent states

For a one-dimensional quantum system, $\Omega = dp \wedge dx = idz \wedge d\bar{z}$, where $(p \pm ix)/\sqrt{2} = z, \bar{z}$. Choose

$$\mathcal{A} = \frac{i}{2}(z \, d\bar{z} - \bar{z} \, dz) \tag{20.59}$$

The covariant derivatives are $\partial_z - \frac{1}{2}\bar{z}$ and $\partial_{\bar{z}} + \frac{1}{2}z$. Holomorphic polarization corresponds to $P = \partial/\partial\bar{z}$, leading to the condition

$$(\partial_{\bar{z}} + \tfrac{1}{2}z)\Psi = 0 \tag{20.60}$$

This is equation (20.46) for this example. The solutions are of the form

$$\Psi = e^{-\frac{1}{2}z\bar{z}} \, \varphi(z) \tag{20.61}$$

where $\varphi(z)$ is holomorphic in z.

The vector fields corresponding to $z, \bar{z}$ are

$$z \leftrightarrow -i\frac{\partial}{\partial\bar{z}}$$
$$\bar{z} \leftrightarrow i\frac{\partial}{\partial z} \tag{20.62}$$

These commute with $P = \partial/\partial\bar{z}$ and so are polarization-preserving. The prequantum operators corresponding to these are

$$\mathcal{P}(z) = -i(-i)\left(\frac{\partial}{\partial\bar{z}} + \tfrac{1}{2}z\right) + z = -\frac{\partial}{\partial\bar{z}} + \tfrac{1}{2}z$$
$$\mathcal{P}(\bar{z}) = -i(\,i)\left(\frac{\partial}{\partial z} - \tfrac{1}{2}\bar{z}\right) + \bar{z} = \frac{\partial}{\partial z} + \tfrac{1}{2}\bar{z} \tag{20.63}$$

In terms of their action on the functions $\varphi(z)$ in (20.61) corresponding to Ψ's obeying the polarization condition, these can be written as

$$P(z)\,\varphi(z) = z\,\varphi(z)$$

$$P(\bar{z})\,\varphi(z) = \frac{\partial\varphi}{\partial z} \tag{20.64}$$

The inner product for the $\varphi(z)$'s is

$$\langle 1|2\rangle = \int i\frac{dz \wedge d\bar{z}}{2\pi}\Psi_1^*\Psi_2$$

$$= \int i\frac{dz \wedge d\bar{z}}{2\pi}\,e^{-z\bar{z}}\,\varphi_1^*\,\varphi_2 \tag{20.65}$$

What we have obtained is the standard coherent state (or Bargmann) realization of the Heisenberg algebra.

20.6.2 Quantizing the two-sphere

We consider the phase space to be a two-sphere $S^2 \sim \mathbf{CP}^1$ considered as a Kähler manifold. Using complex coordinates $z = x + iy, \bar{z} = x - iy$, the standard Kähler form is

$$\omega = i\,\frac{dz \wedge d\bar{z}}{(1 + z\bar{z})^2} \tag{20.66}$$

The metric is given by $ds^2 = e^1 e^1 + e^2 e^2$ where the frame fields are

$$e^1 = \frac{dx}{1 + r^2}, \qquad\qquad e^2 = \frac{dy}{1 + r^2} \tag{20.67}$$

where $r^2 = z\bar{z}$. The Riemannian curvature is $\mathcal{R}^{12} = 4e^1 \wedge e^2$, giving the Euler number

$$\chi = \int \frac{\mathcal{R}_{12}}{2\pi} = 2 \tag{20.68}$$

The phase space has nonzero $\mathcal{H}^2(\mathcal{M})$ given by the Kähler form. As we have discussed, the symplectic two-form must belong to an integral cohomology class of $\mathcal{M}$ to be able to quantize properly. We take

$$\Omega = n\,\omega = i\,n\,\frac{dz \wedge d\bar{z}}{(1 + z\bar{z})^2} \tag{20.69}$$

where n is an integer. In this case, $\int_M \Omega = 2\pi n$ as required by the quantization condition. The symplectic potential can be taken as

$$\mathcal{A} = \frac{in}{2}\left[\frac{z\,d\bar{z} - \bar{z}\,dz}{(1 + z\bar{z})}\right] \tag{20.70}$$

The covariant derivatives are given by $\partial - i\mathcal{A}$. The holomorphic polarization condition is

$$(\partial_{\bar{z}} - i\mathcal{A}_{\bar{z}})\Psi = \left[\partial_{\bar{z}} + \frac{n}{2}\frac{z}{(1 + z\bar{z})}\right]\Psi = 0 \tag{20.71}$$

This can be solved as

$$\Psi = \exp\left(-\frac{n}{2}\log(1+z\bar{z})\right) f(z) \tag{20.72}$$

Notice that $n\log(1+z\bar{z})$ is the Kähler potential for Ω. The inner product is given by

$$\langle 1|2\rangle = i\alpha \int \frac{dz \wedge d\bar{z}}{2\pi(1+z\bar{z})^{n+2}} f_1^* f_2 \tag{20.73}$$

Here α is an overall constant, which can be absorbed into the normalization factors for the wave functions. Since $f(z)$ in (20.72) is holomorphic, we can see that a basis of nonsingular wave functions is given by $f(z) = 1, z, z^2, \cdots, z^n$; higher powers of z will not have finite norm. The dimension of the Hilbert space is thus $(n+1)$. We could have seen that this dimension would be finite from the semiclassical estimate of the number of states as the phase volume. Since the phase volume for $\mathcal{M} = S^2$ is finite, the dimension of the Hilbert space should be finite.

It is interesting to see this dimension in another way. The polarization condition (20.71) is giving the $\bar{\partial}$-closure of Ψ with a $U(1)$ gauge field $\mathcal{A}$ and curvature $\mathcal{R}_{12}$. The number of normalizable solutions to (20.71) is thus given by the index theorem for the twisted Dolbeault complex, i.e.,

$$\text{index}(\bar{\partial}_V) = \int_{\mathcal{M}} \text{td}(\mathcal{M}) \wedge Ch(V) \tag{20.74}$$

where, for our two-dimensional case, the Todd class $\text{td}(\mathcal{M})$ is $\mathcal{R}/4\pi$ and the Chern character $Ch(V) = \text{Tr}(e^{iF/2\pi}) = \text{Tr}(e^{\Omega/2\pi})$ is $\int \Omega/2\pi$ for us. We thus have

$$\text{index}(\bar{\partial}_V) = \int_{\mathcal{M}} \frac{\Omega}{2\pi} + \int_{\mathcal{M}} \frac{\mathcal{R}}{4\pi}$$
$$= n+1 \tag{20.75}$$

Notice that, semiclassically, we should expect the number of states to be $\int \Omega/2\pi = n$. The extra one comes from the Euler number in this case.

An orthonormal basis for the wave functions may be taken to be

$$f_k(z) = \left[\frac{n!}{k!\,(n-k)!}\right]^{\frac{1}{2}} z^k \tag{20.76}$$

with the inner product, for two such functions f, g,

$$\langle f|g\rangle = i(n+1) \int \frac{dz \wedge d\bar{z}}{2\pi(1+z\bar{z})^{n+2}} f^* g \tag{20.77}$$

Notice that this is the same as (20.73) but with a specific choice of $\alpha = n+1$. This is the value which gives $\text{Tr}1 = n+1$ as expected for an $(n+1)$-dimensional Hilbert space.

Classically the Poisson bracket of two functions F and G on the phase space is given by

$$\{F, G\} = \Omega^{\mu\nu} \partial_\mu F \partial_\nu G$$
$$= \frac{i}{n}(1 + z\bar{z})^2 \left(\frac{\partial F}{\partial z} \frac{\partial G}{\partial \bar{z}} - \frac{\partial F}{\partial \bar{z}} \frac{\partial G}{\partial z} \right) \qquad (20.78)$$

Consider now the vector fields

$$\xi_+ = i \left(\frac{\partial}{\partial \bar{z}} + z^2 \frac{\partial}{\partial z} \right)$$

$$\xi_- = -i \left(\frac{\partial}{\partial z} + \bar{z}^2 \frac{\partial}{\partial \bar{z}} \right)$$

$$\xi_3 = i \left(z \frac{\partial}{\partial z} - \bar{z} \frac{\partial}{\partial \bar{z}} \right) \qquad (20.79)$$

It is easily verified that these are the standard $SU(2)$ isometries of the sphere. The Lie brackets of the ξ's give the $SU(2)$ algebra, up to certain factors of $\pm i$ compared to the standard form, having to do with how we have defined the ξ's; we have put in certain multiplicative constants so that the quantum algebra has the standard form. Further, the ξ's are Hamiltonian vector fields corresponding to the functions

$$J_+ = -n \frac{z}{1 + z\bar{z}}$$

$$J_- = -n \frac{\bar{z}}{1 + z\bar{z}}$$

$$J_3 = -\frac{n}{2} \left(\frac{1 - z\bar{z}}{1 + z\bar{z}} \right) \qquad (20.80)$$

The prequantum operators $-i\xi \cdot \mathcal{D} + J$ corresponding to these functions are

$$\mathcal{P}(J_+) = \left(z^2 \partial_z - \frac{n}{2} z \frac{2 + z\bar{z}}{1 + z\bar{z}} \right) - i\xi_+^{\bar{z}} \mathcal{D}_{\bar{z}}$$

$$\mathcal{P}(J_-) = \left(-\partial_z - \frac{n}{2} \frac{\bar{z}}{1 + z\bar{z}} \right) - i\xi_-^{\bar{z}} \mathcal{D}_{\bar{z}}$$

$$\mathcal{P}(J_3) = \left(z \partial_z - \frac{n}{2} \frac{1}{1 + z\bar{z}} \right) - i\xi_3^{\bar{z}} \mathcal{D}_{\bar{z}} \qquad (20.81)$$

Acting on the polarized wave functions, $\mathcal{D}_{\bar{z}}$ gives zero. Writing Ψ as in (20.72), we can work out the action of the operators on the holomorphic wave functions $f(z)$. We get

$$\hat{J}_+ = z^2 \partial_z - n z$$
$$\hat{J}_- = -\partial_z$$
$$\hat{J}_3 = z \partial_z - \tfrac{1}{2} n \qquad (20.82)$$

If we define $j = n/2$, which is therefore half-integral, we see that the operators given above correspond to a unitary irreducible representation of $SU(2)$ with $J^2 = j(j+1)$ and dimension $n+1 = 2j+1$. Notice that there is only one representation here and it is fixed by the choice of the symplectic form Ω.

From the symplectic potential (20.70) and from (20.29) we see that an action which leads to the above results is

$$S = i\frac{n}{2}\int dt\, \frac{z\dot{\bar{z}} - \bar{z}\dot{z}}{1 + z\bar{z}} \tag{20.83}$$

where the overdot denotes differentiation with respect to time. This action may be written as

$$S = i\frac{n}{2}\int dt\, \text{Tr}(\sigma_3 g^{-1}\dot{g}) \tag{20.84}$$

where g is an element of $SU(2)$ written as a (2×2)-matrix, $g = \exp(i\sigma_i\theta_i)$ and σ_i, $i = 1, 2, 3$, are the Pauli matrices. In this action, the dynamical variable is an element of $SU(2)$. If we make a transformation $g \to g\, h$, $h = \exp(i\sigma_3\varphi)$, we get

$$S \to S - n\int dt\, \dot{\varphi} \tag{20.85}$$

The extra term is a boundary term and does not affect the equations of motion. Thus classically the dynamics is actually restricted to $SU(2)/U(1) = S^2$. The choice of parametrization

$$g = \frac{1}{\sqrt{1 + z\bar{z}}}\begin{pmatrix} 1 & z \\ -\bar{z} & 1 \end{pmatrix}\begin{bmatrix} e^{i\theta} & 0 \\ 0 & e^{-i\theta} \end{bmatrix} \tag{20.86}$$

leads to the expression (20.83).

Even though the classical dynamics is restricted to $SU(2)/U(1)$, the boundary term in (20.85) does have an effect in the quantum theory. Consider choosing $\varphi(t)$ such that $\varphi(-\infty) = 0$ and $\varphi(\infty) = 2\pi$. In this case $h(-\infty) = h(\infty) = 1$ giving a closed loop in the $U(1)$ subgroup of $SU(2)$ defined by the σ_3-direction. For this choice of $h(t)$, the action changes by $-2\pi n$. e^{iS} remains single valued and, even in the quantum theory, the extra $U(1)$ degree of freedom is consistently removed. If the coefficient were not an integer, this would not be the case and we would have inconsistencies in the quantum theory. Thus the quantization of the coefficient to an integral value is obtained again, from a slightly different point of view.

So far we have used a local parametrization of S^2 which corresponds to a stereographic projection of the sphere onto a plane. Another more global approach is to use the homogeneous coordinates of the sphere viewed as $\mathbf{CP}^1$. We use a two-component spinor u_α, $\alpha = 1, 2$, with the identification $u_\alpha \sim \lambda u_\alpha$ for any nonzero complex number λ. We also define $\bar{u}_1 = u_2^*$, $\bar{u}_2 = -u_1^*$ or $\bar{u}_\alpha = \epsilon_{\alpha\beta}u_\beta^*$, where $\epsilon_{\alpha\beta} = -\epsilon_{\beta\alpha}$, $\epsilon_{12} = 1$. The symplectic form is

$$\Omega = -in\left[\frac{du\cdot d\bar{u}}{\bar{u}\cdot u} - \frac{\bar{u}\cdot du\, u\cdot d\bar{u}}{(\bar{u}\cdot u)^2}\right] \tag{20.87}$$

where the notation is $u \cdot v = u_\alpha v_\beta \epsilon_{\alpha\beta}$. It is easily checked that $\Omega(\lambda u) = \Omega(u)$; it is invariant under $u \to \lambda u$ and hence is properly defined on $\mathbf{CP}^1$ rather than $\mathbf{C}^2 - \{0\}$. The choice of $u_2/u_1 = z$ leads to the previous local parametrization. The symplectic potential is

$$\mathcal{A} = -i\frac{n}{2}\left[\frac{u \cdot d\bar{u} - du \cdot \bar{u}}{\bar{u} \cdot u}\right] \tag{20.88}$$

Directly from the above expression we see that

$$\mathcal{A}(\lambda u) = \mathcal{A}(u) - d\left(i\frac{n}{2}\log(\bar{\lambda}/\lambda)\right) \tag{20.89}$$

This means that $\mathcal{A}$ cannot be written as a globally defined one-form on $\mathbf{CP}^1$. This is to be expected because $\int \Omega \neq 0$ and hence we cannot have a globally defined potential on $\mathbf{CP}^1$. From (20.32), we see that the prequantum wave functions must transform as

$$\Psi(\lambda u, \bar{\lambda}\bar{u}) = \Psi(u, \bar{u}) \, \exp\left[\frac{n}{2}\log(\lambda/\bar{\lambda})\right] \tag{20.90}$$

The polarization condition for the wave functions becomes

$$\left[\frac{\partial}{\partial \bar{u}_\alpha} - \frac{n}{2}\frac{u_\beta \epsilon_{\beta\alpha}}{\bar{u} \cdot u}\right]\Psi = 0 \tag{20.91}$$

The solution to this condition is

$$\Psi = \exp\left(-\frac{n}{2}\log(\bar{u} \cdot u)\right) f(u) \tag{20.92}$$

Combining this with (20.90), we see that the holomorphic functions $f(u)$ should behave as

$$f(\lambda u) = \lambda^n \, f(u) \tag{20.93}$$

$f(u)$ must thus have n u's and hence is of the form

$$f(u) = \sum_{\alpha's} C^{\alpha_1 \cdots \alpha_n} \, u_{\alpha_1} \cdots u_{\alpha_n} \tag{20.94}$$

Because of the symmetry of the indices, there are $n+1$ independent functions, as before. There is a natural linear action of $SU(2)$ on the u, $\bar{u}$ given by

$$u'_\alpha = U_{\alpha\beta} \, u_\beta, \qquad\qquad \bar{u}'_\alpha = U_{\alpha\beta} \, \bar{u}_\beta \tag{20.95}$$

where $U_{\alpha\beta}$ form a (2×2) $SU(2)$ matrix. The corresponding generators are the J_a we have constructed in (20.81, 20.82).

20.6.3 Compact Kähler spaces of the G/H-type

The two-sphere $S^2 = SU(2)/U(1)$ is an example of a group coset which is a Kähler manifold. There are many compact Kähler manifolds which are of the form G/H, where H is a subgroup of the compact Lie group G. In particular G/H is a Kähler manifold for any compact Lie group if H is its maximal torus. The maximal torus is the subspace of G generated by the mutually commuting generators. Another set of spaces is given by $\mathbf{CP}^N = SU(N+1)/U(N)$. There are many other cases as well.

One can take the symplectic form as proportional to the Kähler form or as a combination of the generators of $\mathcal{H}^2(M)$ and quantize these spaces as we have done for the case of S^2. In general, they lead to one unitary irreducible representation of the group G, the particular choice of the representation being determined by the choice of Ω.

In most of these cases, the Kähler form can be constructed in a very simple way. As an example consider $\mathbf{CP}^2 = SU(3)/U(2)$. A general element of $SU(3)$ can be represented as a unitary (3×3)-matrix. We define a $U(1)$ subgroup by elements of the form $U = \exp(iI^8\theta)$, where

$$I^8 = \frac{1}{\sqrt{6}} \begin{pmatrix} 1 & 0 & 0 \\ 0 & 1 & 0 \\ 0 & 0 & -2 \end{pmatrix} \tag{20.96}$$

We also define an $SU(2)$ subgroup which commutes with this by elements of the form

$$h_{SU(2)} = \begin{pmatrix} h_{2\times 2} & 0 \\ 0 & 1 \end{pmatrix} \tag{20.97}$$

These two together form the $U(2)$ subgroup of $SU(3)$. Consider now the one-form

$$\mathcal{A}(g) = iw\,\mathrm{Tr}(I^8 g^{-1} dg) \tag{20.98}$$

where g is an element of the group $SU(3)$ and w is a numerical constant. If h is an element of $U(2) \subset SU(3)$, we find

$$\mathcal{A}(gh) = \mathcal{A}(g) - wd\theta \tag{20.99}$$

We see that $\mathcal{A}$ changes by a total differential under the $U(2)$-transformations. Thus $d\mathcal{A}$ is defined on $\mathbf{CP}^2$. Evidently it is closed, but it is not exact since the corresponding one-form is not globally defined on $\mathbf{CP}^2$, but only on $G = SU(3)$. Thus $d\mathcal{A}$ is a nontrivial element of $\mathcal{H}^2(\mathbf{CP}^2)$. We can use $d\mathcal{A}$ as the symplectic two-form. The integrals of $d\mathcal{A}$ over nontrivial two-cycles on $\mathbf{CP}^2$ will have to be integers; this will restrict the choices for w. Alternatively, we take the action to be $S = \int \mathcal{A}$; for e^{iS} to be well defined on $\mathbf{CP}^2$, we will have restrictions on the w. The wave functions are functions on $SU(3)$ subject to the restrictions given by the action of $SU(2)$ and $U(1)$. In other words, we can write, using the Wigner $\mathcal{D}$-functions

$$\Psi \sim \mathcal{D}^{\alpha}_{AB}(g) \tag{20.100}$$

Here α is a set of indices which labels the representation; A, B label the states. $\mathcal{D}^{\alpha}_{AB}(g)$ is the AB-matrix element of the group element g in the irreducible representation of G characterized by the labels α.

The groups involved in the quotient can be taken as the right action on g. The transformation law for $\mathcal{A}$ then tells us that Ψ must transform as

$$\Psi(gh) = \exp(iw\theta) \; \Psi(g) \tag{20.101}$$

This shows that the wave functions must be singlets under the $SU(2)$ subgroup, acting on the right of g, and carry a definite charge w under the $U(1)$ subgroup. This restricts the choice of values for the index B in (20.100). Further w must also be quantized so that it can be one of the allowed values in the unitary irreducible representations of $SU(3)$ in (20.100). Once the indices B are chosen this way, the index A is free and so the result of the quantization is to yield a Hilbert space which is one unitary irreducible representation of the group $SU(3)$. w is related to the highest weights defining this representation. (There are many representations satisfying the condition on the charges for the subgroup. But the polarization condition, which we have not discussed for this problem, will impose further restrictions and choose one representation which is a "minimal" one among the various representations allowed by the charges.) To do this in more detail, notice that $I^8 = \sqrt{3/2}\, Y$, where Y is the hypercharge; thus $\mathcal{D}_{AB}(gU) = \mathcal{D}_{AB} \exp(i\sqrt{3/2}\, Y_B\theta)$, Y_B being the hypercharge of the state B, identifying w as $\sqrt{3/2}Y_B$. Unitary representations of $SU(3)$ have hypercharge values quantized in units of $\frac{1}{3}$; $SU(2)$ invariant states have hypercharge quantized in units of $\frac{2}{3}$. Equation (20.101) then tells us that w must be quantized as $n\sqrt{2/3}$, where n is an integer.

An explicit parametrization can be obtained as follows. Evaluating the trace in (20.98), we get

$$\mathcal{A} = -iw\sqrt{\frac{3}{2}}\; u^*_\alpha du_\alpha \tag{20.102}$$

where $g_{\alpha 3} = u_\alpha$. Evidently, $u^*_\alpha u_\alpha = 1$ and we can parametrize it as

$$u_\alpha = \frac{1}{\sqrt{1 + \bar{z} \cdot z}} \begin{pmatrix} 1 \\ z_1 \\ z_2 \end{pmatrix} \tag{20.103}$$

The symplectic two-form corresponding to the potential (20.102) is

$$\Omega = -iw\sqrt{\frac{3}{2}}\; du^*_\alpha du_\alpha$$
$$= -iw\sqrt{\frac{3}{2}} \left[\frac{d\bar{z}_i\, dz_i}{(1 + \bar{z} \cdot z)} - \frac{d\bar{z} \cdot z\, \bar{z} \cdot dz}{(1 + \bar{z} \cdot z)^2} \right] \tag{20.104}$$

This Ω is proportional to the Kähler two-form on $\mathbf{CP}^2$. Requiring that Ω should integrate to an integer over closed nontrivial two-surfaces in $\mathbf{CP}^2$ will lead to the same quantization condition on w. This is similar to the case of the two-sphere or $\mathbf{CP}^1$. The polarization condition will tell us that the states are functions of u_α's only, not u_α^*'s, and the condition on the $U(1)$-charge will fix the number of u_α's to be n. The states are thus of the form

$$\Psi \sim u_{\alpha_1} u_{\alpha_2} \cdots u_{\alpha_n} \tag{20.105}$$

This corresponds to the rank n symmetric representation of $SU(3)$. One can get other representations by other choices of H.

More generally one can take

$$\mathcal{A}(g) = i \sum_a w_a \text{Tr}(t^a g^{-1} dg) \tag{20.106}$$

where t^a are diagonal elements of the Lie algebra of G and w_a are a set of numbers. H will be the subgroup commuting with $\sum_a w_a t^a$; if w_a are such that all the diagonal elements of $\sum_a w_a t^a$ are distinct, then H will be the maximal torus of G. $\mathcal{A}$ will change by a total differential under $g \to gh$, $h \in H$ and $d\mathcal{A}$ will be a closed nonexact form on G/H. If some of the eigenvalues of $\sum_a w_a t^a$ are equal, H can be larger than the maximal torus. Upon quantization, for suitably chosen w_a, we will get one unitary irreducible representation of G, and w_a will be related to the weights defining the representation.

20.6.4 Charged particle in a monopole field

The symplectic form for a point particle in three dimensions is

$$\Omega = dp_i \wedge dx_i \tag{20.107}$$

The usual prescription for introducing coupling to a magnetic field involves replacing p_i by $p_i + eA_i$. The Lagrangian has a $A_i \dot{x}_i$ added to it, giving the canonical one-form $m\dot{x}_i dx_i + eA_i dx_i$. Writing $p_i = m\dot{x}_i$, the symplectic two-form is found to be

$$\Omega = dp_i \wedge dx_i + eF \tag{20.108}$$

where $F = dA = \frac{1}{2} F_{ij} dx^i \wedge dx^j$ is the magnetic field strength. Thus in terms of the symplectic two-form the minimal prescription for introducing electromagnetic interactions amounts to adding eF to Ω. We can use this to discuss a charged particle in the field of a magnetic monopole which has a radial magnetic field $B_k = gx_k/r^3$, where g is the magnetic charge of the monopole and $r^2 = x_i x_i$. The monopole is taken to be at the origin of the coordinate system. Thus Ω is

$$\Omega = dp_i \wedge dx_i + \frac{1}{2} eg\epsilon_{ijk} \frac{x_k}{r^3} dx_i \wedge dx_j \tag{20.109}$$

We can identify the basic Hamiltonian vector fields. Contraction of $X_i = -\partial/\partial p_i$ with Ω gives $-dx_i$ identifying it as the vector field for x^i. The contraction of $P'_i = \partial/\partial x_i$ gives

$$i_{P'}\Omega = -dp_i + eg\epsilon_{ijk}\frac{x_k}{r^3}dx_j \qquad (20.110)$$

P' is not a Hamiltonian vector field, but since we get dx_i from contraction of $\partial/\partial p_i$ with Ω we see that the combination

$$P_i = \frac{\partial}{\partial x_i} - eg\epsilon_{ijk}\frac{x_k}{r^3}\frac{\partial}{\partial p_j} \qquad (20.111)$$

is a Hamiltonian vector field and corresponds to p_i. The Poisson brackets are thus

$$\{x_i, x_j\} = -i_{X_i}dx_j = 0$$
$$\{x_i, p_j\} = -i_{X_i}dp_j = \delta_{ij}$$
$$\{p_i, p_j\} = -i_{P_i}dp_j = eg\epsilon_{ijk}\frac{x_k}{r^3} \qquad (20.112)$$

It is also interesting to work out the angular momentum. Under an infinitesimal rotation by angle θ_k the change in the variables x_i, p_i are

$$\delta x_i = -\epsilon_{ijk}x_j\theta_k$$
$$\delta p_i = -\epsilon_{ijk}p_j\theta_k \qquad (20.113)$$

The corresponding vector field is therefore

$$\xi = -\epsilon_{ijk}\theta_k\left(x_j\frac{\partial}{\partial x_i} + p_j\frac{\partial}{\partial p_i}\right) \qquad (20.114)$$

Upon taking the interior contraction of this with Ω, we find

$$i_\xi\Omega = -\epsilon_{ijk}\theta_k\left(-x_j dp_i - p_i dx_j\right) - \theta_k eg(\delta_j^m\delta_k^n - \delta_k^m\delta_j^n)\frac{x_j x_n}{r^3}dx^m$$
$$= \theta_k d(-\epsilon_{ijk}x_i p_j) + \theta_k eg\left(\frac{dx_k}{r} - \frac{x_i x_k}{r^3}dx^i\right)$$
$$= \theta_k d\left(-\epsilon_{ijk}x_i p_j + eg\frac{x_k}{r}\right) \qquad (20.115)$$

The angular momentum which is the generator of rotations is thus

$$J_i = \epsilon_{ijk}x_j p_k - eg\frac{x_i}{r} \qquad (20.116)$$

This shows that the charged particle has an extra contribution to the angular momentum which is radial. In fact $\hat{x} \cdot J = -eg$.

By converting the Poisson brackets to commutators of operators we can set up the quantum theory. The only unusual ingredient is the following. The symplectic two-form is singular at $r = 0$. Thus we need to remove this point for a nonsingular description. If we do so, then the space has noncontractible two-spheres and so there is a quantization condition for Ω. Integrating Ω over a two-sphere around the origin (which is the location of the monopole), we get $4\pi eg$. The general quantization condition (20.58) becomes

$$eg = \frac{n}{2} \tag{20.117}$$

This is the famous Dirac quantization condition stating that the magnetic charge must be quantized in units of $1/2e$. This value $1/2e$ is the lowest magnetic charge corresponding to one monopole. The argument for quantization also follows from noting that in the quantum theory, the eigenvalues of any component of angular momentum have to be half-integral. $\hat{x} \cdot J = -eg$ then leads to the same quantization; this was first noted by Saha.

20.6.5 Anyons or particles of fractional spin

We consider relativistic particles in two spatial dimensions. In general, they can have arbitrary spin, not necessarily quantized, and hence they are generically referred to as anyons (or particles of any spin).

We will work out some of the theory of anyons starting with a symplectic structure. A spinless particle may be described by a set of momentum variables p^a and position variables x^a, $a = 0, 1, 2$. The canonical structure or symplectic two-form is given by

$$\Omega = g_{ab} \, dx^a \wedge dp^b \tag{20.118}$$

where $g_{ab} = diag(1, -1, -1)$ is the metric tensor. For a charged particle, as discussed above, the coupling to the electromagnetic field A_a by the minimal prescription is equivalent to $\Omega \to \Omega + eF$.

The motion of the relativistic charged particle is given by the (classical) Lorentz equations

$$\frac{p^a}{m} = \frac{dx^a}{d\tau}$$
$$\frac{dp^a}{d\tau} = -\frac{e}{m} F^{ab} p_b \tag{20.119}$$

τ is the parameter for the trajectory of the particle (with mass m). We have chosen a specific parametrization or equivalently a gauge-fixing for the gauge freedom of reparametrizations of the trajectory and so the equations (20.119) are not invariant under reparametrizations. Equations (20.119) tell us that the infinitesimal change of τ is given, on the phase space, by a vector field

$$V = \frac{p^a}{m} \frac{\partial}{\partial x^a} - \frac{e}{m} F^{ab} p_b \frac{\partial}{\partial p^a} \tag{20.120}$$

The canonical generator of the τ-evolution, say, G, is defined by $i_V \Omega = -dG$. This gives $G = -p^2/(2m) +$ constant. Anticipating the eventual value of the constant, we choose it to be $m/2$. Basically this is the *definition* of the mass. Thus

$$G = -\frac{1}{2m}(p^2 - m^2) \qquad (20.121)$$

Since we need reparametrization invariance, the τ-evolution must be trivial. Thus we must set G_0 to be zero for the classical trajectories. Quantum theoretically, this can be implemented by

$$G\Psi = 0 \qquad (20.122)$$

This will be the basic dynamical equation of the quantum theory.

The symplectic form may be written as

$$\Omega = \frac{1}{2}\Omega_{AB}\, d\xi^A \wedge d\xi^B$$

$$\Omega_{AB} = \begin{pmatrix} 0 & -g_{ab} \\ g_{ab} & eF_{ab} \end{pmatrix} \qquad (20.123)$$

where $\xi^A = (p^a, x^a)$ denotes both sets of phase space variables. This symplectic form leads to the commutation rules

$$[x^a,\ x^b] = 0$$
$$[p^a,\ x^b] = ig^{ab}$$
$$[p^a,\ p^b] = ieF^{ab} \qquad (20.124)$$

These relations are solved by

$$p_a = i\partial_a + eA_a \qquad (20.125)$$

The condition of trivial τ-evolution, namely, (20.122), becomes

$$[(\partial_a - ieA_a)^2 + m^2]\Psi = 0 \qquad (20.126)$$

This is the Schrödinger equation (in this case, the Klein-Gordon equation) which describes the quantum dynamics of the particle. One can easily show that the Lorentz equations (20.119) are quantum mechanically realized by

$$i\frac{\partial \xi^A}{\partial t} = [\xi^A, G] \qquad (20.127)$$

In a more general situation, one can obtain the Schrödinger-type equation as follows. We start with the symplectic two-form. From the equations of motion, we find the generator of the τ-evolution. Setting this generator to zero on the wave functions gives us the equation we are seeking. To realize this as a differential equation we must solve the commutation rules in terms of a set of coordinates and their derivatives (namely, canonical variables).

For a free anyon with spin $-s$, the symplectic structure is given by

$$\Omega = dx^a \wedge dp_a + \frac{1}{2} s \, \epsilon_{abc} \frac{p^a dp^b \wedge dp^c}{(p^2)^{3/2}} \qquad (20.128)$$

The commutation rules are given by

$$[x^a, \, x^b] = is \, \epsilon^{abc} \frac{p_c}{(p^2)^{3/2}}$$
$$[p^a, \, x^b] = ig^{ab}$$
$$[p^a, \, p^b] = 0 \qquad (20.129)$$

Consider the Lorentz generator J^a defined by

$$[J^a, \, p^b] = i\epsilon^{abc} p_c$$
$$[J^a, \, x^b] = i\epsilon^{abc} x_c \qquad (20.130)$$

It is easy to see that J^a is given by

$$J^a = -\epsilon^{abc} x_b p_c - s \frac{p^a}{\sqrt{p^2}} \qquad (20.131)$$

This shows that the particle has a spin $-s$, easily seen in the rest frame with $p^0 = m$, $p^1 = p^2 = 0$. The expression for J^a is analogous to (20.116) except for the different signature for spacetime. Because of this change of signature, there is no quantization of the coefficient of the second term in the symplectic structure (20.128). The value s is not quantized. Alternatively, there is no closed two surface which is not the boundary of a three-volume. We can solve the commutation rules (20.129) in terms of canonical variables as

$$x^a = q^a + \alpha^a(p)$$
$$\alpha^a(p) = s\epsilon^{abc} \frac{p_b \eta_c}{p^2 + \sqrt{p^2} p \cdot \eta} \qquad (20.132)$$

where $\eta^a = (1, 0, 0)$ and $[q^a, q^b] = 0$, $[p^a, p^b] = 0$, $[p^a, q^b] = ig^{ab}$ or $q^a = -i\frac{\partial}{\partial p_a}$.

Using (20.132) for x^a in (20.131), we can write

$$J^a = -i\epsilon^{abc} p_b \frac{\partial}{\partial p^c} - s \frac{p^a + \sqrt{p^2}\eta^a}{\sqrt{p^2} + p \cdot \eta} \qquad (20.133)$$

We see that $p \cdot J + s\sqrt{p^2} = 0$. With $p^2 = m^2$, we see that the spin is indeed $-s$. Thus the symplectic structure Ω of (20.128) is indeed appropriate to describe the anyon.

The symplectic structure Ω for anyons in an electromagnetic field is now obtained by $\Omega \to \Omega + eF$. Thus

$$\Omega = dx^a \wedge dp_a + \tfrac{1}{2}s\epsilon_{abc}\frac{p^a dp^b \wedge dp^c}{(p^2)^{3/2}} + \tfrac{1}{2}eF_{ab}dx^a \wedge dx^b + \mathcal{O}(\partial F) \quad (20.134)$$

With the introduction of spin it is possible that Ω has further corrections that depend on the gradients of the field strength F. This is indicated by $\mathcal{O}(\partial F)$ in the above equation. We shall not discuss the case of anyons in an electromagnetic field in any more detail here. One can actually obtain a wave equation as indicated and show that the gyromagnetic ratio for anyons is 2.

20.6.6 Field quantization, equal-time, and light-cone

We consider a real scalar field φ with the action

$$S = \int d^4x \, \frac{1}{2}\left[\frac{\partial\varphi}{\partial x^0}\frac{\partial\varphi}{\partial x^0} - \frac{\partial\varphi}{\partial x^i}\frac{\partial\varphi}{\partial x^i}\right] - U(\varphi) \quad (20.135)$$

where $U(\varphi) = \tfrac{1}{2}m^2\varphi^2 + V(\varphi)$. By considering a general variation and identifying the boundary term at the initial and final time-slices, we find, using (20.31),

$$A = \int d^3x \, \dot{\varphi} \, \delta\varphi \quad (20.136)$$

where we use δ to denote exterior derivatives on the field space and $\dot{\varphi} = \partial\varphi/\partial x^0$. The time-derivative of φ must be treated as an independent variable since A is at a fixed time. By taking another variation we find the symplectic two-form

$$\Omega = \int d^3x \, \delta\dot{\varphi} \wedge \delta\varphi \quad (20.137)$$

Recalling that the Poisson brackets for the coordinates on the phase space are the inverse of the symplectic two-form as a matrix, we find

$$\begin{aligned}
\{\,\varphi(\boldsymbol{x},x^0),\varphi(\boldsymbol{x}',x^0)\,\} &= 0 \\
\{\,\varphi(\boldsymbol{x},x^0),\dot{\varphi}(\boldsymbol{x}',x^0)\,\} &= \delta^{(3)}(\boldsymbol{x}-\boldsymbol{x}') \\
\{\,\dot{\varphi}(\boldsymbol{x},x^0),\dot{\varphi}(\boldsymbol{x}',x^0)\,\} &= 0
\end{aligned} \quad (20.138)$$

Upon replacing the variables by operators with commutation rules given by i times the Poisson brackets, we get the standard equal-time rules for quantization.

This phase space also has a standard Kähler structure. Consider the fields to be confined to a cubical box with each side of length L and volume V. With periodic boundary conditions, we can write a set of mode functions as

$$u_k(\boldsymbol{x}) = \frac{1}{\sqrt{V}} \exp(-i\boldsymbol{k}\cdot\boldsymbol{x})$$

$$k_i = \frac{2\pi n_i}{L} \quad (20.139)$$

Here n_i are integers. The fields can be expanded in modes as

$$\varphi(x) = \sum_k q_k u_k(\boldsymbol{x})$$

$$\dot{\varphi}(x) = \sum_k p_k u_k(\boldsymbol{x}) \tag{20.140}$$

The reality of the fields requires $q_k^* = q_{-k}$, $p_k^* = p_{-k}$. Substituting the mode expansion and simplifying, Ω becomes

$$\Omega = \sum_k \delta p_k \wedge \delta q_{-k}$$

$$= i \sum_k \delta a_k \wedge \delta a_k^* \tag{20.141}$$

where we define

$$a_k = \frac{1}{\sqrt{2}}(q_{-k} - ip_k), \qquad a_k^* = \frac{1}{\sqrt{2}}(q_k + ip_{-k}) \tag{20.142}$$

This shows the Kähler structure and we can carry out the holomorphic quantization as in the case of coherent states.

A somewhat more interesting example which illustrates the use of the symplectic structure $\omega_{ij}(\boldsymbol{x}, \boldsymbol{x}')$ is the light-cone quantization of a scalar field. We introduce light-cone coordinates, corresponding to a light-cone in the z-direction as

$$u = \frac{1}{\sqrt{2}}(z + t)$$

$$v = \frac{1}{\sqrt{2}}(z - t) \tag{20.143}$$

Instead of considering evolution of the fields in time t, we can consider evolution in one of the the light-cone coordinates, say, u. The other light-cone coordinate v and the two coordinates $x^T = x, y$ transverse to the light-cone parametrize the equal-u hypersurfaces. Field configurations $\varphi(u, v, x, y)$ at fixed values of u, i.e., real-valued functions of v, x, y, characterize the trajectories. They form the phase space of the theory. The action can be written as

$$S = \int du\, dv\, d^2 x^T \left[-\partial_u \varphi \partial_v \varphi - \tfrac{1}{2}(\partial_T \varphi)^2 - U(\varphi) \right] \tag{20.144}$$

Again from the variation of the action S, we can identify the canonical one-form $\mathcal{A}$ as

$$\mathcal{A} = \int dv\, d^2 x^T \left(-\partial_v \varphi\, \delta\varphi \right) \tag{20.145}$$

(This was denoted by Θ in Chapter 3.) The symplectic two-form is given by

$$\Omega = \frac{1}{2} \int d\mu d\mu' \; \Omega(v, x^T, v', x'^T) \; \delta\varphi(v, x^T) \wedge \delta\varphi(v', x'^T)$$

$$\Omega(v, x^T, v', x'^T) = -2 \, \partial_v \delta(v - v') \delta^{(2)}(x^T - x'^T)$$

$$d\mu = dv \; d^2 x^T \tag{20.146}$$

From this point on, the calculation is identical to what we did in Chapter 3. The fundamental Poisson bracket can be written down from the inverse to Ω. Writing

$$\delta(v - v') \; \delta^{(2)}(x^T - x'^T) = \int \frac{d^3 p}{(2\pi)^3} e^{-i p_u (v - v') - i p^T \cdot (x^T - x'^T)} \tag{20.147}$$

we see that

$$\Omega^{-1}(v, x^T, v', x'^T) = \frac{1}{2} \int \frac{d^3 p}{(2\pi)^3} \frac{1}{i p_u} \; \exp(-i p_u (v - v') - i p^T \cdot (x^T - x'^T))$$

$$= \frac{1}{4} \epsilon(v - v') \; \delta^{(2)}(x^T - x'^T) \tag{20.148}$$

Here $\epsilon(v - v')$ is the signature function, equal to 1 for $v - v' > 0$ and equal to -1 for $v - v' < 0$.

The phase space is thus given by field configurations $\varphi(v, x^T)$ with the Poisson brackets

$$\{\varphi(u, v, x^T), \varphi(u, v', x'^T)\} = \frac{1}{4} \epsilon(v - v') \; \delta^{(2)}(x^T - x'^T) \tag{20.149}$$

The Hamiltonian for u-evolution is given by

$$H = \int dv \; d^2 x^T \; \left[\frac{1}{2}(\partial_T \varphi)^2 \; + \; U(\varphi) \right] \tag{20.150}$$

The Hamiltonian equations of motion are easily checked using the Poisson brackets (20.149).

Quantization is achieved by taking φ to be an operator with commutation rules given by i-times the Poisson bracket.

20.6.7 The Chern-Simons theory in 2+1 dimensions

The Chern-Simons (CS) theory is a gauge theory in two space (and one time) dimensions. The action is given by

$$\mathcal{S} = -\frac{k}{4\pi} \int_{\Sigma \times [t_i, t_f]} \mathrm{Tr} \left[AdA + \frac{2}{3} A^3 \right]$$

$$= -\frac{k}{4\pi} \int_{\Sigma \times [t_i, t_f]} d^3 x \; \epsilon^{\mu\nu\alpha} \, \mathrm{Tr} \left[A_\mu \partial_\nu A_\alpha + \frac{2}{3} A_\mu A_\nu A_\alpha \right] \tag{20.151}$$

Here A_μ is the Lie-algebra-valued gauge potential, $A_\mu = -it^a A_\mu^a$. t^a are hermitian matrices forming a basis of the Lie algebra in the fundamental representation of the gauge group. We shall consider the gauge group to be $SU(N)$ in what follows, and normalize the t^a as $\text{Tr}(t^a t^b) = \frac{1}{2}\delta^{ab}$. k is a constant whose precise value we do not need to specify at this stage. We shall consider the spatial manifold to be some Riemann surface Σ and we shall be using complex coordinates. The equations of motion for the theory are

$$F_{\mu\nu} = 0 \qquad (20.152)$$

The theory is best analyzed, for our purposes, in the gauge where A_0 is set to zero. In this gauge, the equations of motion (20.152) tell us that $A_z = \frac{1}{2}(A_1 + iA_2)$ and $A_{\bar{z}} = \frac{1}{2}(A_1 - iA_2)$ are independent of time, but must satisfy the constraint

$$F_{\bar{z}z} \equiv \partial_{\bar{z}} A_z - \partial_z A_{\bar{z}} + [A_{\bar{z}}, A_z] = 0 \qquad (20.153)$$

This constraint is just the Gauss law of the CS gauge theory.

In the $A_0 = 0$ gauge, the action becomes

$$S = -\frac{ik}{\pi} \int dt d\mu_\Sigma \, \text{Tr}(A_{\bar{z}} \partial_0 A_z) \qquad (20.154)$$

For the boundary term from the variation of the action we get

$$\delta S = -\frac{ik}{\pi} \int_\Sigma \text{Tr}(A_{\bar{z}} \delta A_z) \Bigg]_{t_i}^{t_f} \qquad (20.155)$$

We can now identify the symplectic potential as

$$\mathcal{A} = -\frac{ik}{\pi} \int_\Sigma \text{Tr}(A_{\bar{z}} \delta A_z) + \delta\rho[A] \qquad (20.156)$$

where $\rho[A]$ is an arbitrary functional of A. The freedom of adding $\delta\rho$ is the freedom of canonical transformations. As in the case of the scalar field, δ is to be interpreted as denoting exterior differentiation on $\tilde{A}$, the space of gauge potentials on Σ. $\tilde{A}$ is also the phase space of the theory before reduction by the action of gauge symmetries. (The space of gaueg potentials was denoted by $\mathcal{A}$ in Chapter 16; here we use $\tilde{A}$ to avoid confusion with the symplectic potential.)

The symplectic two-form Ω is given by $\delta\mathcal{A}$, i.e.,

$$\Omega = -\frac{ik}{\pi} \int_\Sigma \text{Tr}(\delta A_{\bar{z}} \delta A_z) = \frac{k}{4\pi} \int_\Sigma \text{Tr}(\delta A \wedge \delta A)$$

$$= \frac{ik}{2\pi} \int_\Sigma \delta A_{\bar{z}}^a \delta A_z^a \qquad (20.157)$$

(We will not write the wedge sign for exterior products on the field space from now on since it is clear from the context.)

The complex structure on Σ induces a complex structure on $\tilde{\mathcal{A}}$. A_z, $A_{\bar z}$ can be taken as the local complex coordinates on $\tilde{\mathcal{A}}$. Indeed we have a Kähler structure on $\tilde{\mathcal{A}}$; Ω is k times the Kähler form on $\tilde{\mathcal{A}}$ and we can associate a Kähler potential K with Ω given by

$$K = \frac{k}{2\pi} \int_\Sigma A_{\bar z}^a A_z^a \tag{20.158}$$

Poisson brackets for $A_{\bar z}$, A_z are obtained by inverting the components of Ω and read

$$\{A_z^a(z), A_w^b(w)\} = 0$$
$$\{A_{\bar z}^a(z), A_{\bar w}^b(w)\} = 0$$
$$\{A_z^a(z), A_{\bar w}^b(w)\} = -\frac{2\pi i}{k} \delta^{ab} \delta^{(2)}(z - w) \tag{20.159}$$

These become commutation rules upon quantization.

Gauge transformations are given by

$$A^g = gAg^{-1} - dgg^{-1} \tag{20.160}$$

Infinitesimal gauge transformations are generated by the vector field

$$\xi = -\int_\Sigma \left[(D_z\theta)^a \frac{\delta}{\delta A_z^a} + (D_{\bar z}\theta)^a \frac{\delta}{\delta A_{\bar z}^a} \right] \tag{20.161}$$

where D_z and $D_{\bar z}$ denote the corresponding gauge covariant derivatives. By contracting this with Ω we get

$$i_\xi \Omega = -\delta \left[\frac{ik}{2\pi} \int_\Sigma F_{z\bar z}^a \theta^a \right] \tag{20.162}$$

which shows that the generator of infinitesimal gauge transformations is

$$G^a = \frac{ik}{2\pi} F_{z\bar z}^a \tag{20.163}$$

Reduction of the phase space can thus be performed by setting F to zero. This is also the equation of motion we found for the component A_0. Notice also that

$$\Omega(A^g) - \Omega(A) = \delta \left[\frac{k}{2\pi} \int_\Sigma \mathrm{Tr}(g^{-1}\delta g\, F) \right] \tag{20.164}$$

(In the second term F is the two-form $dA + A \wedge A$.)

The reduced set of field configurations are elements of $\tilde{\mathcal{A}}/\mathcal{G}_*$ where $\mathcal{G}_*$ denotes the group of gauge transformations, $\mathcal{G}_* = \{g(x) : \Sigma \to G\}$.

The construction of the wave functionals proceeds as follows. One has to consider a line bundle on the phase space with curvature Ω. Sections of this bundle give the prequantum Hilbert space. In other words, we consider functionals $\Phi[A_z, A_{\bar{z}}]$ with the condition that under the canonical transformation $\mathcal{A} \to \mathcal{A} + \delta\Lambda$, $\Phi \to e^{(i\Lambda)}\Phi$. The inner product on the prequantum Hilbert space is given by

$$\langle 1|2 \rangle = \int d\mu(A_z, A_{\bar{z}}) \, \Phi_1^*[A_z, A_{\bar{z}}] \, \Phi_2[A_z, A_{\bar{z}}] \qquad (20.165)$$

where $d\mu(A_z, A_{\bar{z}})$ is the Liouville measure associated with Ω. Given the Kähler structure, this is just the volume $[dA_z dA_{\bar{z}}]$ associated with the metric $||\delta A||^2 = \int_\Sigma \delta A_{\bar{z}} \delta A_z$.

The wave functionals so constructed depend on all phase space variables. We must now choose the polarization conditions on the Φ's so that they depend only on half the number of phase space variables. This reduction of the prequantum Hilbert space leads to the Hilbert space of the quantum theory. Given the Kähler structure of the phase space, the most appropriate choice is the Bargmann polarization which can be implemented as follows. With a specific choice of $\rho[A]$ in (20.156), the symplectic potential can be taken as

$$\mathcal{A} = -\frac{ik}{2\pi} \int_\Sigma \text{Tr}\big(A_{\bar{z}}\delta A_z - A_z\delta A_{\bar{z}}\big) = \frac{ik}{4\pi} \int_\Sigma \big(A_{\bar{z}}^a \delta A_z^a - A_z^a \delta A_{\bar{z}}^a\big) \qquad (20.166)$$

The covariant derivatives with $\mathcal{A}$ as the potential are

$$\nabla = \Big(\frac{\delta}{\delta A_z^a} + \frac{k}{4\pi}A_{\bar{z}}^a\Big), \qquad \overline{\nabla} = \Big(\frac{\delta}{\delta A_{\bar{z}}^a} - \frac{k}{4\pi}A_z^a\Big) \qquad (20.167)$$

The Bargmann polarization condition is

$$\nabla \, \Phi = 0 \qquad (20.168)$$

or

$$\Phi = \exp\Big(-\frac{k}{4\pi} \int A_{\bar{z}}^a A_z^a\Big) \, \psi[A_{\bar{z}}^a] = e^{-\frac{1}{2}K} \, \psi[A_{\bar{z}}^a] \qquad (20.169)$$

where K is the Kähler potential of (20.158). The states are represented by wave functionals $\psi[A_{\bar{z}}^a]$ which are holomorphic in $A_{\bar{z}}^a$. Further, the prequantum inner product can be retained as the inner product of the Hilbert space. Rewriting (20.165) using (20.169), we get the inner product as

$$\langle 1|2 \rangle = \int [dA_{\bar{z}}^a, A_z^a] \, e^{-K(A_{\bar{z}}^a, A_z^a)} \, \psi_1^* \, \psi_2 \qquad (20.170)$$

On the holomorphic wave functionals,

$$A_z^a \, \psi[A_{\bar{z}}^a] = \frac{2\pi}{k} \frac{\delta}{\delta A_{\bar{z}}^a} \, \psi[A_{\bar{z}}^a] \qquad (20.171)$$

As we have mentioned before, one has to make a reduction of the Hilbert space by imposing gauge invariance on the states, i.e., by setting the generator $F^a_{z\bar{z}}$ to zero on the wave functionals. This amounts to

$$\left(D_{\bar{z}} \frac{\delta}{\delta A^a_{\bar{z}}} - \frac{k}{2\pi} \partial_z A^a_{\bar{z}} \right) \psi[A^a_{\bar{z}}] = 0. \tag{20.172}$$

Consistent implementation of gauge invariance can lead to quantization requirements on the coupling constant k. For nonabelian groups G this is essentially the requirement of integrality of k based on the invariance of e^{iS} under homotopically nontrivial gauge transformations. We now show how this constraint arises in the geometric quantization framework. Consider first the nonabelian theory on $\Sigma = S^2$. The group of gauge transformations $\mathcal{G} = \{g(x): \ S^2 \to G\}$. Obviously $\Pi_0(\mathcal{G}) = \Pi_2(G) = 0$ and $\Pi_1(\mathcal{G}) = \Pi_3(G) = \mathbf{Z}$. Correspondingly, one has $\Pi_1(\tilde{\mathcal{A}}/\mathcal{G}) = 0$ and $\Pi_2(\tilde{\mathcal{A}}/\mathcal{G}) = \mathbf{Z}$. The nontriviality of $\Pi_2(\tilde{\mathcal{A}}/\mathcal{G})$ arises from the nontrivial elements of $\Pi_1(\mathcal{G})$. Therefore, consider a noncontractible loop C of gauge transformations,

$$C = g(x, \lambda), \qquad 0 \le \lambda \le 1$$
$$g(x, 0) = g(x, 1) = 1 \tag{20.173}$$

We can use this to construct a noncontractible two-surface in the gauge-invariant space $\tilde{\mathcal{A}}/\mathcal{G}$.

We start with a square in the space of gauge potentials parametrized by $0 \le \lambda, \sigma \le 1$ with the potentials given by

$$A(x, \lambda, \sigma) = (gAg^{-1} - dgg^{-1})\sigma + (1 - \sigma)A \tag{20.174}$$

The potential is A on the boundaries $\lambda = 0$ and $\lambda = 1$ and also on $\sigma = 0$. It is equal to the gauge transform A^g of A at $\sigma = 1$. Since A^g is identified with A in the quotient space, the boundary corresponds to a single point on the quotient $\tilde{\mathcal{A}}/\mathcal{G}$ and we have a closed two-surface. This surface is noncontractible if we take $g(x, \lambda)$ to be a nontrivial element of $\Pi_3(G) = \mathbf{Z}$. We can now integrate Ω over this closed two-surface; for this calculation, we may even put A equal to zero and use $A(x, \lambda, \sigma) = -\sigma dgg^{-1}$. We then have $\delta A = -\delta\sigma dgg^{-1} - \sigma d(\delta gg^{-1}) - \sigma[\delta gg^{-1}, dgg^{-1}]$. Using this in the expression for Ω and carrying out the integration over σ, we get

$$\int \Omega = \frac{k}{4\pi} \int \mathrm{Tr}(dgg^{-1})^2 \delta gg^{-1}$$
$$= \frac{k}{12\pi} \int \mathrm{Tr}(dgg^{-1})^3$$
$$= -2\pi k Q[g] \tag{20.175}$$

where, in the second step $dgg^{-1} = \partial_i gg^{-1} dx^i + \partial_\lambda gg^{-1} d\lambda$; we include differentiation with respect to the spatial coordinates and with respect to the internal coordinate λ. $Q[g]$ is the winding number (which is an integer) characterizing the class in $\Pi_1(\mathcal{G}) = \Pi_3(G)$ to which g belongs; it is given by

$$Q[g] = -\frac{1}{24\pi^2} \int \mathrm{Tr}(dgg^{-1})^3 \qquad (20.176)$$

From general principles of geometric quantization we know that the integral of Ω over any closed noncontractible two-surface in the phase space must be an integer, see (20.58). Thus (20.175) and (20.176) lead to the requirement that k has to be an integer. This argument can be generalized to other choices of Σ.

The situation for an Abelian group such as $U(1)$ is somewhat different. Consider the case of $G = U(1)$ and with Σ being a torus $S^1 \times S^1$. This can be described by $z = \xi_1 + \tau\xi_2$, where ξ_1, ξ_2 are real and have periodicity of $\xi_i \rightarrow \xi_i+$ integer, and τ, which is a complex number, is the modular parameter of the torus. The metric on the torus is $ds^2 = |d\xi_1 + \tau d\xi_2|^2$. The two basic noncontractible cycles (noncontractible closed curves) of the torus are usually labeled as the α and β cycles. Further the torus has a holomorphic one-form ω with

$$\int_\alpha \omega = 1, \qquad\qquad \int_\beta \omega = \tau \qquad (20.177)$$

Since ω is a zero mode of $\partial_{\bar{z}}$, we can parametrize $A_{\bar{z}}$ as

$$A_{\bar{z}} = \partial_{\bar{z}}\chi + i\frac{\pi\bar{\omega}}{\mathrm{Im}\tau}a \qquad (20.178)$$

where a is a complex number corresponding to the value of $A_{\bar{z}}$ along the zero mode of ∂_z. This is the Abelian version of (17.60).

For this space, $\Pi_0(\mathcal{G}) = \mathbf{Z} \times \mathbf{Z}$, by virtue of gauge transformations $g_{m,n}$ with nontrivial winding numbers m, n around the two cycles. Consider one connected component of $\mathcal{G}$, say, $\mathcal{G}_{m,n}$. A homotopically nontrivial $U(1)$ transformation can be written as $g_{m,n} = e^{i\lambda}e^{i\theta_{m,n}}$, where $\lambda(z,\bar{z})$ is a homotopically trivial gauge transformation and

$$\theta_{m,n} = \frac{i\pi}{\mathrm{Im}\tau}\left[m\int^z(\bar{\omega} - \omega) + n\int^z(\tau\bar{\omega} - \bar{\tau}\omega)\right] \qquad (20.179)$$

With the parametrization of $A_{\bar{z}}$ as in (20.178), the effect of this gauge transformation can be represented as

$$\chi \rightarrow \chi + \lambda$$
$$a \rightarrow a + m + n\tau \qquad (20.180)$$

The real part of χ can be set to zero by an appropriate choice of λ. (The imaginary part also vanishes when we impose the condition $F_{z\bar{z}} = 0$.) The physical subspace of the zero modes is given by the values of a modulo the transformation (20.180), or in other words,

$$\text{Physical space for zero modes} \equiv \mathcal{C}$$
$$= \frac{\mathbf{C}}{\mathbf{Z} + \tau \mathbf{Z}} \tag{20.181}$$

This space is known as the Jacobian variety of the torus. It is also a torus, and therefore we see that the phase space $\mathcal{C}$ has nontrivial Π_1 and $\mathcal{H}^2$. In particular, $\Pi_1(\mathcal{C}) = \mathbf{Z} \times \mathbf{Z}$, and this leads to two angular parameters φ_α and φ_β which can be related to the phases the wave functions acquire under the gauge transformation $g_{1,1}$. The symplectic two-form for the zero modes can be written as

$$\Omega = \frac{k\pi}{4} \frac{d\bar{a} \wedge da}{\text{Im}\tau} \int_\Sigma \frac{\bar\omega \wedge \omega}{\text{Im}\tau}$$
$$= -i\frac{k\pi}{2} \frac{d\bar{a} \wedge da}{\text{Im}\tau} \tag{20.182}$$

Integrating the zero-mode part over the physical space of zero modes $\mathcal{C}$, we get

$$\int_{\mathcal{C}} \Omega = k\pi \tag{20.183}$$

showing that k must be quantized as an even integer for $U(1)$ fields on the torus due to (20.58). (The integrality requirement on Ω arises from the use of wave functions which are one-dimensional, i.e., sections of a line bundle. If we use more general vector bundles, this quantization requirement can be relaxed; however, the probabilistic interpretation of such wave functions is not very clear.)

The symplectic potential for the zero modes can be written as

$$\mathcal{A} = -\frac{\pi k}{4} \frac{(\bar{a} - a)(\tau d\bar{a} - \bar\tau da)}{(\text{Im}\tau)^2} \tag{20.184}$$

The polarization condition then becomes

$$\left[\frac{\partial}{\partial \bar{a}} + i\frac{\pi k}{4} \frac{(\bar{a} - a)\tau}{(\text{Im}\tau)^2} \right] \psi = 0 \tag{20.185}$$

with the solution

$$\psi = \exp\left[-i\frac{\pi k}{8} \frac{(\bar{a} - a)^2 \tau}{(\text{Im}\tau)^2} \right] f(a) \tag{20.186}$$

where $f(a)$ is holomorphic in a. Under the gauge transformation (20.180) we find

$$\psi(a + m + n\tau) = \exp\left[-i\frac{\pi k(\bar{a} - a)^2 \tau}{8(\text{Im}\tau)^2} - \frac{\pi k n(\bar{a} - a)\tau}{2\text{Im}\tau} + i\frac{\pi k \tau n^2}{2} \right] f(a + m + n\tau) \tag{20.187}$$

Under this gauge transformation $\mathcal{A}$ changes by $d\Lambda_{m,n}$ where

$$\Lambda_{m,n} = i\frac{\pi kn(\tau\bar{a} - \bar{\tau}a)}{2\,\mathrm{Im}\tau} \tag{20.188}$$

The change in ψ should thus be given by $\exp(i\Lambda_{m,n})\psi$; requiring the transformation (20.187) to be equal to this, we get

$$f(a + m + n\tau) = \exp\left[-i\frac{\pi kn^2\tau}{2} - \pi ikna\right] f(a) \tag{20.189}$$

This shows that $f(a)$ is a Jacobi Θ-function. On these, $\bar{a}$ is realized as $(2\,\mathrm{Im}\tau/k\pi)(\partial/\partial a) + a$. The inner product for the wave functions of the zero modes is

$$\langle f|g\rangle = \int \exp\left[-\frac{\pi k\bar{a}a}{2\,\mathrm{Im}\tau} + \frac{\pi k\bar{a}^2}{4\,\mathrm{Im}\tau} + \frac{\pi ka^2}{4\,\mathrm{Im}\tau}\right] \bar{f}g \tag{20.190}$$

It is then convenient to introduce the wave functions

$$\Psi = \exp\left[\frac{\pi ka^2}{4\,\mathrm{Im}\tau}\right] f(a)$$

$$= \exp\left[\frac{\pi ka^2}{4\,\mathrm{Im}\tau}\right] \Theta(a) \tag{20.191}$$

On these functions, $\bar{a}$ acts as

$$\bar{a} = \frac{2\,\mathrm{Im}\tau}{\pi k}\frac{\partial}{\partial a} \tag{20.192}$$

20.6.8 θ-vacua in a nonabelian gauge theory

Consider a nonabelian gauge theory in four spacetime dimensions; the gauge group is some compact Lie group G. We can choose the gauge where $A_0 = 0$ so that there are only the three spatial components of the gauge potential, namely, A_i, considered as an antihermitian Lie-algebra-valued vector field. The choice $A_0 = 0$ does not completely fix the gauge, one can still do gauge transformations which are independent of time. These are given by

$$A_i \rightarrow A_i' = gA_ig^{-1} - \partial_i g\, g^{-1} \tag{20.193}$$

The Yang-Mills action gives the symplectic two-form as

$$\Omega = \int d^3x\,\, \delta E_i^a\,\, \delta A_i^a$$

$$= -2\int d^3x\,\, \mathrm{Tr}\,(\delta E_i\,\, \delta A_i) \tag{20.194}$$

where E_i^a is the electric field $\partial_0 A_i^a$, along the Lie algebra direction labeled by a. The gauge transformation of E_i is $E_i \rightarrow gE_i g^{-1}$. The vector field generating infinitesimal gauge transformations, with $g \approx 1 + \varphi$, is thus

$$\xi = -\int d^3x \left[(D_i\varphi)^a \frac{\delta}{\delta A_i^a} + [E_i, \varphi]^a \frac{\delta}{\delta E_i^a} \right] \tag{20.195}$$

This leads to

$$i_\xi \Omega = -\delta \int d^3x \, [-(D_i\varphi)^a E_i^a] \tag{20.196}$$

The generator of time-independent gauge transformations is thus

$$G(\varphi) = -\int d^3x \, (D_i\varphi)^a E_i^a \tag{20.197}$$

For transformations which go to the identity at spatial infinity, $G(\varphi) = \int \varphi^a G^a$, $G^a = (D_i E_i)^a$. $G^a = 0$ is one of the Yang-Mills equations of motion; it is the Gauss law of the theory. In the context of quantization, this is to be viewed as a condition on the allowed initial data and enforces a reduction of the phase space to gauge-invariant variables.

As discussed in Chapter 16, the spaces of interest are

$$\tilde{\mathcal{A}} = \left\{ \text{space of gauge potentials } A_i \right\} \tag{20.198}$$

$$\mathcal{G}_* = \left\{ \text{space of gauge transformations } g(\boldsymbol{x}) : \mathbf{R}^3 \rightarrow G \right.$$
$$\left. \text{such that } g \rightarrow 1 \text{ as } |\boldsymbol{x}| \rightarrow \infty \right\} \tag{20.199}$$

The transformations $g(\boldsymbol{x})$ which go to a constant element $g_\infty \neq 1$ act as a Noether symmetry. The states fall into unitary irreducible representations of such transformations, which are isomorphic to the gauge group G, up to $\mathcal{G}_*$-transformations. The true gauge freedom is only $\mathcal{G}_*$. The physical configuration space of the theory is thus $\mathcal{C} = \tilde{\mathcal{A}}/\mathcal{G}_*$. In Chapter 16, we also noted that $\mathcal{G}_*$ has an infinity of connected components so that

$$\mathcal{G}_* = \sum_{Q=-\infty}^{+\infty} \oplus \, \mathcal{G}_{*Q} \tag{20.200}$$

Q is the winding number characterizing the homotopy classes of gauge transformations. The space of gauge potentials $\tilde{\mathcal{A}}$ is an affine space and is topologically trivial. Combining this with $\Pi_0(\mathcal{G}_*) = \mathbf{Z}$, we see that the configurations space has noncontractible loops, with $\Pi_1(\mathcal{C}) = \mathbf{Z}$. Our general discussion shows that there must be an angle θ which appears in the quantum theory. We can see how this emerges by writing the symplectic potential.

The instanton number $\nu[A]$ for a four-dimensional potential is given by

$$\nu[A] = -\frac{1}{32\pi^2} \int d^4x \; \text{Tr} \left(F_{\mu\nu} F_{\alpha\beta} \right) \epsilon^{\mu\nu\alpha\beta}$$

$$= \frac{1}{16\pi^2} \int d^4x \; E_i^a F_{jk}^a \epsilon^{ijk} \tag{20.201}$$

The density in the above integral is a total derivative in terms of the potential A, but it cannot be written as a total derivative in terms of gauge-invariant quantities. $\nu[A]$ is an integer for any field configuration which is nonsingular up to gauge transformations. It is possible to construct configurations which have nonzero values of ν and which are nonsingular; these are the instantons, also considered briefly in Chapter 16.

We may think of configurations $A(\boldsymbol{x}, x_4)$ as giving a path in $\tilde{A}$ with x_4 parametrizing the path. $\nu[A]$ can be written as

$$\nu[A] = \oint K[A]$$

$$K[A] = \int d^3x \; F_{jk}^a \delta A_i^a \epsilon^{ijk} \tag{20.202}$$

The integral of the one-form K around a closed curve is the instanton number ν and is nonzero, in particular, for the loop corresponding to the instanton configuration. We can also see that this one-form is closed as follows:

$$\delta K[A] = -2 \int d^3x \; \delta\text{Tr} \left[F_{jk} \delta A_i \right] \epsilon^{ijk}$$

$$= -4 \int d^3x \; \text{Tr} \left[(D_j \delta A_k) \delta A_i \right] \epsilon^{ijk}$$

$$= -4 \int d^3x \; \text{Tr} \left[\partial_j \delta A_k \; \delta A_i + [A_j, \delta A_k] \delta A_i \right] \epsilon^{ijk}$$

$$= 0 \tag{20.203}$$

In the last step we have used the antisymmetry of the expression under permutation of δ's, cyclicity of the trace, and have done a partial integration. We see from the above discussion that $K[A]$ is a closed one-form which is not exact since its integral around the closed curves can be nonzero.

The general solution for the symplectic potential corresponding to the symplectic two-form in (20.194) is thus of the form

$$\mathcal{A} = \int d^3x \; E_i^a \delta A_i^a + \theta \; K[A] \tag{20.204}$$

Use of this potential will lead to a quantum theory where we need the parameter θ, in addition to other parameters such as the coupling constant, to characterize the theory. The potential $\mathcal{A}$ in (20.204) is obtained from an action

$$S = -\frac{1}{4} \int d^4x \; F^a_{\mu\nu} F^{a\mu\nu} + \theta \, \nu[A] \tag{20.205}$$

Thus the effect of using (20.204) can be reproduced in the functional integral approach by using the action (20.205). Since the relevant quantity for the functional integral is $\exp(iS)$, we see that θ is an angle with values $0 \leq \theta < 2\pi$. Alternatively, we can see that one can formally eliminate the θ-term in $\mathcal{A}$ by making a redefinition $\Psi \to \exp(i\theta\Lambda)\Psi$, where

$$\Lambda = -\frac{1}{8\pi^2} \int \mathrm{Tr}\left(AdA + \frac{2}{3}A^3\right) \tag{20.206}$$

Notice that $2\pi\Lambda$ is the Chern-Simons action (20.151) for $k = 1$. Λ is not invariant under homotopically nontrivial transformations. The wave functions get a phase equal to $e^{i\theta Q}$ under the winding number Q-transformation, showing that θ can be restricted to the interval indicated above. This is in agreement with our discussion after equation (20.52).

20.6.9 Current algebra for the Wess-Zumino-Witten (WZW) model

The WZW action was introduced in Chapter 17 in the context of evaluating the two-dimensional Dirac determinant. One can think of the WZW action in its own right as defining a field theory in two dimensions. For this one uses the Minkowski signature; the dynamical variables are group-valued fields $g(x^0, x^1)$. The action is given by

$$S(g) = -\frac{k}{8\pi} \int_{\mathcal{M}^2} d^2x \; \mathrm{Tr}(\partial_\mu g g^{-1} \partial^\mu g g^{-1}) + \frac{k}{12\pi} \int_{\mathcal{M}^3} \mathrm{Tr}(dg g^{-1})^3 \tag{20.207}$$

The first term involves integration over the two-dimensional manifold $\mathcal{M}^2$;the second term, the Wess-Zumino (WZ) term, requires extension of the fields to include one more coordinate, say, s, and corresponding integration. We can take $\mathcal{M}^3$ as a space whose boundary is the two-dimensional world, or we can take $\mathcal{M}^3 = \mathcal{M}^2 \times [0, 1]$ with fields at $s = 1$ corresponding to spacetime. Different ways of extending the fields to $s \neq 1$ will give the same physical results if the coefficient k is an integer. This quantization requirement arises from the single-valuedness of the transition amplitudes or wave functions. This result was also shown in Chapter 17; we just note here that the WZ term, being a differential form, is not sensitive to the signature of the metric and so the argument presented in Chapter 17 will be valid in the Minkowski case as well.

In the Minkowski coordinates we are using here, the Polyakov-Wiegmann identity becomes

$$S(hg) = S(h) + S(g) - \frac{k}{4\pi} \int d^2x \; \mathrm{Tr}(h^{-1}\partial_i h \partial_j g \; g^{-1})(\eta^{ij} + \epsilon^{ij}) \tag{20.208}$$

where $\eta^{ij} = diag(1, -1)$ is the two-dimensional metric and ϵ^{ij} is the Levi-Civita tensor. By taking small variations, $h \approx 1 + \theta, \theta \ll 1$, this identity gives the equation of motion

$$(\partial_0 - \partial_1)\left(\partial_0 g\, g^{-1} + \partial_1 g\, g^{-1}\right) = 0 \qquad (20.209)$$

This is also equivalent to

$$(\partial_0 + \partial_1)\left(g^{-1}\partial_0 g - g^{-1}\partial_1 g\right) = 0 \qquad (20.210)$$

There are two commonly used and convenient quantizations of this action which correspond to the equal-time and lightcone descriptions. There is a slight difficulty in obtaining the symplectic potential from the surface term resulting from time-integration. This is because the expression for the WZ term is written for a spacetime manifold which has no boundary, so that it can be the boundary of a three-volume. The variation of of the WZ term can be integrated to give

$$\delta\Gamma_{WZ} = \frac{k}{4\pi}\int_{\mathcal{M}^2} \mathrm{Tr}(\delta g g^{-1}I^2) \qquad (20.211)$$

where $I = dgg^{-1}$. Reintegrating this over the parameter s, we get the form of the WZ term written on $\mathcal{M}^2 \times [0,1]$,

$$\Gamma_{WZ} = \frac{k}{4\pi}\int_0^1 ds \int_{\mathcal{M}^2} \mathrm{Tr}\left[\partial_s g g^{-1}I^2\right] \qquad (20.212)$$

We will use this form to identify the symplectic potential.

First we consider the equal-time approach. The action, separating out the time derivatives, is

$$\mathcal{S}(g) = -\frac{k}{8\pi}\int_{\mathcal{M}^2} d^2x\, \mathrm{Tr}(\partial_0 g g^{-1})^2 + \Gamma_{WZ} \qquad (20.213)$$

If we vary this and look at the surface term from the integration over time, we get the symplectic potential as

$$\mathcal{A} = -\frac{k}{4\pi}\int dx\, \mathrm{Tr}(\xi I_0) + \frac{k}{4\pi}\int \mathrm{Tr}(\xi I^2) \qquad (20.214)$$

Here $\xi = \delta g g^{-1}$, and the last term still has integration over s as well as x. Exterior derivatives are given by

$$\delta\xi = \xi^2$$
$$\delta I_1 = \partial_1 \xi + \xi I_1 - I_1 \xi \qquad (20.215)$$

Upon taking exterior derivatives, the second term in $\mathcal{A}$ becomes a total derivative and can be integrated over s as follows. Using $I^2 = dI$,

$$\delta \frac{k}{4\pi} \int \mathrm{Tr}(\xi I^2) = \frac{k}{4\pi} \int \mathrm{Tr}\left[\xi^2 I^2 - \xi d(d\xi + \xi I - I\xi)\right]$$

$$= \frac{k}{4\pi} \int \mathrm{Tr}\left[-d(\xi^2 I)\right]$$

$$= \frac{k}{4\pi} \int \mathrm{Tr}\left[-\partial_1(\xi^2 I_s) + \partial_s(\xi^2 I_1)\right]$$

$$= \frac{k}{4\pi} \int \mathrm{Tr}\left[\xi^2 I_1\right] \tag{20.216}$$

The symplectic two-form is now obtained as

$$\Omega = -\frac{k}{4\pi} \int dx \, \mathrm{Tr}\left[\xi^2 I_0 - \xi^2 I_1 - \xi \delta I_0\right] \tag{20.217}$$

Notice that I_0 must be considered as an independent variable, as is usually done in equal-time quantization.

Consider a vector field $V_1(\theta)$ whose interior contraction has the effect of replacing ξ by θ. If we expand $\xi = \delta g g^{-1} = (-it^a)E_{ab}\delta\chi^b$, we can explicitly write

$$V_1(\theta) = \int dx \, (E^{-1})_{bc}\theta^c \frac{\delta}{\delta\chi^b} \tag{20.218}$$

The action of $V_1(\theta)$ on g is to make the left translation $V_1(\theta)g = (-it^a)\theta^a g$. We also define

$$V_2(\theta) = \int dx \, f^{abc}\theta^b(I_0 - I_1)^c \frac{\delta}{\delta I_0^a} \tag{20.219}$$

which has the effect of replacing δI_0 by $[\theta, I_0 - I_1]$ upon taking a contraction with Ω. The contraction of $V = V_1 + V_2$ with Ω gives

$$i_V \Omega = -\delta \int dx \left[\frac{k I_0^a \theta^a}{8\pi}\right]$$

$$\equiv -\delta J_0(\theta) \tag{20.220}$$

Thus V is a Hamiltonian vector field corresponding to J_0.

Another vector field of interest is

$$W(\theta) = \int dx \, \left(\partial_1\theta^a + f^{abc}\theta^b I_1^c\right)\frac{\delta}{\delta I_0^a} \tag{20.221}$$

This has the effect of replacing δI_0 by $\partial_1\theta + [\theta, I_1]$. Contraction with Ω gives

$$i_W \Omega = -\delta \int dx \left[\frac{k}{8\pi} I_1^a \theta^a\right]$$

$$\equiv -\delta J_1(\theta) \tag{20.222}$$

Thus W is a Hamiltonian vector field corresponding to J_1.

The currents of interest are, once again,

$$J^a_\mu = \frac{k}{8\pi}(\partial_\mu gg^{-1})^a \tag{20.223}$$

For the Poisson brackets, we find

$$\{J_0(\theta), J_0(\varphi)\} = -i_V \delta J_0(\varphi) = i_V \int dx \; \text{Tr} \; \delta\left(\frac{k}{4\pi}I_0\varphi\right)$$

$$= \frac{k}{4\pi} \int dx \; \text{Tr} \; [\theta, I_0 - I_1]\varphi$$

$$= J_0(\theta \times \varphi) \; - \; J_1(\theta \times \varphi) \tag{20.224}$$

$$\{J_0(\theta), J_1(\varphi)\} = i_V \int dx \; \text{Tr} \left(\frac{k}{4\pi}\delta I_1\varphi\right)$$

$$= J_1(\theta \times \varphi) - \frac{k}{8\pi}\partial_1\theta^a \; \varphi^a \tag{20.225}$$

$$\{J_1(\theta), J_1(\varphi)\} = -i_W \int dx \; \text{Tr} \left(\frac{k}{4\pi}\delta I_1\varphi\right)$$

$$= 0 \tag{20.226}$$

In these equations $(\theta \times \varphi)^a = f^{abc}\theta^b\varphi^c$. These can be combined to yield

$$\{J_+(\theta), J_+(\varphi)\} = J_+(\theta \times \varphi) - \frac{k}{4\pi}\partial_1\theta^a\varphi^a \tag{20.227}$$

where $J_+ = J_0 + J_1$. The classical Hamiltonian is given by

$$H = \frac{4\pi}{k} \int dx \, [J^a_0 J^a_0 + J^a_1 J^a_1] \tag{20.228}$$

In the lightcone quantization, we introduce coordinates u, v,

$$u = \frac{1}{\sqrt{2}}(t - x), \qquad\qquad v = \frac{1}{\sqrt{2}}(t + x)$$

$$\partial_u = \frac{1}{\sqrt{2}}(\partial_0 - \partial_1), \qquad\qquad \partial_v = \frac{1}{\sqrt{2}}(\partial_0 + \partial_1) \tag{20.229}$$

This is different from (20.143), but given the structure of the equations of motion (20.209), these are easier. The action becomes

$$S = -\frac{k}{4\pi} \int \text{Tr}(\partial_u gg^{-1}\partial_v gg^{-1}) + \Gamma_{WZ} \tag{20.230}$$

We take v as the analog of the spatial coordinate and consider evolution in u. The surface term for u-integration will arise from variations on the $\partial_u gg^{-1}$-term. This gives

$$\mathcal{A} = -\frac{k}{4\pi} \int_v \text{Tr}(\xi\partial_v gg^{-1}) + \frac{k}{4\pi} \int_{v,s} \text{Tr}(\xi(dgg^{-1})^2) \tag{20.231}$$

Once again, the last term still involves the s-integration, but upon making another variation, we can integrate this as before and get

$$\Omega = \frac{k}{4\pi} \int dv \ \mathrm{Tr} \left[\xi \partial_v \xi + 2\xi^2 \ I_v \right] \tag{20.232}$$

The equation of motion is $\partial_u I_v = 0$. The contraction of $V_1(\theta)$ as defined in (20.218) gives

$$i_{V_1} \Omega = \frac{k}{4\pi} \int dv \ \mathrm{Tr} \left[2\theta \partial_v \xi + 2\theta \xi I_v - 2\xi \theta I_v \right]$$

$$= \frac{k}{2\pi} \int dv \ \mathrm{Tr} \left[\theta (\partial_v \xi + [\xi, I_v]) \right]$$

$$= -\delta \int dv \ \left[\frac{k I_v^a \theta^a}{4\pi} \right]$$

$$\equiv -\delta J_v(\theta) \tag{20.233}$$

The current J_v is given by

$$J_v^a = \frac{k}{4\pi} (\partial_v g g^{-1})^a \tag{20.234}$$

The Poisson brackets are given by

$$\{ J_v(\theta), J_v(\varphi) \} = i_{V_1} \frac{k}{2\pi} \int \mathrm{Tr}(\delta I_v \varphi)$$

$$= \frac{k}{2\pi} \int \mathrm{Tr} \left[\partial_v \theta \varphi + \theta I_v \varphi - I_v \theta \varphi \right]$$

$$= \frac{k}{4\pi} \int I_v^a (\theta \times \varphi)^a - \frac{k}{4\pi} \int \partial_v \theta^a \varphi^a$$

$$= J_v(\theta \times \varphi) - \frac{k}{4\pi} \int \partial_v \theta^a \varphi^a \tag{20.235}$$

The algebra of currents given by (20.227), or the light-cone version (20.235), is an example of a Kac-Moody algebra.

References

1. Two general books on geometric quantization are: J.Sniatycki, *Geometric Quantization and Quantum Mechanics*, Springer-Verlag (1980); N.M.J. Woodhouse, *Geometric Quantization*, Clarendon Press (1992).
2. For the discussion of symplectic structure and classical dynamics, see V.I. Arnold, *Mathematical Methods of Classical Mechanics*, Springer-Verlag (1978); V. Guillemin and S. Sternberg, *Symplectic Techniques in Physics*, Cambridge University Press (1990); J.V. José and E.J. Saletan, *Classical Dynamics: A Contemporary Approach*, Cambridge University Press (1998). The proof of Darboux's theorem can be found in these books.

3. A different proof of Darboux's theorem is outlined in R. Jackiw, *Diverse Topics in Theoretical and Mathematical Physics*, World Scientific Pub. Co. (1995).

4. Coherent states are very useful in diverse areas of physics, see, for example, J.R. Klauder and Bo-Sturé Skagerstam, *Coherent States: Applications in Physics and Mathematical Physics*, World Scientific Pub. Co. (1985).

5. The quantization of the two-sphere and other Kähler G/H spaces is related to the Borel-Weil-Bott theory and the work of Kostant, Kirillov and Souriau; this is discussed in the books in reference 1. In this context, see also A.M. Perelomov, *Generalized Coherent States and Their Applications*, Springer-Verlag (1996).

6. Our treatment of the charged particle in a monopole field is closely related to the work of A.P. Balachandran, G. Marmo and A. Stern, Nucl. Phys. **B162**, 385 (1980).

7. The argument relating Dirac quantization to the quantization of angular momentum is given in M.N. Saha, Indian J. Phys. **10**, 141 (1936).

8. Anyons have been around in physics literature for a while; there is also evidence that the quasi-particles in quantum Hall effect can be interpreted as anyons. For early work, see E. Merzbacher, Am. J. Phys. **30**, 237 (1960); J. Leinaas and J. Myrheim, Nuovo Cimento **37**, 1 (1977); G. Goldin, R. Menikoff and D. Sharp, J. Math. Phys. **21**, 650 (1980); *ibid.* **22**, 1664 (1981); F. Wilczek, Phys. Rev. Lett. **49**, 957 (1982); F .Wilczek and A. Zee, Phys. Rev. Lett. **51**, 2250 (1983). For a recent review, see F. Wilczek, *Fractional Statistics and Anyon Superconductivity* (World Scientific, Singapore, 1990). For anyons in the quantum Hall system, see R.B. Laughlin, Phys. Rev. Lett. **50**, 1395 (1983); various articles in *Physics and Mathematics of Anyons*, S.S. Chern, C.W. Chu and C.S. Ting (eds.) (World Scientific, Singapore, 1991).

9. Our geometric description is based on R. Jackiw and V.P. Nair, Phys. Rev. **D43**, 1933 (1991). This symplectic form is also related to the discussion of spinning particles in A.P. Balachandran, G. Marmo, B-S. Skagerstam and A. Stern, *Gauge Symmetries and Fibre Bundles*, Springer-Verlag (1983). There are other approaches to anyons; some early works are C.R. Hagen, Ann. Phys.(NY) **157**, 342 (1984); Phys. Rev. **D31**, 848, 2135 (1985); D. Arovas, J. Schrieffer, F. Wilczek and A. Zee, Nucl. Phys. **B251**, 117 (1985); M.S. Plyushchay, Phys. Lett. **B248**, 107 (1990); Nucl. Phys. **B243**, 383 (1990); D. Son and S. Khlebnikov, JETP Lett. **51**, 611 (1990); D. Volkov, D. Sorokin and V. Tkach, in *Problems in Modern Quantum Field Theory*, A. Belavin, A. Klimyk and A. Zamolodchikov (eds.), Springer-Verlag (1989).

10. The result $g = 2$ for anyons is in C. Chou, V.P. Nair and A. Polychronakos, Phys. Lett. **B304**, 105 (1993).

11. The Chern-Simons term is due to S.S. Chern and J. Simons, Ann. Math. **99**, 48 (1974). It was introduced into physics literature by R. Jackiw and S. Templeton, Phys. Rev. **D23**, 2291 (1981); J. Schonfeld, Nucl. Phys. **B185**, 157 (1981); S. Deser, R. Jackiw and S. Templeton, Phys. Rev. Lett. **48**, 975 (1982); Ann. Phys. **140**, 372 (1982). By now it has found applications in a wide variety of physical and mathematical problems. In a brilliant paper, Witten showed that the Chern-Simons theory leads to the Jones polynomial and other knot invariants, E. Witten, Commun. Math. Phys. **121**, 351 (1989). Some of the early papers on the Hamiltonian quantization are: M. Bos and V.P. Nair, Phys. Lett. **B223**, 61 (1989); Int. J. Mod. Phys. **A5**, 959 (1990) (we follow this work, mostly); S. Elitzur, G. Moore, A. Schwimmer and N. Seiberg, Nucl. Phys. **B326**, 108 (1989); J.M.F. Labastida and A.V. Ramallo, Phys. Lett. **B227**, 92 (1989); H. Murayama, Z. Phys. **C48**, 79 (1990); A.P. Polychronakos, Ann. Phys. **203**, 231 (1990); T.R. Ramadas, I.M. Singer and J. Weitsman, Comm. Math. Phys. **126**, 409 (1989); A.P. Balachandran, M. Bourdeau and S. Jo Mod. Phys. Lett. **A4**, 1923 (1989); G.V. Dunne, R. Jackiw and C.A. Trugenberger, Ann.Phys. **149**, 197 (1989).

12. Geometric quantization of the Chern-Simons theory is discussed in more detail in S. Axelrod, S. Della Pietra and E. Witten, J. Diff. Geom. **33**, 787 (1991).

13. References to the θ-parameter have been given in Chapter 16, C.G. Callan, R. Dashen and D. Gross, Phys. Lett. **B63**, 334 (1976); R. Jackiw and C. Rebbi, Phys. Rev. Lett. **37**, 172 (1976). If the spatial manifold is not simply connected one may have more vacuum angles; see, for example, A.R. Shastri, J.G. Williams and P. Zwengrowski, Int. J. Theor. Phys. **19**, 1 (1980); C.J. Isham and G. Kunstatter, Phys. Lett. **B102**, 417 (1981); J. Math. Phys. **23**, 1668 (1982).

14. The WZW model was introduced by E. Witten in connection with non-abelian bosonization, E.Witten, Commun. Math. Phys. **92**, 455 (1984). This model has become very important because it is a two-dimensional conformal field theory and various rational conformal field theories can be obtained from it by choice of the group and level number. For a survey of some of these developments, see P. Di Francesco, P. Mathieu and D. Senechal, *Conformal Field Theory*, Springer-Verlag (1996). The Poisson brackets we derive go back to Witten's original paper.

Appendix:Relativistic Invariance

A-1 Free point-particles and the Poincaré algebra

In this appendix we shall briefly discuss the relativistic wave functions of free particles. The symmetry operations relevant to relativistic invariance are translations (in space and time) and Lorentz transformations (including spatial rotations). To obtain the wave functions, we need a Hilbert space on which we can realize translations and Lorentz transformations as unitary transformations. In other words, we need a unitary irreducible representation (UIR) of the algebra of translations and Lorentz transformations. The requirement of unitarity is clear since all physical observables generate unitary transformations on the Hilbert space in quantum mechanics. The qualification of irreducibility is a little more subtle, so we take a moment to recall that in nonrelativistic quantum mechanics we have a similar situation. For one-particle dynamics in one dimension, the relevant operators are $\hat{x}$ and $\hat{p}$ with the commutation rules

$$[\hat{x}, \hat{x}] = 0, \qquad [\hat{p}, \hat{p}] = 0, \qquad [\hat{x}, \hat{p}] = i \tag{A-1}$$

We need a unitary representation of this algebra of observables. One way to represent the operators $\hat{x}, \hat{p}$ is as follows. We can consider complex-valued functions $\psi(x, p)$ with the normalization condition

$$\int dp\, dx\ \psi(x, p)^* \psi(x, p) = 1 \tag{A-2}$$

The operators act by

$$\hat{p}\psi = -i\frac{\partial}{\partial x}\psi$$

$$\hat{x}\,\psi = \left(i\frac{\partial}{\partial p} + x\right)\psi \tag{A-3}$$

Clearly, this gives the wrong quantum mechanics since we are specifying the wave functions as functions of x and p. We can also see that this representation is reducible. In fact, we see that we can impose a condition

$$\frac{\partial \psi}{\partial p} = 0 \tag{A-4}$$

on the wave functions and still obtain a representation of (A-1). In particular, in this case,

$$\hat{p}\psi = -i\frac{\partial}{\partial x}\psi, \qquad \hat{x}\psi = x\ \psi \qquad (A-5)$$

The normalization condition can now be taken as

$$\int dx\ \psi^*\psi = 1 \qquad (A-6)$$

This is the standard Schrödinger representation. Since we are able to obtain a representation on this smaller space of functions obeying the condition (A-4), the former representation (A-3) is reducible. The Schrödinger representation can be shown to be irreducible; i.e., there is no smaller function space on which the algebra (A-1) can be realized. (Properly speaking, one should consider bounded operators obtained by exponentiation.) We thus see that quantum mechanics may be identified as a unitary irreducible representation (UIR) of the algebra of observables. (For the algebra (A-1), there is only one representation, the Schrödinger representation, up to unitary equivalence, a result due to Stone and von Neumann. With an infinite number of degrees of freedom, or on nonsimply connected spaces, there can be many UIR's and the physical consequences can be different depending on which UIR one chooses. This is not an issue of immediate relevance to our discussion; we have discussed some of these issues in Chapters 12, 15 and 20.)

Returning to the question of the relativistic point-particle, we consider the symmetry transformations in some more detail. The action for the motion of a relativistic point-particle is the mass m times the proper distance, or

$$S = \int dt\ L = -m \int \sqrt{\eta_{\mu\nu}dx^\mu dx^\nu} \qquad (A-7)$$

The metric tensor in the above expression is the Minkowski metric with Cartesian components $\eta_{00} = 1$, $\eta_{ij} = -\delta_{ij}$ and all other components being zero. The symmetries of the theory are clearly the symmetries of the proper distance $ds = \sqrt{\eta_{\mu\nu}dx^\mu dx^\nu}$. We first consider the continuous symmetries which can be understood in terms of infinitesimal transformations. We write

$$x^\mu \to x'^\mu = x^\mu + \xi^\mu \qquad (A-8)$$

where ξ^μ is infinitesimal. The requirement that this be a symmetry transformation of the proper distance gives

$$\eta_{\alpha\nu}\frac{\partial\xi^\alpha}{\partial x^\mu} + \eta_{\mu\alpha}\frac{\partial\xi^\alpha}{\partial x^\nu} = 0 \qquad (A-9)$$

The solution to this condition is given by

$$\xi^\mu = a^\mu + \omega^{\mu\nu}x_\nu \qquad (A-10)$$

where $a^\mu, \omega^{\mu\nu}$ are constant parameters and $\omega^{\mu\nu} = -\omega^{\nu\mu}$. $\omega^{\mu\nu}$ are the six parameters of Lorentz transformations, ω^{0i} being the relative velocities of the frames connected by the transformation (A-10). $\theta^i = \frac{1}{2}\epsilon^{ijk}\omega_{jk}$ are the angles of spatial rotations and a^μ are the parameters of translations. For any function of x^μ, we can write the generators of the transformations immediately from (A-8, A-10).

$$\delta f(x) = \xi^\mu \frac{\partial}{\partial x^\mu} f = a^\mu \frac{\partial f}{\partial x^\mu} + \omega^{\mu\nu} x_\nu \frac{\partial f}{\partial x^\mu}$$

$$= \left(i\, a^\mu P_\mu - \frac{i}{2}\omega^{\mu\nu} M_{\mu\nu} \right) f \tag{A-11}$$

where

$$P_\mu = -i\frac{\partial}{\partial x^\mu}, \qquad M_{\mu\nu} = x_\mu P_\nu - x_\nu P_\mu \tag{A-12}$$

(The factors of i are convenient for the sake of hermiticity; these operators will be interpreted later as physical quantities.) The commutation rules of these operators are easily worked out to be

$$[P_\mu, P_\nu] = 0$$
$$[M_{\mu\nu}, P_\alpha] = i(\eta_{\mu\alpha} P_\nu - \eta_{\alpha\nu} P_\mu) \tag{A-13}$$
$$[M_{\mu\nu}, M_{\alpha\beta}] = i\,(\eta_{\mu\alpha} M_{\nu\beta} + \eta_{\nu\beta} M_{\mu\alpha} - \eta_{\mu\beta} M_{\nu\alpha} - \eta_{\nu\alpha} M_{\mu\beta})$$

This algebra of translations and Lorentz transformations is the Poincaré algebra. Although we obtained this algebra by considering the infinitesimal transformations on a scalar function, it is of general validity. One can also characterize particles with spins in terms of this algebra.

For example, for a vector-valued function $A_\mu(x)$, the transformation rule can be worked out by treating $A_\mu dx^\mu$ as a scalar; i.e., $\delta A_\mu dx^\mu = A_\mu(x')\, dx'^\mu - A_\mu(x) dx^\mu$. Explicitly,

$$\delta A_\mu = \xi^\nu \frac{\partial A_\mu}{\partial x^\nu} + \frac{\partial \xi^\nu}{\partial x^\mu} A_\nu$$

$$= \left(ia^\alpha P_\alpha - \frac{i}{2}\omega^{\alpha\beta} M_{\alpha\beta} \right)_\mu^{\ \nu} A_\nu \tag{A-14}$$

where

$$P_\alpha = i\frac{\partial}{\partial x^\alpha}$$
$$M_{\alpha\beta} = (x_\alpha P_\beta - x_\beta P_\alpha) + S_{\alpha\beta}$$
$$(S_{\alpha\beta})_{\mu\nu} = -i(\eta_{\mu\alpha}\eta_{\nu\beta} - \eta_{\nu\alpha}\eta_{\mu\beta}) \tag{A-15}$$

$S_{\alpha\beta}$ is to be thought of as a 4×4-matrix, α, β specifying the type of Lorentz transformation we are interested in and μ, ν specifying the matrix elements. Equation (A-15) is in matrix notation, i.e., expressions like $a^\alpha P_\alpha, x_\alpha P_\beta -$

$x_\beta P_\alpha$ are proportional to the identity matrix δ_μ^ν. One can easily verify that P_α, $M_{\alpha\beta}$ in (A-15) obey the same commutation rules as in (A-13), with matrix multiplication understood for $S_{\alpha\beta}$. $S_{\alpha\beta}$ is the spin contribution to the Lorentz generators.

For a constant vector k_μ, the Lorentz transformation is generated entirely by $S_{\mu\nu}$. The finite transformation is given by the composition of many infinitesimal transformations as

$$k'_\mu = \left[\lim_{N \to \infty} \left(1 - \frac{i}{2} \frac{\omega^{\alpha\beta}}{N} S_{\alpha\beta} \right)^N \right]_{\mu\nu} k^\nu$$

$$= \left[\exp(-\frac{i}{2} \omega^{\alpha\beta} S_{\alpha\beta} \right]_\mu^\nu k_\nu \qquad \text{(A-16)}$$

$$\equiv L_\mu^\nu k_\nu$$

The matrix L_μ^ν obeys

$$L_\mu^\alpha L_\nu^\beta \eta_{\alpha\beta} = \eta_{\mu\nu} \qquad \text{(A-17)}$$

and further $det\ L = 1$, $L_0^0 = 1$. (These are sometimes further qualified as proper orthochronous Lorentz transformations. Improper transformations can have a determinant equal to -1; they can be understood in terms of composing the proper orthochronous transformations with discrete transformations such as parity and time-reversal. The discrete transformations which are also symmetries of the proper distance will not be discussed here; they are most easily understood in the functional integral language and are explained in chapter 12.)

The Poincaré algebra (A-13) is the basic algebra of observables for free point-particles. From our understanding of quantum mechanics, we can thus obtain free-particle wave functions by studying the unitary realizations of the Poincaré algebra. P_μ, being the generator of translations, will be the four-momentum of the particle, $P_0 = H$ being the Hamiltonian. M_{ij} will generate spatial rotations and hence $J^i = \frac{1}{2}\epsilon^{ijk} M_{jk}$ is the angular momentum. $M_{0i} \equiv K_i$ generate Lorentz transformations connecting frames of reference which are in relative motion; these are the so-called boosts. One can decompose the Poincaré algebra in terms of these components as follows:

$$[P_i, P_j] = 0, \qquad [P_i, H] = 0$$
$$[J_i, P_j] = i\ \epsilon_{ijk} P^k, \qquad [J_i, H] = 0$$
$$[K_i, P_j] = i\ \delta_{ij} H, \qquad [K_i, H] = iP_i$$

$$[J_i, J_j] = i\ \epsilon_{ijk} J^k \qquad \text{(A-18)}$$
$$[J_i, K_j] = i\ \epsilon_{ijk} K^k$$
$$[K_i, K_j] = -i\epsilon_{ijk} J^k$$

In arriving at these equations, it is useful to remember that ϵ^{ijk} is the numerical Levi-Civita symbol, so that $\epsilon^{ijk}\epsilon^{iab} = \delta^{ja}\delta^{kb} - \delta^{jb}\delta^{ka}$.

A-2 Unitary representations of the Poincaré algebra

We begin the construction of unitary realizations of the Poincaré algebra by noticing that the P_μ commute among themselves and so can be simultaneously diagonalized. We can thus define a set of states, $|p\rangle$, depending on a four-vector p_μ such that

$$P_\mu|p\rangle = p_\mu|p\rangle \tag{A-19}$$

Now, from the commutation rule (A-13), we can work out the finite Lorentz transformation of P_μ as

$$P'_\mu = L(\omega)^\nu_\mu P_\nu \tag{A-20}$$

(A-13) is the infinitesimal version of this result with $P'_\mu = U(\omega)P_\mu U^{-1}(\omega)$, $U(\omega) = \exp(-\frac{i}{2}\omega^{\mu\nu}M_{\mu\nu})$. For the eigenvalue four-momentum p_μ, we thus have $p'_\mu = L^\nu_\mu p_\nu$.

From the transformation law for P_μ, it is clear that P^2 is invariant under Lorentz transformations, and since the P's are commuting operators, P^2 commutes with all P_μ and $M_{\mu\nu}$; i.e., it is a Casimir operator for the Poincaré algebra. The possible representations can therefore be specified by the value of P^2. The possibilities are

1. $P^2 > 0$,   say, $P^2 = m^2$
2. $P^2 < 0$
3. $P^2 = 0$.

$P^2 > 0$ will describe massive particles of mass m. (In fact the condition $P^2 = m^2$ may be taken as the definition of the mass m.) $P^2 < 0$ is unphysical, corresponding to propagation faster than light. $P^2 = 0$ will describe massless particles such as the photon.

There is also another Casimir operator for the Poincaré algebra. This is given as W^2, where W^μ is defined as

$$W^\mu = \epsilon^{\mu\nu\alpha\beta}P_\nu M_{\alpha\beta} \tag{A-21}$$

W^μ is called the Pauli-Lubanski spin vector. We can use W^2 to characterize the representations further. However, instead of analyzing the possible values of W^2 in general, we shall simply analyze the representations of interest and calculate W^2 for these.

Given that $p^2 = m^2$ or zero, for the states $|p\rangle$, we have $p_0 = \pm\sqrt{\mathbf{p}^2 + m^2}$ (or $p_0 = \pm\sqrt{\mathbf{p}^2}$). It is easy to see from the Lorentz transformation of p_μ that L^ν_μ do not change the sign of p_0. Therefore we further characterize representations by the sign of p_0; for the physically interesting cases (with positive energy), we have

$$p^2 = m^2, \qquad p_0 > 0$$
$$p^2 = 0, \qquad p_0 > 0. \tag{A-22}$$

The states $|p\rangle$ are thus labeled by the three-momentum $\boldsymbol{p}$, with $p_0 = \sqrt{\boldsymbol{p}^2 + m^2}$ (or $p_0 = \sqrt{\boldsymbol{p}^2}$ for the massless case). These states for all of p-space obeying the conditions (A-22) should give us a complete set of states. Because (A-22) are Lorentz-invariant, we choose a Lorentz-invariant measure of p-integration as

$$d\mu(p) = \frac{d^4p}{(2\pi)^3}\delta(p^2 - m^2)\theta(p_0) \tag{A-23}$$

(with $m = 0$ for the massless case). The p_0-part of the integration with this measure is trivial; carrying out this integration, we get, instead of (A-23),

$$d\mu(p) = \frac{d^3p}{(2\pi)^3}\frac{1}{2\omega_p} \tag{A-24}$$

where ω_p is the positive square root solution for p_0, $\omega_p = \sqrt{\boldsymbol{p}^2 + m^2}$; we integrate over all of $\boldsymbol{p}$. The completeness condition for the states $|p\rangle$ can be written as

$$\int |p\rangle \frac{d^3p}{(2\pi)^3}\frac{1}{2\omega_p}\langle p| = 1 \tag{A-25}$$

Since P_i is hermitian, we must have orthogonality of states of different values of $\boldsymbol{p}$. The orthonormality condition corresponding to the completeness relation (A-25), is

$$\langle p|p'\rangle = (2\pi)^3 2\omega_p\,\delta^{(3)}(p - p') \tag{A-26}$$

A-3 Massive particles

So far, we have only specified the action of the momentum operators P_μ. We must now specify how Lorentz transformations act on $|p\rangle$. For this we follow the procedure of Wigner. The strategy is to go to a special frame, suitably chosen, construct a representation there and bring it back to a general frame by appropriate Lorentz transformations. To begin with we must thus choose a special p_μ obeying (A-22). For the massive case, which we shall discuss first, we can take

$$p_\mu = p_\mu^{(0)} = (m, 0, 0, 0) \tag{A-27}$$

This corresponds to the rest frame of the particle. There is a special Lorentz transformation $L(\omega_0) = B(p)$, which gives p_μ from $p_\mu^{(0)}$, i.e.,

$$p_\mu = B(p)_\mu^\nu\, p_\nu^{(0)} \tag{A-28}$$

Actually there are many choices for B. Explicitly, one choice is given by

$$B_\mu^{\ \nu} = \delta_\mu^\nu - \frac{(p_\mu + m\eta_\mu)(p^\nu + m\eta^\nu)}{m(p_0 + m)} + \frac{2p_\mu\eta^\nu}{m} \tag{A-29}$$

where η_μ is a fixed vector with $\eta_0 = 1$, $\eta_i = 0$.

There are still some transformations we can do in the rest frame. The transformations $L(\omega)$ such that $L(\omega)p^{(0)} = p^{(0)}$ form the *little group* or the isotropy group of $p^{(0)}$. (Here $(Lp^{(0)})_\mu = L_\mu^\nu p_\nu^{(0)}$; we use an obvious matrix notation.) In our case, it is evident that these are the spatial rotations $R(\omega) = L(\omega)$, with $\omega^{0i} = 0$. (Clearly such rotations are an ambiguity in the choice of B also.) Unitary representation of these rotations is well known; it is standard angular momentum theory. A representation is characterized by the highest value, denoted s, of a component of angular momentum, say, J_3. There are $(2s + 1)$ states and we have the transformation rule

$$U(\theta)\,|s, n\rangle = \sum_{n'} \mathcal{D}_{nn'}^{(s)}(\theta)\,|s, n'\rangle \qquad (A\text{-}30)$$

Here $\theta^i = \frac{1}{2}\epsilon^{ijk}\omega_{jk}$. The $\mathcal{D}_{nn'}^{(s)}(\theta)$ are the standard Wigner $\mathcal{D}$-matrices of angular momentum theory; they obey the unitarity condition $\mathcal{D}^\dagger \mathcal{D} = 1$, in matrix notation. We may write them out as $\mathcal{D}_{nn'}^{(s)}(\theta) = \langle s, n|e^{i\theta^k J_k}|s, n'\rangle$ where J_k are angular momentum generators and $|s, n\rangle$ are the spin-$s$angular momentum states.

The crucial observation is that the representation for the full Poincaré group can be obtained on states of the form $|p, s, n\rangle$. In other words, a basis for the Hilbert space is given by products of the form $|p\rangle \otimes |s, n\rangle$, where $|s, n\rangle$ provide a unitary representation of the little group, in this case, rotations of $p^{(0)}$. The action of a general Lorentz transformation can be obtained from the action of rotations as in (A-30). For this one constructs, for every Lorentz transformation, a pure spatial rotation, called a Wigner rotation, given by

$$R(\theta_W) = B^{-1}(L(\omega)p)\,L(\omega)\,B(p) \qquad (A\text{-}31)$$

Since $R(\theta_W)p^{(0)} = p^{(0)}$, we see that the combination on the right-hand side of equation (A-31) is indeed a pure rotation. Thus for every $\omega^{\mu\nu}$, we can associate rotation angles θ_W^i given by (A-31). The action of a general Lorentz transformation with parameters $\omega^{\mu\nu}$ can then be defined as

$$U(\omega)|p, s, n\rangle = \sum_{n'} \mathcal{D}_{nn'}^{(s)}(\theta_W(p, \omega))\,|Lp, s, n'\rangle \qquad (A\text{-}32)$$

Since the action of rotations on $|s, n\rangle$ is unitary, i.e., $\mathcal{D}^\dagger \mathcal{D} = 1$, we see that (A-32) defines a unitary realization of Lorentz transformations. This can be explicitly checked with the scalar product (A-26). A representation such as (A-32) is called an *induced representation* since the action of Lorentz transformations is induced from the action of rotations.

The nature of massive one-particle states is now clear. They are characterized by the spin s with $(2s + 1)$ polarization states. s determines the $\mathcal{D}$-matrices to be used. The action of translations on such states is given by $e^{-iP\cdot a}$, which is simply a factor $e^{-ip\cdot a}$ on the momentum eigenfunctions. Lorentz transformations, including rotations, act on the states as in (A-32).

It is instructive to work out the Wigner rotation explicitly for an infinitesimal Lorentz transformation. From (A-31)

$$R \approx 1 - B^{-1}\omega B + (\delta B^{-1}) B \qquad \text{(A-33)}$$

Working out the components, we find that

$$R^0_0 = 1, \quad R^i_0 = 0 = R^0_i, \quad R^j_i = \delta^j_i - \omega^j_i - \frac{1}{(p_0 + m)}(\omega_{0i}p^j - \omega^{0j}p_i) + \cdots \qquad \text{(A-34)}$$

This identifies the Wigner rotation parameter as

$$\theta^i_W(p,\omega) \approx \frac{1}{2}\epsilon^{ijk}\left(\omega_{jk} + \frac{2}{(p_0 + m)}\omega_{0j}p_k\right) \qquad \text{(A-35)}$$

The Pauli-Lubanski spin operator, for $p^{(0)}_\mu$, reduces to the angular momentum operator, $W^0 = 0$, $W^i = mJ^i$. This explains why W^μ is called the spin operator. W^2 is given by $m^2 s(s+1)$ for the representations appropriate to massive particles.

A-4 Wave functions for spin-zero particles

For a particle of spin zero, since $s = 0$, we do not have any nontrivial $\mathcal{D}$-matrices for the transformations. The action of a finite translation by a^μ on the state $|p\rangle$ is given by

$$e^{-iP \cdot a}|p\rangle = e^{-ipa}|p\rangle \qquad \text{(A-36)}$$

This shows that the x-space wave function for a particle of momentum p can be taken as

$$\psi_p(x) = \langle x|p\rangle = e^{-ipx}, \qquad \text{(A-37)}$$

since $x \to x + a$ is reproduced by the action of $e^{-iP \cdot a}$ as in (A-36). (Here px denotes $p \cdot x = p^0 x^0 - \boldsymbol{p} \cdot \boldsymbol{x}$.) $\psi_p(x)$ is a Lorentz scalar. The scalar product for the x-space wave functions is taken as

$$\int d^3x \; [\psi_p(x)^*(i\partial_0\psi_{p'}(x)) - (i\partial_0\psi_p(x)^*)\psi_{p'}(x)] = 2p_0(2\pi)^3\delta^{(3)}(p - p') \qquad \text{(A-38)}$$

The choice is dictated by the requirement of consistency with the normalization condition (A-26). The scalar product

$$\langle 1|2\rangle = \int d^3x \; [\psi_1^*(i\partial_0\psi_2) - (i\partial_0\psi_1^*)\psi_2] \qquad \text{(A-39)}$$

is easily checked to be Lorentz-invariant.

In our calculations in text, for simplicity of interpretation of creation and annihilation of particles, we have considered the particles to be in a cubical box of volume $V = L^3$, with the limit $V \to \infty$ taken at the end of the calculation. (In this limit, we will recover the full Lorentz symmetry as well.) In this case, the wave functions for a particle of momentum $\boldsymbol{p}$ can be taken as

$$u_p(x) = \frac{e^{-ipx}}{\sqrt{2\omega_p V}} \tag{A-40}$$

We shall use periodic boundary conditions on the wave functions. Ultimately, of course, physical results should not be sensitive to the boundary behavior in the limit of $V \to \infty$; so this convenient choice should be fine. With periodic boundary conditions, i.e., $u_p(x + L) = u_p(x)$ for translation by L along any spatial direction, the values of $\boldsymbol{p}$ are given by

$$p_i = \frac{2\pi n_i}{L} \tag{A-41}$$

(n_1, n_2, n_3) are integers. The wave functions $u_p(x)$ obey the orthonormality relation

$$\int_V d^3x \ \left[u_p^*(i\partial_0 u_{p'}) - (i\partial_0 u_p^*)u_{p'} \right] = \delta_{p,p'} \tag{A-42}$$

where $\delta_{p,p'}$ denotes the Kronecker δ's of the corresponding values of n_i's, i.e.,

$$\delta_{p,p'} = \delta_{n_1,n_1'} \delta_{n_2,n_2'} \delta_{n_3,n_3'} \tag{A-43}$$

In the limit of $V \to \infty$, we have

$$\delta_{p,p'} \to \frac{(2\pi)^3}{V} \delta^{(3)}(p - p')$$

$$\sum_p \to \int V \frac{d^3p}{(2\pi)^3} \tag{A-44}$$

The wave functions ψ_p (or u_p) are obviously solutions of the equation

$$i\frac{\partial \psi}{\partial t} = \sqrt{-\nabla^2 + m^2} \ \psi \tag{A-45}$$

The differential operator on the right-hand side is not a local operator; it has to be understood in the sense of

$$\sqrt{-\nabla^2 + m^2} f(x) \equiv \int \frac{d^3p}{(2\pi)^3} \ e^{i\boldsymbol{p}\cdot\boldsymbol{x}} \sqrt{\boldsymbol{p}^2 + m^2} f(p) \tag{A-46}$$

where

$$f(x) = \int \frac{d^3p}{(2\pi)^3} e^{i\boldsymbol{p}\cdot\boldsymbol{x}} f(p) \tag{A-47}$$

One can, of course, define a local differential equation whose solutions are the wave functions (A-40). It is the Klein-Gordon equation

$$(\Box + m^2)\psi = 0 \tag{A-48}$$

where $\Box$ is the d'Alembertian operator, $\Box = \partial_\mu \partial^\mu = (\partial_0)^2 - \nabla^2$. Equation (A-48), however, has, in addition to (A-40), solutions of the form e^{-ipx} with $p_0 = -\sqrt{p^2 + m^2}$. At the level of one-particle wave functions, it is difficult to interpret such solutions. However, in quantum field theory, they can be interpreted consistently and the Klein-Gordon equation becomes the basis for the discussion of spin-zero particles.

A-5 Wave functions for spin-$\frac{1}{2}$ particles

For a spin-$\frac{1}{2}$ particle, $s = \frac{1}{2}$, we have two spin states. Translations act as in (A-36) and the wave functions have two components, viz.,

$$\psi_{p,r}(x) = \langle x|p, s = \tfrac{1}{2}, r\rangle \tag{A-49}$$

($r = 1, 2$.) There are now nontrivial $\mathcal{D}$-matrices in the transformation law. In this case, they are the 2×2 rotation matrices

$$\mathcal{D}^{(\frac{1}{2})}(\theta) = \exp\left(i\frac{\sigma_a}{2}\theta^a\right) \tag{A-50}$$

where σ_a are the Pauli matrices

$$\sigma_1 = \begin{pmatrix} 0 & 1 \\ 1 & 0 \end{pmatrix}, \quad \sigma_2 = \begin{pmatrix} 0 & -i \\ i & 0 \end{pmatrix}, \quad \sigma_3 = \begin{pmatrix} 1 & 0 \\ 0 & -1 \end{pmatrix} \tag{A-51}$$

The wave functions are solutions to the equation

$$i\frac{\partial \psi_r(x)}{\partial t} = \sqrt{-\nabla^2 + m^2}\,\psi_r(x) \tag{A-52}$$

For many examples of spin-$\frac{1}{2}$ particles of mass m, we also have antiparticles of the same mass. The antiparticle wave functions $\phi_r(x)$ obey equation (A-52), with ϕ_r in place of ψ_r. One can then define a combined wave function $\Psi(x)$ by

$$\Psi(x) = \begin{pmatrix} \psi_r(x) \\ \phi_r^*(x) \end{pmatrix} \tag{A-53}$$

$\Psi(x)$ is a four-component column vector. It obeys the equation

$$i\frac{\partial \Psi}{\partial t} = \gamma_0 \sqrt{-\nabla^2 + m^2}\,\Psi \tag{A-54}$$

where

$$\gamma_0 = \begin{pmatrix} 1 & 0 & 0 & 0 \\ 0 & 1 & 0 & 0 \\ 0 & 0 & -1 & 0 \\ 0 & 0 & 0 & -1 \end{pmatrix} \tag{A-55}$$

This is one version of the Dirac equation for spin-$\frac{1}{2}$ particles, the so-called Foldy-Wouthuysen representation. As with spin-zero particles and the Klein-Gordon equation, one can seek a local differential equation for $\Psi(x)$. The local equation is the usual version of the Dirac equation. Instead of transforming (A-54) to a local form, it is easier to show that (A-54) follows from the Dirac equation.

A-6 Spin-1 particles

For spin $s = 1$, we have three components and the wave functions have the form

$$\psi_{p,r}(x) = \langle x | p, s = 1, r \rangle = e^{-ipx} \epsilon_r(p) \tag{A-56}$$

$(r = 1, 2, 3.)$ ϵ_r transforms as a vector under rotations, i.e., infinitesimally

$$\delta \epsilon_r = \omega_{rk} \epsilon_k \tag{A-57}$$

ϵ_r behaves like the spatial components of a four-vector. A local spin-1 analog of the Klein-Gordon equation might look like

$$(\Box + m^2)\psi_r(x) = 0 \tag{A-58}$$

The functions (A-56) are evidently solutions to (A-58) for any $\epsilon_r(p)$. This equation is, however, not manifestly covariant since only spatial components of a four-vector are involved. A local, manifestly covariant equation would be

$$(\Box + m^2)\psi_\mu(x) = 0 \tag{A-59}$$

$(\mu = 0, 1, 2, 3.)$ However, ψ_μ has one more component, namely ψ_0, more than we need. Thus if one would like to have a local manifestly covariant equation, one can use (A-59), but must impose additional constraints on ψ_μ to eliminate the unwanted degree of freedom. Alternatively, one may choose an equation different from (A-59) with extra symmetries which help us to eliminate the unwanted degree of freedom. These extra symmetries are gauge symmetries. They arise, from the particle point of view, because of the mismatch of the number of physical spin states, namely, $(2s+1)$, and the number required for a suitable Lorentz vector or tensor in terms of which a local manifestly covariant equation can be constructed. (Gauge symmetries also have a deep geometric interpretation, as discussed in text.) Gauge symmetries are required for all spins $s \geq 1$.

A-7 Massless particles

Although we derived the Poincaré algebra as the symmetry of the action for a massive point-particle, it holds for a massless particle as well. This can be seen by considering the symmetry of the equations of motion rather than the action or by noting that massless particles obey $ds = 0$. (They follow null geodesics on a general spacetime.) Thus massless particles are also described by the UIR's of the Poincaré algebra. In this case, we have $p^2 = 0$ and $p_0 > 0$. We do not have the possibility of going to the rest frame of the particle. A general solution to $p^2 = 0$, $p_0 > 0$ can be constructed by Lorentz transformations of a special vector $p_\mu^{(0)} = (1,0,0,1)$. In our discussion of massive particles, we considered the transformations which left $p_\mu^{(0)} = (m,0,0,0)$ invariant, viz., rotations. From the action of the rotations, via the use of Wigner rotations, we could obtain the action of a general Lorentz transformation on the states. One can do a similar construction for massless particles by considering the transformations which leave the vector $p_\mu^{(0)} = (1,0,0,1)$ invariant or in other words the isotropy group of this vector. Infinitesimally, these transformations are given by those $\omega^{\mu\nu}$ which obey $\omega^{\mu\nu} p_\nu^{(0)} = 0$. One can check that a general Lorentz transformation which preserves $p_\mu^{(0)}$ is of the form

$$
\begin{aligned}
U &\approx 1 + i\left(\omega^{01}(M_{01} + M_{13}) + \omega^{02}(M_{02} + M_{23}) + \omega^{12}M_{12}\right) \\
&\approx 1 + i\left(\omega^{01}T_1 + \omega^{02}T_2 + \omega^{12}M_{12}\right)
\end{aligned}
\tag{A-60}
$$

$T_i = M_{0i} + M_{i3}$, $i = 1,2$, and M_{12} generates rotations around the direction of the spatial momentum $\boldsymbol{p}$, in this case the z-axis. The commutation rules among these operators are

$$
[M_{12}, T_\pm] = \pm T_\pm, \qquad [T_+, T_-] = 0
\tag{A-61}
$$

where $T_\pm = T_1 \pm iT_2$. We can also check that for the special choice of $p_\mu^{(0)}$, W^2 is given by

$$
W^2 = T_+ T_-
\tag{A-62}
$$

One class of states and representations is obtained by

$$
T_+|p, \lambda\rangle = 0, \qquad T_-|p, \lambda\rangle = 0, \qquad M_{12}|p, \lambda\rangle = \lambda|p, \lambda\rangle
\tag{A-63}
$$

This has $W^2 = 0$. In this case, since both W^2 and p^2 are zero, and $W \cdot p = 0$ from (A-21), so we must have W^μ proportional to p^μ. In fact, $W^\mu = \lambda p^\mu$. This equation gives an invariant definition of λ. We may write

$$
\lambda = \frac{W^0}{p^0} = \frac{\boldsymbol{p} \cdot \boldsymbol{J}}{p^0}
\tag{A-64}
$$

λ is called the helicity of the particle. Since M_{12} generates rotations, invariance under $\omega^{12} \rightarrow \omega^{12} + 2\pi$ gives $\lambda = 0, \pm 1, \ldots$, for single-valued representations and $\lambda = \pm\frac{1}{2},\ \pm\frac{3}{2}, \ldots$, for the double-valued representations. The photon ($\lambda = \pm 1$) and the graviton ($\lambda = \pm 2$) are examples of the realization of these representations. Just as in the massive case, one can construct a suitable B_μ^ν such that $p_\mu = B_\mu^\nu p_\nu^{(0)}$ and Wigner rotations and the explicit realization of a general Lorentz transformation on these states. We do not discuss these matters, since it is a little easier for these cases to obtain the wave functions and the transformations by solving local equations of motion from a field theoretic approach.

One can also construct representations for which $T_\pm$ are not represented by zero. However, such representations have an infinite number of polarization states and do not seem to be of any physical significance.

A-8 Position operators

We have not discussed the position yet as an operator to be included in the algebra of observables. From the point of view of obtaining the representations, this does not make much difference, x^μ appear explicitly in the x-space wave functions. However, x^μ is not appropriate as a position operator. For example, x^0 denotes time. It is, in the case of point-particles, just a variable parametrizing the path of the particle; to consider it as an operator would lead to difficulties of interpretation. Newton and Wigner have defined a proper notion of position operator and calculated it for various cases. We shall not discuss this in detail. For the case of massive spin-zero particles, one can see quite easily that $\boldsymbol{x}$ is not appropriate. It is not self-adjoint with the scalar product (A-26). A modification which gives a self-adjoint operator is

$$\boldsymbol{x} = i\nabla_{\boldsymbol{p}} - i\frac{\boldsymbol{p}}{2E} \tag{A-65}$$

This is appropriate in the sense that it coincides with the position operator of nonrelativistic theory for small velocities; it is the center of mass for localized wave packets. Equation (A-65) is the Newton-Wigner (NW) operator for this case. For particles with spin, the NW position operator has spin-dependent terms in general.

A-9 Isometries, anyons

It is interesting to follow the logic of the previous sections, viz., of understanding the one-particle states in terms of representations of the symmetry algebra, in some unusual situations. As an example, let us consider free particle motion in a spacetime which is not necessarily flat Minkowski space. The action is given by $\mathcal{S} = -m \int \sqrt{g_{\mu\nu}dx^\mu dx^\nu}$, where $g_{\mu\nu}$ is the metric tensor. The requirement that $x^\mu \to x^\mu + \xi^\mu$ be a symmetry gives

$$g_{\alpha\nu}\frac{\partial\xi^\alpha}{\partial x^\mu} + g_{\mu\alpha}\frac{\partial\xi^\alpha}{\partial x^\nu} + \xi^\alpha\frac{\partial g_{\mu\nu}}{\partial x^\alpha} = 0 \qquad \text{(A-66)}$$

We can write this equation as

$$\nabla_\mu\xi_\nu + \nabla_\nu\xi_\mu = 0 \qquad \text{(A-67)}$$

where

$$\nabla_\mu\xi_\nu = \partial_\mu\xi_\nu - \Gamma^\alpha_{\mu\nu}\xi_\alpha$$
$$\Gamma^\alpha_{\mu\nu} = \tfrac{1}{2}g^{\alpha\beta}\left(-\frac{\partial g_{\mu\nu}}{\partial x^\beta} + \frac{\partial g_{\beta\nu}}{\partial x^\mu} + \frac{\partial g_{\mu\beta}}{\partial x^\nu}\right) \qquad \text{(A-68)}$$

$\Gamma^\alpha_{\mu\nu}$ is the Christoffel symbol and $\nabla_\mu\xi_\nu$ is the covariant derivative of ξ_ν. Equation (A-67) is the Killing equation, the solutions ξ^μ give the isometries or transformations which leave the distance ds (and hence the particle action) invariant. The isometries will form a group, the isometry group, of the spacetime of metric $ds^2 = g_{\mu\nu}dx^\mu dx^\nu$. From our general discussion of quantum mechanics, the Hilbert space for one-particle motion on such a spacetime will be given by a UIR of the isometry group. A simple concrete example is provided by anti-de Sitter spacetime, which has the metric

$$ds^2 = dz_0^2 - dz_1^2 - dz_2^2 - dz_3^2 + dz_4^2 \equiv \eta_{\mu\nu}dz^\mu dz^\nu + dz_4^2 \qquad \text{(A-69)}$$

with

$$\eta_{\mu\nu}z^\mu z^\nu + z_4^2 = R^2 \qquad \text{(A-70)}$$

One can solve (A-70) explicitly in terms of local coordinates, valid in some coordinate patch, as

$$z_0 = R\sin t, \qquad z_4 = R\cos t\ \cosh\chi$$
$$(z_1, z_2, z_3) = \cos t \sinh\chi\ (\cos\theta, \sin\theta\cos\varphi, \sin\theta\sin\varphi) \qquad \text{(A-71)}$$

Substitution of these in (A-69) will give a more standard four-dimensional way of writing the metric. Using the presentation of the space in terms of (A-69), (A-70) with the auxiliary variable z_4, we see that the isometries are transformations which leave the quadratic form $(\eta_{\mu\nu}z^\mu z^\nu + z_4^2)$ invariant. These are the pseudo-orthogonal transformations forming the group $SO(3,2)$ (and some discrete symmetry transformations). Thus UIR's of $SO(3,2)$ will describe possible types of particle motion on anti-de Sitter spacetime.

As another example, consider three-dimensional spacetime, with a metric of the form

$$ds^2 = dt^2 - g_{ij}dx^i dx^j, \qquad\qquad i,j = 1,2 \qquad \text{(A-72)}$$

For simplicity the nontriviality of the metric is restricted to the spatial dimensions. For flat space, $g_{ij} = \delta_{ij}$, and the rotation $\delta x_i = \theta\epsilon_{ij}x^j$ is clearly a symmetry. The states are of the form $|p, s\rangle$ with the action of rotations in the rest frame, for which $p_\mu = p_\mu^{(0)}$, given by

$$U(\theta)|p^{(0)}, s\rangle = e^{is\theta}|p^{(0)}, s\rangle \tag{A-73}$$

If we allow multivalued representations, s need not be an integer. Particles with any value of spin are generically called anyons. The quasi-particles relevant to the fractional quantum Hall effect are of this type. (For this system, the physics is essentially planar and so a two-dimensional description is reasonably accurate. Also a nonrelativistic approximation is quite adequate.)

If space is a sphere of radius R,

$$g_{ij}dx^i dx^j = R^2(d\theta^2 + \sin^2\theta d\varphi^2) \tag{A-74}$$

in terms of the usual angular coordinates θ, φ on the sphere. The isometries in this case are

$$\xi_{(1)} = (-\sin\varphi, -\cos\varphi \cot\theta), \quad \xi_{(2)} = (\cos\varphi, -\sin\varphi \cot\theta), \quad \xi_{(3)} = (0, 1) \tag{A-75}$$

The generators $L_\mu = -i\xi^i_{(\mu)}(\partial/\partial x^i)$ obey the angular momentum algebra $[L_\mu, L_\nu] = i\epsilon_{\mu\nu\alpha}L_\alpha$, $(\mu, \nu, \alpha = 1, 2, 3)$. In other words, the isometry group of the two-dimensional sphere is $SU(2)$, the angular momentum group. Unitary representations clearly require integer or half-odd-integer values of spin. Thus a spherical world cannot support anyons, at least with the action $S = -m \int ds$. It is also interesting to investigate what kind of anyons are possible if the world is a two-dimensional torus of metric $ds^2 = d\theta^2 + d\varphi^2$.

References

1. The uniqueness theorem on the representation of the Heisenberg algebra is due to M.H. Stone, Proc. Nat. Acad. Sci. USA, **16**, 172 (1930); J. von Neumann, Math. Ann. **104**, 570 (1931).
2. The representation theory of the Poincaré group is due to E.P. Wigner, Ann. Math. **40**, 149 (1939).
3. The position operator was introduced by T.D. Newton and E.P. Wigner, Rev. Mod. Phys. **21**, 400 (1949).
4. Isometries and the Killing equation are discussed in most books on the general theory of relativity, see, for example, S.W. Hawking and G.F.R. Ellis, *The Large Scale Structure of Space-time*, Cambridge University Press (1973). Anti-de Sitter space is also given here.
5. The Poincaré group analysis for anyons in three dimensions is given in B. Binegar, J. Math. Phys. **23**, 1511 (1982).

General References

1. Lowell S. Brown, *Quantum Field Theory*, Cambridge University Press (1992).
2. C. Itzykson and J-B. Zuber, *Quantum Field Theory*, McGraw Hill Inc. (1980).
3. J. M. Jauch and F. Rohrlich, *The Theory of Photons and Electrons*, Springer-Verlag (1955 & 1976).
4. Michio Kaku, *Quantum Field Theory: A Modern Introduction*, Oxford University Press, Inc. (1993).
5. F. Mandl and G. Shaw, *Quantum Field Theory*, John Wiley (1984).
6. Michael E. Peskin and Daniel V. Schroeder, *An Introduction to Quantum Field Theory*, Westview Press (1995).
7. P. Ramond, *Field Theory: A Modern Primer*, Addison-Wesley Pub. Co. Inc. (1990).
8. Lewis H. Ryder, *Quantum Field Theory*, Cambridge University Press (1985 & 1996).
9. S. S. Schweber, *An Introduction to Relativistic Quantum Field Theory*, Harper and Row, New York (1961).
10. S. S. Schweber, *QED and the Men Who Made It*, Princeton University Press (1994).
11. Steven Weinberg, *The Quantum Theory of Fields: Volume I Foundations*, Cambridge University Press, (1995);
    The Quantum Theory of Fields: Volume II Modern Applications, Cambridge University Press, (1996);
    The Quantum Theory of Fields: Volume III Supersymmetry, Cambridge University Press, (2000).
12. Anthony Zee, *Quantum Field Theory in a Nutshell*, Princeton University Press (2003).
13. J. Zinn-Justin, *Quantum Field Theory and Critical Phenomena*, Clarendon Press (1996).

Index

Graduate Texts in Contemporary Physics